AF574131

Benchmark Papers in Geology

Series Editor: Rhodes W. Fairbridge
Columbia University

A selection from the published volumes in this series

Volume
17 METAMORPHISM AND PLATE TECTONIC REGIMES / *W. G. Ernst*
18 GEOCHEMISTRY OF IRON / *Henry Lepp*
19 SUBDUCTION ZONE METAMORPHISM / *W. G. Ernst*
29 METALLOGENY AND GLOBAL TECTONICS / *Wilfred Walker*
34 CRYSTAL FORM AND STRUCTURE / *Cecil J. Schneer*
37 STATISTICAL ANALYSIS IN GEOLOGY / *John M. Cubitt and Stephen Henley*
44 MINERAL DEPOSITS, CONTINENTAL DRIFT AND PLATE TECTONICS / *J. B. Wright*
57 MEGACYCLES: Long-Term Episodicity in Earth and Planetary History / *G. E. Williams*
62 OROGENY / *John G. Dennis*
67 ECONOMIC EVALUATION OF MINERAL PROPERTY / *Sam L. VanLandingham*
69 MINING GEOLOGY / *Willard C. Lacy*

Benchmark Papers in Geology / 70

A BENCHMARK® Books Series

MINERAL EXPLORATION

Edited by

WILLARD C. LACY

James Cook University of North Queensland

Hutchinson Ross Publishing Company

Stroudsburg, Pennsylvania

Benchmark Papers in Geology, Volume 70
Library of Congress Catalog Card Number: 82-969
ISBN: 0-87933-425-8

85 84 83 1 2 3 4 5
Manufactured in the United States of America.

LIBRARY OF CONGRESS CATALOGING IN PUBLICATION DATA
Main entry under title:
Mineral exploration.
(Benchmark papers in geology; 70)
Includes indexes.
1. Prospecting—Addresses, essays, lectures.
I. Lacy, W. C. (Willard C.) II. Series.
TN270.M655 622'.1 82-969
ISBN 0-87933-425-8 AACR2

Distributed worldwide by Van Nostrand Reinhold Company Inc.,
135 W. 50th Street, New York, NY 10020.

CONTENTS

PART III: REGIONAL ANALYSIS

PART IV: RECOGNITION

SERIES EDITOR'S FOREWORD

The philosophy behind the Benchmark Papers in Geology is one of collection, sifting, and rediffusion. Scientific literature today is so vast, so dispersed, and, in the case of old papers, so inaccessible for readers not in the immediate neighborhood of major libraries that much valuable information has been ignored by default. It has become just so difficult, or so time consuming, to search out the key papers in any basic area of research that one can hardly blame a busy person for skimping on some of his or her "homework."

This series of volumes has been devised, therefore, as a practical solution to this critical problem. The geologist, perhaps even more than any other scientist, often suffers from twin difficulties—isolation from central library resources and immensely diffused sources of material. New colleges and industrial libraries simply cannot afford to purchase complete runs of all the world's earth science literature. Specialists simply cannot locate reprints or copies of all their principal reference materials. So it is that we are now making a concerted effort to gather into single volumes the critical materials needed to reconstruct the background of any and every major topic of our discipline.

We are interpreting "geology" in its broadest sense: the fundamental science of the planet Earth, its materials, its history, and its dynamics. Because of training in "earthy" materials, we also take in astrogeology, the corresponding aspect of the planetary sciences. Besides the classical core disciplines such as mineralogy, petrology, structure, geomorphology, paleontology, and stratigraphy, we embrace the newer fields of geophysics and geochemistry, applied also to oceanography, geochronology, and paleoecology. We recognize the work of the mining geologists, the petroleum geologists, the hydrologists, and the engineering and environmental geologists. Each specialist needs a working library. We are endeavoring to make the task of compiling such a library a little easier.

Each volume in the series contains an introduction prepared by a specialist (the volume editor)—a "state of the art" opening or a summary of the object and content of the volume. The articles, usually some twenty to fifty reproduced either in their entirety or in significant extracts, are selected in an attempt to cover the field, from the key papers of the last century to fairly recent work. Where the original works are in foreign languages, we have endeavored to locate or commission translations. Geologists, because of their global subject, are often acutely aware of the oneness of our world. The selections cannot therefore be restricted to any one country, and whenever possible an attempt is made to scan the world literature.

To each article, or group of kindred articles, some sort of "highlight commentary" is usually supplied by the volume editor. This commentary should serve to bring that article into historical perspective and to emphasize its particular role in the growth of the field. References, or citations, wherever possible, will be reproduced in their entirety—for by this means the observant reader can assess the background material available to that particular author, or, if desired, he or she too can double check the earlier sources.

A "benchmark," in surveyor's terminology, is an established point on the ground that is recorded on our maps. It is usually anything that is a vantage point, from a modest hill to a mountain peak. From the historical viewpoint, these benchmarks are the bricks of our scientific edifice.

RHODES W. FAIRBRIDGE

PREFACE

Mineral exploration owes much to the mineral prospector, the mine geologist, and the earth scientist. It entails theoretical aspects of mineral deposition that assist in the generation of targets for search as well as the practical aspects of recognition, testing, evaluation, and reporting. This volume brings together a collection of papers that laid the foundation for effective search and development. It should be used in conjunction with the companion volume, *Mining Geology*.

Literature for this volume has been grouped into seven parts. Part I presents a historical perspective in terms of techniques, discoveries, expenditures, people, and the nature of effective management that resulted in successful exploration efforts. Part II considers the basic factors regarding probability of successful search, while Part III illustrates the variety of regional approaches to the generation of targets and the screening of prospects.

Recognition of alteration haloes and oxidized mineral deposit surface exposures are considered in Part IV, while Part V includes contributions and procedures that assist in testing and evaluation.

Clear and accurate reporting and presentation of data are critical to those who make the financial decisions. Suggestions regarding the writing of reports and map illustrations are included in Part VI. Part VII includes important, but often overlooked, literature that considers aspects of ethics and environmental responsibilities.

Many of the papers included have been severely edited in order to permit broader coverage of the subject matter by reprinting a variety of articles. However, the basic concepts and philosophy of approach have been retained in order to illustrate the differences in search philosophy between geologists in different countries.

I would like to thank James Cook University of North Queensland and the University of Arizona for the use of their libraries and facilities and Ann Quadroy for her typing and editing.

WILLARD C. LACY

CONTENTS BY AUTHOR

INTRODUCTION

HISTORICAL PERSPECTIVE

While the philosophers have continued to debate the origin and genetic processes that form ore deposits, the exploration geologists have continued to make discoveries. Often the discoveries were made on the basis of some personal concept, philosophy or inferred relationship that failed to stand up to the scrutiny of science but, at the time, that found ore deposits. For example, astrologers, following the concepts of Aristotle, proposed that metals were generated in the crust by the action of rays emitted from the heavenly bodies—the sun being responsible for the formation of gold. Thus, quite logically the equatorial zone was the logical place to search for gold since it was subjected to more intense sun rays. This philosophy guided the Spanish conquistadors, and they were successful in their search and discovered some of the world's great gold deposits.

The alchemists of the sixteenth century proposed that deposits of the baser metals were continually undergoing transition or transmutation, under solar influence, to the nobler metals. In the process, a powerful exhalation, *witterung,* passed through the ground, eventually destroying the gold, poisoning the ground, and leaving a reddish earthy residue in which a few flakes of gold might survive, but being underlaid by base metal deposits. The redstained gossans led to discoveries of base and precious metal deposits.

Abraham Werner (Crook, 1933), at the turn of the eighteenth century a kind of scientific pope who ruled over geological thought at the time until challenged by James Hutton, proposed that a universal flood precipitated metals to form ore deposits. His theory fitted many of the central European stratiform deposits, enabling the discovery of extensions and new deposits.

Was it luck? Most theories, no matter how wild we consider them to be, are based upon some observed relationship. Thus, often the observed pattern of relationships rather than the truth of the theory really guides the search. Each new concept or observed relationship leads to search in new areas and results in discoveries. We have seen, and still utilize, many pattern relationships such as those found in contacts and cupolas of intrusives, volcanic centers, favored sedimentary units, favored time units, mobile belts, lineament breaks, subduction zones, rift structures, and so on.

The most appropriate response when a revolutionary concept is proposed is: Assuming that this might be true, where would it tend to localize mineral deposits? In my own experience, too often have I scoffed at some new concept or proposal only to find that it was in part true and resulted in occasional concentrations that formed ore deposits.

From a scientific point of view we are concerned with: Why are ore deposits where we observe them to be? However, from a practical point of view we ask: Where are the deposits that have not yet been found or put into production that can be found and/or put into production economically? Both questions are relevant and must be integrated into an exploration philosophy.

Mineral exploration is based upon a number of basic assumptions that deposits are yet to be found; that the deposits can be found by presently available techniques; that the deposits can be developed, if they are discovered; and that the mineral components can be recovered and sold at at profit. If any of these basic assumptions are not valid, exploration is not justified.

Prior to the middle of the nineteenth century, the geologist did not concern himself with the search for ores. Search was in the realm of the prospector. Evaluations were made by the prospector by crude panning. Discoveries were by chance or were made by following a trail of gold back to its source by panning stream sediments—one of the oldest and most effective techniques of the prospector. When the gold disappeared from the stream sediment, he then moved up adjacent slopes, panning the overburden until, hopefully, he found the lode from which the gold was derived. A similar technique was used in the search for tin, tungsten, mercury, and lead.

The prospector worked by rule-of-thumb methods but had a remarkable eye for landscape, vegetation, and color variations as he searched systematically for ore bodies. He seemed to know by instinct where valuable concentrations of minerals were likely to occur, and few modern discoveries are made that do not have the marks of earlier prospector activity.

What his experienced eye was looking for, as he worked in isolation, was the "float" from veins in the outwash gravels of streams or in the hillside soil, and the minute flakes of gold, or perhaps cassiterite, that had eroded from a vein by weathering and stream action.

Ability to recognize significant minerals, as well as the discolored and stained outcrops from which the metal values had been leached by weathering processes, was an important capability of the successful prospector. He had his own methods of identifying minerals by scratching them with his knife, or by tasting them, and he knew the importance of examining the physical conditions of the gold particles that would reveal their distance of travel and mode of formation.

The prospector recognized so-called kindly country where the geological environment was favorable for the discovery of certain types of metals. Greenstone country in western Australia and Canada, for example, is regarded as friendly for gold, and granite is unfriendly for gold but friendly for tin. The experienced prospector goes further than this, however; the presence of certain igneous intrusions in greenstone tells him that it is kindlier country than one devoid of the intrusions, and other things being equal, he knows that soft and schistose greenstone is kindlier than hard and blocky greenstone and in still more detail he willl prefer one schistose greenstone because it has seams of ironstone, quartz veinlets of particular luster, or pyrite of a particular form. The exploration geologist today might put these observations in technical terms of favorable host, localizing structure, and wallrock alteration, but he is looking for the same guides to ore.

In looking for topographic guides to ore, topographic anomalies such as a small knob jutting up from a flat plain of sharp, wall-like ridge would catch the eye of the prospector. In some areas mineralization might be associated with badly fractured soft rock that would form valleys or saddles in the ridges; these more obscure sites would then attract the prospectors' attention—particularly when accompanied by certain vegetation.

As the effectiveness and reliability of the prospector declined, with diminishing abundance of those deposits that commenced at the grass roots, professional geologists were required to make examinations and to detail the geological environments in order to detect clues of hidden deposits, aided by search techniques such as geophysics and geochemistry. Costs of exploration rapidly escalated. Exploration was left as a part-time responsibility of the mine geologist.

Following World War II, three major factors resulted in a review of the mineral exploration function: first, the uranium boom in the United States where thousands of small companies were organized to search for uranium deposits and quick profits; second, growing concern on the part of governments and major operating companies regarding the ability of mineral reserves to meet projected demands; and third, the success that the petroleum companies were experiencing in their worldwide exploration programs.

In 1952, the Kennecott Corporation established an exploration branch (Bear Creek Mining Co.) independent of mining operations. The geologists recruited for this task were given special instruction in the recognition characteristics of the various types of deposits and in the utilization of geochemical and geophysical techniques. The success of this approach rapidly resulted in its adoption by most major mining companies, and independent exploration teams were formed. The field of operation spread throughout the world. This organizational change, with accompanying change in search philosophy, gave great stimulus to the development of geochemical and geophysical techniques and interpretations. It resulted in a change in the mineral reserve picture from one of predicted shortages of nearly all metals to one of universal surplus.

The study of ore deposits, like other branches of geology, had entailed four methods of approach: local, regional, topical, and comprehensive. The local method was carried out to a high degree of perfection by the mine geologist in studies of individual mines or districts. The assignment of the mine geologist was to predict the contortions of the ore body in advance of mining, to find additional ore shoots and ore extensions, and to evaluate adjoining prospects. We have considered this approach in a companion Benchmark volume, *Mining Geology* (Lacy, 1983).

With the advent of the exploration geologist, the regional approach was emphasized (Billingsley and Locke, 1941; Mayo, 1958; Anderson, 1966; Schmitt, 1966; Noble, 1976). Studies established clusters and belts or ore districts as being related to particular geological environments, structural or stratigraphic, and these features served to guide the exploration geologist into virgin areas beyond active mines or known mineralization.

Plate tectonics concepts suggested specific tectonic environments for specific types of deposits (Wright, 1977). Genetic models for different types of deposits were developed, and a flood of age dates of deposits established a pattern of time relationships (Watson, 1973).

These days, the exploration geologist rarely has the opportunity to observe a virgin outcropping of a mineral deposit that has not been examined, broken, and sampled by some previous prospector or geologist. However, opportunities for chance discoveries still exist in most parts of the world. New exposures are brought to light by erosion of a stream's bank or a land slip, and the chance water-well, road-cut, or foundation excavation is still possible and may result in the discovery of new shoots in known mining districts or may establish new districts. However, the demands by society for replenishing depleting reserves does not permit reliance upon such chance discoveries. Organized search programs, utilizing both established and new geological concepts and search techniques, are required.

From pattern analysis and saturation prospecting, exploration has moved in the direction of conceptual models. These are, on the basis of well known deposits, predicted in terms of tectonic, stratigraphic, and structural position and identified under cover by peripheral characteristics or geochemical, geophysical, or geological patterns that match the predictive model (Jerome, 1966; Cook, 1968). This approach has many advantages where surface exposures are limited or deposits are blind. It assists visualization of the target when only limited or remote portions can be observed. It also has disadvantages in that it tends to narrow vision by search for a specific type of deposit that may not correspond in all details with the model, or the model may be erroneous. Thus, the geologist might overlook a deposit that departs from the model. It often leads to high risk and/or ill-advised exploration on the basis of superficial similarities to the model.

This volume is concerned primarily with geological aspects of exploration with only casual references to geophysical and geochemical techniques. These will be the subject of additional compilations.

REFERENCES

Anderson, C. A., 1966, Areal Geology of the Southwest, in *Geology of the Porphyry Copper Deposits, Southwestern North America,* S. R. Titley and C. L. Hicks, eds., University of Arizona Press, Tucson, Ariz., pp. 3–16.

Billingsley, P., and A. Locke, 1941, Structure of Ore Districts in the Continental Framework, *Am. Inst. Mining, Metall. and Petroleum Eng. Trans.* **144:**9–64.

Cook, D. R., 1968, Suggestions for Increasing Metalliferous Ore Discoveries by More Effective Use of Exploration Techniques and Other Considerations, *Am. Inst. Mining, Metall. and Petroleum Eng. Preprint 68-I-72,* 16p.

Crook, T., 1933, *The History of the Theory of Ore Deposits,* Thomas Murby and Company, London, pp. 33–37.

Jerome, S. E., 1966, Some Features Pertinent in Exploration of Porphyry Copper Deposits, in *Geology of the Porphyry Copper Deposits, Southwestern North America,* S. R. Titley and C. L. Hicks, eds., University of Arizona Press, Tucson, Ariz., pp. 75–86.

Lacy, W. C., ed., 1983, *Mining Geology,* Benchmark Papers in Geology, vol. 69, Hutchinson Ross, Stroudsburg, Pa.

Mayo, E. B., 1958, Lineament Tectonics and Some Ore Districts of the Southwest, *Mining Eng.* **9:**1169–1175.

Noble, J. A., 1976, Metallogenic Provinces of the Cordillera of Western North and South America, *Mineralium Deposita* **11:**219–233.

Schmitt, H. A., 1966, The Porphyry Copper Deposits in Their Regional Setting, in *Geology of the Porphyry Copper Deposits, Southwestern North America,* S. R. Titley and C. L. Hicks, eds., University of Arizona Press, Tucson, Ariz., pp. 17–34.

Watson, J., 1973, Influence of Crustal Evolution on Ore Deposition, *Inst. Mining and Metallurgy Trans.* **82:**B107–B113.

Wright, J. B., ed., 1977, *Mineral Deposits, Continental Drift and Plate Tectonics,* Benchmark Papers in Geology, vol. 44, Dowden, Hutchinson & Ross, Stroudsburg, Pa., 397p.

Part I

MINERAL EXPLORATION SINCE 1940

Editor's Comments on Papers 1 Through 5

1 DERRY
Exploration—Targets and the Choice of Weapons

2 DERRY and BOOTH
Mineral Discoveries and Exploration Expenditure—A Revised Review 1966–1976

3 ALBERS
Discovery Rate and Exploration Methods for Metallic Mineral Deposits in the U.S., 1940–76

4 JORALEMON
The Ore Finders

5 MILLER
Excerpts from *Corporations, Ore Discovery, and the Geologist*

The track record of teams of exploration geologists since 1940 has been very impressive. Although the statistics given in Papers 1, 2, and 3 refer primarily to North America, the same exploration methods and resultant discoveries occurred in South America, Africa, Australia, and elsewhere. Many of the discoveries were in reality rediscoveries where changing economics and technology had altered cut-off grades. However, many discoveries were in entirely new districts with no record of previous production. Derry (Paper 1) points out that in spite of the rapid development of exploration techniques in geochemistry and geophysics, application of geological search methods continues to be the dominant and most successful method of search, though assisted by the other techniques. An important aspect that rises from the compilation is the high cost per discovery. This cost amounted to about $40 million per discovery in Canada during the early 1960s as determined by Derry and Booth (Paper 2). Another important factor is the long preproduction times required—up to twenty-three years as determined by Albers (Paper 3) within the United States where access and availability of infrastructure and technological capability are considered to be readily available.

Joralemon (Paper 4) calls upon personal experience to emphasize that there are many ways that an ore deposit may come into existence other than simply being discovered. Further, he emphasizes that in following a set sequence of exploration steps and stages, the deposit may be missed or funds will be unnecessarily expended.

The question is often asked as to why some companies are so successful in their exploration programs, while others with equal budgets are so unsuccessful. There is no easy answer to this query other than to emphasize that discoveries are made by people who have ideas, who are willing to walk that extra mile, who are willing to put their job on the line by making a risky recommendation, and who see more than the geologist who previously examined a prospect. Miller (Paper 5) deals with the exploration geologist and various corporations and the relationships that benefit both.

1

Reprinted from pages 151, 153, 155, 157, 159, 161, and 163 of *Mining Mag.* **120:**151-163 (1969)

EXPLORATION—TARGETS AND THE CHOICE OF WEAPONS

Duncan R. Derry

Fig. 1. Scenery in the high Andes where surface is masked by volcanic ash

In the combination of enthusiasm, exhilaration, frustration and disappointment that makes up the life of those who are deeply involved in searching for new ore deposits, the use and choice of technical aids plays an ever-increasing part. At the same time our ideas on the origins and life histories of orebodies are undergoing radical changes. This article is not an attempt to describe these technical aids in any detail—there is neither the space to do so nor does the writer have the qualifications. The professional in exploration will find little new to him, except some costs to compare with those of his own projects. The object of the article is to assess, from one man's viewpoint, the present relative value of various techniques under different conditions, and their relative cost, for the benefit especially of those who make decisions on mining exploration budgets and programmes. Since some of such people may have operational or financial backgrounds, rather than geological or geophysical, some fairly elementary explanations are included and some generalisations made that may not be scientifically accurate.

SINCE 'Exploration' is a word of rather variable definition it might be well, at the start, to define the type of search that is being considered, as well as some matters concerning costs:

(1) Only 'grassroots' exploration is dealt with, i.e. the search for new deposits, preferably 'big game', in areas that are relatively unexplored or in those where a second attack, using new weapons, seems justified. The search for extensions or nearby repetitions of known orebodies involves a different approach and is *not* included in these discussions.

(2) Non-ferrous metallic deposits are only considered and some, but not all, of the techniques would be applicable only if the metals were in the form of sulphides.

(3) Where costs are mentioned, unless otherwise defined, the *total* cost of a project, or per unit area, is the one given, including follow-up procedures, ground acquisition (other than by purchase from a private owner), administration and overhead.

All costs have been converted to U.S. dollars at the exchange rate prevailing at the time of the operation. It is this overall cost, rather than the cost of operating a particular piece of equipment per line mile or unit area, that is of most interest to the person responsible for instigating, promoting or approving a programme.

In an earlier paper[1] the author pointed out that, in Canada, between 1955 and 1966, 123 non-ferrous metal mines were put in production at an average rate of 10·4 per year, while new discoveries made during the period 1951 to 1966 of sufficient merit to justify immediate plans for production amounted to only 4·7 per year. The balance depended on discoveries made in prior years but which were not considered at the time to warrant immediate exploitation. Doubt was expressed as to whether we were putting 'on the shelf' deposits of possible future value in the same number as we were taking down from the shelf deposits left there by our predecessors. To a large extent the same situation exists in Australia and Africa. These 'shelved' deposits were usually surface showings, noted during the early prospecting of a country, and either too low grade or too inaccessible to be of interest at the time; or subsequent work proved them to be much more extensive or of higher grade than originally indicated. It is doubtful how many more of these remain to be

The table alongside shows the number of Canadian non-ferrous mineral discoveries (resulting in production) for the time intervals given and the proportions in which principal credit was given respectively to conventional prospecting, geological deductions, geophysics or geochemistry

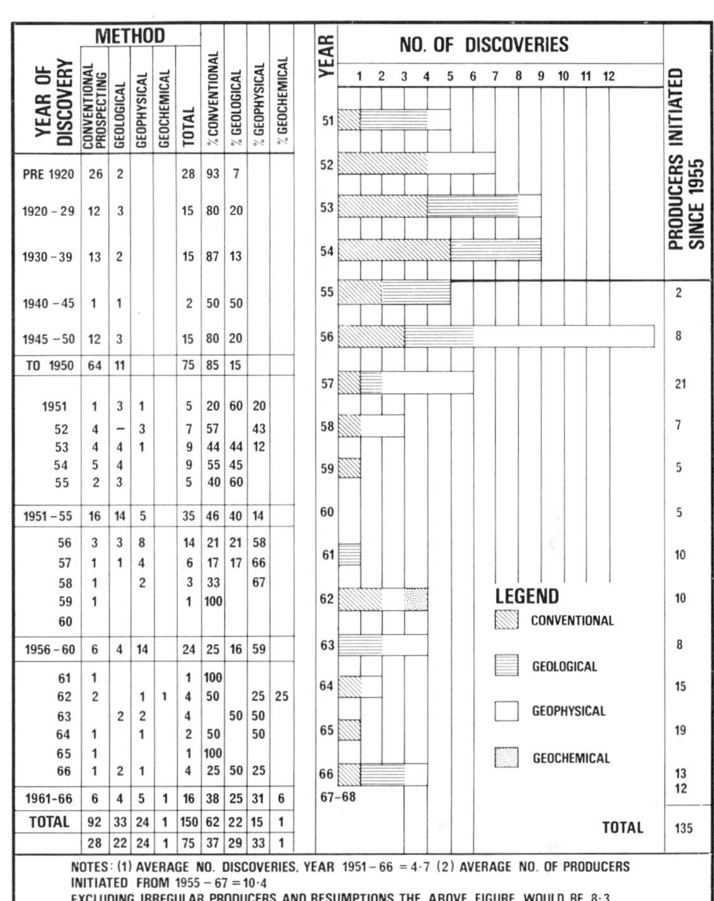

YEAR OF DISCOVERY	METHOD: CONVENTIONAL PROSPECTING	GEOLOGICAL	GEOPHYSICAL	GEOCHEMICAL	TOTAL	% CONVENTIONAL	% GEOLOGICAL	% GEOPHYSICAL	% GEOCHEMICAL
PRE 1920	26	2			28	93	7		
1920 - 29	12	3			15	80	20		
1930 - 39	13	2			15	87	13		
1940 - 45	1	1			2	50	50		
1945 - 50	12	3			15	80	20		
TO 1950	64	11			75	85	15		
1951	1	3	1		5	20	60	20	
52	4	–	3		7	57		43	
53	4	4	1		9	44	44	12	
54	5	4			9	55	45		
55	2	3			5	40	60		
1951 - 55	16	14	5		35	46	40	14	
56	3	3	8		14	21	21	58	
57	1	1	4		6	17	17	66	
58	1		2		3	33		67	
59	1				1	100			
60									
1956 - 60	6	4	14		24	25	16	59	
61	1				1	100			
62	2		1	1	4	50		25	25
63		2	2		4		50	50	
64	1		1		2	50		50	
65	1				1	100			
66	1	2	1		4	25	50	25	
1961-66	6	4	5	1	16	38	25	31	6
TOTAL	92	33	24	1	150	62	22	15	1
	28	22	24	1	75	37	29	33	1

NOTES: (1) AVERAGE NO. DISCOVERIES, YEAR 1951 – 66 = 4·7 (2) AVERAGE NO. OF PRODUCERS INITIATED FROM 1955 – 67 = 10·4
EXCLUDING IRREGULAR PRODUCERS AND RESUMPTIONS THE ABOVE FIGURE WOULD BE 8·3

tested, proven and exploited. More and more we must depend on finding new, hidden deposits—hidden by overburden, heavy vegetation or because they do not reach rock surface. To find these it is necessary to depend increasingly on geophysics and geochemistry, with geological studies and ideas governing the selection of areas and interpretation of the results.

There is another major change in the last 20 years affecting the business of searching for mineral deposits. This is in the type of deposit sought—we are interested much more in the size of the 'game' for which we are hunting, even at the expense of quality. The change is, of course, one of emphasis. We have always, naturally, wanted to find large deposits if possible but, prior to 1950, there had been no major change in the proportion of new mines having a daily capacity of, say, 1,000 tons or more. With the increasing cost of labour relative to metal prices, and hence the cost of mining a ton of ore underground in a small scale operation, the small or moderately sized mine has lost much of its attraction.

While preparing this article a rough count was made, from a list of producing non-ferrous metal mines in Canada in 1967, of those whose production was initiated since 1955, but excluding satellite mines contributing ore to the mill of a parent mine. Out of a total of 55 mines, only 19 were treating less than 1,000 tons a day and these accounted for 7·5 per cent of the total capacity of the 55. This is a fundamental change from the period prior to 1955. The same trend is seen in Southern and Central Africa, in South America and, more than anywhere, in the U.S.A. If it is accepted that interest is limited to relatively large orebodies, we can use a wider mesh fish net—to change metaphor slightly—and forget about the small ones that get away. The trend has an important influence on the techniques employed and the way they are employed.

The parts played by the various techniques can now be considered, under the three general headings of Geology, Geophysics and Geochemistry.

Geology

Geology must be the foundation of the strategy in the selection of areas of interest, whatever techniques are to be

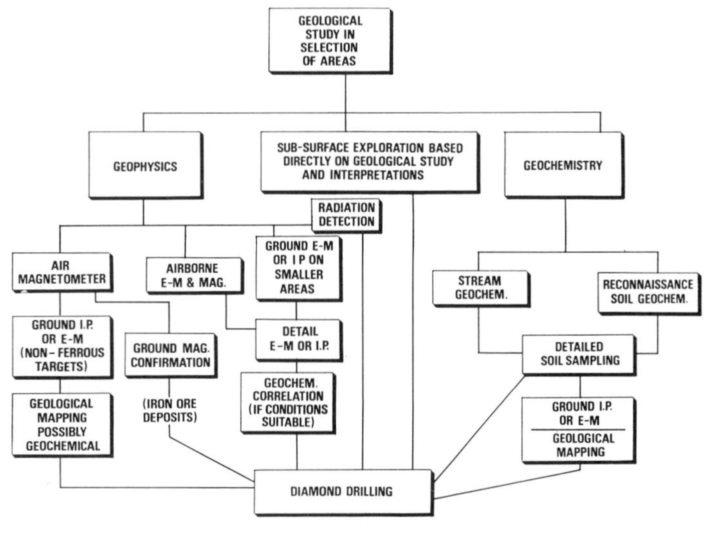

Chart shows combinations and sequences of techniques commonly used in exploration of large areas

used in the actual search for orebodies. However efficient these are, they can only be successful if the area chosen actually contains orebodies—this is a truism surprisingly often overlooked. The more obvious areas by empirical deduction, such as the extension of areas containing producing mines, have already been attacked in most parts of the world, perhaps several times and using various tools. Increasingly in 'grassroots' exploration, the selection of new areas is becoming a matter of developing a working hypothesis on which the selection is based. And increasingly the application of geological theory is dependent on the improvement and use of geophysical and geochemical techniques. The attractiveness of the area, from a geological standpoint, must be balanced against relative cost due to inaccessibility, roughness of terrain, and suitability to the application of the new techniques.

In all references to geological work, it is taken for granted that use will be made, on an increasing scale, of photogeology and the use of air photographs generally in the selection of areas, on the geological mapping of the area selected and the interpretation of geophysical results.

Over the past ten years or so, there has been a major change in the opinions of the majority of mining geologists on the matter of the genesis and life histories of many non-ferrous metal ore deposits. Essentially the trend is to accepting, for a larger proportion of base and precious metal deposits, greater dependence in origin on their enclosing host rocks and less on ore solutions introduced from an outside source at a much later geological period. From the economic standpoint this increases the significance of a detailed study and knowledge of the volcanic and sedimentary history of a region when it is being considered for selection of areas for exploration campaigns.

One of the best recent examples of success in following a working hypothesis was the discovery of the orebody in Kidd Township, Ontario, by the Texas Gulf Company. This remarkable sulphide body of zinc, copper and silver resulted from the persistent pursuit, in several parts of Canada, of the rocks formed during a particular stage in a period of volcanic activity. The basic weapon used was airborne electro-magnetics, followed in time by various ground geophysical surveys and finally drill testing.

Geological knowledge must also be used in the interpretation of the results of geophysical or geochemical surveys, and in planning a drilling programme to test anomalies.

Finally, in relatively rare cases, a 'grassroots' programme may be based solely on geological mapping and deduction, followed by drilling without the intermediary of geophysics or geochemistry. This applies especially to exploration for strata-bound deposits on the basis of structural or stratigraphic interpretation, where the target is at depths too great for the application of present, or easily available, geophysical techniques. Such direct use of geology is likely to become more important in the future as areas of favourable geology untested by geophysical tools become scarcer.

Fig. 4. Drilling a base metal anomaly on an Irish farm

Geophysics

Magnetic methods

The magnetometer, and its predecessor, the dip-needle, is the oldest of the geophysical weapons. It is still widely used on the ground and its use from aircraft has become a major factor in most regional studies with applications to pure science, and to the search for minerals and petroleum.

Increasingly governments are undertaking airborne magnetic surveys to give a basis for sub-surface geological interpretation, and to encourage follow-up surveys of a more specific nature by mining or petroleum companies.

With rather rare exceptions, the only metal that is likely to be indicated *directly* by an airborne magnetic survey *alone* is iron. This limitation makes the method more suitable for government use in a country where the finding, financing and operating of mines is left to private capital. Unless a government is itself in the business of mining exploration, it is something of an embarrassment for it to locate a mine. As a basic study, comparable with the geological mapping of an area, an airborne magnetic survey is invaluable and is increasingly being accepted as a necessity.

A more direct application of airborne magnetic surveys, flying on lines more closely spaced, is its use in areas of intrusive bodies on the contacts of which skarns are developed. Magnetite is a major mineral in skarns and may form deposits of commercial value for the iron alone. Frequently, however, chalcopyrite and other sulphides are associated with magnetite in such skarn or contact deposits. For this reason the combination of air magnetic and ground Induced Polarisation methods is often employed in such areas.

Negative airborne magnetic anomalies have been used, with some success, to pinpoint areas of disseminated mineralisation of the 'porphyry copper' type.

In the follow-up programmes of airborne surveys, ground magnetic surveys, often in concert with electro-magnetic, are routine, being fast and having a relatively low cost per mile of line covered. Where the ore itself is magnetic, e.g. a pyrrhotite-pentlandite ore of nickel, direct targets for drilling may be indicated. More often the ground magnetometer is used: (1) To interpret and interpolate geology in overburden-covered areas, and (2) To aid in the confirmation that a conductor indicated by E-M is caused by sulphides rather than by a graphitic zone, although lack of positive magnetic indications cannot be taken to eliminate the presence of sulphides.

Electro-magnetic methods—airborne

Here, as with magnetic methods, a technique used originally only on the ground has become airborne with remarkable success. In the actual identification of heavy sulphide deposits in Canada, no method has had a greater impact on exploration policies and procedures than have airborne electro-magnetic techniques. First used in the late 1940s by the late Dr. Stanley Davidson, 'Airborne E-M' has branched out into many variations such as the two-frequency method, single frequency with in-phase and out-of-phase measurements, the 'In-put' method of Barringer, etc. No attempt will be made here to describe these variations, but all depend essentially on an airborne transmitter producing a primary electromagnetic field from which a secondary field is produced around conductive bodies. This secondary field is picked up by a receiver in the aircraft or towed in a 'bird' trailed behind. Units have been installed in single and twin-engined fixed-wing aircraft as well as in helicopters.

Most economical method

The most economical method in cost per line mile is probably the single-frequency in-phase/out-of-phase method installed in a relatively large single-engined aircraft, where the transmitting coil is installed on one wing-tip and the receiving coil on the other. Depth of effective penetration, while depending partly on the minimum size of the ore-body sought, is about 300ft from the transmitter which, in practice, means a maximum of 200ft from surface. The cost of flying, plus the processing, interpretation and presentation of data normally varies today between $14 and $25 (U.S.) per line of mile flown, depending on the nearest air base available and the size of the area covered in a single project. With twin-engine aircraft, giving a greater safety factor and capable of carrying equipment that may, in some cases, give greater depth penetration, the cost range would of necessity be higher.

The above, however, is only the beginning of the cost of the complete survey since the follow-up work on the ground, including identification of the airborne anomaly, geological, ground geophysical and sometimes geochemical studies, and drill testing usually costs more than the airborne study—sometimes five times as much or more. Some examples are given below.

Ground electromagnetic methods

Ground electromagnetic methods are normally used on a specific property, or in the ground follow-up of an airborne survey, rather than as an initial reconnaissance tool. It is probably the most essential of the follow-up tools in an airborne E-M survey since the conductor indicated by the airborne work must be confirmed and accurately placed before any drill testing can be undertaken. Again there are many variations of the method which will not be described here.

Induced polarisation

This method, which has not as yet been adapted to airborne use, employs the principle of exciting individual sulphide grains or crystals to produce secondary fields rather than that from a continuous conductive body, as is the case in the original electro-magnetic methods. Since it is applicable to a body containing quite widely disseminated sulphides (as little as 2 per cent by volume), it has been widely used in the search for deposits of the porphyry copper type. It is normally not a reconnaissance tool—at least it is rather expensive for such—but it is widely used in testing property-sized areas that have been selected on geological evidence, and also as a follow-up tool in an airborne magnetic or electro-magnetic survey.

Fig. 5. Inspecting core from a drill hole testing an anomaly on an Irish farm

Various other geophysical methods such as the measurement of spontaneous polarisation, gravity variations, and the use of seismic data have their uses in specific problems but could not be regarded, at present, as basic tools for 'grassroots' exploration.

Measurement of radioactivity

The geiger counter, followed later by the scintillometer and gamma-counter, have been the standard tools of the prospector or geologist searching for uranium since this metal became of economic importance. To the prospector for uranium the geiger counter, or light scintillometer, plays a part similar to the 'pan' to the gold prospector. To the geophysicist or geologist, it is a geophysical tool with which measurements may be made systematically on a grid of cut lines and the results contoured. Few, if any, of the uranium deposits now in production could have been found without such instruments.

Airborne scintillometers, until very recently, have been less successful. They were quite widely used in the uranium boom of the 1950s, but at that time separation of the radioactivity due respectively to uranium, thorium and potassium was not possible, with the result that broad potassium-rich areas tended to mask the more confined effects of uranium. Up until 1968 no exploitable deposits of uranium, so far as the author is aware, had been credited to airborne scintillometers. Late in 1968, however, what appears to be a deposit of some importance (although no details have been released at the time of writing) was intersected in a drill hole on the west side of Wollaston Lake in Northeastern Saskatchewan. It is understood that this discovery was directly due, in the first place, to an airborne scintillometer survey, using a GIS-2 which is a 'discriminating' unit, i.e. separates the radioactivity due to uranium from that due to thorium or potassium.

Over the last couple of years it has become increasingly common in Canada, and probably elsewhere, for radiometric equipment to be used as an ancillary to other airborne surveys whose primary target is base metal. The additional cost, on an overall square mile basis, is relatively minor.

The weakest point to date in any radioactivity-detecting instrument, ground or airborne, is its lack of penetration below rock or soil surface. The detection of radon gas escaping from a uranium source, while still in an experimental stage, may be a partial solution to this problem.

Programmes based on airborne E-M

It is not easy to separate, in importance or credit for discovery, the various elements involved in an airborne geophysical project. Normally magnetics are measured simultaneously with electromagnetics, and often radioactivity as well. Ground follow-up includes confirmation and positioning of the conductor by ground E-M and usually magnetometer, plus the mapping of any geology available on surface. This may be a difficult and expensive stage of the work in rough and inaccessible country. Finally, if ground confirmation is obtained, a diamond drill is brought in to find what is causing the anomaly and to provide samples if sulphide mineralisation is the cause.

This routine is particularly suitable to the Canadian Precambrian Shield and parts of the Appalachians. Glaciation has removed the oxidised zone that is such a handicap in many other countries, and has left a covering of till that severely limits the areas available for surface prospecting. Successes in Canada include the first orebodies of the Thompson nickel area in Manitoba, the Mattagami orebody in N. Quebec, the Poirier and New Hosco bodies in the same general area, the Kidd Township orebody north of Timmins, the Heath-Steele mine in New Brunswick (the larger Brunswick Mining and Smelting orebodies would have shown up well in an airborne E-M survey but had been discovered by drilling on surface indications before the airborne survey was carried out).

The method has not been as successful on the African or Australian continents, nor in the American continent south of Canada. This is natural in view of the much greater depth penetration needed to get below the zone of oxidation, particularly in the arid parts of Africa and Australia. It is also not very suitable in mountainous country, even where oxidation is not a factor, because of the impossibility of keeping the aircraft, and transmitter, at a constant altitude above ground.

Cost of airborne electro-magnetic programme

While the cost of the actual flying varies a certain amount, depending on equipment used, distance from base and total area to be flown, the variation in the *total* cost is much greater. The latter variation depends more than anything on the number of conductors that are considered to warrant ground checking and the accessibility for ground crews. The following four examples, with which the author was closely concerned, give some idea of this variability:

Case P-1

Locality and Conditions. Northern Ontario Precambrian. Rather scarce outcrops, generally level ground covered by moderately thick conifer forest and muskeg. Outcrop under 2 per cent of area. No roads. Ground crew access by helicopter.

Size of Area. 210 square miles (544 km^2).

Programme. Airborne E-M and magnetometer coverage at 1/6th mile intervals. Ground follow-up by geology and ground geophysics in vicinity of selected conductors only. No conductors drilled.

Cost. Total cost U.S. $132,000. Cost per square mile, U.S. $603 ($243/km^2).

Case P-2

Locality and Conditions. Northwest Ontario Precambrian. Rugged but differences in elevation under 300ft. High proportion of lakes. Outcrop about 10 per cent of land areas.

Size of Area. 100 square miles (259 km^2).

Programme. Airborne E-M and magnetometer at 1/8th mile intervals with ground follow-up by geology and geophysics on conductor areas. Diamond drilling on 16 conductors (total 17 holes), which is much higher than average for an area of this size, some holes drilled from lake ice in winter.

Cost. Total cost U.S.$112,000. Cost per square mile, U.S.$1,120 ($433/km^2).

Case P-3

Locality and Conditions. Northern Quebec Precambrian. Quite rugged country but under 600ft difference in elevation. Heavy bush, numerous lakes. Proportion of outcrop under 5 per cent. Road access in parts.

Size of Area. 800 square miles (2,072 km^2).

Programme. Airborne E-M and magnetometer at 1/6th mile intervals and ground follow-up of conductor areas by geology and geophysics. Three conductors drilled with total of nine drill holes. Number of conductors justifying investigation smaller than average for area of this size.

Cost. Total cost, U.S. $175,000. Cost per square mile, U.S. $220 ($84/km^2).

Case P-4

Locality and Conditions. Northern Quebec Precambrian. Generally level peneplane with few ridges. Outcrops less than 2 per cent. Covering of bush and muskeg on glacial till, etc. No roads. Fair availability of lakes and rivers for float plane landings.

Size of Area. 116 square miles (300 km^2).

Programme. Airborne E-M and magnetometer coverage at 1/8th mile intervals. Follow-up by geological mapping, ground geophysics and geochemistry followed by drill testing. Cost includes only the first six diamond drill holes on discovery later developed to copper-zinc producer (Poirier Mine).

Cost. Total cost of programme up to point of indicating a deposit of definite interest, but not including some later investigations of other anomalous areas: U.S. $111,000. Cost per square mile, $950 ($370 km^2).

Note 1: This programme as defined was completed in 1959 and today would probably cost about 30 per cent more.

Note 2: The breakdown of expenses, expressed as the nearest whole number percentage of the total, was: Airborne E-M (company-owned aircraft and equipment)—15 per cent, Staking and option payments—10 per cent, Line-cutting—6 per cent, Geology—10 per cent, Ground geophysics—9 per cent, Geochemistry—5 per cent, Diamond drilling (6 holes)—24 per cent, Assays, Laboratory and sundry expenses—6 per cent, Administration and overhead—15 per cent.

In all four above cases the flying was by a single-engine Otter aircraft equipped with single frequency, in-phase/out-of-phase E-M unit, magnetometer and, in Case P-3, scintillometer. The larger proportion of conductors could only be reached by float plane or helicopter.

Geochemistry

This, like its cousin geophysics, is becoming yearly more specialised and more complex. In its use in 'grass-roots' work, the methods most used can be divided conveniently into stream sampling and soil sampling. The former is more commonly used as a first approach when dealing with a large area. The measurement of metal content in water of streams is used on some occasions but, in general, due to the variations consequent on changes in precipitation, the samples most regularly used in stream sampling are those of silt at the edge of a stream, i.e. deposited during flood conditions, or from the bottom sediment. In dealing with stream sediments, the metal content we are measuring is mainly that which has been in solution and has been adsorbed on the surface of clay particles. Depending on the particular conditions these values may be augmented by actual grains of metalliferous minerals carried mechanically down the stream. The proportion of any metal that has been adsorbed from solution on to the surface of clay particles, as compared with that contained in mechanical grains, gives some indication of the distance from the source of such values. For this reason a combination of hot and cold extraction has been used in many geochemical programmes over the past ten years.

The determination of metal content, regardless of the method of extraction, may be accomplished in a number of ways but, over the past few years, the atomic absorptiometer has been used to an increasing extent.

Soil sampling involves taking samples from just below the surface organic layer at regular intervals in lines that may be cut, measured and marked through the country but in other cases may be done in reconnaissance fashion using air-photographs for control. Soil sampling may replace stream sampling as a reconnaissance tool, particularly where there is no well-marked drainage pattern. The method is also used to check, and study in detail, an area indicated by stream sampling to be of interest. Hot extraction is normally used in soil samples, although in certain cases, where upward or sideways movement of metal values is suspected, cold extraction tests may also be of value.

The sampling of vegetation, e.g. the twigs or leaves from certain trees or bushes that show an affinity for some metals, has been used on an experimental scale. It is, however, not easy to control and supervise when using locally-trained samplers and the author has yet to see a case where an anomaly (of economic significance) shown by vegetation would not equally well be indicated by normal soil sampling.

In a general reconnaissance of an unknown area the three metals most commonly tested are copper, lead and zinc. Of the three zinc is the most mobile and lead the least. Many other metals give quite satisfactory anomalies in stream sediments or soils but, in most cases, they are associated with one or more of the above three metals and it is relatively unlikely that an anomaly of a metal other than these three would show up clearly without a corresponding anomaly from copper, lead or zinc. There are, however, exceptions and more may be shown by statistical studies in progress. Analysing for mercury has recently been found to be of considerable value in searching for certain types of ore because it tends to give a broader aureole around the source and is, therefore, detectable by wider-spaced sample lines. Ores containing sulph-arsenides and antimonides are more likely to be amenable to mercury geochemistry.

The ideal situation, theoretically, for the use of geochemistry as a primary tool in the search for ore is one where a residual soil covers the rocks from which it is derived without subsequent disturbance by glacial action, sharp changes of climate or cover by later alluvium. Thus the Precambrian areas of Canada and Scandinavia, although geologically favourable and already productive, are not particularly suitable for geochemical attack by reason of the intense glacial scouring and covering by till not derived from the nearby underlying rocks. Mainly for this reason geochemistry has, so far, been credited as the primary tool in the discovery of only one producing mine (Mount Newman) in Canada. However, refinement of geochemical techniques is being developed along two lines of particular use in Canada:

1. Sampling of glacial till in an effort to pick 'trains' of values that may be traced back, on the reverse of glacial flow direction, to their source in the same way as the Scandinavians have successfully traced the source of boulders.

2. Efforts to find how far metal values released since glaciation may penetrate a layer of till.

Australia and Africa present difficulties of a different sort and, considering that the covering of rock on both continents is largely residual soil, geochemical methods have not been as successful as might have been hoped. The problem on both continents is mainly that of drastic changes in climate. In Australia a tropical, moist climate during Tertiary times resulted in deep oxidation followed by an arid climate that produced a caliche soil, with the result that metal values, with some exceptions, have been leached out of surface material. In Africa an alternation of extreme pluvial and arid periods during the one million years of Pleistocene history has also resulted in much leaching and distribution of surface metal values.

The last two paragraphs merely touch on the complexities of climate history and soil development with which geochemistry, in its application to prospecting, has to contend. As an exploration tool it is in a very early stage of development and we may expect to see rapid advances in the next decade. The geochemical laboratory at Imperial College, London, under Dr. J. Webb, is probably the most advanced, along the lines discussed, of any similar organisation in the Western World and we may look with confidence for improved techniques both in field practice and computation of measurements, resulting from experimental work now in progress.

Fig. 6. Geochemical sampling in jungle on the slopes of the Andes

The country where geochemical methods (as the primary exploration technique) have shown by far the greatest success, relative to other methods, is the Irish Republic. Out of five discoveries, three of which have resulted in current production and two of unproved viability, four were definitely due to geochemistry in the first place. The fifth (Mogul Mines at Silvermines, Co. Tipperary) resulted from the application of both geochemistry and geophysics to the projection of a geological formation from the locus of former production. In the other cases, geophysics was a valuable tool in the better definition and interpretation of the anomaly, but the first attention was directed by geochemistry.

Geochemistry has evidently been employed with a good deal of success in Russia, particularly in the southern latitudes. The series of papers on this work that many geochemists were looking forward to hearing at the 1968 International Geological Congress in Prague was, unfortunately, curtailed by the Russian occupation of the country.

Projects based on geochemical reconnaissance

Below are given costs of three projects, in which geochemistry was the primary tool, carried out in three different parts of the world during the last six years:

Case C-1

Locality and Conditions. Tropical Andes. Treeless but volcanic-ash covered in higher altitudes. Jungle cover on lower slopes. Extremely rugged. Very limited road access, requiring road-building for follow-up work.

Size of Area. Concession approximately 1,000 square miles but 20 per cent eliminated due to heavy volcanic ash cover so net 800 square miles (2,072 km²).

Programme. Reconnaissance geology and stream sampling over 800 square miles, mainly jungle. Detailed stream and soil sampling of four anomalous areas and limited geophysical follow-up. (A drilling campaign carried out around an old mine has been excluded from the costs given below.)

Cost. Total cost, excluding expenditure on detailed work around old mine and allowing for some credit on used equipment, over a three-year period—U.S.-$350,000. Cost per square mile a (total of 800 square miles)—U.S.$435 ($168 km²).

Case C-2

Locality and Conditions. Central America. Tropical jungle except for some areas of savanna and open pine ridge. Few roads. Transport very difficult except in limited dry period.

Size of Area. Concession 2,020 square miles of which 500 were eliminated by preliminary geology. 1,300 square miles covered by reconnaissance geology and some prospecting. 800 square miles covered by stream sampling and more detailed prospecting.

Programme. Selection of 800 square miles by reconnaissance geology as above on which stream sampling carried out. Follow-up of certain areas by more detailed geochemistry and geological mapping to be completed.

Cost. Total cost of above programme and allowing an estimate for one season follow-up, U.S.$250,000. Cost per square mile (800 square miles), U.S.-$310 ($120 km²).

Case C-3

Locality and Conditions. Irish Republic. Twenty-four separated areas selected on geological grounds. Farmed country, some bog, little outcrop. Excellent road availability.

Size of Area. 24 areas of about 25 square miles each—600 square miles (1,554 km²).

Programme. Stream (or flowing ditch) sampling at rate of approximately 2·6 samples per square mile. Detailed geology and soil sampling in anomalous areas followed in some cases by induced polarisation and test drilling. The I.P. survey was critical in pin-pointing, within the broader area of the copper soil anomaly, the deposit that later became a producer[2].

Cost. Total cost up to the point of a discovery (Gortdrum Mine) including only the first six drill holes—U.S.-$75,000. Cost per square mile—U.S.-$125 ($48 km².)

Note: This programme, as defined, was completed nearly five years ago and would cost approximately 30 per cent more if carried out today.

It may be seen from the above three examples that there is a very wide variation in overall cost per square mile—Case C-1 being about three times Case C-3. This difference is largely a matter of accessibility for the geological and geochemical crews in the initial, and still more the follow-up, work. If Case C-3 had been a single area, as in the other two, the difference in cost per square mile would have been still greater. To date the two more expensive projects have not resulted in a commercial ore body and the least expensive has—but this is a matter of change, and perhaps Irish luck.

Before leaving this brief review of the use of geochemistry, mention should be made of its use as a secondary tool in the assessment of anomalies obtained in a geophysical survey. In the case of an airborne electro-magnetic survey in a country like Canada, where an indicated anomaly may be under a heavy covering of glacial till, a geochemical survey of the soil may have little significance, but samples taken by auger or light drill of the overburden at regular intervals between surface and bedrock may be of considerable value. In fact the method has been used with some success as a primary exploration tool in searching for new veins in the vicinity of known ore deposits by United Keno Hill Mines in the Mayo district, Yukon. As a check on geophysical anomalies the method is still in an experimental stage. It is doubtful if anyone would discard an otherwise promising electro-magnetic anomaly because geochemical testing was negative but, where one has a large number of conductors resulting from an airborne survey, they might be screened and given priorities by such geochemical testing at a lower cost than by conventional diamond drilling.

As in so many cases in geophysical or geochemical surveys once one has a target indicated, the value and cost of confirmatory surveys must be balanced against the likelihood that diamond drilling will thereby be saved, and the cost of such abortive drilling. In some cases a confirmatory geochemical survey, if some type of drill is required, may cost nearly as much to transport and carry out as would a diamond drill hole that is quite likely to be eventually required anyway. Nevertheless, the use of geochemistry as a secondary tool, even in glaciated areas, is likely to increase as methods of its application are further refined.

Conclusions

1. If we hope to find sufficient new orebodies to replace those that are presently being mined we must depend increasingly on the use of existing and improved geophysical and geochemical techniques.

2. The selection of the area, on the basis of geological knowledge and reasoning, is still the most important single step in the road towards mineral discovery in a previously unproved area.

3. The 'choice of weapons', i.e. the geophysical or geochemical techniques, must vary with the climate and topography of the area chosen as well as on geological environment and type of orebody sought. Combinations of methods are often desirable, e.g. airborne E-M and magnetometer - ground E-M follow-up (possibly geochemical sampling or soil or overburden for anomaly grading); geochemical stream-sampling—detail soil sampling—induced polarisation.

Almost all lead to testing by diamond drilling or, in some cases in dry areas, non-coring drilling.

4. Examples are quoted showing costs varying between $125 and $1,120 per square mile in areas of between 100 and 800 square miles. The greater part of the wide variation in costs per unit area is due to the size of the area and to conditions of terrain, weather, transportation and the number of anomalies investigated rather than to the costs of the particular methods employed.

Acknowledgements

The author would like to acknowledge assistance in the preparation of this paper by Mr. Brian Middleton and Mr. Ian Thompson, both of whom are associated with him in consulting practice and who participated in most of the examples given of exploration projects. He would also like to thank Dr. E. L. Evans, Chief Geologist of Rio Algom Mines Ltd., for supplying some figures and comments on one of the airborne electro-magnetic projects on which he and the author collaborated some years ago.

References

[1] Derry, D. R. Contribution to Symposium 'Man and Minerals'. Can. Min. & Met., Feb. 1968.

[2] Thompson, I. S. 'The Discovery of the Gortdrum Deposit, Co. Tipperary, Ireland'. Can. Min. & Met., April 1967.

Fig. 7. Use of mules by a prospecting party on a jungle trail on the slopes of the Andes

2

Copyright ©1978 by Mining Journal Ltd.
Reprinted from *Mining Mag.* **129:**430-433 (1978)

MINERAL DISCOVERIES AND EXPLORATION EXPENDITURE— A REVISED REVIEW 1966-1976

Duncan R. Derry and James K. B. Booth

THIS paper was presented, as an invited contribution, to the session on "Exploration Economics" organised by the Canadian Institute of Mining and Metallurgy and held at its Annual Meeting in April 1977 in Ottawa.

Figures previously presented by one of the co-authors, and published (1), are updated and additions include some figures and graphs comparing exploration expenditure and discovery rate with metal price trends, non-ferrous metal production and items of Federal or Provincial legislation in Canada.

A necessary change from the previous presentation is in the definition of "discovery" in view of the rather unusual ratio today between some metal prices and the capital cost of achieving production and the interest on such capital. This is explained below.

Discovery rate and method

In Table 1 are listed the mineral discoveries made from the year 1965 to 1976. In each case the principal metal or mineral is indicated and a symbol shows the principal technique responsible for the discovery. The guidelines and definitions used in this list are as follows:

(1) Only non-ferrous metallic deposits and asbestos are included. Iron is excluded since most production decisions result from changes in demand and transport facilities rather than by new discoveries. The same applies in most cases to non-metallic minerals such as evaporites or phosphates. Asbestos, on the other hand, has more in common with base metals in occurrence and in the methods of exploring for it. Similarly carbonatites, whether eventually mined for niobium, apatite or other minerals, usually result from specific discoveries.

Definition of "discovery"

In the previous study (*C.I.M. Transactions*, Volume LXXIII, 1970) the definition used was "material of sufficient grade and width to lead to production without deliberate delay. For example, a deposit that was shelved for the future is not included". Under current conditions (1977), when the relation of the metal prices to availability and cost of capital is unusually unfavourable, it seems essential to use a less rigorously limited definition since a deposit of, for example, copper that would normally be placed in production without delay cannot be brought to production under today's conditions. Accordingly the amended definition for a "discovery" as used in this study is "A deposit of sufficient grade and tonnage to justify being brought to production under normal or average conditions of metal prices and interest on borrowed capital".

Date of discovery

As before, the most specific date, and the one used in these tables, is that on which ore was first exposed (usually in a drill hole) of sufficient grade and thickness to lead to production if sufficient work showed sufficient extent. It is *not* the date, for example, of an airborne geophysical survey or a geochemical programme that might have been responsible for the drilling programme, nor is it the date when a commercial orebody was confirmed.

Method of discovery

This is often a difficult decision and leads to considerable comment and criticism from readers of such articles as this. Rather few orebodies are now found by only one method or by one person. If, however, we attempt to give credit to every method involved in an individual discovery we have no basis for classification. Therefore the method that played the most crucial part in the discovery, according to the best written

Table 1 Discoveries 1965-1976

1965	1966	1967	1968	1969	1970
(1) Husky (Keno) Ag, Pb, Zn—GC (2) Anvil Pb, Zn—CV (3) Agnew Lake U—GL	(1) Madeleine Cu—GL (2) Brenda Cu, Mo—GL (3) Pyramid Pb, Zn—GP (4) Jameland Cu—GP (5) Hab (Eldorado) U—CV	(1) Newmont-Ingerbelle Cu—GL (2) Utah's Island Copper Cu—CV	(1) Louvem (Soquem) Cu, Zn—GP (2) South Bay (Uchi) Cu, Zn, Ag—GP (3) Valley Copper Cu—GP (4) Rabbit Lake U—RA (5) Sam Goosly Ag, Au—GC(?)	(1) Ruttan Lake Cu, Zn—GP (2) Mattabi Cu, Zn—GP (3) Centennial Cu-Zn—GP (4) Gibraltar Cu—GL	(1) Sturgeon Lake Cu, Zn—GP (2) Umex-Thierry Cu, Zn—GP (3) Cluff Lake "D" Deposit U—RA (4) Cluff Lake "NRF" Deposit U—RA
1971	**1972**	**1973**	**1974**	**1975**	**1976**
(1) Lyon Lake Cu, Zn—GP (2) J.A. (Bethlehem) Cu—GL(?) (3) Cluff Lake (Claude Lake Deposit) U—RA (4) Ming (Rambler) Cu, Au—GP (5) Afton Cu—GL (6) Canada Tungsten "E" Zone W—GL (7) Lessard Cu—GP	(1) Magusi River Zn, Cu—GP (2) Hebecourt Syndicate Cu—GP (3) Howard Pass Zn, Pb—GC (4) DEF/Minto Cu—GC (5) Creek Zone (Sturgeon L.) Cu, Zn—GP	(1) Cuvier Pb, Zn—GC (2) Lemoine Cu, Zn, Ag, Au—GP (3) Grasset Lake Cu, Zn—? (4) Grum Zn, Pb—GP (5) Reed Lake, Man. Cu, Zn—GP (6) West Arm Cu—GP	(1) Corbet Cu, Zn—GL (2) Brouillon A-1 Zn, Ag, Cu—GP (3) Izok Lake Cu, Zn—GP (4) Hood River Cu, Zn—? (5) Amoco (Burnt Bush) Au—GP	(1) Brouillon B Cu, Au—GP (2) Key Lake (Gartner Orebody), Sask. U—RA:GL (3) Goldstream Cu, Zn—CV	(1) Montcalm (Teck) Ni, Cu—GP (2) X-25 Orebody Pine Point—GL (3) Key Lake—Deilmann Orebody —RA: GL

ABBREVIATIONS: CV —Conventional Prospecting
GL —Geological
GP —Geophysical
GC —Geochemical
RA —Radioactive

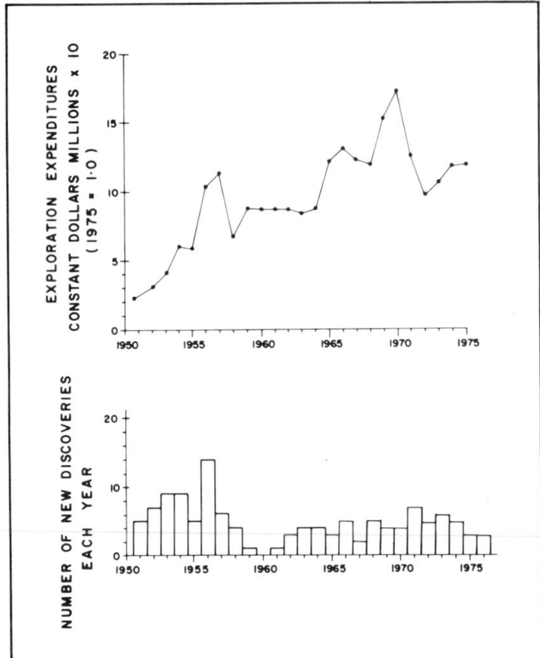

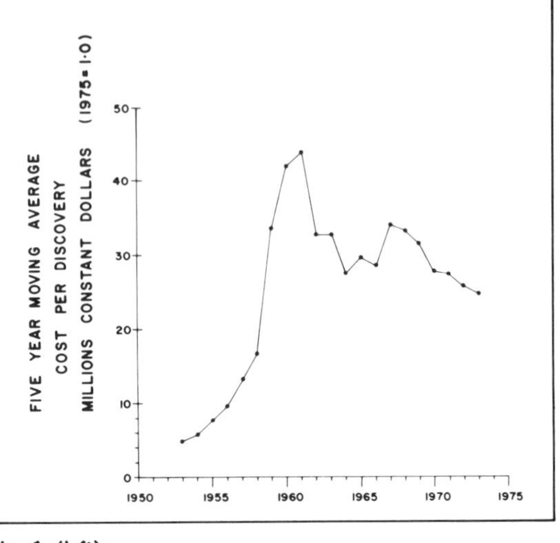

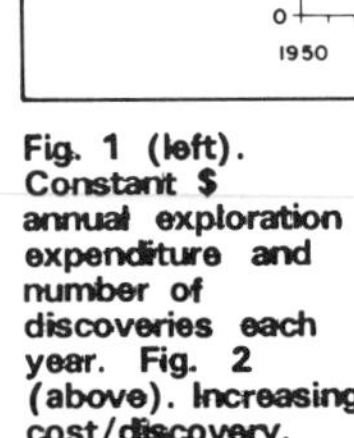
Fig. 1 (left). Constant $ annual exploration expenditure and number of discoveries each year. Fig. 2 (above). Increasing cost/discovery.

or verbal advice available, is the one listed.

Comments on discoveries between 1971 and 1976

This period covers discoveries made since the previous review referred to above. Discoveries are listed in Table 1 which also includes, for comparison, those made in the previous six years 1965 to 1970.

An obvious unsatisfactory aspect of a tabulation of this sort is the wide variation in tonnage and eventual value of "discoveries" each of which is listed as one unit. This disadvantage is, however, unavoidable without undertaking a very complex study involving weighting each discovery by its assumed future production value. In numbers of discoveries in this six-year period it may be seen that 1971 had the highest rate with seven and the lowest was shared equally by 1975 and 1976 with three each.

Three of the discoveries in this six-year period were uranium and made within the Province of Saskatchewan near the edge of the Athabasca Sandstone Basin or in the Cluff Lake structure within it. The Rabbit Lake discovery had already been made in 1968 and two of the Cluff Lake bodies in 1970. In all cases the initial indications, beyond the generally favourable geological environment that instigated the search, was an airborne radioactive anomaly but in no case was this directly over the orebody. The eventual discovery in each case was by drilling on a target selected on the basis of a study of the glacial and bedrock geology and the aid of scintillometry and uranium or radon geochemistry. This has been marked "GL" in the case of Cluff Lake bodies and "RA-GL" at Key Lake but the sequence of events seems to have been similar in both cases.

Non-ferrous base metal discoveries in this six-year period amounted to 25. In sixteen of these the credit for the technique playing the most critical part is given to geophysics, in the majority of cases airborne electromagnetic surveys. Geochemistry was given the chief credit for three of the discoveries and the application of geological reasoning for four. It may be significant that the role played by geochemistry and geological reasoning in new discoveries has increased since the previous period probably relating to the increasing interest in lower grades and greater disseminations of mineralisation.

Factors affecting exploration expenditure and discovery rate

The fact that there were only three clearly-established discoveries (other indications of mineralisation found may subsequently be upgraded to "discoveries") in each of the years 1975 and 1976 is an obvious cause for concern. In attempting to find the reasons for, and possible ways of reversing, this trend it is of interest to compare the following factors and trends:

(1) The Discovery Rate compared with Exploration Expenditure in Canada that was applied to the types of deposits included in this review. This leads to the theoretical average cost per discovery for each year.

(2) The trend of Exploration Expenditures related to the value of production of the same metals and minerals for each year in the period.

(3) The trend of Exploration Expenditures compared with metal prices over the same period.

(4) The effect on Exploration Expenditures, and hence on Discovery Rate, of political actions by Federal or Provincial Governments.

The figures used for Exploration Expenditures and Value of Production of the corresponding minerals are derived from Statistics Canada. *All such figures have been adjusted to constant (1975) dollars using the Consumer Price Index.*

The graph in Figure 1 shows the constant dollar annual exploration expenditure related to the number of discoveries in each year. There is a rough correspondence in the peaks, and in the downturn in the last few years. The increased cost of finding an orebody is seen in the increasing divergence of the Exploration Expenditure curve from the relatively level or declining discovery rate. This increasing cost per discovery, with a later reversal which is commented on below, is seen in Figure 2.

Figure 3 shows the Annual Canadian Non-Ferrous Metal production, in constant dollars, from 1951 to 1975, with the new discovery record for the same period plotted below. This brings out, with depressing clarity, the inadequecy of our exploration effort if Canada is to maintain its position as a metal producer in a world context.

In Figure 4 the 1951-1975 prices of four base metals (1975 dollars) have

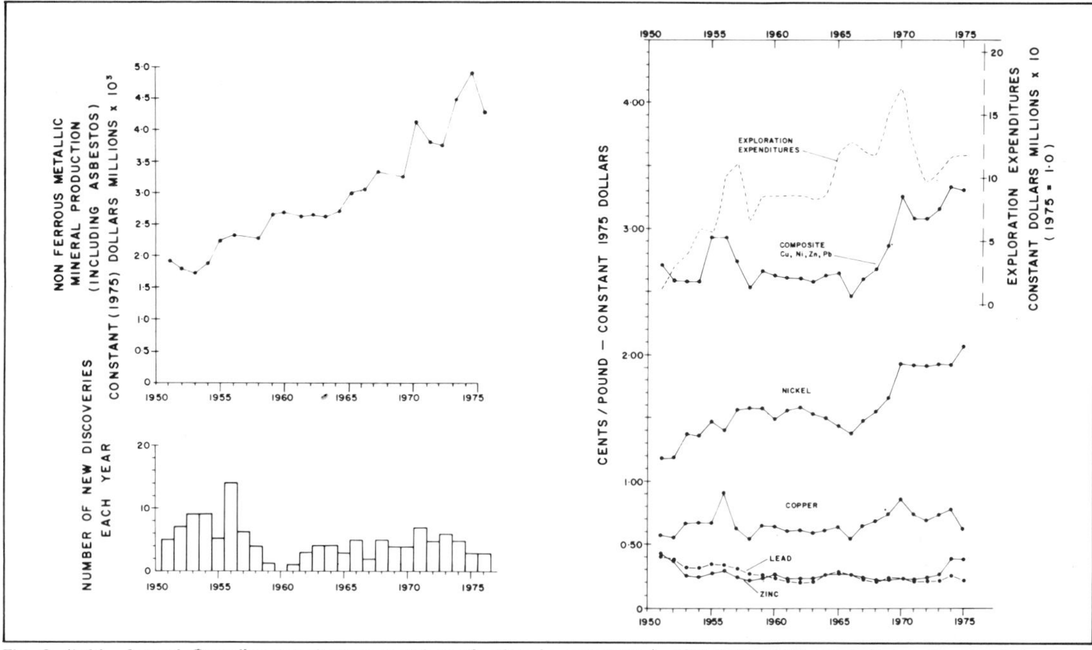

Fig. 3 (left). Annual Canadian non-ferrous metal production in constant $, 1951-1975, with new discovery record for same period. Fig. 4 (right). 1951-1975 prices of four base metals (1975 $) plotted separately and accumulatively. Also shown are exploration expenditures in same period.

been plotted separately and accumulatively. On the same chart is plotted Exploration Expenditures for the same period.

Discussion of graphs

A study of the graphs in Figures 1 to 5 (inclusive) illustrates the following trends:

(1) Annual Canadian Non-Ferrous Metallic Mineral Production from 1951 to 1975, in terms of constant (1975) dollars, increased from $1.9 billion to about $4.5 billion, an increase of 137%.

During the same period, composite metal prices (Figure 4) have only increased from 270c to 330c representing a corresponding increase of only 22% which is almost entirely due to nickel; prices for zinc, lead and copper have remained relatively flat, on the average, in real terms. To look at it another way, the increase in the composite metal price is only 22% over this period, practically all due to nickel, whilst the value of the metal production, and hence consumption of the National inventory, has increased 137%.

Because only one-sixth of the value of metal production has been added by the increase in composite metal prices, the real growth in metal production and therefore inventory consumption has been 115%, representing an average annual rate of increase of 4.6%. Since the rate of discovery of new reserves is falling (Figure 3), the obvious conclusion is that we are drawing from inventory at a rapid rate and consequently suffering a net depletion in the National inventory.

(2) In this same 25 year period from 1951 to 1975, *in terms of* 1975 *dollars*, annual exploration expenditures increased from $22 million in 1951 to a peak of $173 million in 1970, an increase of nearly 700%, and to $120 million in 1975, representing an increase over 1951 of about 450% (Figure 1). During the same period, the discovery rate has declined slightly, meaning that mines are becoming much more expensive to find (Figure 2).

The average cost of finding a mine during each five-year period between 1950 and 1975 increased, *in terms of constant* 1975 *dollars*, from about $5 million per mine in the 1950-1955 period to a peak of around $40 million per mine in the early 1960's and has since declined to an average of about $25 million per mine in the first five years of the '70's. This recent decline is probably not the healthy sign it first appears to be. It reflects to a large degree the current decline in exploration expenditures which will mean a corresponding fall-off in discoveries in the immediately ensuing period since, as might be expected, there is a time lag of a year or two between exploration expenditures and the positive results from that expenditure.

(3) In summary, and viewing all values in terms of constant 1975 dollars, production of non-ferrous metals, and therefore consumption of non-ferrous reserves, more than doubled in the 25 years from 1951 to 1975 and is still rising. At the same time, non-ferrous base metal prices other than nickel remained relatively constant, while exploration expenditures increased at least 450% and the rate of discovery of new reserves showed a declining trend.

The conclusion is inescapable, and merely proves what the Canadian mining profession already knows, that costs are rising whilst prices remain relatively constant and National mineral reserves are being depleted faster than we are adding to them.

(4) Up to 1970 the rising cost of finding a mine and the net depletion of the national inventory was simply a reflection of increasing difficulty of finding new deposits under deeper soil and rock cover; this could have been overcome by technological advances and a corresponding increase in the national exploration budget together with the increases in metal prices which normally follow these trends in a free market. Since 1970, however, the problem has been compounded and the situation worsened by a spate of both Federal and Provincial legislation increasing taxes and at the same time adding difficulties and additional costs by the imposition of well-meaning but often poorly applied environmental and marketing controls. Figure 5 shows the inevitable effect of these negative factors in the drop in exploration expenditures

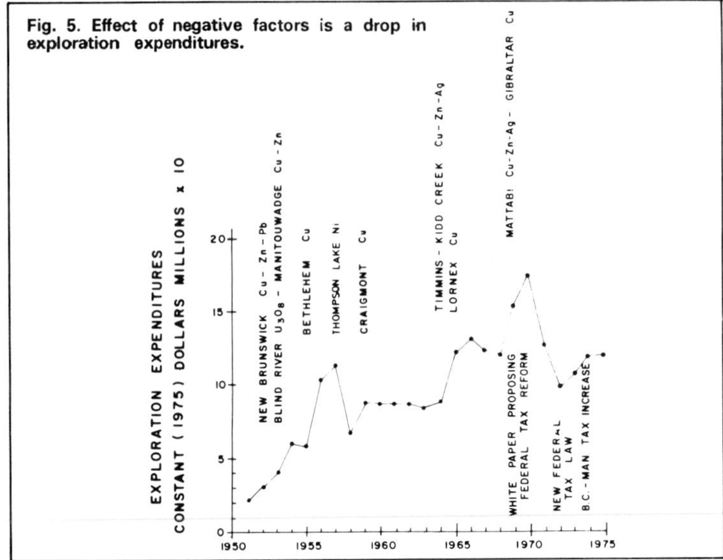

Fig. 5. Effect of negative factors is a drop in exploration expenditures.

which began with the uncertainties created by the initiation of Federal tax reforms in 1969 and the subsequent political harassment as a result of the struggles between the Federal and certain Provincial Governments over royalties and taxes to be imposed on the mining industry.

There has been an upturn in exploration expenditures since 1972, relatively modest in constant dollars, reflecting both the slightly firmer metal prices through 1973 and 1974 and the easing of the more punitive Provincial royalties and mining taxes in 1976. Unfortunately, these positive factors have been offset by depressed copper prices but, on the whole, there has been a slightly improving trend developing recently and it is to be hoped that this trend will continue.

(5) The evidence leaves little doubt, however, that Canada's inventory of ore reserves can only be maintained, and with it Canada's position as a major producer, if economic circumstances and Government policies combine to promote the very significant increases in exploration expenditures which will be needed merely to hold the national mine inventory position in the face of increasing technical difficulty in finding new mineral reserves.

At the risk of confusing an apparently simple conclusion, it must be added that a real increase in world metal prices, that is an increase beyond the normal rate of inflation, would also have the effect of replenishing the national mineral inventory by converting marginal reserves in the ground into ore-grade reserves.

References

Derry, D. R., 1968—Man and Minerals—Exploration. *Can. Min. & Met. Bull.*, February 1968.

Derry, D. R., 1970—Exploration Expenditure, Discovery Rate and Methods. *Can. Min. & Met. Bull.*, March 1970.

Derry, D. R., 1976—The Trend in Mining Exploration Expenditures in Canada, *Can. Min. & Met. Bull.*, August 1976.

3

Reprinted from *Eng. and Mining Jour.* **178:**71-73 (1977)

DISCOVERY RATE AND EXPLORATION METHODS FOR METALLIC MINERAL DEPOSITS IN THE U.S., 1940-76

John P. Albers
Geologist, US Geological Survey

IN 1967, A. H. LANG of the Geological Survey of Canada published a report[1] on discovery methods for mines that began production in Canada after 1955, and in 1970, Duncan R. Derry of Toronto contributed a paper[2] on the rate of new mineral discoveries in Canada, the methods used, and the annual exploration expenditures from 1950 to 1969. The report here presents some of the same type of information for the US during the interval 1940-1976. However, only mines having an annual production of 150,000 tpy or more are included here, and there is no detailed information on annual company exploration expenditures in the US.

The basic data in the report were derived from four principal sources: 1) published literature, 2) US Geological Survey commodity geologists, 3) state geologists, and 4) officials of mining companies. Although the data are thought to be generally reliable, there is a strong likelihood that they are incomplete. There is also a possibility of error because, for some of the mines listed, the discovery date and/or principal exploration method that led to discovery are based on the recollection of an individual having intimate knowledge of a property, rather than on published documentation. Nevertheless, the report gives a reasonably accurate summary that will be useful for educational purposes and possibly also as an aid in planning research and exploration programs by government organizations, companies, or private individuals.

Of the 65 mining properties listed in Table 1, 50 were producing in 1976. Fifteen deposits, mostly large porphyry copper, had not yet been brought into production at the end of 1975. An additional unknown, but small, number of deposits had been discovered by the end of 1975 but were not yet in production. Data are insufficient to permit including this last category of deposits in the discussion.

Of the 50 mines in Table 1 known to be producing in mid-1976, each has a capacity of at least 150,000 tpy of ore. Important deposits such as Ajo, Bingham, Anaconda Vein, and Morenci are not included because they were discovered prior to 1940. The 65 properties listed are chiefly base metal deposits, but they include also major uranium deposits, precious metals, and a few other types. Phosphate, potash, and most iron and titanium deposits are not included because the basis for their exploration and discovery generally depends on technological and economic factors different from those controlling the discovery of concealed base metal and precious metal deposits.

Some discoveries are rediscoveries

Because the minerals search has in the recent past concentrated to an ever-increasing degree on finding concealed orebodies, only mines discovered since 1940 have been listed. However, data were compiled for active mines back to the mid-19th century.

The older data show a period of intense prospecting from the Civil War to around 1890, during which such famous deposits and districts as Butte, Bingham, Homestake, Coeur d'Alene, and Ajo were discovered. Many other deposits also were found during this period by prospectors, who worked the rich surficial oxides for precious metals, only to abandon the deposits when primary sulphide base metal ores were uncovered.

Many of these old, partially worked deposits have in recent years become targets for sophisticated exploration by mining companies, and some have proven to be the surface manifestations of economically important porphyry copper or other types of deposits. Consequently, some of the discovery dates shown in Tables 1 and 2 are, in a sense, "rediscovery" dates for some deposits, rather than the date of original discovery by the prospector of earlier days. However, the interval between discovery and first production for such new mines can be determined meaningfully only if the rediscovery dates are considered to be the time of discovery.

Four methods of discovery are identified in the tables: conventional prospecting, geological inference, geophysical anomaly, and geochemical anomaly. Of the 65 discoveries in Table 1, 46 were made by geologic inference, seven by geophysical anomaly, two by geochemical anomaly, four by conventional prospecting, and six by a combination of methods in which each method played a role of equal importance in discovery.

As Derry points out, identifying the principal method of discovery is often difficult because few ore deposits are found now by only one method or by only one person. Obviously, geologic inference must be a factor in a very high percentage of discoveries. But if credit is given to every method involved in a discovery, there is no basis for classification. Therefore, the method that played the most important part in the discovery, according to the best advice available, is the one listed in the tables.

Years elapsed between discovery and production of 50 metal mines in US

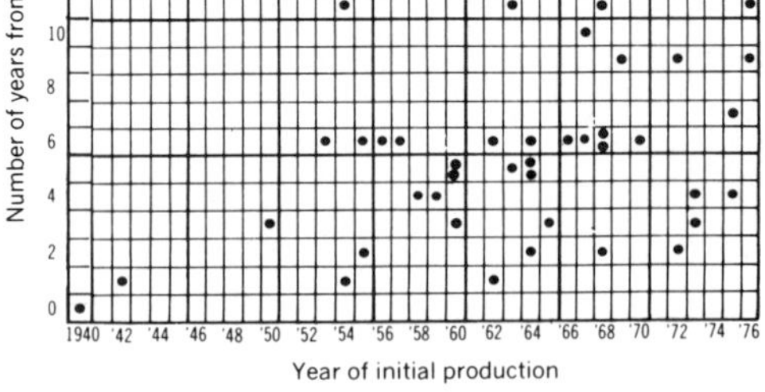

Number of years from discovery to production

0 2 4 6 8 10 12 14 16 18 20 22 24

1940 '42 '44 '46 '48 '50 '52 '54 '56 '58 '60 '62 '64 '66 '68 '70 '72 '74 '76

Year of initial production

Table 1—Discovery and production data for 50 US metal mines, 1940 through 1976[1]

Property	Metal(s) or mineral(s) produced	Date of discovery or rediscovery[2]	Date of first production	Method of discovery*	Years between discovery and first production	Type of operation[3]	Size of operation[4]
Carlin, Nevada	Au	1962	1965	O	3	P	C
Cortez, Nevada	Au	1966	1968	■	2	P	B
Battle Mountain, Nevada	Cu, Ag	1961	1967	O §	6	P	B
Creta, Oklahoma	Cu	1962	1964	O	2	P	E
Lakeshore, Arizona	Cu	1968	1975	O	7	U	A
Mineral Park, Arizona	Cu	1959	1964	■	5	P	A
Nacimiento, New Mexico	Cu	1970	1972	O	2	P	B
Pima, Arizona (Mission)?	Cu, Ag	1951	1957	O	6	P	A
Pinto Valley, Arizona	Cu	1971	1975	O	4	P	B
Twin Buttes, Arizona	Cu, Ag	1957	1968	O	11	P	A
Tyrone, New Mexico	Cu, Ag	1950	1973	O	23	P	A
Lower OxHide, Arizona	Cu	1967	1976	O	9	P	A
San Xavier, Arizona	Cu	1955	1969	O §	14	P	C
Yerington, Nevada	Cu	1945	1961	O	16	P	A
Questa, New Mexico	Mo	1960	1966	O	6	P	A
Henderson, Colorado	Mo	1965	1976	O	11	U	A
Continental, New Mexico	Cu, Zn, Ag	1970	1973	O	3	U	B
Esperanza, Arizona	Cu, Mo	1955	1959	O	4	P	A
San Manuel, Arizona	Cu, Mo, Zn, Ag	1943	1956	O	13	U	A
Sierrita, Arizona	Cu, Mo	1960	1973	X	13	P	A
Copper Cities, Arizona	Cu	1943	1954	O■	11	P	A
Sacaton, Arizona	Cu	1961	1975	O	14	P	B
Lucky Friday, Idaho	Pb, Zn, Ag	1940	1940	X	0	U	E
Blue Hill, Maine	Pb, Zn, Cu	1963	1972	O §	9	U	E
Viburnum, Missouri	Pb, Zn, Cu, Ag	1955	1960	O	5	U	B
Brushy Creek, Missouri	Pb, Zn	1960	1973	O	13	U	B
Buick, Missouri	Pb, Zn, Ag	1960	1969	O	9	U	B
Fletcher, Missouri	Pb, Zn	1958	1964	O	6	U	B
Indian Creek Mines, Missouri	Pb, Zn, Ag	1947	1953	O	6	U	C
Bergen, Utah	Pb, Zn	1950	1963	O	13	U	E
Ozark, Missouri	Pb, Zn, Ag	1962	1968	O	6	U	B
Friedensville, Pennsylvania	Zn	1954	1958	O	4	U	D
Immel, Tennessee	Zn	1954	1968	O	14	U	E
Jefferson City, Tennessee	Zn	1950	1956	O	6	U	D
New Market, Tennessee	Zn	1952	1963	O	11	U	C
Young, Tennessee	Zn	1959	1964	O	5	U	E
Magmont, Missouri	Zn, Pb, Cu, Ag	1962	1968	O	6	U	C
Schullsburg, Wisconsin	Zn, Pb	1947	1950	O	3	U	D
Sunnyside, Colorado	Zn, Pb, Au, Ag	1961	1962	O	1	U	E
McIntyre, New York	Ti	1941	1942	X	1	A	D
Dave Group, Sec. 10, Wyoming	U_3O_8	1955	1960	§	5	P	C
F Group, Wyoming	U_3O_8	1957	1960	O	3	P	B
Highland, Wyoming	U_3O_8	1969	1973	§	4	P	B
Humeca, Utah	U_3O_8	1964	1970	O	6	U	B
Lucky Mc, Wyoming	U_3O_8	1953	1954	§	1	P	B
Paguate, New Mexico	U_3O_8	1956	1962	O	6	P	A
Pay-Aljob, Wyoming	U_3O_8	1953	1955	§ X	2	P	C
Schwartzwalder, Colorado	U_3O_8	1949	1955	X	6	U	B
Sec. 1, T. 13N., R. 9W., New Mexico	U_3O_8	1957	1967	O	10	U	C
Walker Lease, Wyoming	U_3O_8	1958	1963	§	5	P	C
Partial list of new deposits discovered but not yet in production							
Golden Sunlight, Montana	Au	1946?	—	O			
Copper Creek, Arizona	Cu	1966	—	O			
Heddleston, Montana	Cu	1966	—	O			
Helvetia, Arizona	Cu	1970	—	O §			
Hillsborough, New Mexico	Cu	1975	—	O			
Flambeau, Wisconsin	Cu	1968	—	§			
Kalamazoo, Arizona	Cu	1965	—	O			
Poston Butte, Arizona	Cu	1970	—	O			
Red Mountain, Arizona	Cu	1970	—	O			
Safford (Kennecott), Arizona	Cu	1955	—	O			
Safford (Phelps Dodge), Arizona	Cu	1957	—	O			
Vekol, Arizona	Cu	1965	—	§			
Yerington Largo (Phelps Dodge), Nevada	Cu	1970	—	O			
Spar Lake, Montana	Cu, Zn	1962	—	O			
Punkin Hollow, Nevada	Fe	1961	—	§			

***Legend for method of discovery:**

O—Geologic inference
X—Conventional prospecting
§—Geophysical anomaly
■—Geochemical anomaly
O§—Combination of geologic inference and other anomaly(ies)

1) List of mines and their sizes is adapted in part from MINING MAGAZINE, September 1975, pp 188-189, supplemented by additional information from other sources. List includes concealed deposits and deposits whose surficial exposures are of no commercial value. 2) See text for explanation. 3) P—open pit; U—underground; A—alluvial.

4) Legend: A—over 3 million tpy (average 7 million tpy); B—1 million to 3 million tpy; C—500,000 to 1 million tpy; D—300,000 to 500,000 tpy; E—150,000 to 300,000 tpy. (Operations of less than 150,000 tpy are excluded.)

Data indicate reduced discovery rate since 1971

As Table 1 shows, of the 65 deposits considered in this review, 15* have not yet begun operations and are still in our mineral resource bank. While these undeveloped deposits provide a limited cushion of reserves (principally for the copper industry) for the short term, new discoveries must be made at a substantial rate if the cushion is to be maintained.

Data in Tables 1 and 2 suggest that the number of discoveries has dropped markedly since 1971. However, not all reported discoveries have been listed, because of inadequate or conflicting information on dates, and some discoveries have probably not been reported. Furthermore, the lower number of discoveries in the early 1970s could prove to be only a brief aberration in the long term picture. Any one of these factors or a combination of them might explain the apparent dearth of new discoveries. On the other hand, the phenomenon might result from a sharp cutback in exploration activity or from increasing scarcity of geologically favorable prospecting ground, with a consequent increase in the difficulty of finding additional deposits.

Brown[3] observes that statistics compiled by the Canadian Diamond Driller's Association show that exploratory drilling for new metal mines in Canada dropped from 6.2 million ft in 1970 to 4.4 million ft in 1974—a very substantial decline. According to Brown, this and other indications show that, under present conditions, the Canadian mining industry is not spending enough to achieve the discovery and development of the new resources that are necessary to maintain the relative importance of Canadian mineral production beyond the mid-1980s.

Data on annual exploratory drilling for mineral resources and on total annual exploration expenditures are not available for the US. However, if the downward trend observed by Brown for Canada is also applicable to the US, the indicated decline in new domestic discoveries could be the beginning of a long term trend, with unfavorable implications for the US in its ability to satisfy the ever-increasing need for mineral raw materials.

The interval between discovery and first production

The average time between discovery and first production for the 50 mines listed is about seven years. The accompanying graph shows that the time between discovery and production ranges up to 23 years.

Technological factors, economic factors, or a combination of these may affect the discovery-to-production time span for a particular deposit. Economic factors noted by Cranstone and Martin[4] include the following:

- The owners of the deposit may have deposits of higher grade or greater profitability that can supply all the ore they can possibly sell.

- The owners may not be able to raise capital or to afford the financial risk of bringing a new mine into production.

- The markets are not currently available.

- The owners prefer to await higher prices.

Some of the 15 deposits still not in production (Table 1) are in the category of nonproducers simply because they have been discovered too recently to have permitted development. At least five of the deposits have been discovered since 1970, and all but three of the remaining 10 were discovered during the 1960s. Whatever the reason that these properties have not yet been brought into production, the third column in Table 2 indicates that production starts have, in general, fluctuated in a fairly regular manner since the mid-1950s. □

*This is a minimal number, at least for porphyry copper deposits discovered since 1940 but not yet in production.

References

1) Lang, A. H., "Discovery methods of post-1955 new producers," CANADIAN MINING JOURNAL, Vol. 88 (1967), No. 1, pp 47-50.
2) Derry, D. R., "Exploration expenditure, discovery rate, and methods," CIM BULLETIN, Vol. 63 (1970), No. 695, pp 362-366.
3) Brown, R. D., "Taxation and our future resources," CIM BULLETIN, Vol. 69 (1976), No. 769, pp 124-126.
4) Cranstone, D. A., and H. L. Martin, "Are discovery costs increasing?," CANADIAN MINING JOURNAL, April 1973.
5) MINING MAGAZINE, "International mining survey," Vol. 133 (1975), No. 3, pp 185-221.

Table 2—Number of discoveries and production starts each year, 1940-1976

Year	Number of discoveries, by method*	Production starts
1940	X	1
1941	X	—
1942	—	—
1943	O ★	—
1944	—	—
1945	O	—
1946	O	—
1947	OO	—
1948	—	—
1949	X	—
1950	OOO	—
1951	O	—
1952	O	—
1953	§ ★	1
1954	OO	2
1955	★ OO § O	2
1956	O	2
1957	OOOO	1
1958	O §	1
1959	■ O	1
1960	OXOO	3
1961	★ OO §	1
1962	OOOOO	2
1963	★	3
1964	O	4
1965	OO §	1
1966	■ OO	1
1967	O	2
1968	O §	5
1969	§	2
1970	OO ★ OOO	1
1971	O	—
1972	—	2
1973	—	5
1974	—	—
1975	O	3
1976	—	2

*Legend:
O — Geologic inference
X — Conventional prospecting
§ — Geophysical anomaly
■ — Geochemical anomaly
★ — Combination of geologic inference and some type(s) of geophysical and/or geochemical anomaly

4

Reprinted from *Mining Eng.* **27**:32-35 (Dec. 1975)

The Ore Finders

Peter Joralemon

Man has always felt more comfortable breaking his life into categories or classifications. An otherwise bewildering life somehow becomes easier to bear when divided into neat measurable sections, each of which is fitted into its proper niche.

It seems likely that the world's first economic geologist, when faced with the world's first two ore bodies, started his career by classifying them. Early on, he divided his ore bodies into those born of water, of fire, and of air. In retrospect this threefold classification was probably about as useful as the more complex ones that followed.

This division of ore bodies probably did not help him in his search for new deposits, but it made him feel more secure in dealing with those already known.

As geology evolved, the various classifications of ore bodies were developed and expanded in an attempt to arrive at something that might help in the understanding of ore bodies. Some geologists preferred the simplistic approach of classifying ores by the economic materials contained—iron deposits, copper deposits, etc. While this approach left very little room for argument, it did not offer much of an intellectual challenge to the geologic fraternity.

A second school brought forth a descriptive classification based on the form of the deposit—vein, stockwork, etc. This was probably more helpful than the classification by metal, but it tended to emphasize that ore bodies are gradational and strongly resist such attempts at descriptive classification. A "vein" was originally conceived to be a simple tabular body, but Nature is seldom simple and this concept did not fit comfortably with the complex Coeur d'Alene vein systems or the massive tabular deposits such as the Mountain Pass rare earth body.

Then came what was the most useful and certainly the most generally accepted method: the genetic classification—fathered by Lindgren and dressed up by his successors Graton and Buddington. This system told, in a word, where and how the deposit had been formed and how it might be expected to behave at depth and laterally. Those of us brought up in this school found it generally useful and some of us became specialists in certain categories of ores. Trained in the West, the home of epithermal deposits, I focused my efforts on this type while my eastern Canadian neighbors were taking a hard look at deposits formed by magmatic segregation.

This system of classification probably did not find many new mines (although it could), but it helped immeasurably to understand the deposits already and accidentally discovered. Our faith in this system has unfortunately been shaken as younger geologists point out that our theories of ore formation are generally flawed and, in places, are entirely haywire. The old concept of mineralizing solutions rising toward the surface from a magmatic source, depositing their contents as they rose, was apparently too simple. Instead we must now visualize continents bobbing about on a semi-plastic sima "sea" with parts of the continents constantly melting and reconsolidating, all the while concentrating the sparse quantities of metals disseminated through the rocks into economic deposits. We must imagine great piles of submarine volcanics whose gaseous emanations form base metal deposits right there among the fishes. I find myself yearning for the simpler days of Lindgren and Graton and thinking that these old and obsolete hypotheses were after all not entirely unsuccessful as aids in ore finding and ore understanding.

The New Classification

It seems that a new system of classification of ore bodies needs to be introduced, a system that will not contradict the new theories of ore genesis and will still live comfortably with the older geologists. My classification, presented here, involves none of the problems of ambiguity met in earlier systems, nor is it dependent on any particular theory of ore genesis. On the contrary, it seems simple enough to be understood by the most obtuse of us and yet may serve some practical purpose in ore finding. My classification simply divides ores into the following nine categories, according to who is best qualified to discover them: (1) geologist's ore; (2) geophysicist's ore; (3) metallurgist's ore; (4) engineer's ore; (5) miner's ore; (6) assayer's ore; (7) lawyer's ore; (8) politician's ore; (9) prospector's (barroom) ore.

In these days of overspecialization it becomes increasingly difficult for a man of any given persuasion to work on categories of ore other than his own.

Geologist's Ore

Many ore discoveries attributed to geologists properly belong in another of my categories. Others have been explained by geologists only after discovery by men with other skills. Such hindsight discoveries have proven to be of only limited value.

Ideally, geologist's ore is discovered as follows: The geologist starts with a study of structural lineaments which will establish a two- or threefold intersection as the primary target for further study. From here, intensive geologic mapping follows, accompanied by geochemical sampling and as much microscopic work as will be allowed. Some trenching may be required in this phase, but drilling will under no circumstances be considered until all other tools have been exhausted. Our theoretical geologist's ore target will show a number of concentric shells of mineralization increasing in intensity to the core, which is finally selected as the drill target. I have found that in many instances this core of mineralization was visually apparent from the beginning and could have been drilled immediately, making unnecessary the prolonged academic study.

This theoretical case history of developing a drill target can frequently be stretched out into over a year's program before the moment of truth must be met in the drill core. Indeed, I have known some such programs that have become almost lifetime projects of the geologist, forever undisturbed by drilling. This classic approach to geologist's ore seldom bears fruit, despite its popularity.

From time to time an important discovery is made that can be clearly attributed to careful geologic work. The Kalamazoo extension of the San Manuel ore body beyond the great fault is a modern example of geologist's ore. Similar geologic work in the past led to the discovery of the United Verde extension ore body downfaulted from the eroded top of the United Verde deposit, and to the discovery of the deep downfaulted Ruby Hill ore at Eureka, Nev.

Although many geologists have made a stab at ore discovery through a study of regional structural lineaments, I know of no ore body that can honestly be attributed to this type of geologic reasoning.

The study of depth and mineralogic zoning as a tool for ore finding is perhaps also overvalued. I know of a possible porphyry copper occurrence that was turned down by two able geologists; one because erosion had clearly stripped the deposit down to its very roots, and the other, seeing the same evidence, because the ore body was still buried at a depth of over 5000 feet below the surface!

The more effective use of geology in ore finding is as an art rather than an exact science. Those who are sure that they know all about geology will find powerfully little ore.

In summary, geologist's ore at its most successful is an application of descriptive geometry, projecting known ore beyond faults or into different and more favorable formations. The more effective use of geology in ore finding is as an art rather than an exact science. I am afraid that those who are sure that we know all there is to know about geology will find powerfully little ore.

Geophysicist's Ore

Most geophysical "discoveries" are the products of geophysical surveys in or near already known ore occurrences. Geophysicist's ore is more common in Canada than in the United States (Kidd Creek is an outstanding example) because of the scarcity of outcrops in the northland or perhaps because our Canadian neighbors are more willing than we to take chances. Domestic examples of geophysicist's ore are found at Pumpkin Hollow near Yerington, Nev., at the cluster of blind copper deposits at Milford, Utah, and at Pima, Ariz.

It is well known that there are many more geophysical anomalies than there are ore bodies; and there are geologists who are somewhat beyond their depth in evaluating geophysical data and may tend to minimize the importance of geophysicist's ore. For maximum effectiveness, I believe that geophysical work should be evaluated by geophysicists assisted, where necessary, by geologists. If the geophysical anomaly is to be drilled, the geophysicist should certainly be allowed to have the first say in locating the holes. The geologist may be able to log the core.

The recommendation made by Dr. S. E. Ward several years ago that any serious geophysical work be preceded by the drilling of several test drill holes to allow the interpretation of geophysical results has been pretty much ignored by geologists. And yet I have seen so many cases where limited preliminary drilling could have saved tens of thousands of dollars in unnecessary geologic work.

Metallurgist's Ore

My third class of ore may be divided into two subcategories: ores of metals for which no use is presently known; and ores so refractory that satisfactory recovery techniques have not yet been developed. In the former category, the Mountain Pass rare earth deposit comes to mind. When I examined this great and rich deposit over 20 years ago, practically the sole use for rare earths was in cigarette lighter flints—for thorium, in gas lamp mantles. With such very limited consumption this deposit could scarcely be considered ore. But over the past 20 years many new uses of rare earths have been developed, and this deposit has now become one of the greatest ore bodies of its type in the world.

In the other subcategory belong such deposits as Round Mountain, Nev. This deposit had been known and thoroughly tested years ago, but was simply too low grade to respond to the then accepted methods of ore dressing. A short time ago, heap leaching of low-grade gold ores was successfully attempted at Cortez, and with this hitherto untested method the great Round Mountain stockwork deposit became ore. This surely must be classified as metallurgist's ore.

An enormous body of finely intergrown fluorspar, sphalerite, and beryl near Eureka, Nev., so far has defied metallurgical efforts to produce marketable products. When that is accomplished, that interesting deposit will also join the ranks of metallurgist's ore.

My metallurgist's ore classification emphasizes the fact that any unusual concentration of mineral or metal has a good chance of becoming valuable when metallurgists have worked long enough on the problem.

Engineer's Ore

Perhaps engineer's ore first came into prominence when the sharp pencil of D. C. Jackling flew through a huge array of figures to prove that low-grade material could be mined profitably if handled in sufficient volume. This was some 70 years ago, but the effects of Jackling's revolution are still felt everywhere.

It was in 1909 that John Greenway of Calumet and Arizona read of Jackling's work at Bingham and ordered a junior engineer, Ira Joralemon, to find and develop a similar ore body in Arizona. From this order evolved the New Cornelia ore body at Ajo.

Today it seems that many great companies elect to expand the flow of raw material by increasing the scale of existing operations rather than finding new ore bodies. In the short pull this approach may make monetary sense. In the long range, of course, it simply hastens the day when our supply of resources becomes exhausted.

Another type of engineer's ore is the deposit so remote from sources of supply, labor, power, and consumption that it had not been considered ore until engineers planned and put into being a practical system of transportation. Perhaps the most striking example of this type of engineer's ore is Freeport's great Gunung Bijih (Ertsberg) copper mine in Irian Jaya, Indonesia. Underground mining methods such as block caving or longwall mining have made possible profitable mining operations in cantankerous deep-seated deposits. These, too, are examples of engineer's ore.

Miner's Ore

An overwhelming majority of new ore discoveries, particularly in the United States, have simply been extensions or enlargements of ore bodies long known and long mined on a small scale. Such discoveries are of miner's ore.

Narrow, high-grade copper veins at Ajo were worked over 100 years ago with ores shipped by wagon, stern wheelers, and sailing ships to Swansea, Wales. Native copper was being produced at Santa Rita, N. M., by Indians over 200 years before the large open-pit operation began in 1910. The discovery of native silver in 1875 near Superior, Ariz., led to the later development of the Magma copper mine. The rich but complex gold-base metal ores at Oracle, Ariz., were mined for many years before the huge adjoining San Manuel deposit was recognized.

Gold was discovered in the Butte district in 1864 and this led to the discovery of bonanza silver values 10 years later, and even later to the development of the rich copper veins. The Berkeley pit, which began operations in 1955, was the direct outgrowth of prospecting for narrow gold veins.

Similarly, early silver and gold mines at Bingham, Park City, and Eureka, Utah, were the forerunners of the enormous base metal mines of later years. Darwin was born as a precious-metal camp, later was important as a lead-zinc producer, and still later as a tungsten district.

Profitable new mines can and often have grown out of old, small, high-grade mines. The smaller miner who started the original operation laid the groundwork for the later great discovery. This, then, is miner's ore, a very important category indeed. But despite the importance of this class, many exploration people act under strict orders *never* to look at an old, small mine.

Assayer's Ore

This class of ore has gained importance with the recent upward adjustment of the gold and silver prices. When gold was at $35 per oz and silver $0.90 per oz, samples of lode ore analyzed by fire assay gave perfectly reliable results. A variation of 0.01 oz Au was comparatively unimportant and it was generally recognized that the fire assay would give slightly lower results than actual smelter production. At that time, however, the analytical methods for determining the grade of a placer deposit were entirely different. Where the gold content could be as low as 0.01 oz per ton of gravel and still be commercial, it was agreed that the only satisfactory way to determine the actual gold content of the gravel was to measure the weight of the total recoverable gold in a known volume of gravel. "Assaying" of a placer deposit consisted of making a gravity concentrate from the gravel and recovering by amalgamation the total gold.

At the present or anticipated gold price it is possible that certain gold ores containing from 0.02 to 0.03 oz per ton may be considered commercial. In such cases the techniques used in placer testing will be more satisfactory than the long-accepted fire assay of a relatively small sample.

Although an enormous amount of geologic work went into the study of the disseminated copper deposit at Butte, I would be inclined to classify this great ore body as assayer's ore, since the determination of its potential value was based to such a large extent on the assaying of many carefully collected samples.

And finally, the discovery of certain large arsenical gold ore bodies not hitherto recognized by the systematic sampling of old mine dumps would place such deposits in the assayer's ore category.

Lawyer's Ore

Lawyer's ore, while not necessarily an entirely new discovery, can nevertheless be extremely profitable.

The discovery of a small accidental fraction of open ground in the center of an operating open-pit copper mine some years ago is well known. In the eyes of the law and those who profited by it, this constituted every bit as much of a valid discovery as a claim staked in virgin ground, and the negotiated value of the claim probably exceeded the value of most claim fractions staked by geologists.

From time to time, large blocks of privately held ground become open to exploration as a result of extensive negotiations. Portions of Indian reservations may be made available through the action of lawyers, or skillful negotiating may convince some of the electronic and space firms that hold so much land in Nevada to release the mineral rights. Ore found through such negotiations may also be classified as lawyer's ore. This is an outstanding example of ore stemming from skills outside the geologic persuasion.

Politician's Ore

I have established this category to cover ores found and developed in foreign countries whose investment climate has recently become or is about to become favorable. While it is the politician who makes this class of ore, it is generally discovered, developed, or promoted by that unusual and talented breed known as "international lawyer." Those I have met appear to know, on a first-name basis, everyone from the king or prime minister to the best moneylender and the most effective procurer. They are generally able and, remarkably often, right.

Some time ago I came across such a man who was promoting tin in Bolivia, armed with a treasure trove of reliable information. Although the political climate seemed a bit uncertain, my company asked me to have a look because of the persuasiveness of the "international lawyer." The properties looked promising and initial test work bore out our fondest hopes, but the company was dissuaded by the ubiquitous "cooler heads" and abandoned the project. Hindsight now shows that had we heeded our foreign advisor we would have been in production within a year, would have returned our investment in the first year of operation, and would have continued a profitable operation for perhaps three more years. At that time the operation would have been nationalized, but within three years would have been returned to us as operators under virtually the same conditions we had started with.

Politician's ore can be very rewarding, provided that the proper advisor is selected. But choosing the proper exploration man is always a vital part of a successful exploration program.

Prospector's Ore

I put this class of ore at the end of my list because this seems to be the position to which it has been relegated by most geologists. Since nearly all mines operating anywhere in the world today have been the outgrowth in some way of a prospector's discovery, I am inclined to put prospector's ore at the very head of the list and in so doing to accord the prospector the credit due him.

I read in the journals that the prospector is either a dying breed or has already gone to his reward. I am told by exploration geologists that there is nothing left for the prospector to find, even if there was one still alive. Even worse, I am told that those prospectors who may still be able to roam the hills have nothing to teach the geologist. I know of some geologists, mostly under 30, who so dread spending time with the prospector that they will furtively sneak out alone in the evening to examine the property just to avoid having to listen to him.

Now I admit to some puzzlement about "blowouts," or about the process by which the rocks enclosing the ore were "all burned up." But on the whole I have never spent time with a prospector who could not teach me something.

Perhaps prospectors have a poor reputation because, not having the benefit of formal training, they may tend to look for ore in places where it obviously could not occur. A good prospector operates with a sort of sixth sense, an inner dial that lights up when he sees a favorable ore showing. Perhaps this may be a form of ESP, but I like to think it is no more mysterious than a good imagination.

I would like to cite two examples of prospector's ore based on my own experience. In 1953, an untrained prospector put a Geiger counter on some outcrops in the Gas Hills of central Wyoming. Finding a considerable kick, the prospector, Neil McNiece, located some claims and then started his arduous effort to interest a company in his discovery. From his work was born the Lucky Mc Uranium Corp., and a group of Salt Lake City mining men agreed to back him to a limited extent. I was asked to serve as a sort of troubleshooter for this group.

In the words of Josh Billings, "Taint the not knowing that adds so much to the colossal sum of human ignorance as the knowing so much that ain't so."

This was the first, or nearly the first of the Wyoming deposits to be tested and our group was stumbling as we learned of the behaviour of this strange ore occurrence. Early on in our effort, the Atomic Energy Commission was attracted to the property and did some work of its own. After several months the AEC geologist, who might wish to remain anonymous, told me that having drilled the property, he had concluded (1) there would never be any commercial uranium ore developed outside the Colorado Plateau, (2) commercial uranium would never be found in any Tertiary formations, and (3) the Lucky Mc claims in particular would never have a profitable production. I frequently thank my stars that we elected to heed McNiece and his prospector's hunch rather than the AEC geologist and his education.

A few years later I had the opportunity to work with Spud Huestis, who had opened up what he optimistically thought was a porphyry copper deposit in southern British Columbia. Along with thousands of other geologists of my and earlier generations, I knew for a fact that the North American porphyry belt was, with the exception of Butte, restricted to the southwestern part of our country. British Columbia, on the other hand, was known to be the land of small high-grade deposits.

But Spud did not have the benefit of this knowledge, and in his ignorance of geologic principles proceeded to lay the groundwork for the development of one of the greatest prophyry belts in the world. Long after it had been shown that Bethlehem Copper could be properly classified as a porphyry copper deposit, many of the world's outstanding geologists were scorning British Columbia as a site for exploration.

While prospectors are pretty much ignored in the US, they play a very important role in Canada and other more progressive countries where they are granted a participation in any mines they may find. Our exploration people could learn a lot from these practices.

Before settling on the term "prospector's ore," I had considered introducing the term "barroom ore" in its place. My thought was that a geologist could make an in-depth geologic study of a district in order to determine possible exploration targets, or he could come to the nearest saloon and ask the resident prospectors a few intelligent questions. The latter process is generally faster, cheaper, and far more effective than the former.

Mineral Exploration Today

Here, then, are nine categories or classifications of ore, each requiring a different set of skills for discovery and development. It seems apparent that an exploration program guided by a dedicated geologist or an alert geophysicist or a keen metallurgist may well miss all the targets save his own category of ore. And the more inflexible he may be in his approach to ore finding, the more ore bodies he will miss. Perhaps the most valuable single attribute of a successful exploration man is a lively imagination. The past great ore discoveries have been made by men who look on ore finding as an art rather than an exact science. I can see no indication that this pattern has substantially changed.

I like the expression "explorationist," recently proposed by Kenyon Richard. Perhaps I can carry this a step further than Richard intended and consider a successful explorationist as a man who is, if not master of, at least familiar with the nine disciplines of ore finding. By using all skills available, blended with a good dose of imagination, I feel sure the cost of discovery of an important ore body can be greatly reduced from the present absurd estimate of $25 million for each discovery.

To get along with the business of exploration it may be well to keep in mind the deep philosophy of Josh Billings, who summed it all up in his greatest lesson: "Taint the not knowing that adds so much to the colossal sum of human ignorance as the knowing so much that ain't so."

5

Reprinted from pages 836–841, 844–845, and 846–847 of *Econ. Geology*
71:836–847 (1976)

Corporations, Ore Discovery, and the Geologist

LEO J. MILLER

Abstract

Multinational mining corporations produce about 27 billion dollars of mined products annually. Approximately 17 billion dollars is produced by mining corporations with annual mineral sales above 100 million dollars. Their ore deposits are, in most cases, either discovered or acquired as raw prospects by economic geologists within the corporate exploration department. An additional 5 billion dollars of annual mined products in the Western World are controlled by national government organizations. Approximately 3.5 billion dollars of governmental mined products were previously mined by multinational corporations. Therefore, the multinational corporation is the primary producer of the Western World's raw mineral materials, and the fundamental discoverer and/or acquirer of the ore deposits is the economic geologist within the mining corporation.

The structure of mining corporations varies from a group system with a parent house to complete ownership of mines by one corporation. The lack of a stereotype operational system for mining corporations indicates that we are within a period of growth, still searching for the best type of vehicle for mineral discovery.

Mining corporations explore by (1) territorial or (2) specialist group systems. The specialist approach to specific classes of deposits is proving more effective as evidenced by the recent ore discoveries at Thompson, Mattagami Lake, Kidd Creek, Lee Creek, and Rustler Hills.

There are about twenty classes of ore deposits searched for by specialist groups. These deposits represented 37 billion dollars in gross sales in 1973. Base metals account for 34 percent of the total, and it is within the copper-lead-zinc-nickel fields that most of the new specialization has occurred.

Since 1948, major ore discoveries within new districts have been placed in production within every genetic class of deposit except Witwatersrand gold, red-bed copper, and tin. These ore discoveries range from raw subsurface ore deposits using geological and geophysical methods through raw discoveries outcropping in remote areas. In the purest form of discovery, the explorationist appears to "peak out" at a particular knowledge level at the moment of discovery. Thereafter, a period of "expert senility" occurs as to the specific class of mineral deposit. The time has arrived for the explorationist to switch to another class of deposit for discovery.

The successful explorationist is intelligent and creative with an above-average capability for rationalizing numerous variables within a mappable framework. As a rule, compilation of data followed by field observation leads to new concepts for ore discovery.

The successful explorationist, as a rule, receives little credit for an ore discovery, due to dilution within the corporate management system and, therefore, he changes corporations. Since the value and life of a major discovery is so great, corporations have not suffered financially on a short-term basis. However, the graveyard of mining corporations is growing, and this is due to the loss of their unique explorationists.

It is suggested that the Society of Economic Geologists establish an Ore Discovery Committee to publish detailed accounts of major ore discoveries. The value of such reports is both for the benefit of the successful explorationists and for the education of members of our Society on techniques of ore discovery.

The mining corporations must realize that to maintain continued growth, successful explorationists must be financially rewarded. The rewards should be significant stock awards, high salaries, and increased freedom to explore. With the proper credit and sufficient reward to the explorationist, a multinational corporation will have continued growth, limited only by the excellence of competitive corporations.

Introduction

Mining corporate vehicles and ore discoveries are the life blood of all exploration geologists. Together, they form the basic raw material requirements for the advancement of our civilization. The purpose of this paper is to examine the function, character, and impact on our society of these three separate entities and to suggest methods by which our ore discovery rate may be improved.

* This talk is printed as it was presented before the Society of Economic Geologists, Feb. 18, 1975, in New York, N.Y.

TABLE 1. 1973 Annual Production of Raw Minerals and Metals of the World's Mining Companies, in Dollars

Company	Production	Company	Production
Anglo American	$1,866,884,000	Broken Hill Pty Ltd.	$264,000,000
Kennecott	1,370,675,000	Cominco	259,364,000
United States Steel	1,241,185,000	Cyprus	257,096,000
Chilean Government	1,140,901,000	Johannesburg Consolidated Invest.	239,450,000
International Nickel	986,811,000	Moroccan Government	239,064,000
Indian Government	929,112,000	Peruvian Government	233,739,000
Zaire Government	771,059,000	Outokumpu	203,640,000
Asarco	570,025,000	Union Corp.	202,290,000
Rio Tinto—Zinc	532,966,000	Granges AB	195,480,000
AMAX	522,237,000	Freeport	194,842,000
Brazilian Government (iron)	516,000,000	St. Joe	188,067,000
Anaconda	514,120,000	Island Creek Coal	185,778,000
Newmont	500,555,000	Boliden	183,713,000
Zambian Government	498,557,000	Ideal Basic	178,559,000
Bethlehem	496,472,000	Hanna	162,216,000
Phelps Dodge	492,804,000	Mt. Isa	158,791,000
Consolidation Coal	491,073,000	Pittston Coal	152,624,000
Consolidated Gold Fields	468,810,000	International Minerals and Chemical	146,600,000
Noranda	412,934,000	Johns Manville	139,000,000
Texasgulf	406,533,000	Lorraine Mines (France)	136,000,000
Falconbridge	367,750,000	ISCOR	135,094,000
General Mining	357,999,000	Reserve Mining	132,504,000
Utah International	354,155,000	Atlas Consolidated	126,470,000
Kiruna	348,000,000	Mauritania	121,488,000
Pennzoil	328,001,000	Mexican Government Sulphur	120,000,000
De Beers	285,340,000	Kerr-McGee	117,564,000
Gold Fields of South Africa	275,000,000	Copper Range	108,428,000

Note: The bauxite companies are omitted due to lack of data.

I am grateful for being honored by the Society of Economic Geologists as their Distinguished Lecturer in Applied Geology for 1975.

At this moment in my life as an exploration geologist, I am not involved in the search for any specific deposit. My work, rather, is to stimulate and organize geological explorationists within a corporate framework so that a maximum rate of ore discovery is obtained. Therefore, this paper is an outgrowth of my knowledge in a field to which I am totally committed.

Corporations

Mining exploration corporations are evolutionary organizations which are still in a youthful stage of development. They provide the framework within which we exploration geologists discover mines. The leading mining corporations are listed on Table 1 in decreasing order of their raw material gross sales from mined products in 1973. I wish to emphasize that these corporations are listed in order of raw materials sales and no value is given for upgrading of their various products. The reason for portraying the raw material producers in this manner is because, as explorationists, we are primarily concerned with the mined product, and we usually are not involved in upgrading iron to steel, or phosphate rock to phosphoric acid. For example: U. S. Steel, which mines $1,250,000,000 in raw materials has gross sales of $5,400,000,000, due to the processing of their ores to steel; while IMC, which mines $146,000,000 in potash and phosphate rock, has sales of $490,000,000, due to upgrading of phosphate rock to phosphoric acid and dry fertilizer products.

Anglo American and Kennecott top the list of raw material producing companies and are totally different in corporate structure. Anglo American Corporation of South Africa operates by a group system in that the parent house provides a complete range of administrative and technical services to the companies within the group. Anglo American usually holds less than 50 percent interest in the individual companies. The $1,800,000,000 in raw material sales represents only Anglo American's portion of this production, which is primarily derived from 40 percent of the total Witwatersrand gold production; 8 percent of the diamond sales of the Central Selling Organization; 34 percent of Zambian copper; and 23 percent of South African platinum.

Kennecott operates its own mines or through wholly owned subsidiaries, such as the Peabody Coal Company. Kennecott obtained 57 percent of its raw material revenues from metals and 43 percent from coal in 1973.

The success of two such diverse corporate systems, and all variations in between, portrays the lack of stereotypes within our exploration and mining vehicles.

The three leading raw material mining companies in the Western World are Anglo American, Kennecott, and United States Steel. These companies are multinational, although each has its major raw material source within its country of origin:

	Major Raw Material	% of Foreign Production
Anglo American (South Africa)	Gold	20%
Kennecott (United States)	Copper	5%
United States Steel (United States)	Iron Ore	33%

The raw material position of all the multinational companies is constantly being threatened by nationlization. Anglo has lost 51 percent of its Zambian copper production; Kennecott lost 100 percent of Braden; and U. S. Steel lost 51 percent of Cerro Bolivar, in December 31, 1974. As a result of such divestitures, a new breed of national companies has entered the raw material exploration scene in Zambia, Chile, and Zaire (Table 1). If this trend continues, it will have a great effect on the exploration geologist. Western geologists are already restricted from exploration within the communist world, and further territorial limitation may herald an era of stagnation for the explorationists. However, most restrictions at this time have been placed on various commodities unique to the particular country, such as iron, copper, and phosphate. It appears that these new national mining groups will eventually follow the pattern of Outokumpu in Finland and develop their various natural resources through an economic cycle which passes through a profitable peak and falls to uneconomic levels as lower grade ores are mined and processed.

On the production table there are a group of companies that have revitalized their exploration staffs and are raising their raw material production capacity through ore discovery:

Company	Discovery in Production
Rio Tinto-Zinc	Bougainville, Hamersley
Texasgulf	Kidd Creek, Lee Creek
Utah	Blackwater, Island Copper
Pennzoil (Duval)	Rustler Hills

Mining corporations survive on new ore discoveries, the basis of the system. All ore deposits are exhaustible, and the historical past is littered with dead corporations which were unable to satisfy their need for raw materials.

However, mining corporations also change their complexion through a more insidious disease which is expansion into fields outside their historical expertise.

The original ore discovery, as a rule, raised geologists and/or mining engineers to the chief executive position in mining corporations. These men lead their company through the mining, benefication, and final processing of the concentrates to elemental metals, phosporic acid, alumina, etc. At this juncture, the need for the company to continue to grow would be based on a new raw material discovery or expansion into fabrication and therefore into sales of a multitude of products. Since mines are difficult to discover, many mining corporations follow the latter route that leads to lower unit profit but higher sales volume. As a result, the management passes from the hands of the geologists and mining engineers to accountants and salesmen. It is not long before the exploration division becomes a questionable but still tolerable expense.

Finally, high-volume sales of a multitude of conglomerate products brings about immense tax problems which require the chief executive branch to be dominated by lawyers. Under the guidance of this system of management, the corporation may continue to thrive, but it is no longer an attractive institution for exploration geologists. It is significant to note that the previous mining company could have survived as a raw material producer if the explorationist could have had an acceptable discovery rate. However, at that critical moment in time when the mining company desired continued growth, the explorationist failed and a transition commenced into a multitude of finished products which was irreversible.

Exploration geologists should carefully examine the stage of growth of a particular mining corporation and analyze the character of management prior to employment.

One cannot underestimate the importance of style and philosophy within mining corporate management. A successful explorationist will probably succeed irrespective of management, but a high rate of ore discoveries can only be achieved by corporations who support their successful explorationists through "thick and thin".

Corporate Exploration Techniques

Most mining corporations were born by the ore discovery of excellent prospectors. As the earth became less virgin, corporations built exploration departments which, in themselves, are still evolving in their search for the most successful exploration philosophy.

The techniques of ore discovery for corporations lie at two extremes: (1) purchase of an ore deposit and (2) the raw discovery. Successful mining corporations utilize each approach and the geological methods employed in each case are remarkably similar. However, the geological and geophysical methods employed for the exploration of different classes of ore deposits are usually dissimilar. It is within this area of expertise that our mining corporations are still searching for a solution.

Mining exploration corporations are organized within three different frameworks: (1) territorial, (2) specialists, and (3) a mixture of the two.

Under a territorial regime, an exploration district is outlined, such as the western United States, followed by establishment of an office. The territorial group then proceeds to explore for all minerals that may occur within that district. The territorial system of exploration cannot survive into our forthcoming era of totally blind ore deposits. The knowledge required in geology and geophysics for the different classes of deposits is beyond the mnemonic capability of man. Early in this century, the mining corporations drifted into the separation of exploration departments, based upon the degree of symmetry possessed by the particular class of ore deposit. Coal or phosphate or sulfur geologists did not search for porphyry copper or volcanogenic deposits. The territorial system with its regional offices is too restricted in that the geological system which controls the locale of a certain class of ore deposit continues beyond the territorial boundary.

Specialist groups within the corporate system are composed of geologists who devote a minimum of four years to the study and exploration of specific types of ore deposits. At the outset, it must be emphasized that a geologist does not give up his birthright of generality by interim specialty. On the contrary, he becomes a broad generalist, on a strong data base, near the close of his life.

A list of the most important classes of ore deposits is portrayed on Table 2 in order of decreasing annual contribution to our world's economic wealth.

Coal dominates worldwide mining, but due to its symmetrical and sedimentary properties, and, therefore, ease of discovery, it has not required the exploration effort of other raw materials. The importance of base metals is somewhat submerged in this type of class organization, but it is significant to note that porphyry, volcanogenic, red-bed, carbonate, nickel-copper, and Proterozoic base metals have a total annual raw mine value of $12,500,000,000. In fact, the raw material importance of the various mineral groups based on percentage of the total value mined in 1973 is:

Base metals	34%
Coal	28%
Iron and manganese	15%
Precious metals and diamonds	12%
Industrial minerals	6%
Others	5%
	100%

The future of large mining exploration corporations hinges on the expertise they acquire in each class of deposit. As an example of the volume of knowledge required within the various systems, the contrast between exploration for volcanogenic copper-zinc-silver and porphyry copper deposits is described.

TABLE 2. 1973 Annual Production, in Dollars, of Minerals and Metals Separated into Exploration Systems

Coal	$10,467,000,000
Iron	5,246,160,000
Porphyry copper	5,086,046,000
Volcanogenic copper-zinc-silver	2,975,000,000
Witswatersrand gold	2,700,000,000
Red-bed copper	1,400,000,000
Carbonate lead—zinc	1,345,000,000
Diamonds	1,331,760,000
Archean-Proterozoic nickel-copper	988,000,000
Tin	933,000,000
Phosphate	908,328,000
Bauxite	836,550,000
Proterozoic lead-zinc-copper	741,640,000
Potash	506,400,000
Manganese	399,000,000
Uranium	382,000,000
Layered intrusive platinum-palladium	360,000,000
Sulfur (Frasch)	291,000,000
Chrome	208,000,000
Illmenite-rutile	100,000,000
	$37,204,884,000

Over 800 mines occur as volcanogenic massive sulfide deposits and all are confied to eugeosynclinal belts. They are unlike most other economic mineral occurrences, in that they formed throughout geological time, commencing 3,000 million years ago at Mons Cupri in Australia and are still forming today on Hokkaido Island, Japan. Approximately 85 percent of the deposits occur within rhyolite ash-flow tuffs or nuée-ardente fragmentals. As a rule, they border the ultramafic spine of the volcanic arc within the eugeosynclinal system. Therefore, the exploration specialist within the corporate exploration team must have a working knowledge of over 800 deposits plus a complete knowledge of eugeosynclinal ultramafic, volcanic, and sedimentary systems. Lacking this comprehension of massive sulfide deposits, exploration teams have mistakenly searched within continental rhyolitic ash-flow tuff or in the vicinity of continental layered ultramafics. The origin of volcanic magma within modern eugeosynclines is within the ultramafic mantle, as evidenced by seismic measurements. That is, epicenters of magmatic disturbances commence as deep as 700 kilometers, and as shallow as 25 kilometers, along the Benioff zone where the crust is thin, about 10 kilometers. The volcanic products consist of about 85 percent andesite-basalt and 15 percent rhyolite-dacite. In contrast, continental rhyolites, andesites, and basalts originate where the crust is 40 kilometers thick and contains no massive sulfide deposits. Therefore, it is important for an explorationist to realize that an intimate and detailed knowledge of a specific type of volcanology is required for the search for volcanogenic massive sulfide deposits.

Copper which is related to porphyry copper deposits is tied to an entirely different set of exploration parameters than copper related to volcanogenic deposits. In contrast, there are over 100 porphyry copper deposits in the world, all of which are associated with subvolcanic intrusives. The intrusives are within volcanic arcs and also continental volcanic systems. The key to future porphyry copper exploration appears to be defining the pattern of these magmatic systems. This definition may be accomplished by mapping Mesozoic and/or Tertiary volcanic rocks in order to outline the subvolcanic magmas or by direct mapping of the magmatic systems where possible. Porphyry copper deposits have not been generated continuously throughout geological time but are mostly confined to the geological period from late Mesozoic to Recent times. However, the Gaspé porphyry copper occurs within mid-Paleozoic rocks, and uneconomic Archean porphyrylike deposits occur in Quebec.

In order to find hidden ore deposits, the geologist must "peak out" on his knowledge of a specific class of ore deposits. One cannot be engrossed in eugeosynclinal volcanic phenomenon and the multitude of variations of contained massive sulfide deposits and simultaneously be completely knowledgeable in subvolcanic magma of both eugeosynclinal and continental types, plus variations of over 100 porphyry copper deposits. Therefore, successful corporate mining exploration of the future will be guided by a unique class of specialists who confine their restricted specialty to brief periods of time. As a result, the individuals will develop into true generalists who focus periodically on specific ore deposits. Their prime forte will be the ability to organize a multitude of seemingly unrelated data into a logical sequence of facts which leads to an ore discovery.

Ore Discovery

The concept of ore discovery has been interpreted in numerous ways historically. Within the modern corporate system, an ore discovery is a mineral deposit which can be brought into profitable production within a five-year period. A technical success is the discovery of a mineral deposit which is subeconomic into the indefinite future. From a corporation point of view, exploration geologists can be judged successful only on their rate of ore discovery, not on the basis of technical successes. The rate of return on invested capital is much higher on an ore deposit near a railroad than one in a remote area. It also follows that it requires more intelligence and creativity from the geologists to discover ore within reach of our industralized society. Discovery of mineral deposits in the arctic of North America and the deserts of Australia usually requires the perseverance of the prospector without the need for abstract thought. Herein lies the major problem facing exploration geologists today, that is, what human values are to be honored in an ore discovery in contrast to other values that may be purchased on the open market place? The answer to this question must be based upon the human traits which are lauded in particular individuals who contributed to the advancement of our civilization. Can we agree that it was the ability to develop abstract thoughts in such men as Aristotle, Copernicus, Newton, and Einstein that is respected today? Furthermore, to judge their importance within the milieu of their contemporaries, one only has to remove them from the scene and examine the void. And so it is with ore discovery. As an example, examine the components of a recent sulfide discovery. They are:

1. Corporate financial vehicle,
2. A geologist,
3. Geophysical tools, and
4. Drill core with ore.

In the majority of cases, only the geologist is unique. The other ingredients, corporation, geophysical tools, and drillers, may be contracted on the open market. In the minority of cases, a corporation with blanket geophysical coverage may find ore deposits. However, this technique passed its heyday by 1960 and awaits the development of a new geophysical tool, such as an airborne gravimeter.

I conclude that credit for an ore discovery should go to the explorationist who conceives the approximate position of the mineral deposit within his mind and then has the courage of his convictions to follow through to discovery. At this moment in time, the majority of successful explorationists are not contractable but are buried within corporate systems. A partial purpose of this paper is to set into motion a sequence of events which will liberate these unique and valuable individuals in order to increase their rate of discovery.

The value of mineral discoveries varies, due to their economic dependence upon the requirements of various industralized societies. Steaming coal in Alaska is of little value compared to steaming coal near New York. However, this fact is changeable, since time is also an important variable. It requires more intelligence and creativity to find particular mines than others, and therefore the following lists is a priority outline of types of ore discoveries which exemplify the uniqueness and value of the successful explorationists in the field of ore discovery.

Types of Ore Discoveries Arranged in Decreasing Value as to Geological Ability

1. *A raw subsurface ore discovery*
   a. Outside a mining district of its type

b. Not based on blanket geophysical coverage
   Examples:
   - Rustler Hills sulfur, Texas
   - Kidd Creek copper-zinc, Ontario
   - Pine Point lead-zinc, Northwest Territories

2. *A raw subsurface ore discovery*
   a. Within a known mining district of its type
   b. Not based on blanket geophysical coverage
   Examples:
   - Pima copper, Arizona
   - Henderson molybdenum, Colorado
   - Blackwater coking coal, Queensland
3. *A raw subsurface ore discovery*
   a. Based on blanket geophysical coverage
   b. Guided by geological mapping
   Examples:
   - Thompson nickel, Manitoba
   - Mattagami Lake copper-zinc, Quebec
   - Ranger uranium, Northern Territory
4. *A raw surface ore discovery*
   Examples:
   - Mount Tom Price iron, Western Australia
   - Weipa bauxite, Queensland
   - Anvil lead-zinc, Yukon
5. *A raw surface mineral discovery within remote areas which may be mined within five years*
   Examples:
   - Arctic copper-lead-zinc, Alaska
   - Isok Lake copper-zinc, Northwest Territories
   - Agnew nickel, Western Australia
6. *Land put together by "finders" within known mineral districts on fairly obvious mine extensions*
   Examples of this class occur in the Bushveld and western United States coal and uranium. This class of deposit emphasizes the importance of the corporation and minimizes the value of the geologist. That is, the "finder" seeks out the best corporate vehicle so that his fee is guaranteed.

TABLE 3. Major Mine Discoveries of the Last Twenty-Five Years with Gross Mine Production Value for 1973

1948	Knob Lake	iron	$312,000,000
1949	Mexican Sulphur	sulfur	140,000,000
1951	Pima	porphyry copper	120,000,000
1952	Granduc	volcanogenic	29,000,000
1953	Gas Hills-Shirley Basin	uranium	193,000,000
1953	Mission	porphyry copper	63,000,000
1953	Brunswick	volcanogenic	53,000,000
1954	Heath Steele	volcanogenic	12,500,000
1954	Cerro Bolivar	iron	227,000,000
1954	Pine Point	carbonate lead-zinc	143,000,000
1954	Gaspé	porphyry copper	40,000,000
1955	Viburnum Trend	carbonate lead-zinc	143,000,000
1955	Thompson	nickel	150,000,000
1955	Ambrosia Lake	uranium	105,000,000
1955	Bethlehem	porphyry copper	45,620,000
1956	Palabora	Proterozoic	150,000,000
1956	Craigmont	porphyry copper	30,104,000
1956	Mattagami Lake	volcanogenic	64,000,000
1957	Mt. Whaleback	iron	316,000,000
1959	Kidd Creek	volcanogenic	280,314,000
1960	Weipa	bauxite	103,500,000
1961	Lee Creek	phosphate	20,000,000
1961	Lake Dufault	volcanogenic	14,000,000
1962	North Florida	phosphate	28,000,000
1962	Blackwater	coal	74,000,000
1962	Mt. Tom Price	iron	352,800,000
1963	Carlin	gold	20,000,000
1964	Bougainville	porphyry copper	307,530,000
1964	Twin Buttes	porphyry copper	91,000,000
1965	Prieska	volcanogenic	28,600,000
1965	Kambalda	nickel	109,500,000
1955	Anvil	proterozoic lead-zinc	104,000,000
1966	Gilbralter	porphyry copper	83,000,000
1967	Mattabi	volcanogenic	37,444,000
1967	Rustler Hills	sulfur	60,000,000
1968	Lornex	porphyry copper	70,000,000
1968	Island Copper	porphyry copper	63,000,000

[*Editor's Note:* Material has been omitted at this point.]

The Geologist

Intertwined in the discussion of the corporate exploration vehicle and concomitant ore discoveries has been the suggestion that there exists a unique and rare individual entitled an "explorationist". This explorationist forms the "guts" of all successful modern exploration corporations. He is able to move to different corporate vehicles and continue to discover ore deposits. The abandoned corporation, meanwhile, may live 50 years on the fruits of his past discoveries, but without the acquisition of another unique explorationist, the corporate discovery rate may sink to zero, followed by a "hit and miss" expensive mine purchase program.

Within the western world, there are few ore-finding explorationists. Many more could be developed through proper training, but due to the extremely competitive world of the corporation, many young, capable geologists falter, lose self-confidence, and fail, when possibly on the verge of an ore discovery.

The physical and mental traits necessary for a modern successful explorationist are herein listed in decreasing priority:

1. Physical fitness
2. Creativity
3. Intelligence
4. Optimism
5. Persistence
6. Non-meek and non-humble
7. Adventurous

Physical fitness is of first priority because without this characteristic, all other traits cannot be properly utilized. Fatigue undermines one's ability to perform mentally. After a difficult ascent through heavy bush in mountainous terrain, unfit geologists are finished for the day while, in fact, the work has just begun. In Western Australia, one may travel by plane from the relatively cool south to the hot northern desert, and upon alighting from the aircraft be greeted with a blast of ambient air temperatures of between 110°F and 120°F. The unfit may survive at a pace of 50 percent below the fit.

Intelligence is the ability to store data. The more intelligent are selective in the data stored. For instance, the environmental facts covering the carbonate lead-zinc deposits in the world must be selective facts and stored in one's mind. It is only then that the data is instantaneously retrievable and utilizable throughout one's period of time in that particular specialized field.

Creativity is one's ability to recall selected stored data in order to formulate abstract ideas. There are at least four ways in which the all-important creativity characteristic may be brought to bear on ore deposit discovery.

First, in the formulation of a new exploration approach to a specific class of ore deposits, a creative individual works from a broad philosophical consideration of the earth's distribution of these deposits as a whole. As the data become increasingly specific from a global map to a broad regional map and onto an area map, the reliability of the multitude of variables must be constantly kept in mind. Larger and more detailed final maps allow the creative mind to mentally erase, and transmute, variables in microseconds in order to settle on the ore control or location of the hidden ore deposit. The compiler must constantly be aware of how far questionable and possibly erroneous data are influencing his guides to ore.

Secondly, creative geologists place all the data on one map in lieu of transparancies. For example, one map may contain geological outcrops, magnetic, I.P., and geochemical information. The ability to erase mentally part or all of any particular class of data plus keeping in mind the exactness of portions of the information is the mark of creative geologists.

Thirdly, creativity requires creative maps, not "quadrangle" geological maps. Ore control maps (as example, specific itme horizons) only are preserved if proven correct. Otherwise, they are just one more creative effort destined to hit the wastebasket.

And finally, to create a mental map from a high vantage point in some field terrain requires data recall of weeks of field traverse. This type of mental map is the interim or final visual and "computer-type" exercise in the authenticity of the recorded data.

The optimistic trait of a successful explorationist is the most apparent identifiable characteristic. His optimism carries the explorationist through the depths of despair, brought on by erroneous data, poor stratigraphic correlations, lack of monetary support, etc.

Persistence, tied to optimism, is the trait necessary for the discovery of most ore deposits, since the first hole is not always ore.

The lack of humility is a characteristic of almost all successful explorationists and is an identifiable virtue. The corporate system does not tolerate meek individuals. An explorationist, once he has focused on the particular class of deposit, must not only overcome the problem of communication with the rocks, but must also fight a lonely battle through the ramparts of the corporation. However, both the biological and physical sciences have case histories of ruthlessness as evidenced by Wilson and Crick on DNA and Lawrence and Oppenheimer on the first nuclear explosion.

Adventurous explorationists tend to go beyond the normal demands of corporations. These men pack guns for protection in Guerrero, Mindanao, and New Guinea, risk their lives in old mine shafts, and free-climb numerous vertical cliffs during the course of everyday traverses. This spirit of adventure and/or curiosity is not engendered by the corporate system, salary, fringe benefits, or the old-age pension. Rather the explorationist is on the trail of an ore deposit, and its discovery is all that will satisfy his creative mind.

Exploration geologists live within a lonely and frustrating world. His only work happiness is brought about by ore discovery, and discoveries are rare. Individuals within management fields may observe the fruits of their work within hours, weeks, or years, and as a rule this is true of research scientists, but not so the explorationist. His fate is usually obscurity. He is commonly recognized as a threat to other departments within a corporate system and, therefore, through self-preservation changes corporate vehicles. An excellent example are the men who were responsible for the discovery and the development of the great coking coals of the Bowen Basin. They are no longer with the parent company, which has since grown rich through their valiant efforts.

In order to sustain our civilization's need for raw materials, we must cultivate every potentially great explorationist. He must be raised out of obscurity and gain recognition and wealth with or without managerial responsibilities. It must be realized that successful explorationists are a product of their genetic environment and not a product of their academic background.

In order to develop a multitude of successful explorationists, it is necessary to construct avenues by which recognition and reward can be achieved. It is not meant here to contradict the concept of purity of purpose; however, the handful of pure explorationists is insufficient to discover the necessary raw materials to continue the advance of our civilization. The primary ingredients are creativity, intelligence, and courage. Fame and wealth can drive one to optimism, persistence, and fitness.

[*Editor's Note:* Material has been omitted at this point.]

Summary

Mining corporations represent a particular segment of multinational companies which bridge the political and geographical boundaries of nations. They offer the opportunity for the governments and peoples of various nations to participate in raw material wealth of countries other than their own. Movements of nationalism throughout the world are commencing to limit the future of multinational mining corporations. However, this trend could change as the raw material resources of countries diminish and a need for multinational talent, and capitalism, becomes the logical avenue for ore discovery.

Mining corporations age and, as a rule, their rate of ore discovery diminishes. Management passes from the explorationist to engineering to accounting and finally to the legal department. The fault for this transition lies at the feet of the explorationist, due to his lack of ore discovery during those critical periods requiring growth.

Ore discoveries of the past have been made by prospectors and, in remote parts of the world, the prospector remains the dominate discoverer. However, exploration geologists are becoming the prime "ore finders" and this trend will continue on into the future. Ore discoveries are being made by specialists in particular classes of deposits. Specialty in particular class systems should last for brief periods, say, four years, due to the concept of "senile experts".

Successful exploration geologists are unique explorationists and, as such, fame and wealth should be their reward. However, this is not the case since their history of discovery is buried within company records. As a result, the ore discovery is credited to others by hearsay. It is important that the Society of Economic Geologists consider the appointment of an Ore Discovery Committee in order to document the details of each new discovery in our Bulletin.

Our rate of ore discovery is low due to a poor reward system within mining corporations. Unique and successful explorationists depart and, as a rule, have a slow "start-up time" with a new corporate vehicle. Therefore, successful explorationists should be rewarded with freedom, high salary, and generous stock options.

The goal of mining corporations is to increase their rate of ore discovery. In order to accomplish this feat, they must maintain a continuum of exploration logic which is exemplified by their own particular successful unique explorationist. If this can be accomplished, then the multinational mining corporation will continue to be the leading supplier of the world's raw materials.

Acknowledgments

The helpful discussions with all of the geologists of Texasgulf Inc. are gratefully acknowledged, especially in the determination of what are the parameters of a great mining corporation and the characteristics of great explorationists.

Texasgulf Inc.
5934 McIntyre Street
Golden, Colorado 80401

Part II

EXPLORATION PHILOSOPHY

Editor's Comments on Papers 6 Through 10

The search for mineral deposits entails a great deal of risk. Slichter (Paper 6) refers to it as the world's biggest and best gambling business in which the winnings are limited only by the character of the deposits, legal restrictions, competition from others, and the limitations of the exploration group. Slichter suggests the use of vital statistics in the appraisal of expectations of finding ore. The concept was originally introduced by Allais (1957) but is brought into better focus for the exploration geologist in Slichter's paper.

Many refinements to the statistical evaluation of mineral wealth of regions have been proposed, as by Slichter et al. (1962) and Harris (1966). However, the probability of the existence of an ore deposit in an area is very different from the probability of a specific individual or a specific group making a discovery of that deposit. The statistical approach is most useful in giving guidance regarding the spacing of geochemical or geophysical grid lines or the spacing of grid drilling.

Risk in extrapolation of ore potential from intercepts in drill holes or underground workings depends upon a great deal more than simple geometry. It will vary with different geological environments. McLaughlin had this relationship in mind in writing Paper 7. The terminology he proposed for the classification of ore deposits based upon economic expectations has never been widely used, and this important paper and its concepts have drifted into obscurity.

As mineral economists sought to establish criteria for the projection of reserves into potential reserves or resources, as mine geologists sought to determine the potential and probable life of a mining district, and as exploration geologists sought to determine the potential target sizes, attempts were made to establish characteristic models of various types of ore deposits. Lasky (Paper 8) made an important contribution in establishing that, in certain types of ore deposits, a fixed and predictable relationship often exists between grade and cumulative tonnage. Lovering (1969) raises a word of caution in the application of this technique to metals such as lead and silver, which may not follow the exponential relationship shown by copper. There may be gaps in the tonnage/grade curve.

Brant (Paper 9) attempts the very difficult task of predicting the profitability of different order of magnitude deposits that are likely to be discovered within specific provinces and of predicting the probability of making a discovery. Such an approach serves well to aid in the selection of an area in which to concentrate exploration effort. However, neither an ill designed program nor an exploration geologist who lacks an inquisitive nature will locate a deposit.

Bailly (Paper 10) suggests how the explorer can achieve good luck by carefully considering all factors and circumstances that are assessable or controllable and amenable to prediction and action. The explorer is not at the mercy of statistics but can beat the odds by well defined exploration programs.

REFERENCES

Allais, M., 1957, Method of Appraising Economic Prospects of Mining Exploration over Large Territories: Algerian Sahara Case Study, *Management Sci.* **3:**285–347.

Harris, D. P., 1966, A Probability Model of Mineral Wealth, *Am. Inst. Mining, Metall. and Petroleum Eng. Trans.* **235:**199–216.

Lovering, T. S., 1969, Mineral Resources from the Land, in *Resources and Man,* P. Cloud, ed., Freeman, San Francisco, pp. 109–134.

Slichter, L. B., W. J. Dixon, and G. Meyer, 1962, Statistics as a Guide to Prospecting, in *Mathematical Techniques and Computer Applications in Mining and Exploration, Symposium Proceedings,* College of Mines, University of Arizona, Tucson, Ariz., pp. F1:1–27.

6

Reprinted from pages 570–575 of *Mining Eng.* **12**:570–577 (1960)

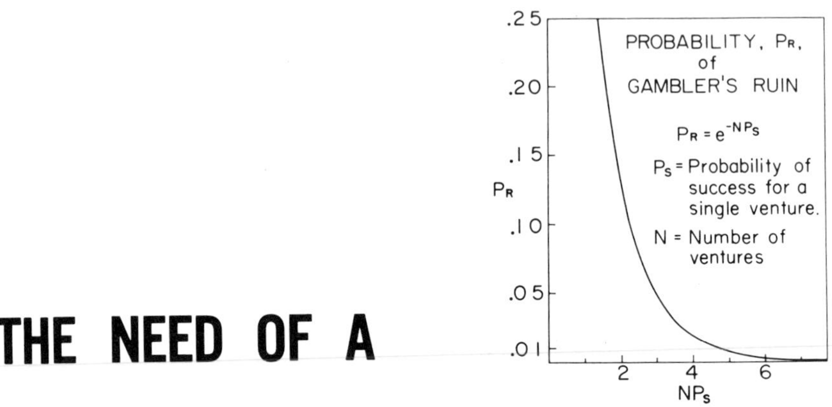

THE NEED OF A NEW PHILOSOPHY OF PROSPECTING

by LOUIS B. SLICHTER

Prospecting is certainly the world's biggest and best gambling business. It is a game where the chips cost many thousands and where many millions, even billions, can be won. An attractive feature of this gamble is the fact that the players are free to rig the odds as favorably as possible. Only the limited bounty of nature, the restrictions of the laws of the land, the competition from other players, the limited sagacity of the player himself—only these, and other factors, restrict the possibilities for large winnings.

Today probably about half the land surface of the earth remains to be searched for signs of ore visible at the surface. It is fortunate that so much land still remains for exploration by relatively economical methods. Looking to the future, it is a good guess that the costs of finding ore will increase in response to increasing technical difficulties. To continue to achieve profits from prospecting, it will become necessary to analyze the business more skillfully, in a more and more sophisticated manner. Whenever feasible, qualitative concepts must be replaced by their quantitative equivalents, improvements, and extensions.

THE GAMBLE

The fact that prospecting is a gambling business reminds us at once of the elementary rule known as the *law of gambler's ruin*. I hesitate to mention so simple a rule in this company of experts. This rule expresses the rather serious chance of going broke when the odds for success are small, merely by a normal run of bad luck regardless of the long-run expectations of gain. The only sure way of avoiding this special risk of *gambler's ruin* is to have enough capital, and the will, to continue the play many times and thus ride out the inevitable runs of bad luck. For example, if the probability of success is one in ten for each venture, there is a 35 pct chance that ten successive ventures will fail in a row. But if one has the capital to continue the play through a run of 100 failures, then the chance of *gambler's ruin* is only 5 in 100,000. Almost any management, I suppose, would be under terrific pressure to cease the play during a long run of bad luck. To continue to prospect under such circumstances requires great confidence in the premises, and a firm conviction in the long-term expectation of profit. There is evidence which I shall mention that in the future even the strongest companies may not have the necessary capital to prospect on the ideal scale. Joint ventures will be very desirable as a means for extending the available venture

L. B. SLICHTER, Member AIME, is Director of the Institute of Geophysics, University of California, Los Angeles.

capital. The essential difference between the fascinating business of prospecting and the more prosaic insurance business lies in the law of *gambler's ruin*. The insurance companies place so many thousands of bets that they are obviously immune from the hazard of *gambler's ruin*. Mining prospecting is *not* thus immune and qualifies as an excellent example of a first-rate gambling business.

STATISTICAL ANALYSES OF OREBODY OCCURRENCES

The use of geophysics and geochemistry in prospecting, and especially the use of the airplane, changes the character of the operation from the bird-dog search for special situations, used by the old-timers, to the wholesale uniform search of large tracts of land. The unit of coverage appropriate for an airplane, say 1,000 to 10,000 sq miles, is so large that the statistics of orebody distributions may become basic in determining rational prospecting plans. As the areas explored become larger, the details of the special situations, strikes, dips, sizes, and other characteristics of the orebodies become submerged in the so-called *law of averages*.

Areal Distribution: A familiar qualitative statement about the statistical distribution of orebodies is the axiom *more ore is to be found near known mines*. This axiom may be given quantitative expression in terms of the observed distribution of known mines throughout large mining areas. Suppose that a large mining area is subdivided into many equal units of a size, for example, of 1000 sq km. On average, how many mines occur per unit, and what is the chance that a unit of area will contain at least n mines where n is 0, 1, 2, 3, etc? A good beginning has been made in the study of such questions. Under suitable conditions, Allais and Blondel[1] find that the areal distribution of orebodies is well represented by the familiar Poisson distribution. The Poisson type of distribution is very commonly encountered. For example, on a long stretch of open highway, the spacing of cars accords with the Poisson distribution. Again, in prospecting for uranium, the random clicks of a geiger counter are spaced in time in this way.

Returning to the distribution of mines, let us divide the area of 125,000 sq miles—centered at Boulder Dam—which was studied by Nolan[2] into 357 squares of 1000 sq km each and examine the distribution among these subdivisions of the 154 mining districts which have produced in excess of $10,000. Similarly, the Precambrian intrusive and sedimentary areas of Ontario were divided into squares of 1000 sq km each (185 in number) and the count of mines, square by square, recorded. (Mines within two miles of one another were counted as a single mine.) The results of such counts for the two regions are shown in Table I. Here the first column represents n, the rank of a 1000-sq km square in respect to the number of mines now known within it. In the second column are listed the number of unit squares containing precisely n mines. The third column lists the percentage of these areas which contains n or more mines. For example, in the Basin and Range region, two or more mines are found in 9.5 pct of the sub-areas. The mean number of mines per 1000 sq km is 0.43 in the Basin and Range area, 0.80 in the Ontario area.

These results are presented graphically in Figs. 1(A) and 1(B). The scale of ordinates which is logarithmic represents the probability $P(n)$ that n or

SUGGESTED READING FOR BASIC CONCEPTS OF STATISTICAL ANALYSIS

1. Bennett, C. A. and Franklin, N. L.: *Statistical Analysis*. John Wiley and Sons, 1954.
2. Hald, A.: *Statistical Theory With Engineering Applications*. John Wiley and Sons, 1952.

more known mines will occur in any randomly chosen unit square. The plotted points which represent the observed counts are well presented by the straight lines $P(n) = 10^{-0.514n}$, $P(n) = 10^{-0.356n}$, respectively, in the two regions. The two associated graphs which are concave downwards represent theoretical accumulative Poisson distributions having the appropriate means, 0.43 and 0.80 respectively. It is clear that the straight lines representing the exponential functions provide a better fit to the data than do the Poisson curves. The exponential probability distribution has important implications. It implies that the chance that one or more additional mines exist in any unit is *independent* of the numbers that are already known to exist. There is no indication that a unit of this size has been appreciably depleted in potential undiscovered wealth because some two, three, or four mines have already been found. The statistics indicate that the mines are there to be found and that the lack of discov-

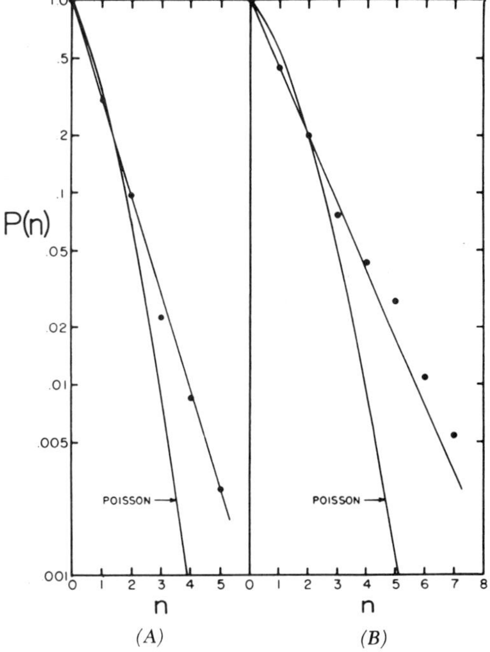

Fig. 1—(A) Comparison of observed distribution of mining districts in Basin and Range area with Poisson distribution. (B) Comparison of observed distribution of mines in Precambrian intrusive and sedimentary area of Ontario with Poisson distribution. Note: Ordinate P is the probability that a unit area (1000 km²) will contain at least n mines, where n is 0, 1, 2, 3, etc.

eries is attributable to difficulties of discovery rather than absence of targets for discovery.

If the Poisson approximation were valid, the axiom *look for mines near known mines* would be bad advice, in a narrow sense, since in such a distribution there is less chance of finding more in areas where the density is already high. No doubt the reason for the truth of the adage is the fact that more adequate geological information is revealed in the known mines. Thus the rule probably means not that mines are more abundant but are easier to find in the neighborhood of known mines. The statistical evidence here presented indicates that the abundance of *undiscovered* mines in these areas so much exceeds the few per 1000 sq km that have already been discovered that these discoveries do not reduce to any appreciable degree the number remaining to be found. Much more extensive studies of mine distributions in many other regions are of course needed. Such statistics may contribute significant evidence relative to the abundance of undiscovered mineral resources. The areal distribution of orebodies is obviously a basic consideration in the rational planning of prospecting.

Distribution of Orebody Values: Another familiar assertion about prospecting is the statement that the rare big orebodies—the so-called *elephants*—furnish the real rewards. This axiom seems to be accurately expressed quantitatively by the statement that the logarithms of the values of orebodies are distributed in accordance with the normal law of errors. Like the Poisson distributions, the log-normal law is also very frequently encountered both in nature and in economics. The production of the 285 mining districts in Nolan's Basin and Range study follows closely the log-normal pattern (Fig. 2). In this area 4 pct of the districts produced 80 pct of the wealth, and 16 pct of the districts accounted for 97 pct of the production. According to Fig. 2, only one district in 1000 would produce over one billion dollars. Clearly, the log-normal distribution in the size of orebodies leads to dramatic implications in the economics of search. In discussing the economics of prospecting an area of one million square kilometers in the Sahara Desert, Allais[1] estimated that the single largest discovery would account for between 35 and 40 pct of the value and that 25 pct of the finds would account for 65 to 80 pct of the values. We shall consider later how greatly the large orebodies should influence the strategy of prospecting.

Physical Dimensions: Associated with the log-normal distribution of orebodies in respect to value, there is a similar log-normal distribution in respect to their size and physical dimensions. Clearly both the absolute dimensions, and to a lesser degree the relative dimensions (namely the geometrical fineness ratios), as well as the associated values of the orebodies, are the chief variables which determine the most profitable plan of blind search. However, it may happen (and it often does) that satisfactory direct information about the sizes of orebodies is not available. In this case, a useful estimate of the size distribution, which is insensitive to the kind of ore involved, can be derived from the known distribution of the orebodies in respect to value. For example, within a factor of about 2, the average ores of ten important metals from gold to iron carry the same values per pound, or per cubic foot.[3] Thus one may adopt a standard factor to convert value to volume, which for convenience I shall assume to be about $1.00 per cu ft.

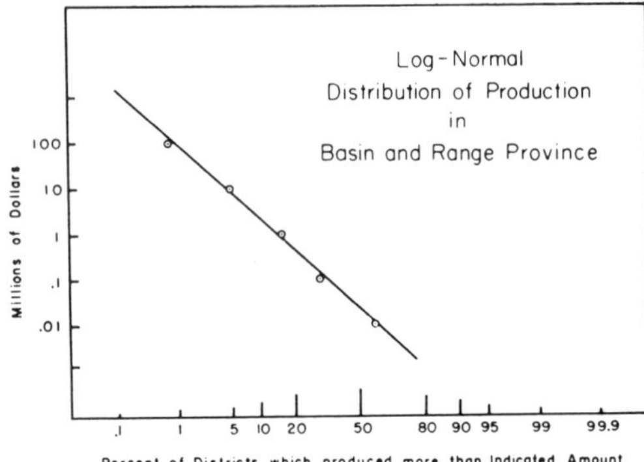

Fig. 2—Log-normal distribution of production in Basin and Range area. (A straight line on this graph represents a log-normal distribution.)

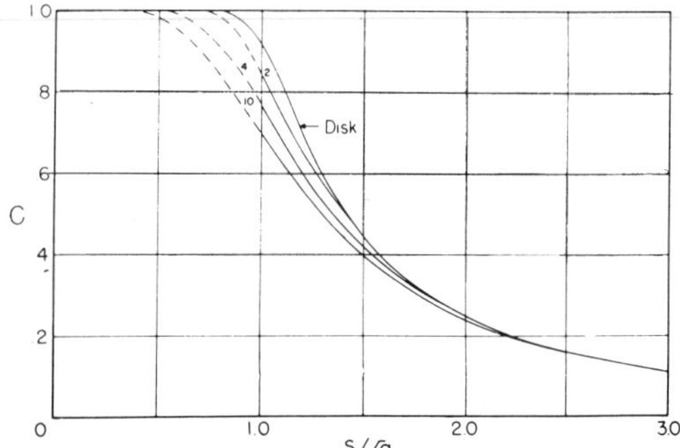

Fig. 3—Completeness of search by drilling grid.

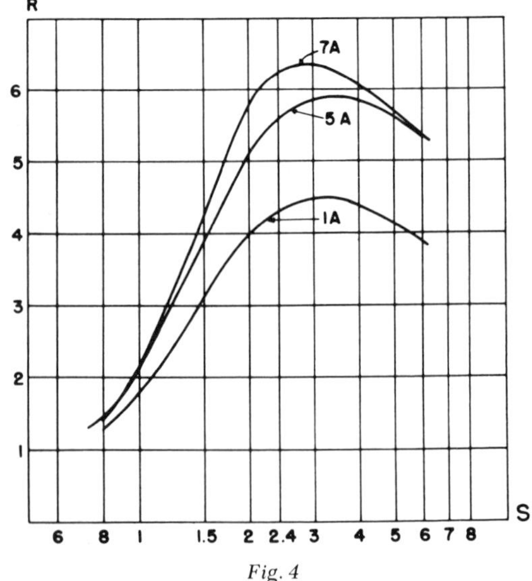

Fig. 4

PROBABILITY OF DISCOVERY

Having briefly considered the distribution of targets in space and in size, I shall next discuss some search problems in prospecting for blind orebodies. In these search problems, the basic compromise dictated by economics is between the completeness or thoroughness of search and the cost of such thoroughness. Usually the desirable compromise is that which maximizes the ratio of profits to cost.

Prospecting by Drilling: In some simple types of exploration problems, notably in exploration by the drill, it is feasible to calculate the probable completeness of search in terms of the size and shape of targets, and in terms of the drillhole spacing. Thus one can determine the best spacing of the drillholes. Fig. 3 shows the completeness of search for randomly oriented orebodies of various fineness ratios as a function of the drillhole spacings. The important item here is the ratio of the area *a* of the target to the area associated with each drillhole of the pattern. Long, thin bodies are somewhat more difficult to find than equi-dimensional bodies of the same area. A computed example appears in Fig. 4 showing the effect of hole spacing on the prospecting profit ratio.*

Prospecting by Geophysics: Let us turn now from prospecting by the drill to prospecting by geophysics, and in particular to prospecting by the airplane. In the geophysical search problem, one of the important questions concerns a suitable meaning for the word *discovery*. If a drill strikes good ore, one does not quibble much about the meaning of *discovery*, but a geophysical survey may reveal many so-called *anomalies*, most of which are not caused by ore. The identification of potential ore anomalies among similar effects caused by other geologic influences remains a first-class problem. Better ways of appraising the significance of geophysical anomalies, and of distinguishing them from the background of geologic noise, are needed and indeed are being continually developed.

Obviously, in the last analysis, the dangerous role of personal judgment or intuition in identifying and appraising anomalies can be eliminated only by performing the entire job mathematically.

For the present, however, let us avoid the complex subject of the interpretation of geophysical maps, however fascinating this may be. Let us agree for present purposes that the word *discovery* means only that the geophysical anomaly is crossed by at least one traverse, or better perhaps, by at least two, or if you prefer by at least three.[5] The point here is that one needs enough data to develop the outlines of a significant geophysical indication, and so the number of *discovery traverses* needed depends upon the local circumstances.

With this definition one may compute the probability of discovering targets of various sizes and shapes by surveys having various traverse spacings. In one case, at least, it is well known that the results may be expressed in a very simple way, namely, when the traverse spacing exceeds the major horizontal dimension of targets of any convex shape, which are randomly oriented, then the probability of discovery is proportional only to the perimeter of the target. In Fig. 5 the perimeter of elliptical targets is shown plotted against the major dimension. When the fineness ratio a/b is greater than 3, the perimeter is essentially proportional to

*These examples are more fully discussed in reference 4.

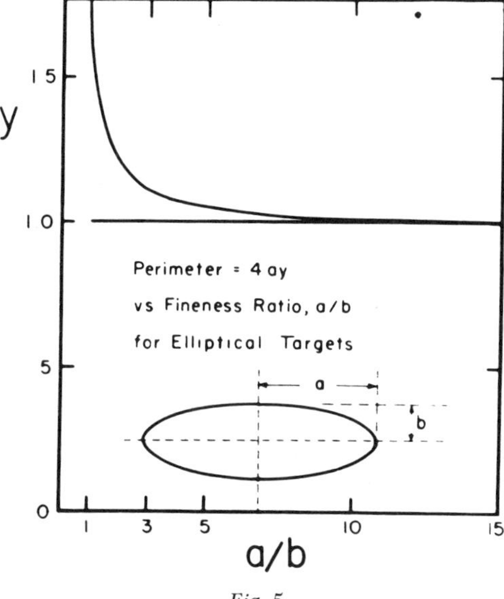

Fig. 5

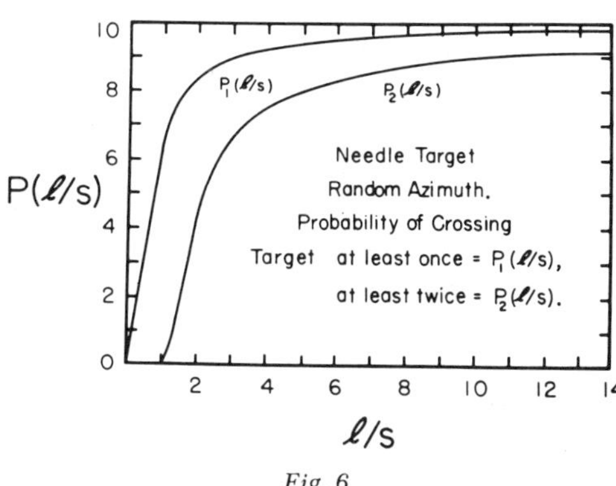

Fig. 6

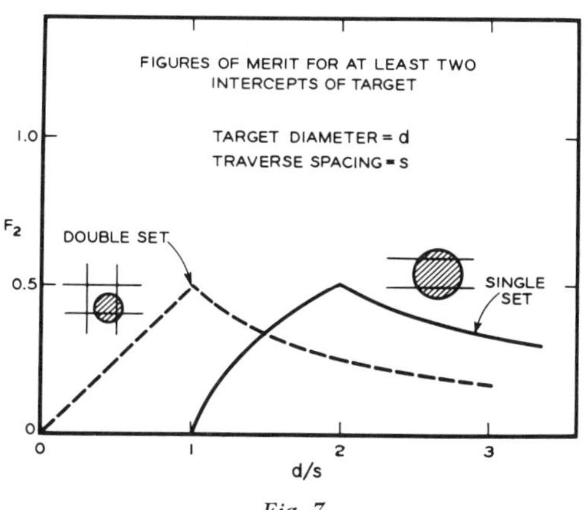

Fig. 7

Table I. Areal Distribution of Mines in Basin and Range Area and in Ontario

A. Basin and Range (154 Mines, Area 357 Units)

n = Areal Density of Mines in Number per 1000 Km²	Number of Units of Area 1000 Km² Each Containing n Mines	Percent of Units Contain-ing n or More Mines	Percent According to Cumulative Poisson Distribution (ε = 0.43)
0	250	100.00	100.00
1	73	29.97	34.88
2	26	9.52	7.01
3	5	2.24	0.81
4	2	0.84	0.08
5	0	0.28	0.00
6	1	0.28	0.00
7	0	0.00	0.00

B. Precambrian Intrusive and Sedimentary Areas of Ontario (147 Mines, Area 185 Units)

n = Areal Density of Mines in Number per 1000 Km²	Number of Units of Area 1000 Km² Each Containing n Mines	Percent of Units Contain-ing n or More Mines	Percent According to Cumulative Poisson Distribution (ε = 0.8)
0	104	100.00	100.00
1	45	43.78	55.07
2	22	19.46	19.12
3	6	7.57	4.74
4	3	4.32	0.91
5	3	2.70	0.141
6	1	1.08	0.0184
7	1	0.54	0.00207
8	0	0.00	0.0000

BASIN AND RANGE PROVINCE

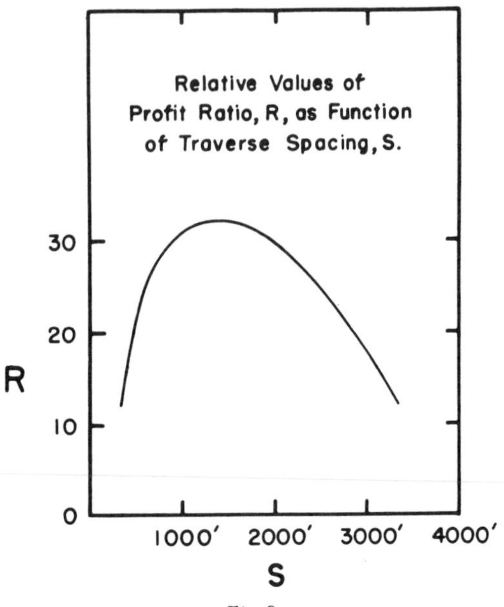

Fig. 9

the major dimension; the maximum deviation from this rule, for the circle, is 57 pct.

In the extreme case of long needle-like anomalies randomly oriented, the probabilities of detection by at least one or by at least two traverses are indicated by Fig. 6. Probability curves of this type are basic in determining the best spacing of traverses either in aerial surveys or in ground surveys where essentially continuous data along parallel lines are taken.

To express a *factor of merit* in such prospecting operations, one should divide the probability of discovery by the cost of discovery. Neglecting for the moment overhead charges, it will be assumed that this cost is merely a flying cost, at a fixed charge per mile flown. In this case, the factor of merit F is proportional to the probability of discovery times the spacing between traverse lines.

In the important limiting cases when the orebody is near its maximum detectable depth, there is a well-known tendency for the geophysical anomaly to become much more equi-dimensional than is the causal orebody. For this reason the other extreme case, namely the problem of detecting a circular disk, is of interest. For this case of the disk, Fig. 7 shows the figure of merit F as a function of traverse spacing on the hypothesis that two crossings over the target constitute discovery. In Fig. 7, in addition, the corresponding figure of merit is shown for the case of a double set of traverses in a criss-cross checkerboard pattern. If two crossings are required to insure discovery, it is clear that the criss-cross flight pattern may be the better. It provides higher values of the figure of merit whenever the traverse spacing exceeds two-thirds of the disk's diameter.

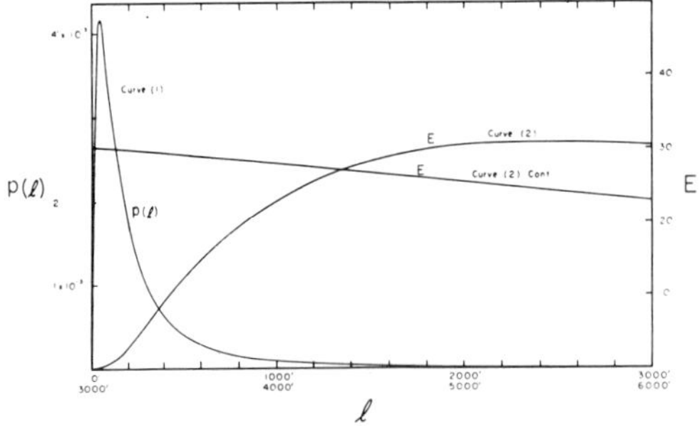

Fig. 8—Distribution of Basin and Range-type ore occurrences as to length (Curve 1) and value (Curve 2).

Determination of Flight Pattern: We have considered some chief items which affect the efficiency of search. The following example will show how these items may be adjusted to arrive at the *best* flight pattern. The Basin and Range type of ore distribution will be assumed, together with a minor hypothesis that the ratios of dimensions, namely the shapes of orebodies, are not affected by their absolute size. The dimensions of the orebodies are then also distributed log-normally. Fig. 8 shows this distribution as to length for the Basin and Range orebodies, on the assumption that the lengths are ten times the mean width. The chief feature is the predominance of small orebodies. Half the bodies are less than 220 ft long, and bodies of 40 ft length are the most abundant. They occur 330 times more frequently than those of 2200-ft length. The second curve in this diagram represents the distribution in value of orebodies as a function of their length. Despite the scarcity of large bodies, the maximum values are represented by those of 2200 ft length. These have an individual value of $50 million each, but they occur in the Basin and Range area only at the rate of 1 per 50,000 sq miles. *In toto* they have a value

500 times greater than those in the most numerous 40-ft category. If we further assume that the bodies are randomly oriented and that two traverses across any orebody suffice for its discovery we find that the most ore is discovered per dollar of flight cost when the traverses are spaced 1300 ft apart. This result is depicted in Fig. 9, which shows the effect upon the profit-ratio of varying the traverse spacing. This represents the solution of the problem of finding the best flight pattern under the special conditions postulated.

COMPETITIVE ASPECTS

As I have remarked on a previous occasion, prospecting plans should also "be governed by the practical requirements of competitive prospecting. The company that sees the favorable opportunity early and acts promptly may achieve a major advantage over others that delay too long in improving their analysis of a situation. The competitive aspects of prospecting obviously add another interesting dimension to the range of considerations which should guide the decisions of management about prospecting ventures. One sometimes wonders whether these managers are being paid enough."

SUMMARY

A few of many kinds of statistical data have been mentioned which are needed for guiding geophysical prospecting. Without belittling in the least the importance of improving all the tools for finding ore—geological, geophysical, geochemical—I have urged the primary importance of appraising the *expectations* of finding ore. There is room, I suggest, for improvement in the business of maximizing the returns from prospecting. The vital statistics in prospecting are just as vital to prospectors as are the corresponding vital statistics in the insurance business. Actuarial experts of a new type are needed to compile the necessary information and to appraise the risks in prospecting blind. We need to improve the art of comparing prospecting ventures in terms of the odds—in just one word, in terms of *numbers.*

As a student I recall a colorful professor who used to admonish his class with the famous dictum of Pythagoras: "The nature of things consists in Numbers." I have never been able to understand these mystic words, but I suppose that our modern million-dollar computing machines would be very pleasing to Pythagoras. When suitably interpreted, his dictum probably contains much wisdom. For future prospectors who will search boldly and rationally for all-important giant orebodies concealed in vast areas, perhaps he has furnished a suitable motto—*The nature of things consists in numbers.*

REFERENCES

[1] M. Allais: Method of Appraising Economic Prospects of Mining Exploration Over Large Territories: Algerian Sahara Case Study. *Management Science*, 1957, vol. 3, pp. 285-347.

[2] T. B. Nolan: The Search for New Mining Districts. *Economic Geology*, 1959, vol. 45, pp. 601-608.

[3] L. B. Slichter: Some Aspects, Mainly Geophysical, of Mineral Exploration. *Natural Resources*, McGraw-Hill Book Co., Inc., Chapter 15, p. 407.

[4] L. B. Slichter: Geophysics Applied to Prospecting for Ores. *Economic Geology*, 50th Anniversary Volume, 1955, pp. 967-968.

[5] W. B. Agocs: Line Spacing Effect and Determination of Optimum Spacing Illustrated by Marmora, Ontario, Magnetic Anomaly. *Geophysics*, 1955, vol. 20, pp. 871-885.

R. M. Ellis and J. H. Blackwell: Optimum Prosepcting Plans in Mining Exploration. *Geophysics*, 1959, vol. 24, pp. 344-358.

D. G. Krige: A Statistical Analysis of Some of the Borehole Values in the Orange Free State Goldfield. *Journal of the Chemical, Metallurgical and Mining Society of South Africa*, 1952, vol. 53, pp. 47-70.

[*Editor's Note:* The appendix of formulas referred to or utilized in the text has been omitted.]

7

Reprinted from pages 589-590, 595-601, 605, 617-618, and 620-621 of *Econ. Geology* **34:**589-621 (1939)

GEOLOGICAL FACTORS IN THE VALUATION OF MINES.*

DONALD H. MCLAUGHLIN.

INTRODUCTION.

SOME years ago, the manager of a mining company, with which I was associated, had a habit of annoying his geological staff by making uncomplimentary distinctions between what he called "real ore" and "geological ore." By the first term, he meant, of course, ore that could be estimated by rule-of-thumb engineering methods with the precision that passes for certainty in some quarters, in contrast to ore that required a higher order of imagination to recognize. After some reflection, I am now willing to accept the distinction as a valid one, but with the comforting observation that in most successful mines there have been periods when the value that could properly be attached to "geological ore," as just defined, considerably exceeded that calculated from strictly defined ore reserves.

Among the many factors that determine future profits and that provide the means of estimating present value, the one that above all others gives mining ventures their own particular quality is *ore supply*. On it, the whole enterprise depends; and when the ore is exhausted, the period of profits or of hope of profits is over.

Mining, as investments go, is usually regarded as a more speculative undertaking than the average industry, but once the existence of an adequate ore-body is established, the dangers and difficulties in modern practice are surely not enough to put it in

* Presidential Address read before the Society of Economic Geologists, New York Meeting, Feb. 14, 1939.

an especially high risk category. Engineering and technical problems in mining and metallurgy may at times be of unconventional sort and require more than ordinary ingenuity in their solution, but, given the raw material in proper quantity and grade, either in the ground or in a bin, one can confidently expect, these days, that some successful way will be found to recover the metal. Indeed, apart from questions involving ore supply, the hazards that have to be faced in mining are merely those that plague industrial enterprises in general. Wage scales, cost of materials, and taxes; fluctuations in output and changing needs of consumers; metal prices; all introduce variables that may be more difficult to estimate and predict than the vagaries of ore deposits. But these uncertainties are by no means unique to mining.

Differences in the degree of probability with which the quantity and composition of the ore in a deposit can be appraised create great contrasts between various mining ventures from the standpoint of financial risk. In some, where all the available orebody can be easily drilled or tested by other reliable means, the danger of loss through lack of ore can be practically eliminated except with regard to the initial expenditure on exploration. At the same time, however, the prospects for something better are likewise eliminated, or at least are very much reduced, and the enterprise loses much of its appeal to those who hope for more spectacular benefits. At the other extreme are mines on deposits in which, for one reason or another, development cannot be pushed far ahead of actual operations, where tonnage or grade cannot be precisely measured, and where expectation of ore depends largely upon geologic evidence.

[*Editor's Note:* Material has been omitted at this point.]

CLASSIFICATION OF ORE DEPOSITS ON BASIS OF ASSURANCE OF ORE SUPPLY.

The striking differences in the nature of the evidence on which estimates or appraisals of ore supply depend are of such vital concern to those who risk their funds in mining ventures that geologists may very properly be called upon to review their assemblage of facts and fancies about ore deposits and to express their opinions concerning the degree of confidence in the existence of ore that is justified in deposits of various types. As a step in this direction, an attempt is made on the following pages to classify deposits according to the assurance of ore supply that they are ordinarily capable of providing with the amount of exploration that is customary in good current practice in active producing mines developed on them. With this object in mind, three major groups, as defined below, are set up, and the common types of ore deposits are assigned to them as nearly as possible by the features that determine the category of risk to which they belong, when appraised as sources of ore.

I. Plenemensurate ore-bodies—those capable of being fully measured and sampled at an early stage in the operations;

II. Partimensurate ore-bodies—those in which prospects for ore in addition to proved reserves remain a substantial element of value until the later stages of the life of the mines based on them;

III. Extramensurate ore-bodies—those difficult to explore and measure much in advance of mining, in which the value of prospects for ore based on geologic evidence exceeds the value based on proved reserves throughout most of the life of mines supported by them.

These headings, of course, merely designate successive parts of a gradation in which the various deposits are arranged as nearly as possible according to the ratio between assured ore supply and prospective ore (or "geologic ore") that is usually attained in them in good practice, when mining operations are in full swing. The rough classification given below does not conform strictly to genetic lines, for other considerations, some of which are only indirectly dependent on concepts of origin, must also be taken into account if deposits with similar qualities from a valuation standpoint are to be grouped together. Assignment of particular deposits to one or another of the major groups and to their place in the sequence is determined to a large extent by the following characteristics:

(1) Size, form and attitude of the deposit.
(2) Accessibility to exploration.
(3) Continuity or regularity in metal content.
(4) Persistence of the mineralization, and particularly of the metal-bearing phase, as revealed by mineral species, textures and alteration.
(5) Controls determined by structure and environment.

In the review of various deposits regarded in terms of ore supply, which makes up the bulk of the remaining part of this address, these features will be emphasized. Examples, for the most part, have been selected from districts or mines of major rank as metal producers, or from those that are fairly well known.

Classification of Ore Deposits according to Assurance of Ore Supply Afforded by Them.

I. PLENEMENSURATE DEPOSITS.

Mechanical concentrates not deeply buried by later formations.

Gold dredging gravels.
Sacramento Valley fields,* Cal.
Bulolo field, New Guinea.

Tin dredging gravels.
Malayan fields.

Flat-lying or gently inclined sedimentary ore bodies.

Sedimentary iron ores.
Luxembourg-Lorraine field.
Jurassic iron ores of England.

Sedimentary manganese ores.
Manganese ores of Trans-Caucasia.

Low grade ores or iron formations, locally enriched.
Mesabi iron range, Minn.

Residual ore deposits formed by weathering from present surface.

Lateritic iron ores.
Mayari and Moa fields, Cuba.

Bauxite ores.
Arkansas deposits.

Supergene ores formed from earlier sulphide mineralization, not buried by later deposits.

Oxidized ores above sulphide ore bodies of large horizontal section.
Silver-pacos ores of Cerro de Pasco, Peru.
Copper sulphate ores of Chuquicamata, Chile.
Ajo carbonate ores (now exhausted), Ariz.

Copper carbonate ores in flat-lying limestone beds, or on flat fractures.
Katanga, Belgian Congo.

Chalcocite-covellite disseminated ores of porphyry copper type.
Chino, New Mex.
Inspiration-Miami ore body, Ariz.
Ray, Ariz.

Concordant veins or lodes in bedded rocks, confined to specific structural units within range of complete exploration at an early stage.

* The few examples in this and corresponding subheadings are listed merely for illustration.

Copper sulphide ores of Northern Rhodesian type (in part).
Roan Antelope Mine, N. Rhodesia.

Veins, lodes or pipes, or ore shoots in them, bottomed within reach of relatively easy exploration.

Primary mineralization in downward converging bodies in many porphyry copper deposits.
Ely, Nev.

Downward converging ore pipes.
Platinum ore pipes of the Bushveldt, Transvaal.
Certain massive sulphide or breccia copper ore pipes.

Termination of veins or ore shoots at definite limits established by a flat petrological or structural locus of some sort.
Cobalt, Ont.

Restriction of ore to a shallow hypogene zone in certain epithermal deposits.
Individual ore shoots at Guanajuato, Mex.

Limitations imposed on persistent ore bodies by property boundaries.

Veins or lodes dipping across boundary planes.
Outcrop mines on the Witwatersrand, Transvaal.

Ore shoots or lodes plunging across boundaries.
Ore shoots in the central group of mines at Broken Hill, N.S.W.

Flat ore bodies or ore shoots extending beyond boundary planes.
Specific mines in Tri-State district (Mo., Okla., Kan.).

II. PARTIMENSURATE DEPOSITS.

Sedimentary ore deposits in inclined folded or faulted beds.

Bedded iron ores extending beyond practical range of complete development.
Birmingham iron ores, Ala.
Iron formations of the Old Ranges (Mich., Wis. and Minn.).

Residual enrichments along deep loci.

Iron ore shoots along deep troughs.
Ore bodies in Old Ranges (Mich., Wis., and Minn.).

Widespread sulphide ores (or their oxidized equivalents) in flat or gently inclined limestones remote from known intrusives.

Lead-zinc ores of the Mississippi Valley type.
Southeastern Missouri lead district.
Tri-State lead-zinc district.

Concordant veins, lodes, or reefs, in beds or flows, in inclined structures persisting to great depths.

Native copper lodes of Lake Superior type.
Lodes of Keweenaw Point, Mich.

Copper sulphide lodes of N. Rhodesian type (in large part).
Mufulira, N. Rhodesia.

Gold ores of the banket type.
Witwatersrand gold field, Transvaal.

Ores with close magmatic affiliations.

Platinum-bearing layers in ultra-basic rocks.
Merensky horizon, Bushveldt.

Chromite layers.
Bushveldt and Rhodesian chromite deposits.

Magnetite layers or lenses in igneous rocks, generally syenitic.
Mineville, N. Y.
Kiruna, Sweden.

Copper and copper-nickel sulphide ore bodies, as layers, lenses or pipes of major size, massive or disseminated, in relatively fresh igneous rocks.
Sudbury, Ont.

Massive sulphide replacement lodes, lenses, mantos and pipes of major size in highly altered rocks, persistent as individual bodies or repeated along well defined loci.

Pyrrhotite gold ores.
Morro Velho, Brazil.

Pyrrhotite copper gold ores.
Noranda, Que.

Pyritic copper ores.
Rio Tinto, Spain.
United Verde, Ariz.

"Crater margin" ore bodies.
Cerro de Pasco, Peru.

Hypothermal lead-zinc-silver lodes.
Broken Hill, N.S.W.

Pyrometasomatic deposits of large size.
Fierro, N. M.
Highland Boy Mine, Bingham, Utah.

Copper and copper silver limestone replacements of major size.
Morococha (in part), Peru.
Bisbee (in part), Ariz.

Lead-zinc-silver limestone replacement ore bodies of major size.
Leadville, Colo. (in part).
Santa Eulalia, Chihuahua (in part).

Lodes of discordant and concordant stringers, constituting ore bodies concordant with inclined or folded layered structures persisting to great depths.

Continuous gold-bearing lodes of quartz stringers and mineralized rock along well defined loci.
Homestake ore bodies, S. Dak.

Repetition of separate gold-bearing lodes along well defined loci.
Dome Mine, Ont.

Continuous zones of gold-bearing stringers in barren rock.
Alaska-Juneau Mine.

Steep continuous veins or lodes, discordant wholly or in part to layered structures, if any.

Gold quartz veins of relatively deep origin.
Veins of Porcupine, Ont.
Mother Lode, Cal.

Siliceous gold-bearing replacements along well-defined shears of relatively deep origin.
Kalgoorlie, W. A.
Kirkland Lake, Ont.

Base-metal veins of moderately deep to shallow origin on strong persistent fractures.
Multimetalled veins of Butte, Mont.
Copper-silver veins of Magma Mine, Ariz.
Copper-silver-lead-zinc veins of Casapalca, Peru.
Silver-lead and silver-copper veins of the Coeur d'Alene district.

Gold-silver veins of moderate to shallow depth of origin, on strong and persistent fractures.
San Dimas district, Durango, Mex.
Pachuca, Mex.

III. EXTRAMENSURATE DEPOSITS.

Mechanical concentrates concealed by later formations.

Ancient gold placers.
Tertiary channels of the Sierra Nevada, Cal.
Deep-leads of Victoria, Aust.

Pyrometasomatic deposits.

Copper sulphide ore bodies, irregularly distributed in skarn zone.
Seven Devils, Idaho.

Limestone replacement ores.

Massive sulphide copper mantos, layers, and pipes.
Morococha, Peru (in part).
Bisbee, Ariz. (in part).

Massive sulphide, lead, zinc, silver mantos and pipes.
Santa Eulalia, Chihuahua (in part).
Tintic, Utah.

Siliceous gold ores.
Black Hills, S. Dak., ores in Cambrian dolomite.

Repetition of relatively short veins or other ore bodies in a broad locus.

Successive flat veins reoccurring in depth in intrusive bodies of restricted size.
Morning Star Mine, Woods Point, Victoria.

Small veins or irregular masses of ore at intervals along or parallel to axes of folds.
Bendigo, Victoria.

Small ore shoots, irregularly distributed in extensive veins.

Rich gold ore shoots in quartz veins.
Alleghany district, Cal.

Flat-pitching discontinuous ore shoots repeated in depth.
Norseman district, W. A.

Plenemensurate Deposits.

After the initial risks of the prospecting or exploration stage have been successfully passed in mining enterprises based on deposits listed in this group, uncertainties with regard to ore supply should be largely eliminated. In all of them, the entire extent of the ore-body is within range of exploration generally from the surface, or under some circumstances from mine workings at a lower horizon. In each, the continuity of ore is such that reliable projections can be made between widely spaced intersections. Except for placers that are elongated along river channels, the principal deposits of this group are extensive laterally. Relatively large areas are needed to control adequate tonnage of ore, but once such properties are acquired, the valuable deposit can be traced to its limits or to property lines at an expense that is usually small in proportion to the scale of the proposed or actual investment. The hope of spectacular improvements in deposits in this class is traded for the comforts of security when the exploration stage ends.

[*Editor's Note:* Material has been omitted at this point.]

Partimensurate Deposits.

Expectation of profits from ore still to be developed is a substantial part of the total value of all mines on deposits of the types listed under this heading. Ores of sedimentary origin and ores formed by residual enrichment are again noted; but in this group they possess structural characteristics that make their complete exploration at an early stage impractical.

[*Editor's Note:* Material has been omitted at this point.]

Extramensurate Deposits.

Although a comfortable margin of assured ore exists in most smoothly running mines, there are a few in which the nature of the ore-body practically enforces almost exclusive dependence on geological evidence. Representatives of a number of genetic types may be assigned to this group, but in each case they will be found to be deposits not easily delimited by economically practical means. Ore shoots too erratically distributed throughout a locus that is too large to be systematically explored, those in extremely heavy or wet ground too costly to keep open except immediately in advance of mining, isolated repetitions of ore or recurrent bonanzas that create sudden but short-lived prosperity after intervals of profitless operation may all serve as sources of ore for successful and even long-lived mines; but in them the danger of failure from lack of ore usually justifies so high a risk rate in the valuation formula that their present value is substantially reduced, unless the geologic evidence is unusually persuasive.

Shortage of ore reserves can, of course, arise in mines of many sorts where stoping is allowed to press too closely on exploratory development, but in the deposits classed as extramensurate, this condition is the rule rather than the exception.

[*Editor's Note:* Material has been omitted at this point.]

CONCLUSION.

Except in the clearly defined group in which the available ore supply is completely delimited—in which hazards and opportunities that give zest to mining are almost eliminated—prospects for finding ore in addition to clearly revealed and measurable reserves must be recognized as an element to be taken into account in any fair appraisal of the value of a mine. Although tonnage and grade of ore that is still to be developed cannot be precisely estimated by routine methods, nevertheless a direct or indirect appraisal of them is made whenever a mine is valued at a higher price than can be strictly calculated from formally developed ore. Estimates of this sort are unavoidably made in the majority of mine valuations and involve assumptions with regard to persistence or recurrence of ore that to be trustworthy must conform to geologic probabilities.

Any attempt to predict specific conditions in the solid rock in advance of exploration necessarily involves risks; but they can be kept to tolerable limits if the appraiser possesses a sound scientific philosophy of what ore deposits really are, supplemented by a shrewd appreciation of their individual vagaries in the many different geologic environments in which they are formed.

8

Reprinted from *Eng. and Mining Jour.* **151**:81-85 (1950)

HOW TONNAGE AND GRADE RELATIONS HELP PREDICT ORE RESERVES[1]

S. G. LASKY[2]

Part I—Why Mathematical Laws Apply to Ore Deposits

DISCHARGE A SHOTGUN at a target. The pattern of holes within the circles of the target might be called a standard pattern of dispersion or frequency. Mathematically it can be expressed in several ways. If the number of holes in each circle is compared with the number of holes in adjacent circles, the equation describing the pattern has a definite form, commonly known as the Gaussian or normal equation of probability. If the target is small, so that some of the shot missed it, the number of holes that the outer circles of a larger target would have contained can be easily calculated.

Likewise, nature seems always to operate in an orderly fashion, never haphazardly, even though man has chosen to tag some of her actions as "random." Moreover, in the pleasing phraseology of D'Arcy Thompson,[3] "Nature rings her many changes on a few simple themes. The same expressions serve for different orders of phenomena. The swing of a pendulum, the flow of a current, the attraction of a magnet, the shock of a blow, have their analogues in a fluctuation of trade, a wave of prosperity, a blow to credit, a tide in the affairs of men. The same exponential curve may illustrate a rate of cooling, a loss of electric charge, the chemical action of a ferment or a catalyst."

In a recent paper[4] I implied that these same principles apply to the distribution of mineral in some ore deposits, as follows:

"In many [mineral] deposits in which there is a gradation from relatively rich to relatively lean material, there appears to be a consistent mathematical relation between tonnage and grade, according to the equation . . . Grade $= K_1 - K_2$ log tonnage. The equation seems to apply . . . to such widely diverse kinds of deposits as the low-grade syngenetic manganese deposits in the Artillery Mountains of Arizona, the vanadium and phosphate [deposits] of Idaho and Wyoming, the Falconbridge nickel deposit, the Alaska-Juneau gold deposit. . . . The equation applies also to the porphyry copper deposits. . . ."

With the porphyry copper deposits as an example, the general pattern of this relationship has now been worked out to the point where it can be presented in more detail. There is nothing new or revolutionary either in the pattern itself or in the method of analyzing or expressing it. It represents nothing more than another frequency pattern, similar to those mentioned. What does appear to be new is the fact that more practical information than has heretofore been appreciated—including geologic conclusions—can be deduced from past records of mining operation.

The characteristics of each mineral deposit, of the kind being discussed here, can be expressed in terms of five curves, three of which are classical or standard frequency curves. One of these is what is known among statisticians and biologists as the "curve of diminishing increments"; the other two are the S-shaped or summation curve and its differential, the bell-shaped "curve of first differences," both described by Thompson.[5] The S-shaped curve is now 350 years old and "represents a certain common pattern among Nature's mechanisms. The growth of a child or of a nation, the history of a railway, or the speed between stations of a train; the spread of an epidemic, or the evolutionary survival of a favored type—all these things run their course . . . after the fashion of the S-shaped curve."[6]

The five curves mentioned are empirical curves discovered from available factual data.

As a conclusion from the analysis presented herein and from related studies, it may be stated as a general principle that in many mineral deposits in which there is a graduation from relatively rich to relatively lean material, the tonnage increases at a constant geometric rate as the grade decreases. Or, as the tonnage increases, the mineral content of the cumulative tonnage decreases at a constant declining rate.

I am not prepared to specify all the geologic types to which this principle may apply, nor even to claim it applies to every example of a particular type. Specific application can be determined only by trial. It fits the porphyry coppers, as described herein, and apparently also other deposits of similar type, in which small quantities of ore minerals are scattered through great volumes of shattered rock. It applies also to some kinds of bedded deposits, as illustrated by the Ar-

[1] Published by permission of the Director, U. S. Geological Survey.

[2] Chief, Mineral Resources Section, U. S. Geological Survey.

[3] Thompson, D'Arcy W., On Growth and Form, The Macmillan Co., p. 139, 1942.

[4] Lasky, S. G., Mineral Resource Appraisal by the U. S. Geological Survey, Colorado School of Mines Quarterly, p. 24, Vol. 45, No. 1A, January, 1950.

[5] Thompson, D'Arcy, op. cit., Chap. III, The Rate of Growth, pp. 78-285.

[6] Idem, p. 158.

tillery Mountains manganese deposit[7] and by the Idaho-Wyoming phosphate-vanadium deposits.[8] Calculations equating assay values against drill-hole footage for a well-known limestone replacement deposit suggest that the principle may have still wider geologic application.

Nor am I prepared to explain the geologic reasons why a mineral deposit should conform to normal frequency pattern beyond saying that a mineral deposit is a normal phenomenon of nature. The final distribution of copper in a porphyry deposit, for example, is the net resultant of several individual phenomena; the spacing and size of cracks in cubic miles of shattered rock; the nature, volume, concentration, and velocity of the solutions that brought in the copper; and the set of factors that caused precipitation of the copper. Each of these phenomena may be expected to adhere to an orderly scheme or pattern; it is therefore not surprising that their resultant also adheres to such a scheme.

This Mathematical Approach Offers Useful Information

The kind of study described herein has considerable practical significance. Some practical applications of the porphyry copper study itself (see Part II) are discussed briefly below, as examples designed to suggest to the reader the usefulness of similar studies with respect to the particular commodity, deposit, or problem that may concern him.

The most obvious practical implication is that it is possible, even ahead of actual exploration, to infer first, the tonnage and grade of material remaining to be found in a deposit, and second, the tonnage and range in grade of the material contributing the most metal. The validity of these conclusions patently depends upon the extent to which the curves can be extrapolated beyond the data. Operations on the porphyry copper deposits were begun on ore whose average grade was somewhat more than 2% of copper, whereas the present average grade is about 0.8 or 0.9%. Inasmuch as the curves fairly rigidly define the situation over that range of grade (about 1½%), it would seem reasonable, geologic evidence permitting and as an operating theory, to project them over an additional two- or three-tenths of a percent, which is the maximum possible extension remaining. Expressed another way, it seems practicable to project for 1 in. a curve already about 6 in. long.

A Typical Porphyry Deposit

A hypothetical average porphyry deposit has been calculated as part of the study. The characteristics of this "average" deposit can be said to be the expectable characteristics of any new porphyry copper that may be discovered. When a deposit actually is discovered and factual data are obtained, these characteristics can be modified accordingly to yield a new set of expectable features. As may be noted from Fig. 5, the ten deposits shown may be grouped roughly into three families —one family composed of H and I; a second family composed of E, G and J; and a third composed of A, B, C, D, and F. The expectable characteristics of a new discovery can be analyzed still further by calculating the average deposit for each of these three families and comparing any factual data with them.

These applications have manifest practical importance. For instance, study offers a basis for appraising the national copper position, inasmuch as the porphyry coppers contribute more than two-thirds of domestic copper production. Should the Michigan and Butte deposits also be amenable to this kind of study, the basis would be good indeed. The part played by the porphyry coppers, in such a national appraisal, can be determined by means of summing up the conclusions concerning individual deposits and by reference to the curve in Fig. 5 marked "U. S. total." This curve is the sum of the ore curves for the individual deposits and represents the known porphyry copper deposits as a group. From it can be read the total amount of ore (and therefore the total copper) present in the ground in known deposits for any specified average grade. Since we know the amount of past production, it would seem that we can deduce a good approximation of what remains. Taking into account irreducible technologic losses and costs, we can further deduce the copper remaining to be produced and the economic conditions necessary to achieve this copper production.

Part II—Nature of Data

The data used include, first, production records, and second, history of reserve estimates. It appears that the unit of study must be a geologic entity, as contrasted with the partial geologic entity represented at places by the operations of a single company. For example, Miami and Inspiration together constitute the geologic entity at Miami, and Kennecott and Consolidated Coppermines constitute the entity at Ely.

The well-managed mining company usually has to maintain a continuing estimate of its ore reserves, bringing estimates up to date once or twice a year. The estimate usually includes, first, only that material whose presence has been demonstrated with considerable assurance, and second, only material sufficiently rich to support profitable operation under the conditions of the time. Thus at any particular time the deposit is known to contain more mineral-bearing rock than is included in the estimate, this rock consisting of material still lower in grade than that accepted by the estimator and perhaps also of material of equal or better grade but not demonstrated to his satisfaction. The estimate is expressed in terms of the weighted average grade of all the material included.

With respect to the porphyry copper deposits, constantly improving efficiency in mining and processing has made it possible to mine each year, on the average, lower-grade material than was mined the year before. As a result, each year (again, on the average) the estimate of reserves is based on a lower cutoff than for the year before and has a lower average grade. Thus, over the years there is built up a record of increasing tonnage and decreasing grade. If the record is long enough, one should be able to read from it the relation between this tonnage and grade.

Traditionally, reserves estimates for porphyry copper deposits have been kept equal to 25 or 30 years' production. They are based on a tremendous amount of exploration data, particularly at those mines that have been in operation for many years. The record is, therefore, long and comprehensive. It is the record of a series of samples consisting of production to any date plus the material included in the reserve estimate of that date. The study described herein is based on the assumption that these samples are representative of the deposit; if they are, then it does not matter

[7] Lasky, S. G., idem; Lasky, S. G., and Webber, B. N., Manganese Resources of the Artillery Mountains Region, Mohave County, Ariz., U. S. Geol. Survey Bull. 961, Figs. 3 and 4, 1949.

[8] By inference from personal communication from V. E. McKelvey.

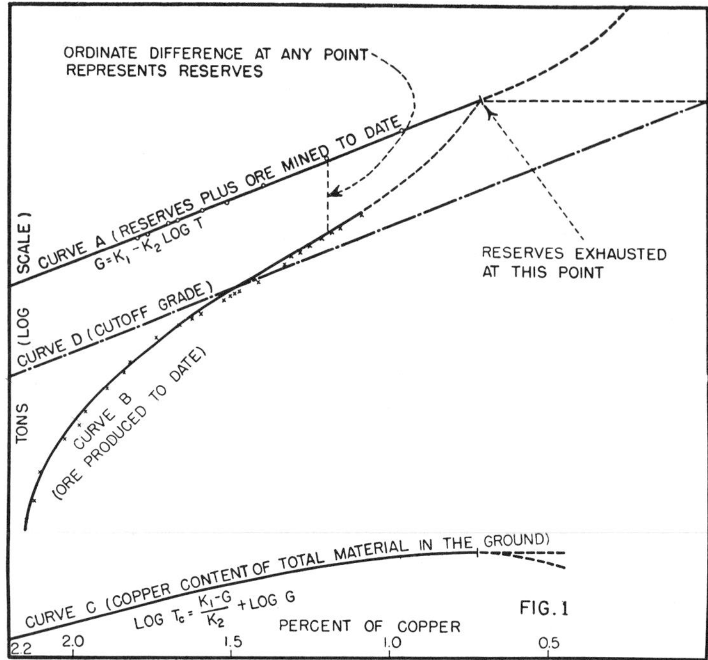

RELATION BETWEEN GRADE AND TONNAGE of ore produced to date, total material in the ground, and contained copper for a typical porphyry copper deposit. Total material in ground in curve A includes reserves plus ore mined to date. Ordinate scale left off in order not to reveal identity of deposit.

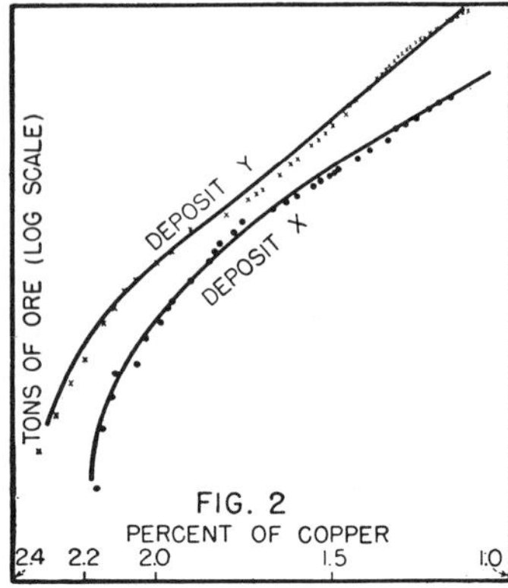

RELATION BETWEEN GRADE and cumulative tonnage of ore mined. Ordinate scale left off in order not to reveal identity of deposit.

Table 1—Sample production record for a hypothetical porphyry copper deposit

(TABLE A, Production by years; TABLE B, Production in order of decreasing grade)

TABLE A			TABLE B			
Year	Tons (000)	Grade (% Cu)	Grade (%Cu)	Tons 000	Cumulative Tonnage (000)	Av. Grade
1912	1,400	2.12	2.16	2,900	2,900	2.16
1913	2,900	2.16	2.12	1.400	4,300	2.14
1914	3,600	1.83	1.83	3,600	7,900	2.00
1915	3,100	1.64	1.82	2,300	10,200	1.99
1916	4,300	1.63	1.76	1,800	12,000	1.92
1917	2,300	1.82	1.70	500	12,500	1.92
1918	1,800	1.76	1.64	3,100	16,100	1.86
1919	500	1.70	1.63	4,300	20,400	1.80
1920	1,900	1.55	1.59	3,200	23,600	1.78
1921	3,300	1.50	1.55	1,900	25,500	1.76
1922	3,200	1.59	1.50	3,300	28,800	1.73
1923	3,500	1.42	1.42	3,500	32,300	1.71
1924	4,000	1.08	1.14	3,600	35,900	1.65
1925	2,500	1.13	1.13	2,500	38,400	1.61
1926	3,600	1.14	1.12	3,000	41,400	1.58
1927	4,000	1.11	1.11	4,000	45,400	1.53
1928	3,000	1.12	1.08	4,000	49,400	1.49
1929	3,100	1.05	1.05	3,100	52,500	1.46
1930	1,700	1.01	1.01	3,300	55,800	1.43
1931	1,600	1.01				

that the deposit may not have been drilled to its ultimate limits. The grade-tonnage "pattern" revealed by the samples is analogous to the shotgun pattern described in the opening paragraph, by means of which the number of holes that would be present in the outer circles of a large target can be calculated from the pattern of holes in the small target actually used.

Table 2—Relative Magnitude of Porphyry Copper Deposits

DEPOSIT	According to inferred total copper content STANDING	According to production to 1948 and reserves currently estimated STANDING
A	4	3
B	1	1
C	7	7
D	6	6
E	2	4
F	7	7
G	3	2
H	9	8
I	8	9
J	5	5

When the material estimated as reserves at any particular time is added to and averaged with the ore produced to that time, the resulting figure can be said to approximate the total amount of material in the deposit at that average and corresponding cutoff grade. Inasmuch as ever lower-grade material is mined from year to year on the average and is included in succeeding reserve estimates, each succeeding "total" value shows a greater figure for tonnage and a lower figure for average grade. When the tonnage is plotted against the average grade, there results a curve that corresponds to equation:

$$G = K_1 - K_2 \log T \ldots\ldots (1)$$

wherein T is the tonnage produced to any given time plus the estimated reserves, G is the weighted average grade of this tonnage, and K_1 and K_2 are constants to be determined for each deposit. K_2 is always preceded by a minus sign, inasmuch as T increases as G decreases.

Curve A, Fig. 1, is an example of a plot for an actual deposit. The plotted points are so tightly spaced that the curve does not have to be fitted to them but, on the contrary, is closely defined by them. According to this curve, the tonnage in this deposit increases at a compound rate of 14.9% for each 0.1% decrease in grade.

Production Records Show Tonnage-Grade Relationship

The grade of ore mined during any given period is adjusted by the operators to the economics of the time and thus reflects the influence of an independent non-geologic factor. The effect of this factor can be modified by statistically reassembling the annual production figures according to grade of ore produced. Table 1 illustrates this operation. Further, in order to be able to compare this record with that obtained from combining production figures with reserve estimates, the values of the declining grades are summed up into cumulative form, and the corresponding average grades are calculated, as in Table 1-B. When

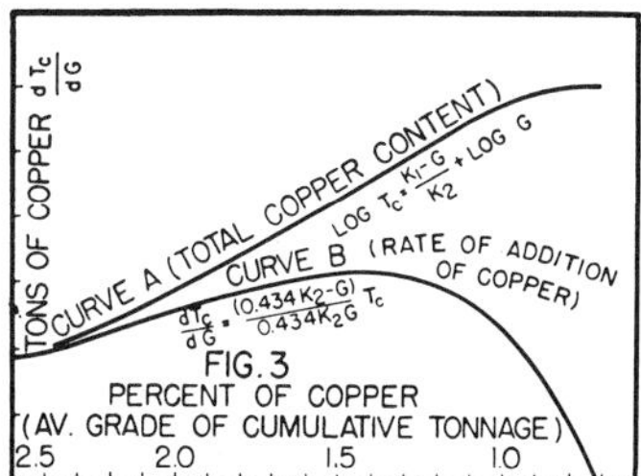

RATE OF ADDITION of copper and total copper content with respect to grade of ore for a typical porphyry copper deposit. Ordinate scale omitted

the cumulative values are then plotted against the corresponding average grades, the resulting plot for porphyry copper deposits typically has the form shown in Fig. 2, in which plot X is for the same deposit as Fig. 1, and plot Y is for a second actual deposit. Plot X is duplicated as curve B in Fig. 1.

The left-hand part of this curve represents the years of operation during which higher grade parts of the deposit were selectively mined, and the right-hand part represents the years after bulk mining was established, rich and lean ore then being mined together. In the phraseology of palmistry, this curve is the typical "life-line" of a porphyry copper.

Curve A, Fig. 3, which represents the same deposit as Fig. 1, shows the form of the curve that is obtained when copper content is plotted against the grade of the ore that contains it. Curve A is replotted as curve C, Fig. 1. The equation of this curve is

$$\text{Log } T_c = \frac{K_1 - G}{K_2} \log G \ldots (2)$$

or

$$T_c = 10^{\frac{K_1 - G}{K_2}} G$$

Equation 2 has been derived algebraically from equation 1 by first changing equation 1 to read as follows:

$$G = K_1 - K_2 \log T + K_2 \log G - K_2 \log G \ldots\ldots\ldots\ldots\ldots (3)$$

and then developing equation 3 to the form of 2 by converting log T + log G to its equivalent, log T_c, where T_c is the copper content in tons.

For any particular deposit, the values of K_1 and K_2 are the same for both equation 1 and equation 2.

This is Thompson's S-shaped curve mentioned earlier.

If cumulative ore tonnage is plotted against the cumulative copper content, the resulting curve has the form shown in Fig. 4. This is the "curve of diminishing increments" mentioned in the introduction. Its equation is

$$y = a(1 - e^{-kx})$$

$$\text{or} \quad T_c = a(1 - e^{-kT_0}) \ldots\ldots\ldots (4)$$

where T_0 is the cumulative tonnage of ore at any average grade, T_c is the tonnage of copper in this ore, and *a* is the total copper in the deposit.

The maximum shown by the curve for equation 2 (Curve A, Fig. 3, and curve C, Fig. 1) is the point at which copper ceases to be added; as this is a cumulative curve, at that point all copper-bearing material in the deposit has been included. From there on, only barren material could be added. In other words, the zero cutoff has been reached. Thus, practically, the curve should stop at that point, or continue as a horizontal line. The total copper content (and thus the limit of the copper curve) can be determined also by solving for *a* in equation 4; the differential of equation 4 is

$$\frac{dy}{dx} = k(a - y) \ldots\ldots\ldots\ldots (5)$$

and *a* is the point at which the plot of this equation intersects the ordinate.

The point of zero cutoff determines also the practical limit of the curve for equation 1 (curve A, Fig. 1), for, practically speaking, the curve cannot extend beyond the point at which all the metal-bearing material is included. If barren material were averaged in, the curve would extend along the dotted path shown. It must be pointed out here that all rock contains a little copper, but as this hardly exceeds a few parts per million, zero cutoff can be considered synonymous with zero content.

The zero cutoff point indicates the extreme limit of the curve, but for some deposits the practical end of the curve is at some point to the left of the zero cutoff. Curve A and B, Fig. 1, converge, as logically they must. Since curve A represents reserves plus ore produced to date, and curve B represents only ore produced to date, the ordinate difference at any point represents reserves; when the deposit is mined out, the reserves are exhausted and the two curves end at a common point.

For some deposits these two curves intersect at a point to the left of that determined by the zero cutoff, and for those particular deposits the extreme limit of curve A

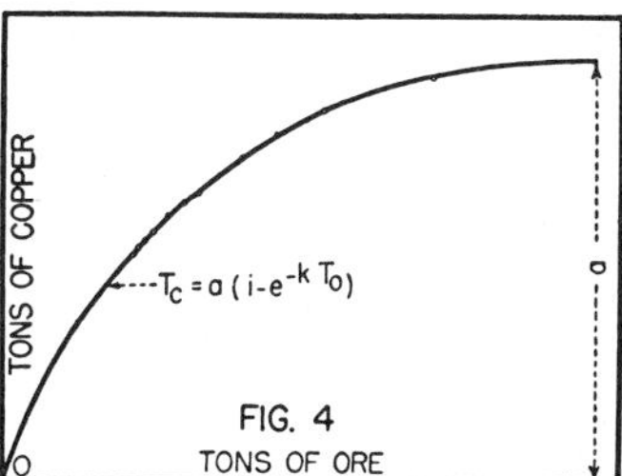

RELATION BETWEEN CUMULATIVE ore tonnage and copper content for a typical porphyry copper deposit. Ordinate scale omitted.

is taken at the point of intersection. This may or may not be a true mathematical point, but on the basis of present available data it seems to be a practical one.

Curve B, Fig. 3, is the form of the curve for the differential of equation 2. The differential equation is

$$\frac{dy}{dx}\left(\text{or } \frac{dT_c}{dG}\right) = \frac{(0.434\,K_2 - G)\,T_c}{0.434\,K_2G} \ldots\ldots\ldots (6)$$

This curve is the "bell-shaped curve of first differences" mentioned in the introduction. It is one way of expressing the frequency of the various grades of ore, from rich to lean, that compose the deposit. For the deposit illustrated here, most copper is contributed by the particular material—100 million tons—that pulls the average grade down to the range 1.7 to 1.2%. By reference to the cutoff curve, explained below, it appears that this 100 million tons ranges in grade between 0.5 and 1.0% of copper.

Cutoff Grade

Consideration of cutoff grades leads to uncertain results. The cutoff grade used in calculating reserves is usually considered a "trade secret" by most mine operators. Little factual information is available, and what is available does not check with the mathematical analysis.

The relation between cutoff and average grade can be deduced mathematically by two methods:

One method depends on the fact that for every value for tonnage and corresponding average grade there is a corresponding cutoff grade, and on the fact that every tonnage value includes all material at and above the corresponding cutoff grade. Any additional material

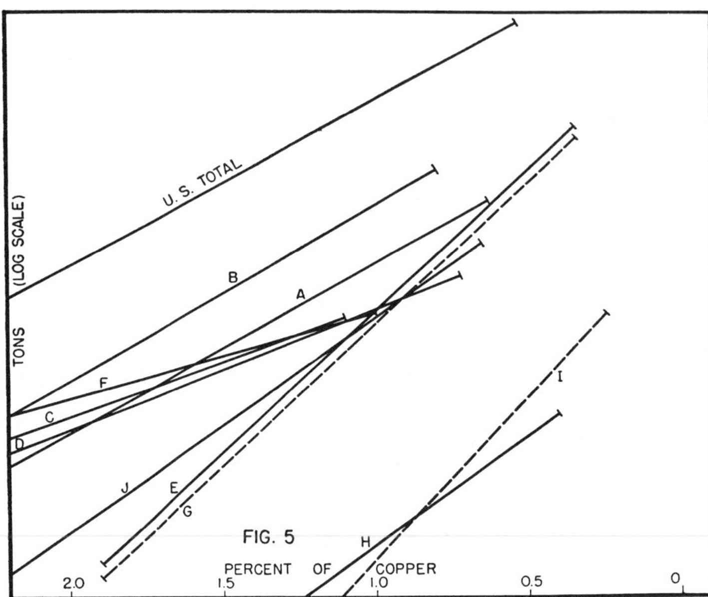

GRADE-TONNAGE RELATIONS AND RELATIVE MAGNITUDE of ten American porphyry copper deposits. Groups A, B, C, D, F; J, E, G; and H, I are deposits with similar characteristics. Ordinate scale left off in order not to reveal identity of individual deposits.

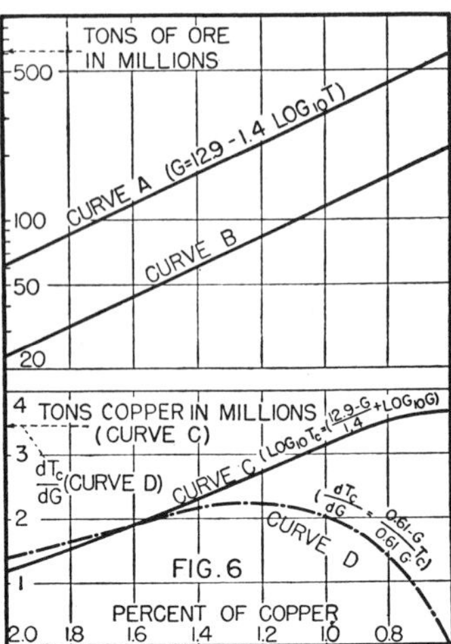

CHARACTERISTICS of hypothetical average porphyry copper deposit. Curve A, total material in the ground plotted against average grade; Curve B, total material plotted against cutoff grade; Curve C, copper content; Curve D, rate of addition of copper.

(ΔT) would be below the existing cutoff grade, and the minimum grade of the additional material would then become the cutoff grade for the new total. If ΔT is taken as the tonnage corresponding to a particular grade bracket (ΔG), the lower value of this grade bracket is the cutoff grade applying to the entire deposit at the point of the corresponding ΔT. Both ΔT at any point and the average grade of ΔT can be calculated readily from equations 1 and 2.

As ΔT is made smaller and smaller, its average grade approaches the cutoff value; if it is made small enough, its grade may be accepted as equal to the cutoff. When the cutoff values so obtained are plotted against their corresponding tonnages, they define a curve of the same type as curve A, Fig. 1, and parallel to it.

The second method is equally short. The determination of zero cutoff point has already been described. This constitutes one point on the cutoff curve. A second point can be determined from the fact that for the highest-grade ore in the deposit the cutoff grade is the same as the average grade. This grade is roughly equal to K_1 of equation 1, as becomes evident by taking T at one ton. Connecting the two points gives the cutoff curve.

The curves obtained by the two methods are not precisely the same, one being parallel to the curve for average grade and the other diverging slightly from it; but for the range of values applicable to the deposits, the two methods give nearly identical results. Curve D, Fig. 1, is the cutoff curve, as thus determined, companion to curve A. The actual cutoff values used by the company engineers in estimating reserves during past years are higher than those indicated by the curve, but available data are insufficient to analyze the discrepancy. The reason for it may lie in the fact that the cutoff grade used by these engineers is not actually the grade of the leanest material included but is the average grade of the leanest minable block.

There is no necessary relation between curve D and B. Whereas curve D relates to the entire deposit as a geologic phenomenon, curve B relates to scattered grade blocks incompletely mined out; in effect, curve B represents only skeleton data.

Fig. 5 shows the grade-tonnage relations and relative magnitude of all the porphyry copper deposits of the United States except for a few of the smallest.

Table 2 compares the relative magnitude (on the basis of contained copper) as deduced from the mathematical analysis described in this paper, with the relative magnitude as would appear to be true on the basis of reserves plus production to date.

Fig. 6 shows the characteristics of the hypothetical average porphyry deposit calculated from the eight major deposits by the least-square solution. The equation for it is as follows:

$$G = 12.9 - 1.4 \log T \quad \ldots\ldots (7)$$

The hypothetical average deposit contains a total of somewhat more than 600 million tons of ore averaging about 0.6% of copper, and in it the tonnage increases roughly at the rate of 18% for each unit decrease in grade of 0.1%. Most of the copper in it is contained in the 175 million tons having a copper content between 0.5 and 0.9%.

Of the many persons, mostly friends and geologists, with whom I have discussed the subject of this paper, I am especially and sincerely indebted to the following: R. M. Hernon, James Grunig, and W. W. Rubey of the U. S. Geological Survey; Julian W. Feiss of the U. S. Bureau of Mines; Walter A. Hendricks of the Department of Agriculture; Donald H. Menzel of Harvard College Observatory; and James P. Pollock. I am indebted above all to V. E. McKelvey of the Geological Survey, who called my attention to the logarithmic relationship on the Wyoming-Idaho phosphate-vanadium deposits.

9

Reprinted from *Mineralium Deposita* **3**:1–17 (1968)

The Pre-Evaluation of the Possible Profitability of Exploration Prospects

Third Hugh Exton McKinstry Memorial Lecture, Harvard University, April 25, 1967

ARTHUR A. BRANT

Geophysical Department, Newmont Exploration Ltd., Danbury, U.S.A.

The need for exploration and discovery is apparent from the doubling of tonnage mined about each 12 years. The dependency of exploration expenditure on profits and cash flow is indicated. The capital outlays for selected U.S. and Canadian mines are listed, Tables 1 and 2. The increasing exploration risk, i.e., money spent per discovery, necessitates that criteria be evolved to evaluate exploration projects on the basis of their profitability potential. This involves the chance of occurrence, the chance of actual discovery and the chance of sufficient economic value. These factors are discussed and their evaluation is outlined. An operational equation is set up, and the return on an exploration venture, and a series of exploration ventures, versus the odds against involved, are presented as a possible means of pre-evaluating exploration projects.

Die Notwendigkeit, Exploration zu treiben, ergibt sich aus der Tatsache, daß sich die Tonnage des Fördergutes im Bergbau alle zwölf Jahre verdoppelt. Die Abhängigkeit der Explorationskosten vom Profit und Kapitalfluß wird aufgezeigt. Tabelle 1 und 2 stellen die Kapitalauslagen für ausgewählte Grubenbetriebe in den U.S. und in Canada dar. Das zunehmende Explorationsrisiko, d.h. die Ausgaben pro Entdeckung, verlangt die Entwicklung von Kriterien zur Beurteilung von Explorationsprojekten auf einer Erfolgsaussichts-Basis, die den Zufall (Chance), daß das Vorkommen vorhanden ist, daß es auch wirklich entdeckt wird und daß es genügend abbauwürdig ist, einschließen. Diese Faktoren werden erörtert und ihre Ermittlung und Beurteilung wird erläutert. Es ergibt sich eine praktische Gleichung, mit der der Gewinn eines und einer Serie von Explorationswagnissen ermittelt und den negativen Faktoren gegenübergestellt wird. Es besteht somit die Möglichkeit, Explorationsprojekte im voraus zu beurteilen.

I hope to show:

1. The need for exploration;
2. The increasing cost of exploration;
3. The necessity for venture selectivity;
4. A yardstick for decision judgement.

To the extent that the subject has been studied, the complexities have become increasingly apparent. The presentation, therefore, will be sketchy.

1. The Need for Exploration

A ten year running average shows world copper consumption increasing at 4% per annum cumulatively, a typical growth curve, which provides a straight line between log of tonnage and the increasing time in years, Fig. 1. Tons increase per year $\simeq$ i x present tons where i is the growth rate.

The equation of such an uptrending curve with semi-log straight line plot is:

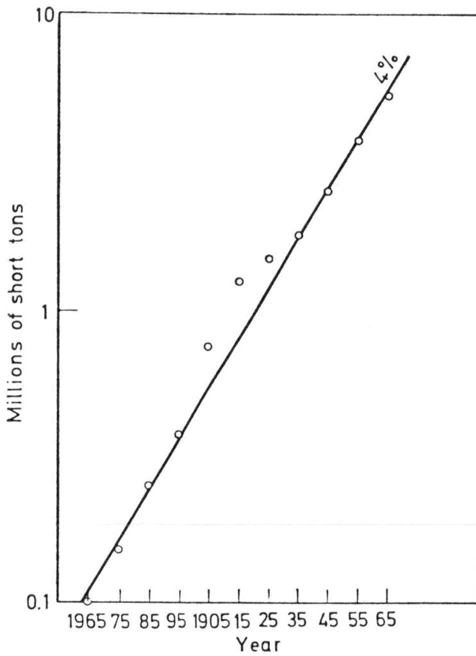

Fig. 1. Ten year running average of world copper consumption

$$\frac{y}{y_0} = e^{(n-n_0)\log(1+i)} = (1+i)^{n-n_0}$$

i.e., $\log y - \log y_0 = (n - n_0) \log (1+i)$

where y_0 is the initial or reference value
n_0 is the initial or reference date
y is the present or future value at year n and
i the cumulative rate or growth rate.

For such a law $\frac{dy}{dn} = \log(1+i)\, y \simeq iy$ where i is small

and $\frac{d}{dn} \log y = \log (1+i)$.

Please note that the above is merely the growth equation over the increasing slope part of the curve. In life and nature, growth is usually followed by a flattening off, and frequently by a decay at a more rapid rate than the original growth.

The statistical data presented are fairly short term, usually of a generation's duration, and projection will be to the immediate future, so the simple growth law may apply.

U.S. copper grade mined shows a cumulative decrease of 1.8% per annum, Fig. 2, at an average increasing price trend of 2.4% per annum. If this is taken as world-wide then the tonnage mined per annum is increasing at $4+1.8=5.8\%$ per annum, implying a doubling of tonnage mined every twelve years. So twelve years from now we should be capable of mining twice as many tons at an average grade 25% below today's.

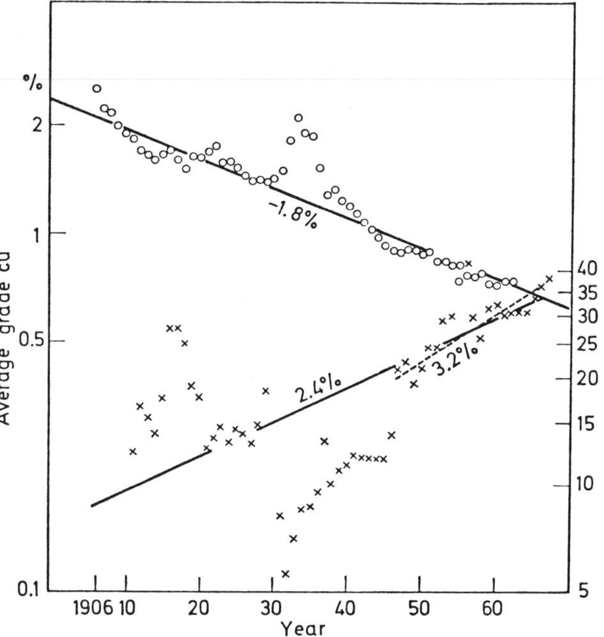

Fig. 2. Average grade of copper mined in U.S. (oo) and average New York price per pound (× ×)

The lead time between discovery and production is four to eight years.

If trends continue, the need for exploration, more precisely discovery, is apparent.

2. The Increasing Cost of Exploration

To-date, exploration expenditures have fluctuated with the price of copper, i.e., with mining profits. Fig. 3, for example, shows the almost exact correlation between Canadian metal exploration expenditures after Dr. D. DERRY and the New York price of copper.

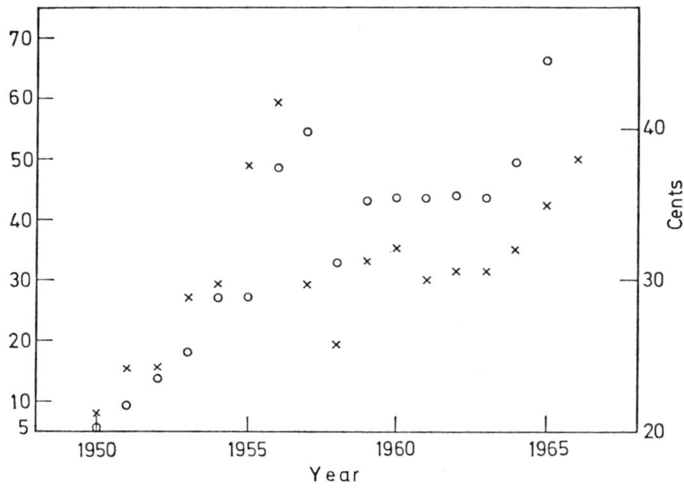

Fig. 3. Dollar exploration in millions, Canada (oo) compared to average New York price for copper (× ×)

Nevertheless, a growth in exploration expenditures at a rate of 11% cumulative is apparent for Canada, Fig. 4, lower half. Ours at Newmont Mining Corporation have increased 10% per annum in the last ten years, as is true for many U. S. mining companies. About 5% is increased costs, so net growth rate has been 5—6% per annum.

Exploration costs as a fraction of gross mining revenue have been increasing at 5% per annum in the Canadian case and certainly similarly for world-wide U. S. based companies; as seen on Fig. 4, upper half.

This suggests that Canadian mining gross has been increasing at 11—5 or 6% per annum as is apparent in Fig. 5, upper line. For the American coppers, Fig. 5, lower line, the increase hasbeen nearly the same.

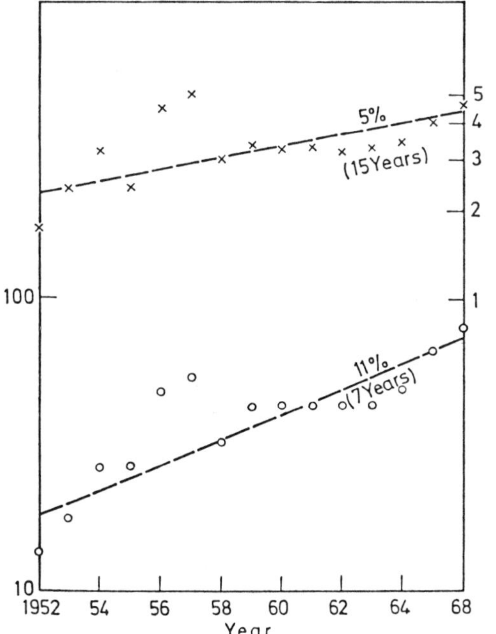

Fig. 4. Dollar exploration in millions, Canada (below), dollar exploration as % of production (above)

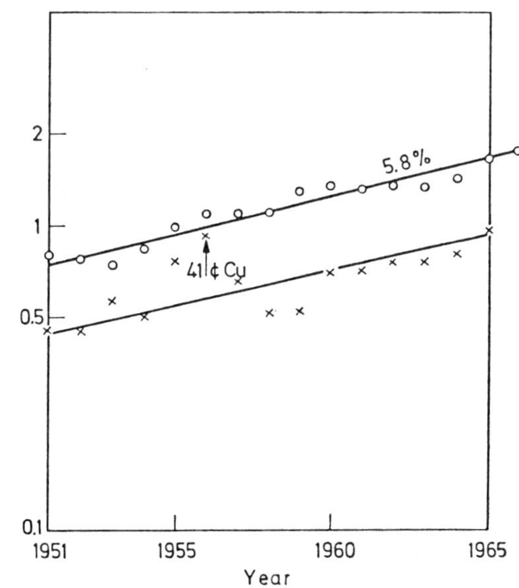

Fig. 5. Dollar production Canadian mines (above), dollar production U. S. copper mines (below) in billions

World-wide, however, a 4% increase in Cu production at a 1.8% lower grade, but 2.4% higher price, suggests an increase in gross revenue of about 4.6%. When the South African, South American, and Australian contributions are recognized, it leaves little increase outside the U. S., Canada, and the countries mentioned.

One should also work in the decreasing purchasing power of the vanishing American dollar. Fig. 6 shows the growth in the presumed cost of living index undoubtedly slanted toward a flatter slope since the element of quality of item has long since ceased to be considered. This indicates a 1.4% per annum cumulative decrease in purchasing power or labor equivalent. Thus, the world-wide copper gross revenue is increasing at no better than 4.6—1.4 or 3% per annum, a negligible amount, for the C. L. E. index and population is rising as fast or faster in most other countries.

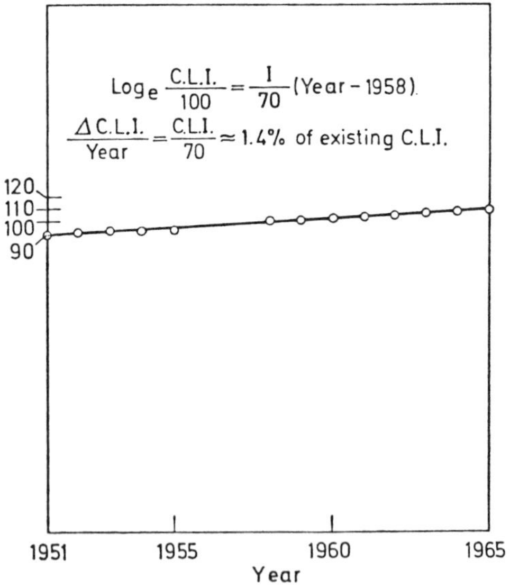

Fig. 6. Cost of living index

Presuming the increase in demand for ore tons of Cu, what is the limit to be expected in exploration expenditures? We have seen, Fig. 3, that exploration expenditures fluctuate with the price of copper, i. e., are highly dependent on profit.

Based on 1951—56 inclusive, for Cu, Pb, and Zn mining in the U. S. A., the average rate of cash flow in percents of gross were, profits after taxes 13%, depletion 10% and depreciation 2 1/2 % = a cash flow on average of approximately 25% of gross revenue.

Dividends, priority 1, account for disbursement of 50% of net income or 25% of average cash flow; capital and new venture investments, priority 2, for 50% of cash flow, leaving 25% of cash flow or 6% of gross for debt retirement, priority 1, capital reserve, and for exploration, drilling and research, priority 3. It looks, therefore on average, as if about 5% of gross, providing the present average level of profits maintain, represents an upper limit for exploration, drilling and research costs.

For selected, active, profitable mining companies in the U. S. A., exploration drilling and research expenditures as of 1966 were running at about 15% of cash flow. Our expenditures in this direction at Newmont are about 6—7% of gross revenue which is near the upper limit in relation to our income.

The recent level of exploration in Canada at 4.6% of gross, Fig. 4, upper half, reflects the influx of outside money, and represents a level toward the limit of the monetary resources available. It is, however, dangerous to relate exploration to gross revenue rather than cash flow.

Bringing exploration, research and drilling expenditures to a lb. of Cu in the U. S., indicates a present level of about 1.4 ct. per lb., with an upper allowable limit of 1.8 ct. per lb. at current prices. Our exploration work at O'okiep, South Africa in the earlier sixties was at a 0.9 ct./lb. Cu rate, and is now about 1.2 ct.

Mine labor wages are increasing at 6.7%/annum and costs per lb. of Cu mined at 4.2%/annum, see Fig. 7. With many fluctuations, Fig. 2, lower plot, the price of copper, uncorrected for C. L. E. is increasing at 2.4% per annum longer term, and 3.2% since 1947. Gross copper revenue in the U. S. then should have increased at the 5.8% increased production and 2.4% increased price less the 1.8% decrease in grade less increased labor costs and other costs which conservatively can be equated to the C. L. E. of 1.4%. Thus, increase in gross copper revenue is 5.8% + 2.4% — 1.8% — 1.4% = 5.0% approximately, as is apparent in Fig. 5, the lower plot.

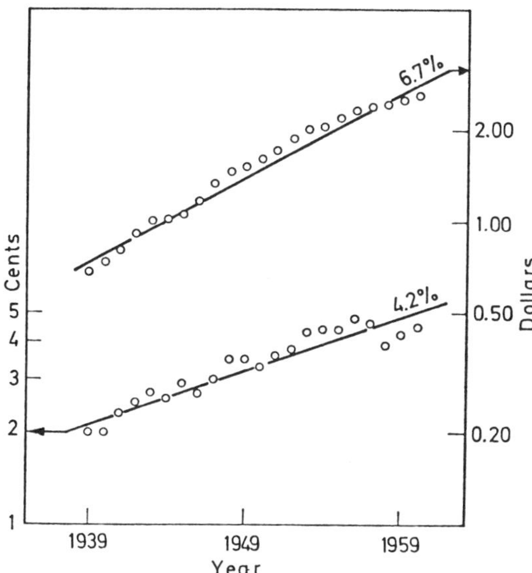

Fig. 7. Mine labor per hour (above), labor per pound copper (below)

It is interesting that in terms of constant dollars, the price of copper has actually been decreasing by about 0.4% per annum (Fig. 8). Thus, if present trends continue, exploration + research + exploratory drilling expenditures

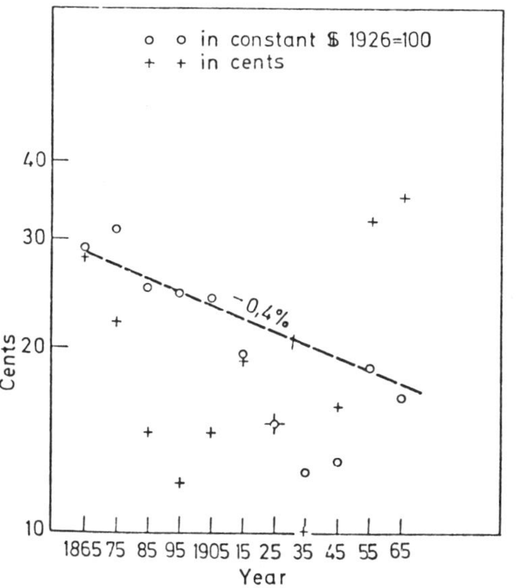

Fig. 8. Ten year running average of price of copper

could increase to a limit of about 5% of gross mining revenue, but are and will be highly contingent on the profit margin. Since we are now at a level of 4% gross, a flattening out is expected with any further rise paralleling that in cash flow or gross.

By way of comparison, the case for thirty outstanding petroleum companies is as follows: Geology, geophysics and land leases, some 8% of cash flow, 15% of net income.
Research some 3% of cash flow.
Exploratory drilling some 8% of cash flow. (Dry holes were considered exploratory, pay holes development).
Together, these total 19% of cash flow or 3.6% of gross. So mining, in its small way, is spending almost as much proportionally on exploration and research. Compare the 44 billion gross and 8.3 billion cash flow of these thirty companies to copper mining's 1 billion gross and 250 million cash flow.

Briefly, let us consider the capital investment in a mining venture, for this is the great funnel into which the large share of cash flow is poured, leaving exploration as the younger brother.

Tables 1 and 2 list the capital involved in the start-up of a series of open pit copper mines, and several underground higher grade ventures. When grades are considered, a good horseback figure for start-up capital is derived: = $ 200 × net lbs. Cu per ton × tons per day to be mined, or without smelter, $ 150 instead of the $ 200 figure.

Now, $ 30,000 of investment is necessary on average to create one tax paying, not tax based, job. Thus, the capital investment of 101 million at San Manuel should have created 3400 jobs. Present payroll is 2050 persons, which in turn gives rise to some 300 outside, or a total of 2350 resulting jobs.

The capital investment necessary to increase production highlights a simple consideration. U. S. level of copper consumption for 1966 is 16.5 lbs. per capita. World consumption, excluding United States, is 3.1 lbs. per capita per annum. Indian consumption is less than 0.5 lbs. per capita. To bring India say to the U. S. level of copper consumption would require 500 million × 16 × 200 ·/. 300 days or 5 billion dollars, on the present basis of costs and technology. Foreign export of capital (the rest is

Table 1. *Capital investments large tonnage-low grade*

Mine	Tons/Day	Capital in millions	$/ton Capital	Comment
San Manuel	36,000	101	2800	Smelter, townsite Cu, porphyry Block caving
Palabora	33,000	99	3000	Smelter townsite Cu, porphyry type Open pit
Pima	10,000 initial 18,000 present	41.6	4160 2300	No smelter or townsite Cu, porphyry type Open pit
Esperanza	8,00 initial 14,000 present	20	2500	No smelter or townsite Cu, porphyry type Open pit
Mineral Park	10,000 initial 15,000 present	33	3300	No smelter or townsite Cu, porphyry type Open pit
Bethlehem	4,000	13	3250	No smelter Cu, replacement and disseminated Open pit
S. Peru	36,000	228	6330	Smelter, two townsites 150 mile rail mine to smelter Cu, porphyry type, open pit
Endaco	8,000 initial 12,000 present	21.2	2650	No smelter, townsite, MoS_2, open pit
Mission	10,000 initial	25	2500	No smelter or townsite Cu, porphyry, open pit

Open pit types = Without smelter $ 2750/ton/day capital
— With smelter and townsite $ 4000/ton/day capital
Average grade = .8 or 16 lbs. Cu/ton recoverable
Investment about $ 200.00 per lb. Cu recoverable per day

used internally) is roughly 10 billion per year world-wide. In short, it would take all the world's effort for $^1/_2$ year to bring about such material equality. Looked at another way, if the U. S. copper consumption were divided equally to the rest of the world's population, it would result in 1 lb. of copper per person, or about twice present Indian level.

It appears then that the great socialistic wave of enforced equality that will engulf us in a generation's time will do little to alleviate material deficiency. In fact, outside of Sweden, where Stockholm is rapidly taking on the appearance of a mediterranean city, I know of no socialist country that has ever shown the capability to feed itself. Charter membership in the United Nations seems to require an avowal of socialism, inability to generate internal capital, inability to feed your people and an increasing debt status. Perhaps this is why South Africa is so alienated for it qualifies in none of these aspects.
This leads to a little history, the growth, ascendency and decay of one time pre-eminent countries as outlined in Table 3. It is noted that the period of

Table 2. *Capital investments underground replacement and miscellaneous*

Mine	Tons/Day	Capital in millions	$/Ton Capital	Comment
Mattagami	4,000	28	7000	Cu, Pb, Zn replacement Townsite, no smelter
Fox Lake	4,000	29	7250	Cu, Pb, Zn replacement Townsite, no smelter
Tsumeb	1,000 initial	8.6	8600	Cu, Pb, Zn pipe Townsite, initially no smelter
O'Okiep	2,000 initial	5.6	2800	Cu in noritoids initially no smelter, but townsite
Granduc	8,000	75	9450	Cu in shears in altered seds Townsite, 30 mile road, 11 mile tunnel
Carlin	2,500	10	4000	Gold in silicified ls. no townsite, open pit
Imc K. Potash	7,200	65	9030	Potash Underground mining Townsite

Underground replacement types $ 7000/ton/day
Average grade Cu equivalent recoverable = 2–1/2% = 50 lbs. Cu/ton
Investment = $ 150/lb. Cu recoverable

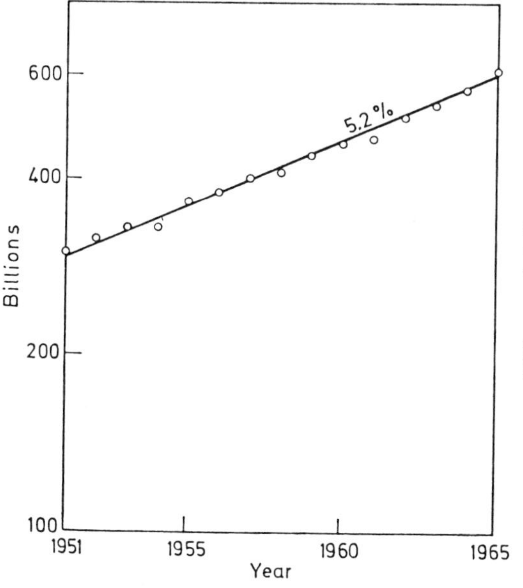

Fig. 9. Net national product, U.S.A.

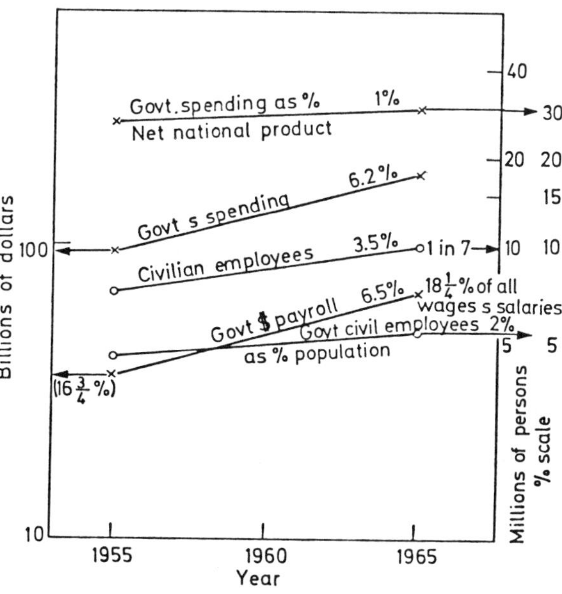

Fig. 10. U.S. Federal & State Expenditures & Employees; Government spending

ascendency never exceeds the period of growth, struggle and sacrifice.
So the growth curves do not rise forever but, man being man, flatten out and turn down. The decisions we make today determine whether our ascendency is 100 or 150 years, a short 5–8 generations, and popular decisions do much to shorten the interval.
Before we leave growth curves, a word about one of the fastest growing endeavors in the U.S.A., namely government, federal and local.
Fig. 9 shows the recent increase in the U.S. net (about 80% of gross) national product at a 5.2% rate less 1.4 for C.L.E. = 3.8%.
Government spending is rising at 6.2%, government payroll at 6.5%, and government civilian employees by 3½%, Fig. 10. These figures suggest that in 60 years (3 generations), all employables (40% of population), could be on government payrolls. By 1980, for example, the Department of Agriculture will have more employees than there are farmers in the U.S. One in seven civilian employees are now in government.
Personal income taxes are 125 billion out of gross wages and salaries of some 500 billion, roughly 25%. Thus, each man spends the first 90 days of each year working for the government, i.e., ¼ of all our labor, representing a first claim before family or any other responsibility. The pyramids indentured 6% of the Menes dynasty labor force for the full year, i.e., 24% of all man-hours, while the great wall of China required a ⅓ year's labor of each man or 33%. The 20th century angle is that you may serve your indentureship at home.
The point I wish to make in its simplest form is the extent to which this curtails investment and increase in tax paying jobs. Seventy billion of governmental payroll if cut in half could provide the investment for 35 billion ·/. 30,000, the investment per job, or 1 million new tax paying jobs per annum.

Table 3. *Growth, ascendency and decay of pre-eminent countries*

Country	Period of growth, struggle and sacrifice	Period of power ascendency, wealth	Decay
Athens	Cleisthenes, Solon, Pisistratus 600–490 b.c. 110 years	Marathon Salamis To end of Peloponnesian War 490–404 b.c. 85 years	404 b.c. To Chaeronea 338 b.c. 70 years
Rome	Federation of Italy 265 b.c. To Actium 31 b.c. 230 years	Actium 31 b.c. to death of Commodus 192 a.d. 220 years Depletion of monetary reserves 30–115 a.d. Loss of export markets 100 a.d. Urban Renewal Augustus on Welfare, Augustus; Maintenance grants, Trajan Bureaucracy increase, Hadrian on Increasing social anarchy	192 a.d. to Alaric 410 a.d. 220 years
Spain	Union of Castile and Aragon 1479 Final overthrow of moors 1492 Conquest of Mexico and Peru by 1533 65 years	1533 to end of Philip II 1598 65 years	1598 to end of Philip IV 1665 65 years
Great Britain	1588 Armada 1755 India 1763 Canada 180 years	Treaty of Paris 1763 to First World War 1918 155 years	1918 on
U.S.A.	1783 to 1945 160 years	End of Second World War 1945 Add 100 years? = 2045	

The oversimplified arithmetic is as follows: Let us take and invest $ 30,000 annually and presume we are yearly successful in generating one tax paying job of $ 12,000 income. Tax accrual to government is then $3000 + 2 \times 3000 + 3 \times 3000 + \ldots\ldots$.

The same $ 30,000 invested annually in government will but sustain $2^1/_2$ jobs of $ 12,000 income from which the tax input is simply $7500 + 7500 + 7500 + \ldots\ldots\ldots$

The first case should result in stable prices, the second case in a rise in prices by $2^1/_2$/total working force, since the proportionate increase of goods will not have taken place.

The sustenance of government is 18% of the price of copper or 7 ct. for every pound.

We worry about inflation, both price and monetary. Governmental overhead always exerts pressure toward price inflation, yet the problem is never brought into the open.

A worker receives, let us say $ 1.25 for a fraction of an hour's work. If taxation takes 25 ct. of this, he has $ 1.00 left. But the fabricator or merchant pays the $ 1.25 as labor and in his turn faces a 50% levy on profits, up to 5% on sales, plus other forms of state, federal and municipal taxes which he must add to the $ 1.25 to reach his marketing price if he is to break even, and for which the laborer must pay if he chooses to buy the item. In essence, the tax level deducts from real wages and adds to prices. Higher taxes then always exert pressures toward price inflation, even though they may restrict monetary inflation. The scientific revolution and associated high productivity has done much to alleviate, disguise and defer the consequences.

Slowly and surely excessive overhead strangles any country as adamantly as it bankrupts a business. But tax incentives, a powerful weapon that increases tax paying jobs and therewith tax, seems scarcely thought of in the 20th century, outside of perhaps Russia who having tasted the bitter fruits of socialist society can move only in one direction — away from it.

3. The Necessity for Venture Selectivity

Having wandered far afield let us now consider the necessity for exploration selectivity. One, is the increase in basic costs, engineering salaries 4%, costs of drilling 6%, associated overheads 4%, labor 6%, in all some 5% per annum so that our 10 or 11% increase in exploration of Fig. 4 is only some 5 or 6% actual. Two, the decrease in direct outcrop and geologic, i. e., prime prospects, or said another way the increase of risk. With recent discoveries in depth south of Banner, Arizona (Anaconda), east of San Manuel, Arizona (Quintana), and near Empire, Colorado (Climax), we are realizing the potentials at depth.

Therefore, with increasing cost and complexity of the exploration phases, criteria must be developed to select the prospects with the best chance of discovery and economic return.

4. A Yardstick for Decision Judgement

Let us consider the factors which may permit the evaluation of a prospect from a statistical or operational point of view.

Let P_1 be the probability of an economic occurrence within the prospect area;

P_2 the probability of actual discovery, if in fact the occurrence is present;

P_3 the probability of the deposit being of sufficient total value to more than repay the exploration ventures, capital necessary, etc.

All exploration steps will be multiplied by $(P_1\,P_2\,P_3)^{-1}$ since any stage requires having passed through the previous, as well as the necessity of passing the subsequent.

If we consider everything brought to the starting time T_0, and express all factors in present values at say 6% then $\sum E\,(P_1\,P_2\,P_3)^{-1} + D + C \leq \overline{V} + A + L$ where E are the exploration costs including preliminary drilling, D are development drilling, plus underground or other development costs,

C is the capital for plant,

$\overline{V}$ is the anticipated profit value (after taxes) of the deposit,

A is the capital amortization,

L is the depletion,

$\sum$ expresses the series of exploration ventures that are generally necessary for discovery to result.

Amortization is simply $0.1\,C\,(1.06)^{-1} + 0.09\,C\,(1.06)^{-2}$, etc. for 9 years with a present value relative to C of 0.53 C. $C - A = 0.47$ C. (You never really recover your capital investment through allowable amortization.)

Depletion in most cases is 15% of net profit before taxes, so that with the prevailing 50% profits tax $\overline{V} + L = 1.30\,\overline{V}$. If the mine runs 15 or 20 years with a 4 year interval between exploration and start, the present value of the gross production V is respectively 51% and 45%, so let us take 48%.

Average profit $\overline{V}$ is 13% of gross and $\overline{V}+L=$ 17% of gross. Thus, the present value of the deposit is $0.48 \times 0.17\ V = 0.08\ V = V/12.5$.

Development costs vary widely. Those in the U. S. A. can be written off against other operating profit. In this case, we shall choose to capitalize development and amortize with the plant costs C.

D may vary from 10 to 25% of C with 15% an average value. Thus, 0.47 C becomes 0.47 (C+D) or $0.47 \times 1.15\ C = 0.55\ C$.

Let us now try to evaluate $P_1\ P_2\ P_3$ respectively. Fig. 11 shows the number of mines per unit area of 1000 square km = 400 square miles for Ontario (after SLICHTER 1960) and the basin and range (after NOLAN 1959). These are clearly straight lines on a semi-log plot. Hence, the probability of at least one or more mines being present in a selected single venture area is $e^{-kn\ (400/\text{area sq. miles})}$. Now k for the two cases shown is respectively 0.82 and 1.18.

The log law is important. It shows that chances are just as good where mines have been found, the chance of a new mine being independent of the number already existing.

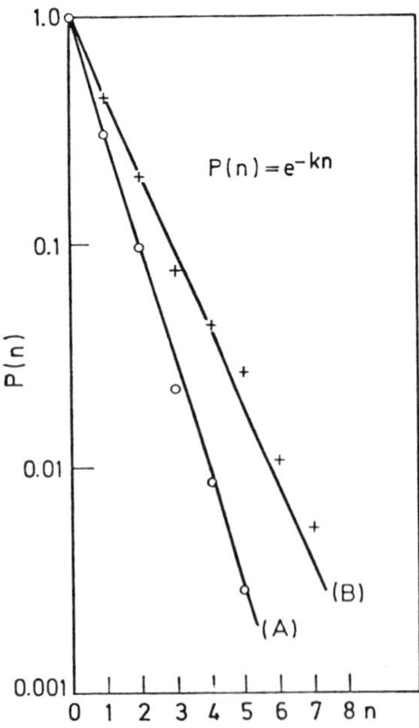

Fig. 11. Distribution of mines per 1000 km² in Ontario (B) and Basin and Range (A)

In the areas considered, these are not all the eventual mines, so that the probability of occurrence has a flatter slope, say $P_n = e^{-0.7n\ 400/\text{area in sq. miles}}$.

No matter how large an area you select, the chances of finding one or more mines is never better than one, i. e., 100%. The probability decreases rapidly with decreasing size of area if the choice is purely random. This is apparent, for the chance of a mine decreases with area, and leaves you with the increased number of area units from which to make selection, so that the chance certainly decreases as at least the square of the factor 400/area.

From a purely random point of view:

P_1 for an area of 40 sq. miles is $e^{-0.7 \times 10}$ or 1 chance in 1000.

This brings up the case for geological and geophysical sieving toward area selection. First, you select geological and structurally favorable areas. Then you narrow these down to a prime area of endeavor.

Suppose we scan 400 sq. miles and select therefrom 10 sq. miles. Chance of occurrence within 400 sq. miles is $e^{-0.7} \simeq 1/2$. Chance of nonoccurrence within the 390 sq. miles eliminated is also approximately $1/2$. So chance of occurrence within the selected 10 sq. miles is now $1/2 \times 1/2$ or $1/4$, a reasonable figure, much increased over the 1 in 1000 chance for a randomly selected 40 sq. mile block. Thus, geology has amply justified itself.

This consideration, however, cannot be extended. Suppose we have scanned 1200 sq. miles, reducing it to a 10 sq. mile block. Chance of occurrence is $e^{-0.23n}$ about 0.8, chance of nonoccurrence 0.2 and chance of occurrence in our 10 sq. miles is now 0.16, less than $1/4$. We have lost ground because of the low chance of nonoccurrence, i. e., high value of occurrence in the area discarded. In short, if area scanning and elimination is to bear fruit, the chance of nonoccurrence in ground discarded must be maintained about $1/2$. If we could accomplish this, our chance in the selected 10 sq. mile unit out of the 1200 sq. miles would be 0.4 compared to 0.25 based on the scanning of 400 sq. miles.

It is apparent that scanning of large areas may be self-defeating, if they are reduced to but a single prime target.

Fig. 12 shows probability of at least a single occurrence in area scanned and reduced to a single block of just several square miles. The probability flattens out while costs rise with increase of area scanned.

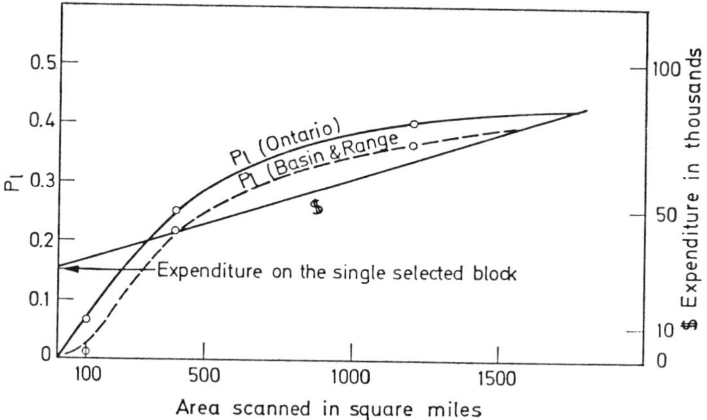

Fig. 12. Probability of occurrence within a single selected area derived from a larger scanned area

However, with a larger scanned area, more than one prime area is apt to result, so that we have a chance to make successive plays, which will be considered later.

Now, most of our exploration selections are based on geologic and other information, which in effect is a scanning. Coveragewise, 100 square mile units might be considered as having been evaluated in reaching decisions in the average case. The exploration block directly attacked is usually of several square miles, and may in part have been further narrowed by earlier work. For such a case, the chance of occurrence is about $^1/_{16}$. For e^{-1n} rather than $e^{-0.7n}$, this factor would be $^1/_{50}$. This is merely saying, look where you have mines. For discussion purposes then we could take $P_1 = {}^1/_{25}$.

P_2 is the probability that if the ore body is there, you will find it.

This is very difficult to evaluate.

Factors to be considered are:

Occurrence distribution with depth.

Depth and nature of cover.

Variation in response of geophysical, drilling and other technological procedures with depth to target, and size and shape of target.

Fig. 13 shows the geophysical anomaly signal in units of geologic plus instrumental noise versus depth below surface on a log log basis. Depending on the method used and shape of body sought, the anomaly may drop off anywhere from a first to 7th power of depth. The thickness of cover generally reduces your effective penetration into bed rock, i. e., reduces the volume surveyed. If the cover is also conducting, e. g., much of Australia, Cyprus, Cuba, the effect for electrical methods is again to further reduce depth penetration by shifting the straight lines to the left and also increasing the geologic noise.

Your chance of detection is given by the dotted curve which drops off sharply as the noise level is reached.

The depth scale is somewhat arbitrary and could be shifted left or right depending on method, body size, body shape and cover conditions.

Fig. 14 provides an interesting operational study of New Brunswick by R. GORE (1958) after the discoveries. The anomalies are linear

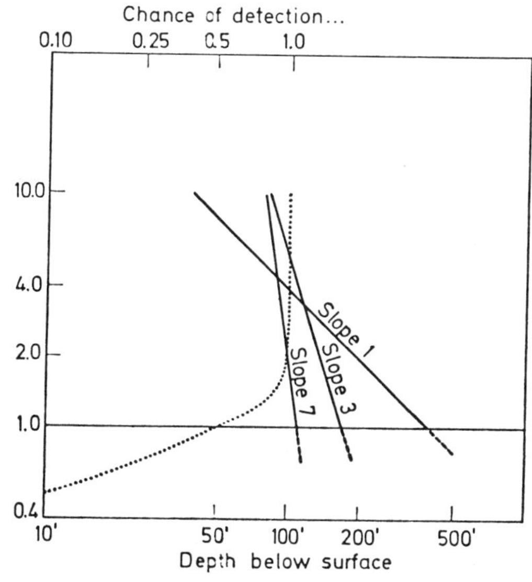

Fig. 13. Ratio of anomaly signal to geologic & other noise versus depth

with a range of lengths. The geophysical line spacing is varied. Probability of detection is considered with all anomalies being drilled. The exploration costs are ratioed to the return expected from the bodies detected. The profitability related to line spacing is in fact only positive for a narrow range of line spacings approximately equal to the occurrence lengths. The variation of ore occurrence with depth is likely amenable to formulation. Four typical ore occurrence distributions are shown on Fig. 15 with their type of formulation indicated. The problem then is to develop mathematical expressions and evaluations for the probability of detection for the occurrence that may be present.

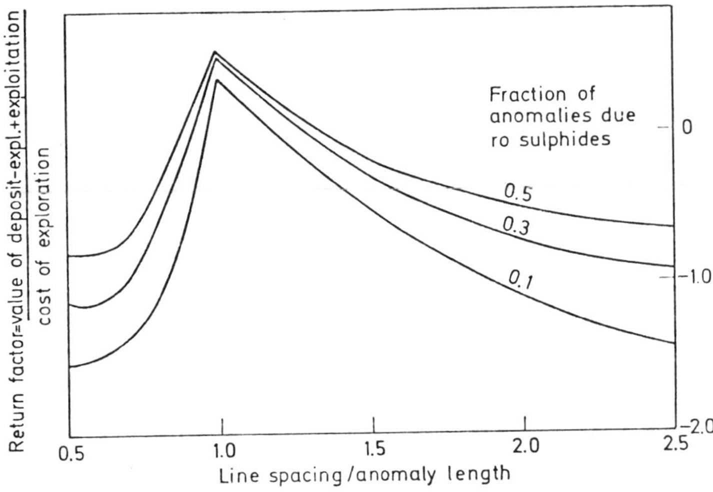

Fig. 14. Monetary return as a function of geophysical line spacing to anomaly length

The Canadian Precambrian case is simple. Geologic noise, i.e., unwanted anomalies are frequently present and of similar magnitude to the ore body signal. Depth penetration is sufficient. Going to Fig. 13, the probability of detection and recognition for anomaly/noise = 1 is perhaps $^1/_2$. For the other types of ore distribution depending on depth, size, shape, etc. detection or recognition ranges from 1—0. Taking into consideration the likelihood of deep covered ore and the limitations of geophysics and geology, one would say that the chances of finding ore, if it is present, might be on the average of $^1/_2$, i.e., we miss about $^1/_2$ of the ore bodies that actually exist.

Thus, P_1 P_2 together, for our example, could be taken = $^1/_{50}$. The next factor to consider is P_3, the probability that the discovery has a gross value V or greater. NOLAN (1959), and ALLAIS (1957) have clearly shown that the distribution of ore deposit gross values, and/or tons, follows a log normal law. This means that if the frequency, i.e., number of occurrences, with gross value of given stepped amounts are plotted as ordinates against the log of the corresponding values, a characteristic symmetrical distribution curve results.

Fig. 16 shows the gross production values of a series of West Australian gold mines after D. CAMPBELL.

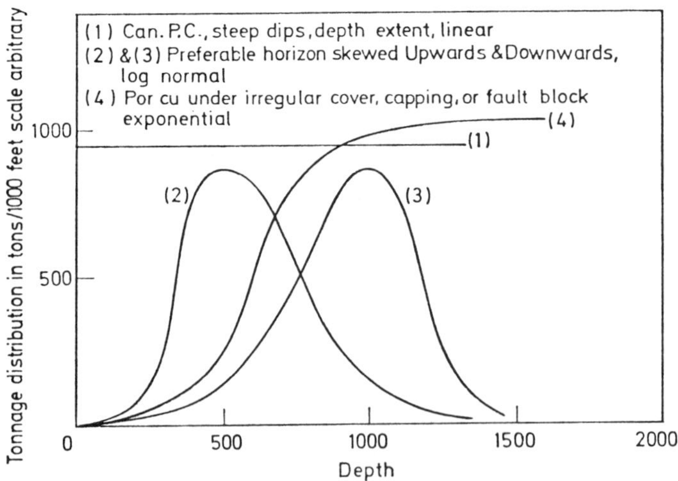

Fig. 15. Functionalized variation of ore occurrence with depth

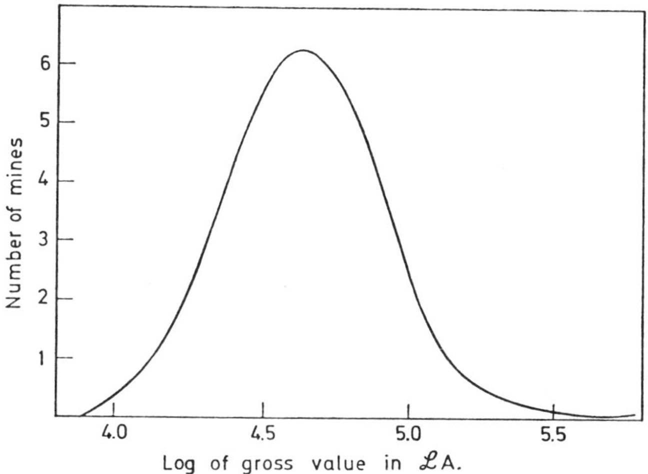

Fig. 17. Log frequency distribution of gross values for a series of W. Australian Gold Mines

Fig. 17 is the same curve plotted as a log frequency distribution with the number of mines having a given plateau or step of value plotted against the log of the value. Actually, a histogram results but this has been smoothed. The percent having values equal or greater than a given step can now be set up and a log probability plot developed as in Fig. 18, the dashed curve. The number of mines having values equal to, or greater than, any step log value have been counted, from Fig. 17, and expressed as a % of the total number of mines considered. This % is then plotted as ordinate versus the log value of the middle of the step. From the resulting plot, the probability that a discovery will be equal to or greater than a given gross value can be read off.

The solid curve of Fig. 18 is for western U.S. coppers taken from ALLAIS (1957), for the early nineteen fifties, with gross values adjusted to present dollars. It shows that for the geologic conditions of such an area, the chance that a discovery will gross 100 million dollars or more is about 33%, i.e., 1 in 3; 300 million or more is 1 in 6, and 1 billion dollars or better about 5%, or 1 in 20.

The value V corresponding to 50%, Fig. 18, is the median value. $LV_{16\%} - LV_{50\%} = \sigma$ where σ^2 is the variance, and where logs are to base e. The mean value $V_m = LV_{50\%} + \sigma^2/2$ and is obviously greater than the median because of the few richer mines in the distribution.

For plot 1, Fig. 18, the median value is 40 million and $\sigma = 2$, whence the mean value is 300 million and the most frequent value $= LV_{50\%} - \sigma^2 = 750{,}000$ gross, very low. Since σ is high, we know immediately that many low valued mines have been included in the statistics.

Let us return to our example for the western U.S.A.:

$P_1 P_2 = 1/50$ for the prospect based on selection from a 100 square mile evaluation.

$P_1 P_2 P_3$ for 100 million dollars gross or better $= 1/150$.

$P_1 P_2 P_3$ for 300 million dollars gross or better $= 1/300$.

$P_1 P_2 P_3$ for 1000 million dollars gross or better $= 1/1000$.

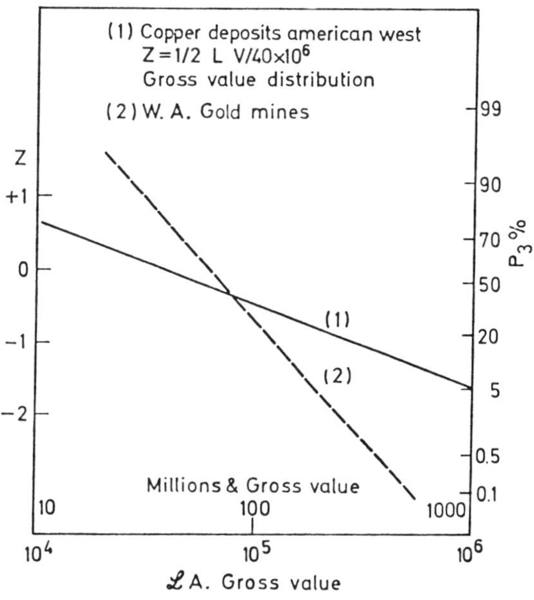

Fig. 18. Log probability plot for a series of W. Australian Gold Mines (2) and for Western U.S. Coppers (1)

These figures are reasonable, for experience has shown that about 1 tested prospect in 100 makes a viable mine.

Now consider a single exploration venture. $ 50,000 is a reasonable initial expenditure; 40% drilling, 20% geophysics, 20% geology and 10% miscellaneous.

Compare 50,000 $(P_1\ P_2\ P_3)^{-1}$ + 0.55 C vs. V/12.5. C based on daily production over a 15—20 year period expressed as lbs. of copper works out to 10, 30 and 100 million respectively. The first two values are too low and should be about 20 and 40 million of capital investment.

Hence we compare

50,000 × 150 + 11 million vs. 100/12.5 or 8 million,
50,000 × 300 + 22 million vs. 300/12.5 or 24 million, and
50,000 × 1000 + 55 million vs. 1000/12.5 or 80 million.

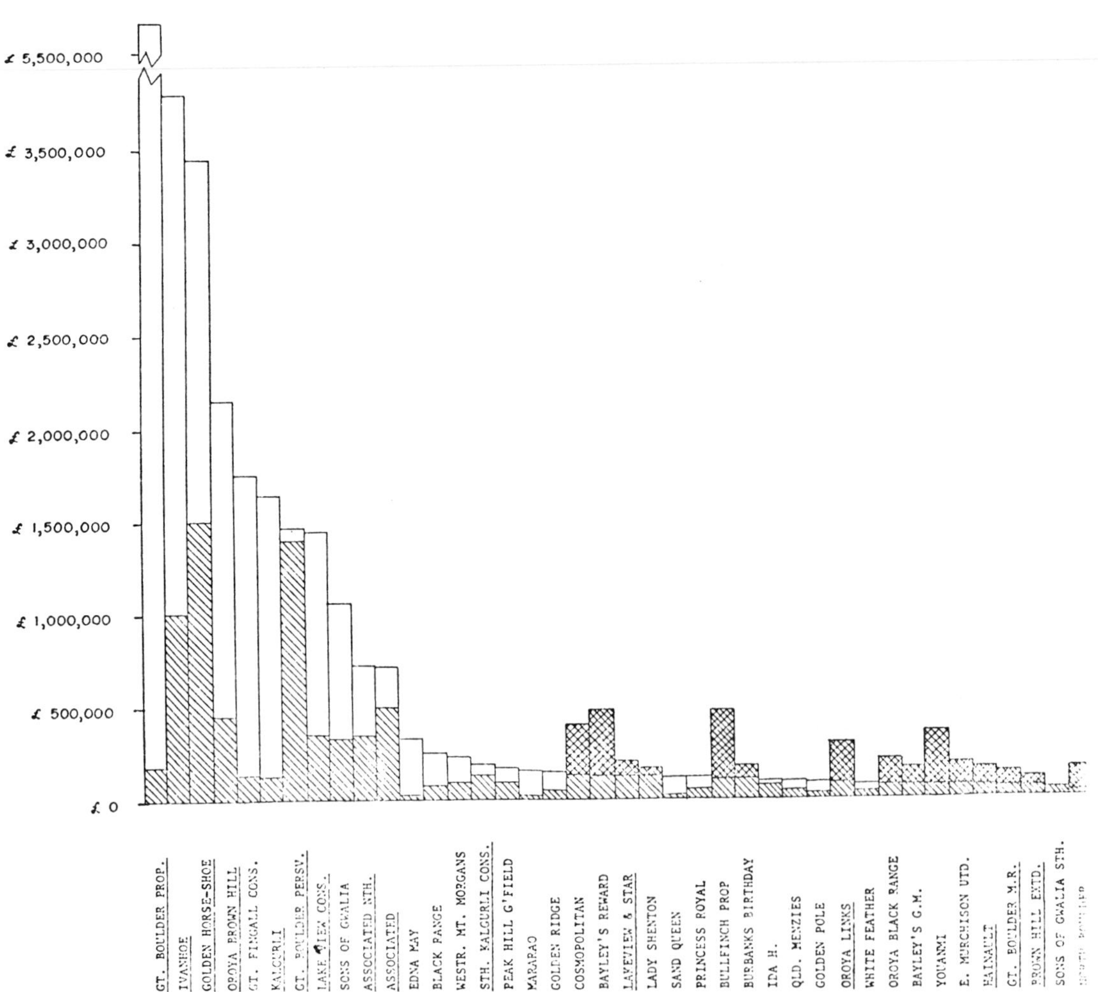

Fig. 16. Gross production values for a series of W. Australian Gold Mines

Let us consider the following table:

Initial Risk	Odds Against	Return = V/12.5—0.55 C	Return / Risk × Odds	Return / Investment
50,000	150	(8—11) millions	Loss	Loss
50,000	300	(24—22) millions	$\frac{2\times 10^6}{300\times 50{,}000} = 0.13$	$\frac{2}{22} = 11\%$
50,000	1000	(80—55) millions	$\frac{25\times 10^6}{1000\times 50{,}000} = 0.4$	$\frac{25}{55} = 45\%$

This points up the obvious truth of looking for the big ones.
Actually, these are minimum return/risk × odds or profitability ratios, since the probability P_3 is based on the value of V the present value of gross, being equal to, or greater than, those taken.
If Fig. 18 is considered for the sections or segments equal to and above the V values selected, the mean expected gross values related to at least 100, 300 and 1000 million, are 400, 750 and 1800 million respectively. The associated reasonable capital investments now are 40, 75 and say 120 million.

The table now becomes:

Initial Risk	Odds Against	Return = V/12.5—0.55 C	Return / Risk × Odds	Return / Investment
50,000	150	(50—22) millions	$\frac{28\times 10^6}{150\times 50{,}000} = 3.75$	$\frac{28}{22} = 1.3$
50,000	300	(94—40) millions	$\frac{54\times 10^6}{300\times 50{,}000} = 3.6$	$\frac{54}{40} = 1.4$
50,000	1000	(225—66) millions	$\frac{159\times 10^6}{1000\times 50{,}000} = 3.2$	$\frac{159}{66} = 2.4$

The considerations are by no means as simple as the above. Considerable drilling may be done after discovery, toward evaluation, without resulting in a potential operation. Roads, town sites, power, wages, transport all relate in the potential return. The risk is thus considerably more than the 50,000 prospecting venture. Computerized flow sheets exist permitting all the various factors to be evaluated in assessing a presumed tonnage, grade, operation and market.

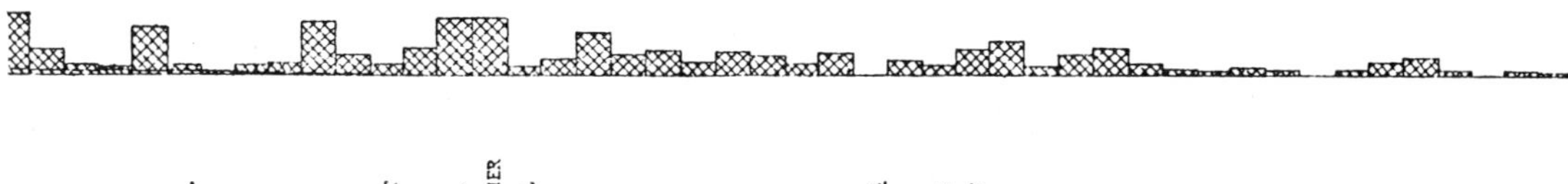

Fig. 16 (continued)

The above approach does, however, permit a means of prospect pre-evaluation. In view of the known geology, location, operating conditions, etc., is the proposed prospect possible of providing a profitability ratio of 3 or better, and a return ratio of 1 or better?

The associated chance of loss in all cases is $1—P_1 P_2$ or 49/50. Odds against are hence 49 to 1.

The chances are then high for a series of no luck ventures, which cannot be sustained, in short-gambler's ruin.

If P_s is the chance of detection $= P_1 P_2 = 1/50$, then the chance of n successive failures is e^{-nPs} and the chance of at least once success is $1—e^{-nPs}$, Fig. 19.

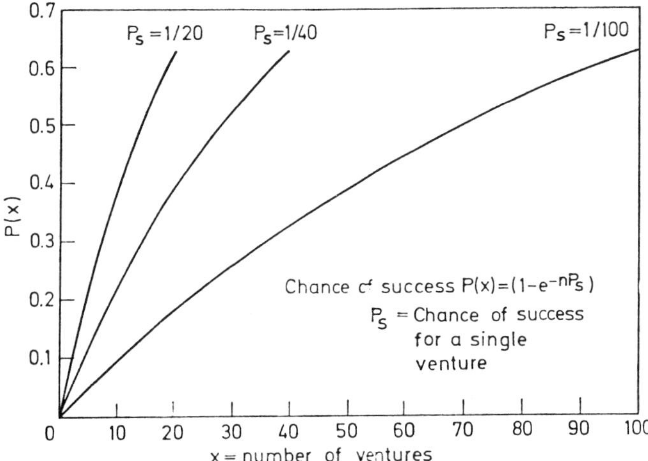

Fig. 19. Chance of success for a series of ventures

If we make 50 ventures, the chance of 50 successive failures is $e^{-1} = 0.37$ and the chance of at least one success is 0.63. Chance of loss is now only 0.37 or odds against are only 3 to 2 as compared to 49 to 1 for the single venture. The profitability ratio likewise goes down but not so rapidly.

We now compare:

$50 \times 50{,}000 \times 3/2 \times 3$ vs. 50—22 million
$50 \times 50{,}000 \times 3/2 \times 6$ vs. 94—40 million
$50 \times 50{,}000 \times 3/2 \times 20$ vs. 225—66 million

resulting in return/risk $\times$ odds ratios of:

$$\frac{28 \times 10^6}{2.5 \times 10^6 \times 4.5} = 2.5$$

$$\frac{54 \times 10^6}{2.5 \times 10^6 \times 9} = 2.4$$

$$\frac{159 \times 10^6}{2.5 \times 10^6 \times 30} = 2.1$$

and return/investment ratios of:

$$\frac{28 \times 10^6}{(2.5 + 22) \times 10^6} = 1.1$$

$$\frac{54 \times 10^6}{(2.5 + 40) \times 10^6} = 1.3$$

$$\frac{159 \times 10^6}{(2.5 + 66) \times 10^6} = 2.3$$

Thus for 50 ventures, the profitability ratio is 2.3 and odds against discovery are 3 to 2.

Pay off ratio is $2.3/1.5 = 3/2$.

In a single venture, the profitability ratio is 3.5 but odds against discovery are 49 to 1.

Pay off ratio is then $3.5/49 = 1/14$.

Thus, by a series of exploration ventures you can drastically reduce the odds of loss, and can place yourself in a position where the pay off ratio is actually favorable, i.e., greater than 1. This explains why joint or group ventures are so frequent. Providing you have the means to sustain the effort, it looks as if intelligent exploration can be profitable.

Does exploration pay? In the last 15 years, Newmont and associates have spent perhaps 40 million in exploration, research and exploration drilling. Our return has been a stake in Southern Peru Copper and Palabora, several new ore bodies at O'okiep, two small mines near Tsumeb, Carlin Gold, and what will likely be a small 100 million tons at 0.65% porphyry type copper in Arizona.

HADDON F. KING of Consolidated Zinc-Rio Tinto of Australia, in a private summary, showed that basic exploration expenditures of some 25 million in the last 12 years resulted in discoveries of bauxite on the east side of the Gulf of Carpentaria, their stake in the Hammersley Iron Range and the recent copper find in Bougainville of to-date some 100 million tons of 0.75 copper.

As long as demand and cash flow maintain, exploration will be active, and most companies are now working to the limit of their financial resources.

Sketchy as this presentation has been, it does present a possible mechanism on which to judge the merits of a prospect, namely its profitability ratio or potential for profit. Reasonable statistics must be available. The crucial ones are those dealing with the frequency of occurrence, Fig. 16, on which P_1 is based, and that dealing with the distribution of values, Fig. 18, involving P_3, which varies from geologic to geologic province. Several of the conclusions evolved are followed in empirical practice such as go where the ore is, look for the big ones, spread your bets.

Not so greatly emphasized is the necessity to make sure your sieving of the larger studied area to select the bet, does in fact eliminate to a minimum level of certainty of 1/2, else you may well be defeating your purpose. It is here that geology and geophysics may make an increasingly vital contribution. Our inadequate ability to reach depth, Fig. 13, and our inability to diagnose geophysical anomalies with a higher degree of certainty means we may be missing perhaps half of the existing ore bodies.

To me it is an honor to have assisted, albeit inadequately, in the memory of Professor McKinstry. I remember the quiet, silver haired man with the shy smile taking our then small, delighted children to the movie in Prescott and his quiet courage on the three hour ride out of the Shasta hills after his horse, being led along the trail, balked at a creek, then suddenly plunged forward on to him. His thoughtful work in the seemingly simple but so complex Victoria Goldfields, and in the ochre colored scrub gum areas of Kalgoorlie is still referenced to in the awe due it.

This quiet man, who knew a man has to do what he must, not what he wishes; this gracious man with the fast disappearing respect and deference for others. May his type long live.

References

Allais, M.: Method of Appraising Economic Prospects of Mining Exploration Over Large Territories: Algerian Sahara Case Study, Management. Science **3**, 285–347 (1957).

Campbell, J. D.: From unpublished company files.

Derry, D. R.: "Man and minerals" and "The Promises and Challanges at C + 100". Canad. Inst. Mining & Metallurgy, Annual Gener. Meet. at Ottawa, 1967.

Gore, R. C.: A Mathematical Model of Geophysical Exploration for Elongated Ore Deposits. Submitted in partial fulfillment of the requirements for the Degree of Master of Science, M.I.T. Department of Geology and Geophysics, September 1958.

Nolan, T. B.: The Search for New Mining Districts. Econ. Geol. **45**, 601–608 (1959).

Slichter, L. B.: The Need of a New Philosophy of Prospecting. Mining Engineering, 1–7, (June 1960).

Received December 4, 1967

Dr. A. A. Brant, Director
Geophysical Department, Newmont Exploration Ltd., 44 Briar Ridge Road Danbury, Connecticut 06810, U.S.A.

10

Reprinted from *Mining Cong. Jour.* **58:**31–37 (April 1972)

Mineral Exploration Philosophy

By PAUL A. BAILLY
President
Occidental Minerals Corp.

MINERAL EXPLORATION OBJECTIVE:
DISCOVER AND ACQUIRE NEW ORE RESERVES AND MINERAL RESOURCES IN ORDER TO:

1. ...MAINTAIN & INCREASE PROFITABILITY OF PRODUCTION ASSETS:		2. ...CREATE NEW ASSETS & PROFIT CENTERS
A. MINE/MILL OPERATIONS	B. RAW MATERIAL SUPPLY TO TREATMENT PLANTS	A. PRODUCT DIVERSIFICATION B. GEOGR. DIVERSIFICATION

ANALYSIS AND APPRAISAL OF AVAILABLE RESOURCES

PLANNING GRID WHAT RISK? WHAT EXPENDITURES?	WHERE? POLIT. - SOCIAL - FISCAL CLIMATE? WHERE? MINERAL PRODUCTION/POTENTIAL? U.S.A.	CANADA	AUSTRALIA	ETC...
ACQUIRE PRODUCER				
ACQUIRE DEVEL. RESERVE				
DEVELOP KNOWN DEPOSIT				
EXPLORE KNOWN TARGETS				
EXPLORE FOR NEW TARGETS				
- NEAR KNOWN DEPOSIT				
- IN MINING DIST.				
- IN MINERAL BELT				
- IN FAV. VIRGIN AREA				

EVALUATION CRITERIA

Fig. 1. Example of a format for defining exploration goals.

IN A RECENT DISCOURSE on the Mining Law of 1872, a distinguished mining lawyer admitted: "As a practical matter, exploration must precede discovery of minerals." This startling statement reminds us that exploration is only the means of discovery and finds its justification in its end product, not in its activities.

We frequently hear that there are only three components in exploration successes: money, people and luck; and it has often been said that luck is the most important ingredient. Luck is best defined as the leverage of unforeseeable circumstances on the evolution of an undertaking. In this sense, luck plays a part in shaping exploration results.

For instance, it is not possible to foresee, before drilling, whether the grade of a chalcocite blanket below a large area of leached capping will be 0.06 or 0.04 percent copper and the thickness 100 or 200 ft. But let's say that during land negotiations you lose a big part of the large, leached capping target to a competitor who heard of your work and dealings in the area, ran a quick check and grabbed the opportunity. Let us say further that your competitor finds the thicker, higher grade part of the blanket on his land and announces a discovery, whereas you end up dropping the acreage you optioned because of your submarginal results. In such circumstances, luck was a factor regarding the grade found, but luck had no part in his success and your failure at land acquisition. Remember, you could and should have acquired all the land. Anyone who mistakenly labels as unforseeable circumstances which can humanly be foreseen and controlled erroneously attributes to bad luck unfavorable events resulting from inexperienced personnel, bad preparation and bad performance.

All exploration factors not foreseeable

An explorer should always attempt to put good luck on his side by trying to forecast and take into account all those factors and circumstances assessable, or controllable, by man and, therefore, amenable to prediction and action. By the same token, completely discounting luck as influencing exploration results assumes that all exploration factors are foreseeable. Such blindness, or conceit, can lead to sad disappointments. Luck may help or deter us; belief in it should never become a substitute for exploration thinking.

The successful exploration process consists of decisions and activities which transform available resources into a new resource: mineral deposits. Exploration philosophy is the intellectual discipline which investigates, defines, explains and improves the body of principles which guide the explorer in achieving this transformation of resources. Nature, including man, is indeed complex. Human, corporate and national resources and goals are varied and changing. As a result, the principles that guide each explorer toward discoveries are also varied and changing. But if we look below the exploration motions, we detect certain unifying principles.

Goal definition precedes exploration philosophy

To begin with, all explorers have a goal more or less clearly defined and they have an array of similar resources in varied quality and quantity. The discovery goal need be defined before one can define an exploration philosophy. After all, if we don't know what we want to find, we don't need a philosophy to look for it.

Fig. 1 is an example of a format that can be used in defining goals. For instance, a copper-producing company with fast depleting ore reserves in the U.S.A., overcapacity at its inland U.S. smelter and refinery, some excellent exploration talent, deep conviction in the future of copper and excellent banking relations may decide to have the following discovery goals:

1-A—Discover additional copper reserves near its existing mines.

1-B—Discover new deposits of copper within 1000 miles of its smelter in North America.

2-A—Diversify its product through discovery of nickel sulfide deposits in Canada, U.S.A. and Australia, entering exploration at the grass roots reconnaissance level if necessary.

2-B—Diversify geographically by discovering new copper deposits in Australasia.

Recently, a new company announced it was organizing a mineral exploration group in a multi-country area, with an initial capital of $1,000,000. I quote: "Included in the broadly based program will be grass roots exploration and initial prospect evaluation leading to the selection of projects for full-scale evaluation. There will be no restriction on the type of minerals in which the new company will be interested." My guess is that such a group will either luckily start at a few good prospects, develop them and considerably reduce the scope of its exploration objectives, or it will try to look briefly at everything, in an effort to make the best selection possible; but not having enough time and funds to analyze any thing at depth, it will soon be quitting exploration. Such program appears to illustrate the definition of philosophy proposed by Ambrose Bierce in his Devil's Dictionary: "Philosophy is a route of many roads leading from nowhere to nothing."

Scarcest resource should control use of other resources

The goals often have to be scaled down, or realistically modified after the available resources are evaluated. There is no point in doing anything unless it is done with high quality resources; in addition, the scarcest resource should always control the use of the other resources. The explorer can mobilize five main resources to make discoveries:

—**First resource:** Human beings actively dedicated to discovery and fully acquainted with the nature, uses, availability and limitations of all the other resources. Ability to perceive and project in three dimensions is an absolute requirement of the successful explorer. In exploration, the quality of human resources is the key ingredient.

—**Second resource:** Knowledge of the earth's surface, subsurface and their relations.

(A)—As to the subsurface, this includes knowledge of rock units, their compositional, depositional and structural relations, knowledge of ore deposits, their settings, their exploitation and their economics, and knowledge of mineral ownership, legislation, regulations.

(B)—As to the surface, this includes knowledge of ownership, climate, vegetation, fauna, human developments and other possible uses, laws, regulations and trends affecting the mineral industry and degree of human interest including attitude of environmental protection activists toward the areas under consideration.

(C)—In each area of interest, the explorer must appreciate the relations of surface and subsurface or, if you prefer, the Ecosystem-Geosystem interface, since this interface often affects exploration with regard to the choice, legality, practicability and performance of geochemical, geophysical and geological surveys and drilling.

—**Third resource:** Mineral detection methods and related techniques, equipment, supplies, contractors and consultants.

—**Fourth resource:** Time.

—**Fifth resource:** Money.

Human resource is key control

Each of these resources is necessary, but none of them is sufficient by itself to make discoveries. Human resources are, however, the key control. But human resources, as well as time, money and knowledge of surface, are resources needed for practically any and all endeavors. The principles which guide our use of the other two resources define our exploration philosophy. These two resources are:

—*knowledge of the earth's subsurface, and*
—*mineral detection methods.*

Ideally, our knowledge of the subsurface should include laws of nature, or concepts which allow us to predict reliably where ore is to be found. Unfortunately, the location of deposits is the result of the interaction of many variables, the relative importance of which in time and space can rarely be even guesstimated. Therefore, when he decides what part of the earth's crust to investigate, the discovery-oriented explorer makes his initial selection according to broad categories as shown in fig. 2 (see p. 35).

He assumes the increasing risks of moving farther from known deposits as he chooses for his hunting ground:

First, areas near producing mines;
Then areas near known deposits;
Then areas near known mineral showing, and
Then areas within known mineral belts.

As he ventures into virgin areas not known to contain deposits of the commodity he seeks, he enters into high risk target areas derived from a geological concept of working hypothesis as favorable to the occurrence of a deposit. He may even assume greater risk and start hunting in a region, or a large area, where he believes he will be able to define smaller target areas that will meet the criteria of his working hypothesis.

On the vertical axis of fig. 2 is the explorer's selection of field procedure. From top to bottom, they are listed in order of decreasing chance of detection for the money spent. Admittedly, this is a personal current rating, not valid historically and not necessarily shared by all explorers, especially if they can claim some success for their favorite procedure.

Much of exploration a multistage procedure

Much of today's exploration normally follows a **multistage procedure** involving properly ordered steps using increasingly discriminating methods which focus progressively more sharply on a target. This sequential progress varies with different types of deposits, but normally involves areas decreasing in size and methods increasing in cost per unit area.

Conventional prospecting consists of search for directly observable mineralization and natural features commonly associated with mineralization. Before many physical and chemical mineral detection systems were developed, conventional prospecting was the only reasonable and discriminating approach. Historically, it claims at least 90 percent of the ore produced to date, but its present value has decreased, except in a few poorly prospected parts of the world.

Next comes **massive multimethod coverage** of the area selected. This procedure is exemplified by multimethod airborne surveys performed with the underlying assumption that all parts of the area selected have an equal chance of containing an ore deposit.

The last procedure mentioned here is used by the explorer who has fallen in love with a **"unique technique"** and uses it exclusively. It may be some kind of counter; some form of geochemical sampling; some specific characteristic of the ground; a certain tingling in his boots, or just simply using a drill anywhere, any time, anyhow.

The combination of initial area selection and choice of field procedure is the discovery philosophy of the explorer at a given time, for a given type of deposit. Some of the many possible combinations, or philosophies, are common enough to have received specific names worth a brief mention here.

The **"Elephant Country"** approach describes only the initial selection of area regardless of field procedure. It assumes that deposits of a certain type tend to occur in areas, or regions, where similar deposits are already known.

"Grass Roots Exploration" usually refers to programs starting with the reconnaissance of areas not known to be mineralized, regardless of the field procedures.

Scientific ore-finding is sometimes used to describe the multiphase discriminating procedure in scientifically selected hunting ground.

The **"Prospector and his Burro"** approach, as illustrated by the three pack burros, refers to conventional prospecting in mineralized or large, favorable areas. Sometimes prospectors employed by companies are sent to conceptually defined target areas in virgin territory, where otherwise the prospector would have gone only by chance.

Some explorers continue to limit their work to **conventional examination** of prospects submitted to them. An apparently active variant of this passive approach is the **"Fortune-teller"**—the explorer simply announces to anyone interested: "Come to us (with your samples) and we'll tell you your fortune."

The **"Grasshopper"** approach characterizes the explorer who limits his work to visual examination of mineral showings. It is often a cop-out for the explorer who spends all his time examining only prospects turned down many times by others; as he continues this feverish merry-go-round, he optimizes his chance of failure. This approach is sometimes used by companies not truly interested in discoveries; they employ explorers for cosmetic public relations effect only.

The **"Throw-the-book-at-it"** approach has been used sometimes to obtain assurance that nothing has been left undiscovered in mineralized areas. Every technique available is used.

"Saturation Exploration" defines massive multimethod coverage of sizable areas in mineral belts or virgin areas. It is used effectively in the search for massive sulfide deposits in covered areas.

The **"Bulldog"** approach describes the obstinate and consistent use of only one unique method, on any part of a mineralized area. This trial and error probing is certainly a dog's life.

The **"Black Box"** approach can be that of the explorer in possession of a unique gadget, or instrument, measuring some physical or chemical characteristic and which is claimed to give near magical indications of the presence of ore. It is unlikely that this approach will be used in conceptually defined areas, since such concentration on one gadget precludes the use of any mental energy to seriously analyze where to look.

"Drillitis" is the philosophy, or affliction, of the explorer who believes the only way to explore is systematic drilling in large, favorable areas. One must realize, however, that in some cases drilling is the only possible way to proceed after initial area selection.

Each approach has its successes

This catalog may give the impression that the exploration community is like a menagerie with its various animal groupings and that there are so many philosophies in this zoo that utter chaos prevails. But, as fig. 3 shows, this catalog can also be used as a score card to record modern discoveries attributable to different approaches. The discoveries listed are only selected examples; there was no attempt to make a complete list.

Important copper discoveries (shown against a blue background) result mostly from scientific ore finding with the possible exception of Pima, found through the massive application of geophysical methods within a mineral belt.

The discovery history of the Blind River District started with Pronto (shown against an orange background in upper part of figure), discovered by prospecting an area selected because of its similarity with South African gold-uranium occurrences. Radioactive surveys, detailed mapping, followed by drilling, resulted in the discovery of Stanrock and Denison. The concepts derived from the Blind River experience were later conceptually extended to the east, where Agnew Lake was discovered. In the last six years, many parts of the Blind River area have received the attention of groups suffering from "drillitis" without any new ore discoveries being announced.

During the last 20 years, multimethod airborne surveys with variable penetration have contributed many new discoveries in areas initially selected within recognized mineral belts, or through a geological working hypothesis. The famous Manitoba Nickel Belt deposits and Timmins illustrate the successes of this approach.

Deep, nonoutcropping deposits of lead-zinc in carbonate strata, and uranium in detrital formations, are not, to this date, amenable to remote detection through geophysical or geochemical techniques. After the outcropping deposits have been discovered through conventional prospecting, the surrounding area and similar basins, rated as favorable on geologic criteria, are then reconnoitered through drilling. To insure against loss of

[*Editor's Note:* In the original, color was used in Figures 2, 3, 4, 6, 7, and 8 primarily for the oral presentation of the paper. It was more to demonstrate a method of approach than to convey detailed information.]

discovery, large acreages have to be tied up. This explains why discoveries of lead in the new Missouri Lead Belt, zinc in mid-Tennessee, and many deep uranium deposits in western basins result from considerable drilling over large areas.

Fig. 3 illustrates that each approach is specially suited for certain types of deposits and that each approach has its successes.

The explorer's ideal is to develop a perfect understanding of the earth's subsurface that will allow him to make reliable predictions about the occurrence of mineral deposits and to select areas that will prove to contain ore deposits. Unfortunately, one is reminded that exploration directors, asked by their management why another company discovered Timmins, while they were frittering money away, could sincerely reply that their own exploration crews were working in "more favorable areas," as related to their exploration concepts. This explains why exploration geologists have usually been dubbed as blind in foresight. However, one perceives modern trends toward improvements in geologic foresight which cannot be ignored.

Fig. 4, inspired from a paper by Bradley [1] shows how geology, as a scientific discipline, relates to exploration. Our knowledge of the geology of mineral deposits usable in exploration consists of the following, in order of increasing generalization:

(1) Primary observations recorded as innumerable descriptions of deposits and their setting and as many ore occurrence patterns expressed by recognized associations of valuable minerals with certain mineralogical, structural, petrological and stratigraphic features;

(2) Concepts and principles derived from many occurrence models or types of deposits without genetic connotations and from some genetic models which give an intellectually satisfactory explanation of the source, transport and deposition of the components of a type of deposit;

(3) A few laws of nature of general validity.

Occurrence models improve predictive powers

As we progress from primary observations to laws of nature, our understanding and intellectual power increase. Our predictive power increases also, as long as we generalize, but this predictive power does not seem to increase in parallel when we try to use these generalizations to predict the existence of specific deposits. Our predictive power improves considerably with the formulation of occurrence models but does not improve further with genetic models and laws. A good example of the power of occurrence models has recently been given by Al Hoagland; he suggests a model for strata-bound zinc deposits in Appalachia and its use in exploration.[2]

The target hypotheses derived from such occurrence models are generated in the mind of the explorer who makes a plausible case for his contention that ore may be found where he predicts it to be. This is in marked contrast to the philosophy of despair expressed by the traditional statement "ore is where you find it."

In recent years, there has been considerable research resulting in a much improved understanding of the transportation in solutions of some of the valuable constituents of mineral deposits. The resulting transportation model has certainly increased in validity, but this improved generalization has not given us any new specific predictive power.

K. B. Krauskopf, in his 1970 Presidential Address to the Geochemical Society, discussed "the source of metals."[3] He argues that a common sequence of events to produce the ore metals in adequate concentrations is slight enrichment in large volumes by sedimentary agents followed by strong local concentration during igneous activities. He concludes, however, that his generalized hypothesis on the source of ore metals tends to give theoretical support to the old defeatist conclusion that "ore is where you find it."

Avoid pitfalls of genetic concepts and general theories

Thus, our exploration philosophy can and should take advantage of geological concepts and principles directly derived from primary observations and occurrence models. It must, however, avoid the pitfalls of genetic concepts and general theories which are intellectually attractive and perhaps correct but neither usable nor productive in selecting specific target areas.

After discussing our knowledge of the earth's subsurface as the principal of two resources used in planning for discoveries, let us now say a few words about the other resource: mineral detection methods and related techniques and equipment. The panoply of available gadgetry and schemes promoted for use in exploration is increasing considerably every year. There are, to date, several dozen service organizations promoting and selling services, procedures, equipment and supplies and the inevitable data processing which did not exist 10 years ago. There are always inventors ahead of all others working on improving an old method, or on a new method or technique which, if successful, may upset the game and make other techniques and equipment obsolete. As part of his philosophy, the explorer must therefore be a wise purchaser and a thrifty consumer of mineral detection techniques, services and equipment. This involves:

1) Encouraging innovators.

2) Keeping up with progress in instrumentation and techniques.

3) Avoid getting stuck with a large investment in obsolete equipment.

4) Insisting on the best possible knowledge and evaluation of capabilities and limitations of each technique before using it.

5) Insisting that all plausible interpretations of any anomaly are objectively developed, evaluated, and integrated with all previous results and interpretations.

6) Using a given method only if it can truly give discriminant results, improving on the knowledge of the mineral potential of the area, at a cost estimated to be not greater than the extra costs incurred in the next exploration phase if the method had not been used at all.

One may well ask the question: Is modern exploration successful? To illustrate my resounding affirmative answer, I have prepared four illustrations which show

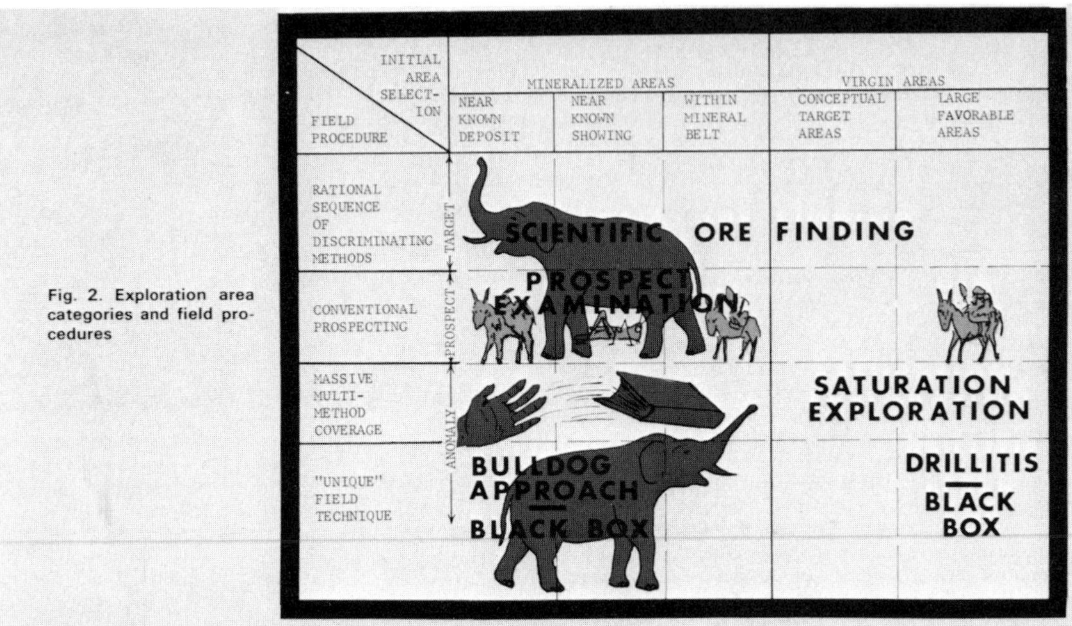

Fig. 2. Exploration area categories and field procedures

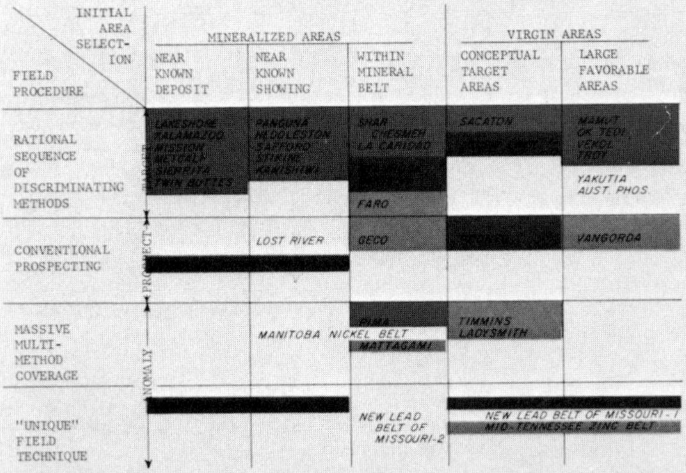

FIELD PROCEDURE / INITIAL AREA SELECTION	MINERALIZED AREAS: NEAR KNOWN DEPOSIT	NEAR KNOWN SHOWING	WITHIN MINERAL BELT	VIRGIN AREAS: CONCEPTUAL TARGET AREAS	LARGE FAVORABLE AREAS
RATIONAL SEQUENCE OF DISCRIMINATING METHODS (TARGET)	*LAKESHORE KALAMAZOO MISSION METCALF SIERRITA TWIN BUTTES*	*PANGUNA HEDDLESTON SAFFORD STIKINE KAWISHIWI*	*SHAR CHESMEH LA CARIDAD FARO*	*SACATON*	*MAMUT OK TEDI VEKOL TROY YAKUTIA AUST. PHOS.*
CONVENTIONAL PROSPECTING (PROSPECT)		*LOST RIVER*	*GECO*		*VANGORDA*
MASSIVE MULTI-METHOD COVERAGE (ANOMALY)		*MANITOBA NICKEL BELT*	*PIMA MATTAGAMI*	*TIMMINS LADYSMITH*	
"UNIQUE" FIELD TECHNIQUE			*NEW LEAD BELT OF MISSOURI-2*	*NEW LEAD BELT OF MISSOURI-1 MID-TENNESSEE ZINC BELT*	

Fig. 3. Selected examples of mineral discoveries

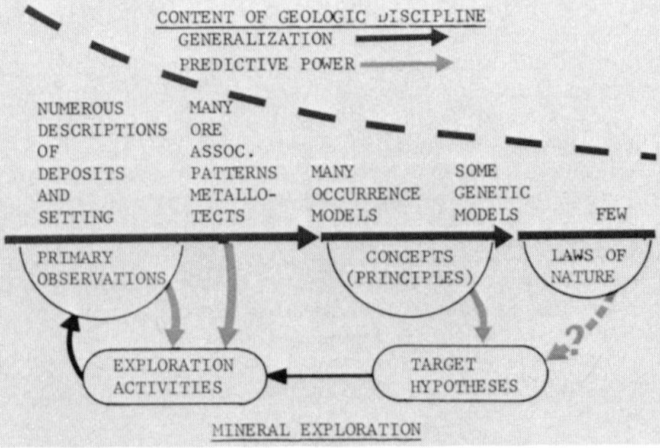

CONTENT OF GEOLOGIC DISCIPLINE
GENERALIZATION
PREDICTIVE POWER

NUMEROUS DESCRIPTIONS OF DEPOSITS AND SETTING
MANY ORE ASSOC. PATTERNS METALLOTECTS
MANY OCCURRENCE MODELS
SOME GENETIC MODELS
FEW

PRIMARY OBSERVATIONS
CONCEPTS (PRINCIPLES)
LAWS OF NATURE
?
EXPLORATION ACTIVITIES
TARGET HYPOTHESES

MINERAL EXPLORATION

Fig. 4. Relationship of geology to exploration

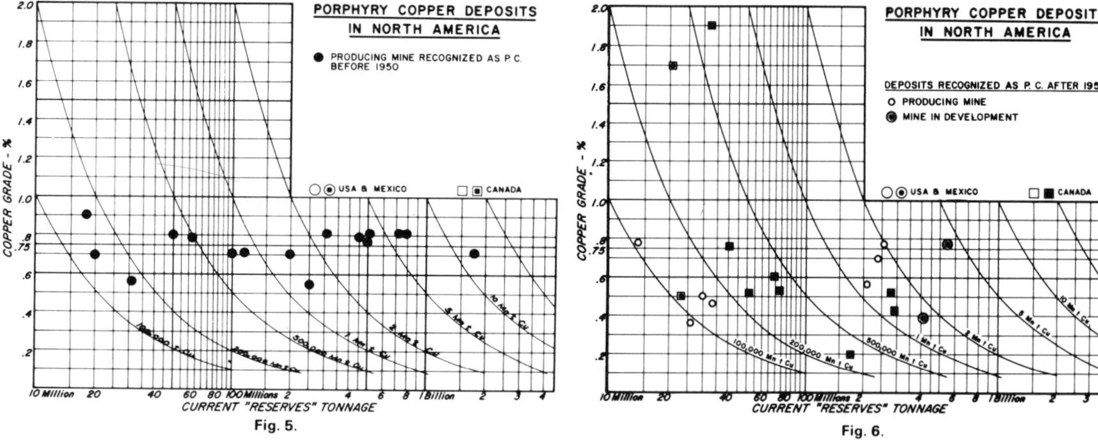

Fig. 5.

Fig. 6.

the results of modern porphyry copper exploration in North America. Similar industry success stories could be presented for other commodities and other types of deposits.

Fig. 5 depicts ore reserves, excluding past production, as best as I could ascertain from publications, releases and prospectuses. The copper grade is shown on the vertical axis and tonnage on the horizontal axis. The tonnage of copper metal contained is shown by the curves. The solid black dots show reserves at 17 currently producing porphyry coppers located in U.S.A., Mexico and Canada, which were recognized as porphyry coppers before 1950.

On fig. 6, the uncolored open circles and squares show 10 mines currently producing, but discovered or recognized after 1950. The open and solid black symbols represent nine deposits, discovered after 1950, currently under development.

Fig. 7 shows, as blue symbols, 14 deposits discovered as suitable for production and at various stages of mine planning. The round symbols with dots in them represent six porphyry copper discoveries made in recent time, which are either being drilled in detail, or shelved because not immediately suitable for production.

The final figure, fig. 8, shows, grouped together, all of these 56 porphyry copper deposits in North America, 39 of which were discovered after 1950. Seven out of 17 old discoveries contain more than two million tons metal in reserves, but only 8 out of 39 new discoveries contain more than two million tons metal. Ninety percent of these new discoveries are in areas well known for their copper mineralization before 1950; this includes 80 percent in close proximity to known deposits, or known showings. Eighty-seven percent of these discoveries are the result of multiphase-sequential discriminating field procedure.

In using all resources at his disposal, the explorer should integrate in his exploration philosphy several strategic principles as follows:

1) **Intelligence.** This includes intelligence about nature which, as a passive enemy to conquer, keeps its secrets well hidden, and intelligence about capabilities and vulnerabilities of all active participants in the game, including one's self, landowners, government agencies and competitors.

2) **Tight security.** The counterpart of intelligence. This avoids leaks of critical plans and ideas to the competition at critical times.

3) **Readiness and a sense of urgency.** Establishing priorities is not enough. A sense of urgency about all high priority items is critical; the low priority items can wait, or should be entirely discarded.

4) **Surprise.** Effective implementation of a good plan is not guaranteed, but is helped through surprise which includes the elements of audacity, originality, speed and secrecy and catches all competitors with their pants down.

5) **Concentration.** Using the best available talent in force at the proper place, at the critical time, is also a key element of success. Concentration on one, or a few commodities, is better than diffusion of effort on many commodities; the same principle applies to areas.

6) Excellent **communications** between all concerned is a prerequisite for effective control of the exploration program and for immediate exploitation of all current results.

7) **Mobility,** or the ability to swiftly adapt plans to results.

8) **Objectivity and hard-mindedness** necessary to abandon a failing project without any regrets and without further use of talent, time and money.

Exploration group shouldn't be isolated in company

Among these strategic principles, good communications and the resulting good morale characterize the human relations aspect of the successful exploration subculture, i.e., the internal operating style of a discovery group. The exploration group within a company should be tuned in with the rest of the company and not isolated, so that its discoveries will meet company

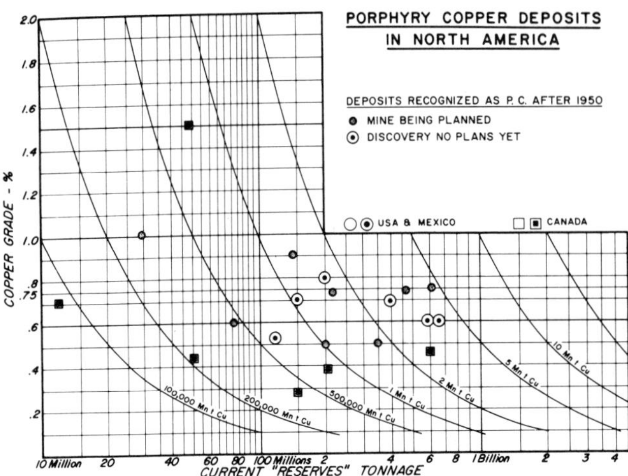

Fig. 7.

goals and can be appreciated and exploited as promptly as feasible. Also, for an exploration group to remain successful, the management of the company must be psychologically prepared to accept and support a long-term exploration effort requiring much staying power. After all, the birth of a mineral deposit and its growth into profitable maturity always takes many years and follows a long gestation preceded by many miscarriages and abortions. An exploration group, to be viable and successful, must feel confident that it does not exist because of a current fad for mineral resources and that the company truly has plans to profit from its discoveries.

In conclusion, as the explorers continue their search for the proverbial pot of gold at the end of the rainbow, and keep striving for new discoveries and for improvements in the effectiveness of their exploration philosophy, they should not forget Shakespeare's admonition:

> "There are more things in Heaven and **Earth** than are dreamed of in your philosophy."

References

[1] BRADLEY, W. H. Geologic Laws, in The Fabric of Geology, Claude C. Albritton, Jr., Editor, 372 pp., Addison-Wesley Publishing Co., 1963.

[2] HOAGLAND, Alan D. Appalachian Strata-Bound Deposits: Their Essential Features, Genesis and the Exploration Problem, *Economic Geology*, Vol. 66, No. 5, 1971.

[3] KRAUSKOPF, Konrad B. The Source of Metals, Geochimica et Cosmochimica Acta, Vol. 35, 1971.

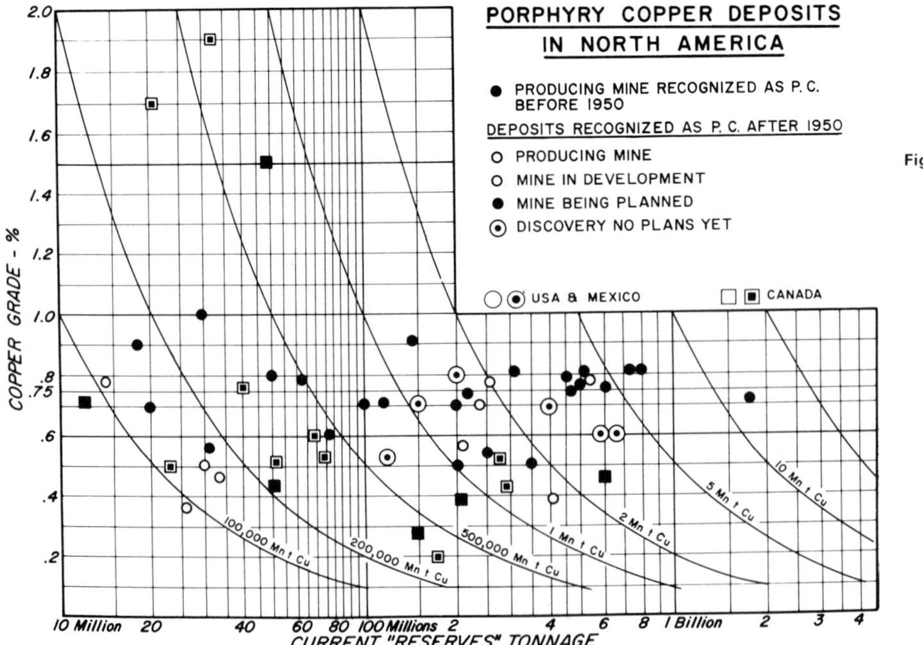

Fig. 8.

Part III

REGIONAL ANALYSIS

Editor's Comments on Papers 11 Through 19

Correlation of the distribution of ore deposits with regional geological structures was first recognized by Élie de Beaumont in his paper "Note sur les emanations volcaniques et metalliferes" (1847). He recognized the significance of broad belts of mountain chains and deformed rocks and intrusives, as well as parallel systems of crustal fractures that bore geographical relations to the distribution of metalliferous deposits. The first comprehensive attempt to correlate local mine and mineral district structures and regional structures of continental magnitude was presented by Billingsley and Locke (1941). The abstract from their paper is presented here as Paper 11, and a longer excerpt is reprinted in Walker (1976). This paper was pretty much neglected for some seventeen years until Mayo (Paper 12) suggested an alignment of major mineral districts in the southwestern United States along or at the intersections of fundamental crustal breaks. Lineament tectonics became a popular exploration technique, though the definition of what constituted a lineament became quite clouded. Wertz (Paper 13) carries fracture analysis into much greater detail, recognizing not only regional fracture trends but also fracture patterns that reflected domal pyramid structures that appeared to be spatially related to mineralized districts.

Availability of satellite imagery gave the lineament-chasing technique a big boost, and any linear feature exhibited on the imagery was presented as a lineament. Gilluly (Paper 14) voices the opinion of the skeptics regarding the unbridled utilization of line drawing on geological maps and aerial photographs or satellite imagery. He rightfully points out that crustal lineaments produce broad zones of fractures and that their intersections encompass large areas—very large in comparison to the size of an ore district.

Reduction of a broad prospective area to establish a target of a size amenable to detailed exploration techniques may be achieved through application of the Connolly (1936) contouring method to establish patterns of zoning or variations in fracture intensity or direction. Taylor and Steveson (Paper 15) demonstrate utilization of this approach in the derivation of potential targets.

Strata-bound deposits in limestones present a challenge to the exploration geologist since they seldom crop out and since their controls are very subtle as is the envelope of wallrock alteration. Therefore, a compilation of a series of exploration case histories is most helpful. Agnew (Paper 16) ties together stratigraphic, structural (local and regional), and chemical parameters in the search for zinc-lead deposits in limestone.

Domination of the igneous-hydrothermal philosophy of ore genesis has given way to wide acceptance of sedimentary and submarine volcanogenic environments of ore deposition as important processes. European, Australian, Canadian, and Japanese geologists have been leaders in the understanding of the nature and environments of deposition of these deposits and in guiding successful exploration for their discovery. King (Paper 17) presents some of the concepts and methods used in establishing stratigraphic relationships.

The realization of an evolving earth and atmosphere/hydrosphere has taken a long time to make an impact upon the search for ore deposits. Watson (Paper 18) introduces the time dimension to the search pattern.

Revived interest in placer-type deposits has raised questions as to the tectonic, basin characteristics and channel processes that result in concentrations of the heavy minerals. A model study under controlled conditions serves to indicate areas and conditions of concentration, and Adams, Zimpfer, and McLane (Paper 19) provide such a model approach to mineral exploration.

REFERENCES

Billingsley, P., and A. Locke, 1941, Structure of Ore Districts in the Continental Framework, *Am. Inst. Mining, Metall. and Petroleum Eng. Trans.* **144:**9–64.

Connolly, H. J. C., 1936, A Contour Method of Revealing Some Ore Structures, *Econ. Geology* **31:**259–271.

Élie de Beaumont, L., 1847, Note sur les emanations volcaniques et metalliferes, *Soc. Geol. France Bull.* **4:**1249 (July 5).

Walker, W., ed., 1976, *Metallogeny and Global Tectonics,* Benchmark Papers in Geology, vol. 29, Dowden, Hutchinson & Ross, Stroudsburg, Pa., 382p.

11

Reprinted from page 9 of *Am. Inst. Mining, Metall. and Petroleum Eng. Trans.* **144:**9–64 (1941)

Structure of Ore Districts in the Continental Framework

By Paul Billingsley* and Augustus Locke,† Members A.I.M.E.

Certain adequately developed mining districts give complete three-dimensional patterns of ore bodies as clusters rising from roots in basement rocks with details controlled by structure of cover rocks. Examples are Goodsprings, Tintic, Bisbee, Leadville, Hedley, Eureka, Ely and Goldfield. In these and others the controlling structures in both basement and cover rock are elements of the neighboring regional structure and have been produced by rock movements in harmony with continental adjustments, which are usually steep tear faults in basement and thrust crumples in cover. Invariably the structures and movements in the districts have long geologic history and result from intersecting or superimposed deformations of different ages. Salt Lake districts show intersecting Uinta and Cordilleran elements; Colorado districts show intersecting pre-Cambrian and Tertiary elements, and Bisbee shows Tertiary superimposed on final Paleozoic folding. Thus the position of districts has been determined by the patterns of North American orogenic belts, which make crossroads in Rocky Mountains and blocks of superimposed deformation in Sierra Nevada, northwest Coast and Canadian Coast Ranges. Essential function of repeated deformation is to strengthen rock, as from shale to slate to schist to gneiss, and thus make it competent to carry channels to depth with or without addition of intrusives. The essence of a mining district is the presence of such competent rocks with long-lived, deep, penetrating breaks reopened to permit passage of heat and associated products from depths to surface. It is suggested that melts such as magmas; metamorphism up to and including granitization; alteration, such as orthoclase-quartz or quartz-sericite or garnet, and also mineralization, may all be by-products of such heat escape along channels made by tectonic forces at specific foci.

[*Editor's Note:* Only the abstract has been reproduced here.]

* Consulting Mining Geologist, Still Waters, Burton, Washington.
† Consulting Mining Geologist, San Francisco, California.

12

Reprinted from *Am. Inst. Mining, Metall. and Petroleum Eng. Trans.* **211:**1169-1175 (1958)

LINEAMENT TECTONICS AND SOME ORE DISTRICTS OF THE SOUTHWEST

Evans B. Mayo

This is a controversial topic. The author, in presenting his analysis, is aware that the study of lineament tectonics and relation of ore districts to regional structure is complicated by insufficient data and, unavoidably, by personal bias.

The development of lineament tectonics has been summarized by Umbgrove.[1] Early attempts to fit ore districts into the Cordilleran framework were made by Billingsley and Locke,[2,3] who do not refer to lineament tectonics, although their approach is similar; they observe that heat and fluids, including the ore-depositing fluids, are most likely to rise at or near intersections of major structures where the crust is fractured, or weakened, to great depth.

As a result of studies distributed over the earth—including ocean basins as well as continents—some tectonists recognize four dominant structural trends: 1) northwest; 2) northeast; 3) nearly east-west, or equatorial; and 4) nearly north-south, or meridional. Baker[4] proposed a theory to account for these trends and Sonder[5] called their world-wide arrangement the regmatic shear pattern. Moody and Hill[6] proposed a much more complicated shear network which, although fascinating and perhaps ultimately useful, will not be followed here. In a recent review of deformation within the Cordillera, Wisser[7] mentioned the four fundamental directions.

The fact that many geologists deny the existence of the regmatic shear pattern implies that the fundamental structures are far from obvious. It may mean, also, that some geologists are not accustomed to examine regional and world maps analytically. The maps require much study, and certain features should be isolated on overlays. Even so, with the present limited knowledge, uncertainties remain. The following analysis is a qualitative experiment, subject to change as information accumulates, and should be supplemented by the western sheet of the Tectonic Map of the United States.[8]

E. B. MAYO is Professor of Geology, University of Arizona, Tucson, Ariz. TP 4776 I. Manuscript, April 8, 1958. New York Meeting, February 1958.

LINEAMENT FRAMEWORK, SOUTHWESTERN U.S.

Many have probably gained the impression that the Cordillera of the West is oriented northwest-southeast and from this, unless experience rules otherwise, it is natural to assume that the structure likewise trends northwest-southeast. To an important extent this is true, yet anyone tracing off the western sheet of the Tectonic Map all the recorded northwest-southeast structures may be surprised to see what a small part of the entire area they occupy. In southwestern U. S. (Fig. 1) the northwest-southeast structures are mostly restricted to the eastern, southern, and western parts. The eastern margin of the Cordillera from northern Colorado to the International Boundary near El Paso, Tex., is obviously determined by some structure other than northwesterly ones. In eastern Nevada and far southward toward the Gulf of California many mountain ranges, valleys, and faults are meridional. In the Rocky Mts. of Colorado, and at many places in southern Arizona, the crystalline Pre-Cambrian is foliated northeast-southwest. The Uinta Mts. of Utah trend approximately east-west, in much the same way as a broad belt of transverse, west-northwest structures—the Texas lineament of Hill[5,9] and Ransome[10] in southern Arizona, southwestern New Mexico, and southern California.

It seems, then that there are four regmatic shear directions in the Southwest, but at many places they are discontinuous, and their projections must be inferred. To clarify these trends the four sets have been isolated into two systems: 1) northwest-northeast and 2) east-west-north-south (Figs. 2 and 3).

Northwest-Northeast System: A number of prominent northwest-trending zones of structure are easily recognized. They are designated by circled Roman numerals, the northeast-trending structures by circled capital letters.

Perhaps no two geologists would agree completely on the positions of all these belts. Names given below are for convenience only, and may be discarded where other names have priority.

(I) The Sierra Nevada-Lower California belt contains the Jurassic-Cretaceous granitic massifs of the Sierra Nevada and Lower California. These

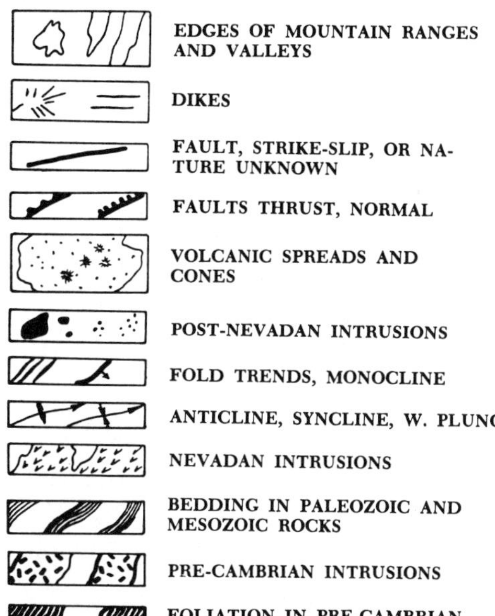

Symbols used on Figs. 1-3

crystalline masses form a barrier between the Pacific coastal region, dominated by northwest trends, and the interior where other structural trends are prominent.

(II) The Walker zone seems to extend from the northern end of the Sierra Nevada, southeastward into northwestern Sonora. Billingsley and Locke,[3] who proposed the name *Walker line,* suggested that this zone curved into southeastern Arizona. It is believed, however, that the line they defined includes several different structural belts.

(III) The southwest Arizona belt is parallel to the Walker zone and reaches from the Patagonia and Santa Rita Mts. of southern Arizona northwestward into southern Nevada.

(IV) The central Arizona belt extends northwestward from the Chiricahua and Mule mountains of southeastern Arizona to the San Francisco volcanic field.

(V) The southwestern New Mexico belt reaches from the International Boundary near Columbus, N. M., toward the Grand Canyon in Arizona. This belt is obscure in the Datil volcanic field and Colorado Plateau and possibly does not cross these features.

(VI) The New Mexico-Utah zone extends from the Guadalupe Mts. in trans-Pecos Texas and New Mexico, past Mt. Taylor, through the high Plateaus volcanic field to the Gold Hill district in Utah. The continuity of this zone, although locally interrupted, seems obvious.

(VII) The southwestern Colorado zone, beginning west of Tucumcari, N. M., extends past Taos and along the southwest side of the San Juan volcanic field and beyond to Bingham, Utah.

(VIII) The central Colorado belt begins in the area of the Spanish Peaks in southern Colorado and trends through Leadville and Climax toward the Washakie Basin in southern Wyoming.

(XI) The Laramie zone trends acutely across the Front Range of Colorado and along the western side of the Laramie Basin, Wyoming.

The following zones appear to be complementary to the above northwesterly trends:

(A) The northeast-trending Santa Rita belt is vaguely shown by faults, aligned intrusions, and the southeastern margin of the Datil volcanic field. It begins southwest of Bisbee, Ariz., and reaches past Tyrone and Santa Rita, N. M., to the northern end of the Tularosa Basin.

(B) The Morenci belt is also vaguely defined and may not extend as far as indicated (Fig. 2). It passes through the Datil volcanic field, its course revealed by alignments of the San Agustin plain, New Mexico, and the modifications at the northwestern ends of valleys at Mogollon, N. M., and Morenci, Ariz. The extension of this zone east of the Rio Grande is marked by faults and by the northeast trend of foliation in Pre-Cambrian rocks.

(C) The Jemez zone, beginning with a southwest-trending trough in southern Colorado, passes through the Jemez caldera and the Mt. Taylor area and along the northwestern border of the Datil volcanic field. The existence of this zone in the Globe-Miami district is suggested by the alignment of post-Nevadan intrusions and other structural features. Between Globe and Ajo, the Jemez zone seems concealed by the overprint of the Texas lineament (1-1, Fig. 3) and other structures. It may extend across southern Arizona to the Pinacate volcanos of northwestern Sonora, and beyond to the little cluster of cones on the Pacific Coast.

Zones (A), (B), and (C) form a broad northeast-trending belt that includes most of the Datil volcanic field as well as many other volcanic features. The broad belt, therefore, seems to have been tectonically active in later geologic time.

South of zone (C) the eastern front of the Cordillera trends somewhat east of north, whereas north of zone (C) the trend is about north-south, or even a little west of north. In the Spanish Peaks region near zone (C), therefore, there is a definite turn in the Cordilleran front, and apparently zones (A), (B), and (C) have had some part in modifying the course of the border between the Great Plains and the Rocky Mts. The broad belt of zones (A), (B), and (C) is obviously of great importance in the regional structure; should it cross some equally important belt of structures, intersections of great interest would be expected.

(D) The Front Range zone is named for the Front Range porphyry belt of Colorado, which forms the eastern part of the zone. The zone appears to extend southwestward along the northwest edge of the San Juan volcanic field to the Carrizo Mts. in northeastern Arizona. It may continue farther, through the San Francisco volcanic field to the area of northeast-trending foliation in the Pre-Cambrian rocks east of the Colorado River.

(E) and (F), two northeast-trending zones, occupy essentially the position of the Wyoming lineament of Ransome (Ref. 10, p. 294). There may exist in the southwestern part of this Wyoming belt a

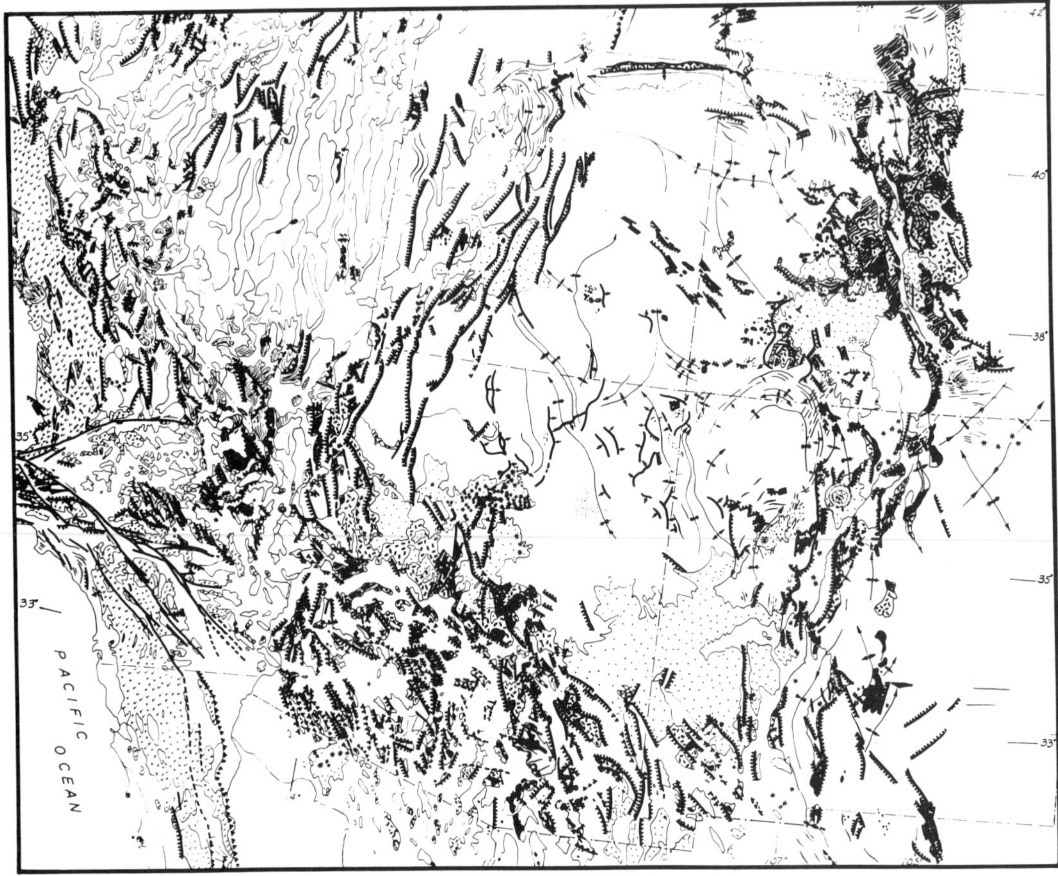

Fig. 1

very broad area of northeasterly trends, partly indicated by foliation in Pre-Cambrian rocks and partly by bedding trends in Paleozoic sediments. Traced northward out of Colorado, the Rocky Mt. ranges end in the northeastern part of this belt, in southern Wyoming. Broad passes intervene between these ranges and the more northern Rockies of Wyoming. The southwestern end of the belt is aimed toward the gap between the granitic massifs of the Sierra Nevada and Lower California.

(G) The southern Nevada zone is mostly shown by the trends of mountains and intervening valleys, and to a lesser extent by aligned intrusions. It should impinge on the eastern front of the Sierra Nevada between Lone Pine and Independence, Calif. Possibly this zone reaches northeastward to the vicinity of Bingham, Utah.

(H) The central Nevada zone is definitely indicated by aligned faults, yet exact position and orientation are uncertain. The structures seem to impinge on the Sierra Nevada in the Mammoth embayment, northwest of Bishop, Calif. This trend is considered largely responsible for the great reentrant in the Sierra front and may be partly responsible for the abrupt broadening of the granitic core southwest of the embayment.

(J) The Carson zone meets the Sierra front near Carson City, Nev. Toward the northeast the course of this belt of structures is suggested by the trends of valleys and mountains, by aligned intrusions, and by faults. To the southwest it may be the reason for some of the granitic prongs along the western edge of the Sierra Nevada batholith.

Although only two sets of lineaments have been examined, it already seems obvious that important geologic features are located at places where these two sets of trends cross. Some of these features, indicated in Fig. 2, are as follows:

1. (1) and (G, H, J), not exclusive—Sierra Nevada batholith.
2. (I) and (C,D,E,)—broadest part of the batholith of Lower California.
3. (II) and (C)—Pinacate volcanos.
4. (II) and (E,F)—large intrusions in Mohave Desert of California.
5. (II) and (G,H,J)—granitic intrusions east of Sierra Nevada.
6. (IV, V) and (D)—San Francisco volcanic field.
7. (V, VI) and (A,B,C)—Datil volcanic field and associated intrusions.
8. (VI) and (F)—High Plateaus volcanic field and intrusions.
9. (VII) and (C)—Jemez Caldera.
10. (VII) and (D)—San Juan volcanic field and intrusions.
11. (VII) and (G?)—Intrusions at and near Bingham, Utah.

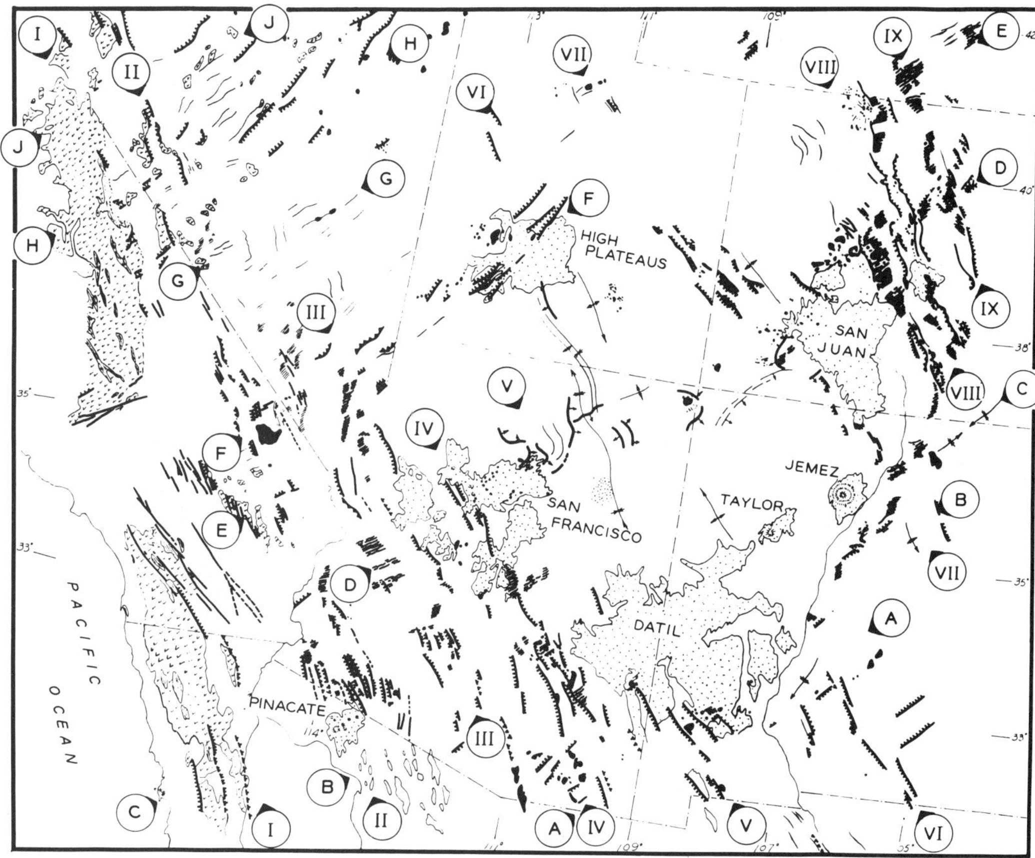

Fig. 2

12. (VIII) and (D)—large area of porphyry intrusions north of San Juan volcanic field, Colorado.
13. (VIII) and (E)—group of intrusions along Colorado-Wyoming border.
14. (IX) and (D)—Front Range porphyry belt, Colorado.

These examples suggest how intersecting weak zones in the crust may have localized the phenomena of intrusion and volcanism. The suggestion is further strengthened by consideration of the east-west-north-south lineament system.

East-West-North-South System: The structures belonging to this system are locally obvious, but in large areas they are either lacking or have not yet been mapped. This is particularly true of the east-west set. As a rule the two sets only approximately follow the east-west or north-south directions. The north-south set is usually oriented somewhat east of north, and the east-west set somewhat north of west. On the basis of information traced from the Tectonic Map,[8] there appear to be the following nearly east-west structures (Fig. 3):

(1) Texas lineament (with a possible branch, 1′). This great belt of transverse structures forms the southern border of the tectonic framework of the western U.S. In the region of Fig. 3, the Texas lineament is by far the most pronounced transverse belt. It can be traced from trans-Pecos Texas on the east to the Transverse Ranges of Southern California on the west. Strands within the belt, as in southern Arizona, are marked by nearly east-west faults and elongated or aligned intrusions. The entire zone, in southern Arizona, may be more than 150 miles wide.

(2) The San Francisco zone is not satisfactorily revealed by the assembled data. The best evidence for its existence is in eastern California and southern Nevada, where trends of intrusive contacts, elongated intrusions, west-northwest faults, and curves in thrusts and in the trend of bedding suggest a transverse disturbance. Some of the cones in the San Francisco volcanic field are arranged in transverse rows; possibly this trend extends into east-central New Mexico where foliation in Pre-Cambrian is oriented west-northwest. If this is the case, there are broad gaps with no evidence, and the supposed San Francisco zone is poorly defined. Perhaps it is valid only where supporting evidence has been mapped.

(3) The Spanish Peaks belt trends from southern Colorado past the San Juan volcanic field and into southern Utah. Groups of intrusions seem to continue this trend across Utah, and faults, intrusions and a belt of thrusts indicate its path across Nevada. This zone apparently reaches the Mother Lode of California near Plymouth.

(4) The Cripple Creek zone reaches from the southern end of the Front Range, Colorado, west-

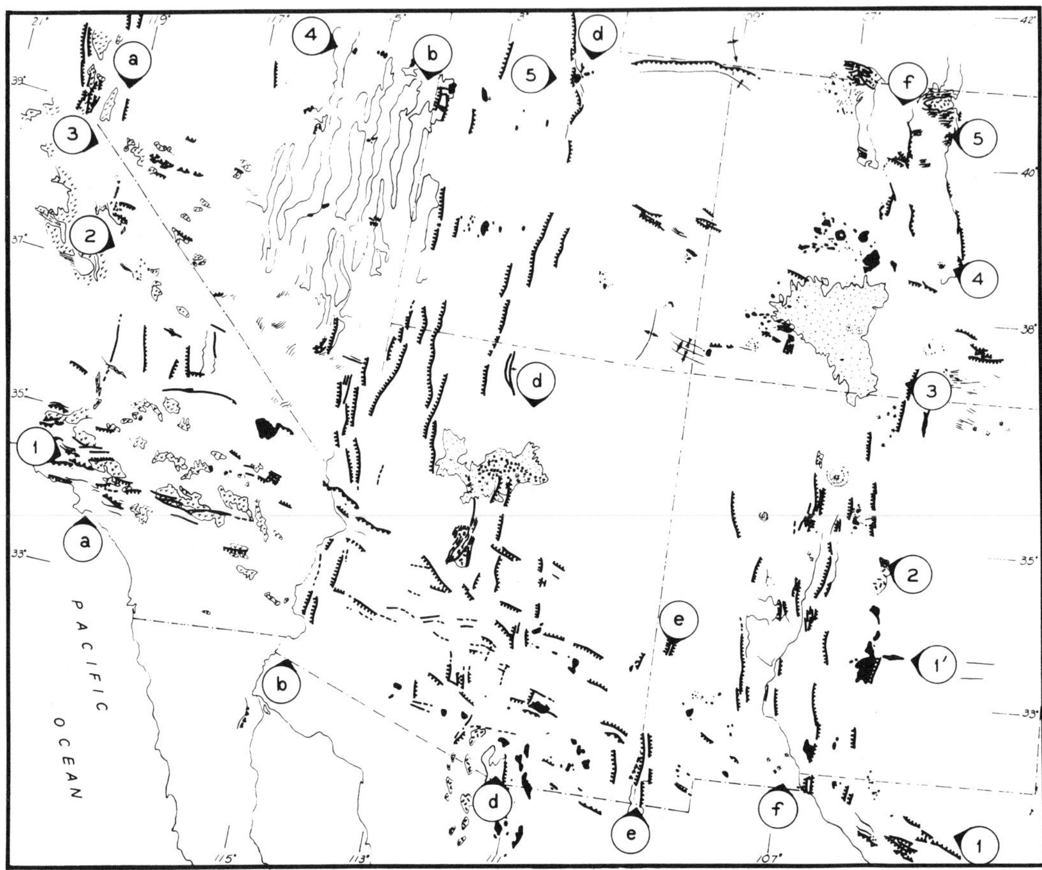

Fig. 3

ward to the great group of porphyry intrusions, and on into eastern Utah. This structure may extend across Utah to the Gold Hill district and westward into eastern Nevada.

There is some evidence (Fig. 3) in eastern Nevada that a transverse structure exists midway between zones (3) and (4).

(5) The Uinta belt begins in northern Colorado and follows along the Uinta axis to Bingham, Utah. Perhaps this is a branch structure, related to the Cripple Creek zone as (1′) is related to the Texas lineament.

Crossing the above structures are the nearly north-south lineaments, of which the following are recognizable:

(a) The Sierra Nevada belt is indicated at the north by north-south faults that at first merely border the Sierra, then slice obliquely into it and across it (e. g., Kern Canyon fault). Perhaps this meridional zone does not cross the Transverse Ranges of southern California.

(b) The Colorado River zone is very broad at the north, in eastern Nevada, and narrows southward along the Colorado River. It reaches the northern end of the Gulf of California.

(d) The Utah-Arizona belt trends north-south through central Utah and Arizona. In Utah and in northern Arizona it forms a boundary between the plateaus to the east and the Basin and Range province to the West.

(e) The Pelonicillo belt parallels the interstate boundary in southeastern Arizona and southwestern New Mexico.

(f) The Cordilleran Front belt, a very broad zone of meridional structures, parallels the eastern margin of the Rocky Mts. in Colorado and New Mexico.

Comparison of Figs. 2 and 3 suggests that most if not all the 14 features listed above are localized not only on intersections of structures belonging to the northwest and northeast sets, but also on north-south-east-west crossings. For example, the Sierra Nevada batholith seems somewhat related to transverse zones (2) and (3) and to north-south zone (a), as well as to the structures already mentioned; the San Francisco volcanic field occupies the intersection of zones (2?) and (d); Datil volcanic field is on the northern margin of the Texas lineament where this is crossed by the north-south structures (e) and (f); the intrusions near Bingham, Utah, occupy the intersection of lineaments (5) and (d), and so on.

It has been noted that some of the structural trends are marked by aligned intrusions, and possibly these intrusions occupy minor intersections on the trends. If this is so, the evidence quite eludes the present analysis. It is otherwise, however, with certain groups of intrusions, with granite masses of regional dimensions, and with some of the most arresting volcanic phenomena. These features occupy the intersections indicated, or at least suggested, by data on the maps.

The above relation suggests that some of the major intersections, having been unusually weak places in the crust, were the sites of especially pronounced and long-continued igneous activity. This repeats Billingsley's and Locke's idea[11] that escape of heat and fluids is particularly favored on the structural intersections.

THE STRUCTURAL INTERSECTIONS

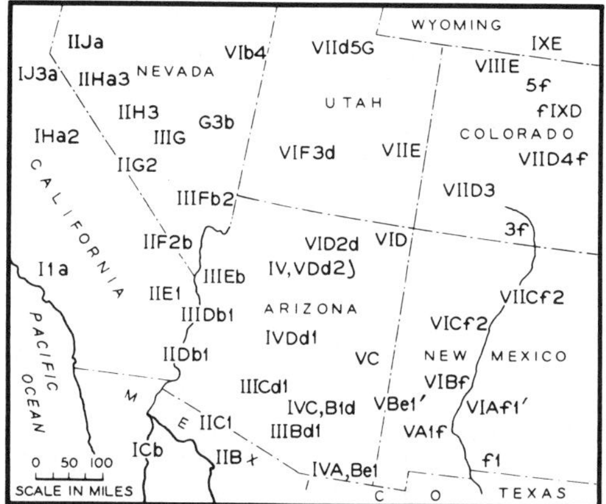

The *geologic crossroads* were located on a tracing (Fig. 4, above) first placed over Fig. 2, then over Fig. 3. There are many intersections, and it is disconcerting to realize that the entire region will be covered, provided the structural zones selected are numerous enough and broad enough. No doubt every place is an intersection of some kind, but this fact is of no aid in selecting areas of special geologic or economic interest, so it becomes necessary to establish guides to selection. This brief inquiry cannot approach a complete list of guides, but a few seem promising:

1. Crossings of the most pronounced structure belts—such as the Texas lineament and zones (A), (B), and (C)—should be of greatest interest.
2. The structural trends concerned should be fairly obvious, suggesting that the trends have been active in later geologic time.
3. More than two directions should meet at the intersection.
4. One or more of the directions should be accompanied by evidence of igneous activity.
5. Small or moderate-sized post-Nevadan intrusions should occupy the intersection, or:
6. If Nevadan granites of regional dimensions are present, the areas of interest should be marginal to the granite.

It is realized that areas of great importance from the viewpoint of exploration may be concealed and may therefore lack some or all of the above characteristics. This is of no importance in the present study, in which the objective is correlation with known ore districts.

With the guides listed above, selections can be made from the intersections in Fig. 4. These intersections are considered according to the transverse belts on which they are located.

INTERSECTIONS, TEXAS LINEAMENT

This broad, transverse belt is crossed by structures that follow the other three regmatic shear directions. Most important of these, apparently, is the northeast bundle of zones (A), (B), and (C). Next in order seem to be northwest zones (III), (IV), and (V); northeast zones (D), (E), and (F); and northwest zones (I) and (II). All the north-south belts reach or cross the Texas lineament.

Structural crossings were arranged into four classes. Intersections that involve the Texas lineament and zones (A), (B), or (C) and at least one other direction, and likewise have small or moderate-sized intrusions, are rated first class. An intersection involving four directions plus intrusions is also rated first class. If four directions are present but there are no intrusions, the intersection is second class, and if there are three directions plus intrusions but zones (A), (B), or (C) are not represented it

Table I. Intersections On and Near Texas Lineament

I	Ia	(2nd class)	II	El	(2nd class)
III	Eb	(3rd class)	III	Dbl	(2nd class)
II	Dbl	(2nd class)	I	Cb	(3rd class)
II	Cl	(1st class)	II	B	(4th class)
IV	DdI	(1st class)	III	Cdl	(2nd class)
III	Bdl	(1st class)	IV	C,Bld	(1st class)
V	C	(4th class)	IV	A,Bel	(1st class)
V	Bel'	(1st class)	VI	Bf	(2nd class)
V	Alf	(1st class)	VI	Afl'	(1st class)
	fl	(4th class)			

Table II. Lineament Intersections and Mining Districts

First Class Intersections		Mining Districts
II	Cl	Ajo
IV	Ddl	Jerome-Bradshaw-Bagdad
III	Bdl	Twin Buttes-Silver Bell
V	Bel'	Morenci-Mogollon
V	Alf	Santa Rita-Tyrone
VI	Afl'	Capitan iron—no important district known
IV	A,Bel	Bisbee-Tombstone-Courtland-Gleeson
IV	C,Bld	San Manuel, Ray, Miami, Superior, Globe, Mammoth, etc.
Second Class Intersections		**Mining Districts**
I	la	No important district known
II	El	Eagle Mountains, iron
II	Dbl	No important district known
III	Dbl	Artillery Mountains, manganese
III	Cdl	No district known
VI	Bf	Magdalena

Table III. Intersections On and Near Transverse Zone (2)

First Class Intersections		Mining Districts
I	Ha2	Southern end of Mother Lode on west; Bishop tungsten on east
VII	Cf2	Pecos
VI	Cf2	Mt. Taylor-Grants region, uranium
II	Fb2	Goodsprings-Searchlight, Nevada; Mountain Pass, California
VI	D2d	Monument No. 2—White Mesa
Second Class Intersections		**Mining Districts**
II	G2	Silver Peak-Goldfield-Rhyolite-Bullfrog
VI	D	Carrizo Mountains, uranium

Table IV. Intersections on Transverse Zones (3), (4) and (5)

First Class Intersections		Mining Districts
I	J3a	Mother Lode
II	Ha3	Aurora—Candelaria
VII	d5G	Bingham-Park City-Tintic, etc.
VI	F3d	Marysvale-Silver Reef-Iron Springs
VII	D3f	San Juan districts
VII	4Df	Cripple Creek
Second Class Intersections		**Mining Districts**
II	Ja	Virginia City (Comstock Lode)
II	H3	Tonopah
	G3d	Pioche
VI	b4	Gold Hill
	f IX D	Front Range Mineral Belt, Leadville, Climax

is second class. If three directions are present without intrusions the intersection is third class; if only two directions are represented it is fourth class. The localities and structural trends represented are shown in Fig. 4 and listed in Table I. The first and second class intersections only are listed, together with the associated mining districts, in Table II.

The contrast between the first and second class intersections is revealed in the relative number and importance of the *hits* scored in each class—seven out of eight in the first class, but no better than three out of six, and probably no major ore district, in the second class.

Intersections On and Near Transverse Zone (2): This vaguely defined, apparently weak or inactive zone compares very unfavorably with the Texas lineament, and its intersections have had to be classified somewhat differently. Where four trends cross and intrusions are present, the crossing is considered first class, but if intrusions are absent it is considered third class, even though important volcanic phenomena are present. Where two or three trends cross and intrusions are present the crossing is second class. All other cases are relegated to third or fourth class and eliminated. Locality VI Dd2 attains to first class because small, ultrabasic intrusions not shown on the map are present, whereas locality IV, V Dd2 is assigned to third class because the surface is mostly covered by volcanic rocks. Even so, uranium has been mined near Cameron, Ariz.

Few places in this transverse zone can be considered major ore districts. Most important, perhaps, are Bishop tungsten in California and Monument No. 2 in Arizona.

Intersections in Zones (3), (4), and (5): Parts of these zones are not satisfactorily revealed by the data on the Tectonic Map, but in general these structures are more obvious than transverse zone (2). It may be that over certain intervals, particularly in Utah and Nevada, the continuity of zones (3) and (4) will never be established. The first and second class intersections are listed in Table IV.

As selected above there are 19 first class intersections and 13 second class intersections. Ore districts are found on 18 of the first class, 10 of the second class. Furthermore, most—but not all—the major ore districts are on the first class crossings.

At first the correlation between intersections and ore districts seems remarkable—about 94 pct for the first class, 77 pct for the second. But these figures take no account of districts that were missed. Excepting some uranium-bearing areas, most of these are of lesser importance. If there is an additional transverse zone between zones (3) and (4), Ely and Eureka may occupy crossings on this zone.

The personal factor is, of course, a disturbing source of error. With all good intentions, the structural trends and their crossings were probably not selected without bias. In spite of many uncertainties, however, it is believed that a systematic relation does exist between the crossings of certain elements of the regional framework and the most important occurrences of ore.

IMPROVING CONTROL

Better maps will provide one answer to present imperfections in correlating ore faci with structure. Fortunately the maps continue to improve, and it should gradually become possible to deal more confidently with two-dimensional geometry. But two dimensions are not enough.

If hypogene ore has come from *below*, something must be known of conditions at depth, and probably the greater the depth the better. Three-dimensional control of ore is best disclosed by mine workings and natural sections. Billingsley and Locke in their second paper[2] offered a few examples, and many more are available. Whatever indirect knowledge is gained of the third dimension—beyond the depth of mines, drillholes, and erosion—must be won by the geophysicist.

Every geological problem, however, is four-dimensional, that is, it involves time. The structural history of any region under investigation holds the answer to the relative importance of various structural trends. Which structures were the longest active and when were they active? Where were the sedimentary basins of the past and when were they deformed? What was the sequence of intrusion, extrusion, differentiation,[10] and alteration? These features have been discussed and shown on maps, but the writer knows of no regional study involving all of them. If such complete data were available the mineral districts might be fitted into the regional framework like jewels into their settings. In that case a proper setting without its jewel would be subject to the most earnest inquiry.

This generalized account can only suggest a direction for further work. Obviously, regional structure is the essential framework for all geological studies, including investigations of ore deposits. It seems, then, that the methods and theory of lineament tectonics should be important in future explorations for economic minerals. The structural framework of the Southwest may prove to be different in many respects from the one proposed here, yet that framework should become more obvious as information accumulates. As the framework becomes more definite it should clarify the positions of known ore districts, reveal localities where new ore might reasonably be searched for, and outline those places where exploration would probably be wasted.

This article was written at the invitation of W. C. Lacy, who generously aided in its preparation. Although the author is acquainted with the geology of parts of the region discussed, the material presented here is mostly compilation. The American Association of Petroleum Geologists kindly granted permission to use information shown on part of the Tectonic Map of the United States.[8] Information was taken, also, from Vincent Kelley's tectonic map of the Colorado Plateau.[13]

REFERENCES

1 J. H. F. Umbgrove: *The Pulse of the Earth*, pp. 304-308, Martinius Nijhoff, The Hague, 1947.

2 P. Billingsley and A. Locke: *AIME Trans.*, 1941, vol. 44, pp. 9-64.

3 R. T. Hill: *AIME Trans.*, 1902, vol. 32, pp. 163-178.

4 C. L. Baker: *Pan American Geologist*, 1933, vol. 59, pp. 19-32.

5 R. A. Sonder: American Geophysical Union, *Trans.*, 1947, vol. 28, pp. 939-945.

6 J. D. Moody and M. J. Hill: Geological Society of America, *Bull.*, 1956, vol. 67, pp. 1207-1246.

7 E. Wisser: *Quarterly of the Colorado School of Mines*, 1957, vol. 52, no. 3, pp. 54-73.

8 P. B. King and Map Committee: *Tectonic Map of the United States*, American Association of Petroleum Geologists, Tulsa, 1944.

9 R. T. Hill: Geological Society of America, *Bull.*, 1928, vol. 39, p. 265.

10 F. L. Ransome: *Problems of American Geology*, pp. 287-376. Yale University Press, 1913.

11 P. Billingsley and A. Locke: *AIME Trans.*, 1933, vol. 501, pp. 2-12.

12 W. C. Lacy: *AIME Trans.*, 1957, vol. 209, pp. 559-562.

13 V. C. Kelley: University of New Mexico Publication in Geology, 1955, No. 5.

14 V. C. Kelley: *USGS Prof. Paper No. 300*, pp. 171-178.

15 C. L. Baker: University of Texas, *Bull.*, 1934, No. 3401, pp. 206-214.

13

Reprinted from *Western Miner,* October 1968, pp. 1-5

EXPLORATION GUIDES IN SOUTHERN ARIZONA: FRACTURE INTERSECTIONS AND DOMES

Jacques B. Wertz

Even with the best geologic maps, there is a constant need for further information and a constant demand for new clues, particularly in exploration work. New ideas and new thoughts are a necessity for progress.

In the structural study of an area, be it Arizona, North Africa, or British Columbia, the basic fracture pattern has to be known first. There is, however, no assurance that the trends shown on the map are really meaningful because they often are secondary in nature and have occurred as a consequence of an intrusion, of doming, or of gravity faulting. Such trends are a very weak basis for assuming the presence of ancient breaks that could constitute any useful guide to exploration.

It is the object of this study to bring out a few concepts that may lead to those zones of weakness that were sufficiently long and, most of all, sufficiently deep to allow either intrusives to be emplaced or mineralized solutions to be tapped.

Even in those places where one is lucky enough to possess accurate and detailed maps, a thorough interpretation is essential in order to establish the actual breaks through the upper surface of the crust.

1. FRACTURES are considered first. For instance:

(a) Where only fractures are recorded or given (Figure 1), their spatial arrangement and more particularly the number and types of downthrows may give a first clue as to the structure. The example of figure 1 indicates, for instance, that the fractures are secondary in nature, being relatively parallel and downthrown toward a common axis. This axis can be safely inferred to represent a major or primary break, a graben in this case.

(b) Oftentimes, a number of fractures are related to or concentrated around an intrusive body (Figure 2). It can be clearly observed here that these fractures are only a resultant of the emplacement of the intrusive. This is especially true in the case of the radial faults which can be categorized as definitely secondary. The elongation axis of the intrusive is really the ancient and important structural trend that reflects a primary fracture. All evidence has long been concealed by later formations or has been erased by the effects of metamorphism or by later intrusions.

The fact that another primary fault (occurring at right angle to the first one) can also be inferred is reflected

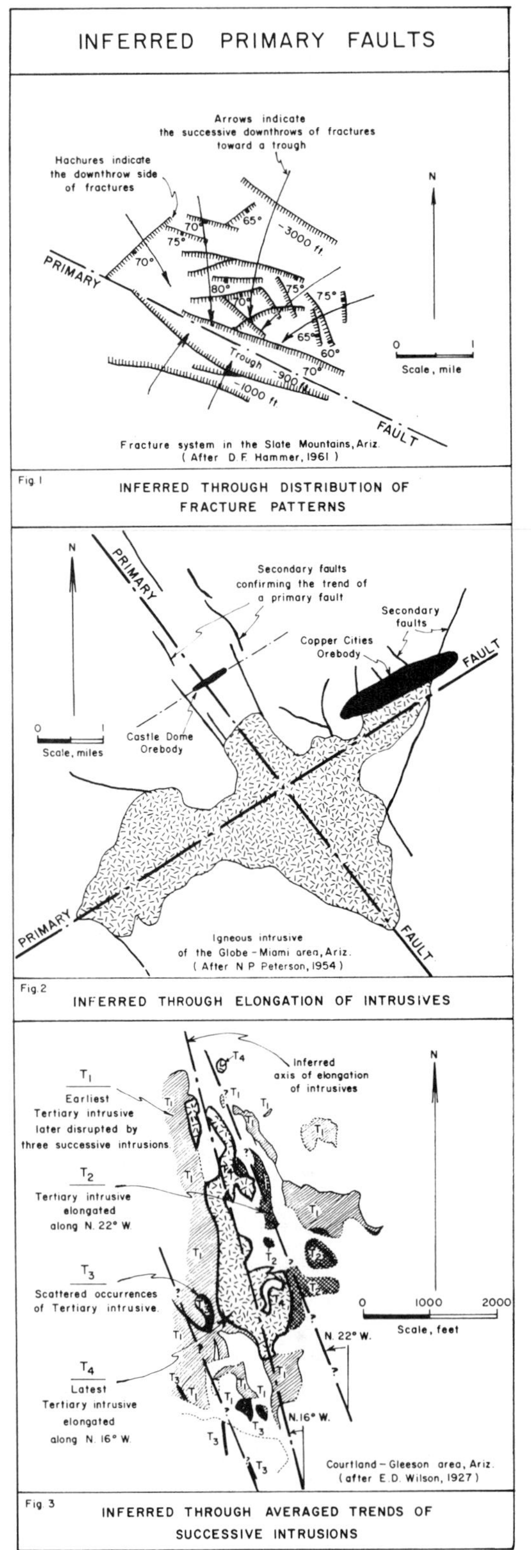

Fig. 1 INFERRED THROUGH DISTRIBUTION OF FRACTURE PATTERNS

Fig. 2 INFERRED THROUGH ELONGATION OF INTRUSIVES

Fig. 3 INFERRED THROUGH AVERAGED TRENDS OF SUCCESSIVE INTRUSIONS

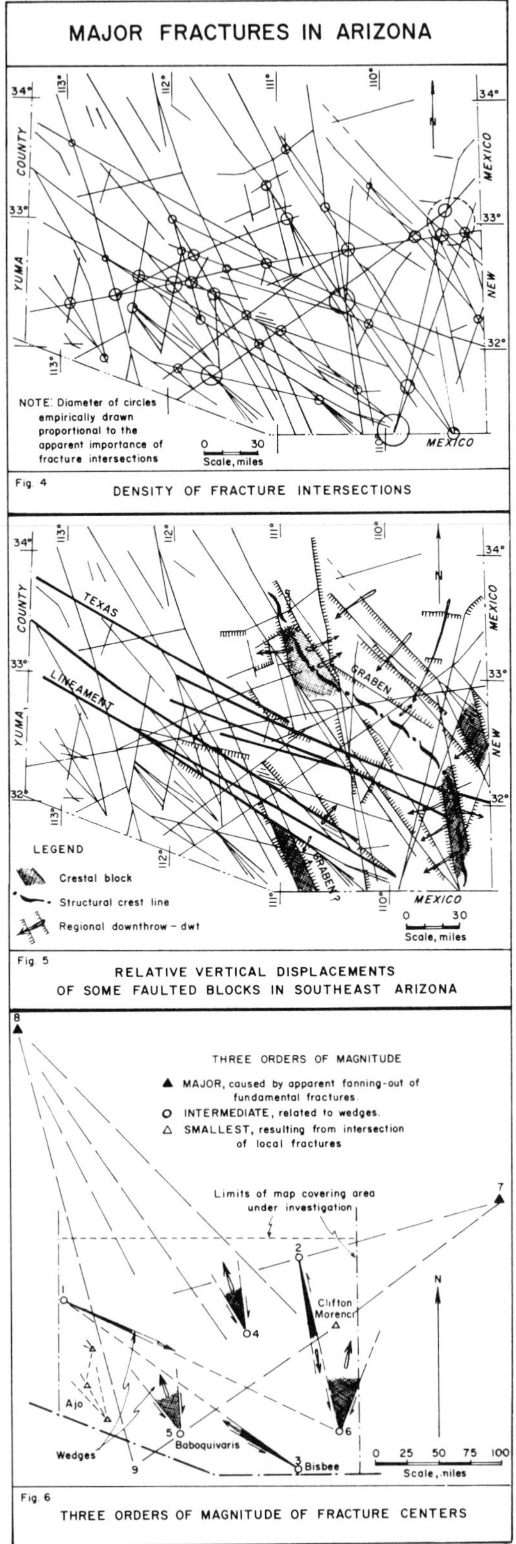

Fig. 4 DENSITY OF FRACTURE INTERSECTIONS

Fig. 5 RELATIVE VERTICAL DISPLACEMENTS OF SOME FAULTED BLOCKS IN SOUTHEAST ARIZONA

Fig. 6 THREE ORDERS OF MAGNITUDE OF FRACTURE CENTERS

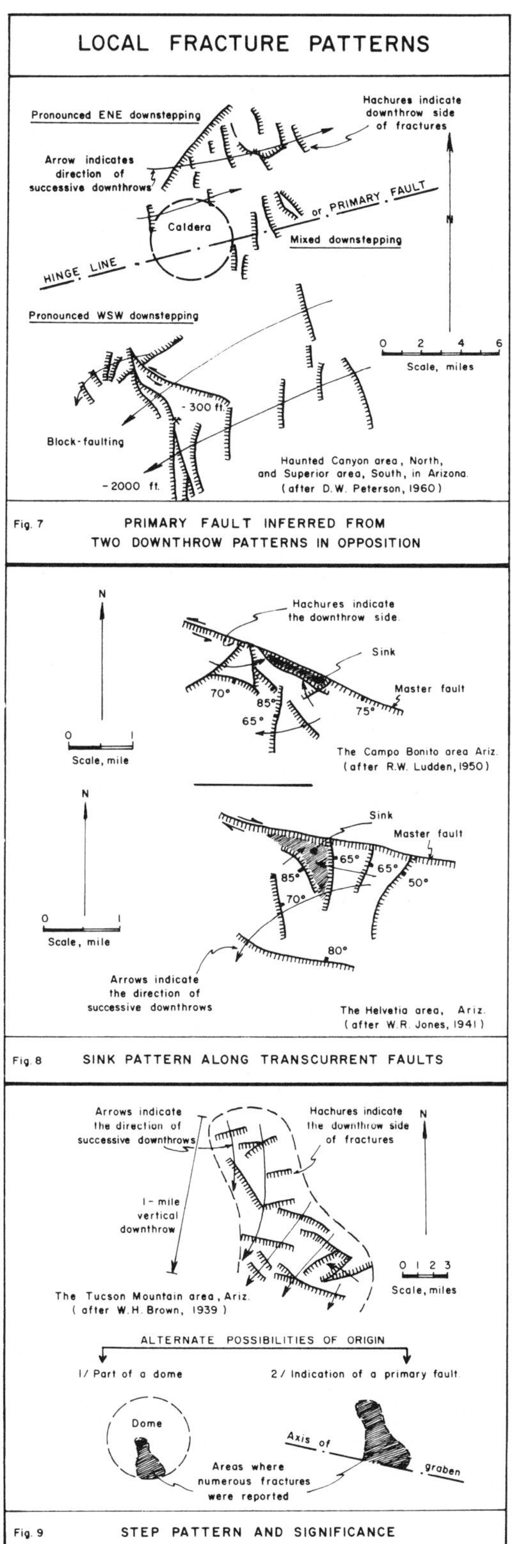

Fig. 7 PRIMARY FAULT INFERRED FROM TWO DOWNTHROW PATTERNS IN OPPOSITION

Fig. 8 SINK PATTERN ALONG TRANSCURRENT FAULTS

Fig. 9 STEP PATTERN AND SIGNIFICANCE

and confirmed by several parallel, secondary faults to the northwest of the intrusive shown on figure 2. The importance of the main primary fault is corroborated by the evidence that the two orebodies — the Castle Dome and the Copper Cities — are oriented perfectly parallel to the axis of the fault. In addition, it is known that the contacts between the several differentiated phases of the intrusive are nearly all oriented parallel to the main fault axis.

(c) In places, one may only observe a complex series of differentiated intrusives, with practically no recorded fracture (Figure 3). The primary fault can safely be inferred as being the averaged elongation trend exhibited by the successive intrusions.

These three structural situations are examples that illustrate the type of inference process to be used in order to systematically interpret geologic maps.

On a regional scale next, one can extrapolate and interpolate a large number of these inferred, primary fractures considered as first-order evidence, with numerous less obvious features of second- and third-order evidence revealed to us mainly by stratigraphic and geomorphologic means.

First-order criteria for recognition of primary breaks are:

(a) **known tension and shear faults;**

(b) **graben (established as such, or inferred);**

(c) **large shear zones;**

(d) **elongation trends of Tertiary dikes;**

(e) **elongation trends of Laramide and Tertiary intrusives;**

(f) **successive igneous intrusions along a same direction.**

Some second-order criteria, which give supporting weight to primary faults when encountered along their trend, are:

(a) **boundary of a fault block;**

(b) **linear juxtaposition of formations belonging to radically different ages (such as Precambrian and Tertiary);**

(c) **relatively straight contacts between elongated areas of metamorphic rocks next to non-metamorphosed sediments;**

(d) **linear trends from aerial photographs or photographic mosaics;**

(e) **small circular intrusives;**

(f) **volcanic craters, cones and calderas.**

Some third-order criteria are:

(a) **topographic and hydrographic trends (cliffs, rivers, elongated lakes) which in places (such as in British Columbia) may become of large importance in reflecting deep structure;**

(b) **frayed embayments (now filled with alluvium) indenting a massive structural block.**

Of great interest is the fact that, empirically, tear faults and tension faults are repeatedly found to be in alignment and also satisfactorily correlatable. This could possibly be accounted for by successive changes in direction of major stresses through time.

Regionally, a detailed study of this sort, with all data compiled from Government and private reports, official maps and University theses, insures a reasonable continuity of trends across a large area. Results consist of a net of primary fractures characterized as follows:

(a) **the most persistent and longest fractures (Figure 4) as well as the apparent density and importance of fracture intersections are empirically determined;**

(b) **the Texas lineament is quite conspicuous, cutting obliquely through the area, and so are the northwest and the northeast trends (Figure 5);**

(c) **definite regional downthrows of fractures are brought out, locating various graben, horsts, and even establishing the structural crest line of southeast Arizona;**

(d) **several important tear faults, some with a mile or more of lateral displacement, are to be further scrutinized.**

DOME, PYRAMID, DOMAL PYRAMID

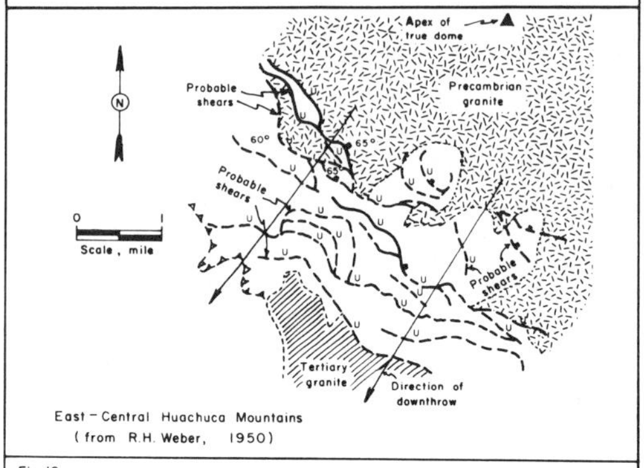

East – Central Huachuca Mountains (from R.H. Weber, 1950)

Fig. 10 RESULT OF A DOMAL PUSH

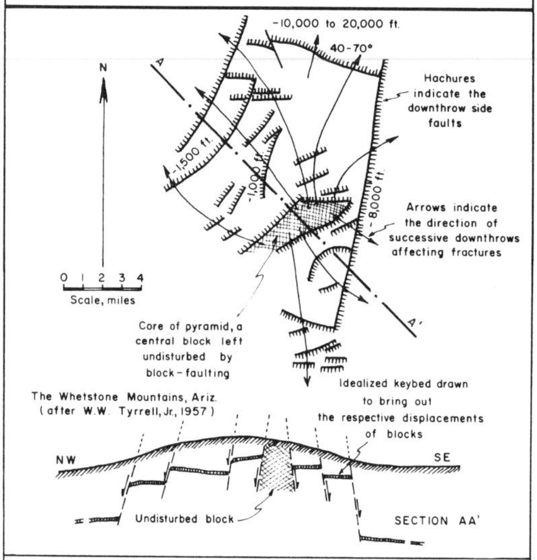

The Whetstone Mountains, Ariz. (after W.W. Tyrrell, Jr., 1957)

Fig. 11 TYPICAL PYRAMID PATTERN RESULT OF BLOCK-FAULTING

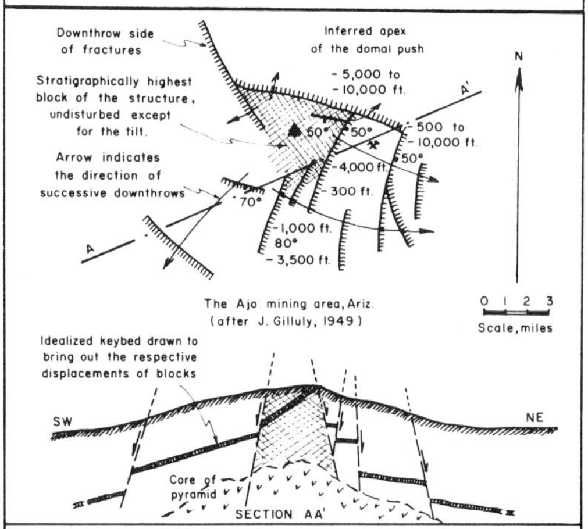

The Ajo mining area, Ariz. (after J. Gilluly, 1949)

Fig. 12 TYPICAL DOMAL PYRAMID, OR BLOCK-FAULTED DOME

One can speculate further:

(a) why do these tear faults apparently occur in pairs, forming the boundaries of at least six blocks or "wedges" that once were the subject of relative displacement;

(b) these wedges are all six moving away from the intersections at which they originate;

(c) why are three of these wedges (the wider ones) roughly oriented parallel to the anticlinorium axis of Arizona, and why are the three others (the thinner ones) oriented seemingly at random?

(d) there are clearly three orders of magnitude in the intersections (Figure 6), from the smallest as near the Ajo mine, for instance, to those where wedges originate, then finally to the regional, inferred points of meeting from which broad trends seem to fan out into neighboring states;

(e) the interest wanders next to these neighboring states and particularly in Sonora, Mexico, where several sets of broad fractures come together.

Locally, there are also some interesting features that come to light and which have to be examined carefully:

(a) two pronounced patterns of downstepping (Figure 7), parallel but trending in opposite directions like two contiguous escalators meeting halfway, reflect the action of a couple that once affected the area. A caldera occurring at the hinge line supports the evidence that this line must have been sufficiently deep to be now considered a primary break correlatable with other primary features;

(b) alongside major transcurrent faults that showed much lateral movement (Figure 8), a small structural "low," elongated parallel to the fracture, is often to be observed on the downthrow side. This "sink" pattern is of interest because it often is the site of minor copper occurrences;

(c) a pattern presenting successive downthrows of fractures in a general direction (Figure 9) may in places be construed to be leading toward a graben, as in figure 1. However, this pattern can also represent the reflection of a dome, observable only over a small portion of it. This difference in interpretation is not easily resolved, although in a dome the fractures should show, preferably, an arcuate arrangement. If this is the case, the apex might approximately be located.

2. **DOMES** are considered next.

In the highly complex geology of Arizona, some of the Tertiary domes were blockfaulted in Miocene time and their recognition can be difficult. Where observable in Arizona, a dome is nearly always characterized by concentric fracturing (Figure 10). Two additional structural patterns, akin to domes, are to be considered also:

(a) a pyramid pattern (Figure 11) may resemble a dome at first but it is caused solely by blockfaulting as indicated by the arrangement of fractures. Caused by slumping of the sides by gravity, this pattern is a barren structure, and no mineralization is to be expected in it.

(b) a hybrid pattern (Figure 12), called a domal pyramid, consists of a dome that has been blockfaulted later on. Mineralization may often coincide with this more complex structure.

In southeast Arizona, a large number of domes have been recognized. They are found to be more numerous to the south, a fact that could at first glance be explained by the lower elevation of the area. Considering a possible south-central Arizona batholith (Figure 13), as first suggested by I. A. Ettlinger in 1928, erosion and truncation of this pluton would have created a peneplaned surface during a pre-Paleozoic interval estimated to have lasted 600 million years. Later reactivation of this Precambrian mass at a number of critical spots in Tertiary times might have been the cause for these domal disturbances. As the surface of the ground slopes from north to south, there are naturally more domes to be expected uncovered by erosion in the south, assuming that the peneplaned surface remained untilted in the meanwhile.

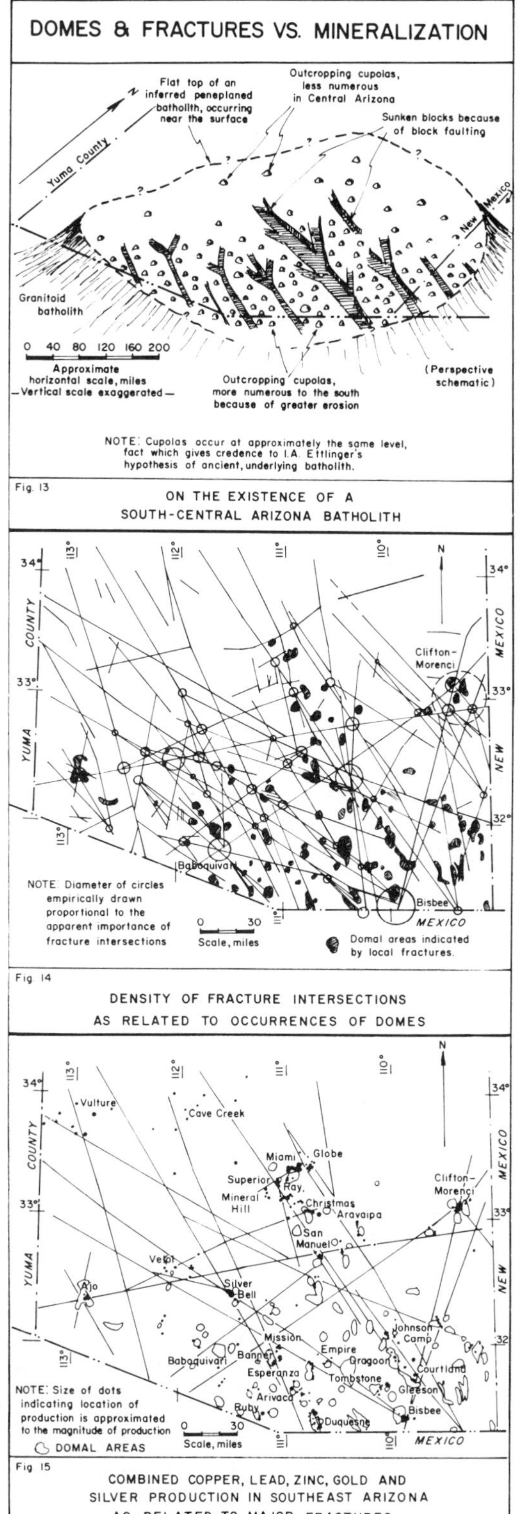

Fig. 13 ON THE EXISTENCE OF A SOUTH-CENTRAL ARIZONA BATHOLITH

Fig. 14 DENSITY OF FRACTURE INTERSECTIONS AS RELATED TO OCCURRENCES OF DOMES

Fig. 15 COMBINED COPPER, LEAD, ZINC, GOLD AND SILVER PRODUCTION IN SOUTHEAST ARIZONA AS RELATED TO MAJOR FRACTURES

3. **SUPERPOSED FRACTURE INTERSECTIONS AND DOMES** (Figure 14) reveal that the domes generally occur near, but not necessarily at, the intersections. Elongated domes are parallel to or occur between parallel fractures but they do not generally straddle these.

The spatial coincidence of the two structural parameters examined so far seems related to the density and importance of intersections but this is not a general rule.

4. **RELATIONSHIP BETWEEN FRACTURE INTERSECTIONS-DOMES WITH MINERALIZATION.**

When comparing the structural conjunctions with the map that shows the mineralization occurrences in southeast Arizona (Figure 15), it can clearly be seen that mineralized areas often occur at the edge of a dome, with sometimes two or several mining areas cluttered at the periphery. Some large centers, like Ajo, Clifton-Morenci, or the agglomeration of small centers, as around Duquesne, may be found almost entirely surrounded by domes. As a whole, there is quite a strong relationship between conjunctions of fracture intersections and domes with mineralized areas.

The two parameters examined here are only local parts of a large puzzle. They are not sufficient to explain all mineral occurrences, partly because many fractures and domes are not detectable at present and partly because there certainly will be other parameters necessary for the determination of potential ore deposits. A sequel to this study will endeavor to further analyze these additional parameters. A conjunction of domes and fracture intersections should indeed be considered in a broader, more complex frame of reference encompassing intrusions as well as ancient stratigraphic basins in order to perhaps better understand the localization of mineralized areas.

CONCLUSIONS

The ideas developed in the foregoing text can be used with profit in other parts of the world. In northern countries, such as British Columbia, the remarkable elongation of numerous lakes as well as the linear stretches shown by many rivers are useful additional features to help in a structural study of this sort. Combined with aerial photographic investigations this structural analysis can become a very powerful tool in the search for ore deposits.

BIBLIOGRAPHY

1. Brown, W. H. Tucson Mountains, an Arizona Basin and Range Type. Geol. Soc. of Am. Bull., v. 50, 1939, p. 697.
2. Ettlinger, I. A. Ore Deposits Support Hypothesis of a Central Arizona Batholith. AIME Tech. Pub. 63, 1928, 23 pp.; (abst), Min. Met., V. 9, 1928, p. 77.
3. Gilluly, J. The Ajo Mining District, Arizona. U.S. Geol. Survey Prof. Paper No. 209, 1946, 112 pp.
4. Hammer, D. F. Geology and Ore Deposits of the Jackrabbit Area, Pinal County, Arizona. Univ. of Arizona Master of Science Thesis, 1961, 156 pp.
5. Jones, William R. The Geology of the Sycamore Ridge Area, Pima County, Arizona. Univ. of Arizona Master of Science Thesis, 1941, 89 pp.
6. Peterson, D. W. Haunted Canyon Quadrangle Map. U.S. Geol. Survey Geologic Quadrangle GQ-128. 1960.
7. Peterson, N. P. Geology and Ore Deposits of the Globe-Miami District, Arizona. U.S. Geol. Survey Prof. Paper 342, 1962, 151 pp.
8. Tyrrell, W. W., Jr. Geology of the Whetstone Mountain Area. Yale Univ. Doctoral Dissertation. 1957.
9. Weber, Robert. The Geology of the East-Central Portion of the Huachuca Mountains, Arizona. Univ. of Arizona Dissertation, 1950, 191 pp.
10. Wertz, J. B. L'Entaille et la Capture d'Arroyo. Accidents Géomorphologiques survenant aux Vallées habituellement sèches du Sud-Ouest des Etats-Unis (Arroyo Cutting and Arroyo Piracy: Geomorphologic Accidents happening to usually Dry Valleys of the southwestern United States.) Revue de Géomorphologie Dynamique, France, v. 15, 1965, pp. 145-157.
11. ———. Relationship of Past Geologic Structural Processes to the Rock Masses Associated with Mining Districts. Open File Report of Investigation (Denver, Colo., offices), U.S. Bureau of Mines Research Center, 1966, 131 pp.
12. ———. Structural Elements of Ore Search in the Basin and Range Province, Southeast Arizona: Domes and Fracture Intersections. SME Transactions (AIME), v. 243, 1968, pp. 276-290.
13. Wilson, E. D. Geology and Ore Deposits of the Courtland-Gleeson Region, Arizona. Arizona BuMines Bull. No. 123, 1927, 79 pp.

14

Reprinted from pages 1507-1508 and 1512-1514 of *Econ. Geology* **71:**1507-1514 (1976)

Lineaments—Ineffective Guides to Ore Deposits

JAMES GILLULY

Abstract

For many years some geologists have been much impressed by real or imagined linear arrangements of various geologic features: drainage lines, faults, joints, volcanoes, ore deposits, and even anticlines. Such real or fancied lineaments have commonly been attributed to deep-seated wrench (strike-slip) faults. Some seem to regard the San Andreas fault as typical; others refer to zones as much as 100 miles wide as lineaments, but surely no one would suggest that a zone that wide would be a guide to ore. In this paper no zones wider than ten miles are considered lineaments.

Some geologists consider lineaments to be the record of worldwide stress fields embracing the entire crust. This paper reviews some of the literature in an attempt to evaluate the significance of lineaments. I conclude that, although sea-floor lineaments—marked as they are by both magnetic and sounding evidence—are indeed important structures, the significance of lineaments in the far more heterogenous continental crust has been greatly exaggerated; the value of lineaments as guides to ore is virtually negligible. Most orebodies, even the greatest such as Butte or Bingham, are so small compared to, say, a ten-mile-wide zone, that their discovery would not be significantly facilitated even though exploration is restricted to the zone. Most postulated ore zones are more than ten miles wide, and many that have been postulated can be shown to be nonexistent.

Lineaments

THE suggestion that the earth's crust is divided into segments by rectilinear faults dates back at least to Elie de Beaumont's "somewhat fantastic pentagonal theory of the arrangement of mountain chains" (quotation from W. H. Hobbs, himself a supporter of the tetrahedral theory of the earth). Hobbs (1904, 1911, 1912) was much impressed by linear arrangements of various topographic and geologic features in New England. The lineaments he recognized are composite: the line of a topographic scarp may be continued by a straight stretch of stream, this by a geologic contact or a fall line.

The most prominent lineaments he recognized trend about N 5° W, a second, less prominent, set is oriented N 50° E, a third is about N 45° W, with less well-defined sets trending N 35° W, N 44° W, N 65° E, N 75° W, and N 85° W. Clearly almost any fault trend drawn at random would lie close to one or another of these trends. The spacing of the nets Hobbs recognized ranged from 40 to 125 miles; their width varied as the scaled width of a pencil line varies with the scale of the map on which it is drawn. Hobbs conceded that lineaments are best recognized on more generalized small-scale maps; they tend to become obscure or zigzags on larger scale maps.

On a much smaller scale, Hobbs found joints in the granite of Cape Ann to fall into sets striking N 20° to 25° W, N to N 5° W, about N 15° E, N 30° to 35° E, about N 45° to 50° W, due E, and N 44° E. He attached regional significance to the principal trends. It would be hard to find a New England lineament trend, however, that differs greatly from one of those he listed.

As Hobbs' Connecticut work was carried out under the auspices of the U. S. Geological Survey, a panel of five field men was appointed to examine the field evidence. The panel unanimously reported that it could see no significance in any of the trends recognized by Hobbs (Arthur C. Spencer, pers. commun., ca 1936). No one denied the individual observations but none could agree with their regional significance or continuity. The varying properties of the rocks from place to place were considered responsible for the local structures.

Sederholm (1913) is sometimes referred to as an advocate of lineaments. He did, indeed, point out the fault control of many topographic features in the landscape of Fennoscandia, but he did not think that the faults and joints were regionally consistent in orientation; in fact he noted that many of the faults curved as much as 30° in strike and from 60° dips to the vertical. Obviously, he considered local, not regional. stress patterns responsible for the faulting.

Raisz (1945) noted a striking topographic lineament extending from the Olympic to the Wallowa Mountains of the Pacific Northwest. But no geologic boundaries are offset along this line, so that even though strikingly linear, the alinement must be considered fortuitous.

Vening-Meinesz (1947) computed the global shear pattern to be expected if the North Pole had migrated from the latitude of India to its present position. Since the polar flattening would thus have traversed about 70° of latitude, he thought that the entire crust would have been subjected to a systematic shear pattern. He was able to convince himself that many continental shelves, mountain ranges, and other major features of topography conform to the deduced shear pattern—so many, in fact, as virtually to prove the reality of polar wandering. He noticed a few N–S and E–W lineaments but considered them aberrant; most lineaments, he thought, contrary to the ideas of many others, lie oblique to both meridians and parallels. He concluded that the worldwide shear patterns he had recognized sufficed to abolish the hypothesis of continental drift.

Sonder (1938, 1939) considered the Atlantic to be bounded and traversed by lineaments. He thought he could identify lineaments following loxodromic curves (making constant angles with the meridians) in the structure of the Swiss Alps. (I have been unable to find other reference to such structures in the alpine literature available to me.) Sonder (1947) accepted Vening-Meinesz' lineaments completely, though he attributed the shears to earth contraction rather than to polar wandering.

Three highly important papers in the literature of structural geology (Anderson, 1951; Hubbert, 1951; and Hafner, 1951) appeared almost simultaneously. Each of these authors emphasized that the load applied to the rock-air interface is merely atmospheric pressure, so that in the shallow crust the vertical is nearly always the smallest compressive stress vector. The medium and maximum stress vectors are therefore in a nearly horizontal plane. This orientation of stresses strongly favors either normal, reverse, or thrust faults to the exclusion of strike-slip. A stress orientation favoring strike slip, that is, one with the intermediate stress vector at a high angle to the horizon, is thus unlikely in the top few kilometers of the crust.

The shallow crust along wrench faults is almost entirely passively transported by the underlying rocks in which the effective wrench-faulting stresses are operating. Faults extending to depths of several kilometers are the only ones likely to be wrench faults, and even these are equally likely to be normal or reverse.

It is not at all surprising, then, that geologic maps show far fewer strike-slip faults than dip-slip varieties. Hubbert (1951), citing Sax, reports that of 2,100 faults mapped in the Netherlands 79 percent were normal and 21 percent reverse. No dominantly strike-slip faults were recognized, though many of those classified must have had a component of strike slip. Moody and Hill (1956) rightly state that the recognition of strike-slip faults has lagged behind that of normal, reverse, and thrust faults, but the most important reason is that there are far fewer of them.

In the oceanic crust, under the load of several kilometers of water and riding passively on currents in the low-velocity zone, strike-slip faults are second only to the normal faults of spreading ridges.

Lineaments and Ore Deposits

Many geologists have suggested that a series of ore deposits are aligned in a direction determined by some deep-seated feature, not necessarily a strike-slip fault. For example, Spurr (1923) thought that a rough alignment of several mining districts in southern Utah and adjoining Nevada was caused by a deep-seated stream of magma flowing eastward.

Gilliland (1962) and Kutina (1969) thought that the Mendocino Fracture Zone, so prominent on the floor of the Pacific (Menard, 1960), extends beneath the continent near the 40th parallel. They suggested that this deep-seated fracture localizes the Bingham and Park City mining districts and accounts for the east-west trend of the Uinta Range. The hidden structure they named the Mendocino Trend. Porter (1924) and Beeson (1927) also considered the Bingham and Park City districts to be controlled by the western extension of the Uinta uplift. The ores of the two districts are very different and there are no known orebodies of consequence in an extension of the range 150 miles to the east.

Park City is indeed on a structural high very near the extension of the Uinta axis, but Bingham Canyon is more than 20 miles to the south and in one of the deepest synclines in the region. The southern Arizona copper districts have been localized, according to Mayo (1958) and Schmitt (1966), by the intersections of cross structures with the Texas Lineament. These postulated structures will be considered in detail.

[*Editor's Note:* Material has been omitted at this point.]

Great Faults without Ore Deposits

If ore deposits were indeed localized by great faults one might expect to find great concentrations along such through-going structures as the San Andreas and Hayward faults of California, the Wasatch fault of Utah, and the Hurricane fault of Utah and Arizona. Yet none of these well-exposed structures is associated with ore deposits, even though they must extend deep into the crust, if not, indeed, into the mantle. If not ore deposits, one might expect to find volcanoes localized by such deep-going structures, yet only a few small ones are locally present. The relatively shallow northeast rift of Kilauea has indeed controlled surface eruptions but not the deep-seated Wasatch fault. It seems, then, that great faults are not especially favorable environments for ore deposits, though it is true that a few deposits are localized by faults.

An Ore-Bearing Lineament

The Mother Lode of California is an outstanding example of ore deposits localized along a through-going reverse fault of regional extent (Knopf, 1929). Scores of profitable gold mines mark the course of this great structure; it is the exceptional lineament that does, indeed, control ore deposition.

Conclusions

Study of geologic and topographic maps shows the extreme rarity of rectilinear features of either landforms or geologic structures. Even the San Andreas fault is somewhat sinuous. Like the island arcs, with their curved foredeeps and subduction zones, the folds and faults of virtually all mountain ranges are sinuous, with axial culminations and depressions, as well as arcuate plan views. It is hardly too much to say that nature abhors straight lines—they are local and very rare on the continents and certainly there are no worldwide stress systems acting throughout the crust. This should be clear when one considers the general prevalence of isostatic equilibrium. An earth incapable of supporting a water load of only 1,000 feet of depth (Lake Bonneville) without yielding surely is incapable of transmitting horizontal stresses over distances of thousands of kilometers. The location of ore deposits, like that of volcanoes, is determined by local or limited regional stresses and rock properties, not by lineaments expressing worldwide stress systems.

Acknowledgments

For critical reading and many helpful suggestions, I am grateful to my friends Paul Averitt, Charles A. Anderson, and J. Fred Smith.

975 Estes Street
Lakewood, Colorado 80215
September 15, December 9, 1975

REFERENCES

Affleck, J., 1963, Magnetic anomaly trend and spacing patterns: Geophysics, v. 28, p. 379–395.
Albritton, C. C. and Smith, J. Fred, 1957, The Texas Lineament: Internat. Geol. Cong., 20th, Mexico City 1956, v. 2, sec. 5, p. 502–518.
Anderson, C. A., 1948, Structural control of copper mineralization, Bagdad, Arizona: Am. Inst. Mining Metall. Petroleum Engineers Trans., v. 178, p. 170–180.
Anderson, E. M., 1951, The dynamics of faulting and dyke formation: Edinburgh, Oliver & Boyd, 206 p.
Baker, C. L., 1934, Major structural features of Trans-Pecos Texas: Univ. Texas Bureau of Econ. Geology Bull. 3401, p. 137–214.
Beeson, J. J., 1927, Mining districts and their relation to structural geology: Am. Inst. Mining Metall. Petroleum Engineers Trans., v. 75, p. 757–792.
Billingsley, P. R., and Locke, Augustus, 1935, Tectonic position of ore districts in the Rocky Mountain region: Am. Inst. Mining Metall. Petroleum Engineers Trans., v. 115, p. 59–68.
Brown, W. H., 1939, Tucson Mountains, an Arizona basin range type: Geol. Soc. America Bull., v. 50, p. 696–760.
Chamberlin, R. T., 1919, A peculiar belt of oblique faulting: Jour. Geology, v. 27, p. 602–613.
Cooper, J. R., 1971, Mesozoic stratigraphy of the Sierrita Mountains, Pima County, Arizona: U. S. Geol. Survey Prof. Paper 658-D, 42 p.
—— and Silver, L. T., 1964, Geology and ore deposits of the Dragoon quadrangle, Cochise County, Arizona: U. S. Geol. Survey Prof. Paper 416, 196 p.
Darton, N. H., 1917, Deming, New Mexico: U. S. Geol. Survey Geol. Atlas, Folio 207.
Drewes, H., 1971, Mesozoic stratigraphy of the Santa Rita Mountains southeast of Tucson: U. S. Geol. Survey Prof. Paper 658-C, 81 p.
Dunham, K. C., 1935, The geology of the Organ Mountains; with an account of the geology and ore deposits of Donna Ana County, New Mefflico: New Mexico Bur. Mines Mineral Resources Bull. 11, 272 p.
Fath, A. E., 1920, The origin of the faults, anticlines and buried "granite ridges" of the northern part of the mid-continent oil and gas field: U. S. Geol. Survey Prof. Paper 128-C, p. 75–84.
Fuller, M. D., 1964, Expression of east-west features in magnetic surveys in parts of the U. S. A.: Geophysics, v. 29, p. 602–622.
Gilliland, W. N., 1962, Possible continental continuation of the Mendocino Fracture Zone: Science, v. 137, p. 685–686.
Gilluly, James, 1956, General geology of central Cochise County, Arizona: U. S. Geol. Survey Prof. Paper 281, 169 p.
Hafner, W., 1951, Stress distribution and faulting: Geol. Soc. America Bull. v. 60, p. 373–398.
Hill, R. T., 1902, The geographic and geologic features and their relations to the mineral products of Mexico: Am. Inst. Mining Metall. Petroleum Engineers Trans., v. 32, p. 163–178.
—— 1928, Transcontinental structural digression [abs.]: Geol. Soc. America Bull., v. 39, p. 265.
Hobbs, W. H., 1904, The lineaments of the Atlantic border region: Geol. Soc. America Bull., v. 15, p. 483–506.
—— 1911, Repeating patterns in the relief and in the structure of the land: Geol. Soc. America Bull., v. 22, p. 123–176.
—— 1912, Earth features and their meaning: New York, McMillan, 506 p.
Hubbert, M. K., 1951, Mechanical basis for certain familiar geologic structures: Geol. Soc. America Bull., v. 62, p. 355–372.
King, P. B., 1965, Geology of the Sierra Blanca region, Texas; U. S. Geol. Survey Prof. Paper 480, 185 p.
—— 1969, The tectonics of North America: U. S. Geol. Survey Prof. Paper 628, 72 p.
—— and Flawn, P. T., 1953, Geology and mineral deposits of pre-Cambrian rocks of the Van Horn area, Texas: Univ. Texas Bureau of Econ. Geology Pub. 5301, 218 p.
Knopf, Adolph, 1929, The Mother Lode system of California: U. S. Geol. Survey Prof. Paper 157, 88 p.
Krieger, Medora, 1968, Geologic map of the Brandenburg Mountain quadrangle, Pinal County, Arizona: U. S. Geol. Survey, Geol. Quad. Map 668.
—— 1969a, Geologic map of the Holy Joe Peak quadrangle, Arizona: U. S. Geol. Survey, Geol. Quad. Map 669.
—— 1969b, Geologic map of the Lookout Mountain quadrangle, Arizona: U. S. Geol. Survey, Geol. Quad. Map 670.
—— 1969C, Geologic Map of the Saddle Mountain quadrangles, Arizona: U. S. Geol. Survey, Geol. Quad. Map 671.
Kutina, J., 1969, Hydrothermal ore deposits in the western states: a new concept of structural control of ore distribution: Science, v. 165, p. 1113–1119.
Landwehr, W. R., 1967, Belts of major mineralization in the western United States: Econ. Geol., v. 62, p. 494-501.
Lasky, S. G., 1938, Geology and ore deposits of the Lordsburg mining district, Hidalgo County, New Mexico: U. S. Geol. Survey Bull. 885.
—— 1947, Geology and ore deposits of the Little Hatchet Mountains, Hidalgo County, New Mexico: U. S. Geol. Survey Prof. Paper 208, 101 p.
Locke, A., Billingsley, P., and Mayo, E. B., 1940, Sierra Nevada tectonic pattern: Geol. Soc. America Bull., v. 51, p. 514–539.
Mayo, E. B., 1958, Lineament tectonics and some ore districts of the Southwest: Mining Eng., v. 10, p. 1169–1175.
Menard, H. W., 1960, The East Pacific Rise: Science, v. 132, p. 1737–1746.
—— and Dietz, R. S., 1952, Mendocino submarine escarpment: Jour. Geology, v. 60, p. 266–278.
Moody, J. D., and Hill, M. J., 1956, Wrench fault tectonics: Geol. Soc. America Bull., v. 67, p. 1207-1246.
Ogle, B. A., 1952, Major shear zone at False Cape, Humboldt County, California [abs.]: Geol. Soc. America Bull., v. 63, p. 1341.
—— 1953, Geology of Eel River Valley area, Humboldt County, California: California Dept. Nat. Resources, Div. Mines and Geology Bull. 164, 128 p.
Peterson, N. P., 1948, Structural control of Globe-Miami district [abs.]: Econ. Geol., v. 49, p. 122.
Porter, C. A., 1924, Intersecting fracture zones and mineral districts: Eng. Mining Jour., v. 118, p. 650–657.
Raisz, E., 1945, The Olympic-Wallowa lineament: Am. Jour. Sci., v. 245A, p. 479–485.
Ransome, F. L., 1904, Bisbee: U. S. Geol. Survey Geol. Atlas, Folio 112.
—— 1915, The Tertiary orogeny of the North America Cordillera and its problems, *in* Problems of American geology: New Haven, Yale University Press, p. 287-376.
Richardson, G. B., 1909, El Paso, Texas: U. S. Geol. Survey Geol. Atlas, Folio 166.
Roberts, R. J., 1960, Alinement of mining districts in north central Nevada: U. S. Geol. Survey Prof. Paper 400-B, p. B17–B19.
—— 1966, Metallogenic provinces and mineral belts in Nevada: Nevada Bur. Mines Rept. 13, p. 47–72.
Sabins, F. F., Jr., 1957, Geology of the Cochise Head and western part of the Vanar quadrangle, Arizona: Geol. Soc. America Bull., v. 68, p. 1315–1341.
Schmitt, H. A., 1935, Structural associations of certain metalliferous deposits in southwestern United States and north-

ern Mexico: Am. Inst. Mining Metall. Petroleum Engineers Trans., v. 115, p. 187–208.
—— 1966, The porphyry copper deposits in their regional setting, *in* Titley, S. R., and Hicks, C. L., eds., Geology of the porphyry copper deposits, southwestern North America: Tucson, Univ. of Arizona Press, p. 17–35.
—— 1959, The copper province of the southwest: Mining Eng., v. 11, p. 597–600.
Sederholm, J. J., 1913, Weiterer Mitteilungen über Bruchspalten mit besonderer Beziehung zur Geomorphologie von Fennoscandia: Comm. géol. Finlande Bull., no. 37, 66 p.
Simons, F. R., 1972, Mesozoic stratigraphy of the Patagonia Mountains, Santa Cruz County, Arizona: U. S. Geol. Survey Prof. Paper 658-E, 23 p.
Smith, J. F., Jr., and Ketner, K. B., 1968, Devonian and Mississippian rocks and the date of the Roberts Mountain thrust in the Carlin-Pinyon Mountains area, Nevada: U. S. Geol. Survey Bull. 1251-I, 18 p.
Smith, J. G., 1965, Fundamental transcurrent faulting in the northern Rocky Mountains: Am. Assoc. Petroleum Geologists Bull., v. 49, p. 1398–1409.
Sonder, R. A., 1938, Die lineamenten Tektonik und ihre Probleme: Eclogae Geol. Helvetica, v. 31, p. 199–238.
—— 1939, Zur Tektonik des Atlantischen Ozeans: Geol. Rundschau, v. 30, p. 28–31.
—— 1947, Discussion of "Shear patterns of the earth's crust" by F. A. Vening-Meinesz: Am. Geophys. Union Trans., v. 28, p. 939–945.
Spurr, J. E., 1923, The ore magmas, v. 2: New York, McGraw-Hill, p. 475–485.
Swan, M. M., 1975, The Texas Lineament—tectonic expression of a Precambrian orogeny [abs.]: Geol. Soc. America Abstracts with Program, p. 1288–1289.
Thomas, G. E., 1974, Lineament block tectonics, Williston-Blood Creek Basin: Am. Assoc. Petroleum Geologists Bull., v. 58, p. 1305–1322.
Tweto, Ogden, and Sims, P. K., 1963, Precambrian ancestry of the Colorado Mineral Belt: Geol. Soc. America Bull., v. 74, p. 991–1014.
Vening-Meinesz, F. A., 1947, Shear patterns of the earth's crust: Am. Geophys. Union Trans., v. 28, p. 1–61.
Wallace, R. E., Griggs, A. B., Campbell, A. B., and Hobbs, S. W., 1960, Tectonic setting of the Coeur d'Alene district, Idaho: U. S. Geol. Survey Prof. Paper 400-B, p. B25-B27.
Wilson, C. W., Jr., 1936, Geology of the Nye-Boulder lineament, Stillwater and Carbon Counties, Montana: Am. Assoc. Petroleum Geologists Bull., v. 20, p. 1161–1188.
Zietz, I., Bateman, P. C., Case, J. E., Crittenden, M. D., Jr., Griscom, Andrew, King, E. R., Roberts, R. J., and Lorentzen, G. R., 1969, Aeromagnetic investigation of crustal structure for a strip across the western United States: Geol. Soc. America Bull., v. 80, p. 1703–1714.

15

Reprinted from *Econ. Geology* **67**:1234-1240 (1972)

An Analysis of Metal Distribution and Zoning in the Herberton Tinfield, North Queensland, Australia

R. G. Taylor and B. G. Steveson

Abstract

Mineralization is localized in a series of emanative centers which trend parallel with, and lie adjacent to the granite-sediment contact. Some areas of the original granite intrusion surface appear to contain series of minor ridges and troughs, whose axial trends also parallel the main contact directions, i.e., N.N.W. and E.N.E. Several major mineralization centers are situated adjacent to E.N.E.-trending ridges, and tend to be sited on the northern flanks. It is suggested that original ridge and cuspate structures on the granite surface played a significant role in localizing the emanative centers.

Two phases of mineralization are suggested, an early tin-tungsten phase associated with greisenization, and a later polymetallic phase. The latter is often zoned, with an inner tungsten zone passing outwards into tin, copper, and lead zones. Development of the zonal sequence varies from center to center, i.e., one or more zones may be absent, and considerable superimposition occurs. It is possible in some areas that a separate lead-zinc-silver mineralization phase may occur.

Introduction

The general geology and mineralization of the Herberton tinfield were reviewed recently in this journal by Blake and Smith (1970), who demonstrated that the mineralization displays a distinct zonation comprising four zones. The latter comprise an innermost tungsten zone followed successively by tin, copper and lead zones, all of which appear to be related to the Elizabeth Creek granite and its metamorphic aureole.

This paper presents a more detailed picture of the metal distributions than was reported by Blake and Smith, and examines relationships among metal distribution, zonation, and various features shown by the granite.

Distribution of The Mineralization

In a recent discussion Taylor (1971) pointed out that, although Blake and Smith (1970) had illustrated the general metal distribution and its zonal arrangement, their interpretation was misleading in that many of the zones shown as being continuous are in fact discontinuous. An inspection of Blake's (1968) detailed plan of the 2,000 mines and prospects indicates that there are many distinct centers of mineralization. This is illustrated in Figure 1, which shows a density contour plot of the mineralization based on the number of known mines and prospects present within chosen unit areas. It is realized, however, that there are many problems involved in producing this type of diagram. For example, the size of the unit area chosen obviously affects the details of the contours, and there is also the problem that a major mine is given a similar weighting to a minor prospect. The unit of area finally selected as a basis for the contour plot was half a mile square which, despite the above limitations, permits adequate display of important features of overall metal distributions.

The contour plot in Figure 1 allows the various centers to be classified into major and minor categories, yields some information as to the differing density of mineralization, and also illustrates general mineralization trends. The major centers are located around the well-known mining districts of Herberton, Irvinebank, Emuford, Coolgarra and Brownsville. Several major centers contain distinct subcenters which often have different salient features.

Two predominant mineralization trends are apparent with axes trending approximately E.N.E. and N.N.W. The importance of these trends appears to vary, with the E.N.E. predominating in the southern regions, and the N.N.W becoming more significant to the north. The majority of the mineralization centers are situated along or adjacent to granite/sediment contacts, and the prominent mineralization trends closely parallel these contacts, suggesting an intimate link between the granite and the mineralization.

The two trend directions outlined by the granite/sediment contact and the mineralization centers appear to intersect in the region of the major Irvinebank tin center. Whether the location of this major

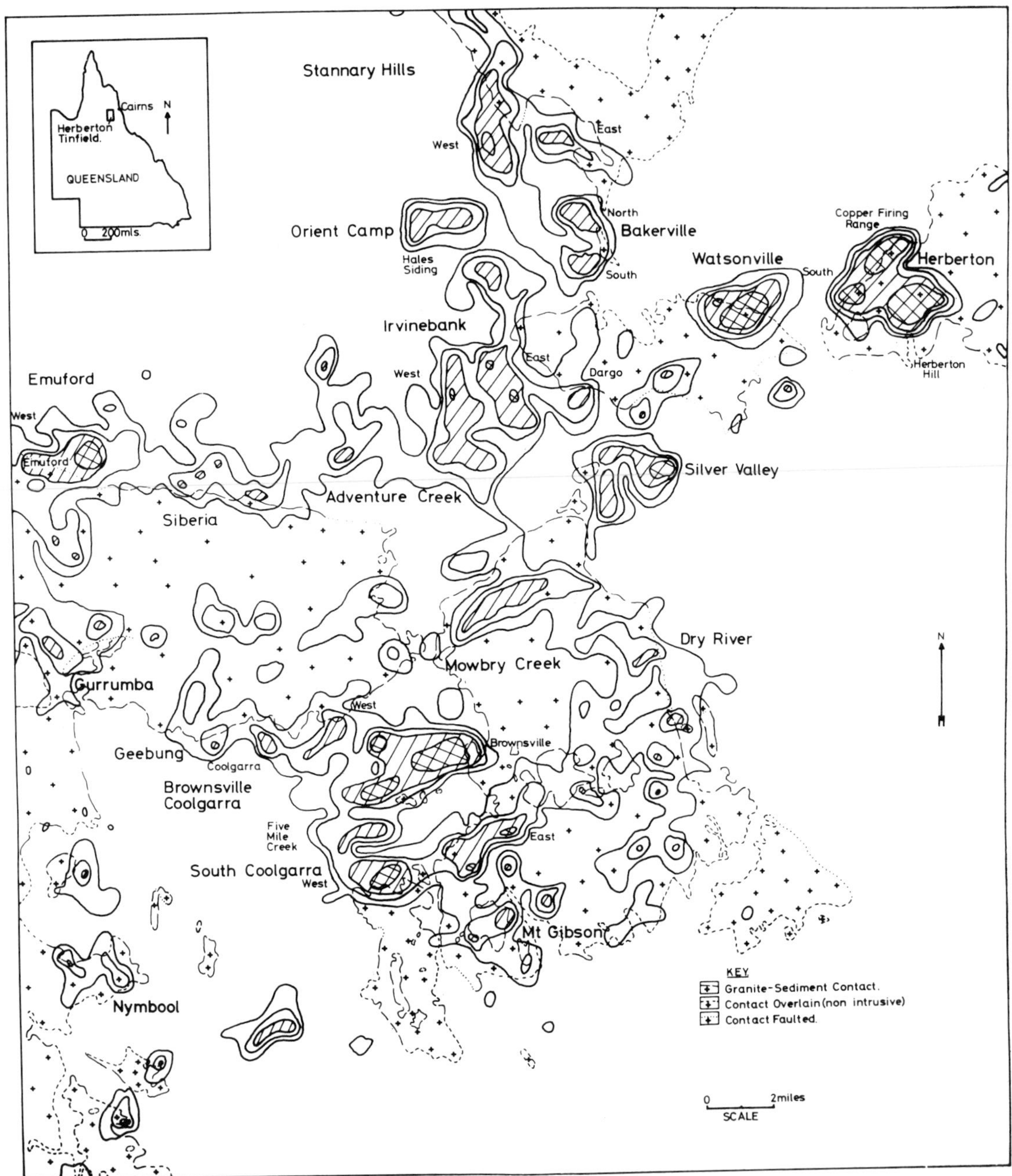

FIG. 1. Herberton tinfield, distribution of mineralization. Contours at 1, 3, 5, 10 mines per unit area of ½ mile square. Main lettering denotes major mineralization centers; minor lettering denotes subcenters.

center is coincidental or is actually controlled by the intersection of the two trends is uncertain. A further feature of interest is that only two major centers are contained entirely within the granite, i.e., Mowbry Creek and the major Herberton subcenter of Herberton Hill.

Factors Controlling the Localization of Mineralization Centers

The Cornish analogy

The localization of isolated mineralization centers is a common phenomenon within major tin fields (Hosking, 1965, 1968), and is particularly well demonstrated in the Cornish mineral field. Dines (1956) outlined the distribution of the major Cornish centers termed "emanative centers" and noted that the majority exhibited a zonal development of mineralization.

Factors controlling the localization of Cornish emanative centers have been discussed by Hosking (1962, 1965, 1968). He concluded that they are associated with well-defined granitic peaks or cusps which tend to develop along granitic ridges. Furthermore, the main mineralization within emanative centers is often disposed along the flanks of ridges rather than directly over them (Hosking, 1965). Recognition of cusps and ridges in Cornwall was based on evidence derived from several lines of approach, namely, detailed collation of underground information, consideration of the outcrops pattern of the granites and their associated metamorphic aureoles, a study of the distribution of dike swarms, and consideration of the distribution of greisen-bordered veins occurring in apical regions of the granite. Both cusps and ridges are thought to develop on the surface of the intruding granitic bodies in response to original fold structures acting as moulds.

Controls of the Herberton mineralization

Cuspate or ridge developments are best observed in regions where granites are barely exposed through erosion. The Elizabeth Creek Granite has not been extensively eroded, and the various isolated outcrops shown in Figure 1 probably all link to an underlying batholith. Indeed, the frequent occurrence of greisen zones, typical of high levels (Shcherba, 1968), and roof pendants of metamorphosed sediments in the southern and western regions indicate that much of the granite surface has not been deeply eroded. A paucity of similar features in the eastern Watsonville-Herberton region suggests a higher level of granitic intrusion and/or a deeper level of erosion.

Assuming that elongated roof pendants and granite-metasediment embayments mark original trough zones on an underlying intrusion surface, it is possible to draw a series of ridge and trough axes in the southern portion of the area, as shown in Figure 2. Axial trends of directions of elongation of roof pendants with regular shapes, and of protruding or embayed features of irregular roof pendants define two very clear trends which parallel the mineralization trends, N.N.W. and E.N.E. This applies generally to similar features shown by the fretted granite-sediment contacts, with those trending N.N.W. being more accentuated and occurring much more frequently. Thus, the original granite surface appears to have been tightly ridged in the N.N.W. direction and gently undulating in the E.N.E. direction. It is uncertain what could have produced this configuration, but original moulding action related to fold structures of the invaded sediments may have been responsible, as was suggested for the Cornish granites (Hosking, 1962). Unfortunately, it is not possible to test this hypothesis for the Elizabeth Creek Granite as the structure of the invaded sediments is insufficiently known (Blake, 1968).

Another approach to outlining potential "high spots" on granite surfaces involves the definition of areas of maximum greisen development (Shcherba, 1968). Several hypothetical ridge structures were outlined in the southern region utilizing the greisen approach (Fig. 2). These also correspond to the main contact and mineralization trends, and a few coincide with axes already delineated by the roof pendant approach.

The parallel existing between directional trends of the greisen veins and the N.N.W.-trending ridge development strongly implies that the two are structurally, and probably genetically related. However, the E.N.E.-trending outline of the greisen zone to the N.E. of Mt. Gibson suggests that greisen development is also related to the E.N.E.-trending surface configuration as well.

The close association between the postulated ridge systems and the positions of the major mineralization centers can be seen from a comparison of Figures 1 and 2. A good correlation exists between the E.N.E.-trending system and the major Brownsville-Coolgarra, South Coolgarra and Mowbry Creek centers. A similar, though weaker correlation is observed around Mt. Gibson. The only major postulated E.N.E. ridge not obviously related to a major mineral center is that occurring around Nettle Creek (Fig. 2). The northern mineralization around Irvinebank, Bakerville, and Stannary Hills probably relates to the N.N.W.-trending ridge system at depth, as suggested by the shape of the northern end of the Brownsville-Coolgarra granite intrusion, and of granite/sediment contact trends around Bakerville, and the fact that the metamorphic aureole, usually less than half a mile wide in the Herberton region, expands to a width of around three miles in the vicinity of the Irvinebank center.

Strong correlation, particularly between E.N.E.-trending ridges and mineralization centers in the southern region, suggests, as in Cornwall, that the

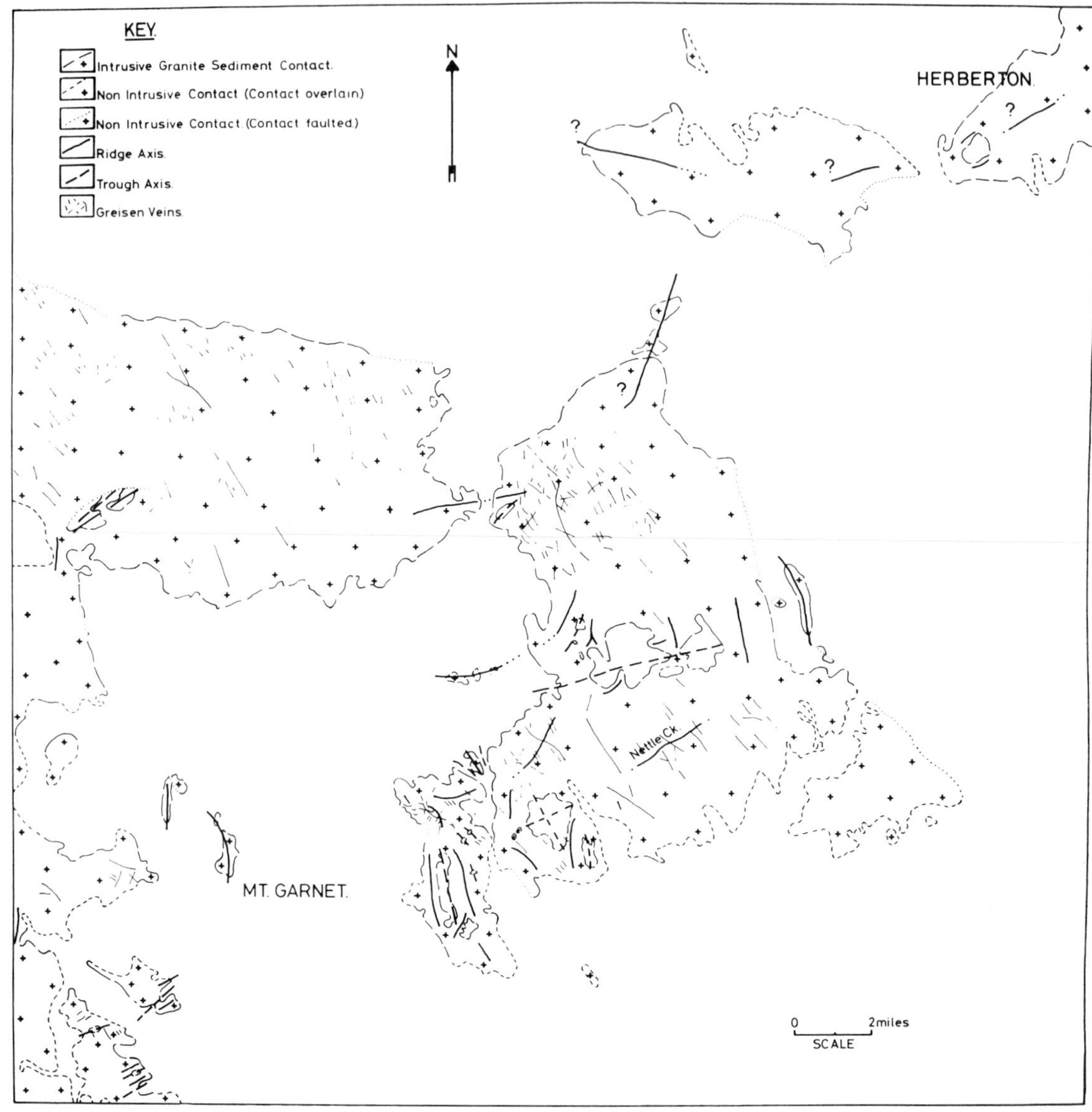

FIG. 2. Interpretation of ridge and trough structures on the surface of the Elizabeth Creek Granite.

location of emanative centers is largely controlled by the detailed structure of the granite surface.

As noted previously, many of the emanative centers in Cornwall lie along the flanks of ridges rather than directly over them. An interesting parallel is seen within the Herberton field, with the Coolgarra-Brownsville, South Coolgarra, and Mowbry Creek centers all showing a tendency toward strongest development on the northern flank of an E.N.E.-trending ridge. The feature is particularly well shown by the Coolgarra-Brownsville center where the south flank of the postulated ridge is devoid of mineralization (Figs. 1–2). The reason for this apparently preferential development along northern ridge flanks is unknown, but it may relate to an asymmetric configuration of ridge profiles. There also appears to be a tendency for major centers to favor the northern flanks of E.N.E.-trending intrusions, e.g., the Herberton, Watsonville, Adventure Creek, Siberia and Emuford centers.

Metal Zoning

Although it is not possible to illustrate all the zonal details of each mineralization center, an approximate picture can be obtained from Figure 3. The detailed picture is extremely complex, and shows many variations from the simple tungsten-tin-copper-lead sequence outlined by Blake and Smith (1970). For instance, no one center displays a clear development of all four zones, although Herberton South, Bakerville, Stannary Hills, Emuford, and Brownsville West lack only a clearly defined lead zone. On the other hand, several other centers display only one zone, e.g., Herberton Hill (tin), and most of the large Irvinebank center (tin).

Salient features of the various metal zones are considered below.

The tungsten zone

Tungsten mineralization reaches maximum development in western and southern regions where it is associated with areas of greisenized granite. Tungsten is also represented in northern and northeastern regions where it occurs only in minor proportions in association with tin zones, e.g., Watsonville, Herberton Hill, and Irvinebank. An anomalous situation is seen in the Copper Firing Range subcenter where tungsten mineralization is apparently associated with the copper-lead zone.

A detailed inspection of tungsten zones associated with greisen in granite reveals that, although normally associated with tin mineralization, they are generally devoid of peripheral copper-lead ores, e.g. South Coolgarra, Mt. Gibson, Gurrumbah, and many smaller centers within the Coolgarra-Brownsville granite mass. Analogous situations have been observed in many tinfields throughout the world (Shcherba, 1968). The tin-tungsten-greisen association is generally considered to represent an early high temperature phase of a continuing mineralization process which may yield later polymetallic phases comprising, for example, tungsten-tin-copper-lead-zinc veins adjacent to earlier greisen development.

The tin-tungsten-greisen association appears to be more extensive in the Herberton district than in Cornwall and, although it rarely contains important ore bodies, has been of considerable economic significance in contributing to important alluvial deposits in the Mt. Garnet region.

The tin zone

The tin zones also show considerable variation in their relationships with succeeding copper and lead zones. Some tin zones are strongly associated with succeeding copper or lead zones (Watsonville, Hales Siding, Stannary Hills and Adventure Creek), while others (Herberton Hill, and most of the Irvinebank center) lack peripheral copper or lead zones. There is, in fact, a continuous gradation between these extremes. The variation is difficult to interpret: it may arise from erosional sectioning of the zonal sequence at different levels, or it could be real, indicating that the zonal sequence does not always develop. It is probable that the present metal distributions are the result of an interplay of many complicating factors which even detailed work may never fully explain.

The copper-lead zone

Apart from association with tin zones, the development of a distinct lead zone is relatively rare (at Orient Camp and Mt. Albion), the lead generally occurring together with copper, e.g., Copper Firing Range, Silver Valley. In view of the fact that the Orient Camp and Mt. Albion Centers are relatively isolated from other centers, it could be argued that they are separate centers unrelated to earlier tin-copper mineralization.

Vertical depth of metal zones

Very few of the 2,000 or so mines and prospects have been worked to depths exceeding 200′. Consequently, it is difficult to comment on the vertical depth of the various zones. However, it is noteworthy that very few of the greisen zones contain workings in excess of 100′, and that deeper workings are more common in non-greisen centers, e.g., Irvinebank, Herberton Hill, Coolgarra, Watsonville, etc. The deepest mine in the tinfield is the Vulcan mine, Irvinebank, which finished operations in a cassiterite-bearing pipe at a depth of 1,462′ below surface. This suggests that the thickness of the zone is probably in excess of 1,000′ in the Irvinebank region. None of the other ore bodies are known to be continuous over this range, although the Great Northern mine situated at the base of the extensively mineralized Herberton Hill reached 600′. Thus, the tin zone here may also exceed 1,000′ in depth.

The copper and lead silver mines were worked only to shallow depths, and the deepest known workings are in the Empress Mine (approximately 350′) of the Copper Firing Range subcenter.

Acknowledgments

The authors extend their thanks to people of the area who gave assistance during the study. Particular thanks are due to Mr. E. Cohen and Mr. G. Greaves of Loloma Mining Corporation for their advice, encouragement and general counsel.

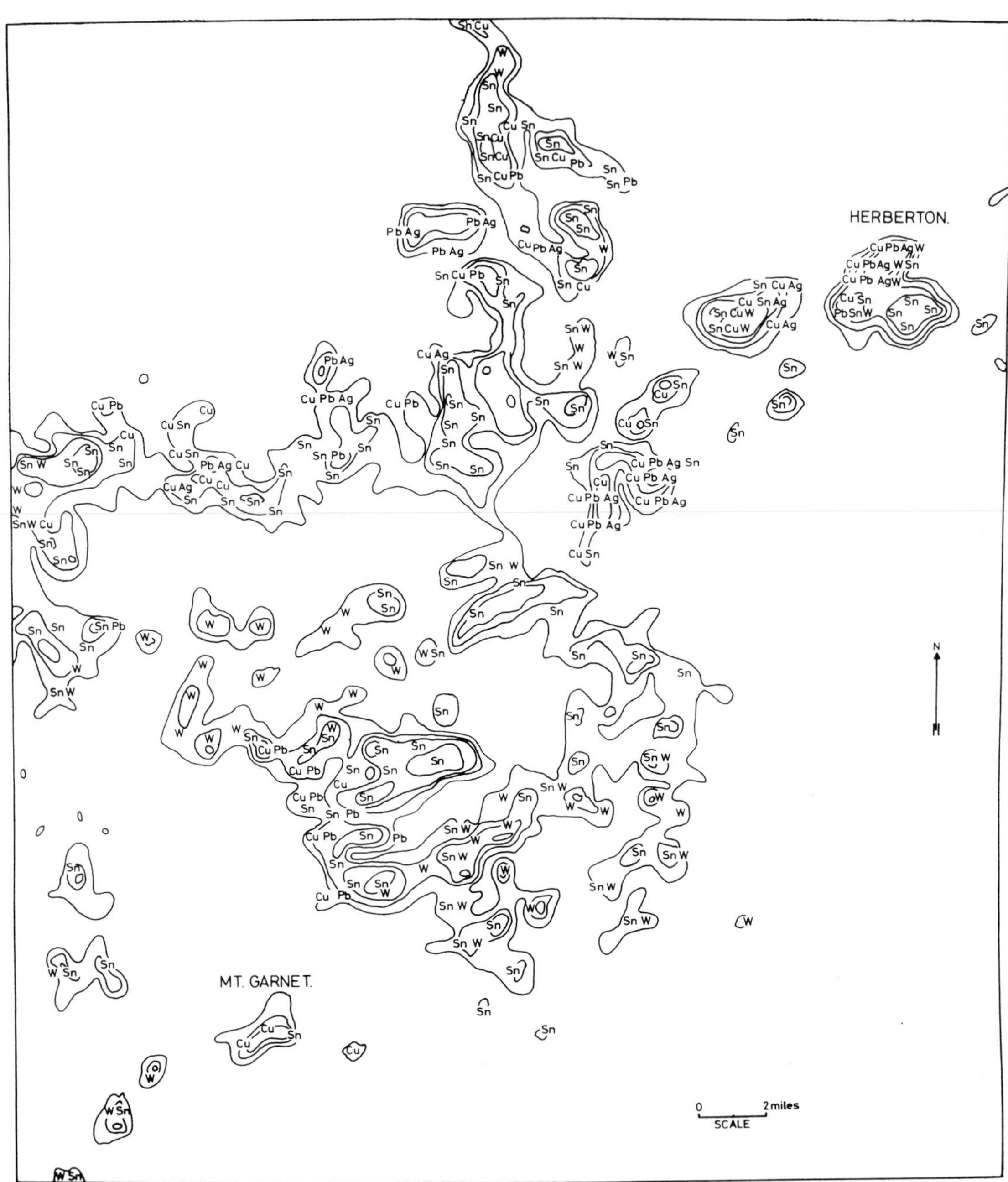

FIG. 3. Herberton tinfield. General distribution of metals. Sn–tin, W–tungsten, Cu–copper, Pb–lead, Ag–silver. Zinc not shown. Contours denote distribution of mineralization—see Figure 1.

Acknowledgment is due also to the preliminary studies of Mr. D. Blake and his associates of the Bureau of Mineral Resources which laid the foundations for this study. The project was financed by the Australian Research Grants Committee, whose support is also gratefully acknowledged. Associate

Professor P. J. Stephenson of the James Cook University of North Queensland kindly read the manuscript, and his interest and encouragement have been greatly appreciated.

GEOLOGY DEPARTMENT,
JAMES COOK UNIVERSITY OF NORTH QUEENSLAND,
TOWNSVILLE, 4810, QUEENSLAND, AUSTRALIA,
August 11, 1971; April 12, 1972

REFERENCES

Blake, D. H., 1968, Regional and economic geology of the Herberton-Mount Garnet area, Herberton tinfield, North Queensland: Australia Bur. Mineral Resources, Geology and Geophysics Rec. 1968/79.

Blake, D. H., and Smith, J. W., 1970, Mineralogical zoning in the Herberton tinfield, North Queensland, Australia: ECON. GEOL., v. 65, p. 993–997.

Dewey, H., 1948, British regional geology: south-west England, 2nd ed.: London, Great Britain Geol. Survey and Mus., 72 p.

Dines, H. G., 1956, The metalliferous mining region of south-west England: London, Great Britain Geol. Survey Mem., 795 p.

Hosking, K. F. G., 1962, The relation between the primary mineralisation and the structure of South-west England, *in* Some aspects of the Variscan fold belt: Inter-Univ. Geol. Cong., 9th, Exeter, Manchester Univ. Press, p. 135–153.

—— 1965, The search for tin: Mining Mag., v. 113, p. 261–273, 308–383 and 448–461.

—— 1968, The relationship between primary deposits and granitic rocks, *in* Technical Conference on Tin, London 1967, [Papers]: International Tin Council, London, v. 1, p. 267–312.

Shcherba, G. N., 1968, Greisens, *in* Genesis of endogenic ore deposits: Moscow, Nedra Press, p. 378–420 (in Russian). Translated in Internat. Geol. Rev., 1970, v. 12, nos. 2 and 3, p. 114–150, 239–255.

Taylor, R. G., 1971, Discussion of Blake, D. H., and Smith J. W. 1970, Mineralogical zoning in the Herberton Tinfield, North Queensland, Australia: ECON. GEOL., v. 66, p. 813–815.

16

Reprinted from *Am. Inst. Mining, Metall. and Petroleum Eng. Trans.* **202:**781–795 (1955)

APPLICATION OF GEOLOGY TO THE DISCOVERY OF ZINC-LEAD ORE IN THE WISCONSIN-ILLINOIS-IOWA DISTRICT

Allen F. Agnew

GEOLOGIC studies for the Federal and State governments in the Wisconsin-Illinois-Iowa zinc-lead district were begun in 1835, and subsequent surveys were made in the three states at intervals until 1916, see Table I.

The current study of the district, Fig. 1, by the U. S. Geological Survey was begun in 1942 in the hope that a systematic investigation would help increase production of zinc and lead, then in extremely short supply. Major emphasis of the USGS program was on detailed mapping of the geologic structure and ore deposits. Preliminary maps and reports covering localities of intensive mining activity have been published during the course of this study, Fig. 2. By 1950 the objectives of the study by the USGS were revised; 7½-min quadrangles of relatively unprospected localities as well as intensively mined localities are now being mapped in Wisconsin. Since 1945 the investigation in Wisconsin has been made in cooperation with the Wisconsin Geological and Natural History Survey, and since 1951 the geologic mapping of areas of particular interest in Iowa has been performed in cooperation with the Iowa Geological Survey. At intervals since 1943 the Illinois State Geological Survey has mapped the geologic structure and ore deposits of localities in the Illinois part of the mining district, see Fig. 2.

Geology was applied by mining company personnel as early as 1853, but only sporadically, with periods of greater application between 1890 and 1925. Much of the so-called geologic work was in reality mining engineering. Since 1946 the major mining companies in the district have employed geologists who have successfully applied geologic techniques in the search for ore.

A. F. AGNEW, Member AIME, is a Geologist, U. S. Geological Survey, Platteville, Wis.

Discussion on this paper, TP 4045 I, may be sent (2 copies) to AIME before Oct. 31, 1955. Manuscript, Aug. 13, 1954. Los Angeles Meeting, February 1953.

History of Mining

Occurrence of galena in the district was known as early as 1658 or 1659, when French explorers heard of lead mines that were apparently in the vicinity of Dubuque, Iowa. In 1690 a trading post was established near Dubuque to obtain galena. The first significant attempt at mining by white men took place in 1788 when the Indians granted to Julien Dubuque mining rights for 20 miles along the west side of the Mississippi River, including the vicinity of Dubuque, Iowa. By 1805 lead was being mined near Galena, Ill. In 1819 permanent settlement of the region was begun, and most of the important lead-producing areas had been found by 1830, see Fig. 3. Lead production in the Wisconsin-Illinois-Iowa district between 1830 and 1871 far exceeded that of any other district in the U. S.[2]

Smithsonite was first mined in 1859, and sphalerite was mined as early as 1867. After 1873 the production of sphalerite exceeded that of smithsonite.

In 1873 annual zinc production from the district first equaled lead production, and from then until 1893 production of zinc and lead was roughly the same. Since 1893 the ratio of zinc to lead mined has generally ranged between 5:1 and 20:1.

In 1942 the Wisconsin-Illinois-Iowa zinc-lead district ranked 14th among zinc-producing districts of the U. S. From 1946 until February 1953 the dis-

Table I. Former Significant Geological Surveys and Their Major Emphasis

Geologist	Wisconsin	Illinois	Iowa
Owen[3]	First general survey of the mining district (Fig. 3)		
Percival[4,5]	Alignment of lead-producing areas		
Whitney[10,30,32]	Detailed descriptions of lead-bearing crevices		
Strong[33]	Detailed description of mine workings		
Chamberlin[34]	General geologic conclusions		
Calvin and Bain[31]			Detailed study of vertical joint (crevice) deposits—lead and zinc
Grant[26]	Mapped folds that control zinc deposits		
Bain[35]	Mapped folds that control the zinc deposits		
Cox[36]		Showed localization of ore deposits in areas of thick *oil rock*	

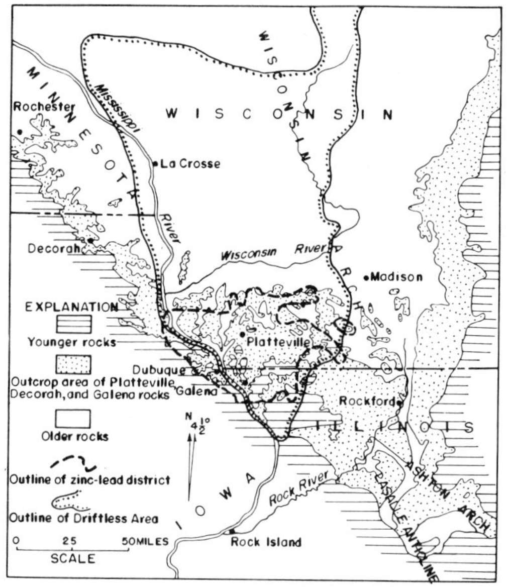

Fig. 1—Map showing relationship of the Wisconsin-Illinois-Iowa zinc-lead district to the outcrop pattern of ore-bearing strata (Platteville, Decorah, and Galena formations) and to the driftless area. (Taken in part from State geological maps, Willman and Templeton.[1])

trict ranked generally as the seventh leading producer of zinc. It ranked 14th among the lead-producing districts in February 1953.

Copper was mined intermittently from 1837 to about 1909, as mixed ore containing copper carbonate and oxide minerals. The principal mines were at Mineral Point, Wis., Fig. 2, where mining ceased in 1880. Copper in the zinc-lead district has been little more than a curiosity, and shipments of the ore probably did not exceed 10,000 tons, of which some lots assayed as high as 20 pct Cu.

Geology

Owen[3] noted that the lead deposits were confined to the area of outcrop of the middle and lower beds of his Cliff limestone, the Galena dolomite of present terminology, see Fig. 4. He subdivided this unit into cherty beds (below) and arenaceous or non-cherty strata (above). Percival[4, 5] subdivided his Blue limestone (now the Decorah and Platteville formations) and since that time his excellent subdivisions have been vaguely or incorrectly used by some authors in an attempt to clarify the stratigraphy. In the course of the present survey a classification of rock units that is strikingly similar to that of Percival was devised for the Decorah and Platteville formations. Reinterpretation of existing subdivisions of the Galena dolomite, and further subdivisions of that formation, resulted in the present classification of that unit.

Pre-Cambrian Rocks: Granite was reported in wells at Platteville and Richland Center, see Fig. 2, at depths of 1714 and 665 ft, respectively. Pre-Cambrian granite was reported at a depth of 1800 ft in city well No. 5, Dubuque, Iowa.

Lower Paleozoic Rocks: Cambrian strata are exposed just north of the principal mineralized area, see Fig. 1, and underlie the mining district. The Cambrian rocks are mainly sandstone and siltstone and aggregate 1000 to 1300 ft thick, but they are not important potential host rocks for zinc-lead deposits.

The dolomite, shale, and sandstone of the Prairie du Chien group (Lower Ordovician) are exposed along the northern fringe of the major mineralized part of the district and in the more deeply incised river valleys in Wisconsin, Fig. 1. They underlie the remainder of the district. Dolomite in the Prairie du Chien is a potentially productive host rock for lead and zinc sulphides.

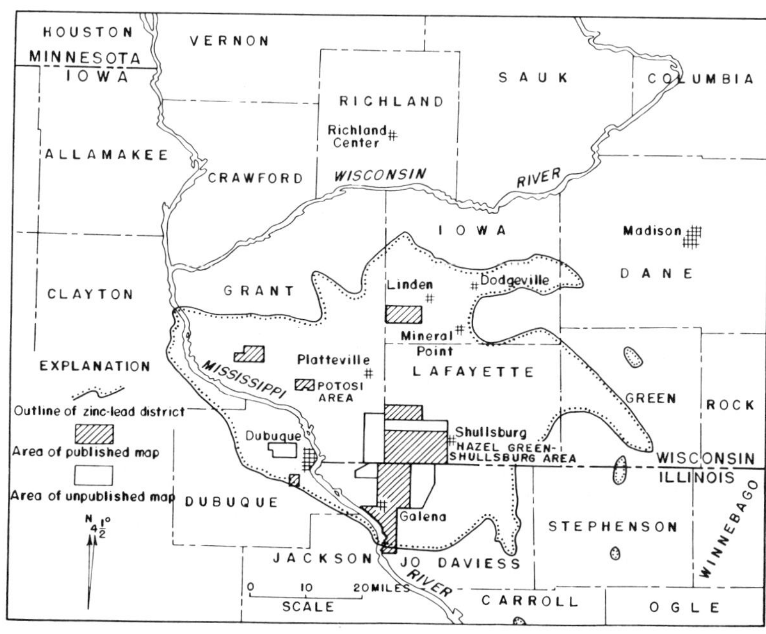

Fig. 2—Map of the Wisconsin-Illinois-Iowa zinc-lead district showing areas covered by published and unpublished geologic maps since 1942.

Areal distribution of the St. Peter sandstone of early Middle Ordovician age is similar to that of the Prairie du Chien. This sandstone is the major aquifer of the mining district, but there is small possibility that it bears zinc-lead ore deposits. The St. Peter and the underlying Prairie du Chien units vary inversely in thickness and together aggregate 270 to 325 ft.

The dolomite, limestone, and shale beds of the Platteville, Decorah, and Galena formations (Middle Ordovician) are at the surface in most of the mining district, Fig. 1; elsewhere in the district, except where erosion exposes the St. Peter and Prairie du Chien rocks, they are present in the subsurface. The Platteville, Decorah, and Galena rocks include all the currently productive strata of the zinc-lead district, and the subdividing of these formations into 18 lithologic units, Fig. 4, was a major step in locating the small folds and faults that controlled the emplacement of the ore. Thicknesses given are for unaltered rocks; alteration due to mineralizing solutions commonly has reduced the Guttenberg and Quimbys Mill members to less than 50 pct of their normal thickness and has likewise affected the McGregor and Ion members. Some of the lithologic units are recognized only locally (as certain subdivisions of the Cherty unit of the Galena formation) and others regionally (as the subdivisions of the Decorah formation). Potentially productive units include the Quimbys Mill member and the upper half of the McGregor member, which have not been prospected in many areas of the mining district despite the fact that ore in these beds was formerly mined in some areas.

Maquoketa strata of Late Ordovician age crop out along the northward- and northeastward-facing cuesta of Silurian rocks in Illinois and Iowa and appear on areal maps as doughnuts around the erosional outliers (locally called mounds) to the north and east, Fig. 1. Maquoketa beds are also present beneath the Silurian rocks south from the cuesta face, and at the center of each mound. The Maquoketa formation, which consists of shale and some dolomite, is not important as a potential ore-bearing unit. On the other hand, these beds can possibly be used to detect mineralized zones beneath, as at places concentrations of iron, lead, and zinc sulphides are found at the Maquoketa-Galena contact because the upward-percolating waters were diverted laterally at this contact.

Dolomite of Silurian age is distributed similarly to the Maquoketa but is slightly more restricted,

Table II. Significant Applications of Geology to Special Problems in the Wisconsin-Illinois-Iowa Zinc-Lead District

► Need for geologic background noted—Daniels (9)*, George (37).

► Potential value of zinc minerals recognized—Owen (3), Daniels (9).

► Relationships between ore deposits and synclines recognized—Whitney (10), Chamberlin (34), Blake (24), Winslow (2).

► Areal mapping of synclines—Grant (26), Bain (35), Grant and Perdue (38), Scott (20). (See Fig. 6).

► Relationship of faults to folds—Jenney (6), Behre, Scott, and Banfield (39).

► Relationship of thick *oil rock* to orebodies—Bain (35), Grant and Perdue (38), Cox (40), Scott (20).

► Production from lower zones ("glass rock," "Trenton") predicted—Daniels (9), Grant (27), Grant and Burchard (41), Cox (40).

► Possibility of production from "Lower Magnesian"—**Pro:** Owen (3), Daniels (9), Percival (4), Murrish (42); **Con:** Whitney (10), Chamberlin (34).

► Mining several crevices by a central shaft—Whitney (30).

* Numbers in parentheses indicate references.

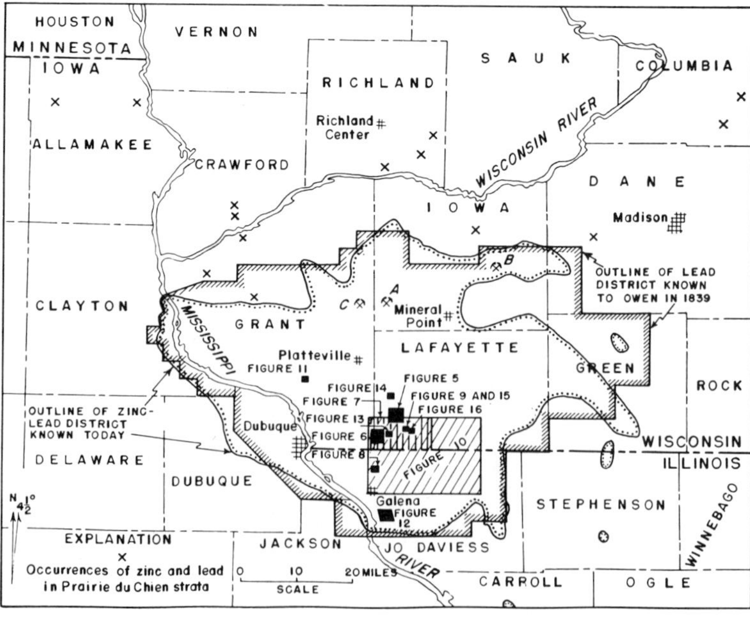

Fig. 3—Map showing lead district known to Owen in 1839, zinc-lead district known today, occurrences of zinc and lead in Prairie du Chien strata, and location of areas shown by Figs. 5-16.

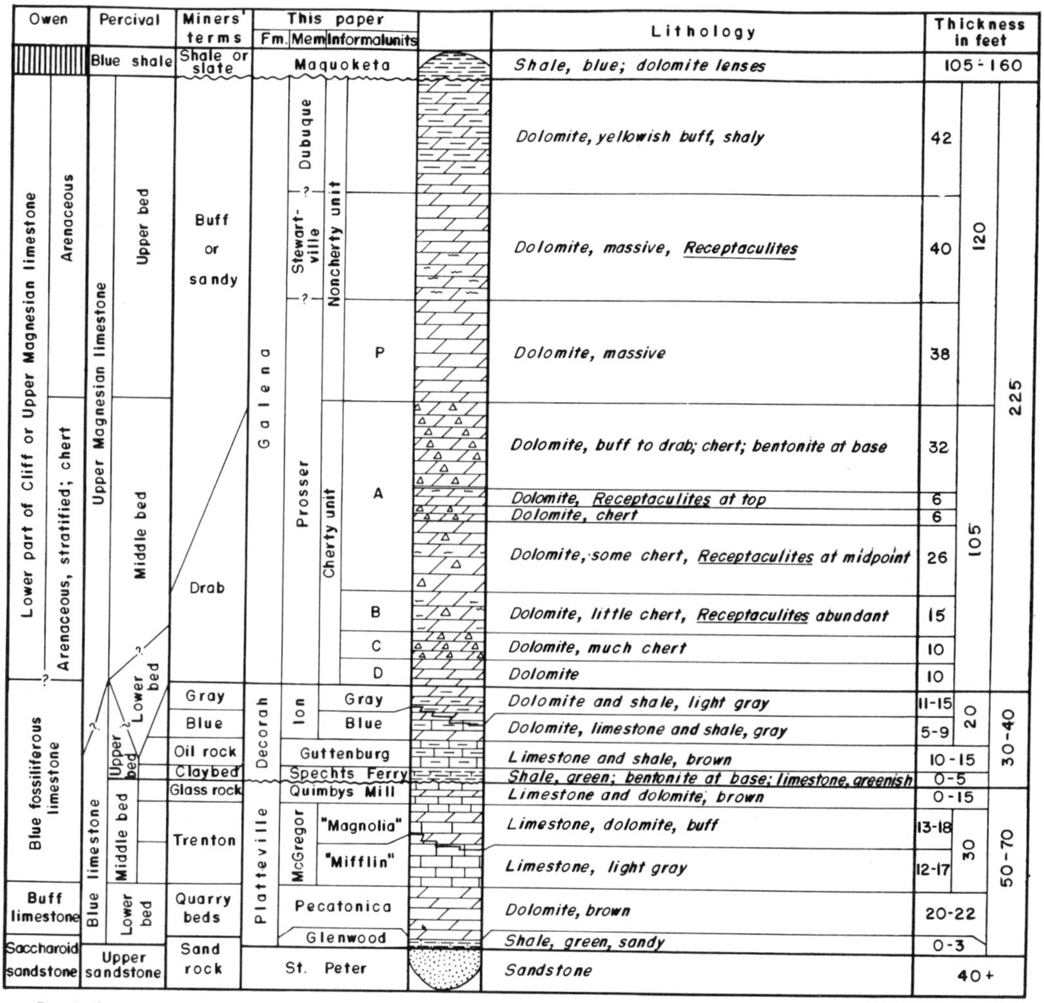

Owen	Percival	Miners' terms	This paper: Fm.	Mem.	Informal units	Lithology	Thickness in feet		
	Blue shale	Shale or slate	Maquoketa			*Shale, blue; dolomite lenses*	105-160		
Lower part of Cliff or Upper Magnesian limestone: Arenaceous	Upper Magnesian limestone: Upper bed	Buff or sandy	Galena	Dubuque		*Dolomite, yellowish buff, shaly*	42	120	225
				Stewart-ville	Noncherty unit	*Dolomite, massive, Receptaculites*	40		
				Prosser	P	*Dolomite, massive*	38		
Arenaceous, stratified; chert	Middle bed	Drab			Cherty unit A	*Dolomite, buff to drab; chert; bentonite at base*	32	105	
					A	*Dolomite, Receptaculites at top*	6		
					A	*Dolomite, chert*	6		
					A	*Dolomite, some chert, Receptaculites at midpoint*	26		
					B	*Dolomite, little chert, Receptaculites abundant*	15		
					C	*Dolomite, much chert*	10		
	Lower bed				D	*Dolomite*	10		
Blue fossiliferous limestone		Gray	Decorah	Ion	Gray	*Dolomite and shale, light gray*	11-15	20	30-40
		Blue			Blue	*Dolomite, limestone and shale, gray*	5-9		
	Blue limestone: Upper bed	Oil rock		Guttenburg		*Limestone and shale, brown*	10-15		
		Claybed		Spechts Ferry		*Shale, green; bentonite at base; limestone, greenish*	0-5		
	Middle bed	Glass rock	Platteville	Quimbys Mill		*Limestone and dolomite, brown*	0-15		50-70
		Trenton		McGregor	"Magnolia"	*Limestone, dolomite, buff*	13-18	30	
					"Mifflin"	*Limestone, light gray*	12-17		
Buff limestone	Lower bed	Quarry beds		Pecatonica		*Dolomite, brown*	20-22		
Saccharoid sandstone	Upper sandstone	Sand rock	Glenwood			*Shale, green, sandy*	0-3		
			St. Peter			*Sandstone*	40+		

Fig. 4—Stratigraphic column of Middle Ordovician strata of the Wisconsin-Illinois-Iowa zinc-lead district showing Owen's classification,[3] Percival's classification,[4, 5] miners' classification, author's classification, graphic and descriptive lithology, and thicknesses.

Fig. 1. Vugs in the Silurian dolomite contain isolated crystals of sphalerite or galena, but the possibility of commercial deposits in the Silurian beds of this district appears to be small.

Post-Silurian Deposits: A mantle of Pleistocene loess covers most of the mining district and is thickest near the Mississippi River. This loess and the residual products from the bedrock constitute the surficial material.

Structure: The Wisconsin-Illinois-Iowa zinc-lead district lies on the south slope of the Wisconsin structural high of pre-Cambrian rocks. The district is just west of the southward extension of this structural high (Wisconsin arch) toward the LaSalle anticline in Illinois. The regional dip, first noted by Owen[3], is about 16 ft per mile south-southwestward across the major mineralized area.

The strata of the district have been subjected to compression that produced a pattern of folds and faults. Although the folds, mentioned first by Owen,[3] are generally of small magnitude (from crest to trough about 30 ft vertically and 1000 to 1500 ft horizontally) and most are local, many fold axes can be traced for several miles. The trends of these axes are principally northeastward, northwestward, and eastward.

Faults, whose potential economic value was noted by Jenney,[6] are of relatively small magnitude and of three general types—reverse, normal, and tear. Dip-slip displacement on the reverse faults is commonly less than 6 ft, but strike-slip displacement is greater; on one of the tear faults it is at least 35 ft.

A well-developed joint pattern is recognizable. Major joint trends are northwestward and northeastward. Subsidiary joints trend eastward, and a very few trend northward. Control of streams by fractures, noted near Richland Center (Fig. 2) by Hobbs,[7] is well displayed in certain areas, Fig. 5.*

* See also Agnew, Flint, and Crumpton.[8]

Rock Alteration: Limestone strata, particularly in the Guttenberg, Quimbys Mill, and McGregor members, Fig. 4, have been locally dolomitized by the ore-bearing solutions; the result is a sugary, more coarsely granular rock. Dolomitic limestone strata illustrate effects of dolomitization in two ways: 1) by bleached areas that show a change from calcareous to dolomitic cement, without apparent change in grain size, and 2) by thin veinlets of fine-grained milky dolomite.

Leaching, especially of limestone strata in the Guttenberg and Quimbys Mill members, has locally resulted in thinning to less than 50 pct of the original thickness. More rarely, as little as 15 pct of the original thickness has been preserved, as shale and argillaceous residuum. McGregor and Ion strata have been similarly leached.

Silica was introduced locally by the ore-bearing solutions and resulted in two products: 1) silicified Guttenberg or Quimbys Mill beds that have the megascopic appearance of unaltered limestone and shale and 2) chert nodules particularly in the Guttenberg, but also in the Quimbys Mill member. Chert in the cherty unit of the Galena dolomite seems to be more abundant near areas that have been mineralized, although quantitative data have not been obtained.

Ore Deposits

Minerals: The ore minerals are sphalerite and galena. Minerals of possible economic value are smithsonite, pyrite, marcasite, and barite. Principal gangue minerals are calcite and limonite.

Lead Orebodies: Lead ore occurs as gash veins in nearly vertical joints (crevices of local terminology) in the noncherty unit and in the upper part of the cherty unit of the Galena dolomite, to depths of 70 ft or more below the topographic surface. The galena is present 1) as isolated crystals or clusters in a matrix of vuggy dolomite or dolomite sand and 2) as veins commonly less than an inch thick. Podlike masses of ore and rock, several feet wide, are present in leached rock at favorable stratigraphic horizons (openings of local terminology) and in a particular opening several podlike enlargements may occur along a crevice. Chimneys or pipes of galena are in some places found at the intersections of crevices. Individual crevices have been mined along their strikes for as much as 1700 ft. The great number of shafts along any particular mineralized crevice was due partly to the early miners' aversion to tramming long distances underground and partly to the small area of each claim. A striking example of the regularity of pattern of these mineralized joints is seen in Fig. 5.

Zinc Orebodies: Most of the zinc ore, with minor lead ore, has been found in the cherty unit of the Galena dolomite, in the Ion and Guttenberg members of the Decorah formation, and in the upper part of the Platteville formation (Quimbys Mill and McGregor members). The principal ore zones are 1) the lower 40 ft of the Galena dolomite and the Ion member and 2) the Quimbys Mill member. Ore occurrences along inclined fractures are referred to as pitches and those along bedding plane separations are flats. These fractures are on the flanks of folds and, like the folds, are believed to be due to compression. In the footwall area of the pitches, which is locally termed core ground, the minerals replaced the rock and filled vugs. These pitch-and-flat orebodies, which are mined as a unit that includes the pitches and the intervening core ground, are as much as 125 ft high, 450 ft wide, and two miles or more long, although an average length is perhaps 1000 ft.

Economic Application of Geology

For the following reasons the *apparent* limits of the Wisconsin-Illinois-Iowa district, Fig. 1, are not necessarily the *actual* limits of the potentially ore-bearing area: 1) The regional dip to the south is responsible for the absence of Platteville, Decorah, and Galena beds to the north. 2) The eastern limit appears to coincide generally with the boundary of the driftless area. Platteville, Decorah, and Galena strata extend toward the east, but are overlain by a variable thickness of glacial material, and the reticence of miners to prospect for zinc-lead deposits through this material is understandable. Zinc-lead deposits may be found east of the driftless area. 3) The southern and western limit of the district seems to coincide roughly with the face of the cuesta of

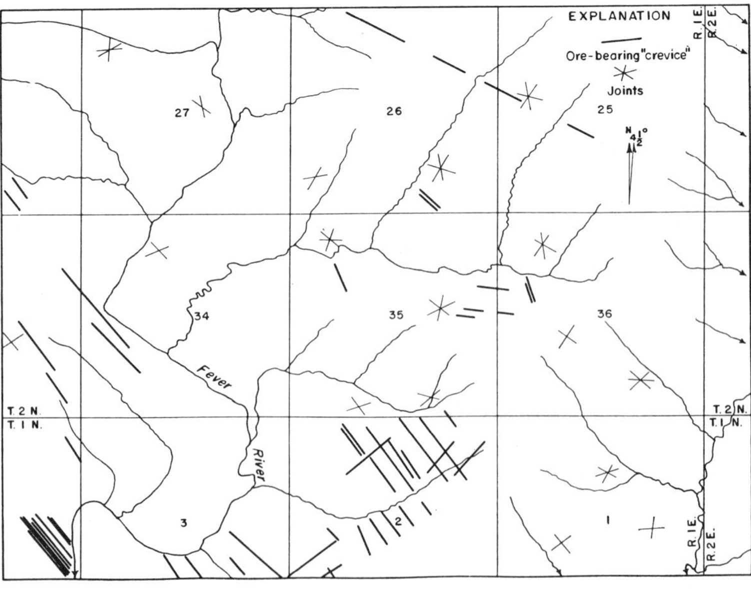

Fig. 5—Map of area southeast of Platteville, Wis., showing drainage control by joints and regularity of ore-bearing joint (crevice) pattern. (From Agnew, Flint, and Crumpton[8] and Heyl and Agnew.[48])

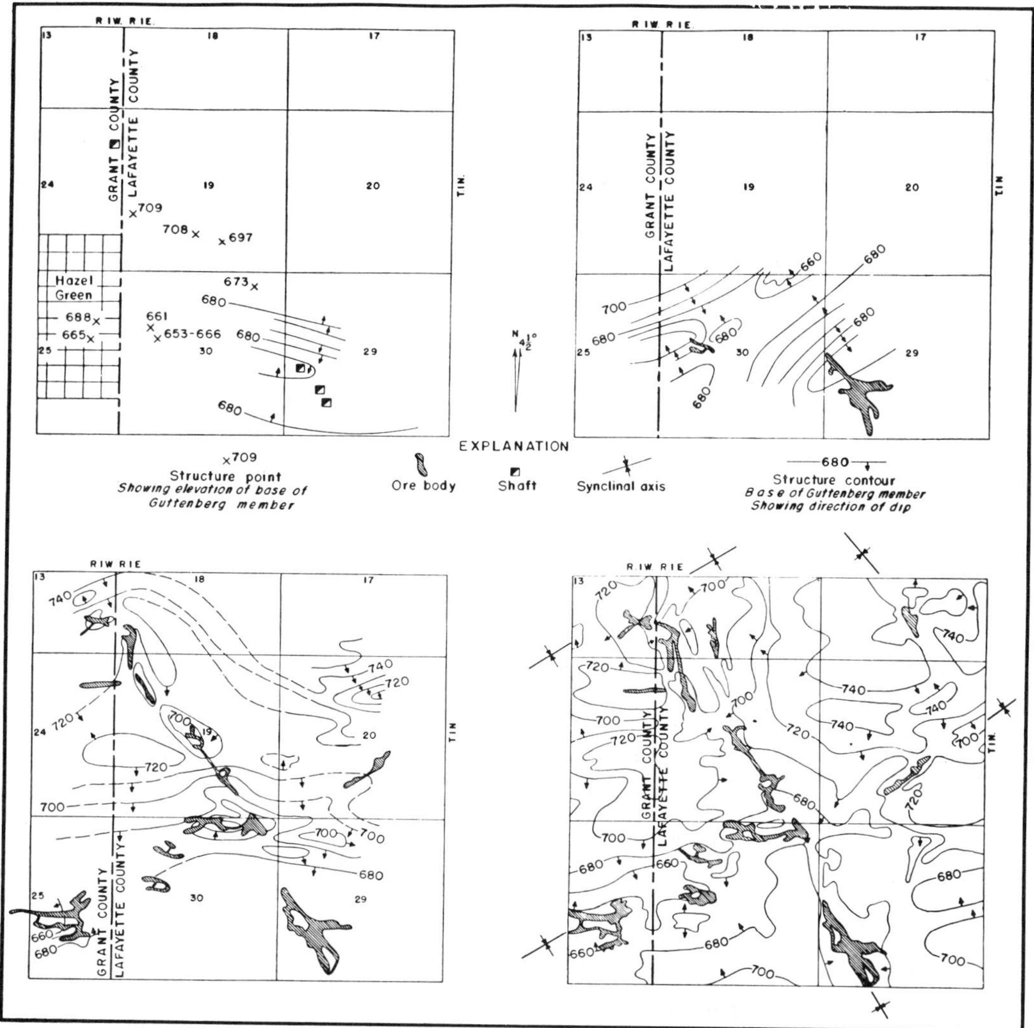

Fig. 6—Evolution of detailed mapping of structure as a result of greater density of data, southwestern Lafayette County (Fig. 3). A. Grant[26]; 10-ft contour interval. B. Wisconsin Geological and Natural History Survey and Wisconsin State Mining Trade School, Hazel Green Special Sheet[11]; 10-ft contours. C. Scott[20]; 10-ft contours. D. Agnew, Heyl, Behre, Lyons[15]; 20-ft contours.

Silurian rocks. The Platteville, Decorah, and Galena strata may bear lead or zinc orebodies to the south and west of this cuesta face. Economic considerations, therefore, presently limit the district on the east, south, and west, because of a) the greater depth of the ore-bearing strata, b) the lack of surface indications of ore because of overlying material, and c) the greater cost of pumping the water from the orebodies that lie to the south and west, owing to the greater depth.

Most of the principal areas of lead mines had been discovered by 1830. After that date lead mines were located within these general areas either on newly discovered crevices or on those already worked. Because of this, Daniels[9] felt it reasonable to expect that "rich discoveries will yet be made upon [the] unexplored grounds [between these general productive areas]." On the other hand, Whitney believed that this "seems to indicate that a considerable portion of the Lead Region is quite destitute of deposits of ore."[10]

Surface indications of lead and zinc deposits were numerous and obvious. This accounts for the discovery of many lead-bearing areas in the few years of prospecting before 1830. Most of these areas are well-dissected by streams, and the minerals were found in valley walls or on slopes. Other areas of mining, however, have been found on uplands, where plowing showed reddish-brown streaks in the wet soil, indicative of mineralized zones, or where float lead or iron minerals were brought to the surface by farming operations. With the boom in zinc mining that began during the 1890's, and the accompanying practice of prospecting by churn drills, exploration expanded into these upland areas, and a number of lead and zinc ore deposits were found.

During the first 70 years of prospecting in this district (1820's to 1890's) lead and zinc bodies were

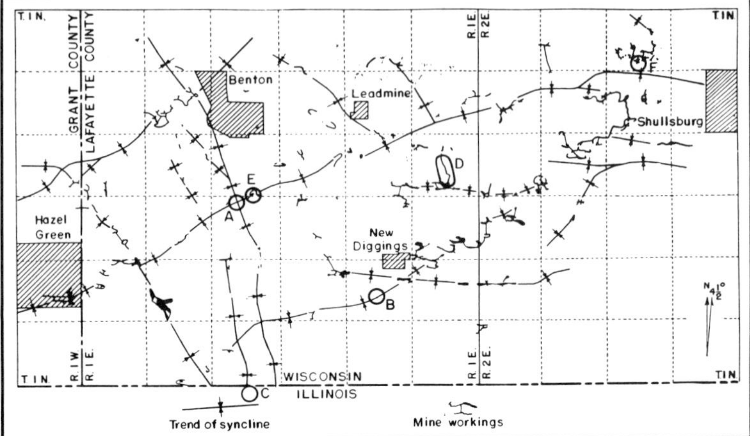

Fig. 7—Map of Hazel Green-Shullsburg area, Wisconsin, showing trends of synclines, zinc orebodies, and localities (A-F) discussed in text. (From Agnew, Heyl, Behre, and Lyons.[15])

found with relative ease. In the next 30 years increasingly greater difficulty was encountered in discovering zinc orebodies, and some geologic techniques were employed by the mining companies to increase the chances for a strike.

Pitch-and-Flat Zinc-Lead Deposits

Synclines: Owen[3] noted that the regional dip of the beds is "subject to undulations," Whitney[10] was the first to record the relationship between the ore deposits and synclines, and Grant was first to map synclines areally,** Fig. 6.

** Strong may have prepared such a map in 1873, for Lapham[11] stated that "Mr. Strong has prepared a map of the Blue Mounds . . . with the Brigham mines, showing geological formations . . . and also the topography by contour lines, and the depth at which each rock would be reached from any point."

Faults: The relationship of faults to the folds in the Wisconsin-Illinois-Iowa district and thus to the control of ore deposits was discovered by Jenney.[6]

Oil Rock: The Guttenberg member of the Decorah formation normally consists of 10 to 15 ft of brown thin-bedded limestone with intercalated brown shale partings, Fig. 4. Locally, as Kay first noted,[12] the limestone component has been leached, resulting in a concentration of brown shale and other argillaceous material; in such places the Guttenberg member is as thin as 2 ft. The *oil rock* of the miners is the brown shaly material. Therefore, in areas of thin Guttenberg, where the shale element is concentrated, miners recognize a thick oil rock; in areas of normal Guttenberg thickness, on the other hand, where the shale component is scattered and obscure, the miners do not recognize oil rock.

The association of many ore deposits with the leached Guttenberg member (or thick oil rock) is a feature useful in prospecting.

Areal Structure Mapping: In 1942 and 1943 the USGS began areal mapping of the folds that control the zinc-lead orebodies. The first locality mapped, selected for its large number of mines and closely spaced drillholes, was the Hazel Green-Shullsburg area in Wisconsin, Fig. 2. At first the folds were mapped on the basis of drillhole records, but because the persons who logged these early holes did not interpret the stratigraphic subdivisions consistently, it was necessary to restudy the stratigraphy in detail. In 1943 and 1944, with information obtained from mapping the geology of more than 40 mines in all parts of the district and from studying outcrops throughout an even more extensive area, the USGS determined the stratigraphic relations of the larger units and subdivided these units, Fig. 4, so that the small folds and faults could be more accurately mapped. During the next few years the Illinois State Geological Survey applied and extended this type of mapping in Illinois in the area just south of the Hazel Green-Shullsburg area, Fig. 2.

In Wisconsin the areal geology was mapped in detail. Owing to dissection and thin surficial deposits, outcrops are numerous, and as most of the holes were drilled in upland areas, geologic information was available in most parts of the Hazel Green-Shullsburg area. The detailed subdivision of the stratigraphic sequence made most exposures useable as data points for structure contouring, which in turn was used to establish the relation of the orebodies to the local folds and to delimit potential areas for exploration. Horizontal and vertical control was obtained by plane table and telescopic-alidade surveys, scale 1:7920. The mapping in Illinois differed in that vertical control was obtained by transit and telescopic-hand-level surveys; horizontal control was derived from the Galena quadrangle.

In 1946 Calumet & Hecla Inc. geologists began mapping the folds with information from prospect holes drilled on a grid pattern.[13] The grid is based on subdivisions of land survey measurements and is a rectangular net 660x1320 ft or 660 ft square. The Eagle-Picher Co. has used a square grid of 660 ft,[14] whereas The New Jersey Zinc Co. drills prospect holes on a grid pattern based generally on 1320-ft spacing. This drilling was designed to obtain uniformly spaced geologic data. Folds disclosed by this means, like those disclosed by outcrop mapping in other areas, are then prospected for ore deposits with more closely spaced drillholes.

Other companies have used a modification of the grid system. In places only a simple grid is used; in other places lines of drillholes are spotted to cross the folds or trends of orebodies. The mineralized areas are then explored with more closely spaced holes, and holes are drilled along the trend of the ore. Geologists of the Eagle-Picher Co. and the Vinegar Hill Zinc Co. are currently using this method with success.

Mapping in the Hazel Green-Shullsburg area[15] showed three general trends of synclinal axes—northwestward, northeastward, and eastward, Fig. 7.

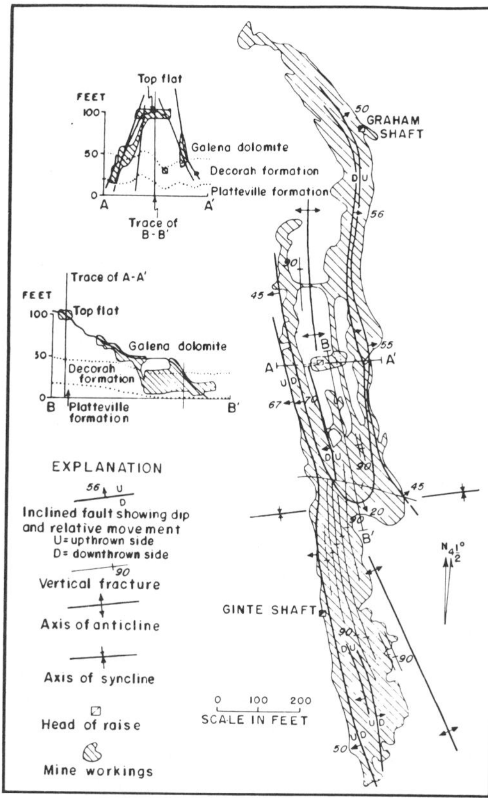

Fig. 8—Plan and cross-sections of Graham-Ginte mine, Illinois, showing opposed pitches and associated crenulated area and cross-syncline and spoon-shaped pitch. (Geology by A. F. Agnew, A. V. Heyl, Jr., E. J. Lyons, C. H. Behre, Jr., and R. M. Hutchinson, 1944-1945.)

In the northwest-trending synclines orebodies trend northwest. They are double-pitch† bodies that lie along a tightly crenulated zone, see Fig. 8; the central anticline is accentuated by steep normal faults and solution-thinned Guttenberg and Quimbys Mill members, see p. 784. The structure is believed due to tectonic compression and was modified by solution. These northwest-trending orebodies are as much as 3000 ft long, 450 ft wide, and 125 ft high, although the average orebody is about 1000 ft long, 200 ft wide, and 60 ft. high.

† *Double-pitch* faults are opposed outward-dipping reverse faults containing ore. One or more ore-filled reverse faults dipping in the same direction are known as *single-pitch* faults.

In the northeast-trending synclines, east-trending horseshoe-shaped orebodies, open to the west, occur *en echelon*. They are single-pitch bodies wrapped around broad folds, Fig. 9. The central part of the fold is broken by weak vertical joints. As very little evidence of solution is visible, the structure appears to have been the result of tectonic compression alone. These east-trending *en echelon* orebodies are as much as 2 miles long, 180 ft wide, and 85 ft high, although the average orebody is about 1200 ft long, 120 ft wide, and 50 ft high.

In the east-trending synclines, east-trending linear orebodies occur. They are single-pitch or double-pitch bodies, each controlled by crenulations that contain a gentle central anticline, which is broken by poorly developed vertical joints. If the orebody is double-pitch, one pitch is commonly much better developed than the other. As is shown in some mines, single-pitch structures have a complementary pitch-type fracture that does not bear ore. Solution effects are visible, but they are not so important as in the northwest-trending orebodies; the tectonically formed structures therefore have not been greatly modified by solution in these east-trending bodies. This type of orebody may be as much as 2000 ft long, 250 ft wide, and 80 ft high, although the average orebody is about 1000 ft long, 125 ft wide, and 50 ft high.

Four examples of the application of geology to the finding of orebodies in this area are:

1) The area southeast of Shullsburg, Wis., had been unprospected because of its cover of Maquoketa strata, as mine operators thought that ore did not occur in areas where the ore zone was overlain by Maquoketa shale, and because the cover of shale necessitates deeper drilling. Success of the exploration program of Calumet & Hecla Inc.[13] in the area is due in large part to the combination of grid drill-

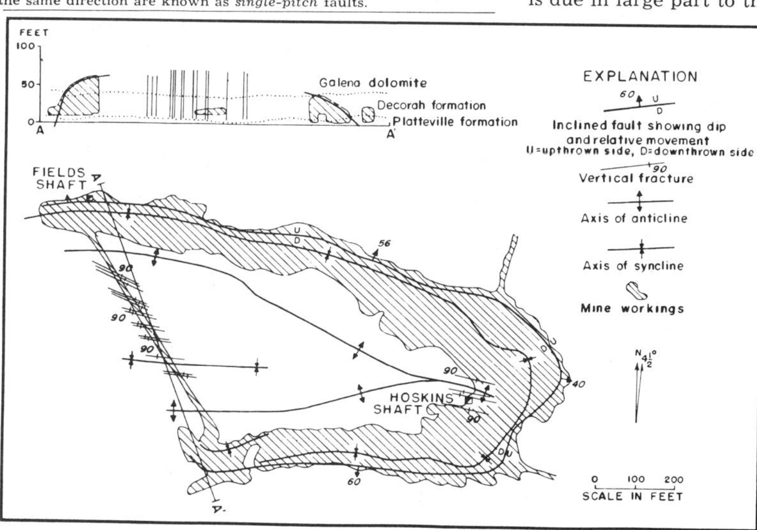

Fig. 9—Plan and cross-section of Hoskins mine, Wisconsin, showing horseshoe-shaped pitch wrapped around a broad fold. Geology by A. F. Agnew and A. V. Heyl, Jr., 1944.)

Fig. 10—Map of area including Galena, Ill., and Shullsburg, Wis., showing area covered by Maquoketa shale and general synclinal axes that could be projected eastward from the Hazel Green-Shullsburg area and northeastward from the Galena area. (From Agnew, Heyl, Behre, and Lyons[15]; Willman and Reynolds[16]; Shaw and Trowbridge[45]; and Grant and Burchard.[41])

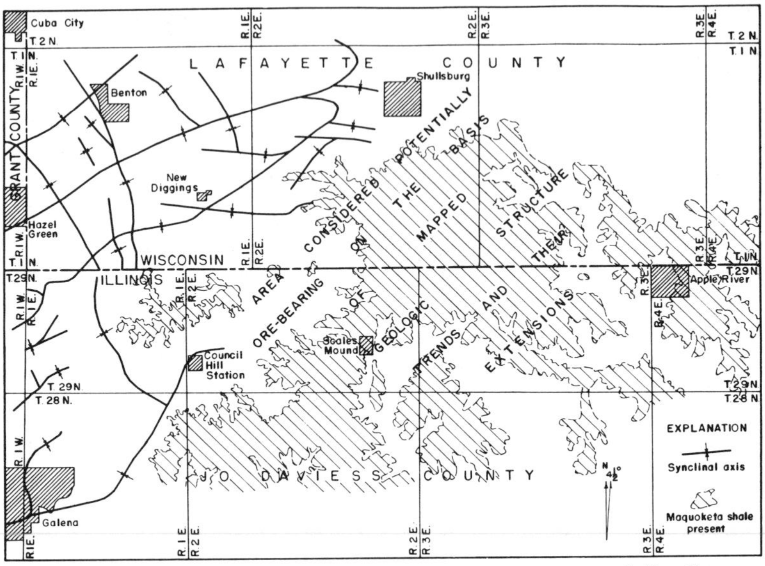

ing methods and outcrop mapping. This company projected to the east and to the northeast the synclinal axes and ore trends shown, Figs. 7 and 10, by outcrop mapping.[15,16] Carrying out its grid drilling program on a prediction of the continuity of structural trends and a repetition of structural pattern, Calumet & Hecla was successful in locating several synclines, some of which contained orebodies now being mined.

2) In the northwest part of Sec. 21, T. 1 N, R. 1 E, the juncture of northeast-trending and northwest-trending synclines was mapped, Fig. 7, *A*. In this vicinity several small patches of ore have been drilled, and recently one of the mining companies discovered an orebody at the juncture which was not known before the structure mapping.

3) In the south-central part of Sec. 26, T. 1 N, R. 1 E, a syncline was mapped, Fig. 7, *B*, showing a bend toward the west. This syncline is the southwestward continuation of the general synclinal trace that extends northeastward for 5 miles and contains at least a dozen zinc-lead mines. Recently a mining company discovered an orebody at the bend in the southwesterly projection of this synclinal trace; the alignment of mines had previously been projected inaccurately because the bend in the syncline had not been recognized.

4) In the center of fractional Sec. 14, T. 29 N, R. 1 E, south of the state line, Fig. 7, *C*, an orebody was discovered in 1947 by one of the mining companies. This ore occurs along the southward projection of one of the mapped synclinal axes.

The outcrop mapping in the Potosi area,[17] Fig. 2, showed the two general trends of synclinal axes, east-southeastward and northeastward, see Fig. 11.

In the east-southeastward synclines zinc-lead orebodies have a similar strike. These orebodies are said[18] to be controlled by single-pitch and double-pitch faults. Outcrop geologic mapping showed a pitch zone along the limb of a fold in a ravine almost a mile to the east of these orebodies, and a westerly continuation of this fold was mapped.

Not long ago one of the mining companies discovered an orebody by drilling along this westerly projection. Active mining in the Potosi area has not been carried on for many years, and the discovery lends impetus to further development of that area.

Late in 1942, in the course of a cooperative exploratory venture with the USGS, the U. S. Bureau of Mines drilled the Gray property in Illinois, because several earlier drillholes had penetrated lean zinc-lead bearing material 40 to 50 ft thick, on an extension of the Black Jack-Pittsburg ore trace, Fig. 12. As a result of this drilling in 1943 the 479,500-ton[19] Gray orebody was discovered. Mining of the ore began late in 1944.

During the early months of drilling the Gray orebody, structure contours based on the drilling on this property and on old records from the Black Jack and the Pittsburg mines to the north, coupled with the discovery of a pitch-type fault in an exposure just north of the Pittsburg mine, showed that a syncline and associated pitches controlled the trend of the orebodies, Fig. 12. In May 1943, therefore, the USGS recommended that a line of five holes be drilled across the projection of this syncline and associated pitches about 2000 ft southeast of the Gray orebody. In the summer of 1943 the Illinois State Geological Survey mapped the outcrops in this area in detail, corroborated the presence of this syncline, and showed its regional setting.[16] In June 1944 the USBM penetrated the Bautsch orebody with the fourth hole of the proposed five-hole cross-section, and by 1945 the drilling had disclosed more than 916,500 tons of ore.[19] The Bautsch orebody was therefore found by drilling on the projection of a syncline and the accompanying pitch zones.

The above illustrations of ore discovery are cited as examples of the application of areal mapping of geologic structure. Many additional ore discoveries by mining company geologists are known.

Spacing of geologic data in great part governs the accuracy of any structural map. Thus, orebodies that lie between synclines shown by outcrop mapping or grid drilling would be overlooked by either of these methods. On the other hand, it should not be inferred that all synclines or all leached areas are ore-bearing. Many of the folds will be found to contain weak evidences of zinc-lead mineralization, or only iron minerals; others may show only the

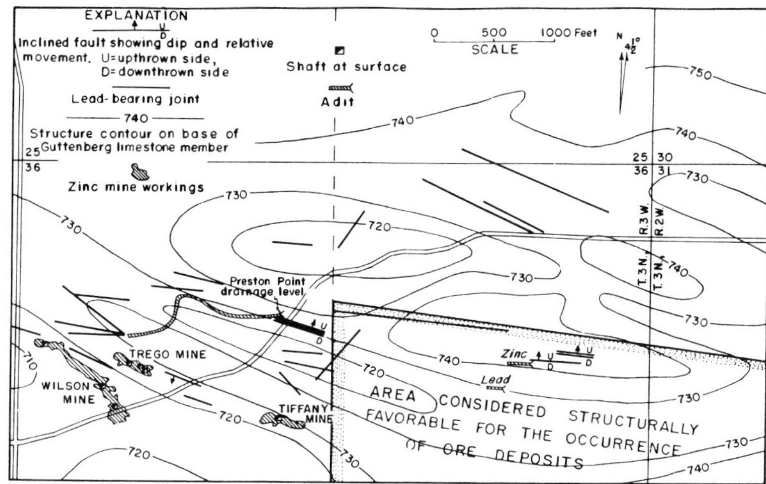

Fig. 11—Map of part of Potosi area, Wisconsin, showing structure trends, orebodies, and potentially ore-bearing area. (From Heyl, Lyons, and Agnew.[17])

leaching effects of the mineralizing solutions, whereas some folds may show no evidence of having been mineralized. The final answer is of course found by prospect drilling.

Structural generalizations are not necessarily valid in all areas. For example, it is common knowledge that north-south orebodies are very rare in the district, yet note the northerly trend of the New Hoskins-Cottingham No. 2 orebody (Fig. 7, *D*) which

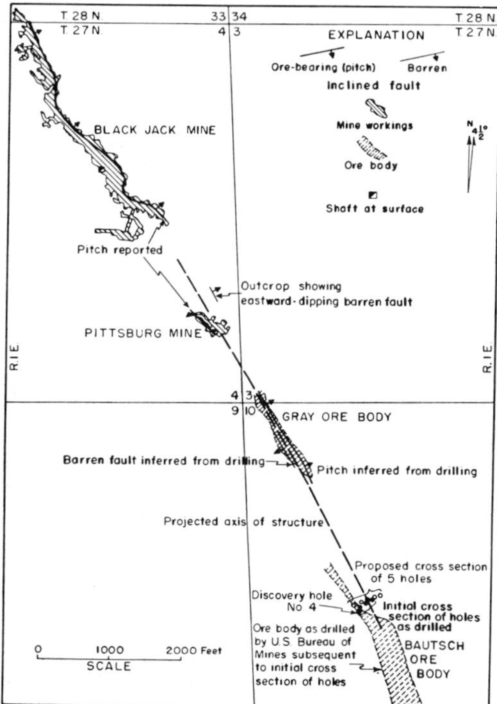

Fig. 12—Map showing Black Jack mine workings, Pittsburg mine workings, Gray orebody, projected geologic structure, proposed and drilled initial cross-section of holes, and the resulting Bautsch orebody discovery, Illinois. (Data in part from Zinner and Lincoln.[46])

was extremely rich. Furthermore, the structural relations of the orebodies south of Shullsburg (Calumet & Hecla Inc. and Vinegar Hill Zinc Co. leases) are not completely known as yet; detailed study of the mines developed in this area may show additional geologic controls or modifications that can be applied in the potentially productive territory farther south and east. Nevertheless, the type of geologic mapping cited on the previous pages, if data are adequate, can rule out areas that are potentially unproductive, effecting a saving in prospecting cost.

Lastly, although all geologists do not agree fully on the source of the ores, or on the precise origin of the pitches, these considerations do not materially affect the use of the surface and subsurface manifestations and geometry of the geologic structure in prospecting.

Structure Mapping in the Mines: Miners can usually follow the ore without difficulty because the ore in most places is easily recognizable underground and the pitches are relatively smooth and continuous. On the other hand, a knowledge of the geology, together with detailed geologic mapping underground, has proved valuable in several places to locate ore shoots that otherwise would have been missed.

The Martin mine, Fig. 13, has developed a horseshoe-shaped orebody in a northeast-trending syncline, Fig. 7, *E*. The orebody is controlled by a north-northeast-directed horseshoe-shaped single-pitch zone.

Mining was proceeding in late 1943 near the south end of the west limb of the orebody easterly toward some drillholes that showed lean ore. The geologic map of the mine prepared at that time indicated that the ore found in these holes was part of a north-northeast-directed core ground[20] body that would continue at least to the end of the crosscut 400 ft to the north, where a drift driven easterly from the shaft had intersected vertical fractures that contained breccia zinc ore; drillholes showed this core ground ore to continue 200 ft beyond the drift. As a result of this geologic mapping the mining company did not cease its operations in this mine, but continued mining the ore in this core ground for 2½ years.

Along the northwest and north side of the orebody the geologic mapping showed several places

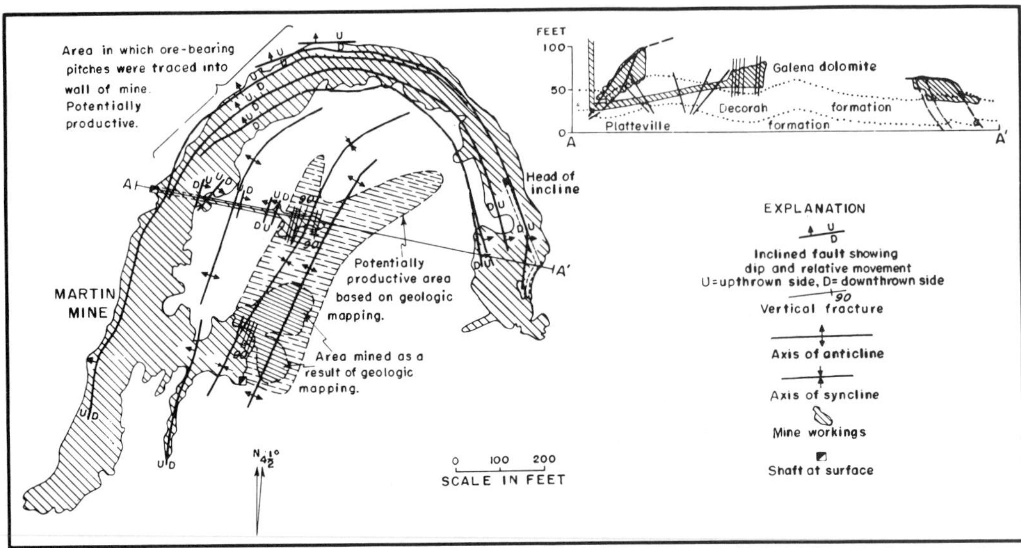

Fig. 13—Plan and cross-section of Martin mine, Wisconsin, showing geologic structure, areas indicated by geologic mapping as potentially productive, and area subsequently mined. (Geology by A. V. Heyl, Jr., A. F. Agnew, C. H. Behre, Jr., and E. T. McKnight, 1943.)

where pitches were traced into the outside wall of the stope; their traces were discovered again in the stoped area, 100 to 200 ft farther along the wall. This suggested that additional ore could be obtained by mining the intervening area.

The Graham-Ginte mine, Fig. 8, is in an elongate northwest-trending double-pitch orebody, controlled by a gentle northwest-trending syncline that has a well-developed central anticline broken by steep normal faults. In 1944 mining was progressing southward from the old Graham mine along drilled-out ore. Geologic mapping by the USGS showed that at the south end of the Graham mine an arcuate pitch connected the east and west pitches at their lower level (Guttenberg and lower part of Ion beds), where an east-trending syncline crossed the north-trending structure. Geologic interpretation suggested that this pitch connected the east and west pitches at higher levels as well, thus forming an inverted spoon-shaped pitch that would be continuous upward until the *top flat* was reached. On the basis of this interpretation the mining company drilled five holes in the core ground area, which was considered barren from evidence in the stopes along the east and west pitches; these holes corroborated the geologic interpretation, and thousands of tons of ore thus found in the core ground area were mined during the next two years.

Interpretation of the geologic map also suggested that south of the cross-syncline the east pitch diverged eastward; this suggests that it might be profitable to prospect east of the southern extension of the mine workings, which are along the west pitch zone and in the core ground.

The Liberty mine, Fig. 14, was developed in an east-trending horseshoe-shaped single-pitch body. Geologic mapping showed that during mining the main pitch had been lost in an area from 200 to 400 ft along strike east of the shaft. On the basis of the geologic mapping a 40-ft cross-drift was driven to

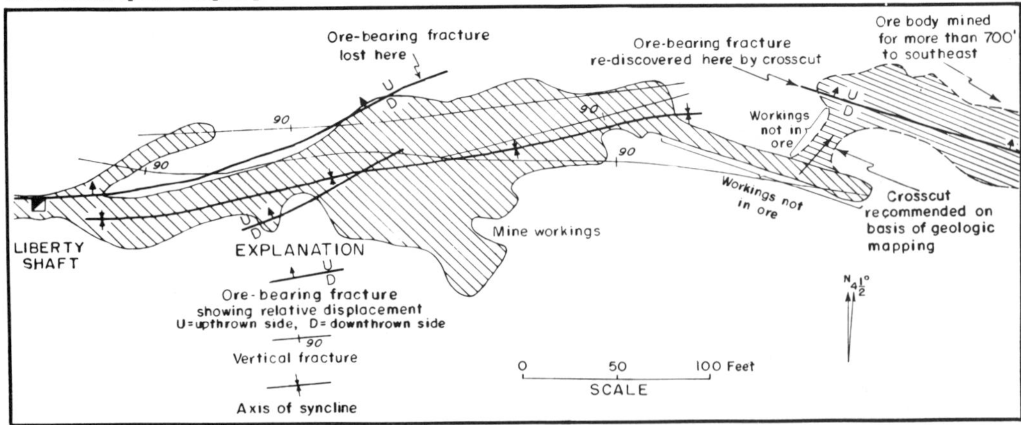

Fig. 14—Map of part of Liberty mine in Wisconsin showing mine workings, crosscut driven on the basis of mapped geologic structure, and part of the orebody discovered as a result. (Geology by A. F. Agnew, A. V. Heyl, Jr., and C. H. Behre, Jr., 1943-1944.)

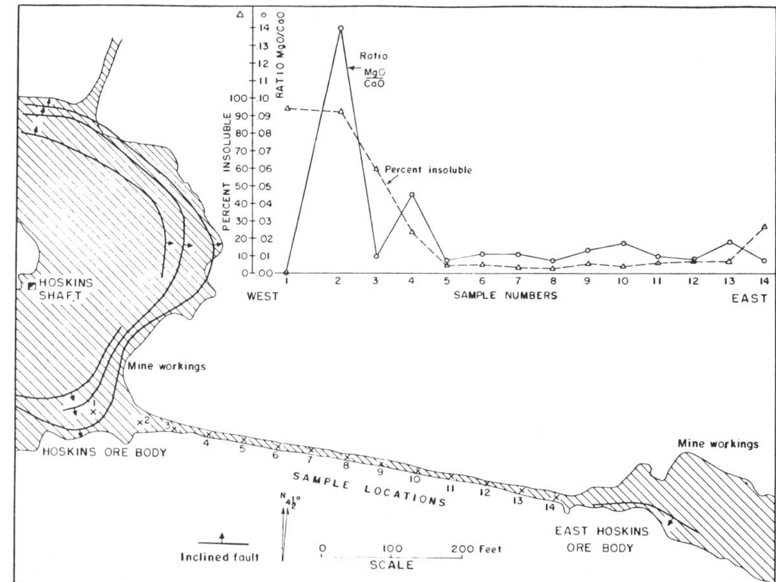

Fig. 15—Map of part of east end of Hoskins mine in Wisconsin showing locations sampled in crosscut. Graph indicates MgO/CaO ratio and percent insoluble material in Guttenberg limestone member. (Geology by A. F. Agnew and A. V. Heyl, Jr., 1944. Analyses by Marie L. Lindberg, USGS, Feb. 13, 1945.)

the north, the pitch zone was rediscovered and was then followed for 700 ft to the east, and the ore thus found was mined during the next two years.

In the Wisconsin-Illinois-Iowa zinc-lead district some of the ore potentialities of mines are missed as mining proceeds. Detailed geologic mapping underground by the USGS during World War II was of value in discovering ore shoots and in noting the economic potentialities of fractures that were neglected as probably unproductive; it is suggested that mining companies may find it profitable to have their geologists do similar mapping.

Recently block caving was introduced into this district. Where this method is used, parts of the faces of the stopes are inaccessible unless geologic mapping is continued each day as mining advances. Together with the height of the open stope (up to 125 ft) and the haze caused by the humidity,‡ this makes accurate geologic mapping difficult. Nevertheless, geologic staffs can contribute to economic mining of the ore and should do stope mapping.

‡ Most of the large mines pump 2000 to 3000 gal of water per min.

Crevice or Vertical Fissure Deposits

Owen[3] cited the following "symptoms of lead ore in Wisconsin": 1) A bench in the outline of a hill. 2) "A small longitudinal depression, or miniature ravine, on a hillside." 3) Sinkholes "in an east and west or in a north and south course." 4) "Rank growth of vegetation in a linear direction, especially of plants with deep-reaching radicals." 5) "Red appearance of the surface." 6) Fragments of calcite at the surface. (If found in large quantities, however, as in the southern and western parts of the district, it is an unfavorable sign.) 7) The presence of *Receptaculites*, which is "a good indication of lead." 8) " 'Gravel mineral' (small pieces of lead ore) in connection with the crumbling and arenaceous appearance of the adjacent magnesian limestone."

Daniels[6] noted that "in the vicinity of veins the rocks are often so changed, as nearly to obliterate their usual characters." Percival[4] was more specific when he wrote that the rock which is now called the Guttenberg member is much decomposed for 6 to 8 ft on each side of a vertical crevice, with normal fossiliferous limestone beyond.

Some of the symptoms of Owen, with additions, are still used by prospectors and by the USGS in outcrop mapping. Sinkholes, vegetation marking traces of joints, and ocher red streaks in the surface material are all useful criteria.

Other Applications of Geology

Other applications of geologic knowledge and techniques to the discovery of ore deposits have been attempted. Some of the significant ones, or those that appear to show promise, are cited below.

Geochemical Prospecting: In 1943 and 1944 mapping by the USGS disclosed halos of SiO_2 and MgO in the Guttenberg and Quimbys Mill members around several orebodies, especially the east-trending horseshoe-shaped single-pitch type. These halos were detected in the field by the hammer and acid bottle. The presence of the halos suggested that chemical analysis of the rock might be revealing, as the following four examples show.

1) In the Hoskins mine, Fig. 9, samples from the Guttenberg member in the drift eastward from the main horseshoe-shaped orebody to the eastern orebody showed a high percentage of insoluble material and a high MgO/CaO ratio near the main body, Fig. 15. As geologic mapping underground showed only slight leaching of the Guttenberg member but did illustrate that the rock was silicified, the insoluble material is believed to be mainly SiO_2.

2) A similar increase in insoluble material near ore was noted in the B. A. T. mine, half a mile east of the Hoskins mine. In the B. A. T. mine the unit sampled was the Quimbys Mill member, Fig. 16. As in the Hoskins mine, an increase in the MgO/CaO ratio near ore was seen at the west end of the crossdrift.

3) At the recommendation of the USGS in 1944 the USBM sampled for magnesium content the soil over part of Coker No. 1 orebody (Fig. 3, *A*) 80 ft below the surface. Analysis of the soil samples showed an increase in the MgO content above the

ore.[21] This increase of MgO over the orebody is even more significant than that in the Hoskins and B. A. T. mines because the underlying bedrock at the Coker mine is dolomite rather than limestone. Thus the dolomitization here had affected dolomitic rocks.

4) In at least two localities in Wisconsin, halos of lead and zinc have been found in soil samples taken above zinc-lead orebodies. At the Porter's Grove and Crow Branch diggings (Fig. 3, *B, C*) soil samples taken by the USGS showed high zinc and lead values across the projection of the alignment of mineralized areas where the zinc and lead minerals are first found as much as 150 ft below surface.[22, 23]

The geochemical indications described above appear significant enough to warrant further research along these lines. Analyses for heavy metals and MgO/CaO ratio are particularly promising and can be determined by microchemical analysis of drill-hole cuttings from so-called barren holes (barren of zinc-lead and even of iron minerals), from cross-cuts underground, and also from analysis of soil samples. In essence, the geochemical methods increase the size of the target.

Geobotanical Prospecting: Owen[3] observed that "rank growth of vegetation in a linear direction, especially of plants with deep-reaching radicals," was a "symptom" of lead ore. This concept was reiterated by Daniels,[9] who noted that "a certain plant, known as the 'masonic' or 'lead weed,' grows only where its roots are fed by lead ore," and he stated that roots of this plant are often found 40 to 60 ft below the surface. Daniels reports that this indication seems to have been learned from Indians.

Recent observations in this district have shown that some old lead pits are marked by the *Indian leadweed* (large-bracted wild indigo), which has a large root. Cottonwoods, and in some places white birches, are characteristically found either in pits or on waste piles of old lead and zinc diggings; the greater acidity in the soil is probably responsible, in part, together with the characteristic deep root growth of these trees.

Geophysical Prospecting: In 1923 a method using "spontaneous polarization" was applied in the Shullsburg, Wis., and the Galena, Ill., areas, see Fig. 2, in an attempt to discover ore. In 1928 self-potential, applied potential, and magnetometer surveys were all made in the Linden and Dodgeville areas of Wisconsin. Not until the middle 1940's, however, was any concerted attempt made to experiment with geophysical methods in the hope of establishing controls that could be employed economically as guides to ore.

Since that time several private companies, the Illinois State Geological Survey, the USGS, and the Geological Society of America, have supported surveys by one or more methods, which have been applied locally in the search for specific orebodies. Methods that have been tested include self-potential and induced-potential, resistivity, electromagnetic, gravity, seismic, radioactivity, and geothermal. Some of these methods have shown promise, but insufficient testing has been done, and the results of most of the surveys have not been released for publication. An integrated program of research in geophysical methods might provide conclusive data regarding the economic value of one or more methods.

A preliminary aerial magnetometer survey has been made by the USGS over part of the mining district, but results have not yet been published. It is believed that such a survey would give information relative to the general topography of the surface of the pre-Cambrian rocks but would not show such relatively small features as the folds controlling the individual zinc-lead orebodies, which are some 1500 ft above the pre-Cambrian surface.

Lower Potentially Productive Zones: Daniels[9] noted that the veins in the Galena dolomite are not present in the underlying "blue limestone" (Guttenberg member and lower part of the Ion member) but "resume again in the buff-colored rock" (*glass rock,* or Quimbys Mill member).

In 1894 Blake[24] withdrew attention from the glass rock as a potential ore host, for he held that the downward-percolating waters that carried the lead and zinc were stopped at the "impervious layers of shale or clay [Spechts Ferry member] at the top of the Trenton limestone." He continued with the statement that ore deposits in the glass rock are in places from which the overlying Galena dolomite has been largely removed by erosion, implying that these deposits are secondary and should not be considered characteristic of the glass rock.

Behre[25] revived the glass rock question by reiterating Grant's[26] views that in the northern part of the district the ore is largely below the glass rock and noted that "mineralized faults cross the 'oil rock' and Spechts Ferry shales." These statements were an attempt to dispel the belief then held by many geologists that the ores, thought to be deposited from descending waters, had been dammed above the glass rock.

Although some mining men knew of mines that had been developed in glass-rock strata, prospecting and mining of these beds in the early 1940's was not systematic; this was due partly to the additional water that they knew would be encountered during mining in these beds. Geologists stressed that zinc-lead production could come from these beds, particularly below strong pitch-zone ore. As a result several mines were deepened to work ore in the glass-rock unit, and a few were developed solely to work glass-rock ore. At one period during World War II one-seventh of the production of this district came from ore in the glass rock.

Since that time prospecting the glass rock has been standard procedure with most operators, and many of the current mine operations are being conducted in orebodies in the glass rock.

Grant[27] noted that at the Little Giant mine (Fig. 7, *F*) ore similar to that in the glass rock had been intersected in a shaft for a distance of 22 ft below the base of the glass rock. This ore, the stratigraphic sequence tells us, occurred in the so-called *Trenton* beds, or McGregor member. As the McGregor strata are argillaceous nodular limestones similar to the Guttenberg and Quimbys Mill members, these strata were recommended by geologists in the 1940's as possible ore hosts in other areas. During World War II ore was mined from the McGregor member as much as 15 and 22 ft below the base of the glass rock, a mile north and a mile south of the Little Giant mine, respectively. Ore in these beds in the areas south of Shullsburg, south of New Diggings, and south of Galena has been intersected by recent drilling. Current prospecting technique includes drilling into McGregor strata in areas that appear favorable because of indications in overlying rocks.

Owen[3] suggested that the *Lower Magnesian* (Prairie du Chien) limestone, if extensive north of the Wisconsin River, "may there yield veins of lead ore." Galena has been mined from these Prairie du Chien strata at several localities north of the district, see Fig. 3. Recent drilling by the USGS[28, 29] has

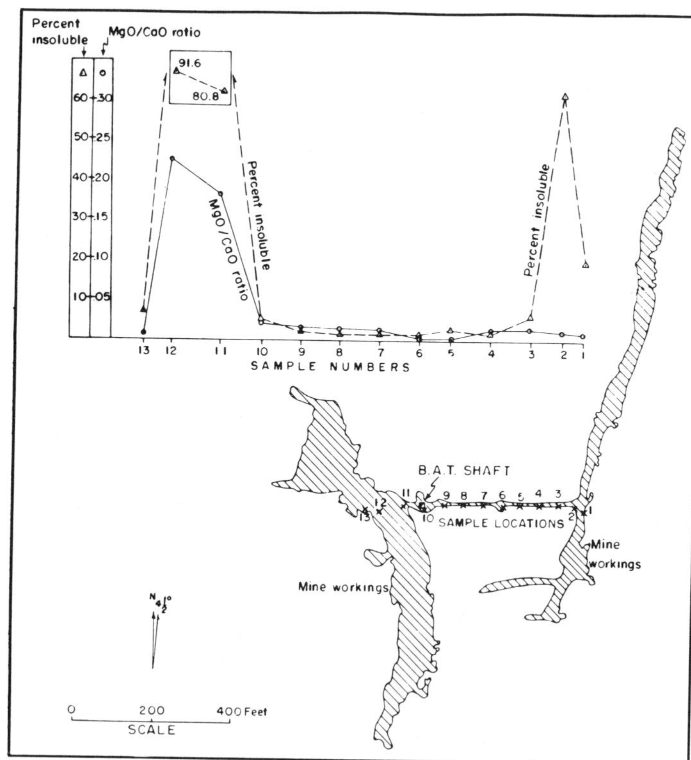

Fig. 16—Map of B. A. T. mine, Wisconsin, showing locations sampled in cross-cut. Graph indicates MgO/CaO ratio and percent insoluble material in Quimbys Mill member. (Geology by A. F. Agnew, A. V. Heyl, Jr., and C. H. Behre, Jr., 1943. Analysis by Marie L. Lindberg, USGS, Feb. 13, 1945.)

shown that at least three zones of mineralized rock exist in this stratigraphic unit. The dolomite and cherty dolomite of the Prairie du Chien are extremely tough, and the cost of breaking this rock during mining operations might be high. Furthermore, the unit is characterized by rapid lateral and vertical facies changes from dolomite to red and green shales and sandstone, and this, together with the disconformable relationship between the Prairie du Chien and the overlying St. Peter sandstone, makes the type of lithology to be expected in even a local area uncertain. As a result many areas will be unproductive at the Prairie du Chien stratigraphic position because of unfavorable rock type. The dolomite of the Prairie du Chien does constitute a possibly productive lower zone, nevertheless, and its potentialities should be explored further.

Mining: Owen noted[3] that the lead shafts were usually abandoned as soon as they reached or penetrated a short distance below the static water level. Even today this fact is mentioned by the old-timers, so it is logical to assume that water has been a factor in prospecting since Owen's time. Potential lead-zinc ore still exists below the lowest workings in many of the crevice mines. Furthermore, zinc-lead ore remains below the floor of some of the pitch-and-flat mines, as was shown during World War II when many orebodies were re-entered and the ore below the floor was taken out.

As was mentioned earlier, p. 785, because of the southward regional dip and consequent increase in the height that water must be pumped, orebodies in the southern part of the mining district and to the south and southwest will contribute more water than those to the north; 2500 gpm is not uncommon in the south, whereas 500 gpm or less is all that need be pumped from mines in the northern part of the district.

Whitney, recognizing the stratigraphic correlation of the openings in the Dubuque area, recommended that "horizontal drifts or excavations are the proper means of exploration, and not vertical ones or shafts."[30] Such drifts or levels, he reasoned, would prospect several crevices at an opening zone and would also provide gravity drainage for the water. He later extended this recommendation to the mines in Wisconsin,[10] citing in the Shullsburg area a level that had been so opened and had drained water from numerous crevices.

Since Whitney's time several mines in many parts of the district have been opened in this way, both in crevice orebodies and in pitch-and-flat orebodies, and water has been drained and ore trammed to the surface by gravity. These orebodies could not have been mined economically otherwise.

Calvin and Bain[31] agreed with Whitney that a series of crevices could be worked as one development by crosscuts at the levels of the openings, but they applied this suggestion to areas where topography would require a shaft, which they recommended be centrally located. They suggested further that raises be put up to mine upper openings.

During recent years this method has been successfully applied to the pitch-and-flat orebodies by several companies. Its application to crevice deposits might be a means of extracting bodies of lead or zinc that can not be mined economically otherwise.

Conclusions

Cooperative geological investigations by the USGS, the Wisconsin Geological and Natural History Survey, and the Iowa Geological Survey, together with studies by the Illinois State Geological Survey, have shown that detailed stratigraphic work and detailed mapping of geologic structure have been important factors in the renewal of interest in the Wisconsin-Illinois-Iowa zinc-lead district. Certain facets of the geologic methods discussed in this paper had their origins in earlier surveys of this district, and after a period of disuse or even discredit, these ideas have been revived and are currently factors in the economic potential of the district.

Detailed mapping of stratigraphy and structure has revealed the following: 1) Pitch-and-flat orebodies are controlled by fracture zones along small folds arranged along larger folds. 2) Using the detailed stratigraphic units established by this cooperative study, these folds and faults can be mapped from outcrops and supplemental drillholes. 3) Detailed geologic mapping underground can help prevent losing ore shoots and misinterpretation of quirks of the orebodies. 4) Geochemical techniques show promise, especially heavy metals and MgO/CaO analysis.

Applied geology has given this district an important tool for distinguishing areas of more promise from less promising ones. As a result 1) the amount of ineffectual prospecting in potentially unproductive areas (structurally) has been decreased, 2) the amount of ineffectual prospecting and development in potentially productive areas has been decreased, and 3) the effectiveness of mining of individual orebodies and parts of orebodies has been increased. Furthermore, the recognition of geologic controls and the application of this knowledge to the discovery of orebodies in recent years has attracted large companies (with geologists) to this district, which had long been considered mined out.

Acknowledgments

The current USGS project was motivated by C. H. Behre, Jr., then at Northwestern University, who since 1933 had been supervising studies in this district by his students and had recognized the potential value of a thorough geologic study of the structure of the ore deposits. Allen V. Heyl, Jr., of the USGS, and the author mapped the mines cited in this paper and discussed many of the concepts presented. Paul Herbert, Jr., now with Tri-State Zinc Inc., has made numerous contributions to the knowledge of stratigraphy of the district. R. R. Reynolds, of Calumet & Hecla Inc., and H. B. Willman, of the Illinois State Geological Survey, contributed to the knowledge of the structure and stratigraphy of the Illinois part of the district. Representatives of mining companies, especially the Vinegar Hill Zinc Co., The New Jersey Zinc Co., and the American Zinc, Lead and Smelting Co., generously allowed the USGS access to their drilling and mining records. These data, as well as the information and counsel given by C. W. Stoops, were of great assistance. Water-well and prospect-hole drillers and mine operators supplied useful records.

References

1 H. B. Willman and J. S. Templeton: Cambrian and Lower Ordovician Exposures in Northern Illinois. *Illinois Academy of Science Trans.*, 1951, vol. 44, pp. 109-125.

2 Arthur Winslow: Lead and Zinc Deposits. Missouri Geol. Survey, 1894, vol. 6, sec. 1, p. 254.

3 D. D. Owen: Report of a Geological Exploration of Part of Iowa, Wisconsin, and Illinois . . . in 1839. *Congressional Documents*, 26th Congress, 1st session, H. Ex. Doc. 239, 1840, pp. 18, 27, 29, 31, 32, 41-42.

4 J. G. Percival: *Annual Report of the Geological Survey of the State of Wisconsin* [*for 1854*], 1855, 101 pp.

5 J. G. Percival: *Annual Report on the Geological Survey of the State of Wisconsin* [*for 1855*], 1856, 111 pp.

6 W. P. Jenney: The Lead and Zinc Deposits of the Mississippi Valley (with discussion). *Trans. AIME*, 1894, vol. 22, p. 642.

7 W. H. Hobbs: Examples of Joint-Controlled Drainage from Wisconsin and New York. I. The Richland Center Quadrangle of Wisconsin. *Journal of Geology*, 1905, vol. 13, p. 366.

8 A F. Agnew, A. E. Flint, and R. P. Crumpton: Geology and Zinc-Lead-Barite Deposits in the Area East of Cuba City, Wis. USGS Mineral Inv. Ser. Field Studies Map MF, 15, 1954.

9 Edward Daniels: *First Annual Report on the Geological Survey of the State of Wisconsin* [*for 1853*], 1854, pp. 22, 23-24, 25, 34, 35, 51.

10 J. D. Whitney: Economical and Mining Geology, *in* James Hall and J. D. Whitney: *Report of the Geological Survey of the State of Wisconsin*, 1862, vol. 1, pp. 221-420.

11 I. A. Lapham: Annual Report for 1873, *in* T. C. Chamberlin: *Geology of Wisconsin*, 1877, vol. 2, p. 12.

12 G. M. Kay: Stratigraphic Setting, Wisconsin-Illinois District, *in* E. S. Bastin and others: *Contributions to a Knowledge of the Lead and Zinc Deposits of the Mississippi Valley Region. GSA Special Paper 24*, 1939, p. 27.

13 H. B. Ewoldt and R. R. Reynolds: Exploration for Lead and Zinc: Procedures Used in the Illinois-Wisconsin District. MINING ENGINEERING, 1951, vol. 3, pp. 230-231.

14 C. O. Dale and W. J. Rundle: Illinois Operations of the Eagle Picher Mining and Smelting Co. MINING ENGINEERING, 1952, vol. 4, pp. 72-76.

15 A. F. Agnew, A. V. Heyl, Jr., C. H. Behre, Jr., and E. J. Lyons: *Zinc-Lead Deposits of the Hazel Green-Shullsburg Area, Lafayette and Grant Counties, Wis.* USGS *Strategic Minerals Investigation Preliminary Report 3-216*, 1948.

16 H. B. Willman and R. R. Reynolds: Geological Structure of the Zinc-Lead District of Northwestern Illinois. Illinois Geol. Survey *R.I. 124*, 1947, plates 4, 7.

17 A. V. Heyl, Jr., E. J. Lyons, and A. F. Agnew: *Geologic Structure Map of the Potosi Lead-Zinc Area, Grant County, Wis.* USGS *Strategic Minerals Investigation Preliminary Map 3-221*, 1948.

18 A. J. Bleser: *Geology and Ore Deposits in the Potosi District*, pp. 16-17. University of Wisconsin thesis, 1904.

19 F. C. Lincoln: New Zinc Mines in the Old Galena District of Illinois. *Mining Congress Journal*, 1945, vol. 31, p. 30.

20 E. R. Scott: *The Structural Control of Ore Deposition in the Upper Mississippi Valley Lead-Zinc District*, fig. 5, p. 29. Northwestern University thesis, 1934.

21 H. R. Joesting: *Experimental Geophysical Surveys in the Coker-Bickford Area of the Wisconsin Lead-Zinc District, Iowa County, Wis.*, 1945. Unpublished USBM report.

22 Lyman Huff: Abnormal Copper, Lead, and Zinc Content of Soil Near Metalliferous Veins. *Economic Geology*, 1952, vol. 47, pp. 524-525, fig. 1, table 3.

23 V. C. Kennedy: *Geochemical Studies in the Southwestern Wisconsin Lead-Zinc District.* In preparation.

24 W. P. Blake: *Discussion of* W. P. Jenney: The Lead-and-Zinc-Deposits of the Mississippi Valley. *AIME Trans.*, 1894, vol. 22, pp. 583, 632, 634.

25 C. H. Behre, Jr.: The Geology and Development of the Wisconsin-Illinois Lead-Zinc District. *Guidebook 9th Annual Field Conference*, Kansas Geological Society, 1935, p. 381.

26 U. S. Grant: *Report on the Lead and Zinc Deposits of Wisconsin.* Wisconsin Geological and Natural History Survey *Bull. 14*, 1906, 100 pp.

27 U. S. Grant: *Preliminary Report on the Lead and Zinc Deposits of Southwestern Wisconsin.* Wisconsin Geological and Natural History Survey *Bull. 9*, 1903, pp. 62, 68, 72, 88.

28 A. V. Heyl, Jr., E. J. Lyons, and A. F. Agnew: *Exploratory Drilling in the Prairie du Chien Group of the Wisconsin Zinc-Lead District by the U. S. Geological Survey in 1949-1950. USGS Circ. 131*, 1951.

29 A. F. Agnew, A. E. Flint, and J. W. Allingham: *Exploratory Drilling Program of the U. S. Geological Survey for Evidences of Zinc-Lead in Iowa and Wisconsin, 1950-51. USGS Circ. 231*, 1953, pp. 7-11, 21-37, plates 3, 4; fig. 4.

30 J. D. Whitney: Lead, *in* James Hall and J. D. Whitney: *Report on the Geological Survey of the State of Iowa* [*for 1855-1856*], 1858, vol. 1, pt. 1, pp. 422-468, figs. 44-52.

31 Samuel Calvin and H. F. Bain: Geology of Dubuque County. *Iowa Geological Survey* [*Annual Report 1899*], 1900, vol. 10, pp. 381-622.

32 J. D. Whitney: Geology of the Lead Region. Illinois Geol. Survey, 1866, vol. 1, pp. 153-207.

33 Moses Strong: Geology and Topography of the Lead Region, *in* T. C. Chamberlin: Geology of Wisconsin, 1877, vol. 2, pp. 645-752.

34 T. C. Chamberlin: Ore Deposits of Southwestern Wisconsin. *Geology of Wisconsin*, 1882, vol. 4, pp. 367-568.

35 H. F. Bain: *Zinc and Lead Deposits of the Upper Mississippi Valley.* Illinois Geol. Survey *Bull. 294*, 1906, 155 pp.

36 G. H. Cox: *Lead and Zinc Deposits of Northwestern Illinois.* U. S. Geol. Survey *Bull. 21*, 1914, 120 pp.

37 H. C. George: The Wisconsin Zinc District. *AIME Bull.*, 1917, vol. 132, p. 2057.

38 U. S. Grant and M. J. Perdue: *Millbrig Sheet of the Lead and Zinc District of Northern Illinois.* Illinois Geol. Survey *Bull. 8*, 1907, p. 340.

39 C. H. Behre, Jr., E. R. Scott, and A. F. Banfield: The Wisconsin Lead-Zinc District. (Preliminary paper) *Economic Geology*, 1937, vol. 32, p. 800.

40 G. H. Cox: *Elizabeth Sheet of the Lead and Zinc District of Northern Illinois.* Illinois Geol. Survey *Bull. 16*, 1910, pp. 36, 40.

41 U. S. Grant and E. F. Burchard: *Lancaster and Mineral Point Quadrangles.* USGS Geological Atlas, *folio 145*, 1907, 14 pp.

42 John Murrish: *Report on the Geological Survey of the Lead Regions* [*for 1863*], 1871, 65 pp.

43 A. V. Heyl, Jr., and A. F. Agnew: *Meekers Grove (Jenkinsville) Zinc-Lead Area, Lafayette County, Wis. USGS Strategic Minerals Investigation Preliminary Map, 1945.*

44 Wisconsin Geological and Natural History Survey, and Wisconsin Mining Trade School: *Hazel Green Special Sheet.* Unpublished map, 1914.

45 E. W. Shaw and A. C. Trowbridge: *Galena and Elizabeth Quadrangles.* USGS Geological Atlas, *folio 200*, 1916, 13 pp.

46 Paul Zinner and F. C. Lincoln: *Exploration of the Gray and Bautsch Lead and Zinc Deposits*, Jo Daviess County, Ill. *USBM R.I. 3847*, 1946, figs. 2-4.

17

Reprinted from pages 25–33 of *Exploration and Mining Geology, with Particular Reference to the Commonwealth of Australia,* L. J. Lawrence, ed., Eighth Commonwealth Mining and Metallurgical Congress Publications, vol. 2, Australasian Institute of Mining and Metallurgy, Parkville, Victoria, 1965, 365p.

THE SEDIMENTARY CONCEPT IN MINERAL EXPLORATION

By

HADDON F. KING[1]

INTRODUCTION

Deposits of both hydrothermal and sedimentary origin have been fully accepted for many years. Bedded iron, phosphate and manganese would be typical of those recognized as sedimentary whilst gold-quartz veins would be representative of the hydrothermal category. The present interest in sedimentary deposition of ore centres around the extension of the sedimentary concept to certain layered ore deposits in which there is strong conformability in attitude between the ore boundaries and the layering of the enclosing rocks.

A transgressive attitude as in Fig. 1(a) establishes the deposit as indubitably younger than the wall rocks. Owing to the common lack of unequivocal evidence on whether a conformable deposit as in Fig. 1(b) is a special case of (a) or is contemporaneous with the enclosing rocks, the hydrothermal theory originally designed for transgressive orebodies was (with some notable exceptions) generally accepted as extensible to conformable bodies until about 1950 when extensive detailed mapping of surface and mine exposures began to make the hydrothermal theory unacceptable to some people as an explanation of this type of ore occurrence.

A sedimentary concept of the origin of sulphide bodies is not new. The first inkling, known to this author, that "transgressive" and "conformable" attitudes were genetically significant may be inferred—if one is so inclined—from the writings of John Woodward (1665-1722) and sedimentary hypotheses have been maintained for certain deposits ever since. A full coverage of the subject would involve examination of the long history of ideas about the Kuperschiefer; the origin of the Rammelsberg deposits; the recent changes in ideas about the origin of Rio Tinto which has been worked and no doubt thought about since pre-Roman times; the development of sedimentary thinking in the Copperbelt of Northern Rhodesia and about the copper deposits of White Pine, Michigan; the interesting exploratory as well as geological effects of acceptance of a stratigraphic habit for the lead-zinc deposits of Bathurst, New Brunswick; and the acutely unresolved conflict of ideas about the origin of the gold deposits of the Rand.

In the differing views published on these subjects there is a suggestion that some of the authors have overlooked the fact that for any one deposit there can be only one mode of origin and consequently that where two opposing interpretations have been reached, one of them, at least, *must* be wrong. To those who are interested in ideas of ore genesis primarily for the use that can be made of them, change is not only acceptable, it is welcome if the change offers improvement in understanding of the problem and in efficacy of application of genetic concepts. It is from this point of view, and especially on the basis of Australian work and personal experience that the subject is discussed here.

SEDIMENTARY IDEAS AND THE HYDROTHERMAL THEORY

The earliest suggestion that a major Australian sulphide ore deposit might be sedimentary was made by Dr. E. D. Peters in 1893 in a report on the Mount Lyell copper deposit, discussed by J. W. Gregory (Gregory, 1905). Montgomery also thought the Mount Lyell deposits to be sedimentary (Gregory, 1905) and compared them to those of Rammelsberg and Rio Tinto. This is interesting in view of the ideas of the origin of Rammelsberg and of the latest thought that the pyrite deposits of Rio Tinto are exhalative-sedimentary (Williams, 1961). Gregory considered the sedimentary hypothesis untenable because of the evidence of molecule-by-molecule metasomatic replacement of schist by ore minerals and because the ore minerals are later than (i.e. occur in) adjacent faults.

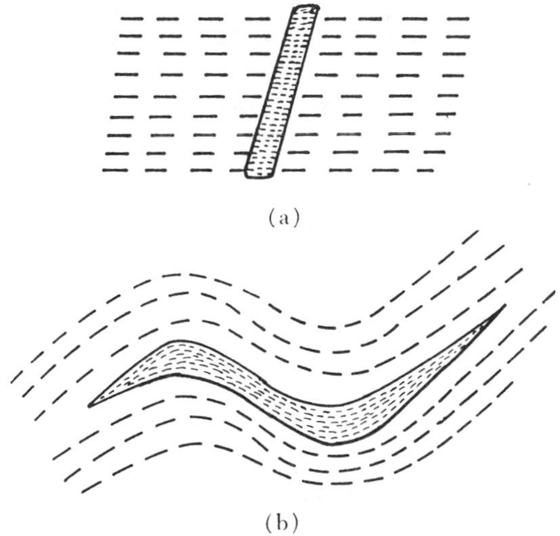

FIG. 1—General structural relationships of (a) discordant orebodies and (b) conformable orebodies.

[1] Conzinc Riotinto of Australia Limited.

Sixty years later, and three years after the Broken Hill lead-zinc deposit had been raised to the position of a classic example of hydrothermal replacement (Gustafson *et al.*, 1950) a small group of geologists of The Zinc Corporation Limited published their dissatisfaction with the hydrothermal explanation of this ore deposit and suggested that it might be sedimentary. It is interesting to look back at the ideas on ore genesis prevailing at that time.

The theory of hydrothermal-magmatic origin of ores arose in its modern form about 1910, although it should be noted that the concept of metasomatic replacement was employed by Gregory in 1903-05 without explanation, implying that it had already attained acceptance as a mode of formation of sulphide orebodies. By about 1950 metallic ore deposits in hard rock had become generally accepted as having been formed, beyond question, by the action of hot fluids (water, vapour, emanations) on solid rocks of the crust, the fluids being residuals in the last stage of crystallization of magmas (usually granitic) and being rich in metals which were deposited in receptive structures or layers principally by substitution of the ore minerals for pre-existing rock forming minerals. Because the presumed source-rock would, by definition, have given up its ore-making constituents in the process of consolidation, it was useless to look to it even if exposed (and its existence was usually conjectural) for signs that it had been the source of an ore deposit; because it had come to be accepted that a replacement deposit could be formed without leaving any trace of the entry or exit of the mineralizing solutions, absence of an observable channel of entry and of signs of the displaced material were regarded as normal. Because of these, and with the passage of time, the hydrothermal-magmatic theory developed a unique invulnerability. Source, mode of concentration, of transport and of emplacement did not need to be demonstrated, not even argued; they could all be taken for granted. Complete reliance on deep seated genetic agencies and on omnipotent fluids denied the enclosing rocks any influence in the ore-making process other than that of passive host. Hence the old Cornish saying "where she be, there she be".

A variation of this general theme of the passivity of the enclosing rocks attained wide acceptance during the two decades 1935-1955: strong genetic influence was assigned to structure (fractures, folds) and to "favourable beds", these being supposedly more amenable to replacement than less stressed environments or adjoining layers. This trend of thought, at its climax (which occurred in Australia about 1939) treated structure as the key to ore occurrence, and all other aspects of the ore geology as relatively unimportant in exploration.

Because mining geology in the English-speaking part of the world has been influenced far more, during the last thirty years, by North American thinking than by the geological ideas of the United Kingdom and the Continent, the attachment to hydrothermal-magmatic replacement and to structural theories (both developed principally in North America) was unaffected by European doubts about the general applicability of these theories. (For example, it was 1952 when this writer learned about Thomas Crook's *History of Theories of Ore Deposition* (1933). An interesting alternative to the hydrothermal theory was, however, proposed by J. S. Brown (1948) and his postulated source is very much like that being invoked today for exhalative-sedimentary ore deposits.) And so it was 1951 when doubts began to develop regarding the adequacy of conventional genetic theory as applied to Broken Hill (King and Thomson, 1953) (Burrell (1942) suggested that the principal gangue minerals were syngenetic but did not extend his argument to the sulphides); 1952 when an evolutionary concept linking conformable and transgressive ore was first mentioned in Australia (King, 1952); 1954 when an association between base metal deposits and volcanic-sedimentary sequences was postulated in eastern New South Wales (Stanton, 1954); and 1955 when an association between uranium occurrences and (a) black shales, or (b) algal limestones was recognized in the Northern Territory (Condon and Walpole, 1955). In all these places a suspicion had arisen that the ore constituents were not late arrivals introduced from some unknown deep-seated source but were contemporaneous with the enclosing rocks. Similar ideas on the genesis of the copper bodies in the Rhodesian Copper Belt began to take shape about 1949 (published in 1953-54 but verbal communication to C. L. Knight in 1950). *The Source Bed Concept* (Knight, 1957) expanded this thinking about specific deposits into a general hypothesis.

Thus we come to the crux of the problem. A late magmatic replacement deposit was by the theory of its origin denied any genetic relationship to its environment. Therefore, unless accompanied by wall rock alteration (conspicuous by its absence from some of the largest deposits) there could be no observable lithological criteria of ore occurrence and it had come to be accepted that even for some major deposits there was no criteria of ore occurrence other than ore itself. A sedimentary deposit being contemporaneous with and related to its environment, could however conceivably have some observable lithological environment which might offer criteria of ore occurrence.

THE TRANSITION

This stage was not reached at one bound. Even those who have been most closely identified with the development of ideas of sedimentary origin started out from a hydrothermal base and at one time either accepted—or would have accepted—conformable sulphide orebodies as hydrothermal.

For the writer and some of his colleagues, the transition to thinking that a sedimentary concept was preferable to the hydrothermal for certain orebodies, was brought about by an increasing awareness of the importance of stratigraphy in the form and, to some extent, the location of sulphide orebodies. It is not merely Broken Hill and Mount Isa that are conformable with the enclosing sediments, almost every sulphide deposit in layered rocks—other than transgressive veins—shows a strong stratigraphic influence and/or is associated with a distinctive rock type (or sequence) of sedimentary origin.

Thus as mapping revealed instance after instance of this stratigraphic influence, it seemed that conformability rather than being an unusual feature to be explained as attributable to the influence of "favourable beds" (as in Mount Isa and Broken Hill) was a normal condition of ore occurrence in layered rock.

This persistently stratiform character of the mineralization accentuated also the ordinary difficulties of fitting these conformable deposits into the hydrothermal theory. For example, "favourability"—never well understood—had been invoked already to explain stratigraphic persistence of mineralization over miles in length. It had now to be extended to tens of miles. Since mineralization was, in many instances, clearly related *spatially* to a time or a condition of the sedimentary sequence, favourability was necessarily a quality of the original sediment. In some instances, the mineralization was explicitly acknowledged to be the only observable criterion of favourability. Could "favourability" and "mineralization" (in the context of conformable ore deposits) be fully synonymous? If so, the extremely difficult problem of ore emplacement in "favourable" beds would disappear and the way would be open for a solution of the problem of concentration.

Additional encouragement for thinking of the "conformable" deposits as genetically different from "vein" deposits has been received from the studies of lead isotope distribution which have been in progress for some years (e.g. Stanton and Russell, 1959). The vein deposits, which are known to have migrated through crustal rocks, have isotope ratios different from those of conformable deposits. Whilst there is much still to be learnt about the significance of these ratios, the observed differences have led Stanton and Russell to conclude that there is a fundamental genetic difference between the character of the leads in the two types of deposits and that this may be attributable to the difference in geological history of the leads, the conformable lead having been shielded from crustal migration by its mode of origin.

It was, however, necessary to recognize that both hydrothermal and sedimentary hypotheses of origin were held by different people for certain deposits, even where there was substantial agreement as to the basic facts. For example, quoting from King and Thomson (1953) on Broken Hill: "There is therefore doubt of the hydrothermal mode of emplacement" (p. 570); "An alternative mode of emplacement . . . must therefore be considered, even sought" (p. 569); "The absence of evidence in support of a hydrothermal introduction of the ore into its present position suggests either that the ore was introduced in some other way or that the ore bearing beds were already orebearing before they were deformed" (p. 571); "Admission of certain constituents as syngenetic reduces the extent and number of components which must be assumed to have undergone a fantastically persistent segregation" (p. 568); "It is preferable to think of them (*the orebearing beds*) as having acquired these characteristics independently prior to being involved in the folding and metamorphism. Finally, Ramdohr's mineragraphic study lends support to the view that the ore was not introduced as it now is" (p. 572); and from King (1958): "It therefore seems appropriate to substitute for the hydrothermal-magmatic assumptions . . . two new assumptions as follow:

(1) The ore constituents were all deposited in the sediments at the time of formation. Sedimentation then provides a single answer to questions of source, transport, concentration and deposition.
(2) The present character of the orebodies is attributable to successive modification by changing conditions in the course of the long geological history of these Precambrian regions" (pp. 165-6). (See also Condon (1959) and Segnit (1961).)

By contrast, from Stillwell and Edwards on Broken Hill: ". . . the ore cannot have been recrystallized by high temperature metamorphism and its textures . . . must be primary features of deposition" (Stillwell and Edwards, 1956, p. 230); "the conception of the origin of the lode advanced by King et alia (1953) cannot be sustained; the interpretation of the Broken Hill ore textures by Ramdohr (1951, 1953) can be rejected and the view advanced by Burrell becomes highly improbable. The evidence of the dolerite dykes requires a return to the older conception of the Broken Hill lode as deposited by high temperature hydrothermal processes, subsequent to folding and metamorphism of the area" (Stillwell and Edwards, 1956, p. 231); "This evidence . . . removes all speculations which connect the lode formation with the metamorphism of the district" (Stillwell, 1959, p. 5).

Just as a new answer to a mathematical problem cannot be adopted safely until the reason for its differing from an old answer has been identified, so in this problem of ore genesis the nature of the "older conceptions" had to be examined. Critical reading of accounts of and arguments for the hydrothermal origin of certain conformable deposits disclosed that the original and depositional character of the ore minerals and textures had been either tacitly assumed[1] or too readily inferred (King and Thomson,

[1] This assumption is a consequence of the emphasis given in conventional genetic studies to the minerals present in ore deposits rather than to the elements. There is an encouraging trend now towards putting the emphasis on the *elements* and recognizing that the mineral suite in any deposit is merely that combination of the elements present which was best suited to the conditions prevailing at the time of (latest) crystallization. The same assumption, if made regarding sillimanite, would require that this common metamorphic mineral should also be a product of hydrothermal replacement.

1953, p. 570; King, 1958, p. 159; Stanton, 1960, pp. 69-70). Acceptance of the ore textures as depositional and original, alone, implies that

1. the host rocks were, prior to the ore-making event, devoid of mineralization;
2. every rock type containing ore minerals must have antedated the arrival of the ore constituents and, conversely, that the ore emplacement must be of later date than the youngest rock type containing ore minerals;
3. neglecting the possibility of open spaces, emplacement of the ore minerals *must* have been by replacement of pre-existing rock-forming minerals;
4. emplacement in certain stratigraphic layers could mean *only* that those layers were more amenable to replacement than others;
5. the ore constituents were relatively late arrivals in relation to other geological events; and
6. the country rocks could play only a receptive role.

Therefore this belief in the depositional and original character of the ore minerals and textures was an integral part of the hydrothermal theory. So long as it was present there was—within the bounds of genetic possibilities so far envisaged—no alternative to a conclusion of late replacement origin. If, however, these textures should be secondary, not primary, the principal objection to an early emplacement of the ore constituents would be removed.

The last decade has brought to light some reasons for not accepting the present character of ore minerals as original and depositional in *all* deposits. (For example, Garlick, 1953-54; Stanton, 1954, 1960; Baas Becking and Moore, 1961; Roberts, 1961; Nixon, 1961; Roberts, 1963.) As a result there is now scientific and experimental support for the earlier surmise that it was anomalous to deny to minerals in ore deposits the same opportunity for growth and change that has long been accepted for minerals in metamorphic rocks.

Once the possibility has been recognized that the present form of the ore minerals may not be original and depositional, the traditional studies of mineral textures (cf. Ramdohr (1951) which presents an interpretation of the mineral textures of the Broken Hill deposit differing from that of Stillwell) cease to have any genetic significance and the way is open for consideration of alternative modes of genesis.

THE SEDIMENTARY CONCEPT

In a genetic hypothesis applicable to conformable deposits, the features which most pressingly demand explanation and accommodation are the following.

1. A stratigraphic habit as precise in some instances, and as invariable under conditions of severe folding, as if the ore constituents had been laid down in the beds in which they were found.
2. Sharp concentration contrasts across the bedding direction but smaller concentration contrasts in the plane of the bedding within the limits of the deposits.
3. Marked and persistent differences in contained metals, in metal ratios and/or in gangue minerals between different layers of certain deposits.
4. Great extent (measured in miles or tens of miles) of the mineralized horizons in which the ore deposits are found.
5. Absence of any indication of a channel of entry of the ore constituents.
6. Absence of any wall rock alteration commensurate with the scale of the ore deposits and the great volume of solution customarily believed to be the means of transport of the ore constituents.
7. Lack of evidence of the material that would have been displaced in an origin by replacement.
8. Association of the ore minerals with iron or other minerals occurring in the same bodies, in the same layers or in adjoining layers in such a way that these minerals, *in a deposit which did not make ore*, would normally be accepted as sedimentary.
9. Certain types of deposits, e.g. fine-grained pyritic Pb-Zn, are known *only* as conformable bodies.

In short, these deposits resemble sedimentary layers in all respects except one—their content of valuable minerals—and one is led to the conclusion that it is their value as "ore" which has denied them dispassionate consideration as being also a layer of "rock". A genetic concept which would allow the ore to be merely one of the many rock types which make up the lithology of the area obviously offers a better fit to the available evidence.

If sedimentary, these deposits are the result of the processes that produced the enclosing rocks, or of events taking place at the same time. Therefore, for any ore deposit in layered rocks, the depositional events—of which the ore deposit is one unusual expression—may be reflected in the enclosing rocks.

What indications do these layered deposits, suspected of being syngenetic, provide of the environment in which they were formed?

1. The mineralized layers which contain the ore deposits commonly consist of a very widespread (often measured in miles) weak mineralization in a single stratigraphic horizon or in a number of horizons within a particular stratigraphic formation, with local areas of much greater than average concentration and thickness of the metallic mineralization.
2. The orebodies are so localized as to suggest either a nearby source, or concentration by post-depositional agencies. Since, however, the degree of concentration is great, even in some of the least disturbed environments, the emphasis is strongly on a local source. The thick-

ness and high degree of concentration of sulphides could result from stagnant depositional conditions with long periods during which no detrital material was deposited.

3. Especially in base metal deposits, an abundance of iron sulphide is common—usually as pyrite—but there are a few important lead-zinc deposits in which iron sulphides are sparse.
4. The enclosing rocks commonly range from shales through limey or dolomitic shales to limestone or dolomite, indicating a moderately shallow to moderately deep offshore sedimentary environment.
5. Carbon in the form of graphite is a common constituent of the ore-bearing rocks and this, together with the calcareous tendency, is indicative of organic activity.
6. Iron formations containing magnetite are also common, though not general, indicating that the base metal deposits are only one of the unusual expressions of the depositional events and that the environmental conditions varied or ranged from oxidizing to reducing.
7. In one lead-zinc-pyrite deposit (Love and Zimmerman, 1960) minute spherical forms in the pyrite have been interpreted as organic. Studies of the production of sulphides by sulphate reducing bacteria have suggested that the possibilities of this process are so wide that it is almost inconceivable that these sulphate reducing (anaerobic) bacteria should not have played a significant role in the formation of sulphide deposits. (It is worth remembering that, compared to the fantastic difficulty of visualizing replacement under "hard rock" conditions as required by the more complex conformable deposits, the depositional and immediate post-depositional stages represent a period when the *whole* of the mineralized bed was accessible for introduction of ore minerals by simple depositional processes if a suitable mechanism should exist. The existence of such a mechanism has long been accepted for iron, manganese, calcium and phosphorus. Baas Becking's early work (continued by K. L. Temple at the C.S.I.R.O. Division of Plant Industry, Canberra) has indicated that lead, zinc, silver and copper may be deposited biogenically and some other elements may now be added to this list.)
8. Volcanic rocks (especially acid to intermediate) are commonly represented in the environment. The relative proportions of volcanic and sedimentary rocks may, however, vary widely from almost all volcanic rocks with thin layers of sediments representing time breaks in the volcanic accumulation, to sedimentary sequences devoid or almost devoid of volcanics of any kind.

Thus the conformable sulphide ore deposits—which are nearly all Precambrian or Palaeozoic—are visualized as having been formed as shallow offshore or basin deposits under reducing conditions in places where there was both an abundant supply of metals and conditions which encouraged the proliferation of sulphate reducing bacteria and where, also, absence of detrital contributions permitted the development of considerable thicknesses of high concentrations of metallic sulphides. The most plausible source of constituents—both sulphur and metals—is volcanic, either as submarine vapours or by river transport of material derived from erosional breakdown of the extensive volcanic surfaces of the Precambrian or the Palaeozoic eras.

Inevitably, this hypothesis about certain Precambrian and Palaeozoic events—though it offers a better fit to the evidence available about conformable ore deposits than did the classical hydrothermal theory—likewise involves assumptions. There appear to be four.

1. Metals were available at the surface of the earth—subaerial or submarine—in the requisite quantities.
2. Metals could be deposited in a sedimentary environment and in the concentrations observed.
3. The absence of indubitable (younger) sediments having the composition of, say, the Broken Hill or Mount Isa *lodes* is attributable to the existence of special conditions during Precambrian or Palaeozoic times.
4. Observed ore textures could be developed *in situ* subsequent to initial deposition of the constituents.

On these the following comments may be offered.

1. The preferred volcanic source remains conjectural, but even at the present time the base metal content of the soil of mineralized areas is sufficient to contribute metals to the sea in quantities which are large by comparison with ore deposits.

   At Northampton, W.A., the lead and copper bodies (of hydrothermal type) are so closely associated with dolerite dykes as to suggest that both ore and dykes may have had a common volcanic origin. If volcanic activity can yield ore as well as silicate rocks (which certainly happens at times) it can yield the constituents of conformable deposits.

   The Oil & Gas Journal, April, 1963, reports the discovery of hot brine solutions at 5,000 feet depth near Niland in the Imperial Valley of California. The brine is described as "containing high quantities of silver, copper and other metals normally concentrated in ore deposits".
2. Deposition in a sedimentary environment is already an accepted mode of formation for some iron (oxide and sulphide), manganese and

phosphate deposits. The geobiological research of Baas Becking and K. L. Temple has revealed a mechanism for biogenic deposition of the sulphides of lead, zinc, copper and other metals. Concentration is conceived by the writer as being brought about chiefly by absence of detrital contributions during the formation of the ore layer. Lead-zinc content of the Mount Isa shales varies from one bed to the next in the same abrupt fashion as does the phosphorus content of the beds of the Phosphoria formation in the western United States.

3. The virtual restriction of major bedded deposits of iron, lead, zinc, copper, manganese and phosphate to Precambrian and to Palaeozoic rocks may be attributed possibly to a greater intensity of volcanic activity during those eras, and to the sedimentary conditions which produced the great thicknesses and extent of acknowledged sedimentary concentration (e.g. banded iron formations).

   These conditions did not exist (in the Pacific zone) during the Mesozoic where major bedded metal concentrations are notably absent. The recurrence of bedded sulphide mineralization at Laloki, New Guinea, in close association with Tertiary volcanics again tends to suggest a genetic link between vulcanism and layered sulphide deposits.

4. The knowledge that biogenic sulphides are deposited as minerals, identifiable by X-ray diffraction; that precipitated lead and zinc sulphides show crystal growth after three weeks at 200°C at normal pressure and at 100°C under pressure and that chalcopyrite is produced in a few hours at the polished surfaces of pyrrhotite and chalcocite in contact with cold water, demonstrates that the birth of the simple ore sulphides, their growth into crystals and the development of complex minerals may all be achieved without the assistance of high temperatures and, therefore, without the influence of molten rock. The ore minerals are not, therefore, necessarily the stable high temperature building units of ore deposits that they are supposed to be. (Thomas Crook (1933, p. 72) records Roman coins converted largely to chalcopyrite and old lead pipes to galena during prolonged burial.)

   With this knowledge, it appears reasonable to assume that deep burial which Precambrian and Palaeozoic rocks have suffered and the very long resulting exposures to elevated temperatures would be sufficient to produce the ore textures observed.

Biogenic, sedimentary and metamorphic processes thus provide a hypothesis for the formation of conformable sulphide deposits which allows significance to their large-scale features and lithological association, appears to meet most of the objections raised on the grounds of small-scale ore textures and can stand up to identification and examination of its basic assumptions.

As yet, however, this hypothesis can only be regarded as firm on these two points of sedimentary emplacement and metamorphic development of texture. Prior to emplacement, between emplacement and the development of a crystalline texture—and perhaps even later—there is a wide scope for processes and events which cannot yet be clearly visualized but which would probably include, in some cases at least, modification, concentration and migration in response to structural and diagenetic influences. What is the source of the constituents—sea water, erosion or submarine volcanic emanations? If non-volcanic, is a deposit the result of a single event of concentration or of more than one stage? If volcanic, how near to or far from the source does deposition occur? Is there a "zoning" of metals away from the source or from the centre of deposition? Have the ancient deposits retained all their original constituents? At this stage there are more questions than answers, but in studies like these, questions are a prerequisite of progress.

Despite the incompleteness of present ideas, so many ore deposits in layered rocks have features consistent with this sedimentary concept that when the still strong bias in favour of a magmatic-replacement origin has been sufficiently overcome or outgrown, the question of early or late emplacement is likely to be asked of every stratigraphically disposed mineral concentration.

In attempting a judgment on the durability of these ideas, the author is well aware that ideas on ore occurrence have been developing for four hundred years and that some of the most firmly held theories have, in time, become of interest as only the quaint beliefs of earlier days. He takes comfort, however, in the history of changes in ideas about the origin of other unusual, but not ore-bearing, rock layers of the Precambrian. For example

1. in Western Australia, in 1926, the dogma that all Archean rocks must be of igneous origin was so strong that F. L. Stillwell (1929) failed to overcome local insistence on a magmatic origin for a banded graphitic rock which he—on field and microscopic evidence—thought to be a sediment;
2. even the banded iron formations continued, until about 1933, to be regarded there as iron-rich shear zones in greenstone;
3. in Broken Hill, before 1892, all the country rocks were considered to have been originally igneous but, stage by stage, as mapping progressed, one rock type after another was accepted (some not without controversy, e.g. Stillwell, 1924) as having been originally sedimentary until today only dykes and obviously intrusive bodies are free from the suspicion of having been originally sedimentary;
4. up to 20 years ago any rock of granitic texture

and composition was almost invariably assigned an igneous intrusive origin. Today most people accept stratiform character in granitic rocks as indicative at least of a hybrid, if not simply sedimentary-metamorphic, origin.

The change from insistence on an igneous origin of layered rocks of unusual composition or texture to, at least, tolerance of the possibility of sedimentary origin has been one of the outstanding features of Precambrian geology of the last thirty years. Reasons for the earlier insistence on a magmatic origin were numerous but one of the principal reasons—advanced even in recent years in this context (Vernon, 1961)—is of special interest: the composition of the rock was inappropriate to, or did not correspond with, any known rock of a sedimentary origin. The significance and inherent weakness of this argument in relation to the genesis of conformable deposits is too obvious to require elaboration.

Looked at in this light, the concept of conformable ore deposits as sedimentary is not the sudden and jarring change of course that it has been represented to be. It is an extension of a well-established trend, yesterday affecting banded iron formations, graphitic layers and granulites, today involving conformable layers of lead, zinc and copper.

In *British Philosophy in the Mid-Century* (kindly lent by E. S. O'Driscoll in the context of this problem) Karl Popper observes that the criterion of whether a theory is scientific is not whether confirmations of it can be found but whether it can be falsified or refuted by "certain possible results of observation".

In the light of his earlier discussion of the hydrothermal theory of origin of conformable deposits, the writer cannot imagine any test that would refute such an origin. The concept of sedimentary origin of conformable deposits could, however, be refuted by finding deposits of comparable size, character and composition in transgressive relationships in the same environment. Since such a relationship would be entirely compatible with the hydrothermal-replacement theory, it must be considered a possible result of observation. The situation appears to justify an attempt to raise the sedimentary concept to the status of a scientific theory by predicting that *deposits* of the same type as the conformable bodies will not be found transgressing the rocks in which the conformable bodies occur.

TOWARDS AN ORE ENVIRONMENT

The suspicion that certain conformable deposits might be sedimentary was closely followed by, or in some instances the result of, the thought that such deposits might be genetically related to a special type of lithological environment. As a result of substantially independent thinking (though all knew of the work by the others), environmental relationships of this kind were discussed by King (1953), Knight (1953), Stanton (1954, 1955) and Condon and Walpole (1955). The rock types identified as associated with mineralization were iron formations, evaporite-like sequences of black shale and pyrite, limestone and volcanics and reef limestones, all of which are today regarded as associated in one way or another with the formation of sulphide deposits. It is of special interest to note, in relation to the following paragraphs, that Stanton writes (Stanton, 1955) "the association is clear only where it is viewed regionally; . . . the correlation (of the rocks and ore) has always been missed in the small scale studies of individual orebodies."

These were followed by *The Source Bed Concept* (Knight, 1957) which, on an even broader view, postulated a sedimentary source for the metals in a wide range of deposits in Africa, North America and Europe as well as Australia, and by *Notes on Ore Occurrences in Highly Metamorphosed Precambrian Rocks* (King, 1958) which suggested that the resemblances between certain ore deposits, their environmental lithology, and of the ore deposits to their environment indicated that the relationship was genetic, not accidental.

Environmental studies of Broken Hill (Thomas, 1960) have shown that all the unusual rock types—granite gneiss, amphibolites, granulite, banded iron formation and the lode, together with the shales (now sillimanite gneisses) which form the bulk of the country rock in the vicinity of the lodes—constitute a geological unit which is as distinctive in the district geology as the lode itself. This unit is twenty miles long and about one mile stratigraphic thickness. There seems now to be no room for doubt that the Broken Hill lodes belong to, rather than happen to appear in, this environment. There is a very strong suspicion that they are genetically related to the unusual rock types in their environment, and that this relationship explains why all the Broken Hill-type mineralization is found in the stratigraphic unit that contains the unusual rock types such as granite gneiss, amphibolite, banded iron formation, etc., and not elsewhere in the district.

In 1960 the Ontario Department of Mines published a report (*Geology of the Manitouwadge Area*, E. G. Pye,. 1957) on the Manitouwadge area in northwestern Ontario, including a geological map of the district. The rock types include granite gneiss, volcanics, shales and banded iron formation as well as acid and basic dykes. The mapping shows a crescentic area, about ten miles long, of shales and volcanics (including banded iron formation). The attitude of the foliation in the granite gneiss on both sides of the shale-volcanic belt is sympathetic with the attitude of the sediments and the sedimentary-volcanic unit thins out both east and west toward the limits of the map. In the legend, the granite gneisses on both sides of the shale-volcanic crescent are assigned the same age, younger than all the other rock types, dykes excepted.

The granite gneisses are interpreted as intrusive but are clearly of the kind that could have been originally sedimentary. Merely by varying the interpretation to this extent, and assigning the greater age to the granite gneiss that lies stratigraphically beneath the sedimentary-volcanic crescent, we reveal another distinctive environmental unit made up of the familiar rock types. It is within this that the Manitouwadge copper-zinc bodies have been found. It is interesting to note that the mapping appears to have been done without any thought of such a relationship. (Stanton and Russell included Manitouwadge as a conformable deposit in their paper on "Anomalous Leads" (Stanton and Russell, 1959).)

In these three regions we have then, for the first time, an inkling of why the base metal deposits occur where they do—they are genetically related to the enclosing rocks and are the result of the same processes that produced the banded iron formation and other unusual rock types. It seems likely that mapping may reveal that other deposits of this kind similarly occupy a place in distinctive environmental units. Possibly some of these units are so large that they have not yet been recognized.

The association of such deposits with observable lithological units provides a great deal of support for the idea that the ore deposits are genetically related to their environment. The units described are still too large to be prospected effectively by sub-surface methods but, for some, reasons can now be advanced (in terms of observable rock types) for these ore deposits occurring where they do. This raises a hope that, with further study and enlightenment, it may be possible to develop criteria, not merely of ore occurrence in this broad sense, but criteria that will discriminate between the kind of environment of mere mineralization and that of mineable ore.

The strong spatial association of volcanic rocks with conformable ore deposits has led to the conclusion that many of them are exhalative-sedimentary, with an implication of a nearby source of constituents. Strong regional fractures, probably deep-seated, have been used already to find conformable ore in the volcanic-sedimentary environment of Noranda and the concept that these structures could give rise to conformable deposits is likely to have the interesting result of identifying the feeder channels so obviously missing in hydrothermal explanations of conformable ore occurrences.

Fractures of this kind may well represent, in some areas, an essential feature of an ore environment. Where they do, the concept of sedimentary emplacement of the ore will have led to a firmer, rather than a weaker belief in the magmatic origin of the ore constituents.

CONCLUSION

From the mapping that has been done by mining companies and other organizations, from the close studies made by some mining companies of particular ore deposits and from related fundamental studies done by a few university and research people, there has emerged a concept of sedimentary emplacement of certain sulphide deposits which—if not entirely new—at least contains the germ of a new idea: a recognizable ore environment. This view of ore occurrence has raised the hope that, at last, some reasons expressible in terms of observable lithological characteristics may be found for sulphide orebodies occurring where they do and, having reached this stage, the next—already in mind in some areas—will be a definition of ore occurrence in terms of geological environment. Unexpectedly, the same relationships which are responsible for the denial of a hydrothermal origin for conformable deposits appear to be leading towards a re-affirmation of the magmatic origin of the constituents.

In the writer's view, this approach promises to lift ideas of ore occurrence out of the realm of miracles and put it, for the first time, where it belongs—in the scheme of natural events. In this view, ore occurrence should become not only understandable but, within the limits imposed by its comparative rarity, perhaps even predictable.

ACKNOWLEDGEMENTS

This paper is one result of a close study of a number of ore occurrences; of company-provided opportunities to see many others; of the searching examination of conventional theory which was stimulated by the strong opposition to the sedimentary concept; and of discussions with a great many geologists in and from four continents, not all of whom are sympathetic to the ideas expressed here. To all of those with whom he has seen and discussed ore deposits, the author feels a sincere indebtedness.

The author is also indebted to Miss F. McIver and Dr. R. L. Stanton for critical reading of an early draft of this paper. S. M. Richards, University of New England, kindly drew the writer's attention to the paper by J. W. Gregory.

REFERENCES

Amstutz, G. C., 1958. *The Genesis of the Mississippi Valley Type Deposits, U.S.A.*, Experentia, Vol. XIV/7, 1958, Basel.

Amstutz, G. C., 1959. Syngenetic zoning in ore deposits, *Proc. Geol. Assoc. Canada*, Vol. II, Dec. 1959.

Baas Becking, L. G. M., and Moore, D., 1961. Biogenic sulphides, *Econ. Geol.*, Vol. 56, No. 2, 1961.

Brown, J. S., 1950. *Ore Genesis: A Metallurgical Interpretation—an Alternative to the Hydrothermal Theory*, Murby & Co., London.

Burrell, H. C., 1942. Ore Types at Broken Hill, Australia, Ph.D. Thesis, Harvard University (unpublished).

Condon, M. A., 1959. Sedimentary structures in the metamorphic rocks and orebodies of Broken Hill, *Proc. Aust. Inst. Min. Met.*, 189, with Discussion, *Proc.* 191, 1959.

Condon, M. A., and Walpole, B. P., 1955. Sedimentary environment as a control of uranium mineralisation in the Katherine-Darwin region, Northern

Territory; Comm. Aust. Dept. National Div., Bur. Mineral Resources Geol. & Geophys. Rep. 4.

Crook, T., 1933. *History of the Theories of Ore Deposition*, Murby & Co., London.

Edwards, A. B., 1955. Cadmium in the Broken Hill Lode, *Proc. Aust. Inst. Min. Met.*, 176.

Edwards, A. B., 1956a. The present status of knowledge and theories of ore genesis, *Proc. Aust. Inst. Min. Met.*, No. 177.

Edwards, A. B., 1956b. Manganese and iron in Broken Hill sphalerites, *Proc. Aust. Inst. Min. Met.*, No. 180.

Edwards, A. B., 1957. Amphibolites from the Broken Hill district, *J. Geol. Soc. Aust.*, Vol. 5, Part 1.

Edwards, A. B., and Carlos, G. C., 1954. The selenium content of some Australian sulphide deposits, *Proc. Aust. Inst. Min. Met.*, No. 172.

Garlick, W. G., 1954. Reflections on prospecting and ore genesis in Northern Rhodesia, *Trans. Inst. Min. Met.*, London, 63 (1953-54).

Gregory, J. W., 1905. The Mount Lyell mining field, Tasmania with some account of the geology of other pyritic orebodies, *Proc. Aust. Inst. Min. Eng.*, Vol. X.

Gustafson, J. K., Burrell, H. C., and Garretty, M. D., 1952. Geology of the Broken Hill ore deposit, Broken Hill, N.S.W.; Bur. Min. Res. Bull. 20; also in *Bull. Geol. Soc. Amer.*, Vol. 61, No. 12, Part 1, Dec. 1950.

King, H. F., 1950. Geological structure and ore occurrence at Norseman, Western Australia, 1941, *Proc. Aust. Inst. Min. Met.*, Nos. 156-157.

King, H. F., 1952. Some thoughts on ore occurrence, Address Broken Hill Branch Aust. I.M.M. Bull. 154.

King, H. F., 1953. *Handbook—Australia and New Zealand*, Vol. VII (Fifth Empire Min. Met. Congr., Melbourne).

King, H. F., 1958. Notes on ore occurrence in highly metamorphosed Precambrian rocks, *Proc. Aust. Inst. Min. Met.* (F. L. Stillwell Anniv. Vol.).

King, H. F., and O'Driscoll, E. S., 1953. The Broken Hill lode in *The Geology of Australian Ore Deposits* (Fifth Empire Min. Met. Congr., Melbourne).

King, H. F., and Thomson, B. P., 1953. The geology of the Broken Hill district, in *The Geology of Australian Ore Deposits* (Fifth Empire Min. Met. Congr., Melbourne).

Knight, C. L., 1953. *Handbook—Australia and New Zealand* (Fifth Empire Min. Met. Congr., Melbourne).

Knight, C. L., 1957. Ore genesis—the source bed concept, *Econ. Geol.*, Vol. 52, No. 7.

Love, L. G., and Zimmerman, D. O., 1960. Bedded pyrite and micro-organisms from the Mount Isa shale, *Econ. Geol.*, Vol. 56, No. 5, August 1961.

Nixon, J. C., 1961. Recrystallization of lead and zinc sulphides, *Nature*, Vol. 192, No. 4802.

Popper, K. R., 1957. *British Philosophy in the Mid-Century*, Geo. Allen & Unwin Ltd., London.

Pye, E. G., 1960. Geology of the Manitouwadge area, Ontario Dept. Mines Vol. LXVI, Part 8 (1957) (66th Anniv. Report).

Ramdohr, P., 1951. The deposits of Broken Hill in New South Wales in the light of new geological knowledge and microscopic ore examination, *Heidelberger Beitrage Zur Mineral u. Petrog.*, Vol. 2, No. 4.

Roberts, W. M. B., 1961. Formation of chalcopyrite by reaction between chalcocite and pyrrhotite in cold solution, *Nature*, Vol. 191, No. 4788.

Roberts, W. M. B., 1963. The low temperature synthesis in aqueous solution of chalcopyrite and bornite, *Econ. Geol.*, Vol. 58, No. 1.

Segnit, E. R., 1961. Petrology of the zinc lode, New Broken Hill Consolidated, Broken Hill, New South Wales, *Proc. Aust. Inst. Min. Met.*, No. 199 (with discussion).

Stillwell, F. L., 1924. Differences of opinion in the recent geological report on Broken Hill, New South Wales, *Proc. Aust. Int. Min. Met.*, No. 53.

Stillwell, F. L., 1929. Geology and ore deposits of the Boulder Belt, Kalgoorlie, Geol. Surv. W.A. Bull. 9A.

Stillwell, F. L., 1959. Petrology of the Broken Hill lode and its bearing on ore genesis, *Proc. Aust. Inst. Min. Met.*, No. 190.

Stillwell, F. L., and Edwards, A. B., 1956. Uralite dolerite dykes in relation to the Broken Hill lode, *Proc. Aust. Inst. Min. Met.*, No. 178.

Stanton, R. L., 1954. Lower Palaeozoic mineralization and its environment near Bathurst, New South Wales. Ph.D. Thesis, Univ. Sydney.

Stanton, R. L., 1955. Genetic relationship between limestone volcanic rocks and certain ore deposits, *Aust. J. Science*, Vol. 17.

Stanton, R. L., 1960. General features of conformable "pyritic" orebodies, *Trans. Canadian Inst. Min. Met.*, Parts 1 and 2, Vol. 63.

Stanton, R. L., 1961. Geological theory and the search for ore, *Min. & Chem. Eng. Review*, Vol. 53, No. 7.

Stanton, R. L., and Russell, M. D., 1960. Anomalous leads and the emplacement of lead sulphide ores, *Econ. Geol.*, No. 54, 4, 588-607.

Thomas, W. N., 1961. Environmental aspects of ore occurrence at Broken Hill—a contribution to a symposium on syngenesis. A.N.Z.A.A.S. Conference, Brisbane, Queensland, May, 1961.

Thomson, B. P., 1953. *Handbook—Australia and New Zealand* (Fifth Empire Min. Met. Congr., Melbourne).

Thomson, B. P., 1959. Discussion on sedimentary structures at Broken Hill, *Proc. Aust. Inst. Min. Met.*, No. 191.

Vernon, R. H., 1961. Banded albite-rich rocks of the Broken Hill district, New South Wales. C.S.I.R.O. Miner. Inv. Tech. Paper No. 3.

Williams, D., 1961. Further reflections on the origin of the porphyries and ores of Rio Tinto, Spain, *Trans. Inst. Min. Met.* (London).

[*Editor's Note:* The source of the King, 1953; Knight, 1953; and Thomson, 1953 references is given incorrectly as the *Handbook*. The articles appear in the *Proceedings*, vol. VII.]

18

Reprinted from *Inst. Mining and Metallurgy Trans.* **82:**B107–B113 (1973)

Influence of crustal evolution on ore deposition*

Janet Watson B.Sc., Ph.D.
Department of Geology, Imperial College, London

The record of mineralization

The abundant evidence relating to the distribution of ore deposits in time and in space provides many indications that the dominant styles of mineralization have not remained constant through the successive stages of earth history. The long-term variations revealed by the record of mineralization attain a particular interest when they are set against the geological evidence of changes in the behaviour and architecture of the earth's crust as a whole which has emerged during the last couple of decades. The object of the present paper is to explore the possible relationships between long-term variations in the style of ore deposition and the evolution of the crust over a time-span of three to four billion years.

At the present day the greater part of the crust is relatively cool, strong and rigid, forming a number of stable plates which move as units in response to movements in the mantle. The plate boundaries, at which new crustal material is added and surplus material is consumed, are tectonically mobile, are characterized by high geothermal gradients and are mechanically weak. Similar contrasts between mobile belts and stable units—which include the cratonic regions of the continental crust—can be recognized at every stage of earth history throughout Phanerozoic and Proterozoic times. Rocks dating from the first billion years of earth history, however, are those characteristic of unstable regimes, and their predominance indicates that the greater part of the early crust was hot, weak and unstable. A major factor in the evolution of the crust appears to have been the emergence, from 2·8 b.y. onwards, of stable massifs in which geothermal gradients were relatively low, and the complementary restriction of many manifestations of tectonic mobility and igneous activity to well defined mobile belts.

It is against this background of crustal evolution that the records of ore deposition are considered. Although quantitative assessments on the lines pioneered by Pereira and Dixon[1] have not been made, nevertheless, the outlines of a pattern which makes sense in the context of crustal history emerge from the data.

Archaean stage (3·8–2·6 b.y.)

The first well documented stage of earth history, lasting well in excess of a billion years, was characterized by widespread mobility. A skin of granitic material broadly comparable with the continental crust of later eras had been formed in some places before the start of this stage. During the Archaean, basic and ultrabasic volcanic rocks, derived from partial melting of the mantle, were repeatedly erupted on to or between the primitive continental rafts and subsided under their own weight to form the infolded tracts commonly called greenstone belts. Granitic material mobilized from the preexisting basement, or differentiated from the mantle, formed buoyant domes and massifs, which tended to rise progressively between the greenstone belts. The hot residuum of the rising granites remained in the lower crust in the form of gneisses, granulites and other high-grade metamorphic rocks. Steep geothermal gradients and chronic instability promoted repeated deformation and metamorphism to such an extent that few rocks remained unmodified.

The occurrence of clastic sediments in the upper parts of many greenstone-belt successions shows that land masses of high relief were subject to erosion, especially near zones of volcanism. Nevertheless, it seems doubtful whether the early granitic crust had the strength to support large land areas standing high above the primitive ocean floors after the manner of the present continents.

The characteristic ore deposits appear to have derived their principal metals directly or by not more than one intervening stage from sources in the mantle, their formation being related to the two dominant geological processes of the early Archaean crust—basic magmatism and granite formation.

The oldest deposits of economic interest are those containing chromium, nickel or copper, which are associated with early basic, ultrabasic or anorthositic intrusions in the greenstone belts (Fig. 1). An unconfirmed age of 3·6 b.y. for the chromite-layered anorthosite of Fiskenaesset in southwest Greenland[2] suggests that ore deposition of this type began within a couple of hundred million years of the formation of the oldest known rocks.[3] At Selukwe in Rhodesia chromite deposits associated with an ultramafic complex predate granites of about 2·9 b.y. in age,[4] and nickel-bearing ores at Kambalda in Western Australia are associated with similar complexes probably little older than 2·7 b.y.

Chromite is the distinctive ore mineral of early layered anorthosites. Chromite or nickel and/or copper sulphides characterize early ultramafic bodies. All these minerals, once segregated from their parent magmas by igneous processes, seem to have remained essentially stable during subsequent phases of deformation and metamorphism. The chromite layers in the Fiskenaesset anorthosite, for example, retained their identity through repeated folding and metamorphism to granulite facies,[5] and chromite masses at Selukwe have similarly survived inversion and low-grade metamorphism.[6]

Manuscript first received by the Institution of Mining and Metallurgy on 29 January, 1973; revised manuscript received on 18 May, 1973. Paper published in August, 1973.

*The substance of this paper formed the Presidential Address to Section C at the Annual Meeting of the British Association held in Leicester in September, 1972.

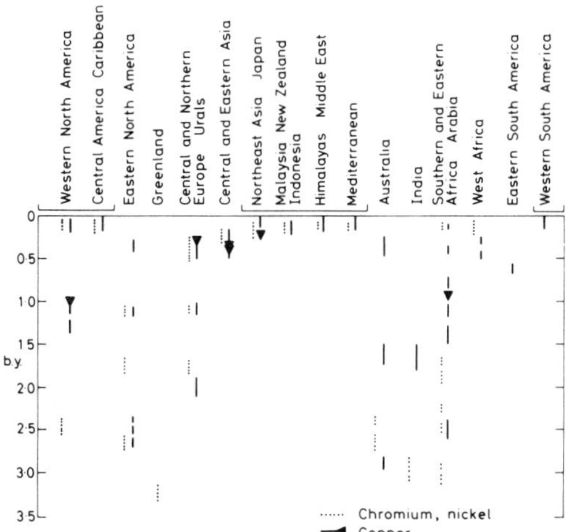

Fig. 1 Deposits characterized by chromium or nickel and copper plotted in terms of age and geographical location. Short vertical lines in Figs. 1–3 indicate approximate time of formation of deposits; inverted triangles show deposits formed mainly by agents of sedimentation and diagenesis; regions which belong to late Phanerozoic mobile belts are enclosed by square brackets

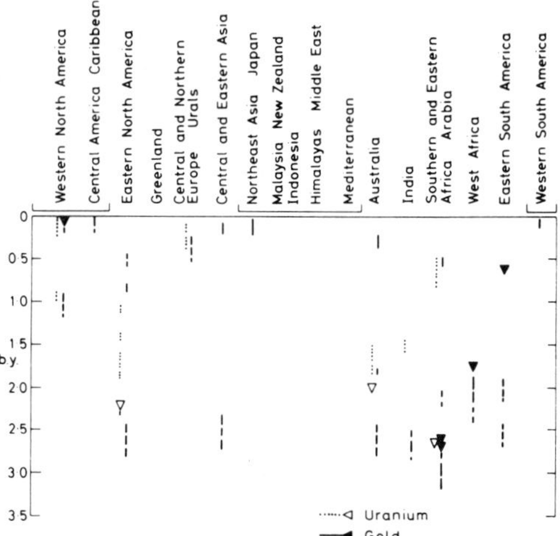

Fig. 2 Deposits characterized by gold and uranium

These very early deposits represent materials segregated from magmas originating in the mantle which took on essentially their present form immediately after their arrival in the crust. The influence of specifically crustal processes on their formation was minimal. It is not surprising, therefore, that rather similar concentrations of chromium, nickel, copper or platinum have been formed in association with mantle-derived intrusive complexes in a variety of crustal settings (Fig. 1). The relative importance of the Archaean deposits is mainly an expression of the scale of Archaean igneous activity fed from sources in the mantle.

The second major group of ore deposits characteristic of Archaean provinces is that dominated by gold, sometimes in association with silver, copper or zinc. Most of these deposits are located in greenstone belts, where they post-date much of the deformation and metamorphism and are generally related, spatially and in time, to granites emplaced in or on the margins of the belts. These relationships formerly led to the inference that gold and associated metals had been introduced into the crust by the rising granites. Recent studies in Canada[7] and in South Africa,[8] however, support MacGregor's early suggestion[9] that mineralization was a two-stage affair, the metals being initially introduced during the formation of greenstone belts but subsequently concentrated into workable deposits by the distinctively crustal processes connected with the mobilization and emplacement of granites.

In the Barberton area of South Africa and in the Archaean provinces of Rhodesia and Tanzania—where greenstone belts well in excess of 3 b.y. in age occur—mineralization began in association with granites dating from at least 2·9 b.y. In the Superior and Slave provinces of Canada and in Western Australia mineralization dates mainly from 2·7 to 2·6 b.y.[7,10,11] In West Africa, where Archaean styles of activity lingered to an exceptionally late date, the Eburnian granites, with which mineralization was connected, range in age down to about 1·8 b.y. Mineralization of the same general style spanned more than a billion years (Fig. 2), coming to an end when the granite–greenstone belt provinces were finally stabilized. The broad mobile belts formed during the succeeding Proterozoic stages are almost devoid of gold deposits of this type, even where mantle-derived basic volcanic rocks are associated with voluminous granites. There seem to have been no further major influxes of gold from the mantle until late Phanerozoic times (Fig. 2).

Aftermath of the Archaean stage

The Archaean stage, characterized by very widespread mobility, petered out over a long transitional period, beginning about 2·8 b.y. ago, during which portions of the continental crust cooled and became more stable. This period—the first episode involving stabilization of large parts of the crust—appears to have established the pattern of geological activities for the remainder of earth history.[12] Its long-term effects on mineralization were shown by the emergence of styles of activity adapted to the newly defined cratonic regimes. A more immediate effect was the accumulation of a unique series of deposits, formed by the recycling of material derived from deposits of the types already discussed.

In several continents the eroded remnants of Archaean granites and greenstone belts are overlain unconformably by sedimentary successions, which began to accumulate soon after the onset of stabilization. Gold and/or uranium is concentrated towards the base of these successions in deposits which, although they have suffered considerable diagenetic or hydrothermal changes, have some of the attributes of placers. Many geologists consider that the gold and uranium of the Witwatersrand System, by far the most spectacular of the ore-bearing formations, were initially introduced by rivers flowing into the Witwatersrand Basin in detrital grains, which were mechanically segregated on account of their high density. The uranium-bearing Huronian deposits at Blind River and other localities in Ontario, the northern Australian deposits at Rum Jungle and the gold-bearing Tarkwaian sediments of West Africa all have relationships broadly similar to those of the Witwatersrand reefs (see, for example, Roscoe[13] and Bowie[14]).

These concentrations, which together supply most of the world's current production of gold and a fair proportion of uranium, represent the first major suite of ore deposits to derive their substance from material recycled at the surface of the earth, largely by erosion and deposition. The oldest deposits in the Dominion Reef Formation of the Witwatersrand date from 2·8 to 2·7 b.y.; the youngest, in the Tarkwaian of West Africa, from about 1·7 b.y. The occurrence of detrital uraninite, pyrite and other minerals which oxidize readily on weathering is in accord with the current view that the Archaean atmosphere was almost devoid of oxygen.

Sources for the gold were available not far from the deposits, since the host successions flank granite–greenstone belt provinces characterized by gold mineralization of the kind already discussed. Uranium mineralization, however, is not recorded on a significant scale in these provinces (Fig. 2). It must be supposed that any detrital uranium minerals were derived from concentrations formed in parts of the granite–greenstone-belt provinces subsequently removed by erosion. This inference is supported by age-determinations of 3·1 b.y. for thorian uraninite and monazite from the Dominion Reef.[15] It seems reasonable to suppose that uranium in the Archaean crust would have settled only in sites near the surface. The relatively high proportions of uranium in the first-formed strata of the Witwatersrand Basin and the preponderance of gold at higher levels in the succession[8] are in keeping with such an inference. That upward migrations of uranium did take place during Archaean times is suggested independently by geochemical studies of the residual granulites and gneisses formed at deep levels in the Archaean crust. In Greenland, Scotland and Norway rocks of these types contain far less than the crustal average amounts of uranium and thorium and are thought to have been impoverished in these elements during high-grade metamorphism.[16,17]

The gold–uranium deposits formed in the aftermath of the Archaean stage were laid down with a comparatively short time-lag after the stabilization of the provinces from which their substance was derived; and although gold-bearing Archaean provinces have been exposed to erosion at many later times, there has been no further episode on a comparable scale. This time-link illustrates what appears to be a general point—major ore deposits which derived their substance from the erosion of mineralized areas have tended to accumulate comparatively soon after the initial episodes responsible for the concentration of the relevant metals in the upper crust. The appearance of numerous gold–uranium deposits soon after the stabilization of the first cratonic massifs suggests that stabilization had made possible the emergence of larger land masses than those which could be supported by the weak early crust, and, hence, had exposed the potential source areas to the forces of erosion.

Late Archaean and early Proterozoic stage (*ca* 2·6–1·6 b.y.)

When activity in the granite–greenstone belt provinces came to a close, systems of mobile belts characterized by persistent instability and high geothermal gradients had been differentiated from a number of cooler continental cratons, usually less than 1000 km in diameter. The igneous and sedimentary assemblages formed under the newly defined cratonic regimes include types which were of little importance in the older Archaean provinces, and a corresponding diversification marks the record of mineralization at this stage.

Basic igneous activity in cratons

The cratons formed towards the end of the Archaean stage were invaded, at various times between 2·6 and 1·9 b.y., by basic magmas derived from partial melting of the mantle. Both dyke swarms of regional extent and large layered intrusions were emplaced as a result of this magmatic activity. The event was comparable in scale with the cratonic basic magmatism of two later periods (*ca* 1·4–0·9 b.y. and *ca* 0·3–0·0 b.y.), which appears to have been genetically related to extension and fracturing of the cratons.

The principal ore deposits associated with cratonic basic magmatism of late Archaean and earliest Proterozoic times are the deposits of chromium, platinum, nickel and iron–titanium, which occur as magmatic segregations in the large differentiated intrusions. The Great Dyke of Rhodesia (*ca* 2·6 b.y.), the Bushveld complex of South Africa (*ca* 1·95 b.y.) and the Petsamo complex in the north of the Baltic shield are among the hosts of these segregations (Fig. 1).

It is worth mentioning that copper mineralization, which was a feature of the two younger basic episodes in the cratons, is of little importance, perhaps because few plateau-lavas (other than the Ventersdorp lavas of the Witwatersrand Basin) have survived.

The Sudbury intrusive complex, which lies close to the eastern border of the principal Archaean craton in Canada, and which is dated at about 1·7 b.y., seems to have no precise analogues among intrusions of other ages. Dietz' invocation of meteor impact[18] and associated suggestions of extraterrestrial sources for the nickel sulphide ores have gained many adherents, and although such a *deus ex machina* seems to raise more problems than it solves, it is difficult to find a place for the complex in the context of crustal evolution.

Kimberlites in cratons

The distinctive pipes and dykes of kimberlite which are the primary sources of diamonds are concentrated almost exclusively in cratonic areas of the continental crust. There seems to be general agreement that kimberlite is, at least in part, derived from the mantle well below the levels from which most basaltic partial melts are generated—the formation of diamonds at a minimum depth of some 200 km being favoured on experimental grounds by some authors.[19] The transmission to the crust of the minute mobile fractions from which kimberlite is derived seems to have been most feasible when and where the upper levels of the mantle were too cool to permit large-scale partial melting, and the appearance of kimberlites, therefore, suggests the existence of a low geothermal gradient. With these considerations in mind, it is perhaps surprising that the first diamondiferous kimberlites made their appearance so early. The kimberlite which supplies Premier mine near Pretoria in South Africa is firmly dated at more than 1·2 b.y., and is probably as much as 1·7 b.y. in age.[20] The occurrence of detrital diamonds in the Roraima Formation of the Guyana shield, which was deposited more than 1·7 b.y. ago, suggests that kimberlites were exposed to erosion in South America even before this date. Cratonic carbonatite–alkaline complexes make their appearance in the geological record at about 1·7 b.y.

Parts of the crust and mantle had, therefore, cooled sufficiently to allow the generation of diamondiferous kimberlites within a billion years of the stabilization of the first cratons. The vast majority of known kimberlites, however, were emplaced during the late Palaeozoic and Mesozoic events connected with the disruption of Laurasia and Gondwanaland. Despite the enormous time-gap of 1·5 b.y. between the first kimberlites and the main swarms, most Phanerozoic kimberlites, especially the diamondiferous ones, are crowded into the Archaean massifs, which represent the first cratons to stabilize.[21] Their preference for long-established cratons, such as those of Tanzania and the Transvaal in Africa and the Anabar massif in Siberia, emphasizes the special characters of these first-formed stable massifs; that such special characters may be traced back to abnormalities in the underlying mantle is suggested not only by the style of igneous activity in them but also by the fact that they tend to form swells in the present continental structure.

Sedimentary deposits of cratons

The formation of relatively strong continental cratons, it was suggested above, facilitated the emergence of moderate-size land masses in which crystalline rocks were soon unroofed by erosion. The metals released by this means became available for recycling at the earth's surface at about the time at which stable environments favourable to the processes of sedimentary differentiation became more extensive. It was in this setting that the gold–uranium deposits already referred to were formed.

The unique banded iron formations of the Lake Superior and Labrador regions of North America, the Hamersley Range of Western Australia, the Ukraine of eastern Europe and several other localities represent segregations formed by chemical or biochemical means. The bulk of these formations consists of rhythmically banded cherts carrying magnetite or hematite. Terrigenous detritus is almost lacking and the purity of the sediments, together with their remarkable lateral consistency,[22] indicate that they accumulated in tranquil conditions.

Banded iron formations of this type were deposited over a short time-range: they came in at about 2·5 b.y., preceded only by thinner jaspery iron formations in certain Archaean volcanic assemblages, and disappeared at about 1·9 b.y. Their limited time-span has been attributed to the effects of organic evolution.[23] The precipitation of iron oxides is seen as a side-effect of the release of oxygen during photosynthesis by primitive organisms incapable of tolerating free oxygen. The subsequent evolution of organisms with a more sophisticated metabolism removed the need for iron as an oxygen acceptor and put an end to the precipitation of banded iron formations. The build-up of oxygen in the atmosphere after 2·0 b.y. is attested by the appearance of red beds and evaporites in the stratigraphical record from about 1·7 b.y. onward.

Deposits of mobile belts

The ore deposits formed in mobile provinces during the late Archaean and early Proterozoic stage were conspicuously different from those of the old granite–greenstone belt terrains. Provinces which consist mainly of regenerated Archaean material tend to be poorly supplied with contemporaneous mineral deposits. The Ubendian belt of East Africa, the Limpopo belt of southern Africa and the Churchill province of Canada, all located at the margins of gold-bearing Archaean terrains and containing reworked greenstone belts, carry only minor amounts of gold. Such gold as survived (for example, at Flin-Flon) in the Churchill province is generally located in areas which largely escaped the effects of regeneration.[24]

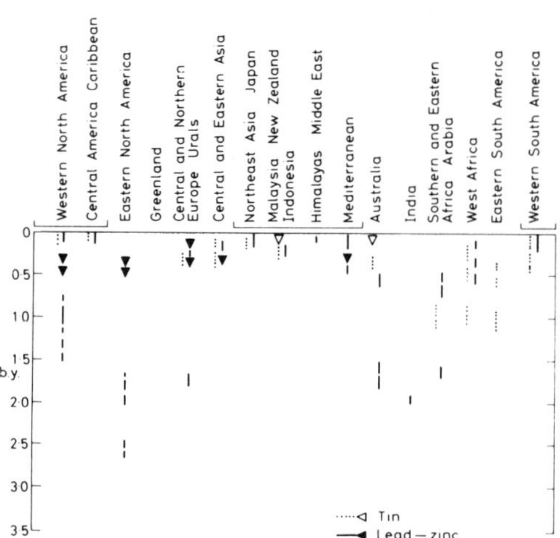

Fig. 3 Deposits characterized by lead and/or zinc and tin

Reworking at the level exposed appears to have dispersed earlier concentrations.

In provinces which incorporate early Proterozoic cover successions, copper, lead, zinc and silver are important.

The copper sulphide ores of the Svecofennide belt in the Baltic shield are associated with, though not always located in, slightly metamorphosed basic volcanic rocks. The lead–zinc (copper) ores of Mount Isa and Broken Hill in Australia also lie in the vicinity of supracrustal groups that include volcanic rocks. These deposits, dating from 1·8 to 1·6 b.y., are of kinds scarcely represented during the Archaean stages of mineralization (Fig. 3). Zinc is indeed an important component of ores in the Abitibi greenstone belt (2·6 b.y.), but the metal is seldom prominent in older greenstone belts, and where it does occur it is rarely associated with significant amounts of lead. The delayed appearance of the typical lead–zinc association may be attributed to the scarcity, during the early stage, of the relatively cool and stable environments which provided favoured sites.

It seems probable that the mobile Archaean crust simply proved inhospitable to mineralization of this type. The many small hydrothermal uranium deposits in the Churchill province of Canada (*ca* 1·8 b.y.) also suggest that conditions suitable for the fixing of fugitive metals had been attained in early Proterozoic times. That such elements shifted repeatedly to new sites is shown by the enormous range of ages obtained from ore minerals in areas of uranium mineralization.[14]

Later Proterozoic and Palaeozoic stages (1·6–0·3 b.y.)

After a terminal phase of abundant granite production many of the early Proterozoic mobile belts became quiescent about 1·8 b.y. ago. The stabilized parts of these belts were welded to older massifs to produce the larger cratons which seem to have dominated the continental crust in mid-Proterozoic times. Extension and fracturing associated with the second major episode of cratonic basic magmatism took place in these cratons over the period 1·4–0·9 b.y. and were accompanied by the development of a new network of mobile belts. The pattern of structures blocked out at this time provided the tectonic framework within which the Atlantic and Indian Oceans were subsequently developed.

The later Proterozoic stage of crustal development saw not only the initiation of a global arrangement of structures recognizably, though remotely, related to the present-day arrangement but also the beginnings of styles of mineralization with a decidedly modern aspect. It will be convenient to concentrate on these new developments rather than to attempt a general survey of mineralization to late Palaeozoic times. Such deposits as the copper ores associated with volcanic rocks in the Caledonian mobile belt of Scandinavia have so much in common with those formed in similar settings in earlier times that comment is superfluous.

Beginnings of tin mineralization

Although cassiterite occurs as an accessory mineral in pegmatites dated at 2·6–2·7 b.y. in West Africa and Rhodesia, forms detrital grains in the Dominion Reef Formation of South Africa (2·8–2·7 b.y.) and is associated with the Bushveld granites (*ca* 1·9 b.y.), tin mineralization of economic importance was largely confined to younger periods (Fig. 3). About a billion years ago mineralization in two rather different settings marked the widespread incoming of tin, tungsten and their associates to the record. The Karagwe–Ankole or Kibaran mobile belt of Africa, which reached the terminal stages of orogenic activity at this time, was characterized by abundant granites and by metamorphism of a low-pressure facies series usually attributed to the operation of a high geothermal gradient. Tin–tungsten deposits are associated with late orogenic (0·9 b.y.) granites at various localities from Uganda to South West Africa. In the Rondônia district of western Brazil[25,26] tin deposits are clustered near a group of high-level anorogenic granites dated at about 0·9 b.y. and emplaced in a basement of granite and gneiss.

These tin-bearing assemblages share many of the characteristic features of younger tin deposits. The Karagwe–Ankole belt, like the tin-bearing Hercynian belt of western Europe,[27,28] was a zone in which a steep geothermal gradient developed; it was not subjected to a major influx of mantle-derived igneous material, and some, at least, of the mineralized granites appear to represent remobilized basement granites. The granites of Rondônia have much in common with the Mesozoic Younger Granites of Nigeria, which are strongly fractionated high-temperature intrusions emplaced in a granitic or gneissose basement. These relationships render it unlikely that the appearance of concentrations of tin, tungsten and allied metals about a billion years ago was connected directly with the influx of material from the mantle.

The well known evidence of repeated mineralization in tin provinces such as those of Nigeria and western South America shows that later phases were commonly richer than the earlier.[29,30] As Schuiling has emphasized in another connexion,[31] rich deposits tend to occur where fields of more than one age overlap. With the remarkably late start to effective tin mineralization, this evidence suggests that recycling within the crust was an inefficient process, which worked best in terrains previously enriched in the metals.

Cratonic basic magmatism and copper mineralization

The exceptionally widespread emplacement of basic dykes and of larger differentiated basic intrusions which took place between about 1·4 and 0·9 b.y. was associated with the extrusion of plateau basalts and, more locally, with rifting of the crust. Magmatic ore deposits of various kinds are associated with the mantle-derived intrusive bodies—for example, the Fe–Ti deposits of the Duluth intrusion in the northern U.S.A. The plateau lavas to which the regional dyke swarms served as feeders are associated with copper deposits of varying importance: those of the Keweenawan lavas of Lake Superior, located mainly in permeable amygdular or conglomeratic horizons, are a major source of the metal, those of Arctic Canada (Coppermine River lavas) and of East Africa (the Bukoban lavas and their equivalents)[32] are scarcely economic. These occurrences, together with the fact that related dykes extend far beyond the present limits of the lava plateau, suggest that copper was transferred to the upper parts of the crust from sources in the mantle on a considerable scale during the period 1·4–0·9 b.y.

It can hardly be a coincidence that copper deposits of sedimentary origin enter the geological record in force at about the same time, occurring in Africa, in Siberia and in North America (Fig. 1). The cupriferous formations from which the huge ore deposits of Zambia, Katanga and the adjacent territories were formed lie in the lower part of the Katangan System, whose basal members[33] represent a molasse dating from the terminal stages of the Karagwe–Ankolean orogeny at or soon after 0·9 b.y. These formations fringe the southern border of the large late Proterozoic craton on which the Bukoban volcanics were extruded about 0·9 b.y. ago. In North America cupriferous sediments of little economic importance characterize horizons in the Proterozoic Belt Supergroup deposited at the western fringe of the craton on which the Keweenawan lavas were extruded between 1·2 and 1·0 b.y. ago.[34] On the assumption that the early deposits of basins fringing the late Proterozoic cratons were derived largely from erosion of the cratons themselves, it seems reasonable to suppose that the copper they contain was directly or indirectly supplied by the recently erupted lavas. Once again, recycling of a metal by surface agencies followed closely the delivery of this metal to the upper parts of the crust.

Lead–zinc in cratonic environments

Towards the end of the Palaeozoic era, and to some extent from Cambrian times onward, lead–zinc concentrations began to appear in the sedimentary cover successions of the cratons, most notably in North America, where the huge deposits of Mississippi Valley type accumulated (Fig. 3).

Sedimentation and diagenesis appear to have been responsible for the initial concentration of lead and zinc in these environments, though substantial redistribution by circulating fluids preceded the formation of the ore deposits themselves. The richest concentrations are in the North American craton, which was flanked on the west by the terrain of Proterozoic lead–zinc mineralization[35] now incorporated in the Cordilleran orogenic belt. The preferred host rocks in the cratonic sequences—limestones, dolomites and associated evaporites—are types whose abundance increased notably over late Proterozoic times with the establishment of an oxygen-rich atmosphere and evolutionary explosion of the Metazoa. The concentration in a sedimentary environment of lead and zinc supplied to the cratonic basin by the erosion of a mineralized terrain was, no doubt, favoured by the abundance of the appropriate host rocks in the Palaeozoic succession. It is interesting that favourable successions in cratonic areas, such as western Europe and northwest Africa, which were not flanked by zones of extensive mineralization, carry smaller concentrations (as in Britain) or are barren.

The latest stage (<0·3 b.y.)

The latest stage of crustal history has been dominated by the fragmentation of large continental cratons and the consequent growth of the Atlantic and Indian Oceans. Disruption and stretching of the cratons were accompanied by the eruption of plateau basalts and the emplacement of mantle-derived dyke swarms and differentiated intrusions, continuing to the present day. Deposits of copper, nickel and platinum—for example, in the Triassic Siberian traps—are broadly comparable with those associated with the products of earlier cratonic basic magmatism (Fig. 1). The initiation of major fracture systems in the cratons coincided in time with the emplacement of kimberlites, especially in the Archaean portions of the cratons. This event, although not new in kind, was on a quite different scale from previous phases of kimberlite emplacement. It was responsible for the influx of most of the world's diamonds and was followed after a short time-lag by the formation of diamondiferous placers, such as those of South West Africa.

In the mobile belts at the junctions of advancing crustal plates a remarkable range of ore deposits is displayed. Some are evidently derived from sources in the mantle, directly or at only one remove. Most of the characteristic deposits associated with the ultrabasic–basic ophiolite suite—massive sulphide copper ores, such as those of Cyprus, and nickel ores of such regions as New Caledonia—fall in this category. Mineralization related to the ophiolitic assemblages—regarded by many geologists[36] as fragments of oceanic crust produced from sources in the mantle at mid-oceanic ridges and emplaced more or less passively within orogenic fold belts—had a good deal in common with the initial mineralization in the Archaean greenstone belts and was on a scale which had not been attained since Archaean times (Fig. 1). The Tertiary gold–sulphide deposits have more complex relationships not unlike those of the porphyry coppers. Both were connected spatially and in time with granites or high-level, acid–intermediate intrusions emplaced in island arcs or mountain belts above active Benioff zones. The metals themselves may have originated in the mantle-derived, oceanic crustal plates overridden at Benioff zones, the acid magmatism serving mainly as an agent of transport and segregation.[37] In style and scale the gold mineralization bears some resemblance to that of the Archaean provinces (Fig. 2).

In other respects mineralization in the mobile belts reflects crustal influences very different from those which prevailed in Archaean times. The diversity of metals represented reflects the maturity of the continental crust. In the Cordilleran and Andean sectors alone copper, lead, zinc, silver, gold, uranium, tin and mercury are all of major importance at some locality. Evidence that the crust was already well stocked with many of these metals makes it natural to infer that recycling of metals by fluids circulating in the upper crust played a major part in mineralization. Such cannibalistic activity had no real counterpart in the immature Archaean crust. Finally, the variety of metals represented is increased by the short time-lapse since the final stages of mineralization in that near-surface deposits of fugitive elements, such as arsenic, antimony, and mercury, have not yet been dispersed by erosion. If precedent is anything to go by, the metals driven up in the mobile belts during Tertiary orogenic phases should be transferred over the next hundred million years to concentrations in cratonic sedimentary environments.

Retrospect

When examined in the light of the present geological situation, the history of mineralization falls into perspective (Fig. 4). Two principal themes may be distinguished—that connected with the transfer of metals from the mantle to the crust and that connected with the redistribution of metals within the crust itself.

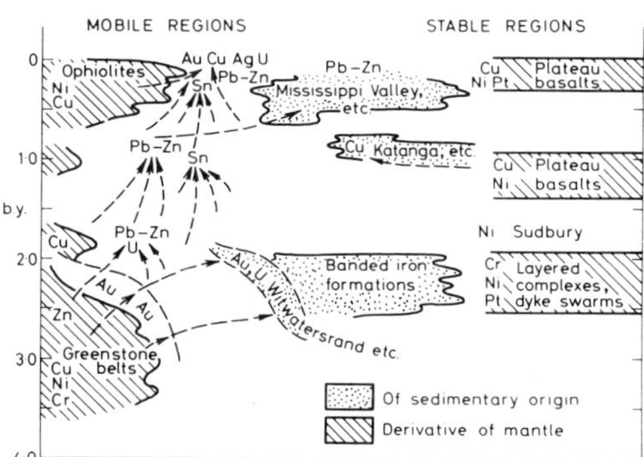

Fig. 4 Summary diagram illustrating possible relationships of ore deposits formed during successive stages of crustal history

Although the most consistent styles of mineralization have been those related to the influx of metals from sources in the mantle, these processes have not continued at a uniform rate. In mobile belts a long Proterozoic period of relative quiescence intervened between the abundant Archaean magmatism of the greenstone belts and widespread formation of Phanerozoic ophiolite assemblages. The accessions of copper, nickel, gold and other metals indirectly or directly associated with the Phanerozoic phases of ophiolite formation were on a scale unmatched since Archaean times; these contrasts suggest that the global tectonics of Proterozoic eras conformed to patterns unlike that of the Phanerozoic model of plate tectonics. In the cratons the major phases of basic magmatism and attendant mineralization coincided with periods of fracturing and crustal extension, the most recent being linked also with continental drift and sea-floor spreading. The emplacement of diamondiferous kimberlites and of alkaline complexes with rare-earth concentrations took place over much the same periods.

The deposits mustered by crustal agencies, more particularly by agencies working at the surface, show conspicuous variations through times which may be attributable to several causes. Deposits of the first billion years show deficiencies which reflect both the immaturity and the instability of the crust. The scarcity of tin and its associates may be put down to the fact that the preliminary recycling necessary for the concentration of these metals had not been completed. The scarcity of lead and uranium seems more likely to have stemmed from the perennial crustal instability which promoted the dispersal rather than the concentration of fugitive elements. The increased importance of uranium and lead–zinc mineralization during the stages which followed the evolution of the first continental cratons illustrates the importance of the appearance of cooler and more stable environments where concentrations of mobile elements might escape subsequent dispersal. The observed changes in tectonic regime appear to provide a more acceptable explanation of variations in the style of mineralization within the crust than do the variations in depth of erosion favoured by Laznicka.[38]

The first suites of ore deposits formed mainly by agents of sedimentation and diagenesis followed rapidly on the emergence of the first land masses in which large mineralized terrains were exposed to erosion. From 2·8 b.y. onward metals which had been transferred to the upper crust by igneous or hydrothermal activity tended to find their way after a short time-lag into the adjacent sedimentary basins. Such metals were concentrated in ore deposits mainly in cratonic environments, which favoured the sorting and reworking of sedimentary materials. The deposits so formed show long-term variations directly or indirectly related to the effects of organic evolution. Some unusual features of the Archaean deposits characterized by gold and/or uranium were connected with anoxygenic conditions of weathering. The early Proterozoic banded iron formations record biologically controlled processes of oxidation; and the majority of younger lead–zinc deposits appear to have depended for their formation on the presence of host rocks formed only in an oxygenic atmosphere. The most recent of all surface concentrations of metals—the dumps of our industrial cities—have been formed by the activities of the newly evolved human species.

Table 1 Incoming of new styles of mineralization

b.y.	
0·0	First deposits formed by human agencies
0·5	First major Pb–Zn deposits of Mississippi Valley type
1·0	First major sedimentary copper ores
1·5	First important tin mineralization
	First evaporites
	First red beds
2·0	First diamonds
	First major banded iron formations
2·5	First stable cratons
	First deposits of cratonic layered igneous complexes (Cr, Pt)
3·0	First ores formed by sedimentary processes (Au, U)
3·5	First gold mineralization in greenstone belts
	First deposits of layered igneous complexes (Cr)

The record of mineralization suggests that a substantial proportion of the metals now fixed in ore deposits had entered the crust well before the end of the Archaean stage. Since these metals have been repeatedly recycled in or at the surface of the crust, the possibility arises that they may have been selectively concentrated in certain crustal environments. It seems probable that such mobile metals as uranium, lead, zinc and mercury, recent concentrations of which are located predominantly in the upper and outer parts of mobile belts or on the cratons, are now recycled mainly by agents operating in the upper crust and are returned to depth only where the collision of crustal plates has forced rock masses bodily down towards the mantle. Such a restriction in the scope of recycling could result in a long-term and large-scale geochemical differentiation in the crust. As Table 1 suggests, there is no reason to suppose that the evolution of mineralizing processes has yet come to an end.

References

1. **Pereira J. and Dixon C. J.** Evolutionary trends in ore deposition. *Trans. Instn Min. Metall.*, **74**, 1965, 505–27.
2. **Evenson M. M. Murthy V. R. and Windley B. F.** In preparation.
3. **Black L. P. *et al.*** Isotopic dating of very early Precambrian amphibolite facies gneisses from the Gothaab district, west Greenland. *Earth planet. Sci. Lett.*, **12**, 1971, 245–59.
4. **Wilson J. F.** The Rhodesian Archaean craton—an essay in cratonic evolution. *Phil. Trans. R. Soc.*, **273A**, 1973, 389–411.
5. **Windley B. F.** Anorthosites of southern west Greenland. *Mem. Am. Ass. Petrol. Geol.*, no. 12, 1969, 899–915.
6. **Stowe C. W.** Intersecting fold trends in the Rhodesian basement complex south and west of Selukwe. *Geol. Soc. S. Afr. Annexure to (Trans.)*, **71**, 1968, 53–77.
7. **Goodwin A. M.** Metallogenic patterns and evolution of the Canadian shield. *Spec. Publs geol. Soc. Aust.*, **3**, 1971, 157–74.
8. **Viljoen R. P. Saager R. and Viljoen M. J.** Some thoughts on the origin and processes responsible for the concentration of gold in the early Precambrian of Southern Africa. *Mineral. Deposita*, **5**, 1970, 164–80.
9. **MacGregor A. M.** The primary source of gold. *S. Afr. J. Sci.*, **47**, 1951, 157–61.
10. **Green D. C. Baadsgaard H. and Cumming G. L.** Geochronology of the Yellowknife area, Northwest Territories, Canada. *Can. J. Earth Sci.*, **5**, 1968, 725–35.
11. **Arriens P. A.** The Archaean geochronology of Australia. *Spec. Publs geol. Soc. Aust.*, **3**, 1971, 11–24.
12. **Sutton J.** Some developments in the crust. *Spec. Publs geol. Soc. Aust.*, **3**, 1971, 1–10.
13. **Roscoe S. M.** Huronian rocks and uraniferous conglomerates in the Canadian shield. *Pap. geol. Surv. Can.* 68–40, 1969, 205 p.
14. **Bowie S. H. U.** Some geological concepts for consideration in the search for uranium provinces and major uranium deposits. In *Uranium exploration geology* (Vienna: International Atomic Energy Agency, 1970), 285–98.
15. **Nicolaysen L. O. Burger A. J. and Liebenberg W. R.** Evidence for the extreme age of certain minerals from the Dominion Reef conglomerates and underlying granite in the western Transvaal. *Geochim. cosmochim. Acta*, **26**, 1962, 15–23.
16. **Heier K. S.** Geochemistry of granulite facies rocks and problems of their origin. *Phil. Trans. R. Soc.*, **273A**, 1973, 429–42.
17. **Lambert, I. B.** The composition and evolution of the deep continental crust. *Spec. Publs geol. Soc. Aust.*, **3**, 1971, 419–28.
18. **Dietz R. S.** Sudbury structure as an astrobleme. *J. Geol.*, **72**, 1964, 412–34.
19. **Kennedy G. C. and Nordlie B. E.** The genesis of diamond deposits. *Econ. Geol.*, **63**, 1968, 495–503.
20. **Allsopp H. L. Burger A. J. and Zyl C. van.** A minimum age for the Premier kimberlite pipe yielded by biotite Rb–Sr measurements with related galena isotopic data. *Earth planet. Sci. Lett.*, **3**, 1967, 161–6.
21. **Clifford T. N.** Tectono-metallogenetic units and metallogenic provinces of Africa. *Earth planet. Sci. Lett.*, **1**, 1966, 421–34.
22. **Trendall A. F. and Blockley J. G.** The iron formations of the Precambrian Hamersley Group, Western Australia. *Bull. geol. Surv. West. Aust.* 119, 1970, 366 p.
23. **Cloud P. E. Jr.** Atmospheric and hydrospheric evolution on the primitive earth. *Science, N.Y.*, **160**, 1968, 729–36.
24. **Davidson A.** The Churchill province. In *Variations in tectonic styles in Canada* **R. A. Price and R. J. W. Douglas eds.** (Toronto: Geological Association of Canada, 1972), 381–433. (*Spec. Pap. geol. Ass. Can.* 11)
25. **Kloosterman J. B.** Granites and rhyolites of São Lourenço: a volcano-plutonic complex in southern Amazonia. *Engenh. Miner. Metal.*, **44**, 1966, 169–71.
26. **Priem H. N. A. *et al.*** Granitic complexes and associated tin mineralizations of 'Grenville' age in Rondônia, western Brazil. *Bull. geol. Soc. Am.*, **82**, 1971, 1095–102.
27. **Zwart H. J.** The duality of orogenic belts. *Geologie Mijnb.*, **46**, 1967, 283–309.
28. **Hall A.** The relationship between geothermal gradient and the composition of granitic magmas in orogenic belts. *Contr. Mineral. Petrol.*, **32**, 1971, 186–92.
29. **Wright J. B.** Controls of mineralization in the older and younger tin fields of Nigeria. *Econ. Geol.*, **65**, 1970, 945–51.
30. **Turneaure F. S.** The Bolivian tin–silver province. *Econ. Geol.*, **66**, 1971, 216–25.
31. **Schuiling R. D.** Tin belts on the continents around the Atlantic Ocean. *Econ. Geol.*, **62**, 1967, 540–50.
32. **Harris J. F.** Summary of the geology of Tanganyika, Part IV: Economic geology. *Mem. geol. Surv. Dep. Tanganyika* no. 1, 1961, 143 p.
33. **Cahen L. and Lepersonne J.** The Precambrian of the Congo, Rwanda and Burundi. In *The Precambrian, vol. 3* **Rankama K. ed.** (New York, etc.: Interscience, 1967), 143–290.
34. **Harrison J. E.** Precambrian Belt basin of northwestern United States. *Bull. geol. Soc. Am.*, **83**, 1972, 1215–40.
35. **Zartman R. E. and Stacey J. S.** Lead isotopes and mineralization ages in Belt Supergroup rocks, northwestern Montana and northern Idaho. *Econ. Geol.*, **66**, 1971, 849–60.
36. **Sillitoe R. H.** Formation of certain massive sulphide deposits at sites of sea-floor spreading. *Trans. Instn Min. Metall. (Sect. B: Appl. earth sci.)*, **81**, 1972, B141–8.
37. **Mitchell A. H. G. and Garson M. S.** Relationship of porphyry copper and circum-Pacific tin deposits to palaeo-Benioff zones. *Trans. Instn Min. Metall. (Sect. B: Appl. earth sci.)*, **81**, 1972, B10–25.
38. **Laznicka P.** Development of nonferrous metal deposits in geologic time. *Can. J. Earth Sci.*, **10**, 1973, 18–25.

19

Reprinted from *Econ. Geology* **73:**416–426 (1978)

Basin Dynamics, Channel Processes, and Placer Formation: A Model Study

JOHN ADAMS, G. L. ZIMPFER, AND C. F. MCLANE

Abstract

The role of basin dynamics, tectonics, and channel processes in the formation of placers has not been well understood. In this study, these problems were investigated in a miniature drainage basin subjected to rejuvenation. Sediment samples collected during the experiment show the response of total sediment discharge and heavy mineral discharge to rejuvenation. Total sediment yield peaked shortly after rejuvenation began and then decayed exponentially. This decay does not characterize magnetite yield which was produced in discrete events separated by periods of little output.

Samples within the source area indicated that magnetite was concentrated and stored on uplands and hillslopes as well as in the floodplain. Valley fill contained horizontal layers deposited as the channel shifted laterally and trough-shaped scour and fill deposits, which marked former channel courses. Within these deposits magnetite was concentrated on bedrock channel bottoms, on alluvial channel bottoms ("false bottoms"), and in similar positions within terrace deposits.

Complex response dominates basin rejuvenation. Initially, placers are destroyed and transported from the basin. Aggradation then overloads channels and heavy minerals are stored. This is succeeded by degradation during which heavy minerals are concentrated and flushed from the basin. Superimposed upon this complex response are shorter periods of heavy mineral transport and storage controlled by internal geomorphic thresholds. In both cases, heavy mineral transport is directly controlled by channel activity. Storage occurs when the channel aggrades, and when the channel degrades heavy minerals are concentrated in the channels and transported from the basin.

Although basin rejuvenation destroys existing placers, it also creates an environment for the formation of new placers. In addition, the complex response of the source basin to a single uplift will cause the multiple reworking of alluvium necessary for placer formation.

Introduction

Two problems relating to placer formation have not yet been resolved. The first is the relationship of placer fields to basin dynamics and regional tectonics, and the second is the relationship of individual placers to channel activity.

Lindgren (1933, p. 225) and Bateman (1950, p. 229) suggest that placer fields are most readily formed in deeply weathered areas of moderate relief that are subjected to progressive small uplifts. These cause stream rejuvenation and successive reworking of stream deposits into rich placers. For the efficient reworking of sediment, a channel gradient must be poised between excessive degradation and aggradation, and Bateman (1950, p. 234) has suggested that a gradient of 0.006 (30 feet per mile) is optimal for placer formation. For similar reasons Sigov et al. (1972) conclude that placer deposits are not formed during periods of major uplift or subsidence because at these times the transport processes are not conducive to the reworking and concentration of low-grade deposits.

Despite the extensive description of bedrock and "false-bottom" placers in the literature, there is no generally accepted mode of formation of these deposits. Cheney and Patton (1967) consider that the concentration of heavy minerals at or near bedrock is axiomatic and that this concentration is caused by unusually large floods that rework the channel alluvium. By contrast Gunn (1968) suggests that the heavy minerals move downward as the gravels of the channel bed are agitated by streamflow. Tuck (1968) considers that heavy minerals are concentrated on the channel bedrock floor during periods of downcutting and that these are later buried by channel aggradation.

The explanation of Tuck is supported by a model study (Shepherd and Schumm, 1974) which indicates that magnetite forms a lag deposit on the channel floor when the channel degrades and that this concentration is buried when the channel later aggrades.

Two concepts recently developed by Schumm (1977) and his students can be used to explain channel and basin processes and relate these to placer formation. The various components of a basin (floodplain, hillslopes, divides, tributaries, and main channel) respond at different times and at different rates to external change (rejuvenation in this case) and this results in a complex response mechanism. The concept of complex response was first developed during experimental work (Schumm and Parker, 1973). It has since been used to interpret valley terraces along Douglas Creek in northwestern Colorado (Womack and Schumm, 1977). Schumm (1973, p. 307) illustrates four stages of complex response (Fig. 1) and describes complex response as follows:

> Incision occurred first at the mouth of the system, and then progressively upstream, successively rejuvenating tributaries and scouring the alluvium previously deposited in the valley. As erosion progressed upstream, the main channel became a conveyor of upstream sediment in increasing quantities, and the inevitable result was that aggradation occurred in the newly cut channel. However, as the tributaries eventually became adjusted to the new baselevels, sediment loads decreased, and a new phase of channel erosion occurred. Thus, initial channel incision and terrace formation was followed by deposition of an alluvial fill, channel braiding, and lateral erosion, and then, as the drainage system achieved stability, renewed incision formed a low alluvial terrace.

The second concept, geomorphic thresholds (Schumm, 1973), shows how channel pattern is controlled by slope thresholds within the fluvial system. It has been known for some time that channel pattern is closely related to channel slope (Leopold and Wolman, 1957). Recent work (Schumm and Khan, 1972) has shown that a critical threshold slope exists above which a straight stream will become meandering and a second critical threshold exists above which the meandering stream will become braided.

During the summer of 1976 research was conducted at the Rainfall Erosion Facility (REF) at Colorado State University into the evolution of the drainage net and the hydrology of an experimental drainage basin. These studies followed the development of the basin from its initial configuration (two sloping planes) to the formation of an "equilibrium" drainage basin. This basin was then rejuvenated by a drop in base level, and the basin was then allowed to develop to equilibrium. Morphological similarities between the REF drainage pattern, drainage density, and stream profile to those of natural drainage basins have been shown by earlier workers (Parker, 1977). Magnetite had been observed to concentrate in the REF channels during an earlier study (Macke, 1977) in an aggrading river channel, but the present experiment provided an opportunity to study placer formation in a drainage basin as it underwent rejuvenation.

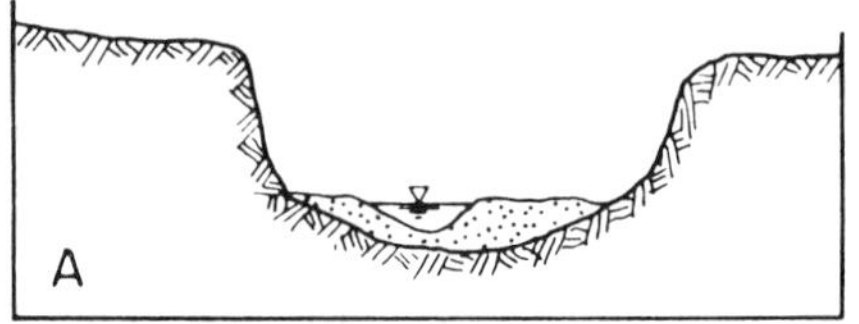

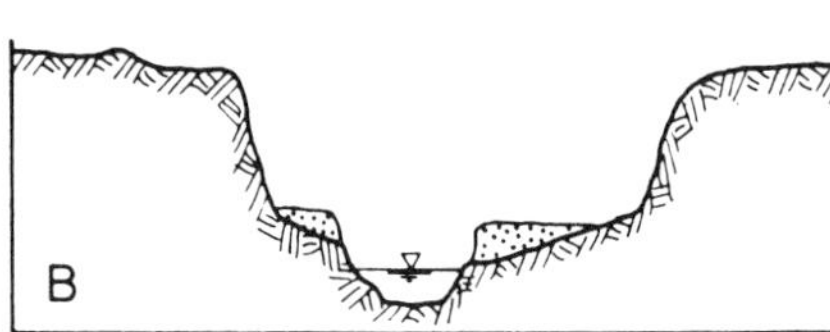

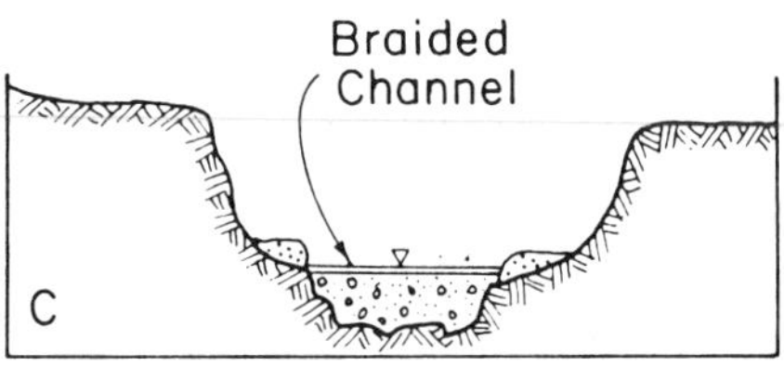

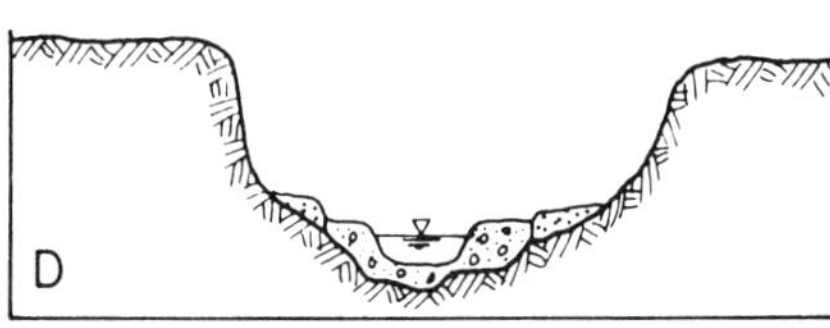

FIG. 1. Diagrammatic cross sections of experimental channel 1.5 m from outlet of drainage system (base level) showing response of channel to one lowering of base level (from Schumm, 1973).

A. Valley and alluvium, which was deposited during previous run, before base-level lowering. The low width-depth ratio channel flows on alluvium.

B. After base-level lowering of 10 cm, channel incises into alluvium and bedrock floor of valley to form a terrace. Following incision, bank erosion widens channel and partially destroys terrace.

C. An inset alluvial fill is deposited as the sediment discharge from upstream increases. The high width-depth ratio channel is braided and unstable.

D. A second terrace is formed as the channel incises slightly and assumes a low width-depth ratio in response to reduced sediment load. With time, in nature, channel migration will destroy part of the lower terrace and a flood plain will form at a lower level.

The time span of tectonic processes makes it impossible to observe the effect of these changes on placer formation in any natural basin. However, the drop in base level in the REF simulates basin rejuvenation and from a study of heavy mineral pro-

FIG. 2. The Rainfall Erosion Facility (REF) before enclosure. Crane shows scale, box is 15 m long.

duction and storage in the basin during this period it is possible to relate placer formation to the rejuvenation processes.

The quantity of heavy minerals stored in the REF allows the formation of placer deposits to be related to the complex response of the basin to rejuvenation. This estimation together with observations of channel activity also allows the recognition of conditions favorable to placer formation. Thus it is possible to use the REF to relate basin and channel processes to the formation of placers and to study the heavy mineral export from a drainage basin through time.

Equipment and Methods

The Rainfall Erosion Facility consists of a 9.1 m by 15.2 m plywood box (Fig. 2) 1.8 m deep that was filled to an average depth of 0.9 m with a 1:1 mixture of local topsoil and plaster sand (mean size 1.55 phi, standard deviation 1.9 phi) which contained 0.4 percent magnetite. This material was compacted by rolling and is referred to as bedrock in contrast with the reworked sediments (alluvium) found in the channels.

Controlled rainfall was applied to the surface at rates of 20, 28, and 56 mm/hr. The highest intensity of rainfall was used to develop the drainage basin and for most of the hydrologic investigations. This resulted in a runoff on the order of 1.3 liters per second. A small flume was attached to the front of the REF to control base level and provide an outlet for runoff and sediment. Sediment production from the model basin was measured only at this outlet flume, and thus information on sediment supply from the individual tributaries to the main channel is not available. During the experiment base level was lowered by removing a plank from beneath the outlet flume and lowering the flume.

During the development of the basin, sediment samples were taken at 15-minute intervals (samples 11–18 and 20–28) for much of the run, and at 30-minute intervals (samples 31–46) for the remainder. Sample intervals during other times (samples 1–10, 19, 20, 29, 30, and 47–50) varied from 8 minutes to 134 minutes with a mean of 31 minutes. Each sample represents the *total* water and sediment discharge from the basin for a measured period that varied between 1.5 and 2.8 seconds. These data were later corrected to a constant sampling period of 2 seconds and then adjusted so that each sample was equivalent to 2 seconds of the 56 mm/hr rain intensity, so that the data would not be biased by the sampling period.

The sediment samples were suction filtered, dried, and weighed to determine the sediment yield. To remove clay and low-density components these samples were washed and panned to about one-eighth of their initial volume in a gold pan. The fine grain size of the magnetite (mean size 3 phi) precluded removal of the fine grains by wet sieving since about one-quarter of the magnetite would also have passed through the sieve. The cleaned samples were subsequently dried and the magnetite fraction separated by a hand magnet and weighed. The error involved in this procedure was approximately 6 percent based on replicate sampling. Magnetite yields were then compared to the total sediment yield, and the deviations from the expected values related to the observations of basin and channel behavior recorded during the experiment.

This experimental drainage basin cannot be considered a scale model. It does not meet the strict rules of scale modeling necessary to maintain geometric, kinematic, and dynamic similitude (Hubbert, 1937; Albertson et al., 1960). Instead, the drainage basin described in this study more closely resembles the analog model described by Chorley (1967) which Hooke (1968, p. 393) calls "a small system in its own right." An analog model, commonly used in experimental geomorphology, does not model an individual drainage basin but models some of the properties of drainage basins in general. This approach is valid if the gross scaling relationships are met, the model reproduces some morphologic characteristics of the prototype (a natural system), and the process which produced the morphologic characteristics in the laboratory can logically be assumed to have the same effect on the prototype (Hooke, 1968). Hooke also demonstrates that a model of this type is necessary because the scale modeling of alluvial systems is not possible at the present state of knowledge. The failure to maintain scaling ratios prevents the direct extrapolation from the model to the prototype, but we consider that the results of experiments of this type may be applied qualitatively to the field, even though the rules of scale modeling were not observed.

Results

The weight of magnetite in the undisturbed material was determined to be 0.408 percent with a standard deviation of 0.025 percent. The magnetite concentration in the sediment discharge samples ranged from 0.068 to 2.01 percent, with a mean value of 0.37 percent. When the magnetite concentration in the samples was plotted against cumulative precipitation (an index of time or stage), the samples showed several distinct peaks (Fig. 3).

Immediately following the base-level drop, sediment yield increased sharply (Fig. 4). After reaching a peak the data show exponential decay similar to the sediment yield curve described by Parker (1977, see also Schumm, 1977) based on more detailed sampling. The sudden lowering of base level produced a knickpoint which migrated up the main channel. As this knickpoint passed tributary junctions, knickpoints formed in the tributaries. These migrating knickpoints defined a "wave of dissection" (Howard, 1971, p. 30) that moved through the drainage system and was one cause of changes in the amount of sediment released to the main channel and the fluctuations superimposed on the general sediment decay curve.

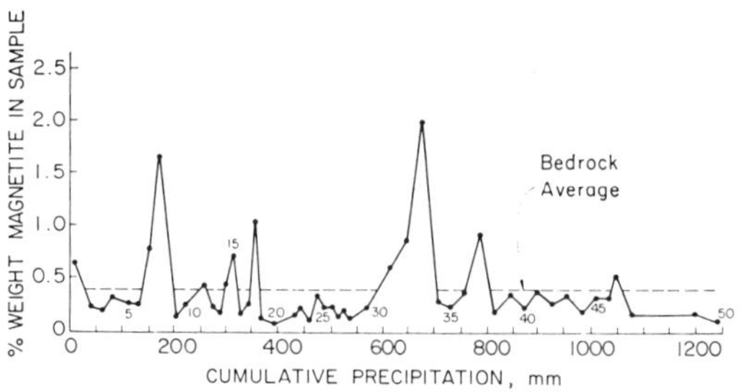

FIG. 3. Variation in magnetite concentration in the samples with cumulative rainfall. The horizontal line represents the average magnetite concentration in the bedrock.

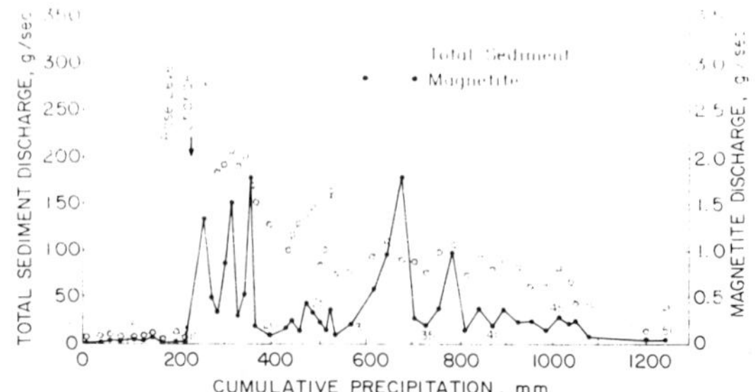

FIG. 4. Production of total sediment and magnetite from the basin. Data have been corrected to a constant sampling period of 2 sec of 56 mm rainfall.

In contrast with the sediment production, exponential decay was not shown by the magnetite production. Instead, magnetite was produced during a series of discrete events that were separated by periods during which little magnetite was moved out of the basin. The data represented by Figure 4 indicate that both sediment and magnetite leave the basin as pulses but that the magnitude of the pulses (relative to the expected values) is relatively greater for the magnetite than for the total sediment production.

The data from Figure 4 were used to plot the cumulative production of magnetite against the cumulative production of total sediment (Fig. 5). The samples indicate that a total of 7,280 kg of sediment and 27.25 kg of magnetite was moved out of the basin during the experiment. Figure 5 shows that there are departures from the straight line of slope 0.00408 that would be expected if the magnetite had been transported from the basin at the same rate as the total sediment.

At three points (samples 11, 15, and 18) slightly more magnetite was removed from the drainage basin than can be explained from the yield of total sediment. This excess magnetite may have been derived from the reworking of alluvial deposits inherited from the preceding cycle of development.

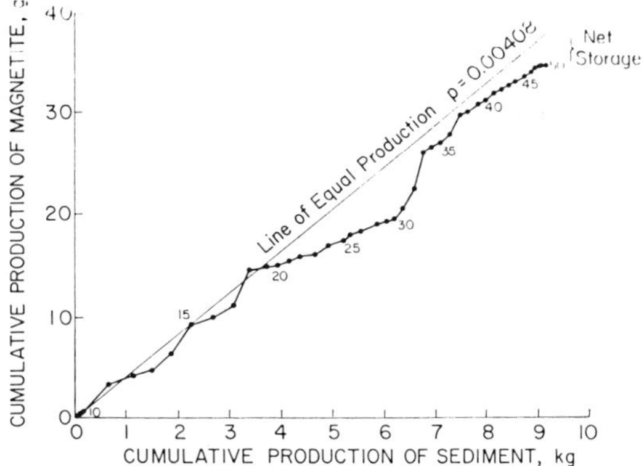

FIG. 5. Cumulative plot of magnetite vs total sediment in the samples. Storage at any point can be calculated from the deviation between the data points and the line of equal production. Points above the line represent transport of magnetite in excess of magnetite freed from bedrock.

An indication of the magnetite stored in the basin can be calculated from the difference between the observed and expected values for each point on the cumulative curve (that is, the vertical deviations of the sample points from the line of equal production). An increase in magnetite storage represents the formation of placers within the basin. By contrast, a decrease in storage represents the flushing of magnetite from the basin and the potential formation of placers downstream of the basin.

The plot of magnetite storage against cumulative rainfall (Fig. 6) shows that magnetite storage in the basin changed dramatically during the experiment. As the end of the experiment was approached the rate of magnetite storage decreased. Figure 4 indicates that sediment and magnetite discharge were decreasing toward their pre-rejuvenation rates and the basin was again approaching equilibrium. Thus the samples discussed here represent changes in the basin as equilibrium was interrupted by a major base-level drop. As this adjustment to rejuvenation occurred, the erosion and reworking of sediment in the basin resulted in the concentration and net storage of 2.5 kg of magnetite.

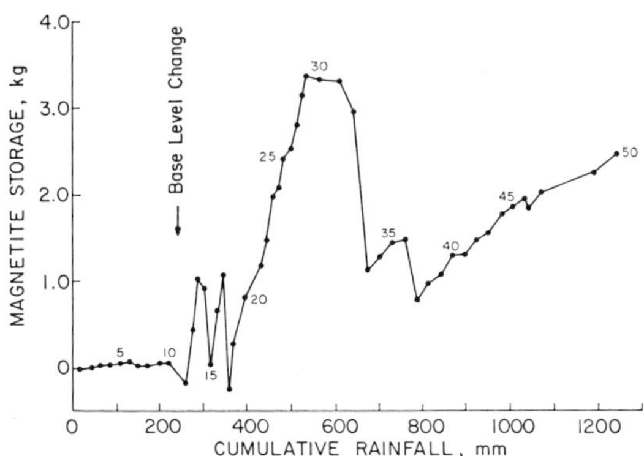

FIG. 6. Storage of magnetite in the basin during the experiment. Data from Figure 5. Zero storage assumed at start of data collection, negative storage represents removal of existing magnetite placers.

Observations after the experiment

After completion of phase two of the experiment, the distribution of magnetite within the basin was sampled. Magnetite was concentrated in the channel alluvium as placer deposits, as indicated below, and was also concentrated on the bedrock surface between the channels where rainsplash had removed much of the lighter alluvium.

Samples of the drainage basin surface were taken at 24 grid locations. At each point a small (20–40 g) sample was scraped from the surface and analyzed for the proportion of weight of magnetite in the sample. These weights ranged from 0.91 to 4.88 percent of the sample weight, with a mean of 1.58 percent. The samples can be divided into three categories: upland samples (13 samples), hillslope samples (4 samples), and floodplain samples (7 samples). Mean proportions of magnetite for each category were calculated. Upland and hillslope concentrations were found to be approximately the same, with means of 1.13 percent and 1.05 percent, respectively. Floodplain samples had a mean of 2.72 percent, more than twice that found in hillslope and upland samples.

The amount of magnetite stored on upland and hillslope surfaces in the basin can be estimated from the product of the drainage area (86.5% of the total basin area of 105 m^2 is upland and hillslope), sample depth (no greater than 0.006 m), bulk density of the surficial material (determined to be 1,390 kg/m^3), and the proportion of magnetite in the samples (1.1%).

The calculation indicates that approximately 8.3 kg of magnetite was held essentially in situ on the bedrock surface. This is an overestimate of the actual magnetite stored because both the actual sampling depth and the thickness of the magnetite surface layer were less than 0.006 m.

Transverse cross sections of the channel were excavated to study the alluvial fill. The valley fill (Fig. 7) contained horizontal layers, deposited as the channel shifted laterally, and trough-shaped scour and fill deposits, which marked former channel courses. Magnetite was frequently concentrated on bedrock channel bottoms, alluvial channel bottoms (false bottoms), and on the bedrock interface of terrace deposits. It was commonly associated with coarser grained alluvium of presumed hydraulic equivalence. A bulk sample of one of these placer deposits contained 15 percent magnetite (38 times background), and locally concentrations of pure magnetite were observed. Eleven nonrandom samples were taken to represent these placer deposits.

These representative samples were combined with the seven grid samples to give a mean concentration of magnetite in the channel fill and terrace deposits of 5.34 percent (13 times background). Valley deposits comprise 13.5 percent (14.11 m^2) of the basin area and have a bulk density of 1,340 kg/m^2. Assuming a mean sampling depth of 0.006 m, this indicates that 6.1 kg of magnetite was stored in these deposits, and, hence, channel storage accounts for slightly less than half of the total basin storage.

The combined estimates indicate 14.4 kg of magnetite were stored in the basin. This is an overestimate of the actual storage because upland sample depths have been overestimated and channel concentrations may be biased toward higher values.

Analysis of the sediment discharge samples suggests a total storage of 4.5 kg, which is in fair agreement with the magnetite estimated by sampling the basin after the experiment.

Magnetite storage and basin activity

The equilibrium of the lower main channel before the base-level drop corresponds to Figure 1A. Little sediment was being transported from the basin (samples 1–10, Figure 4), and both total sediment and magnetite discharge were approximately constant. Very short periods of aggradation and degradation controlled by intrinsic thresholds were observed in the lower main channel, but these had little effect on sediment discharge.

Initial channel response to base-level drop corresponds to Figure 1B. Twenty-seven minutes after sample 10 was collected, base level was lowered 203 mm to rejuvenate the basin. Immediately, the main channel narrowed and downcut through existing alluvial deposits. Total sediment production increased rapidly but sediment export from the basin was not continuous, as shown by samples 11 to 18, and magnetite storage (Fig. 6) went through a period of rapid fluctuation.

Observations made during this period provide an explanation for these fluctuations. Samples 11, 15, and 18 were collected during periods of downcutting. Sample 11 was collected as the first major knickpoint migrated up the main channel. Just before sample 15 was collected, a smaller knickpoint migrated up the main channel. Between samples 17 and 18 the lower main channel narrowed and incised. Magnetite was deposited along the banks of the incised channel,

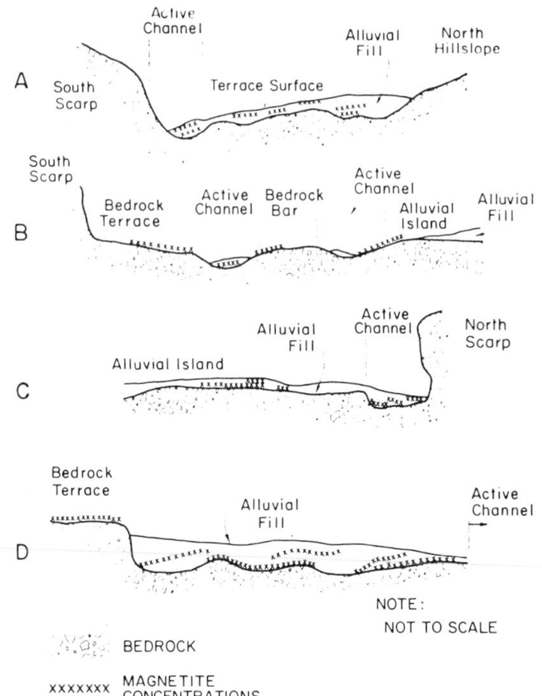

FIG. 7. Sketch cross sections of channel fill. Cross sections (top to bottom) taken at 7.7 m, 5 m, 5 m, and 2.4 m above basin outlet. Cross section C is a continuation of cross section B.

indicating that much magnetite was in transport. Lateral migration of the main channel, and undercutting and failure of the valley scarps, were at a minimum when these samples were collected.

Samples 12, 13, 14, 16, and 17 represent periods of magnetite storage. When these samples were collected, undercutting of the valley scarps and slumping dominated the lower main channel activity. The migration of sediment waves and sediment from the collapse of valley scarps temporarily overloaded the lower channel and resulted in channel braiding, lateral activity, and the storage of magnetite. These periods of magnetite transport and storage (samples 11–18) are attributable to local channel control by internal thresholds.

During the stage corresponding to Figure 1C (samples 19–29) the lower main channel was flooded by sediment from the upper channel and its tributaries. Before the collection of sample 23 response to base-level drop was confined mainly to the lower main channels and the lower parts of its tributaries. After sample 23 was collected, the upper basin responded actively and the lower 5 m of the basin played a more passive role.

This period was dominated by lateral movement in the lower main channel, and much sediment was flushed from storage in upper tributary channels. However, between samples 22 and 23, 26 and 27, and 27 and 28, the lower main channel incised (possibly after waves of sediment migrated out of the basin). This incision was not accompanied by magnetite production because the system as a whole was still dominated by the erosion and transport of stored alluvium from the upper main channel and its tributaries. Between samples 18 and 28 the upper basin was completely flushed of lighter alluvium leaving magnetite deposits exposed on the bedrock channel bottom.

This period of channel aggradation, transport of magnetite-poor sediment, and consequent magnetite storage was the direct result of the complex response of the basin.

No observations of channel behavior could be made during the period of magnetite transport (samples 30–33) because the run was made at night and the REF lacks artificial illumination. However, at the end of the preceding period most tributaries had been flushed of lighter alluvium leaving magnetite as a lag deposit lying upon the bedrock channel bottoms. We postulate that declining sediment supply to the tributary channels caused scour into these magnetite-rich lag deposits. This pulse of high magnitite production was probably brief because these channels would rapidly cut into the underlying "barren" bedrock where magnetite had not yet been concentrated. Some of the magnetite from this pulse was probably stored briefly in the alluvium of the main channel, but the declining sediment supply would progressively affect the channel downstream so that any deposition would be temporary and most of the sediment would be moved completely out of the lower channel.

Magnetite storage returned to lower channel control as the upper tributaries became adjusted to the base-level change and the lower channel shifted from the stage of complex response represented by Figure 1C to the stage represented by Figure 1D (samples 34–50). Samples 34 and 35 represent a period of lateral activity, undercutting of the scarps, and major bank caving. A braided channel filled most of the lower main channel valley. Between the collection of samples 36 and 37, lateral activity ceased and the main channel first narrowed and incised and then became braided again.

During the final period (samples 38–50) the channel was again dominated by lateral activity and remained braided. At this time the whole drainage basin was adjusted to the rejuvenation, and magnetite storage was again controlled by local overloading of the channels.

The formation of bedrock placer deposits and false-bottom placers in particular can be explained by the concepts described above (Schumm, 1977). Channel behavior is strongly influenced by the response of the drainage basin to external change (in this case rejuvenation) and to the crossing of internal thresholds. During periods of low sediment supply the channel can scour and winnow the channel fill to bedrock and so concentrate any heavy minerals present in the channel fill. These concentrations may be on the bedrock surface or at an intermediate depth in the alluvium. Less extreme scour, as expected from the concepts of complex response and geomorphic thresholds, would imply only partial scour of the channel fill and the formation of false-bottom placers above the bedrock surface (e.g., Fig. 1D). Assuming some preexisting distribution of heavy minerals within the channel fill, the deeper the scour, and the more sediment reworked, the richer the deposit concentrated on the scour bottom. Hence, the concentration of heavy minerals in a channel-fill deposit will generally increase downward to the bedrock, as observed in the field examples.

The interpretive framework presented here shows that channel response to a single rejuvenation event results in a period of complex response. This complex response and the adjustment of the channel to internal thresholds result in the formation of multistory placers. Similar multistory placers have been observed during channel aggradation following a rise in base level (Macke, 1977). A degrading channel is more favorable to the formation of rich placers, though the concepts of complex response and geomorphic thresholds ensure that deposition in an aggrading channel will be interrupted by periods of degradation and the formation of false-bottom placers. This may be the origin of placer deposits that occur high above the bedrock surface, such as those in the Ross gold field.

Discussion

Placer morphology

In the present work, similarities between REF heavy mineral concentrations and natural placers were observed. For example, residual placers formed on the model uplands are comparable to the dispersed residual deposits that Krook (1968) considers might be held in situ by vegetation and later remobilized by a climatic change to enter the channels as a workable deposit. Four other placer types were also observed.

Active channel placers: During pauses in the experiment when the channel was drained while it was degrading, magnetite concentrations were observed to form elongate pay streaks on the bedrock. During aggradation these deposits were covered by alluvium

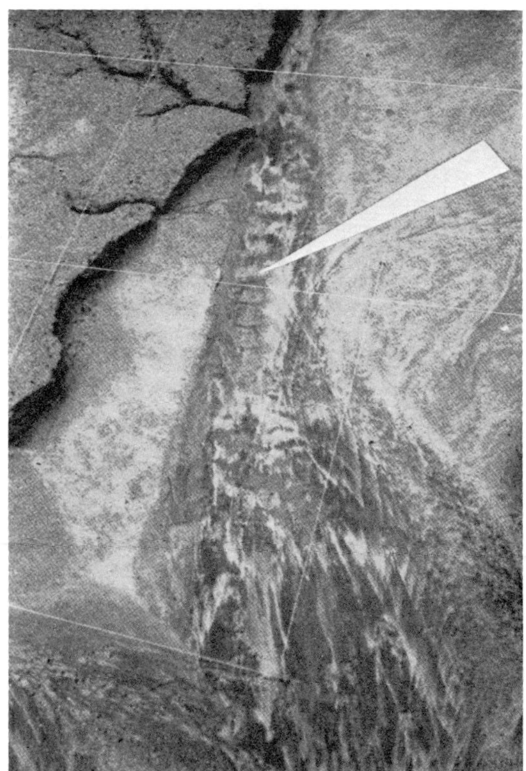

Fig. 8. Magnetite concentrations in the active lower main channel with channel drained for mapping. Flow is toward viewer; grid spacing 1 m. Arrow indicates regularly spaced scours.

to form buried concentrations, which were observed in the channel cross sections.

Figure 8 shows several features of areal heavy mineral concentrations in the model channel. The photograph shows bedrock terraces with scant coarse alluvial cover (left center), rhomboidal magnetite streaks are present at bottom center, and a series of regularly spaced scours and associated magnetite concentrations are present at top center. The regular pattern of scours was caused by the standing waves in the channel that resulted from the shallow water depths and steep slopes in the model. Thus they, like those of Wertz (1949), are probably confined to model studies and should not be considered to represent field examples.

Scour hollows occur at the confluence of tributaries in the model basin (Fig. 9). Mosley and Schumm (1977) suggest that these scours may be a preferred location for placer deposits because scours are deepest at the confluence, and the position of the scour remains in approximately the same location. This promotes the reworking of sediment and concentration of heavy minerals. Their work is supported by the occurrence of magnetite in the scours at a few confluences in the present work (Fig. 9). The presence of scour hollows does not in itself ensure the accumulation of heavy minerals, because the turbulence generated may be sufficient to move both sediment and heavy minerals.

Channel bedrock placers: Almost all channel fills examined in the model basin showed magnetite concentrations on the bedrock-alluvium contact (Fig. 7). These deposits ranged from a thin veneer on bedrock to comparatively thick deposits that formed in the thalweg.

Channel bedrock concentrations have been extensively worked and are the most widely cited in the literature. In the La Cholla area of Arizona, gold was found at 13 and 5 m above bedrock, but "the richest gold-bearing gravel occurs within a few inches of the bedrock" (Wilson, 1961, p. 30). In a second Arizona example, Schrader (1915) described placers in the Greaterville area in which the pay streaks were confined to narrow channels 2 to 4 m wide on bedrock that widened and became more deeply buried

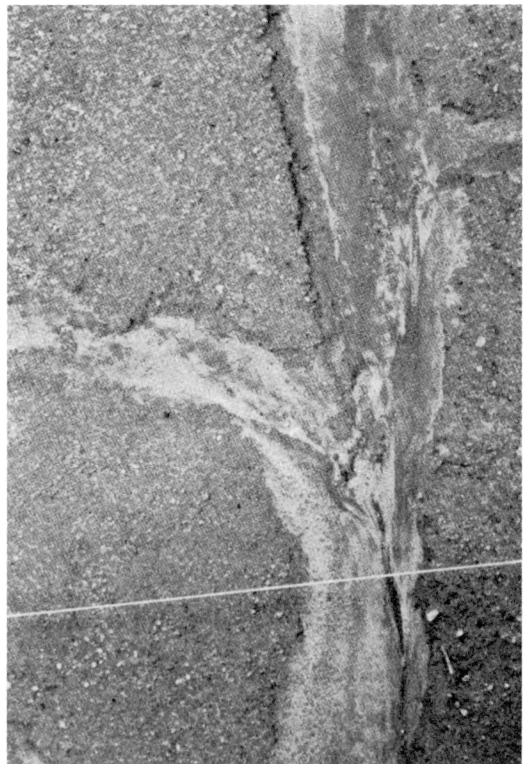

Fig. 9. Scour and associated heavy mineral concentration at tributary confluence.

downstream. Cheney and Patton (1967) conclude from field evidence that the concentration of heavy minerals on or near the bedrock surface is axiomatic.

False-bottom placers: One feature observed in both this experiment and in studies by Macke (Schumm, 1977, fig. 7–32) is the presence of magnetite concentrations within the channel fill. These have been widely recognized in natural deposits as false-bottom placers. Two placer deposits above the bedrock concentration were recognized in the Excelsior claim in the Sierra Nevada by Lindgren (1911, fig. 15) at 8 and 18 m above the bedrock. Eight payable levels separated by barren gravels were present in a 120-meter shaft that failed to reach bedrock in the Ross gold field of New Zealand (Acheson, p. 119, *in* Galvin, 1906). These false-bottom placers may interrupt the rapid increase in value as the bedrock is approached, as is shown in a Yukon terrace gravel (Koch and Link, 1971, fig. 15.2).

Terrace placers: During response to the base-level drop the main channel formed many bedrock terraces that were later abandoned when the channel downcut. These terraces were covered by alluvium and contained heavy mineral concentrations similar to those described from the channel fill deposits. These concentrations indicate that multiple terrace levels may have been formed by a single base-level drop. These terraces and placers correspond to the bench or terrace deposits worked in many gold fields such as in the Sierra Nevada of California (gravels to 900 m above stream level) described by Lindgren (1911) and at Nome, Alaska (200 m above stream level; Lindgren, 1933, p. 230).

Process model for placer formation

The observations and results presented in this paper indicate that placer formation can be dominated by processes occurring in the fluvial system (Schumm, 1977). These processes may be classified into basin processes (dominated by complex response) and channel processes (dominated by geomorphic thresholds).

Complex response characterizes the rejuvenation of the experimental drainage basin. Initially, the lower main channel downcuts while the upper basin remains unaffected by the drop in base level. When knickpoints migrate into the upper basin, the lower main channel is flooded with excess sediment and aggrades. Later, when the upper basin comes into equilibrium with the new base level, the lower main channel will again downcut until it is also in equilibrium. As this response occurs, magnetite is concentrated in the channels and transported from the drainage basin. In general, when the lower main channel is degrading, magnetite is in transport; when aggrading, the magnetite is stored in the alluvial fill.

Superimposed upon this sequence of magnetite transport and storage are shorter periods of transport and storage that are controlled locally by internal thresholds. Channel behavior in the lower main channel can be divided into periods dominated by lateral cutting and periods dominated by vertical downcutting. These alternating periods can be explained by the concept of intrinsic thresholds (Schumm, 1973). During periods of lateral activity the channel is braided or meandering (sinuous), is aggrading, and magnetite is being stored in the drainage basin. During periods dominated by downcutting the channel is straight or meandering (less sinuous), is degrading, and magnetite is being exported from the drainage basin.

Three sources (response to rejuvenation, normal bedload transport, and valley scarp collapse) cause fluctuation in the sediment transport rate, control channel aggradation and degradation, and determine channel slope, pattern, and the type of channel activity. Channel activity, in turn, controls the storage or transport of magnetite from the experimental basin and, hence, the formation or destruction of placers.

The model presented here suggests that the complex response to a single uplift will supply the multifold reworking of channel alluvium that is required for placer formation. Thus, although basin rejuvenation may destroy existing placers, it creates a suitable environment for the formation of new placers.

The conceptual model may be used to explain placer deposits related to the Southern Alps of New Zealand. Despite rapid uplift and modern erosion of gold-bearing schists, few of the short, steep rivers of South Westland carry workable placer deposits. Where present, these placers consist of small pockets of coarse "shotty" gold. Here channel slopes are too steep, and uplift too rapid, to allow the accumulation of much gold in the active channel.

Rapid deposition of the erosion products occurs in the fiords left by the glacial retreat, but again no workable deposits have been found in these sediments. Economic deposits are found only where sediments have been reworked into marine placers, for example, at Okarito and Gillespie Beach.

The older sediments of the Ross gold field were deposited in a similar environment, but here continuous deposition was interrupted by periodic channel degradation and economic multistory placers were formed. These false-bottom placers, which are at least 120 m above bedrock, are most easily explained by the concepts developed here.

Basin rejuvenation caused by uplift is similar to that caused by a drop in base level (as in the model), therefore, channels in coastal areas cut during glacial periods of low sea level might contain higher concentrations of heavy minerals than alluvium, which filled the channels during interglacial periods. Such submerged fluvial deposits have been worked for tin ore in southeastern Asia, and some Alaskan coastal placers may also be of a similar origin.

Conclusions

We hope that the results presented here will be of general interest to geologists and others involved in the understanding of fluvial placers. It should be stressed that the results of the model experiment must be interpreted qualitatively, and that some processes in the REF may not be comparable to natural situations. This experiment has considered only a degrading, rejuvenated basin and has not studied the associated depositional environment. The deposition of heavy minerals in an alluvial-fan environment was the subject of earlier experimental work (Macke, 1977) in the REF.

In this paper both field and model evidence have been used to establish a proposed interpretive framework (conceptual model) for the formation of placers. We lack the opportunity to test the model ourselves, and therefore it is our hope that this framework will be criticized, discussed, and tested.

We conclude that:

1. The REF provides an acceptable analog model for placer formation in a natural system; in both the REF and the natural system deposit morphology is similar.

2. Complex response and the crossing of internal geomorphic thresholds can explain REF channel behavior following basin rejuvenation.

3. Total sediment discharge after basin rejuvenation reaches a peak and then follows an exponential decay curve with superimposed fluctuations. Magnetite discharge does not follow a decay curve but is released as a series of pulses.

4. Only a portion of the magnetite released during the rejuvenation remained in the basin, but much of this was concentrated in the channel deposits.

5. Basin rejuvenation may initially destroy existing placers, but conditions following rejuvenation are favorable to the formation of new placers.

6. Heavy mineral transport and storage and, thus, placer formation can be related to channel behavior.

7. Placers form during alternate periods of mild aggradation and degradation that result from the exceeding of the internal geomorphic thresholds which may follow rejuvenation.

8. Multistory placers with valuable mineral concentrations are favored by degrading channels but may also form in aggrading systems. Multistory placers are formed as the result of complex response and the crossing of internal geomorphic thresholds.

Acknowledgments

We wish to thank Drs. Schumm, Ethridge, and Thompson, and members of the 1976 geomorphology seminar for helpful comments on an earlier draft. The work was supported by a grant from the U. S. Army Research Office and was completed while Adams was supported at Colorado State University by a scholarship from the New Zealand University Grants Committee.

J. A.
VICTORIA UNIVERSITY
WELLINGTON, NEW ZEALAND
G. L. Z. AND C. F. McL.
DEPARTMENT OF EARTH RESOURCES
COLORADO STATE UNIVERSITY
FORT COLLINS, COLORADO 80523
March 30, August 30, 1977

REFERENCES

Albertson, M. L., Barton, J. R., and Simons, D. B., 1960, Fluid mechanics for enginers: Englewood Cliffs, New Jersey, Prentice-Hall, 562 p.

Bateman, A. M., 1950, Economic mineral deposits, 2nd ed.: New York, John Wiley and Sons, 916 p.

Cheney, E. S., and Patton, T. C., 1967, Origin of the bedrock values of placer deposits: ECON. GEOL., v. 62, p. 852–853.

Chorley, R. J., 1967, Models in geography, chap. 3, *in* Chorley, R. J., and Hagget, P., eds., Models in geography: London, Methuen, 816 p.

Galvin, P., ed., 1906, The New Zealand mining handbook: Wellington, New Zealand Government Printer, 589 p.

Gunn, C. B., 1968, Origin of the bedrock values of placer deposits [disc.]: ECON. GEOL., v. 63, p. 86.

Hooke, R. L., 1968, Model geology: Prototype and laboratory streams [disc.]: Geol. Soc. America Bull., v. 79, p. 391–394.

Howard, A. D., 1971, Stimulation of stream networks by headward growth and branching: Geog. Analysis, v. 3, p. 29–50.

Hubbert, M. K., 1937, Theory of scale models as applied to the study of geologic structures: Geol. Soc. America Bull., v. 48, p. 1459–1520.

Koch, G. S., and Link, R. F., 1971, Statistical analysis of geological data, vol. 2: New York, John Wiley and Sons, 438 p.

Krook, Leendert, 1968, Origin of bedrock values of placer deposits [disc.]: ECON. GEOL., v. 63, p. 844–846.

Leopold, L. B., and Wolman, M. G., 1957, River channel patterns: braided, meandering, and straight: U. S. Geol. Survey Prof. Paper 282-B, 85 p.

Lindgren, Waldemar, 1911, The tertiary gravels of the Sierra Nevada of California: U. S. Geol. Survey Prof. Paper 73, 226 p.

—— 1933, Mineral deposits, 4th ed.: New York, McGraw-Hill, 930 p.

Macke, D. L., 1977, Stratigraphy and sedimentology of experimental alluvial fans: Unpub. M.S. thesis, Colorado State Univ.

Mosley, M. P., and Schumm, S. A., 1977, Stream junctions—a probable location for bedrock placers: ECON. GEOL., v. 72, p. 691–694.

Parker, R. S., 1977, Experimental study of basin evolution and its hydrologic implications: Unpub. Ph.D. dissert., Colorado State Univ.

Schrader, F. C., 1915, Mineral deposits of the Santa Rita and Patagonia Mountains, Arizona: U. S. Geol. Survey Bull. 582, 373 p.

Schumm, S. A., 1973, Geomorphic thresholds and complex response of drainage systems, *in* Morisawa, Marie, ed., Fluvial geomorphology: SUNY, Binghamton, N.Y., Publications in Geomorphology, p. 299–310.

—— 1977, The fluvial system: New York, Wiley Interscience, 352 p.

Schumm, S. A., and Khan, H. R., 1972, Experimental study of channel patterns: Geol. Soc. America Bull., v. 83, p. 1755–1770.

Schumm, S. A., and Parker, R. S., 1973, Implications of complex response of drainage systems for Quaternary alluvial stratigraphy: Nature, Phys. Science, v. 243, p. 99–100.

Shepherd, R. G., and Schumm, S. A., 1974, Experimental study of river incision: Geol. Soc. America Bull., v. 85, p. 257–268.

Sigov, A. P., Lomayer, A. V., Sigov, A. V., Storozhenko, K. Y., Khrypor, V. N., and Shub, I. F., 1972, Placers of the Urals, their formation, distribution, and elements of geomorphic prediction: Soviet Geography, v. 13, p. 375-387.

Tuck, Ralph, 1968, Origin of bedrock values of placer deposits [disc.]: Econ. Geol., v. 63, p. 191–193.

Wertz, J. B., 1949, Logarithmic pattern in river placer deposits: Econ. Geol., v. 44, p. 193–209.

Wilson, E. D., 1961, Gold placers and placering in Arizona: Arizona Bur. Mines Bull. 168, 124 p.

Womack, W. R., and Schumm, S. A., 1977, Terraces of Douglas Creek, northwestern Colorado: An example of episodic erosion: Geology, v. 5, p. 72–76.

Part IV

RECOGNITION

Editor's Comments on Papers 20 Through 23

20 BLANCHARD
Use of Ore Guides

21 WISSER
Excerpts from *Oxidation Subsidence at Bisbee, Arizona*

22 BLANCHARD
Excerpts from *Interpretation of Leached Outcrops*

23 BLAIN and ANDREW
Sulphide Weathering and the Evaluation of Gossans in Mineral Exploration

LEACHED OUTCROPS

Little is to be gained if the exploration geologist encounters a mineral deposit but fails to recognize it. As long as ore mineralization crops out on the surface and is marked either by fresh sulphides or a typical gossan cap, recognition is relatively simple. However, where the deposits are hidden beneath alluvium, capping sediments, or volcanics, or where it fails to leave a traditional gossan cap such as with the porphyry copper deposits, greater discernment is required.

Blanchard (Paper 20) outlines the nature and requirements of ore guides. Positive guides can be most helpful, but negative guides are extremely hazardous and have resulted in many lost opportunities.

In limestone host rocks, oxidized sulphide masses are commonly marked by lack of outcrop and evidences of subsidence. Wisser (Paper 21) describes some of the topographic features related to blind oxidizing deposits in limestone host that give evidence of their location.

One of the great contributions to exploration geology was the realization that it is possible, by careful observation, to recognize the character of the original sulphide minerals from an oxidized and leached gossan cap. Augustus Locke (1926) recognized that the presence of disseminated copper deposits was revealed by characteristic leached limonitic capping. The nature of this capping not only indicated that copper minerals had been present but also the copper content. Further investigations of leached sulphide outcrops were carried out by Blanchard and Boswell (1928, 1930, 1934) who established criteria for the recognition of boxworks of the principal sulphide minerals. This pioneer work is summarized by Blanchard (Paper 22). His concepts of the chemistry were simple, but his powers of observation were keen. He established that one could read the residue left behind.

Much of Blanchard's work and illustrations remained unpublished as a consequence of his ill health. This material, fortunately, was collected and published by the Nevada Bureau of Mines (Blanchard, 1968) and its publication has served to again stimulate attention to gossan studies.

Blain and Andrew (Paper 23) carry on Blanchard's concepts to include a much larger assemblage of sulphide minerals and furnish the chemical background for the changes that take place.

REFERENCES

Blanchard, R., 1968, Interpretation of Leached Outcrops, *Nevada Bur. Mines Bull.* 66, 196p.

Blanchard, R., and P. F. Boswell, 1928, Status of Leached Outcrop Investigation, *Eng. and Mining Jour.* **125:**373-377.

Blanchard, R., and P. F. Boswell, 1930, Limonite Products Derived from Bornite and Tetrahedrite, *Econ. Geology* **25:**557-580.

Blanchard, R., and P. F. Boswell, 1934, Additional Limonite Types of Galena and Sphalerite Derivation, *Econ. Geology* **29:**671-690.

Locke, A., 1926, *Leached Outcrops as Guides to Copper Ore,* Williams and Wilkins, Baltimore, Md., 175p.

20

Reprinted from *Eng. and Mining Jour.* **131:**173–175 (1931)

USE OF ORE GUIDES

Roland Blanchard

ORE GUIDES that are infallible cannot be established in a district until the last of its ore has been found; but in districts where development has been carried on over a period of years, and where parts of the ground have been thoroughly cut up, as in the older and more active mining camps, certain ore controls may be established that have been proved to be nearly invariable for the ground dissected, that leave little to the imagination for that particular ground, and that constitute the firmest basis for guiding exploration into adjacent poorly explored or wholly unexplored areas.

To be practicable, ore guides must conform to the following requirements:

1. They must be based upon some simple relation to the ore, such as the association of orebodies with a certain porphyry dike, fracture zone, or limestone bed. As a general working principle, subject to few modifications, the more involved the reasoning back of an exploration program, the less likely is that program to yield ore.

2. They must be so clearly defined and so easy to grasp that the mine foreman or shift boss will have no difficulty in comprehending and applying them. The day for the highly technically phrased report, learned and mysterious in tone, has passed. The geologist's recognized function today, be he consulting or mine geologist, is to simplify, not complicate for the operating staff, the problem of finding ore. Unless the major ore controls are so simply defined that the more responsible staff members engaged in operations underground are able to recognize them readily and grasp their significance, a good chance exists that factors important in ore discovery will be overlooked.

3. Ore guides must be larger and easier to find than the ore; otherwise it may cost more to find the guide than to find the ore itself. The pegmatite with green feldspar, which serves as forerunner to the great Broken Hill lode (Australia), is an excellent guide, since its presence signifies approach to the lode. The envelopes or "shells" of sericitized porphyry that surround the disseminated porphyry copper orebodies in certain districts likewise are excellent guides in such districts, because usually they are two or three times as large as the ore, and the more important ore occurs entirely within them. The veins of native sulphur associated with the copper orebodies at the Britannia mine, on the other hand, are poor guides, because the volume of ore exceeds the volume of sulphur many hundred times, the sulphur veins do not necessarily exist where the ore exists, and the veins, even when present, frequently are not found until the ore itself has been disclosed.

4. The major guides should be clearly distinguished as long-range and short-range guides, so the operating management may be in position to appraise the risk, and apportion expenditures accordingly. At Bisbee, for example, manganese at the surface in association with fissure silicifications is regarded as a guide to copper ore in the limestone beneath. A large surface area with prominent manganese showings corresponds usually to a large ore area, whereas a surface area with small and spotty manganese showings corresponds usually to small and spotty ore. The manganese showings occur as much as 1,800 ft. above the ore, and in many instances play out in depth before the ore is reached. Since they serve only as general guides to the ore areas, they are termed long-range guides. On the other hand, the larger occurrences of hematite are rarely more than 300 ft. from important ore, and nearly always touch that ore at some point. Since these hematite bodies lead directly to ore within a few hundred feet, they are termed short-range guides.

5. The guides should be demonstrated *facts* of ore occurrence that will stand without the crutch of theory to support them. Many ore occurrences may be better understood in the light of geologic theory. But when an exploration program becomes based to an important degree upon theoretical considerations, it is usually foredoomed to failure. One proved fact of ore association, such as the fact that the ore in a given district actually does occur repeatedly in a certain limestone horizon, or in another district occurs consistently a certain distance beneath the old erosion surface regardless of character of rock involved, is worth a dozen theories of where the ore ought to occur.

Though numerous ore guides exist in every mining district, the dominant guides usually may be reduced to three or four major ore controls so simple that even the miners have no difficulty in recognizing them. In favored districts a single major guide often dominates. Thus, at Christmas and San Pedro most of the important orebodies touch somewhere along the "marble line"; that is, along the imaginary line between the garnet zone and the marbleized and unaltered limestone lying beyond. Prospecting the marble line there would have disclosed most of the commercial ore produced by the mines to date. At Pilares the more important orebodies lie within 75 ft. of the outer edge. or periphery, of the 1,000x2,000-ft. fractured oval; prospecting this outer zone would likewise account for probably 85 per cent of the mine's production. At the O. K. mine, near Milford, Utah, the orebodies in the monzonite are associated with small quartz veins that make out from a central plug or cylinder of quartz; and the ore decreases irregularly but rapidly with distance from that central plug. These examples illustrate the simplicity to which prospecting may be reduced in especially favored districts.

IN some districts the problem is not so simple. In one of the larger garnet copper producers of the Southwest, prospecting the marble line would have yielded only a small percentage of the total garnet ore; some of the richest oreshoots were embedded deep within the garnet without touching at any point the marble line. The explanation is that at Christmas and San Pedro mineralization took place under comparatively quiet conditions, and the sulphide orebodies were deposited without distubance in the reducing zone at the outer edge of the oxidizing zone—in other words, at the outer edge of the garnet-epidote zone, along the marble line. At the larger garnet district referred to, turbulent conditions prevailed throughout the period of ore deposition. Though some of the sulphide ore began depositing along the marble line, disturbances occurred which crushed and broke open the garnet areas; and the ore still in solution penetrated and was precipitated more or less at random within the garnet zone itself.

From the foregoing the fact is evident that ore controls established as dependable in one district may be wholly misleading in another district, even though the same ore mineral occurs at both places and the same type of deposit is involved. Because bends concave to the footwall of a vein are the ore localizers at one place, the assumption is not warranted that bends concave to the hanging wall may not be the ore localizers at some other place, even within a single district. Differences in the character of the faulting movements may readily account for this seeming discrepancy, as has been discussed and illustrated with diagrams by Hulin[1]. Emmons[2] has quoted from some unpublished data of my own, showing the results of checking up the prospector's conviction that vein intersections or junctions are the most favorable places to look for rich oreshoots.

Out of 137 cases that have come under my observation in which both legs of the vein carried commercial ore, 74.45 per cent proved richer at the intersection or junction; 11.68 per cent showed no appreciable change; and 13.87 per cent proved poorer at the junction. In this last class, fifteen deposits, or 10.95 per cent of the total cases under consideration, showed noncommercial values at the junction, and seven of the fifteen, or 5.11 per cent of the total, were entirely barren, even though profitable ore had been extracted from each of the veins near by.

Again, the fact that a formation or type of rock is favorable or unfavorable to ore in a given district does not necessarily suggest that it is similarly favorable or unfavorable in another district. In various parts of western North America limestone is the ore carrier; shales and slates are so barren that often they are not considered worth prospecting. Yet in various parts of Australia the despised shales and slates carry large, rich oreshoots, Mount Isa being the outstanding recently discovered example. Nor can it be argued that the shales and slates are here mineralized because limestone is absent; for in several places in northwestern Queensland the ore solutions have passed by the limestone to deposit almost exclusively within the shales and slates. Even within the favored limestone belt of southwestern United States, at the Lookout property, in the Black Range, in New Mexico, the limestone traversed by the ore solutions shows only low-grade mineralization; whereas, an 18- to 20-ft. bed of the usually inhospitable quartzite, lying beyond, carries the commercial orebodies. These examples illustrate the danger of projecting too far or with too much assurance the ore habits of a given district, deposit, or occurrence.

USUALLY, each district presents examples of individual ore habits which must be in large part mastered before exploration programs may be intelligently and economically laid out. Such effective (though admittedly incomplete) mastery may require from several weeks in simple cases to several years in districts whose geology is complicated, as at Bisbee, Ely, Tintic, or Broken Hill; but during that period progress will be made which will continually reduce the exploration hazards. For guidance of the operating staff, as well as to guard against unsound deductions on the part of the geologist himself, the data on which conclusions are based should distinguish clearly between (1) agencies within the district that have a proved and direct bearing upon the finding of ore; (2) agencies within the district not yet fully proved but which seem likely to have such bearing; (3) agencies which have proved effective as ore guides in other districts under conditions closely resembling those in ground being explored. Manifestly, the third group is least reliable. Its inclusion is justified chiefly because of the resulting tendency to sharpen observation and maintain alertness in all concerned.

IN ferreting out the less obvious guides, the temptation to surrender simply because a problem is difficult and seemingly insoluble must continually be combated. Shortly after the World War a certain valuation engineer declared to me that after many years of familiarity with the Bisbee limestone deposits he had concluded that the only way limestone ground could be appraised in the district was on the basis of proximity to Sacramento Hill; any other method was hopeless. On that assumption, ground would be regarded as decreasing uniformly in value outward into the limestone from the main ore locus, Sacramento Hill. This sort of nonsense implies physical and mental laziness on the part of the person making the statement. To work out detailed ore controls in ground like that at Bisbee so as to permit proper evaluation of the various limestone areas is no simple task; but the fact that it has been done proves that it may be done again.

WHEN compiling for discussion in a report the ore guides of a district, one should first create a perspective, or mental framework, upon which to hang the various details. A private report on an Arizona property which I have seen drew an analogy between the deposit and an ocean liner. Commercial ore, dependent upon secondary enrichment, was represented as the deck and that part of the hull above water line; unenriched primary ore, too low in grade to be profitably extracted, was represented as the hull below water line. Although more than a dozen detailed guides were subsequently discussed, the reader was not confused, because each guide was referred to its respective position above or below water line, and likened to some well-known feature of the ship, such as the smokestack or engine room. Shift bosses are usually able to visualize the commercial aspects of a deposit when thus portrayed.

Discussion of any specific guide should be brief. It should contain a clear, full statement of the relation of that guide to the problem of finding ore, together with pertinent applications. Beyond that everything should be ruled out, however tempting the occasion for the author to air his ideas or knowledge. Ordinarily, for such discussion two hundred to five hundred words suffice; rarely are more than a thousand words justified, including references to illustrative cases. If facts alone are presented, as distinguished from ideas and theories, and brevity of style is cultivated, prolongation of the discussion of a single feature beyond six to eight hundred words is usually difficult. Should the author feel that discussion of geologic theory is essential, the appendix constitutes an appropriate cemetery for its interment. That the more reports one writes on the subject of ore finding, the less need does there seem to be for discussion of theory, is a striking fact.

The number of guides to be discussed varies with the scope of the report. Fewer usually will be available or required for an isolated, moderately small deposit than where a major mining district is involved. In a certain silver-lead district in the andesite belt of New Mexico which I have examined, oreshoots occur wholly within a 2-mile length of the X fault zone, which is variably from 40 to 110 ft. wide. Within that zone commercial ore is restricted to sections which exhibit pronounced concave bends with respect to the footwall. Within those bends ore is still further confined to the more highly brecciated areas. And within the brecciated areas the richest ore occurs where the andesite is strongly altered to chlorite with minor epidote. Leached products above the ore are locally helpful in directing exploration, but this guide is of limited application because (1) oxidation does not penetrate below the 150 level; (2) most of the surface ore has been mined; (3) the ore is lenticular down the dip, so that an oreshoot does not necessarily extend upward to the oxidized zone. The major guides in this case are:

1. X fault zone.
2. Concave bends with respect to the footwall.
3. Brecciated parts of the bends.
4. Highly chloritized areas within the breccia portions.
5. Leached capping (local).

Several minor guides dealing with the beginning and ending of veins within the chloritized areas need to be discussed, but so far as the general operating staff is concerned the guides enumerated suffice. For the Ely district I used the classification delineated in the following:

General Guides—three headings.
Guides Within the Porphyry:
 A. Structural Features—five headings.
 B. Expression of Mineralization Above the Ore—two headings.

[1] Carlton D. Hulin. *Engineering and Mining Journal*, Vol. 127, p. 228, 317. Also, "Structural Control of Ore Deposition." *Econ. Geol.*, Vol. 24, No. 1, January-February, 1929.

[2] W. H. Emmons. Vol. 76, p. 308, *Trans. A.I.M.E.*

Guides Within the Limestone:

A. Structural Features—thirteen headings.
B. Expression of Mineralization Above the Ore—eight headings.
C. Expression of Mineralization Below the Ore—four headings.

Miscellaneous—seven headings.

Work done in the Ely district since my visit in 1928 presumably has added to the guides in certain divisions and probably eliminated others that were tentatively included. A list of ore guides once cataloged does not remain stationary so long as exploration continues; imagination remains active, and application and search for new guides are energetically enforced.

With the current guides established for a given mine or district, their application in mine exploration has been found in some instances most effectively carried out through discussion in conference, at which the presence of all members of the operating staff down to and including mine foremen is desirable. Each member is generally handed, a week in advance, the list of prospects to be discussed, and each is required to submit in writing his conception of the favorable and unfavorable factors affecting a given prospect. Mere verbal discussion is unsatisfactory; it permits shirking, and often degenerates into irresponsible and flippant quibbling. The written statement, on the other hand, fixes responsibility, compels sound reasoning, and conduces to sustained, clear thinking. By the time the favorable and unfavorable aspects of a prospect, as reasoned out by the full operating staff, have been collected and threshed over in conference, each member has obtained a vivid conception of why that prospect is to be run, what its chance is for success or failure, and what guides or ore localizing factors he must watch for as exploration progresses. Because each member has contributed something, each feels a personal interest in the outcome.

At times the most carefully thought-out program will be vitiated by accident or by factors that could not have been reasonably foreseen. In the limestone deposits of one Southwestern copper district, for example, certain surface evidences of mineralization occur that are clearly associated with ore deposition. Either by direct or devious routes those evidences have been found to lead downward to ore. But in several instances the orebodies have been cut out by an intrusion of post-ore rhyolite, which, with true irony, failed to destroy the evidence of vigorous mineralization above the ore. To have exploration encounter such a condition when everything pointed to a large, profitable orebody cannot but be disheartening. But since none of us is omniscient, we must become reconciled to the fact that in carrying out any prospecting campaign, disappointments will ensue—sometimes through oversight on our part, sometimes through the vagaries of mineralization which were not subject to analysis prior to actual exploration. A certain allowance must be made for this, particularly when prospecting limestone areas, where ore is characteristically erratic in behavior and occurrence. But that does not mean we should go to the other extreme, and fail to avail ourselves of the most carefully reasoned program that is possible at the time.

As Augustus Locke once stated to me[3], we must remember that we are not competing with a technique that is 100 per cent perfect. The science of finding ore is probably not more than 35 per cent efficient today. If we improve that efficiency 10, or even 5 per cent, we are accomplishing a great deal.

Mistakes that result from unsound deductions or lack of thoroughness in reasoning and applying existing data may be largely eliminated where proper distinction is insisted upon between ascertained fact, reasonable probability, and the more remote conjecture; and where conscientious attempt is made to recognize and acknowledge the degree of doubt involved in using guides not yet securely established in a district. Mistakes caused by the undisclosed vagaries of mineralization are not as immediately subject to control. Usually they can be reduced only in the proportion that such vagaries reveal themselves as distinct district habits. The important point to be kept in mind in any district or with any deposit is that although a part of the ground may have been rather thoroughly cut up, and its ore habits ascertained, other large, well-mineralized areas have usually been but little explored, and concerning their ore habits little has been definitely established. To assume that most of the guides found dependable in proved areas will apply in adjacent untested ground is reasonable; but so long as ore continues erratic in occurrence, constant watch and diligent search must be maintained, both for local exceptions and for new, broad ore controls.

[3]Augustus Locke, oral communication. Acknowledgment for several ideas here discussed is due Mr. Locke, upon whose staff many years ago I acquired my first training in the use of ore guides.

21

Reprinted from pages 761-762, 772-788, and 789-790 of *Econ. Geology* **22:**761-790 (1927)

OXIDATION SUBSIDENCE AT BISBEE, ARIZONA.[1]

EDWARD WISSER.

SUBSIDENCE of overlying rock during oxidation of sulphide replacement ore bodies in limestone, though mentioned in the literature, has been conspicuously neglected as a guide to ore. In Bisbee the writer and his associates undertook its systematic study and found it to exhibit certain simple characteristics identifiable in unmineralized rock as much as 1,000 feet vertically above an ore body.

1 Subsidence in the overlying rock due to removal of support through shrinkage of ore bodies, and of their walls and floors, during oxidation.

These characteristics resemble those established in subsidence due to mining. In other words, the effects of oxidation shrinkage resemble those of stoping. As the ore, its walls and its floor lose volume during oxidation, the unsupported roof fails, and by the dropping out of blocks from its center, begins to take the form of an arch or dome. Successive doming cracks, one above the other, form curved shells above the ore, and as subsidence proceeds, the cracks open wider and wider and the shells successively fail and break into blocks, which fall from the roof and make a jumble of fragments, large and small, like the filling of a caved stope. These blocks fill the opening, take weight and retard the further opening of the doming cracks above. But the subsidence continues and the stresses caused by it are transmitted higher and higher into the overlying rock. The doming cracks are no longer being strongly developed because the jumbled fill below is supporting the larger part of the weight. But, gradually, cracks of irregular direction extend upward and separate the rock into numerous small blocks. These blocks, though slipping on each other, do not rotate. The slipping is of course downward. It eventually results in a slight subsidence of the whole block overlying the ore and in the development around that block of steep fissures that separate it from the uncracked surrounding rock and outline on the surface the region of subsidence.

The effects more nearly resemble those of mining in which the roof comes down gradually (as in the room and pillar method of coal mining) than they do those of mining in which the roof comes down rapidly (as in the long-wall method) ; for, instead of showing a pronounced " draw " the limiting side fissures are usually steep and the fissured area on the surface is usually contained within the upward projection of the oxidized area below.

Oxidation subsidence, because of its simplicity and its direct usefulness as a guide to ore, deserves attention in many districts, including not only those in which the ore bodies occur in limestone, but also those with other rocks, in which its effects may be less striking and its study may prove more difficult.

[*Editor's Note:* Material has been omitted at this point.]

GENERAL FEATURES OF OXIDATION SUBSIDENCE.

The ore bodies in the area studied at Bisbee are replacements in flat-dipping limestone beds, and are therefore in general flatly tabular in shape, their length and breadth being greater than their height. The primary sulphides were pyrite, chalcopyrite and some bornite. The extent of the replacement, as indicated by primary ore elsewhere in the district, varied from the scattering of sulphide grains and blebs through the country rock to a complete replacement of the limestone.

The oxidation of the ore bodies was accompanied by two distinct but closely related types of subsidence. Where the ore body was relatively close to the surface (that is, where the ratio, width of ore body to vertical distance from ore body to surface, is large), the ground above the oxidized ore body tends to subside as a plug bodily into the weakened block. Shear cracks, rising steeply from the periphery of the ore body to the surface, determine the horizontal limits of the plug. The actual downward movement of the plug may be slight.

The second type of subsidence is by doming, effected by the progressive breaking away of the fractured, loosened material above a weakened block by spalling and arching. This material is jumbled by its downward movement. The upper surface of the jumbled ground assumes the shape of a dome, which grows upward by successive slabbing along doming cracks parallel to the dome itself. The proportions of the dome depend upon the strength and structure of the rock overlying the ore body. Other things being equal, the more competent the rock above to act as a beam, the flatter the dome.

The dome is apparently developed to some extent whether or not the marginal cracks occur. Doming appears to be inevitable so long as the plug does not actually descend and come to rest upon the top of the oxidized ground. Where the effects of subsidence are not transmitted to the surface, the subsidence appears to be limited to the block of ground beneath the dome of equilibrium, and the ground above is competent to support itself, despite the opening below, by virtue of the structural stability of the dome. No data are at hand, however, to prove that there are no marginal cracks which die out before reaching the surface, as might be the case with very deep ore bodies.

Where the effects of subsidence are transmitted to the surface

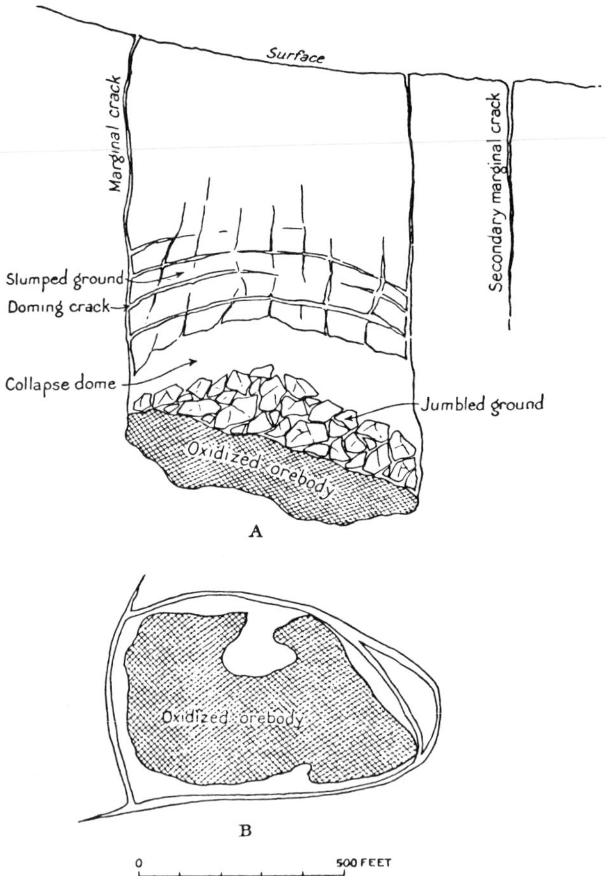

Fig. 5. Subsidence features in massive limestone, near surface.

through marginal cracks, the doming apparently proceeds upward to a greater distance above the ore body than where no marginal cracks occur. This is to be expected, since where marginal cracks occur the footings of the dome have little or no support, and slabbing upward is facilitated by the presence of free faces to which to break. Tensile strength is no longer an important factor in maintaining the slabs in position.

Secondary marginal cracks are cracks parallel to the true marginal cracks and outside of them. Secondary cracks, as seen in plan, may leave and return to the marginal cracks or may remain outside of them. Their dips appear to be identical with those of associated marginal cracks, namely, vertical or nearly so. Some however appear to join adjacent marginal cracks in depth.

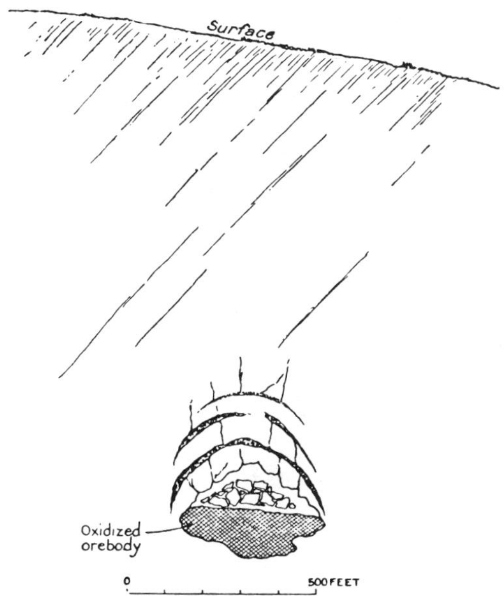

FIG. 6. Subsidence features in massive limestone at depth.

Like those resulting from mining these secondary marginal cracks seem to form most readily where the primary crack runs along a contour of a steep hill. This suggests that allowance must be made in estimating the area of the underlying ore from the surface effects of oxidation subsidence where the effects are manifest on a steep slope.

It has been found that the thickness or thinness of the beds has an important effect on the results. Where the limestone is massive or thick-bedded (Fig. 5) the marginal cracks are well-defined and persistent, with smooth walls. The doming cracks are strong and continuous and the rock above the top of the ore and below the collapse dome is broken into large boulders, which have suffered rotation. Above this "jumbled" ground (Fig. 9) is the "slumped" ground, where the rock is full of minor cracks, most of which carry banded calcite. The cracks are steeply dipping and roughly perpendicular to each other in strike, thus dividing the ground up into irregular prisms. These prisms are cut into blocks by minor doming cracks or bedding planes, whose opening accompanies the upward climb of the dome of collapse.

Where the limestone is thin-bedded (Fig. 7) the marginal

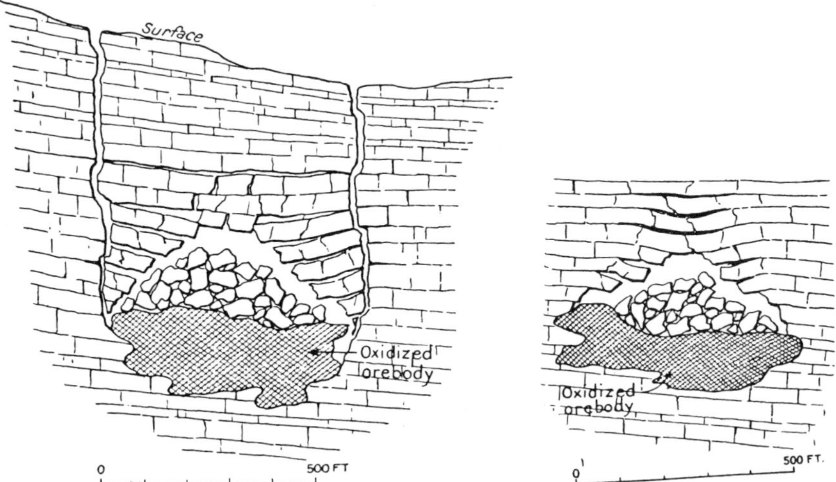

FIG. 7 (left). Subsidence features in thin-bedded limestone, near surface.
FIG. 8 (right). Subsidence features in thin-bedded limestone at depth.

cracks are individually less continuous and are more truly zones of fracturing. The doming cracks are likewise more irregular. The beds are pulled apart to form slump caves in many irregularly spaced places. The collapse dome itself is ill-defined, but the region of slumping as a whole appears to extend higher above the ore, other things being equal, than in massive beds. The thin beds have been separated by sagging for a considerable distance above the ore body, as clearly indicated by interbed fine-banded calcite. In places mud, sand and even conglomeratic material has been deposited between the beds. True doming cracks, convex

FIG. 9. "Jumbled ground." (Photo by H. M. Kingsbury.)

upward, may be lacking, and their places taken entirely by these sag cracks. The greatest sag is directly over the apex of the oxidized ore body, the amount of sag decreasing in all directions toward the margins. It should be remembered in this connection that the thickness or thinness of the beds is relative, and with a very large ore body even thick-bedded limestone may behave in this manner, giving sagging cracks instead of doming cracks. As seen in Fig. 7 however, an irregular dome does occur even with thin-bedded limestone, formed by the ends of the sagging beds breaking off in the manner shown.

THE INFLUENCE OF PRE-OXIDATION FRACTURES.

It is to be expected that where stresses are set up in country above a weakened block of ground, they will find expression along any preëxisting lines of weakness which are at all conformable in attitude and position with the direction of the stresses. As a matter of fact, in the area studied, perhaps 75 per cent. of the breaks mapped run parallel in strike to the three main faults which roughly bound the area, the Czar, Ella and Dividend faults. Such results are certainly to be expected, and should in no way interfere with proper interpretation, for although the oxidation breaks follow fissures established previous to oxidation, they do so only in the country above the ore. That this is so, is proved by the occurrence of banded calcite, indicating reopening of the fissure, only in those parts of the old breaks which traverse the country above the ore.

FEATURES CONFUSED WITH OXIDATION SUBSIDENCE.

It is obvious from the preceeding description of the features due to oxidation subsidence that there is danger of confusion with features resulting from other causes. Thus "weathering" or solution fissures may closely simulate openings due to the withdrawal of support below. Solution of a particular bed, especially where the bed has been warped, gives rise to an opening greatly resembling a doming crack. An ordinary solution cave sometimes attains such size that the roof collapses and "domes" itself. This gives a dome above a mass of rubble in no way distinguishable by itself from a collapse dome caused by oxidation of an underlying ore body. In general the following criteria hold for distinguishing open solution caves from open collapse caves (see Fig. 10):

A. Solution Caves (Fig. 10, *A*).

1. Characteristically pitted and rounded walls, floor and roof.
2. Most of the limestone is removed from the cave, but the latter may be filled with calcite as stalactites and stalagmites, introduced since the formation of the cave.

3. A solution cave usually occurs along a particular soluble bed, seldom crosses the boundaries of this bed, and is commonly associated with a solution fissure.

B. Oxidation Collapse Caves (Fig. 10, *B*).

1. The roof is jagged due to falls of rock.
2. Most of the limestone remains in the cave as fragments, often cemented and partly covered by calcite.
3. The cave is not limited to a particular bed.

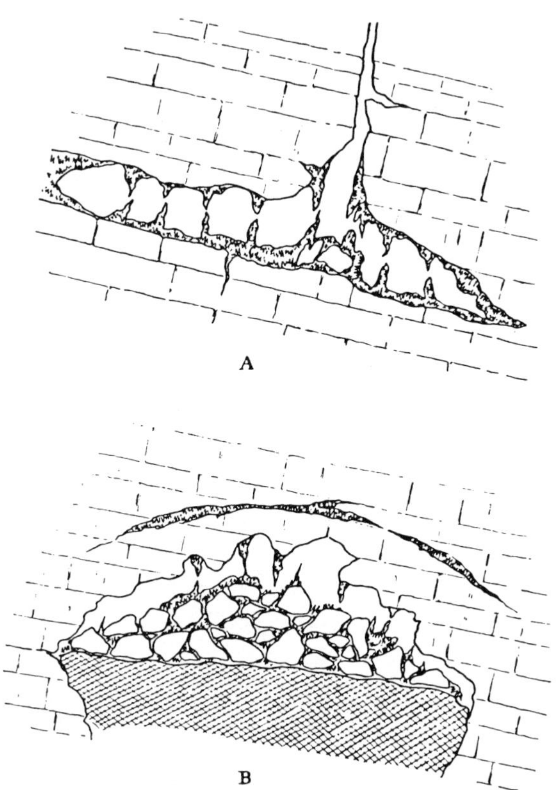

FIG. 10. Solution and collapse caves.

Solution of the limestone has in places been greatly stimulated in the vicinity of oxidized ore bodies. The rock has been intensely eaten along minute fissures, forming a kind of solution breccia superficially resembling collapse caused by the withdrawal of support below. (See Fig. 11.) Open caves are occasionally formed adjacent to oxidized ore bodies by solution alone. While

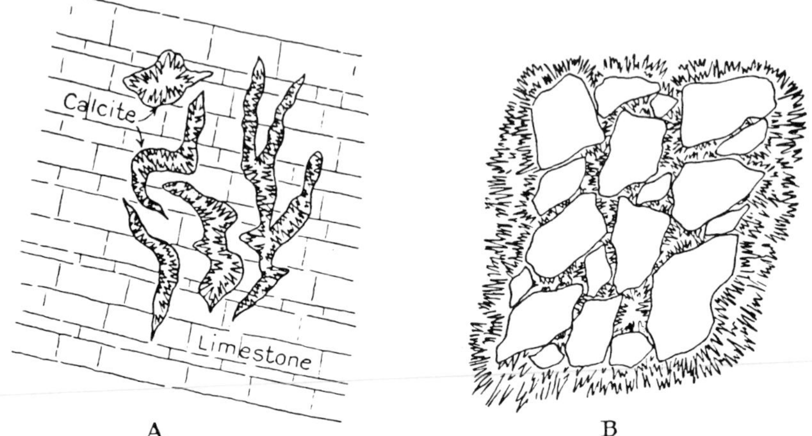

FIG. 11. Solution breccia (*A*) and collapse breccia (*B*).

these caves are usually below the ore body, they may be in almost any position with respect to it. They will seldom be confused with oxidation collapse caves.

THE FIELD APPEARANCE OF OXIDATION SUBSIDENCE AT BISBEE.

A description follows of the characteristics of oxidation subsidence in the ground above the oxidized ore body, from the surface down to the top of the ore body.

Certain small ore bodies at Bisbee, occurring at considerable depth below the surface, show no surface effects of oxidation, but the majority of them do, in the area studied, where detritus does not mantle the overlying surface. The effects upon the surface are obscured, compared with their appearance underground, for several reasons. Ordinary processes of surface weathering tend to obliterate small fissures. Large fissures are often filled by washed-in detrital matter or by the crumbling of the walls. The oxidation fissures of the area studied are typically filled with banded calcite. This material, weathering more rapidly than the enclosing limestone, produces a trough or ditch which makes an efficient trap for the detritus constantly transported by rain-water and gravity down the sides of the hills. The result is that oxidation fissures are difficult to recognize and map for any distance unless they have been reopened and accentuated by mining operations.

Upon the surface, then, whether or not the oxidized ore body below has been mined, the outline of the ore body is often projected by marginal cracks characteristically filled with banded calcite. These cracks are vertical or nearly so, and either located without reference to any faults nearby, and solely with reference to a margin of the ore body below, or are reopened portions of faults which happen to parallel a margin of the ore body. Also "regional" cracks may occur, which have reference to major groups of ore bodies but not to individuals of the group, so that, where an area of oxidation subsidence has been mapped by means of the marginal cracks, we cannot be sure that such an area is continuously ore-bearing.

Where the ore bodies underlie the surface at shallow depth, not only the marginal cracks, but also the zone of slumping may reach the surface, and an area will be mapped which is traversed by innumerable minor fissures with no dominant orientation, but rather sharply cut off from the surrounding country by the marginal cracks. With ore at still shallower depths, the collapsed ground may be seen at the surface, the rotated fragments usually being well-cemented, so that the result is a hard breccia easily recognized.

Subsidence may be indicated on the surface by topographic depressions. These are of two kinds, one caused by the bodily descent of the plug of ground bounded by marginal cracks, and the other caused by the sagging of the beds above an oxidized ore body—a synclinal trough. The former is characteristic of massive rock, the latter of moderately thick-bedded to thin-bedded limestone.

Underground the effects of subsidence are more readily visible, but the limited opportunities for mapping that are afforded by mine workings often counteract this advantage. Upon entering a region of slumped ground, the first thing noticed is the increased fracturing of the ground, which is cut up into blocks by minor fissures, most of which are steeply dipping. They may be open, but most of them are filled with banded calcite. Where mining has occurred below, the cracks have generally been reopened or widened, but the calcite still clinging to their walls shows their pre-mine origin. The contact between the slumped ground and that unaffected is often marked by the marginal crack, vertical or nearly so, with smooth walls bearing thick-banded calcite. Movement along the crack is sometimes shown by slickensiding. Mining operations nearly always reopen such cracks, and mud, sand, gravel, etc. have often been washed into them, although in many cases this material also is pre-mine. Where the limestone is thin-bedded, the separation of the beds due to sag is very striking, and banded calcite usually fills the gaps. In massive ground, doming cracks, gently convex upward, show that the workings are not far above the dome of collapse; the wider the doming crack, the nearer it lies to the dome. These doming cracks, in virgin ground, are often completely filled with calcite. In the cracks which were never completely filled, the calcite rests in great mammilary masses upon the floor, giving a very characteristic appearance.

Upon descending further toward the top of the ore body, the collapse dome may be seen. Its roof resembles that of a doming crack, but beneath the roof is the jumbled ground which directly overlies the ore body. This jumbled ground (Fig. 9) furnishes the most striking of the oxidation subsidence effects, and once seen is not likely to be confused with other breccia types. Here the structure of the country has entirely disappeared. The ground has been broken into fragments which have suffered pronounced movement and rotation. Calcite may have completely cemented these boulders so that the story of collapse must be read upon a solid wall, or the fragments may rest upon each other loosely, ready to tumble into any opening below. Whether or not the collapsed ground has been knit again by calcite into strong rock which stands well in mine workings, the appearance of the whole is strikingly indicative of the removal of support below.

EXAMPLES OF OXIDATION SUBSIDENCE.

The Surface.—(Fig. 12.) The southwest margin of the whole

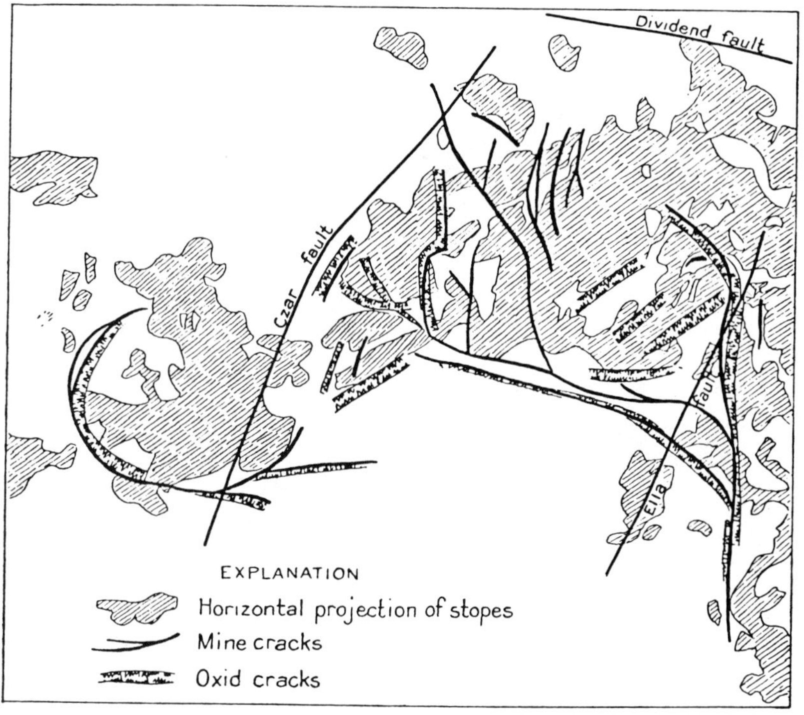

FIG. 12. Plan of surface.

group of ore bodies in this area is defined on the surface by a strong zone of cracks trending northwest-southeast. Roughly perpendicular to this zone of "regional" cracks is a group of minor marginal cracks which closely follow the eastern or western margins of individual ore bodies. These north-south cracks are very numerous in the area, while east-west cracks are rare, except for the above-mentioned regional cracks. There are two reasons for this. First, the individual ore bodies of the area are replacements of favorable beds along their intersections with fractures sympathetic with the Czar fault, which strikes about N. 30° E. The longest dimensions of the ore bodies have also this strike,

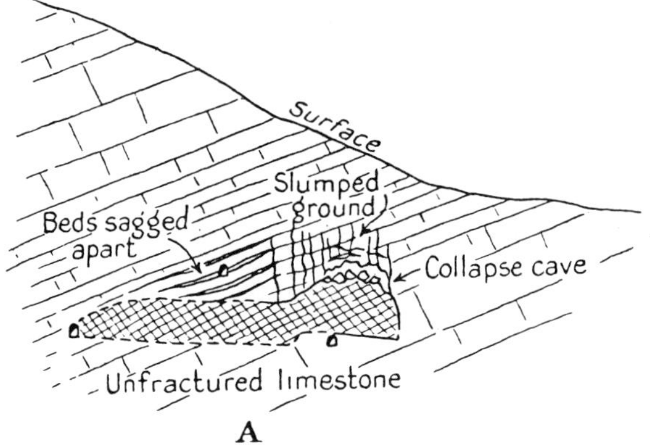

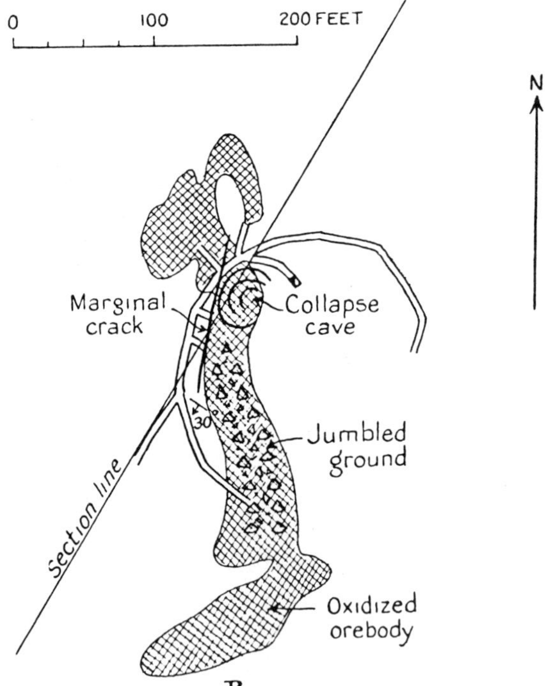

FIG. 13. Example of oxidation subsidence in thick-bedded limestone.

and the ore bodies are often pointed at their northeast and southwest ends. The longest, most important marginal cracks would therefore strike roughly north-south. Second, as the pre-oxida-

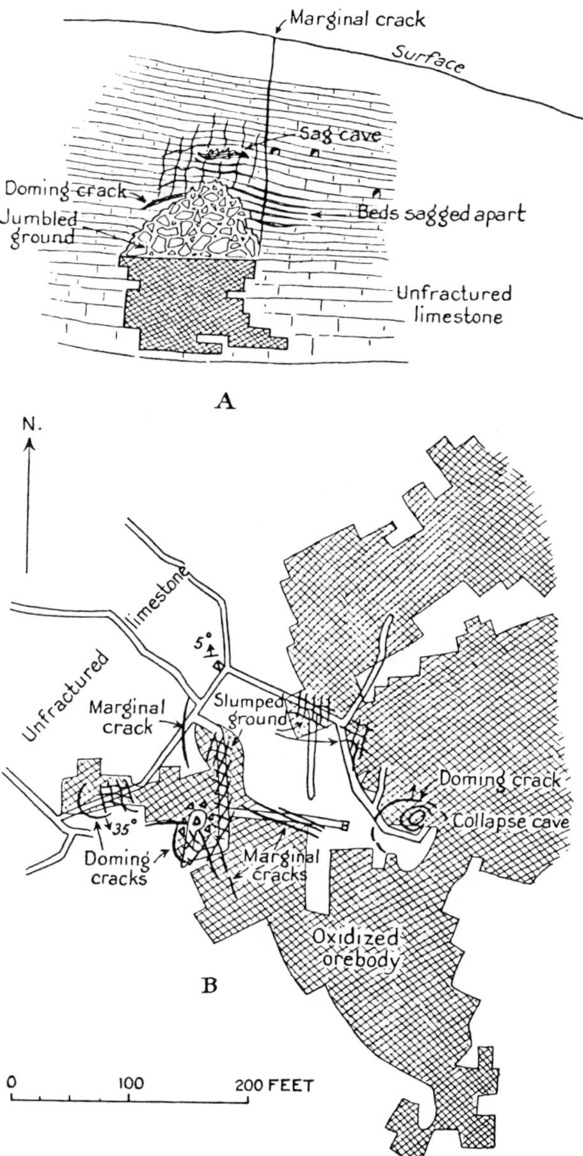

FIG. 14. Example of oxidation subsidence in thin-bedded limestone.

tion fractures of the area are in most cases sympathetic with the Czar fault, stresses set up by oxidation subsidence naturally followed these fractures more readily than along directions transverse to them.

Fifteen separate pre-mine cracks were mapped which were coincident with the projections of stope margins upon the surface. Those not obviously related to stope margins are either regional cracks, related to the whole group of ore bodies but not to individuals, or cracks which run within the horizontal projections of the ore bodies but are parallel to the edges of their thickest portions.

Subsidence in Thick-bedded Limestone.—Fig. 13 shows subsidence effects connected with an oxidized ore body in thick-bedded limestone. A collapse cave surmounts the thickest part of the ore body and beneath the cave lies the jumbled ground. Other manifestations shown are the sagging apart of the beds and a well-defined marginal crack. The sagging apart of the beds is of local occurrence and immediately over one edge of the ore body.

Subsidence in Thin-bedded Limestone.—Fig. 14 shows subsidence effects mapped above extensive ore bodies in thin-bedded limestone. While collapse caves and doming cracks exist immediately above the ore bodies, sag cracks or sag caves occur fully 90 feet above the top of the ore, illustrating the tendency of the sagging effects to climb high in thin-bedded limestone. The close association of the marginal cracks with the ore boundaries is well shown in Fig. 14.

SUMMARY OF RESULTS OF MAPPING SUBSIDED GROUND.

Nineteen areas were mapped as showing oxidation subsidence which were found to be directly above known ore bodies and clearly outlining them.

In two areas above known ore bodies oxidation subsidence was not detected. In one case the limestone is altered to kaolinic material, making heavy ground in which fractures tend to seal and become lost. In the other case close timbering obscured the evidence. In no case was there failure to recognize subsidence above a known ore body where the workings were untimbered, and the structure was not obscured by such alteration as that described above.

Four areas were mapped as showing oxidation subsidence which are not over known ore. In three of these cases lessees told the writer that ore underlay or had underlain the areas. The country below is now inaccessible, so that this could not be verified. The other case was close to the margin of a stope below, and here apparently the subsidence jumped across barren ground between two ore-bearing areas.

APPLICATION OF OXIDATION SUBSIDENCE TO UNDEVELOPED GROUND.

Once the presence of oxidation subsidence has been established, certain conclusions can be drawn from the evidence which serve to point out in a rough way the location of the ore bodies.

The marginal cracks enclose the ore-bearing area and point downward vertically to the margins of the ore body. Marginal cracks alone tell little as to the depth below the point of observation at which the ore body lies. They extend in known instances 300 feet above an ore body 200 by 200 feet in areal extent, and at least 700 feet as regional cracks connected with a great group of ore bodies, and in each of these instances they formerly went higher, since their tops have been removed by erosion. Some judgment may be used here. If the area enclosed is quite small, the ore body probably does not lie far below, since with small oxidized ore bodies at great depth the effects of subsidence would not be transmitted to the surface and no marginal cracks would appear. The possibility of depth increases with the size of the area enclosed by the marginal cracks.

Doming cracks likewise vary in meaning with their size. Small, acute domes indicate small ore bodies not far below. Other things being equal, the larger the ore body in areal extent the larger the dome and the higher it extends above the ore body. The ore body lies beneath the apex of the dome, so that where a doming crack is mapped on a mine level, its horizontal trace is obtained, and this is circular and concentric with the ore body; that is, the center of this circle is vertically above the center of the ore body.

Sagging of the beds in formations of alternating thick and thin beds indicates an ore body at shallow depth, since such sagging would not be transmitted through the massive, competent beds for any great distance above the ore body. Where all the limestone is thin-bedded, sagging may occur at much greater distances above the ore. Slumping, indicated through the division of the ground into blocks by open or formerly open fissures, means comparative closeness to the top of the ore body, and the distance varies inversely with the degree of slump. Collapsed or jumbled ground indicates nearness to the top of the ore and below the dome of collapse. The area of slumping or collapse mapped will usually give the minimum area of the ore body below, since the subsidence seldom extends out beyond the margins of the ore-block; and inasmuch as the jumbled area is bounded above by a dome, the limits of the ore may be outside those of the jumbled area mapped.

As a rough rule, which may be applicable only in the area here studied, it may be noted that the ground at vertical distance greater than 100 feet above ore bodies less than 50 x 50 feet in area is unaffected by slumping. Marginal cracks however may indicate the ore, provided the body lies at shallow depth below the surface.

No rule was worked out regarding the relation of height reached by collapse or slumping to thickness of ore body (the dimensions of the ore body being considered as length, breadth and thickness). The following considerations are offered:

1. With a very thin, interbed ore body, sagging of the beds, as shown in Fig. 15, occurs for a short distance above the ore, the amount of sag decreasing upward and finally dying out.

2. A vertical pipe or vertical tabular ore body has a structurally strong roof, so that, for a given volume of ore, the effects of subsidence (the doming and slumping) will be at a minimum if the ore body has either of these shapes. Furthermore, for a given

volume of ore in a given rock, there is a shape of ore body to give maximum slumping effects. Stated in another way, of two ore bodies having equal horizontal areas of identical outline, but hav-

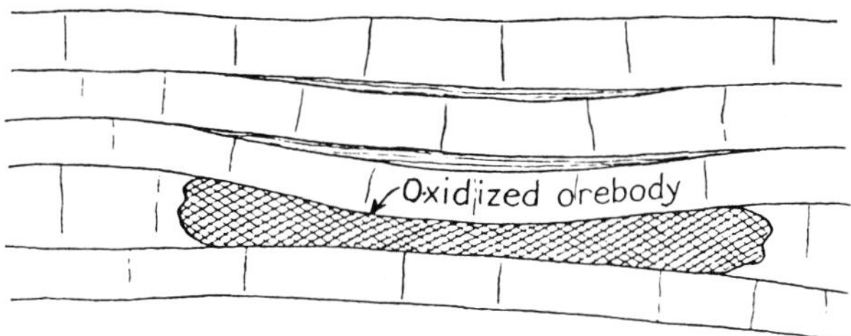

FIG. 15. Sag above thin ore body.

ing different thicknesses, and therefore different volumes, the doming and slumping go higher over the thicker ore body, with one limitation. The theoretical reasons for this follow.

3. The doming above a cavity tends to extend upward until that height is reached at which the dome is competent to carry the overlying load of rock; so that, were the spalled fragments which drop off from below the dome removed, the two ore bodies under consideration would have domes of the same height. We can consider the rock above the cavity as a beam, and this beam must, of course, have a certain thickness not to fail. If it should fail, the rock between the cavity and the surface would subside into the cavity. But if the beam should be too thick and strong to fail, its under side would spall off until a dome were produced. At a given depth below the surface, in a given rock, over a cavity of a given size, there is a certain height which the dome must reach to attain stability, just as structural arches must have a given spring in a given material.

It is probable that this theoretical height of the dome of equilibrium is seldom reached. The material from beneath the dome will, if the dome rise high enough, completely fill the original cavity as well as most of the dome itself, due to the " swell " of broken ground. This support will tend to buttress the dome and give it stability at heights under the theoretical height for the given span. In such a case, the theorem that the thicker the ore body the higher the dome follows very simply (Fig. 16).

[*Editor's Note:* Material has been omitted at this point.]

In applying the ideas of oxidation subsidence to undeveloped ground, it must be borne in mind that bodies of massive pyrite undergoing oxidation have a tendency to shrink and to cause the failure of the overlying ground just as do the copper ore bodies. There exists every gradation between ore bodies consisting chiefly of chalcopyrite and bornite, with comparatively little pyrite, and bodies of pyrite containing practically no copper. The study of the oxidation subsidence features by themselves affords no means of ascertaining the copper content of the underlying oxides. A knowledge of the habit of the nearby ground, together with any other lines of evidence possible to bring in is necessary for the final interpretation of oxidation subsidence in terms of ore.

ACKNOWLEDGMENTS.

Grateful acknowledgment is made to Augustus Locke and F. L. Ransome, for criticism and other aid, and to the Copper Queen Branch, Phelps Dodge Corporation, and the Ray Mines, Nevada Consolidated Copper Co., for permission to publish data relating to their mines.

22

Reprinted from pages 344–345, 346, 351–363, 365, and 372–373 of *Chem. Metall. and Mining Soc. South Africa Jour.* **39:**344–374 (1939)

INTERPRETATION OF LEACHED OUTCROPS.

By Roland Blanchard.

[*Editor's Note:* In the original, material precedes these excerpts. Also, Figures 1, 3, 4, 5, 8, 11–14, 17, and 18 have been omitted.]

Interpretation of leached outcrops is a comparatively recent development in economic geology. As a concrete method for translating such outcrops into terms of ore the technique has been evolved wholly during the past fifteen years. Except as regards several common types of copper ore, the progress has been achieved mainly during the past twelve years. Theretofore the possibility of distinguishing a limonite left by the leaching of pyrite from one left by the leaching of chalcopyrite, chalcocite, bornite, sphalerite, or galena either had been accorded scant consideration, or had been looked upon, in general, as both impracticable and visionary.

To-day not only is a limonite derived from the leaching of pyrite clearly distinguishable in most cases from one derived from the other sulphides named, but where the copper, lead or zinc minerals are involved the grade of the material, prior to leaching, may be in many instances judged almost as accurately as if the original sulphides were still visible in the outcrop. That fact, perhaps, serves as the truest yardstick by which to measure the progress achieved.

Basis of Leached Outcrop Interpretation :

Interpretation of leached outcrops is based upon the observed fact that limonite products derived from any given oxidised mineral, such as chalcopyrite, usually differ in minute physical characteristics from corresponding products derived from other minerals. The differences involve cellular structure, texture, pulverulency, size and arrangement of the limonite grains, and a host of other factors. Of the various distinguishing characteristics the cellular structure is the most outstanding, and is the easiest to recognise and identify. It derives its shape or pattern from the cleavage, crystal form, or fracture pattern of the sulphide or other mineral undergoing oxidation, and results from a webwork of limonite or limonitic jasper "eating" its way during oxidation along cleavage or fracture planes. Thus, galena often yields a cubic boxwork, patterned after the cubic cleavage of galena.

Individual characteristics that serve to establish a given limonite product's origin have been described in numerous published papers. To those not directly engaged in the interpretations the descriptions are tedious : they will not be elaborated upon here : for those who desire to pursue the matter in more detail, sketches and photographs are presented, and a bibliography is appended. The present paper confines inself (1) to outlining briefly the early history of the investigation, (2) to describing the general processes by which the judging of a leached outcrop is accomplished, (3) to furnishing several representative examples of the technique's application in the search for orebodies.

[*Editor's Note:* Material has been omitted at this point.]

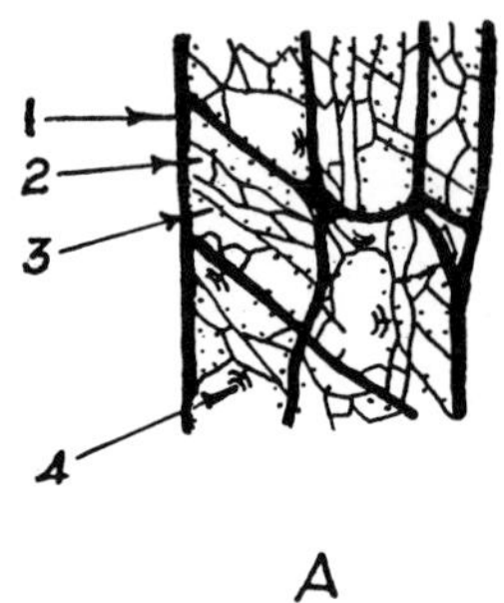

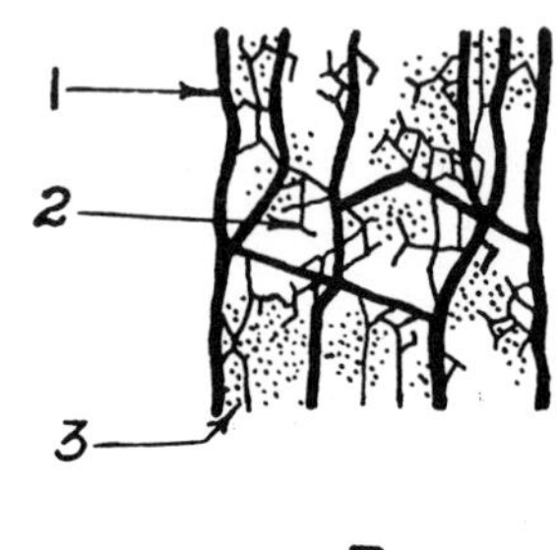

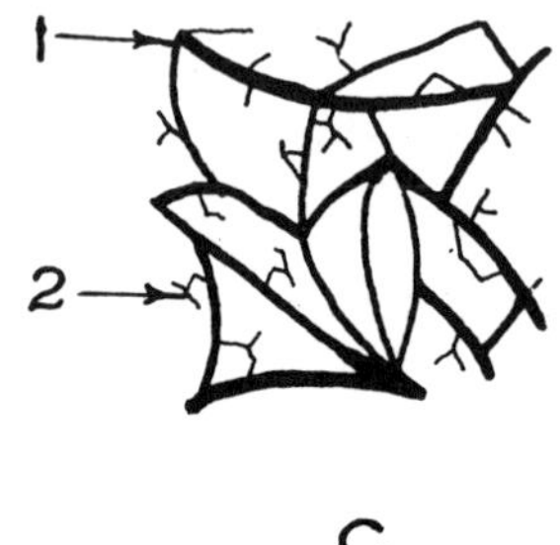

FIG 2.

A.—Sketch showing characteristic coarse cellular boxwork of sphalerite derivation. Carlisle mine, New Mexico. ×5. 1. Coarse cellular boxwork. 2. Well connected cellular webwork. 3. Sandy limonite grains coating cell walls. 4. Limonite rosette. The coarse cell structure is referred to as boxwork; the finer, as webwork. Note well connected cell walls and nearly empty cells; also parallel cell walls crosswise of the main structure. The sandy limonite grains are sparse, and adhere closely to cell walls rather than filling the cell cavities. This is a "key" structure for sphalerite.

B.—Sketch showing characteristic coarse cellular boxwork of chalcopyrite derivation. Rocher de Boule mine, British Columbia. ×5. 1. Coarse cellular boxwork. 2. Generally broken webwork. 3. Granular and pulverulent limonite. Note the less well defined cross structure, generally broken character of the webwork, and scarcity or absence of limonite rosettes. Also that the granular and pulverulent limonite (slightly coalesced in surface exposures) not merely coats cell walls but frequently partly fills intercellular space. Absence of pulverulent limonite in some cells is probably due to admixed pyrite at such places in the sulphide ore. This is a "key" structure for chalcopyrite. Compare with Fig. 3.

C.—Sketch showing characteristic cellular boxwork of bornite derivation. Engel mine, Plumas County, California. ×5. 1. Coarse cellular boxwork. 2. Fragmentary webwork. Note the conspicuous spherical triangular and trapezoidal cell pattern, a "key" structure for bornite. The webwork is fragmentary and sparse. Intercellular space is usually well filled, however, with "partially sintered" limonite crusts or with "relief" limonite. (Neither of these is shown in the sketch.)

The term "Limonite" Defined :

To avoid misunderstanding, it may be well in this connection also to define use of the term "Limonite." As most of us know, Posnjak and Merwin twenty years ago proved that no such mineral as limonite exists.* But most of us, when we hear the term "limonite" used, think of it as denoting the reddish, yellowish or brownish deposits from decomposing iron-yielding minerals in nature. In this paper the term therefore will be used in its commonly accepted sense, even though such use may not be scientifically correct.

Processes by which a Leached Outcrop is Interpreted.

In describing processes by which the judging of a leached outcrop is accomplished it becomes necessary to present a summarised re-statement of much of what has been said before in the various published papers. Justification for the re-statement lies in the fact that without a clear exposition of the underlying principles, intelligent understanding of the steps employed in judging a leached outcrop is not possible.

Three Fundamental Factors :

In interpreting a leached outcrop the three essentials are : (1) A knowledge of what is present in the particular limonite specimen or in the particular area of leached outcrop under investigation ; (2) a knowledge of that type of ore occurrence in all its variations ; (3) a knowledge of the physics and chemistry of the processes needed to produce that particular limonite specimen, or that particular area of leached outcrop, as it is to-day.

The second requirement, namely, a knowledge of ore occurrences in all their variations, never can be attained, but may be to some extent approached through a sufficiently broad field experience.

Because the prospector so frequently possesses at best only the first, and the inspecting engineer or geologist only the first and second, of those requirements, mistakes are made in selecting new areas for prospecting ; whereas opportunities for finding ore-bodies in older districts are sometimes overlooked. Because the academic geologist so often is deficient in the first two requirements, that is, in field experience and a first-hand, intimate knowledge of ore occurrences, he often commits equal or greater errors.

An accurate inter-balance of the three factors is needed. Superficial knowledge of any of the three destroys the balance, and frequently invalidates the conclusions.

Underlying Chemistry of the Technique :

Need of Iron.—Before limonite can form in any case where the oxidation of a sulphide or other mineral is involved, iron salts must be present in the solutions, and the iron must be precipitated from those solutions. It follows that sulphides which do not carry iron, such as chalcocite, covellite, tetrahedrite, molybdenite, sphalerite, or galena, cannot yield limonite unless

*The mineral turgite ($2Fe_2O_3{\cdot}H_2O$) exists to some extent, frequently merely as a transitional stage, but the normal product is the ferric oxide monohydrate, geothite ($Fe_2O_3{\cdot}H_2O$), or its equivalent, lepidocrocite ($Fe_2O_3{\cdot}H_2O$), with variable amounts of absorbed and capillary water. See E. Posnjak and H. E. Merwin, "The Hydrated Ferric Oxides," *American Journal of Science*, Vol. 47, p. 311, 1919.

iron is supplied to them during the oxidation. Field evidence shows, however, that in most instances they do yield limonite. The source of the iron in such cases is various. It may be derived from associated oxidising pyrite within the orebody. It may be derived from a point entirely outside the orebody and be brought in by ground waters during the oxidation. It may be, as often happens in the case of sphalerite, iron that is isomophous with the sulphide itself. From whatever source derived, the iron must be supplied, or no limonite results. Most commonly the iron in ground water is carried in solution as bicarbonate or as ferric sulphate. When carried as ferric sulphate, its most common source is oxidising pyrite, either within or outside the orebody.

reason the Kennecott orebody was so little oxidised is because the region has been recently glaciated, and the processes of weathering effective in many other copper districts have not had time to bring about important oxidation there. That is one reason, and a valid one. But it may be properly asked whether the Kennecott orebody contained sufficient sulphur to have put itself completely into solution. even if it had been exposed, since its formation, to the long-continued, arid weathering conditions of South-western United States.

Covellite, sphalerite, and galena contain just enough sulphur to dissolve the respective sulphides, and to carry away in solution the freed copper, zinc and lead; but, as previously stated, unless iron is

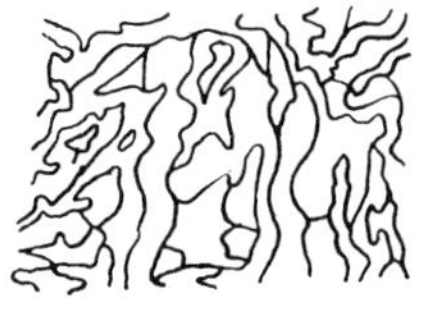

A

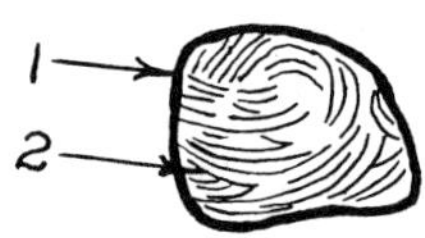

B

FIG. 7.

A.—Sketch showing characteristic Contour boxwork pattern of tetrahedrite derivation. World's Fair mine, Patagonia, Arizona. $\times 5$. In the sketch have been omitted the sandy and dully resinous granules that usually coat cell walls and partly fill cell cavities of tetrahedrite-derived limonite. With cellular structure alone showing, resemblance to contour map of steeply mountainous country is striking. Contour boxwork is the outstanding "key" structure for tetrahedrite.

B.—Foliated boxwork of molybdenite derivation. Hodgkinson District, Queensland. $\times 20$. 1. Outline of cavity. 2. Pseudomorphous replacement by limonite of foliations of the molybdenite. This is a "key" structure for disseminated or small nodular molybdenite. Molybdenite, particularly when massive, frequently oxidises to molybdic ochre without formation of limonite either at the time or as a later decomposition product of molybdic ochre.

Since neither tetrahedrite nor molybdenite contains iron, the iron entering into formation of the limonite in these two cases had to be imported, from an external source, by ground waters. In both cases adjacent oxidising pyrite was the source.

(Courtesy: *Economic Geology.*)

Need of Adequate Sulphur.—Just as iron is needed for the formation of limonite, so adequate sulphur is needed before the sulphide itself can oxidise completely. If insufficient sulphur is present, as with chalcocite and bornite, complete oxidation does not take place unless acid solutions are introduced from an outside source. Theoretically it is possible for half the copper in chalcocite to go into solution without an external sulphur supply; but unless additional acid is imported, complete oxidation does not occur. Geologists have told us that the

introduced during the oxidation, no limonite is formed.

Chalcopyrite contains a slight excess of sulphur, so that all the copper and all the iron go into solution, all the copper and most of the iron are exported, but a small proportion of the iron remains, to be precipitated within the vacated cavity as limonite. Under ideal conditions the amount of iron remaining as limonite amounts to one-third of the total iron originally present. Chalcopyrite is one of the few sulphides thus far discussed that is competent, upon oxidation, to yield limonite without importation of iron from an external source.

Effects of Pyrite.—In the case of pyrite a large excess of sulphur exists. Consequently, where pyrite oxidises adjacent to and contemporaneously with any of the other sulphides, not only is the oxidation of those other sulphides usually hastened, but a source of iron also becomes directly available for the formation of limonite.

In the disseminated chalcocite porphyry deposits a greater or less amount of pyrite usually is present, because in such deposits. chalcocite usually replaces pyrite. There thus generally exist, in those deposits, all the conditions needed not only for bringing about complete oxidation and leaching of the ore minerals, but for the formation of limonite as well.

The chemistry of formation of limonite in disseminated chalcocite deposits is so simple that that type of deposit will be used in this paper as the type example for interpretation of leached outcrops. First. however, it should be noted that since a high ratio of pyrite to other sulphides causes high acidity in the oxidation solutions, and since high acidity favours exportation of iron from the point where the sulphide is oxidising, it follows that the higher the ratio of pyrite to any of the other sulphides, the less likely is limonite to be precipitated at the point of sulphide oxidation. In other words, where the ratio of pyrite to chalcocite is 5 : 1, or 10 : 1. there exists much less likelihood of limonite being precipitated at the point of sulphide oxidation than where the ratio is only 1 : 1.

These three fundamentals of the chemistry of sulphide oxidation, (1) that unless sufficient sulphur is present the sulphides cannot oxidise completely, (2) that even though the sulphides oxidise, limonite cannot form unless iron salts are present in the solutions and the iron is precipitated from the solutions, and (3) the higher the ratio of pyrite to any of the other sulphides named, the less likely is limonite to be precipitated at the point of sulphide oxidation,—these three points are the vital considerations in the interpretation of leached outcrops; for the amount of admixed or associated pyrite with respect to any of the other sulphides named, determines largely the type of limonite produced, its distribution with respect to the cavity or the point where the sulphide particles is oxidising, and, through those two factors, the accuracy with which a given leached outcrop may be interpreted.

Nature of Miami Outcrops:

Leached outcrops over the disseminated copper orebodies at Miami (Arizona) show a characteristic limonite product with thin-walled cellular structure and sharply defined angular pattern, similar to the limonite of Fig. 5 (though in disseminated form), with the limonite precipitated almost wholly within the cavities that have been vacated by the sulphides. At Miami the ratio of chalcocite to pyrite in the ore is approximately 2 : 1. The same type of cellular limonite, precipitated largely within the cavities, is found in most quartz-mica-schist and in most monzonite gangues where that ratio of chalcocite to pyrite prevails.

Nature of Tyrone Outcrops:

Leached outcrops over the disseminated copper orebodies at Tyrone (New Mexico) show a limonite with slightly less angular and slightly more rounded cellular structure. But there, instead of the limonite being precipitated within the cavities, it is precipitated mainly about outer edges of the cavities; and much of the limonite, instead of being cellular, consists of granular and pulverulent products. At Tyrone the ratio of pyrite to chalcocite in the ore is approximately 2 : 1, or just the reverse of the ratio existing at Miami. This type of limonite, precipitated almost wholly about the outer edges of the cavities, is found in most feldspar gangues where that ratio of pyrite to chalcocite prevails.

The Difference Explained.—The explanation is that where the ratio is 2 c.c. : 1 py,* as at Miami, just enough sulphur is present for all the copper and all the iron to go into solution; for all the copper to be exported, but for none of the iron, or very little of it, to be exported. The iron consequently remains, where it is precipitated within

*2 chalcocite : 1 pyrite

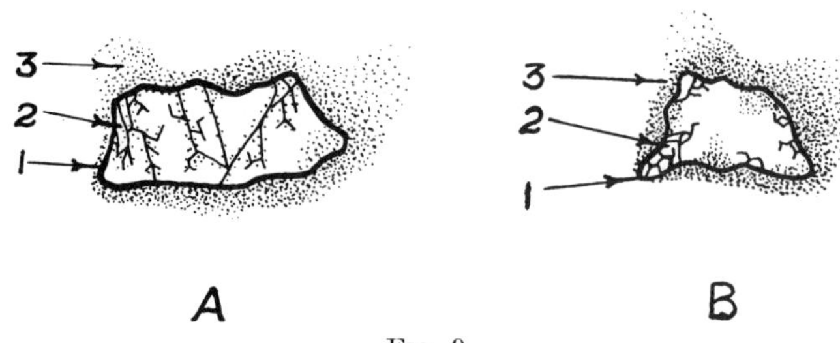

Fig. 9.

A.—Sketch showing characteristic oxidation of essentially pure, disseminated chalcopyrite in a gangue of moderate neutralizing power, such as an orthoclase feldspar rock. Longamundi, Queensland. ×10. 1. Outline of cavity. 2. Fine cellular boxwork, or webwork. 3. Granular and pulverulent limonite. The cellular structure is composed of limonite jasper that has "eaten" its way along fracture planes during incipient oxidation. The granular and pulverulent limonite represents iron precipitated from the oxidation solutions during leaching of the intercellular sulphide residuals ; that outside the cavity has been precipitated largely through the agency of gangue neutraliser. That the neutraliser is moderately efficient as a precipitant is proved by absence of limonite "fog." (See Fig. 16, "A.")

B.—Sketch showing characteristic oxidation of a disseminated mixture of approximately 3 cp : 2 py. Nevada Consolidated Copper Company's pit, Ruth, Nevada. ×10. 1. Outline of cavity. 2. Cellular boxwork, or webwork. 3. Granular and pulverulent limonite. Note that the cellular structure is less angular and more rounded than that of "A," denoting corroding and dissolving action by acid solutions, which in this case have been derived from oxidising pyrite subsequent to formation of the cellular structure. Note also that central portion of the cavity has been more thoroughly cleaned of its iron than in "A," pointing to probable concentration of the pyrite in this portion of the nodule. Concentration of pyrite there explains absence of cellular limonite persistence in central portion of the nodule ; the cellular structure has been formed and is preserved only in the portions that were more dominantly composed of chalcopyrite. Denser precipitation, than in "A," of limonite immediately surrounding the cavity, points to stronger neutralising power of the gangue than in "A." The gangue in this case is monzonite.

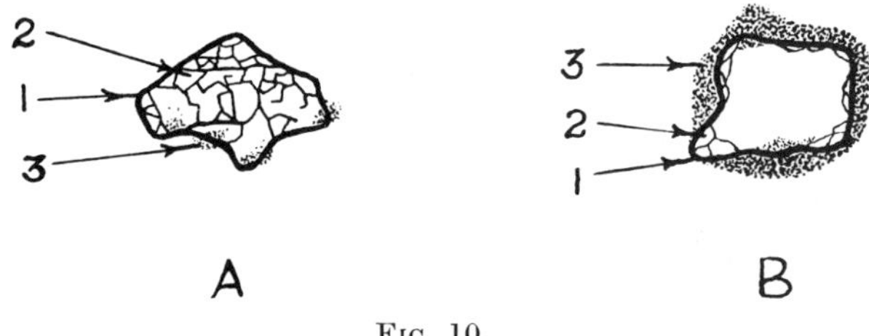

Fig. 10.

A.—Sketch showing characteristic oxidation of a disseminated mixture of approximately 2 c.c. : 1 py. Miami, Arizona. ×10. 1. Outline of cavity. 2. Cellular boxwork of sharply defined, angular pattern. 3. Granular and pulverulent limonite. For the most part the cellular structure is coated with the minute "pin point" nodular crusts or films shown in Fig. 14. These cannot well be reproduced in the sketch. Compare the cellular structure with that of Fig. 5. Less densely formed and less well connected cellular structure in bottom half of specimen than in upper half, together with local occurrence of pulverulent limonite at, and locally outside, edge of the cavity, denotes higher proportion of pyrite in bottom half of the specimen prior to oxidation. The gangue is quartz-sericite schist, of weak neutralising power. Precipitation of the locally exported iron as limonite grains immediately outside the border of the cavity in such a gangue corroborates the other testimony that c.c. predominated markedly over py in the sulphide specimen.

B.—Sketch showing characteristic oxidation of a disseminated mixture of approximately 2 py : 1 c.c. Tyrone, New Mexico. ×10. 1. Outline of cavity. 2. Limonitic crusts. 3. Granular and pulverulent limonite. The pseudo-cellular structure around inner edge of cavity represents largely limonitic crusts of the type illustrated in Fig. 14, rather than cellular boxwork formed along fracture or cleavage planes of the sulphide particle. The gangue is monzonite, as in "B" of Fig. 9. Denser precipitation of limonite around outer edge of the cavity here than in "B" of Fig. 9, although in both cases the gangue is monzonite, points to stronger neutralising power of the gangue at Tyrone. The explanation is that at the Nevada Consolidated pit the monzonite is more sericitized than at Tyrone, thus slightly reducing its neutralising power.

the cavity, usually through hydrolysis, as limonite.

Where the ratio is 2 py : 1 c.c., as at Tyrone, enough sulphur likewise is present for all the copper and all the iron to go into solution ; but with that ratio an excess of sulphur exists, which produces in the solutions an acidity so high that not only is all the copper exported, but all the iron, or essentially all of it, likewise is exported.* The iron is carried outward until the high acidity of the solutions is overcome by the feldspar neutraliser of the gangue ; where it is precipitated as limonite, immediately about outer edges of the cavity.†

Estimating Grade of the Ore Formerly Present: By determining the percentage of voids or cavities in the rock that have been left by leaching of the sulphides, and knowing from the type of limonite produced and the pattern of its distribution that in one case the ratio of the ore is 2 c.c. : 1 py, and in the other case 2 py : 1 c.c., the grade of ore represented by the leached outcrop, in either case, may be readily calculated—quite as readily as if no leaching had occurred, and an experienced observer were making a careful visual estimate of grade of the sulphide ore itself.

Difficulties Involved.—All that is simple. And it thus may be thought that estimating the grade of ore to be expected beneath a leached outcrop is a comparatively easy matter. To dispel that illusion, a few complications may be introduced.

Suppose that, instead of the ratio being 2 py : 1 c.c., as at Tyrone, the ratio is 3 or 3½ or 4 or 5 py : 1 c.c. In that case, all the iron would be exported and an empty cavity would remain, the same as where the ratio is only 2 : 1 ; and, if sufficient neutraliser were present in the gangue, all the iron likewise would be precipitated as limonite immediately about the outer edges of the cavities, the same as where the ratio is only 2 : 1. But in these latter cases a much smaller amount of copper would be represented by a given percentage of voids in the rock than where the ratio is only 2 : 1, even though the arrangement or pattern of limonite distribution in both cases is the same. It thus becomes evident that the arrangement or pattern of limonite distribution with respect to the cavity does not by itself tell the whole story.

Assume, again, that instead of a chalcocite-pyrite mixture being involved, a chalcopyrite-pyrite one is under consideration. In that case all the iron would be exported, and an empty cavity would remain,* where the ratio of py : cp is only 1 : 1, instead of 2 : 1, as in the case of chalcocite. The reason is that in the case of chalcopyrite there exists a higher ratio of sulphur to bases than in the case of chalcocite : consequently less admixed pyrite, or other sulphur from an external source, is needed to produce the same effects as are produced in the case of chalcocite.

Sulphur-Iron Ratio the Important Factor.—It thus becomes evident that the arrangement or pattern of limonite distribution with respect to the cavity is a function primarily, not of the particular sulphides involved in the oxidation, but of the sulphur-iron ratio of the solutions derived from oxidation of the sulphides. In other words, if a covellite-arsenopyrite-pyrite mixture (that freak combination of minerals occurs at one place in Lower California, Mexico) upon oxidation yielded the same relative proportions of sulphur and iron available for the limonite-making reactions as was yielded by a chalcocite-pyrite mixture, or a chalcopyrite-bornite mixture, or a sphalerite-chalcopyrite mixture, or any other combination of sulphides—precisely the same pattern of limonite distribution would be formed in all cases, irrespective of the sulphides involved.

*TUNELL has shown that for the cavity to be completely vacated of iron the ratio of py : cc must be well above 2 : 1. The latter ratio, however, yields a cavity essentially free of iron, as in Fig. 10, "B".

†Not every specimen of leached outcrop picked up over the Miami or Tyrone orebodies shows the characteristics described. A great many exceptions exist in every district, for reasons that are explained later in the paper. The types described are the characteristic and predominant ones.

*Under ideal conditions, of course ; which are not always encountered in the field.

Other Difficulties.—From this it will be seen that the conditions may become more complicated than was the simple case at Miami and Tyrone. Later, when effects of the various types of neutraliser in the gangue upon the formation of limonite are discussed, it will be seen that the conditions may become even more complex. When it is considered further, that the type of limonite produced, and its distribution with respect to the cavity, are determined not alone by the particular sulphides involved, by the sulphur-iron ratio of their oxidation solutions, and by the character of neutraliser in the gangue, but are determined in part also by other factors, such as hydrolysis, adsorption, and dilution of the solutions, it will be found that the problem may become quite formidable. If matters are carried a step still further, and instead of dealing only with disseminated deposits in which each small sulphide speck is surrounded by a large area of neutraliser in the gangue—if, instead of dealing with that type of deposit, there is introduced the massive sulphide type in which nothing but a solid mass of mixed sulphides a hundred feet or more in diameter is involved in the oxidation—the condition, then, may constitute a full challenge to the interpreter's analytical powers.

The Ideal Case.—The ideal case is one like Miami, where the ratio is 2 c.c. : 1 py ; so that (1) all the limonite produced is the most outstandingly characteristic and the most easily recognised chalcocite type, (2) all, or essentially all, the limonite is precipitated within the cavities, (3) just enough neutraliser is present in the gangue to precipitate as limonite immediately about outer edges of the cavities any iron that may be exported where locally an excess of pyrite occurs ; but still not enough to precipitate the copper as copper carbonate, copper sulphate or copper silicate ; so that all dissolved copper becomes available to be carried downward to the zone of secondary chalcocite enrichment below.

Under such ideal conditions, estimating the grade of ore represented by a leached outcrop is a comparatively simple matter. And when dealing with that type of deposit it has been possible to make close estimates of the grade of ore to be expected beneath the outcrops ; and those estimates have been checked by later drilling, and by ordinary underground exploration, and found to be correct within a few tenths of one per cent.

Ordinarily when dealing with leached disseminated copper deposits, the following classification for grading is used :

A. Ore : Material that averages more than 2 per cent. Cu.

B. Doubtful Ore : Material that averages between 1·25 per cent. and 2 per cent. Cu.

C. Low Grade : Material that averages less than 1·25 per cent. Cu but still carries appreciable amounts.

D. Barren Rock : Material essentially barren of commercial copper ; rock that ordinarily carries less than one half per cent. Cu, and may be entirely barren.

It is only fair to state that most leached outcrops are more difficult to interpret than are those at Miami and Tyrone. The reason is that in most of them more factors are involved in the oxidation, and in the formation of their limonites ; and consequently more experience and greater skill are required in the interpretation.

But it likewise is only fair to state that even the complicated cases generally yield to persistent and determined attack.

The Massive Sulphide Type :

In the massive type of sulphide deposit, where the ratio of py : c.c. or other sulphide may be 10 : 1 or 20 : 1, greater difficulty is experienced in interpreting the leached outcrops. The reason is that, where a large excess of pyrite occurs, the smeary crusts and other limonite types characteristic of oxidation under strongly acid conditions, tend to obscure or obliterate the types derived from the other less dominant sulphides to such an extent that the latter may not be present, or may not be visible, in the outcrop in their true proportions. Yet it is worth noting that in the leached outcrops that formerly overlay the massive sulphide deposits of Sacramento Hill at Bisbee (Arizona), of the United Verde Orebody at Jerome (Arizona), and of certain of the Hidden Creek orebodies of the Granby mine at Anyox (British Columbia), evidences of former chalcocite or chalcopyrite, (or bornite in the case of Sacramento Hill), were by no means obliterated, even though the ratio of pyrite to combined copper sulphides was in some cases greater than 10 : 1. Although an accurate estimate of the grade of copper ore leached from such outcrops would not have been possible, enough evidence of former chalcocite, chalcopyrite and bornite was present in the leached outcrops to have demanded exploration on an extensive scale.

Greater Complications with Sphalerite and Galena :

One point should be particularly noted. When pyrite, pyrrhotite, chalcocite, chalcopyrite, bornite, or tetrahedrite go into solution, their limonites, under normal conditions, are precipitated directly from the reactions involved. When sphalerite or galena go into solution, their limonites, under normal conditions, are not pre cipitated directly from the reactions involved ; they do not ordinarily form until the secondary sulphates or carbonates of the zinc and lead minerals have been produced, and these in turn have gone into solution.

The formation of limonite from the zinc and lead minerals therefore is a much more complicated process than is its formation from the more common iron and copper sulphides. This greater complication tends somewhat to obscure the physical features by which the respective limonite products may be recognised, particularly in the case of galena. And that explains the statement made in several of the previously published papers to the effect that of the various limonite products derived from the more common useful sulphides, those of galena derivation are the most difficult of all to recognise with assurance and to interpret accurately. But though the distinguishing characteristics may be largely obliterated, few instances have been found in the field in which they could not be to some extent distinguished, under the hand lens, if a really careful search for them was made.

Complex Massive Sulphide Mixtures :

In the massive type of sulphide deposit complex mixtures of the various sulphides sometimes occur. From what has been said it might be expected that in a mixture of, say, sphalerite, chalcopyrite, galena and pyrite, the solutions derived from oxidation of the different sulphides would mix to such an extent that none of the resulting limonite products characteristic of the individual sulphides would be preserved.

To a limited extent that is true. It is frequently, though not invariably, true where the sulphides are fine-grained and intimately intergrown. But field evidence shows that where the mixed sulphides are massive, with individual sulphide nodules upward of $\frac{1}{4}$-inch (often upward of $\frac{1}{16}$-inch) in diameter, oxidation within a given nodule commonly proceeds more or less independently of oxidation within surrounding nodules of different composition ; and that, if the entire limonite mass be broken into and examined, portions of it often will be found to contain, in place, the limonite products characteristic of the particular sulphide nodules that have been leached.

A

B

Fig. 15.

A.—Sketch showing "cleavage" boxwork, derived from galena of the orthodox, cubic-cleavage type ; as it appears when divested of the partially sintered limonite crusts shown in Fig. 1.

The long, minutely thin, parallel ribs, and the cubic "boxes," are replicas of galena's cubic cleavage. They prove that the limonitic jasper of which they are composed penetrated along the galena's cleavage planes during incipient oxidation,—before the sulphide residuals had been converted into anglesite and cerussite from which the "partially-sintered" limonite crusts of Fig. 1 later were derived. A characteristic of galena-derived limonite is thinness of the cellular walls, they being often not more than ·005 mm. thick. That is one reason why the boxwork is seldom well preserved in outcrops. Lawn Hill, Queensland. ×3.

B.—Sketch showing Diamond Mesh boxwork, derived from "steel" galena. Note the irregular diamond mesh structure, with only occasional "cubic boxes" preserved. Note more particularly the pairs of closely spaced, rigidly parallel, minutely thin webs that constitute the structure's major ribs. The empty space between these pairs of thin parallel webs has not been satisfactorily accounted for. Usually the boxwork is coated with crusts of the "partially-sintered" limonite shown in Fig. 1, which in part obscure the diamond mesh pattern. Close conformance in pattern of this structure to certain crystallisations of cerussite suggest that the pattern may be derived in some instances from such crystalline product ; but no examples have been found to date which show Diamond Mesh boxwork grading directly into crystalline cerussite masses. Eureka, Nevada. ×3.

"Cleavage" boxwork is the "key" structure for galena-derived limonite of orthodox cubic-cleavage galena. It may be a derivative of "steel" galena also. Diamond mesh boxwork has been observed to date only as a derivative of "steel" galena.

(Courtesy : *Economic Geology.*)

That is by no means an invariable rule, but is common enough to serve as a general guide in field work. It explains why at Sacramento Hill, at the United Verde Orebody, and at Anyox, evidence of former copper sulphides was prominent in the leached outcrops, even though an overwhelming amount of pyrite had been present in the sulphide ore prior to leaching.

Effects of Gangue Neutraliser :

Effects of the various types of gangue neutraliser upon formation of limonite is a subject so full of ramifications as to merit a separate paper. Only a few of its outstanding features will be emphasised here.

In the leached outcrops work three standard types of gangue are recognised : (1) gangue with no neutraliser, such as quartz or the highly sericitised ones ; (2) gangue of moderate neutraliser, such as the feldspar rocks ; (3) gangue of strong neutraliser, such as limestone.

A given sulphide, such as pyrite or chalcopyrite, oxidising in a gangue of no neutraliser, commonly yields a different type of limonite than when it oxidises in a gangue of moderate neutraliser. It yields a different limonite product still, when oxidising in a gangue of strong neutraliser, such as limestone.

"Fluffy" Limonite.—If that statement told the whole story, the problem would be simple. Unfortunately, when oxidation occurs in a limestone gangue, or wherever strong neutraliser is present, there is produced a particular type of limonite, called by us the "fluffy" type, because under the hand lens it resembles powdered sugar sprinkled over a surface ; and this

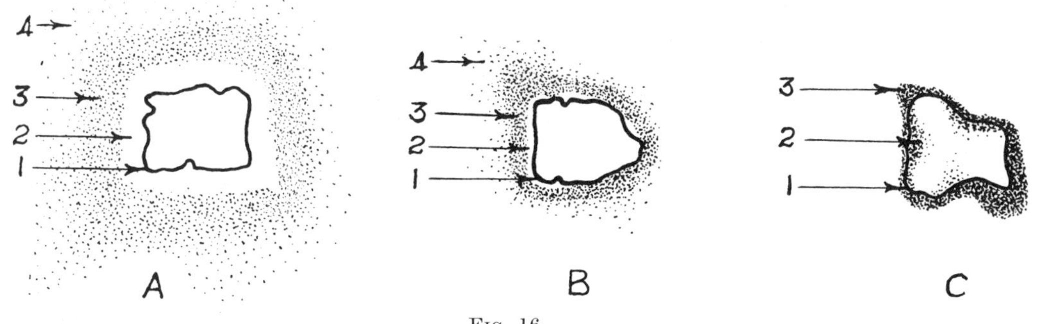

Fig. 16.

A.—Sketch showing characteristic oxidation of pyrite in strongly kaolinised abaskite porphyry. Silverbell, Arizona. ×10. 1. Outline of cavity. 2. Inner bleached zone. 3. Limonite halo. 4. Outer zone of limonite "fog." All iron has been exported from the cavity, and has been carried a substantial distance beyond its borders before the weak neutralising power of the gangue, due to strong kaolinisation, could bring about the iron's precipitation as limonite. This has given rise to the inner bleached zone, or area within which the highly acid solutions, derived from oxidation of the pyrite, travelled outwards before their acidity was overcome by the gangue's neutralising power. Weak neutralising power of the gangue, relative to acidity of the oxidation solutions, is attested further by (1) failure of the gangue to precipitate the limonite more densely within the halo, and 2) persistence of the zone of limonite "fog," or feeble limonite precipitation, for a substantial distance outward from the halo or zone of maximum limonite precipitation. Adapted from *Engineering and Mining Journal.*

B.—Sketch showing characteristic oxidation of pyrite in a gangue of moderate neutralising power, such as slightly sericitised quartz monzonite. Cactus mine, Utah. ×10. 1. Outline of cavity. 2. Inner bleached zone. 3. Limonite halo. 4. Outer zone of limonite "fog." All iron has been exported from the cavity, as in "A" ; but, because of higher neutralising power of the gangue, the inner bleached zone is imperfectly developed and is much narrower, where present, than in "A," the limonite halo is denser and more concentrated about the cavity, and the outer zone of limonite "fog" is but feebly developed.

C.—Sketch showing characteristic oxidation of pyrite in a gangue intermediate between moderate and strong neutralising power, such as slightly limy shale. Mutton Gully diggings, near Ukalunda, Queensland. ×10. 1. Outline of cavity. 2. Pulverulent limonite precipitated within the cavity. 3. Pulverulent and "fluffy" limonite precipitated immediately outside the cavity. Dense precipitation of the limonite in a narrow zone about the cavity's border, together with its "fluffy" texture, and with precipitation of pulverulent limonite within the cavity itself, attests to moderately strong neutralising power of the gangue. In a gangue of pure limestone—provided neutralising solutions remained in contact with the pyrite during its oxidation—the iron would have been precipitated largely or wholly within the cavity as "fluffy" limonite.

"fluffy" limonite has the same texture and physical appearance in all cases, irrespective of whether the iron that went into the formation of its limonite was derived from pyrite, chalcocite, chalcopyrite, siderite, magnetite, garnet, or any other mineral. By itself the "fluffy" limonite signifies nothing more than that an iron-bearing solution of indefinite source came in contact with strong neutraliser.

That condition would seem to make nearly hopeless the attempt to interpret limonites precipitated in limestone gangues, and it must be admitted that formidable difficulties have been encountered in attempting to unravel the complexities of such products.

But the situation is not hopeless. Often there is preserved within the "fluffy" mass some structure of the original mineral that serves as a specific clue. In the case of siderite or other iron carbonate, the rhombohedral structure frequently is dimly preserved. Often, too, where the sulphide nodule is moderately large, the neutralising effects of the lime manifest themselves only about outer edges of the nodules, where the sulphide came in contact with the gangue ; if the entire mass be broken open, the central portion frequently will be found to contain the limonite products characteristic of the particular sulphide that has been leached—just as so often is the case with massive sulphides of the mixed type.

This, too, is by no means invariable, and the interpretation of such limonites is not quite as simple as has been made to appear ; but it does not seem wise to introduce further complications into the discussion in a paper of this length.

One point, however, merits emphasis. When ground waters that carry in solution strong neutraliser, such as calcium or magnesium bicarbonate, flow over sulphides that are oxidising in a quartz or in a feldspar gangue, they may precipitate within such gangues the same "fluffy" type of limonite as would be precipitated if the oxidation had occurred in limestone itself. When that takes place, the difficulties of correct interpretation are obvious. But even such cases, though taxing the inspecting engineer's ingenuity, nevertheless often become subject to intelligent diagnosis.

Further Influence of Neutraliser.—The influence of neutraliser in the interpretation of leached outcrops is important beyond its effects upon precipitation of limonite. At Miami, neutraliser in the gangue is so low that none of the dissolved copper is precipitated as copper carbonate or other oxidised copper mineral. Virtually all of the dissolved copper becomes available to be carried downward to the zone of secondary chalcocite enrichment below.

At Ajo (Arizona) neutraliser in the gangue is so high that essentially all of the dissolved copper is precipitated as copper carbonate or other oxidised copper mineral, either at, or immediately adjacent to the point where the sulphide particle oxidised. That explains why grade of the oxidised ore at Ajo is the same as grade of the primary ore : secondary enrichment has not taken place ; there has merely been change from primary sulphide copper to secondary oxidised copper, without addition or subtraction of copper content.

With the combination of Ajo's gangue and Miami's primary sulphide deposit, the secondary chalcocite orebodies at Miami could not have been formed, and the Miami district, as a major copper producer, probably would not have become known to the world. With the combination of Miami's gangue and Ajo's primary sulphide deposit, and with an amount of erosion equal to that at Miami, there would have been produced a secondary chalcocite orebody, rivalling in richness and extent that of United Verde Extension, or possibly that of Kennecott itself.*

From these considerations it is clear that determination of the amount and character of neutraliser in the gangue, and knowledge of its effects upon the oxidation solutions, is as important in the interpretation of leached outcrops as is the ability to recognise the particular limonite types produced.

Significance of Limonite Colour :

Much has been said about significance of colour in limonite as a guide to prospecting ; and much of what has been said has been irresponsible and profitless talk.

Nevertheless, certain colours of limonite have been found characteristic of particular sulphides in some districts. As nearly as we have been able to check up in the field, the brick red colour of limonite found in

*Kennecott chalcocite is, of course, for the most part primary.

many disseminated copper districts occurs only over the pyritic portions too low in copper content to be commercial.* Chalcocite limonites, precipitated in a feldspar gangue, commonly show a maroon or seal brown colour ; though when precipitated in a limestone gangue the colour is more likely to be a Bismarck brown. Limonites derived from chalcopyrite frequently show a deeply ochreous colour, and at other times an Indian red. Galena limonites, not exposed to surface weathering, often show an orange-ochreous colour ; but that colour is characteristic also of certain limonite products derived from bornite.

Colour in limonite is determined largely by (1) amount of adsorbed or capillary water present in the limonite, (2) state of aggregation of the limonite particles, (3) size of the limonite grains. A red limonite ground sufficiently fine becomes orange to ochreous ; ground still finer, it becomes yellow.

A great many other conditions, such as acidity of the solutions, rapidity of neutralisation, or adsorption in the case of well kaolinised gangues, affect grain size, and govern the amounts of iron and water that enter into the composition of the limonite. For these reasons constancy of limonite colour, either in the same or in different districts, is not to be rigidly expected. In one district in Arizona chalcocite yields a certain colour of limonite in a gangue of fine grained alaskite porphyry, whereas directly beside it pyrite yields precisely the same colour of limonite in a gangue of the more coarsely grained but closely related alaskite. In still another district, closely inter-disseminated chalcocite and molybdenite yield limonites of precisely the same maroon colour.

Limonite colour, though it may serve in some instances as a general clue to origin of the limonite, must be regarded as a clue only. Accurate interpretation always will depend upon recognition of the texture and type, and upon knowledge of the chemistry and physics of the processes needed to produce that particular type, under the conditions existing in the outcrop during oxidation.

*Except insofar as it may represent limonite of copper carbonate derivation,—mainly restricted to limestone districts.

[*Editor's Note:* Material has been omitted at this point.]

Acknowledgments :

Acknowledgments are due *Economic Geology* and *Engineering and Mining Journal* for permission to reproduce figures, and to republish data, contained in certain of the previous leached outcrop papers.

BIBLIOGRAPHY.

WHITE, CHARLES HENRY, "Prospecting for Disseminated Copper by a Study of Leached Croppings," *Eng. & Min. Journ.*, Mar. 22, 1924.

MORSE, H. W. and LOCKE, AUGUSTUS, "Recent Progress in Leached Ore Capping," *Econ. Geol.*, April-May, 1924.

LOCKE, AUGUSTUS, "Leached Outcrops as Guides to Copper Ore," Williams & Wilkins, Baltimore, Md., 1926.

BOSWELL, P. F. and BLANCHARD, ROLAND, "Oxidation Products Derived from Sphalerite and Galena," *Econ. Geol.*, Aug., 1927.

"Cellular Structure in Limonite," *Econ. Geol.*, Dec., 1929.

BLANCHARD, ROLAND and BOSWELL, P. F., "Notes on the Oxidation Products Derived from Chalcopyrite," *Econ. Geol.*, Nov., 1925.

"Status of Leached Outcrops Investigation," *Eng. & Min. Jour.*, Feb. 18 and Mar. 3, 1928.

"Limonite Products Derived from Bornite and Tetrahedrite," *Econ. Geol.*, Sept.-Oct., 1930.

"Additional Limonite Types of Galena and Sphalerite Derivation," *Econ. Geol.*, Nov., 1934.

" 'Limonite' of Molybdenite Derivation," *Econ. Geol.*, May, 1935.

WISSER, EDWARD, "Oxidation Subsidence at Bisbee, Arizona, *Econ. Geol.*, Dec., 1927.

ANDERSON, A. L., "The Incipient Oxidation of Galena," *Econ. Geol.*, Aug., 1930.

"Notes on the Oxidation of Jamesonite, Sphalerite and Tetrahedrite," *Econ. Geol.*, Dec., 1932.

TRISCHKA, CARL, ROVE, OLAF N. and BARRINGER, D. MOREAU, "Boxwork Siderite," *Econ. Geol.*, Nov., 1929.

BLANCHARD, ROLAND, "Significance of the Iron Oxide Outcrop at Mount Oxide, Queensland," to be published.

Also the following related, more strictly technical papers :

POSNJAK, E. and MERWIN, H. E., "The Hydrated Ferric Oxides," *Amer. Jour. Sci.*, 47, p. 311, 1919.

"The System Fe_2O_3-SO_3-H_2O." *Jour. Amer. Chem. Soc.*, 44, p. 1965, 1922.

POSNJAK, E. and TUNELL, GEORGE, "The System CuO-SO_3-H_2O," *Amer. Jour. Sci.*, 18, p. 1, 1929.

TUNELL, GEORGE and POSNJAK, E., "The Stability Relations of Geothite and Hematite," *Econ. Geol.*, May and Dec., 1931.

"A Portion of the System Fe_2O_3-CuO-SO_3-H_2O," *Jour. Phys. Chem.*, 35, p. 939, 1931.

GREGORY, J. W., POSNJAK, E., MERWIN, H. E., and SOSMAN, R. B., "Equilibrium Relationship of Fe_3O_4, Fe_2O_3 and Oxygen," *Amer. Jour. Sci.*, 30, p. 239, 1935.

MERWIN, H. E. and LOMBARD, R. H., "The System Cu-Fe-S," *Econ Geol.*, Mar.-April, 1937.

MERWIN, H. E. and POSNJAK, E., "Sulphate Incrustations in the Copper Queen Mine, Bisbee, Arizona," *Amer. Mineral*, 22, p. 567, 1937.

[*Editor's Note:* A discussion of the article has been omitted.]

23

Reprinted from pages 119–124, 126–127, 129–138, 141–142, and 144–150 of *Minerals Sci. and Eng.* **9:**119–150 (1977)

SULPHIDE WEATHERING AND THE EVALUATION OF GOSSANS IN MINERAL EXPLORATION.

C. F. BLAIN, B.Sc. (Hons.), M.Sc., D.I.C., Ph.D. and R. L. ANDREW, B.Sc. (Hons).
Mining Geology Division, Geology Department, Imperial College of Science and Technology, London SW7 2BP.

INTRODUCTION

Ancient mine workings, four to five thousand years old, in the weathered zones of sulphide orebodies suggest that ore discovery has been linked to the evaluation of leached, iron oxide cappings since earliest mineral history. Several well-documented examples[1,2] remind us that before modern times prospectors recognized the iron oxide-rich, weathered cappings over massive sulphide bodies, or gossans as they are termed*****, as loci where ores of the base and precious metals commonly occur. Even within the past decade, numerous mineral deposits have been discovered by the recognition of gossan outcrops. In modern mineral exploration, therefore, the methods of assessing the economic significance of gossans and ironstones remain a key area of the technology, because in many mineralized terrains, gossans provide the only visible indication at surface of potential ore at depth.

In the near-surface weathering regime, most hypogene sulphide mineral assemblages are unstable in the presence of weathering agents, particularly water-dissolved oxygen, carbon dioxide, and ionic species. The action of these weathering agents causes the sulphide body to re-equilibrate electrochemically and give rise to more stable secondary sulphide and oxide mineral assemblages. The weathering agents, being derived mainly from the atmosphere, generally decrease in concentration with increasing depth below the surface, and therefore, sulphide bodies commonly form vertically zoned, weathered profiles. Ideally then, the hypogene orebody grades upwards into a more oxidized secondary sulphide zone, which may be enriched in ore metals. This zone is overlain by an oxide zone, leached of ore metals, which is capped by gossan at surface.

Distinctive suites of minerals generally characterize the supergene sulphide zones derived from differing hypogene ore types. The most critical step in the overall gossan-forming process is the transformation of sulphide assemblages to stable oxidate assemblages. At this point, where sulphide oxidizes to sulphate, the intrinsic metal-sulphur bonds are broken and the released metal cations are either dissolved in the co-existing groundwaters, or are precipitated as insoluble oxidate minerals. Substantial amounts of iron, associated with many types of base and precious metal deposits in sulphide form, are fixed above the water table, largely as goethite or hematite. Other metals are partly leached away in the groundwater, and partly precipitated nearby; variously as oxides, carbonates, sulphates, halides, phosphates, or even vanadates, depending to a large extent upon the geological environment. In extreme cases, this type of transformation may obscure the genetic links between gossan and parent sulphide. The residues of iron-bearing minerals, together with varying amounts of introduced silica, are commonly the most abundant constituents of the gossans, or 'iron hats', which cap sulphide mineral deposits.

The weathering environment, however, being characterized by the availability and behaviour of the weathering agents, is complex and governed by many factors, such as climate (temperature and rainfall), geology (the physical

C. F. Blain graduated from the University of Western Australia in 1965, and subsequently worked with Pickands Mather & Co. International in several mineral exploration programmes in Australia before leaving in 1968. In 1969, he was awarded the degrees M.Sc., D.I.C. (Mineral Exploration) from the Royal School of Mines, London, and in 1972 the degree Ph.D. (Geological Sciences) from Queen's University, Kingston, Canada. Since 1972, he has been lecturing in mining geology at the Royal School of Mines, London. His main research interests include the geochemistry of mineral deposits and their oxidized cappings.

R. L. Andrew received his B.Sc. degree from the University of Glasgow in 1970, when he joined the R.T.Z. Group as exploration geologist, working in the U.K. and on foreign aid in Burma. In 1973, he moved to South Africa where he worked on the exploration staff of Palabora Mining Co. and as project geologist with the Hunting Group. Following preparatory studies, he commenced research towards a Ph.D. in 1976 at the Royal School of Mines, London. The research project primarily concerns the evaluation of Cu–Zn–Pb gossans and other ironstones in southern Africa.

*This term was probably first applied to the iron-rich oxidized cappings over the sulphide-bearing lodes in Cornwall. In Cornish usage[3], the term was defined as 'a kind of imperfect iron ore, commonly of a tender rotten substance, and red or rusty iron colour. It is often found shallow in tin, copper and lead mines, and is the proper nidus or matrix for the two latter. It is an upper covering to the ore, levels above thirty fathoms, and is very abundant; whence those lodes are called gossan lodes'. In current usage, the term refers to the oxidized cap rocks over massive or semi-massive sulphide mineralization.

TABLE 1
MINERAL ABBREVIATIONS USED IN FIGURES AND TABLES THROUGHOUT THIS PAPER.

ac : aurichalcite	cp : chalcopyrite	id : idaite	po : pyrrhotite
ag : silver	cr : chrysocolla	ip : iron-poor	pr : pyromorphite
al : anglesite	cs : cerussite	ir : iron-rich	ps : pharmacosiderite
an : acanthite	ct : cuprite	jr : jarosite	pt : polybasite
ar : argentite	cu : copper	js : jasper	py : pyrite
at : atacamite	cv : covellite	kl : kaolinite	qz : quartz
au : gold	dg : digenite	ma : manganosiderite	sc : sauconite
az : azurite	dj : djurleite	mc : marcasite	sd : siderite
bb : blaubleibender	dl : dolomite	mg : magnesite	si : silica
bd : beudantite	ds : descloizite	ml : malachite	sl : sulphur
bh : bindheimite	em : embolite	mm : mimetite	sm : smithsonite
bl : boulangerite	fr : freibergite	mn : minium	sp : sphalerite
bn : bornite	gl : goslarite	mr : millerite	sr : scorodite
br : brochantite	gn : galena	ms : massicot	ss : silicates
bt : bournonite	gp : gypsum	mt : melanterite	st : sulphates
bv : bravoite	gs : gaspeite	nl : nickel-rich	su : sulphur-rich
ca : calcite	gt : goethite	np : nickel-poor	tn : tenorite
cb : carbonates	hl : hawleyite	pa : pyrargyrite	vl : violarite
cc : chalcocite	hm : hematite	pj : plumbojarosite	vn : vanadinite
ch : chalcophanite	hz : hydrozincite	pl : polydymite	wd : wad
cl : chlorargyrite	ia : iodargyrite	pm : psilomelane	wf : wulfenite
cn : chalcanthite	ib : iodobromite	pn : pentlandite	wm : willemite

and chemical nature of the ores and host rocks), geomorphology (land forms, drainage), and groundwater movement. In arid regions of deep water table, for example, oxidized ores are known to occur as deep as 800 m below the present surface[4]. Generally, the phenomenon of supergene alteration of massive sulphide orebodies is confined to zones nearer surface than this. Nevertheless, many sulphide deposits in the weathering regime have been subjected to complex geomorphological, hydrological, and climatic changes over long periods of geological time. The resultant profiles, therefore, differ considerably.

In many cases these complex processes of supergene alteration have enriched the initial ore grades[5,6] below surface. Commonly, however, particularly in mineralized terrains where deep, penetrative weathering has resulted in considerable near-surface mobilization of iron and silica, the supergene alteration has obscured the identity of sulphide deposits at surface. Recently, interest has been renewed in the problems of characterizing gossans that occur in these types of terrain. This interest is manifest, at least to some degree, by several notable contributions to the published literature[7–9]. Early gossan evaluation was largely based on the recognition of diagnostic textures in hand specimen[10,11]. With the more recent advances in the development and application of methods of mineragraphic examination and of rapid chemical analytical methods, considerable information now exists on the textural and geochemical characterization of gossans. With these advances in mind, we consider it timely in this paper to review the development and nature of existing techniques of evaluating gossans in relation to what is known about the processes giving rise to gossan formation.

FIELD RELATIONS

Weathering profiles of massive sulphides vary a great deal in the nature of their development and in their maturity. From a review of the principal massive sulphide ore associations, a suite of type profiles emerges (Figure 1) and the field relations of these are described below by discussion of several well-documented examples from each class. Abbreviations of mineral names (Table 1) are used where appropriate in figures and tables throughout this paper.

Many *nickel–copper sulphide orebodies*, associated with ultramafic rocks, typically occur in the greenstone belts of stable cratonic regions that have been subjected to protracted, penetrative weathering over long periods of time. The supergene alteration profile developed at Kambalda[12] and Mount Windarra[13], Western Australia and Pikwe[14], southern Africa persists well below the contemporary water tables (Figure 1). The hypogene ores at depth are connected by transitional zones (90 to 110 m vertical interval), where pentlandite and some of the pyrrhotite are replaced by violarite, to overlying zones (35 to 45 m vertical interval) of pyrite–(marcasite)–violarite–chalcopyrite ore. Above the water table, the sulphides grade sharply into oxide zones (commonly 30 m deep) of goethite–hematite–silica, which are capped by jasperoidal gossans at surface.

The total depth of supergene alteration of the nickel orebodies around the Kambalda dome varies between 30 and 200 m[12]. The uppermost portions of the violarite–

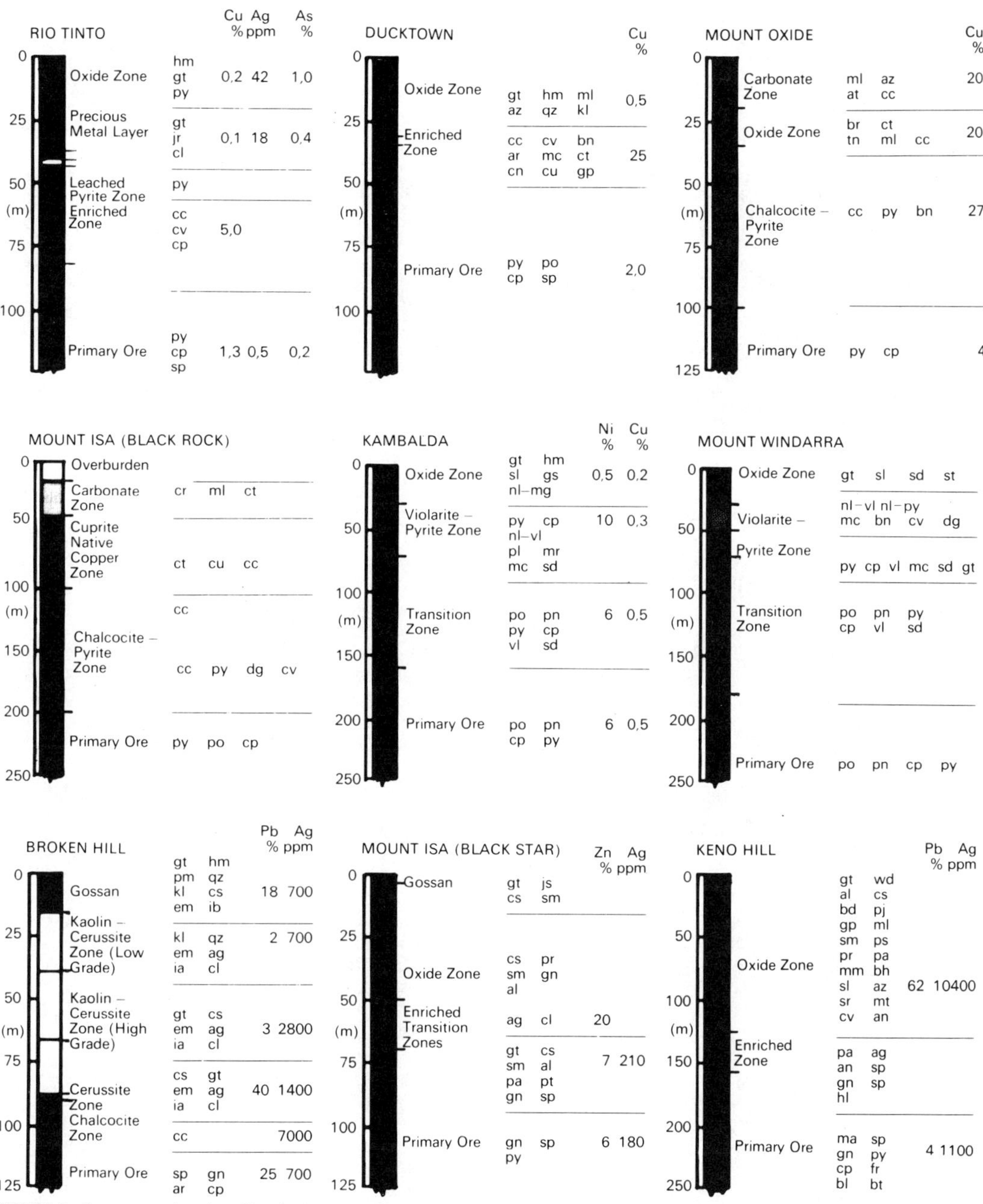

FIGURE 1 Supergene-altered profiles of principal massive sulphide ore types.
Abbreviations for the dominant minerals in each zone are listed in Table 1. Geochemical data are derived from a number of sources[1,2,12,13,18,19,20,22,23,32,33]

[*Editor's Note:* In the original, Figure 1 appears with various colors representing the ore types.]

pyrite zone commonly show evidence of supergene enrichment of nickel in violarite and in some cases, such as Otter Shoot, by the occurrence of supergene millerite[15]. Many of the profiles also contain nickeliferous carbonate minerals[12,16] as products of supergene alteration.

The oxidized profiles of *copper–zinc sulphide orebodies associated with acid volcanic rocks* ideally consist of a basal supergene sulphide enrichment zone, overlain by a narrow earthy precious metal-oxide layer and an upper carbonate-rich oxide zone. This interpreted sequence has been synthesized from the examples illustrated (Figure 1), although none of the examples exhibits the complete progression. In the Salomon-Cerro Colorado area of *Rio Tinto, Spain*[1,17], (Figure 1) the base of the oxide zone is 10 to 40 m deep and the contact is generally sharp. The earthy, precious metal layer (1 to 2 m vertical interval) below the oxide zone, overlies a thin horizon of leached pyrite. A well-developed zone of secondary sulphide enrichment (30 to 40 m vertical interval) grades imperceptibly into the hypogene ore. The copper ores at *Ducktown, Tennessee*[18–20] comprise 60 per cent pyrrhotite and have oxidized in a humid climate producing a shallow, immature weathering profile. Incised valleys have exposed the profile down to the thin chalcocite–suphate zone, where the discovery outcrop contained native copper.

The weathering profile of the pyritic deposits at *Minas Carlota, Cuba*[21], illustrates the strong development of the earthy layer. There the oxide zone (20 to 45 m deep) developed over steeply dipping pyritic bodies, is underlain by a cavernous earthy layer, (typically 8 m vertical interval) which Hill[21] regards as characteristic of the the weathering of pyrite-rich ores.

The well developed, mature profiles at *Mt. Isa*[22] and *Mount Oxide*[23], *Queensland*, however, lack only the near-surface ferruginous cappings. Strong leaching of iron from the oxide zones of these two profiles is apparent in the copper silicate–carbonate–oxide assemblage. At Mount Oxide (Figure 1), 60 m of massive chalcocite was stranded above the present water table, which was marked by a 15 m zone of sooty chalcocite with accessory bornite. The oxidized profiles of *lead–zinc sulphide deposits, associated with carbonate rocks*[24,25] are generally capped by a gossan (silica–hematite–manganese) overlying an oxide zone containing zinc carbonates and sulphates (Table 2). In places, a layer of enriched zinc–lead carbonates containing residual sulphides, grades into the underlying hypogene ore. Galena is more resistant to weathering than pyrite and sphalerite and may persist to surface[121]. No continuous zones of supergene sulphide enrichment occur in the profiles and the extent to which these zones are developed is extremely variable. At *Leadville, Colorado*[26], for example, the oxide zone irregularly overlaps the sulphide zone over a vertical range of 250 m. Supergene sulphide enrichment of copper and silver occurs in economic concentrations. At *Balmat-Edwards, New York*[27], fresh sulphide is commonly overlain by a few centimetres of gossan but strong hematization persists to 16 m below surface. Minor supergene sphalerite and galena have been recognized[28]. Similar near-surface features occur in the profiles of *Austinville, Virginia*[29] and *southwest Sardinia*[30] where karst hollows and pockets in clay overburden are infilled by oxidate minerals. The oxide zone at Austinville is dominated by hemimorphite and smithsonite whereas that of the Sardinian deposits comprises a smithsonite–hematite–hydrozincite assemblage with much residual galena.

TABLE 2
PRINCIPAL MINERALOGICAL ASSOCIATIONS OF SUPERGENE ALTERED PROFILES, LEAD-ZINC SULPHIDE DEPOSITS IN CARBONATE ROCKS (ADAPTED FROM HEYL AND BOZION[24])

Zone	Principal Mineralogy	Accessory Mineralogy
Gossan	sl hm	gt
Oxide–Sulphate	hm gt sl cs pj ca	ml az sm hz
Carbonate–Oxide	hz hm cs ac sm	mn mm ds vn wf ct sc gt
Oxide–Carbonate	hm hz ch sm	ac cs al wf pa
Enriched Carbonate	sm hm ml cs	at pa wm cc ct gn cp ma
Carbonate–Sulphide	sm hm ma	gl sc ca cp gn sp py
Primary Ore	py gn sp cp	ca dl

Other types of lead–zinc–silver sulphide deposit, particularly those not enclosed by carbonate rocks, commonly have well-developed profiles, consisting of a cerussitic siliceous gossan below surface (5 to 10 m), an oxide zone (40 to 50 m vertical interval) dominated by a cerussite–smithsonite–silver halide assemblage and a cerussite-enriched zone (20 to 30 m vertical interval) with varying amounts of smithsonite, anglesite, silver sulphosalts, and residual sulphides. The depth to primary sulphides generally varies from 70 to 100 m below surface. At *Mount Isa, Queensland*[31], over the Black Star orebody (Figure 1), pyromorphite is common in the oxide zone, which has native silver–chlorargyrite–smithsonite lenses in its lower portion. This weathering profile is only half the depth of the adjacent copper orebody profile[22] (Figure 1). The oxide zone at *Broken Hill, New South Wales*[32], essentially is composed of a kaolin–cerussite–silver halide assemblage that may extend to depths of 65 to 140 m below surface. In places, the zone of enrichment wedges out below a thickened oxide zone. A thin horizon of chalcocite enrichment (5 cm to 1 m vertical interval) immediately overlies the primary lead–zinc–silver ore. The oxidized zones at the *Keno Hill – Galena Hill*[33] vein-type, lead–silver–zinc deposits in the Yukon (Figure 1) display an exotic and diverse mineralogy within a wide depth range of 6 to 160 m. A distinct zone of silver enrichment (2 to 20 m vertical interval) immediately overlies the complex primary ore.

In geomorphologically stable areas the depth zonation commonly relates to an even land surface and a present-day water table, but in deeply dissected terrain the zonal development is independent of the contemporary topography and water table. The importance of ancient water tables in controlling the development of sulphide weathering profiles is apparent from the field relations.

THE GEOCHEMICAL ENVIRONMENT OF SULPHIDE WEATHERING

Weathering is largely a consequence of the chemical interaction between mineral phases and co-existing, near-surface groundwaters charged with oxidizing agents. In the pervasive chemical weathering of silicate rocks, hydrolysis-type reactions, involving the release or consumption of hydrogen ions and hydroxyl groups, are commonly regarded as the most important type of chemical change[34,35]. Hydrolysis of silicate minerals promotes the

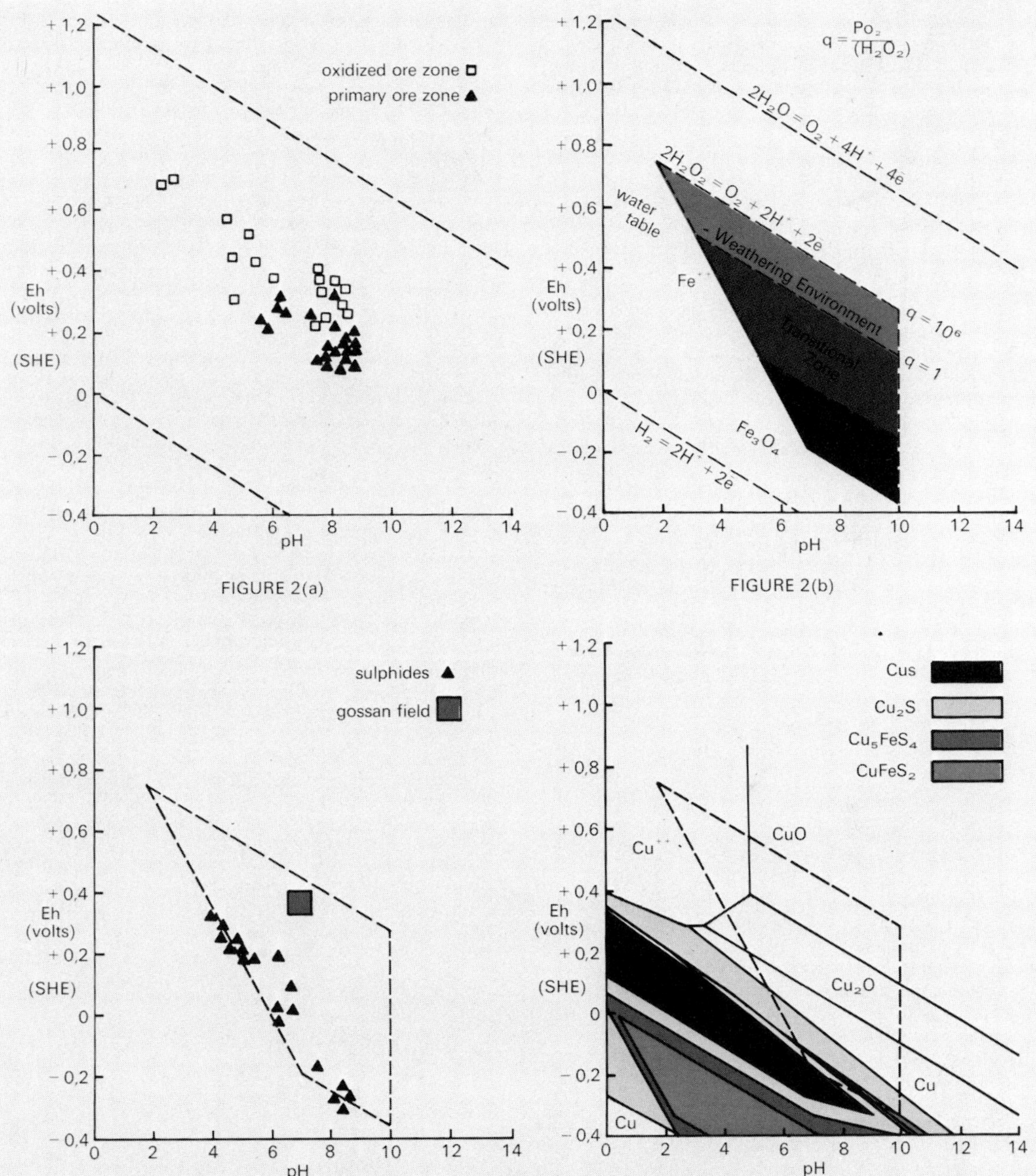

FIGURE 2 Eh–pH diagrams relating to the oxidation of sulphide bodies.

FIGURE 2(a) shows Eh–pH measurements on mine waters associated with oxidizing copper deposits in the southwestern United States. Dashed lines refer to the stability field of water. Figure 2(b) shows the deduced ranges of Eh and pH for oxidizing sulphide systems. Figures 2(a), 2(b) are adapted from Sato[41]. Eh–pH measurements shown in Figure 2(c) are for samples from the supergene–altered part of the Lunnon Shoot nickel profile at Kambalda, Western Australia and the Eh–pH field of some of the Kambalda gossans. The dashed lines refer to the same field shown in Figure 2(b). Figure 2(c) is adapted from Thornber[42]. By projecting this field onto the mineral equilibrium diagram for the Cu–Fe–S–H_2O system (calculated from thermodynamic data at 25°C, assuming 0,1 M solute activities and 0,01 M Cu^{2+} activity), some general phase equilibria applicable to the supergene alteration zones of copper sulphide deposits are indicated in Figure 2(d). Figure 2(d) is adapted from Peters[43].

production of complex silicic and aluminosilicic colloids with the release of cations that may be absorbed onto, or co-precipitated with, colloid phases[36]. Chemical weathering of this type generally results in the formation of a soil profile overlying bedrock[35].

In contrast to almost all silicate rocks, most massive sulphide bodies are coherent electronic conductors[37], and therefore, the main types of reactions involved in the weathering of massive sulphide bodies are electrochemical reactions that take place between the sulphide conductor and the co-existing aqueous media. Supergene alteration of reactive sulphide profiles commonly takes place, therefore, essentially as an electrolytic corrosive process[38]. The general chemical theory relating to these processes has been well summarized by previous workers[37,39,40].

Electrochemical processes consist of couples of oxidation-reduction, half-cell reactions, each of which involves the exchange of electrons. The oxidation half-cell reaction is electron-producing, and for reactions in aqueous media, this can be expressed in the general form:

$$aA + wH_2O \rightarrow bB + mH^+ + n\bar{e},$$

where $\bar{e}$ is the electron, and the other lower case letters specify the numbers of moles of the participating substances. The oxidation reaction is linked to a complementary electron-consuming reduction half-cell reaction. By applying the second law of thermodynamics to the system, the conditions for equilibrium of the oxidation half-cell at 25°C and one atmosphere pressure are:

$$Eh = Eo - \frac{0{,}0592}{n} pH + \frac{0{,}0592}{n}(-a \log(A) + b \log(B))$$

volts, where (A) and (B) are the activities, Eh is the oxidation potential relative to the standard hydrogen electrode, Eo is the standard oxidation potential of the reaction corresponding to the unit activities of all chemical species involved, and pH is the negative logarithm of the hydrogen ion activity. This treatment reduces the significant variables to Eh, pH, and the activities (related to concentration) of the ionic species. This simplification becomes very useful when applied to a multicomponent system as all three variables thus remain dependent on the overall equilibrium of the system. For this reason, it is convenient to describe the geological environment in which sulphide weathering reactions take place in terms of Eh/pH diagrams at specified ionic activities. If the Eh of the environment exceeds the equilibrium Eh of a valid specified reaction, the reaction will take place in the oxidation sense. It should be stressed, however, that the Eh in the natural environment is controlled by all the chemical species present, and not *vice versa*.

The ranges of Eh and pH in the zones of weathering and sulphide oxidation were first systematically investigated by Sato[41] whose results were based upon direct measurement of both Eh and pH in mine waters in the vicinity of oxidizing orebodies. The results of these investigations, summarized in Figure 2(a), define a flattened, wedge-shaped area on the Eh/pH diagram, which portrays the chemical conditions under which sulphide systems commonly oxidize. Based upon geological observations on natural systems, Sato deduced the limiting conditions of this environment in terms of calculated Eh-pH functions (Figure 2(b)). The postulated, lower limit of deep oxidation corresponds approximately to the function: $Eh = 0{,}221 - 0{,}0592$ pH, the equilibrium value of reducing hematite to magnetite. In the zone of weathering, the postulated upper limit is: $Eh = 0{,}895 - 0{,}592$ pH where the theoretical catalytic decomposition of hydrogen peroxide controls the oxidation potential of an aqueous system in contact with free oxygen. The postulated lower pH limit is specified by: $pH = 5{,}96 - 5{,}64$ Eh where goethite is reduced to ferrous ion. The suggested upper limit: $pH = 10$, is based upon consideration of calcite – water equilibrium. The accuracy of these limiting conditions, of course, is dependent upon the validity of the thermodynamic data with which they are calculated. A more serious problem, however, is that the Eh–pH measurements of Sato[41] were made on waters exposed to atmospheric oxygen by mining, and thus there is some doubt as to whether the measured Eh values, probably being variously contaminated, are not higher than the natural Eh values before mining started.

This problem was alleviated by Thornber[42], who estimated the Eh–pH conditions of oxidizing nickel ores in their pre-mining environment. Accurate data on the supergene environment at Kambalda, shown in Figure 2(c), clearly show the progression from low Eh–high pH near the water table. Above the water table, where sulphides have been converted to oxides, the regime is characterized by high Eh and moderate pH. These data are in approximate agreement with the chemical conditions of supergene alteration as deduced by Sato[41], (Figure 2(b)), although Eh in the depth environment is perhaps even lower than predicted. In relation to the corresponding, vertically zoned, mineralogical changes in the oxidizing profile, the data clearly demonstrate that the primary ores progressively equilibrate with co-existing groundwater solutions, the oxidation potentials of which decrease with increasing depth. Furthermore, the tops of oxidizing sulphide bodies are characterized by low pH, whereas the deep zones are alkaline. The chemical mechanisms that give rise to this phenomenon have several important implications in supergene alteration and gossan formation processes, and are therefore discussed in more detail below.

Within the constraints of the available thermodynamic data it is possible to predict the stability of certain sulphide phases within these anticipated Eh–pH conditions of weathering. This approach[40], applied to equilibria

[*Editor's Note:* Figure 3, color photomicrographs illustrating the mineralogical and textural characteristics of the supergene alteration profile over pyrrhotite—pentlandite ores, has been omitted. Also, in the original, the arrows in Figure 4 appear in color.]

of supergene sulphide and oxidate minerals, can serve to indicate which assemblages of minerals may co-exist under the range of Ep–pH variation in oxidizing ore profiles. The stability fields of supergene copper minerals[43] in the system Cu–Fe–S–H_2O at 25 °C are shown in Figure 2(d). A line defining the chemical conditions of the oxidizing profiles thus shows the mineral assemblages expected to be in equilibrium throughout the profile. This approach may be enhanced[40] by considering also the control of varying activities of participating chemical species, on the same equilibria. At present, however, there are hardly sufficient accurate thermodynamic data on sulphide minerals, nor are there sufficient Eh, pH, chemical and temperature data on naturally oxidizing ore profiles, to apply this approach rigorously to problems of supergene alteration. Moreover, it is quite likely that the kinetics of supergene alteration reactions reflect departures from equilibrium conditions. This type of departure might be greatest under geological conditions (such as those imposed in tectonically active zones by a rapidly falling groundwater table), where a large part of the supergene alteration profile would be considerably disequilibrated with its surroundings. The chemical consequences of mineralogical changes induced by perturbations of this type in the oxidation environment may have an important influence on the mechanisms by which orebodies oxidize in various geological terrains. Moreover, it is conceivable that a single profile may have a complex history where the chemical conditions in the oxidizing environment change significantly through time as the profile evolves. Geological considerations, such as these, impose constraints on the rigour with which the precise mechanisms of supergene alteration and gossan formation can be deduced by applying chemical theory.

THE MECHANISMS OF OXIDATION OF SULPHIDE OREBODIES

Since the early observations of Fox[44], Skey[45], and others[46–49], many workers[50–60] have postulated on the role of electrochemical mechanisms in the near-surface oxidation of sulphide deposits. It is appropriate to review generally the currently accepted electrochemical mechanism by which sulphide bodies respond to supergene alteration.

As orebodies are progressively exhumed, they encounter a zone in the weathering regime where their contained sulphide minerals gradually become chemically unstable with respect to co-existing oxygenated groundwater solutions. Initially, a point is reached where the Eh of the solutions exceeds equilibrium Eh value for incipient supergene alteration of an unstable mineral phase to occur[40]. Given favourable kinetics, the mineral oxidizes as a half-cell oxidation reaction, releasing electrons that are conducted through the ore to be consumed elsewhere by a complementary reduction couple. Continued supergene alteration of primary ore can be largely considered a sequence of progressive half-cell oxidation reactions. These are similarly coupled through the massive sulphide conductor to reduction (electron-consuming) half-cell reactions nearer surface. Oxygen or oxygen-bearing compounds need not be involved directly in the low-Eh oxidation reactions[16], although they invariably participate in the reduction couple[37]. In releasing electrons, the oxidizing ore becomes positively polarized and forms an anode[37]. At the top of the conducting orebody, dissolved oxygen, in continuously replenished supply in the groundwater, is reduced to hydroxyl groupings and the consumption of electrons causes negative polarization in the cathodic region. Current flow through the ore is matched by external ionic current flow through the groundwater system, with net migration of positive ions towards the cathode and negative ions towards the anode. Thus, during supergene alteration, massive sulphide bodies probably behave as very large electrochemical corrosion cells[16].

It may be quite common for the electrochemical corrosion mechanism of orebody weathering to act continuously over long periods of time, although at differing rates, as the overall mechanism is probably related to exhumation of the sulphide body relative to the water table[60]. Each of the supergene alteration reactions reflect the tendency of the sulphide profile to equilibrate with groundwaters, the Eh voltage of which decreases with increasing depth. In general, the supergene reaction rates become faster as the departures in Eh voltage from the Eh of reaction equilibria increase. In nature, overall disequilibrium may be sustained by exhumation and movements of the water table. Certainly, above the water table, most supergene sulphides, under the influence of percolating oxygenated solutions, corrode very rapidly. So long as groundwaters induce electrochemical reactions within the sulphide profile, electrolysis must be accompanied by electronic current flow through the orebody and complementary external current flow through the groundwater system.

In this mechanism, the formation of stable iron oxides (gossans) at the top of the conducting profile is thus the last step in the anodic reaction sequence. For most ore types, the final transformation of sulphides to oxides is kinetically a relatively rapid reaction[70]. This commonly also releases metal cations, some of which may percolate below the water table and react with pre-existing sulphides forming supergene enrichment zones. From the viewpoint of evaluating gossans, however, consideration of the total supergene evolutionary sequence of the profile is extremely significant. This is because the gossans are derived from secondary, supergene sulphide assemblages, the mineralogy, geochemistry, and textures of which may differ considerably from those of the primary ores.

To illustrate the evolution of supergene profiles, first consider the general sequence of supergene sulphide development in *pyrrhotite–pentlandite orebodies* (Figure 3). This sequence has been fully described for the profiles at Kambalda[12] and Mount Windarra[13], Western Australia, and current work shows that the development of this general sequence is quite common throughout Western Australia and southern Africa[14]. The deepest (lowest Eh) supergene alteration occurs where pentlandite is replaced by violarite (Vpn) [Figure 3(g)], accompanied by the release of iron and nickel ions. The released nickel reacts with surrounding pyrrhotite to form additional violarite (Vpo) [Figure 3 (f)], and the released iron precipitates mainly as siderite. These two types of violarite are texturally distinct, because violarite after pentlandite (Vpn) retains the octahedral partings of its host, whereas violarite after pyrrhotite (Vpo) assumes a featherlike, pseudomorphous reaction rim irregularly penetrating the basal partings around the grain margins of adjacent monoclinic pyrrhotite [Figures 3(g), (f)]. At higher Eh, residual pyrrhotite oxidizes to form pyrite or marcasite, which, in places develops a colloform texture [Figure 3(e)]. This assemblage of supergene violarite and pyrite, perhaps also with primary pyrite and chalcopyrite, remains essentially

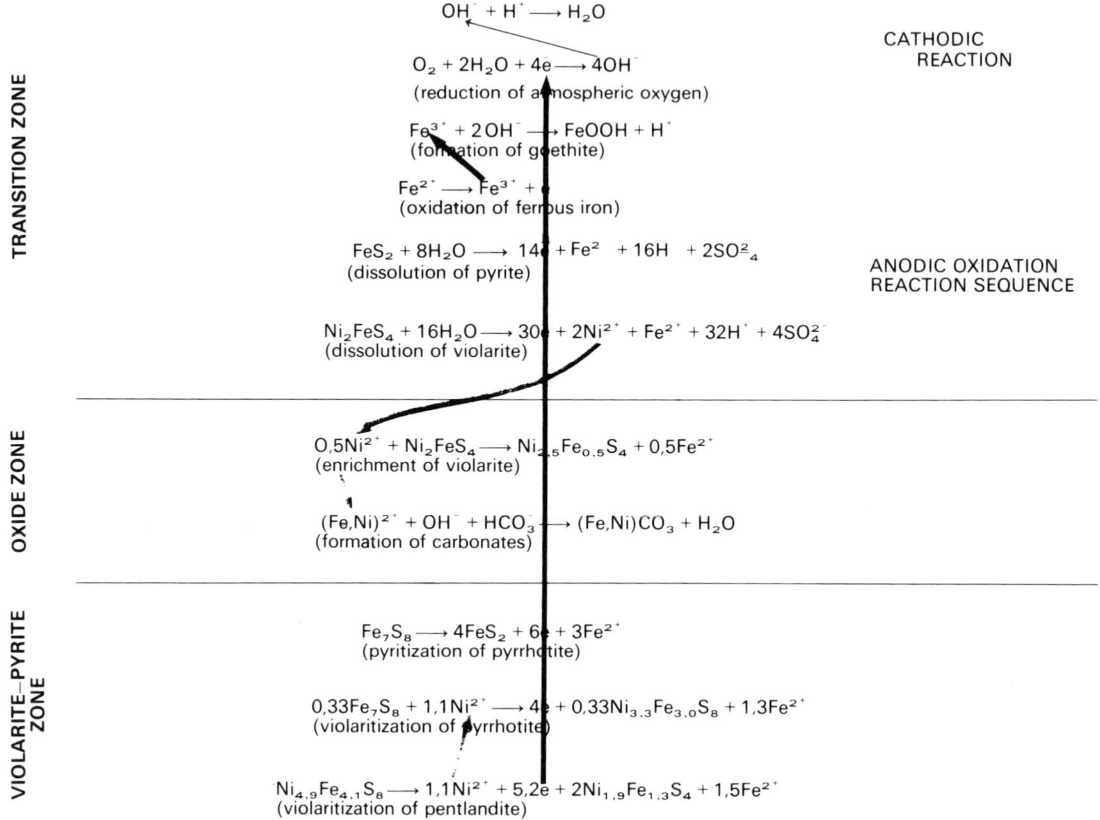

FIGURE 4 Chemical reactions and their inter-relations during the electrochemical alteration of nickel sulphide bodies in the supergene environment. (Adapted from Nickel et al[12].)

stable below the water table. In the higher Eh regime above the water table, all sulphides decompose, and oxides and hydrous oxides of iron are precipitated. Nickel ions, leached during sulphide decomposition, percolate below the water table, where they replace iron in minerals such as violarite[12], or where they form supergene millerite[15] enriching the ore grade. The chemical changes in the overall supergene alteration profile are summarized schematically in Figure 4. The significant point in the case of oxidizing nickel orebodies is that the supergene sulphide assemblages, from which the gossans develop, differ considerably from the primary assemblages.

The general sequence of supergene sulphide development in massive *pyrite–chalcopyrite (bornite) orebodies*, illustrated in Figure 5, has long been recognized to be of widespread occurrence[18,61]. Although the geological controls of these processes are relatively well documented[6,18], the detailed chemical and electrochemical mechanisms are rather complicated. In general terms, the process starts immediately above the water table, as the dissolution of chalcopyrite (with pyrite), releases copper ions, which, in the consequently acidified aqueous medium, are capable of percolating down the ore profile to the water table (Figure 6). There the copper ions react with chalcopyrite, (bornite), and to a lesser extent, with pyrite. These supergene enrichment reactions consume copper, release iron, hydrogen ions, and sulphur species, thus enhancing the overall metal:sulphur ratio of the ores by giving rise to djurleite–chalcocite–anilite assemblages[62], the so-called 'chalcocite' ores. It has been shown experimentally that the replacement reactions of chalcopyrite (and bornite) release 1,45 hydrogen ions for every copper ion consumed, whereas the replacement of pyrite releases 1,71 hydrogen ions for every copper ion consumed[63]. Consideration of the unit cell data indicates that replacement, at these atomic proportions, takes place virtually at constant volume for chalcopyrite (and bornite); the replacement of pyrite, however, is accompanied by a significant volume decrease. Under these conditions, the enrichment reactions take place at electrical neutrality. It is quite probable that these sulphide enrichment reactions, which can take place by ionic exchange up to considerable depths beneath the water table, are in nature marginally oxidational, half-cell reactions, being favoured by high copper activity at moderate pH[64]. Experimentally, the enrichment reaction rates of chalcopyrite and pyrite are retarded by increased hydrogen ion activity, whereas that of bornite is evidently accelerated under the same conditions[63]. In natural ore suites, it is evident that primary bornite is more susceptible to enrichment than chalcopyrite, but, with continued consumption of copper, even the tight sulphur–sulphur bonds of pyrite must break as chalcocite-like copper minerals develop. The thoroughly enriched ores thus formed vary in appearance from massive steely-grey aggregates ('massive chalcocite') to powdery black accumulations ('sooty chalcocite')[62].

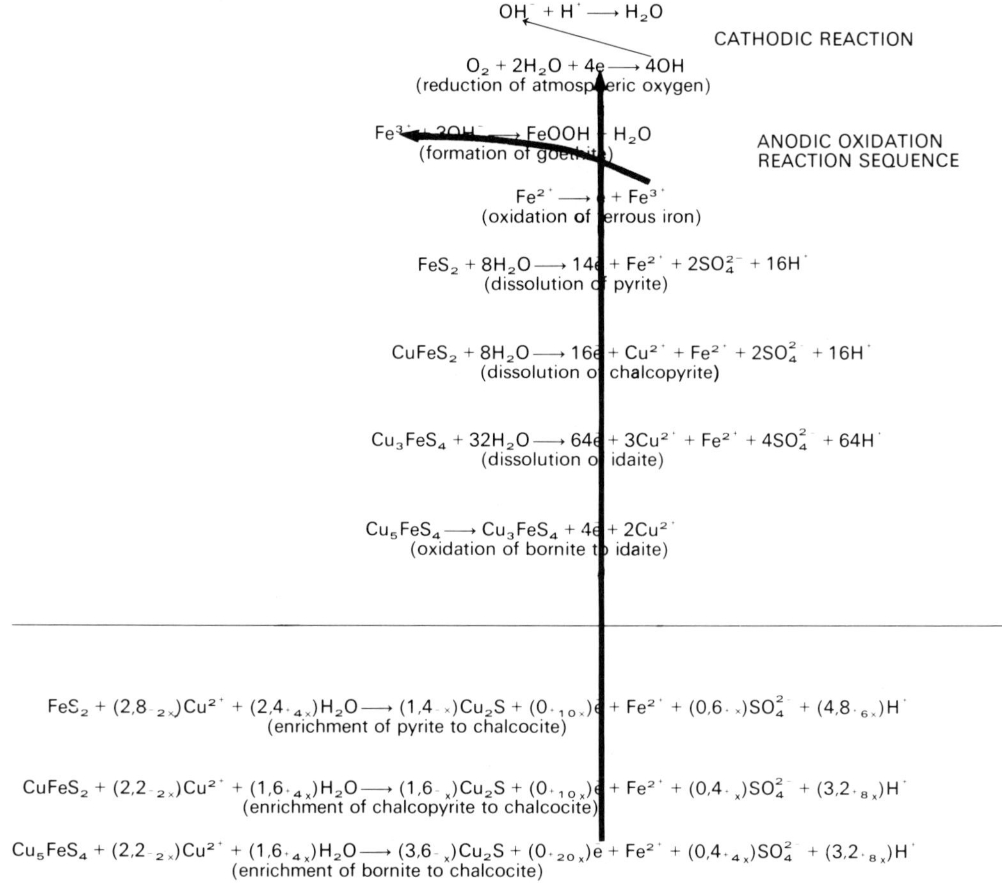

FIGURE 6 Chemical reactions and their inter-relations during the early stages of supergene enrichment of pyrite–chalcopyrite–bornite orebodies.

Note. In the schematic representation of the enrichment reactions, x refers to a very small number which may alter the balance of reactants and products as electrochemical half-cell reactions.

[*Editor's Note:* Figure 5, color photomicrographs illustrating the mineralogical and textural characteristics of supergene enrichment and oxidation of pyrite—chalcopyrite (—bornite) ores, has been omitted. Also, in the original, the arrows in Figure 6 appear in colors.]

Above the water table, the mineral-bound sulphur eventually oxidizes to sulphate. The iron released by dissolution of pyrite, chalcopyrite (and bornite) is generally precipitated eventually as goethite. Under these conditions much of the copper released is soluble and thus promotes the enrichment process. As the enriched ores are oxidized above the water table, however, only part of the copper released is soluble (with the available sulphate anion). Thus enriched copper ores oxidize through complex parageneses, commonly consisting of transitory phases with decreased metal:sulphur ratios (scuh as anilite ($Cu_{1.75}S$), blaubleiblender covellite ($Cu_{1+x}S$), covellite (CuS),[62] to stable native-metal, oxide, or hydrous -carbonate form (Figure 5). In a similar way, bornite (Cu_5FeS_4) oxidizes through the transitory phase idaite (Cu_3FeS_4) to stable assemblages of malachite and goethite[62]. The ore types from which these gossans develop may thus also differ considerably from the primary pyrite-chalcopyrite ores according to the stage in the supergene evolution of the sulphide profile. The resultant oxidate assemblages thus differ considerably.

In contrast to oxidized pyrite–chalcopyrite and pyrrhotite–pentlandite ore associations, *sphalerite–galena* ores

Cathodic Reduction of Dissolved Oxygen Radicals

$$O_2 \xrightarrow{e} O_2^- \xrightarrow{H_2O} HO_2 + OH^- \xrightarrow{e} HO_2^- + OH \xrightarrow{H_2O} H_2O_2 + 2OH^- \xrightarrow{e} OH + 3OH^- \xrightarrow{e} 4OH^-$$

Summed General Reaction

$$O_2 + 2H_2O + 4e \longrightarrow 4OH^-$$

Alteration of Pyrite (Me:S = 1:2) to Goethite)

anode $FeS_2 + 10H_2O \longrightarrow FeOOH + 19H^+ + 2SO_4^{2-} + 15e$

cathode $3{,}75O_2 + 7{,}5H_2O + 15e \longrightarrow 15OH^-$

total $FeS_2 + 17{,}5H_2O + 3{,}75O_2 \longrightarrow FeOOH + 19H^+ + 15OH^- + 2SO_4^{2-}$

net $FeS_2 + 2{,}5H_2O + 3{,}75O_2 \longrightarrow FeOOH + 4H^+ + 2SO_4^{2-}$

Alteration of Chalcopyrite (Me:S = 1:1) to Goethite

anode $CuFeS_2 + 10H_2O \longrightarrow FeOOH + Cu^{2+} + 19H^+ + 2SO_4^{2-} + 17e$

cathode $4{,}25O_2 + 8{,}5H_2O + 17e \longrightarrow 17OH^-$

total $CuFeS_2 + 18{,}5H_2O + 4{,}25O_2 \longrightarrow FeOOH + Cu^{2+} + 19H^+ + 17OH^- + 2SO_4^{2-}$

net $CuFeS_2 + 1{,}5H_2O + 4.25O_2 \longrightarrow FeOOH + Cu^{2+} + 2H^+ + 2SO_4^{2-}$

Dissolution of Chalcocite (Me:S = 2:1)

anode $Cu_2S + 4H_2O \longrightarrow 2Cu^{2+} + 8H^+ + SO_4^{2-} + 10e$

cathode $2{,}5O_2 + 5H_2O + 10e \longrightarrow 10OH^-$

total $Cu_2S + 9H_2O + 2{,}5O_2 \longrightarrow 2Cu^{2+} + 8H^+ + 10OH^- + SO_4^{2-}$

net $Cu_2S + H_2O + 2{,}5O_2 \longrightarrow 2Cu^{2+} + SO_4^{2-} + 2OH^-$

FIGURE 7 Examples of some common gossan-forming reactions showing the effect of metal:sulphur ratio on H^+/OH^- balance at the sulphide–oxide interface.

(even where pyrite is a dominant component), do not characteristically give rise to any significant zones of supergene sulphide development or metal enrichment beneath the sulphide–oxide interface. At Mount Isa, for example, where the copper ores are thoroughly enriched, the 10 m transition zones, marking the upper limits of the pyritic lead–zinc orebodies, contain an assemblage of sphalerite–galena–anglesite–cerussite–smithsonite[22] Similarly, the lead–zinc sulphide ores at Broken Hill, New South Wales, are characteristically altered directly, without significant supergene modifications, to stable carbonate and sulphate assemblages[32]. The same style of alteration has been noted for many of the lead–zinc sulphide deposits in North America[65]. There is, however, a certain amount of evidence to suggest that minor amounts of supergene galena, sphalerite, and associated cadmium and silver sulphides do form in the weathering environment[28,30,66,67]. Their occurrence is mainly restricted to the thorough weathering of either concentrated sulphide residuum in surface[66] and karst[30] hollows, or in subordinate amounts as open-space fillings in leached sulphide zones[28,67]. Solutions enriched in chloride and bicarbonate ions favour the dissolution of galena, thus enhancing the lead content of the solutions from which the secondary sulphides may subsequently precipitate.

As sulphide minerals corrode to stable oxide, carbonate, and sulphate phases near the water table, they become disconnected from the main sulphide ore zone. At this position in the supergene profile, the disconnected sulphides no longer contribute to the major electrochemical corrosion processes on the scale of the orebody as a whole, because the oxidate phases are poor electronic conductors. The chemical environment above the point where the sulphur in massive sulphide ores oxidizes to sulphate also becomes more of an open system for the introduction and removal of chemical species. It is convenient, therefore, to discuss this part of the process separately, as gossan formation, because the rocks formed in this way remain essentially stable as gossan outcrops.

GOSSAN FORMATION

From the foregoing, it is evident that the corrosive decomposition of sulphides at or near the water table is a genetically critical process in the formation of gossans. Whatever the detailed mechanisms, electrolytic oxidation at the sulphide–oxide interface together with the more mobile cations in solution, and the precipitation of iron as oxides or hydrous oxides. Commonly, this is accompanied by the influx of variable amounts of silica. Necessarily, the evaluation of gossans must be based on the aggregates of these rock-forming minerals that are typically precipitated or retained within or near the decomposed sulphide ore.

Early work[62] has shown that the mobility of iron under gossan-forming conditions is dependent upon pH. At extremely low pH, the retardation in the oxidation of ferrous to ferric ion and the low activity of hydroxyl prevent the *in-situ* precipitation of goethite, and the ferrous ion may be transported varying distances before it precipitates as ferric oxides. Conversely, under favourable gossan-forming conditions at moderate pH, goethite, or limonite is precipitated *in situ* pseudomorphing pre-existing sulphide textures. This line of reasoning led Blanchard and Boswell[69] to classify limonite types in terms of the mobility of iron, as:

- indigenous limonite – in the space formerly occupied by the sulphide mineral,
- transported limonite – in spaces a short distance (a few centimetres or so) from the former sulphide mineral, and
- exotic limonite – outside the confines of the orebody, commonly up to a hundred metres from the former sulphides.

The contained iron in gossans is most commonly in the form of goethite (less commonly as its dimorph, lepidocrocite) and hematite, although in some gossans magnetite occurs as a supergene mineral. Where the iron has moved a considerable distance before precipitation, the gossans are generally referred to as 'transported'[17]. These types of gossans are notoriously difficult to evaluate within the mineral exploration context. The pH control on iron precipitation, and on the mobility of the other ore-contained cations, is therefore of fundamental importance in gossan formation. The factors controlling natural pH in the gossan-forming environment are directly relevant to the interpretation and evaluation of almost all gossans.

The association of acidified groundwater with sulphides that are actively oxidizing is a very common phenomenon. For oxidizing nickel sulphide orebodies, Thornber[42] has documented a progressive increase in pH of co-existing groundwaters, from acid near the water table to mildly alkaline at depth. A generally similar, but more variable relation, is recorded by Sato[41] for oxidizing copper sulphide orebodies (Figure 2). It is informative to summarize the geological factors controlling the pH of groundwater solutions in the oxidation regime of sulphide ores.

The mechanisms of oxidation of massive sulphide orebodies by electrolysis suggest that the pH conditions at the sulphide-oxide interface are largely dependent upon:

- the composition of the sulphide ores,
- the nature of the acid-buffering, weathering reactions, and,
- the overall kinetics of the system.

Of these, the composition of the ore has a fundamental control because the hydrogen ion/hydroxyl radical (H^+/OH^-) balance relates inherently to the sulphide-sulphate transition. For every sulphur atom anodically oxidized to sulphate in the aqueous medium, eight hydrogen ions are released. The anodic oxidation of sulphur to sulphate, together with the concomitant release of metal cations electrically balancing sulphate anions, demands the release of electrons. For every electron cathodically consumed, one hydroxyl grouping is formed. Therefore, the resultant H^+/OH^- balance (measured by pH) at the sulphide–sulphate transition is inherently dependent upon the metal to sulphur ratio of the ores. Ores with a low metal:sulphur ratio, such as pyrite-rich deposits, generate a surplus of hydrogen ions during oxidation at the water table, whereas those with a high metal:sulphur ratio, for example chalcocite-rich deposits, generate a surplus of hydroxyl (Figure 7). So long as the sulphide ore continues to oxidize, the relative surplus of hydrogen ions or hydroxyl groups accumulates until either diffusion permits a dilution or the surplus induces a pH buffering chemical reaction. These buffering reactions are probably of two main types: those resulting from precipitation of minerals from the solution and those where the solutions react with the surrounding wall rock. At moderate or high pH, for example, the precipitation of basic oxides (such as goethite), carbonates (such as siderite, smithsonite, cerrussite, and gaspeite) and basic carbonates (such as hydrozincite and malachite) are favoured. These precipitation reactions consume hydroxyl groups directly or indirectly by hydroxyl-bicarbonate exchange, thus tending to enhance the H^+/OH^- ratio of the residual groundwater. Presumably, this is the reason why enriched copper ores with high metal:sulphur ratios are characteristically capped by goethite-malachite assemblages. Conversely, at low pH, reactions between the solutions and the surrounding wall rocks are favoured, and these generally consume hydrogen ions. This 'neutralizing' effect of the wall rocks was noted by the early workers[11], particularly for carbonate rocks. The acid-buffering potential of common, anhydrous, rock-forming silicate minerals such as feldspars, olivines, and pyroxenes which, by hydrolysis-type reactions, give rise to hydrous sheet silicate minerals, notably clays, is also considerable. Probably, however, for massive unfractured wall rocks, the rates of silicate hydrolysis reactions are relatively slow, and there is abundant field evidence of advanced leaching where ores with low metal:sulphur ratios, particularly pyritic ores, were enclosed by silicate rocks. From this point of view, it is evident that H^+/OH^- balance at the water table is further controlled by the overall kinetics of the system. It has been demonstrated that oxidation of sulphide minerals to sulphate form, the main pH-controlling process, is a relatively rapid reaction[70]. Thus, resultant H^+ or OH^- surplus at the sulphide-sulphate interface is also related to the rate of oxidation. If the orebody is rapidly exhumed, or is exposed by a rapidly falling water table, the rate of oxidation increases[60], tending to enhance the H^+ or OH^- surplus. This then tends to be lowered, but is probably limited by the kinetics of diffusion or the buffering reactions, or both.

Probably, all of these factors control the mobility of iron and other chemical species, together with their fixation in gossan-forming minerals. Two of the major components of gossans are goethite and silica. At the oxidation potentials where sulphides oxidize to sulphate, the stability of goethite is certainly in part pH-dependent. It is quite likely that the acid-buffering, hydrolysis reactions of silicate wall rocks account for the release of silica, which, because of decreased solubility at low pH[71], is pervasively precipitated in gossan zones. The general influence of pH on the mobility of the ore metal cations[71] and on the stability of oxidate minerals[40] is relatively well known. Where iron is precipitated *in situ* as observed in pseudomorphous iron oxide (and iron oxide-silica) replacement, textural interpretation may contribute significantly to gossan evaluation.

In the transformation of sulphide mineral assemblages to iron-rich gossans, it is evident that most of the metals,

TABLE 3
CALCULATED VOLUME CHANGES FOR IN SITU GOETHITE AND HEMATITE REPLACEMENT OF COMMON IRON-BEARING SULPHIDE MINERALS AT CONSTANT IRON.

Reactant Mineral	*Molar Volume of Reactant (cm^3)*	*Percentage Volume Reduction to Goethite (per cent)*	*Percentage Volume Reduction to Hematite (per cent)*
Pyrite	23,9	13	37
Marcasite	24,6	15	38
Chalcopyrite	44,1	53	66
Bornite	100,2	80	85
Pentlandite	159,6	31	50
Violarite	63,8	67	76
Arsenopyrite	26,4	21	43

Note: Percentage volume reduction for alteration of an iron-bearing sulphide mineral to goethite or hematite at constant iron is calculated as follows: e.g. for pyrite to goethite

$$FeS_2 + 2{,}5H_2O + 3{,}75O_2 \longrightarrow FeOOH + 4H^+ + 2SO_4^{2-}$$
$$(V_{py} = 23{,}9) \qquad (V_{gt} = 20{,}8)$$

$$\text{Volume reduction} = \frac{V_{py} - V_{gt}}{V_{py}} \times 100\%$$

Molar volumes for goethite and hematite are 20,8 and 30,2 cm^3 respectively.

V = molar volume (cm^3) of the mineral

which are more mobile than iron, are significantly leached from the system. A proportion of the original metal content, however, is commonly fixed in the oxide zone as stable oxidate minerals of the ore metals, or as adsorbed or co-precipitated metal on finely crystalline goethite, metal-rich colloids or even on gel silica. Part of the leached metal must percolate below the water table to be re-fixed in the sulphide zone by supergene enrichment. The remainder is dispersed throughout the groundwater system giving rise to geochemical dispersion haloes. No reliable chemical data quantifying the amounts of leaching, enrichment and dispersion for differing types of orebodies in differing terrains appear to be documented in the published literature. Yet this type of information is directly relevant to understanding the processes involved in gossan formation.

Brotherton[14] has documented bulk chemical changes that exist across the sulphide–oxide interface for a large number of oxidizing nickel sulphide orebodies. The smoothed average, volumetric, elemental abundances*, plotted against profile depth for the Pikwe nickel deposit in southern Africa (Figure 8), emphasize the dramatic chemical changes at the sulphide–oxide interface. Most of the sulphur leaves the system and this is accompanied by a marked influx of silica. It is informative also to note the relative depletion trends for iron, copper, nickel, and cobalt that characterize their relative mobilities during gossan formation at this locality. In addition to the chemical changes shown, considerable amounts of oxygen and hydrogen enter into the gossan-forming system to combine with iron, forming stable oxides and hydrous oxides, such as goethite and hematite, or as oxygen attached to silicon as silica.

*Because of the amount of leaching involved, it is advantageous to express geochemical abundance in terms of weight equivalents per unit volume in order to portray the geochemical migration of elements into and out of the system. The procedure of transforming weight per cent analyses into weight equivalents per unit volume (where the conversion factor is specific gravity of the solid and included pore space divided by atomic weight) eliminates the apparent enrichment/depletion effects related to changing specific gravity, porosity and composition, thus depicting only the spatial geochemical change in the system. The assumption of constant net volume appears to be reasonable because pre-existing sulphide textures are commonly pseudomorphously preserved in gossans.

The apparent reduction in specific gravity, accompanying net leaching, is commonly also accommodated in part by an *in situ* increase in porosity (Figure 8). Evidently, this varies from almost zero where there is a complete influx of silica, to approximately fifty per cent, where leaching of iron and sulphur occurs without complementary silica influx. It is interesting to speculate on the mineralogical controls of porosity change as this has several important implications in the geochemical and textural evolution of gossans. In addition to the genetic factors relating to the dissolution, migration, and precipitation of iron alluded to above, the mineralogy of the ores must inherently control porosity development. Assuming a regime where all the iron of iron-bearing sulphides is precipitated *in situ* as goethite or hematite and all other cations with sulphur are removed, the porosity changes for common sulphide-oxide reactions are summarized in Table 3. These porosity changes are consistent, at least empirically, with those observed in many iron oxide-rich gossans. Microscopic observation shows that much of the goethite, so precipitated, is extremely fine-grained and undoubtedly a certain component of ore metal is co-precipitated with this goethite. The simultaneous or subsequent influx of amorphous silica or externally derived goethite and hematite then even further complicates the mass-volume relations in the geochemical evolution of gossans.

It is evident that the supergene alteration of massive sulphide orebodies gives rise to rocks at surface that differ markedly from their genetically linked hypogene ores. In consequence, the problem for the exploration geologist of characterizing the primary ores by evaluating gossans is formidable. The possible approaches that lend themselves to this type of evaluation, namely, consideration of the field relations, together with textural, mineralogical, and geochemical features of gossans, are discussed below.

THE IRONSTONE PROBLEM IN MINERAL EXPLORATION

In the course of prospecting for base-metal sulphide deposits, it is frequently necessary to evaluate the economic significance of the exposed rocks that could potentially indicate the occurrence of an ore deposit. This requires a distinction to be made between gossans and other iron-rich rocks of similar appearance, but which are not derived

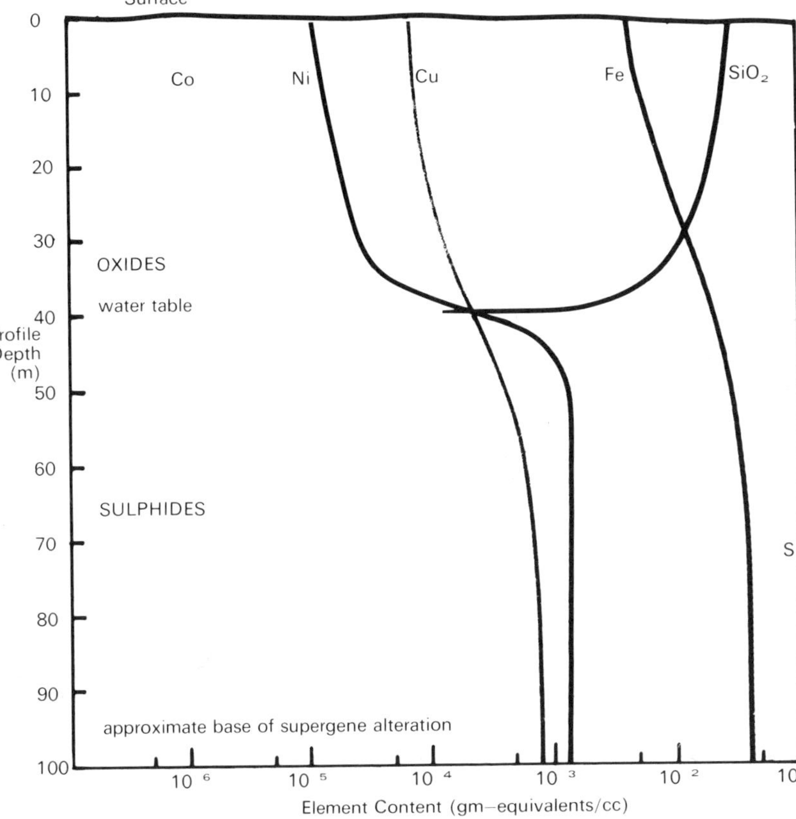

FIGURE 8 Smoothed average geochemical variation trends in the supergene-altered profile of the the Pikwe copper–nickel sulphide body, Botswana.

Note: Element content is given in weight equivalents per unit volume on a logarithmic scale to portray the spatial mobility of the elements within the system. Average specific gravity and porosity values are 4,4 and 3 per cent for the sulphides and 2,4 and 15 per cent for the oxides respectively.

[*Editor's Note:* In the original, the lines representing each element appear in different colors.]

from massive sulphides. This latter group, sometimes with transported gossans and even gossans over uneconomic sulphides, have been variously referred to as 'false gossans'[72,73] or 'pseudogossans'[36,74] to distinguish them from the so-called 'true gossans' over economic massive sulphides. However, 'ore' is defined in complex terms, and this terminology is perhaps unjustified. In the context of this discussion, all weathered rocks derived from massive sulphides are regarded as gossans. Once recognized, it is desirable at an early stage to relate the characteristics of all gossans to the potential size and nature of their parent mineralization, thus effecting efficient follow-up.

Evaluation of iron-rich outcrops (ironstones) is perhaps most difficult in terrains of hot semi-arid climate, such as southwest United States, the Arabian peninsula, southern Africa, and in most of the Australian continent, where Red Desert and Reddish Prairie soil groups[75] testify to the mobilization of iron in the near-surface weathering regime. Pervasive ferruginization, resulting from the upward migration and cumulative accretion of ferric iron near surface, is a common phenomenon in these regions. Also in tropical and equatorial zones of hot wet climate, iron, together with aluminium, titanium, and silica, accumulate as thick, near-surface hardpan layers[76,35] in deeply weathered laterite profiles. In places, polycyclic lateritization has given rise to superimposed or truncated profiles[77]. In both modern and ancient lateritic terrains, the primary lithological features of silicate rocks and sulphides alike are largely obliterated at surface. Where weathering has continued from a lateritic to a semi-arid environment, as in Western Australia[77], the problem of recognition is compounded by the lithological convergence of ferruginous rock units, laterites, and gossans to similar ironstone lithologies.

Because gossans and ironstones are products of surface processes, it is important to equate these units on individual weathering surfaces with specific parent lithologies in the bedrock. Multiple periods of lateritization, locally variable ferruginization, and silicification may also effect lithological divergence by producing widely variable ironstone types from an identical parent, whether silicate, sulphide, or primary ironstone. The convergence and divergence of ironstone types derived from widely differing primary lithologies results in an extremely complex mineralogical, chemical, and morphological problem which cannot readily be resolved in the field.

In mineral exploration, it is sound practice to term all ferruginous rocks 'ironstones' until such time as any features of a true gossan have been recognized. 'Gossan' is a discrete class of ironstone resulting from weathering of sulphides (not necessarily 'ore' in economic terms), whereas the broader term connotes any iron-rich rock of non-sulphide derivation once its identity has been established. In this exploration context, ironstones may be classified on a genetic basis within the following framework:

☐ base metal sulphide gossan,
☐ iron sulphide gossan,
☐ transported gossan,

□ exhalite, sedimentary cherts and iron formation,
□ ferricrete and silcrete,
□ ferruginized rock.

A general discussion of the essential characteristics of each successive class, with specific reference to the ironstone problem, follows below.

Most *sulphide gossans* and other ironstones are variably silicified, chemically stable, and refractory, commonly forming linear ridges or spines of positive topographical expression. Many gossan outcrops, for example those at Broken Hill[32], Australia, Empress Nickel[78], Rhodesia and Otjihase[87], S.W. Africa, reflect fairly accurately the style and configuration of the underlying sulphide body as projected at surface. Elsewhere, gossans may be very poorly expressed at surface. In the Shield areas of Western Australia, gossans overlying nickel–copper sulphide mineralization, for example at the Mt. Edwards deposit, occur as scattered areas of rubble. Elsewhere, gossans over nickel ore shoots may have outcrop dimensions as small as several square metres. Similarly, the gossans are poorly exposed over the copper–zinc sulphide orebodies at Prieska[79] and Areachap[80] in the northwest Cape Province, South Africa. Where exposure permits, diagnostic features of the geological environment of an individual ore type, such as a contiguous alteration halo, a relevant lithological association, or a particular structural setting, may contribute significant evidence towards the characterization of a gossanous outcrop in the field. For example, in southern Africa the association of a gossanous horizon in ferruginous exhalite with stratiform barite and aluminous metasediments[81] is diagnostic of volcanogenic sulphide mineralization.

Based upon field observations, Blanchard[31] noted an exaggeration of the width of sulphide bodies as expressed by the surface gossans. The highly refractory nature of jasperoidal ironstones promotes their residual accumulation and, ultimately, their dispersion as float. This resistant nature also accounts for the isolated persistence of ironstones as large blocks within calcrete. The apparent widening of some gossan outcrops is also attributable to strongly ferruginized wall rocks, as at Woodlawn[131], N.S.W., and various forms of proximally transported gossan (as at Mt. Lyell, Tasmania[82]). Limonite derived from siderite gangue may also contribute appreciably to gossan outcrops at surface[83].

One field criterion used by early workers for recognizing gossans is their 'live' appearance caused by the extreme local variation in red, yellow, brown, and black coloration. This variegated appearance contrasts strongly with the dull, uniform, dark brown limonite that typifies many spurious ironstones. Even jasperoidal gossans tend to retain considerable, residual colour variations on the scale of a hand specimen. Large scale fabrics, such as breccia texture or gross lithological banding, may be discernible on the scale of a gossan outcrop. Other field criteria of a mineralogical or textural nature are described in considerable detail by early workers[31,84].

Iron sulphide gossans, derived variously from 'barren' pyrite and pyrrhotite mineralization, are difficult to distinguish exclusively from base metal gossans in the field. Iron sulphide mineralization (with attendant gossans) is commonly present in a range of base metal sulphide associations, notably with ultramafic nickel–copper, volcanogenic copper–zinc–lead and black, dolomitic shale polymetallic sulphide bodies. The gross configuration, variegated limonite, large scale textural features, and the effects of supergene weathering, discussed above for base metal gossans, are all common to iron gossans.

The field expression of all sulphide-derived gossans is superficially similar and examination of oxidate mineralogy, replica textures and geochemistry affords their correct genetic classification.

Transported gossans are perhaps the most difficult class to recognize in the field because they assume the morphologies of spurious ironstones. Common transported gossan types range from gossan breccias and loosely cemented scree to massive colloform or columnar accumulations of goethite, hematite, or manganese minerals. Breccias commonly formed by the mechanical transport of clasts which were subsequently recemented by an iron oxide matrix at or near the contemporary land surface. Transported gossans are forming in this way at the present time on portions of the hill slopes over the Wadi Wassat pyritic deposits in the arid regions of the Arabian Shield. The gossan breccias near the Wannaway nickel deposit in Western Australia, the Woodlawn[85] base metal deposit in New South Wales, the Mt. Lyell[82] copper deposit in Tasmania and at Minas Carlota[21] in Cuba, may have formed in a similar way. The breccias may be proximal (e.g. Mt. Lyell[82]) or distal (e.g. Minas Carlota[21]) with respect to the *in situ* gossan. Most of the transported massive, colloform and columnar types probably formed by the chemical migration and precipitation of iron as limonite outside the confines of the primary sulphide body. The 'bog iron ores' of Mesa de los Pinos at Rio Tinto,[1,17] Spain, are classical examples of distally transported gossans, comprising rock and gossan clasts in a matrix of sulphide-derived hematite. Proximally transported, massive hematite gossans also occur at Mt. Lyell[82] in Tasmania. In general, apart from the recognition of gossan clasts in an exotic matrix, the direct, positive characterization of transported gossans is relatively uncertain in the field, and diagnosis may be considerably enhanced by geochemical criteria.

Iron-and silica-rich chemical sediments, such as *chert*, *jaspilites*, *taconites*, and their metamorphic derivatives are very oftern recognizable in the field as discrete formations. Similar lithologies, however, are intimately associated with many massive sulphide deposits of volcanogenic affinity, e.g. Broken Hill[86], N.S.W.; Aggeneys and Gamsberg[81], South Africa; Matchless and Otjihase[87], South West Africa; as *exhalite formations*, and their metamorphosed counterparts. Because these exhalite oxide formations are commonly the lateral facies equivalents of the sulphide orebodies, the distinction between the weathered derivatives of both types of facies may in certain terrains pose a considerable exploration problem. Chemical sedimentary or exhalative cherts, with or without a ferruginous component, and cellular iron-rich gossans are lithologically distinct rock types. However, convergence of morphology occurs in this case where gossan has been almost totally flooded by massive supergene silica or meta-jaspilite has been flooded with hematite after magnetite. A distinction between chert and gossan in such extreme situations can only be made in the light of mineragraphic and geochemical evidence.

Other types of ironstone within the broad category of *ferricrete and silcrete* occur in deeply weathered lateritic terrains, where the near-surface accumulation of iron has formed a hard, indurated duricrust[76]. Commonly, the

TABLE 4 COMMONLY OCCURRING OXIDATE MINERALS IN GOSSANS.

METAL	SULPHATE	ARSENATE	OXIDE	CARBONATE	SILICATE	HALIDE	PHOSPHATE	VANADATE	MOLYBDATE
IRON	Jarosite $KFe_3(SO_4)_2(OH)_6$ Melanterite $FeSO_4 \cdot 7H_2O$	Scorodite $FeAsO_4 \cdot 2H_2O$ Symplesite $Fe_3(AsO_4)_2 \cdot 8H_2O$ Pharmacosiderite $KFe_4(AsO_4)_3(OH)_4 \cdot 6H_2O$	Goethite FeOOH Lepidocrocite FeOOH Hematite Fe_2O_3	Siderite $FeCO_3$ Ankerite $Ca(Mg,Fe)(CO_3)_2$			Vivanite $Fe_3(PO_4)_2 \cdot 8H_2O$		Ferrimolybdite $Fe_2Mo_3O_{12} \cdot 8H_2O$
MANGANESE	Mooreite $(Mg,Zn,Mn)_8SO_4(OH)_{14} \cdot 4H_2O$		Pyrolusite MnO_2 Manganite MnOOH Psilomelane $(Ba,Mn)_4O_8(OH)_2$	Manganosiderite $(Mn,Fe)CO_3$	Braunita Mn_2SiO_4		Reddingite – Phosphoferrite $(Mn,Fe)_3(PO_4)_2$ $3H_2O$		
NICKEL	Morenosite $NiSO_4 \cdot 7H_2O$ Retgersite $NiSO_4 \cdot 6H_2O$	Annabergite $Ni_3(AsO_4)_2 \cdot 8H_2O$		Gaspeite $(Ni,Mg,Fe)CO_3$ Regvesite $Ni_6Fe_2(OH)_{16}CO_3 \cdot$ $4H_2O$					
COBALT	Bieberite $CaSO_4 \cdot 7H_2O$	Erythrite $Co_2(AsO_4)_2 \cdot 8H_2O$	Heterogenite CoOOH	Heterogenite					
COPPER Native	Chalcanthite $CuSO_4 \cdot 5H_2O$ Antlerite $Cu_3SO_4(OH)_4$ Brochantite $Cu_4SO_4(OH)_6 \cdot H_2O$	Olivenite $Cu_2AsO_4OH_3$ Clinoclase $Cu_3AsO_4(OH)_3$	Cuprite Cu_2O Tenorite CuO	Malachite $Cu_2CO_3(OH)_2$ Azurite $Cu_3(CO_3)_2(OH)_2$	Chrysocolla $CuSiO_3 \; 2H_2O$ Dioptase $CuSiO_2(OH)_2$	Atacamite $CuCl_2 \cdot 3Cu(OH)_2$	Liebethenite $Cu_2PO_4(OH)$ Pseudomalachite $Cu_5(PO_4)_2(OH)_4$ H_2O		Lindgrenite $Cu_3(MoO_4)_2(OH)_2$
LEAD	Anglesite $PbSO_4$ Plumbojarosite $PbFe_6(SO_4)_4(OH)_{12}$ Lanarkite $PbO \cdot PbSO_4$	Mimetite $PbCl,Pb_4(AsO_4)_3$	Massicot PbO Litharge PbO Minium Pb_3O_4	Cerussite $PbCO_3$ Leadhillite $Pb_4SO_4(CO_3)_2(OH)_2$ Phosgenite $Pb_2CO_3Cl_2$		Phosgenite $PbCO_3 \cdot PbCl_2$	Pyromorphite $PbCl \; Pb_4(PO_4)_3$	Descloizite $Pb(Zn,Cu)VO_4OH$ Vanadinite $PbCl \; Pb_4(VO_4)_3$ Mottramite $Pb(Cu,Zn)VO_4OH$	Wulfenite $PbMoO_4$
ZINC	Goslarite $ZnSO_4 \cdot 7H_2O$		Chalcophanite $(Zn,Mn)Mn_2O_5 \; 2H_2O$	Smithsonite $ZnCO_3$ Hydrozincite $Zn_5(CO_3)_2(OH)_6$ Rosasite $(Cu,Zn)_2CO_3(OH)_2$ Aurichaleite $(Zn,Cu)_5(CO_3)_2(OH)_6$	Hemimorphite $Zn_4Si_2O_7(OH)_2 \cdot H_2O$ Willemite Zn_2SiO_4				
SILVER Native	Argentojarosite $AgFe_3(SO_4)_2(OH)_6$					Chlorargyrite AgCl Embolite Ag(Cl,Br) Iodargyrite AgI			

rocks in this layer have a well-developed, pisolitic concretionary structure, although, in places, they may be massive and blocky, or even cellular. In these terrains, the accumulation of silica in cherty and opaline form, is also relatively common, particularly overlying weathered ultramafic rocks. Both iron- and silica-rich rocks, in this mode of occurrence are a separate genetic class of ironstones that, in certain circumstances, may have a similar appearance to gossans.

In less deeply weathered terrain, such as the eastern Transvaal and southern Cape of South Africa, massive precipitates of goethite, hematite, and manganese oxides are largely derived from iron-leached silicate or carbonate rocks. Jasperoidal ironstones occur as linear bodies capping fractures, as stratabound bodies on lithological contacts or as irregular to tabular accretions adjacent to surface spring deposits. Ferruginous crusts of this type and the ferricretes and silcretes of the more deeply weathered areas can often be distinguished from gossans by the field relations but, failing that, the geochemical and textural properties are generally diagnostic.

The class of ironstone including a range of *ferruginized rocks* results from the pervasive influx and precipitation of iron with variable silica, as goethitic jasper, in weathered rocks. The ironstone inherits the original fabric of the rock and, on all scales, entrained remnants of host rock or discrete relict minerals, commonly amphiboles or micas, may be observed. Ferruginization may traverse bedding or foliation but an overall concordant impression is generally apparent in the field. The geochemical features of this class of ironstone are diagnostic.

Where the genesis of a gossan or ironstone is not apparent in the field, a detailed investigation of the mineralogy, textures or geochemistry may assist in its interpretation.

THE MINERALOGY OF GOSSANS

The mineralogy of most gossans is dominated by 'limonite' and silica in mutually antipathetic abundance. The field term, 'limonite'[31], connotes amorphous, colloidal iron oxides or finely crystalline goethite with subordinate admixed silica, hematite, jarosite[88], lepidocrocite, or manganese oxides (Table 4) in various proportions and states of hydration. The hydrated iron oxide series[89] comprises three mineral species: goethite (αFeO.OH), lepidocrocite (βFeO.OH), and hematite (Fe_2O_3), which show a continuous range of their adsorbed water contents. The poorly crystalline, colloidal nature of goethite also typifies the copreciated gel silica in gossans. Blanchard[69], described this very common association as 'limonitic-jasper', which may be resolved into goethite- or hematite-dominant varieties. Segregations of discrete, finely colloform silica or hyalite commonly occur in jasperoidal gossans. Interpretation of the dominant iron oxide mineralogy is possible in the case of leached cappings over disseminated mineralization, but in sulphide gossans this has little bearing on the determination of the parent sulphide mineralogy[90].

Identification of accessory *residual minerals* in gossans and ironstones by optical microscopy may assist in the interpretation of primary sulphide mineralogy and lithological association. Residual quartz in gossans commonly encloses traces of residual sulphides, perhaps in negative crystal forms. In metamorphic terrains, relict spinels, such as magnetite and gahnite ($ZnAl_2O_4$), from the host rock may point to a protolith of exhalative origin. Residual silicate and clay assemblages, together with gangue minerals, such as calcite, barite, and gypsum, also provide important clues to the nature of the primary ore association or, in the case of ironstones, the nature of the original lithology.

Oxidate minerals, derived from the ore metal sulphides, occur as accessory components in some gossan suites. These are most commonly preserved in poorly silicified gossans, although recognition in hand specimens may be difficult. Carbonates are the predominant oxidates in sulphide gossans, followed by the silicates. The more soluble sulphates, phosphates, and halides are generally only preserved in particularly arid climates. Oxides are strongly represented only as oxidate minerals of manganese and, less commonly, of copper.

Macroscopic recognition of ore-metal oxidate assemblages in gossans may allow immediate, preliminary evaluation of the outcrop in the field. Gossans overlying *nickel–copper sulphides* in places contain light green, crystalline aggregates of gaspeite[12,91] together with nickeloan magnesite[91]. Bright yellow reevesite, the pale- to emerald-green minerals morenosite (fibrous), retgersite[12] (fibrous), and annabergite (earthy), basic nickel oxides and silicates all occur at depth in the oxide zones but are rarely preserved in nickel–copper gossans. Copper carbonates and even sulphates may accompany the nickel oxidates in trace to minor amounts.

Copper–zinc sulphide gossans typically contain the distinctive copper carbonates malachite and azurite, with a bluish-green copper silicate series dominated by chrysocolla. Gossans of particularly arid areas are characterized by the sulphates and basic sulphates of copper[92]; turquoise chalcanthite, green antlerite, atacamite, and blackish green brochantite (Table 4). Cuprite and tenorite are important oxidates derived from iron-poor, copper-rich sulphide assemblages. The zinc oxidate minerals are poorly represented in gossans of this sulphide type because the supergene chemical environment does not favour their formation.

Lead–zinc–silver sulphide gossans contain significant amounts of cerussite[22,24,31,93], smithsonite[24,65] and, less commonly, silver halides[33,93]. Hydrous manganese oxides also make a significant contribution to the assemblages of this gossan type[32,93,94]. Over certain deposits, hemimorphite (e.g. MacArthur River), plumbojarosite (e.g. Gamsberg), hydrozincite, and pyromorphite are diagnostic oxidates. Rarer minerals that characterize this ore type include chalcophanite, argentojarosite[33], descloizite, willemite, anglesite, and goslarite (Table 4).

Diagnostic oxidate minerals of restricted occurrence include scorodite (and green arsenate staining) derived from sulpharsenites[95] or arsenides, pale yellow stibiconite[96] ($Sb_2O_3(OH)_2$) capping antimony sulphides, lindgrenite[92], and ferrimolybdite after molybdenite and black heterogenite derived from cobaltiferous sulphide assemblages.

A wide range of procedures exists to identify specific mineral assemblages in gossans. Residual sulphides may be identified during routine textural investigation of mineragraphic sections whereas the identification of spatially related oxidate and gangue minerals is best made in polished thin section.

Because diagnostic indicator minerals are rare, mineralogy is of limited use in the evaluation of gossans and ironstones. However, the recognition of residual and oxidate minerals, where present, may indicate important features of the primary sulphide, gangue, or lithology.

THE TEXTURAL FEATURES OF GOSSANS

Historical background

The economic implications of supergene sulphide enrichment motivated such eminent geologists as Sales[97], Lindgren[98], Emmons[18], and Locke[99] to investigate this phenomenon in the period 1900 to 1930. Their work dealt principally with the porphyry copper mineralization of southwest North America. The earliest accounts of the interpretative application of leached outcrop textures was produced by Locke[10] in 1926. In the period between 1925 and 1935, interest spread to the interpretation of the gossans overlying massive base metal sulphide mineralization, principally by the studies of Blanchard and Boswell[69,100–105]. Subsequently, Blanchard extended his work on gossans in Australia[11,23,106–108], with particular reference to deposits in New South Wales and Queensland.

Blanchard and his co-workers showed how the leached derivatives of the principal base metal sulphide minerals and associated gangue minerals from numerous deposits commonly exhibit characteristic cellular boxwork textures on a macroscopic scale. In addition, they interpreted a wide variety of crustiform aggregates of limonite as being texturally diagnostic of individual parent minerals. The limitations of these criteria were recognized by their proponents[101], but, within a restricted geographical and metallogenetic province, an experienced prospector could no doubt discern diagnostic textures or even limonite colours in gossanous exposures[31,84].

In the period 1945 to 1965, work generally in the field of prospecting technology turned noticeably toward developments in geochemical and geophysical exploration techniques. Published work on gossans during this period is limited to only a few reviews[109,110,72], most of which draw heavily on the early work.

The discovery of the Kambalda nickel sulphide deposits in 1966 initiated a period of intensive exploration activity throughout the Eastern Goldfields regions of Western Australia. Because of the limited applicability of conventional prospecting techniques in this deeply weathered terrain, the evaluation of gossans was soon recognized as an important exploration tool. Most of the gossans overlying the nickel sulphide deposits in that terrain, however, are massive and jasperoidal, so the cellular boxwork criteria, documented by Blanchard[31], found little application. The need to recognize and evaluate the nickel gossans and other similar rocks thus stimulated renewed interest in gossans. Application of modern, mineragraphic, and analytical techniques to the examination of gossans and ironstones, has since led to considerable advances in this field.

The application of techniques of optical microscopy has shown that morphological features of boxworks are similarly expressed on a microscopic scale as residual outlines in some of the massive jasperoidal gossans[111–113]. Pseudomorphous replacement of supergene sulphide textures, preserved ultimately by goethite–hematite–silica, are readily identifiable on polished mineragraphic surfaces. Many of the supergene alteration textures, illustrated in the literature of 1910 to 1940[10,18,19,114], are known, by mineragraphic examination, to occur as pseudomorphous replica textures in gossans derived from similar ore associations. In many gossans, however, particularly those over iron-poor and pyrite-rich ores, textures of the sulphide minerals are not preserved. Even on outcrops where textures are preserved in the gossans, detailed sampling shows that the preservation may be confined to specific zones of varying size. Nevertheless, the specialized investigation of the mineralogical and textural features of gossans maintains an important function in modern mineral exploration in conjunction with the more versatile roles of field investigation and interpretation of geochemical data.

The basis of textural interpretation

Ever since the pioneering work[10,31] on the interpretation of gossan textures, it has been realized that the applicability of these techniques in mineral exploration is ultimately limited by the nature of the chemical precipitation that takes place at the site where sulphides are oxidized to sulphates. At this site the intrinsic sulphur bonds in the ore are broken and the textures of the oxidation products can only be interpreted in terms of the textures of the ores if precipitation of released or introduced species is crystallographically controlled by the pre-existing ore minerals. This preservation of ore textures is commonly accomplished by *in situ* precipitation of iron or a passive influx of silica as the ore is oxidized. The factors controlling this process, summarized above, are delicately balanced and, if subject to perturbation, may result in conditions that are sufficiently corrosive to destroy all sulphide mineral-related textures as the gossan forms. Where this occurs, clearly the textural approach of evaluating gossans is not applicable as a tool within the scope of mineral exploration programmes. In many gossans, however, there is at least partial preservation of sulphide mineral-replica textures.

Mineragraphic examination of oxidizing sulphide ores commonly shows that many minerals, at the incipient stage of oxidation, preferentially alter along specific crystallographic planes, notably grain boundaries, cleavages, parting planes, and twin boundaries. The alteration products, mainly limonite and introduced silica, accrete laterally and symmetrically forming a network of interlocking septa. Even with subsequent leaching of all the retained sulphides, the early-formed septa may survive as a delicate 'boxwork', classically associated with gossans. The morphology of these boxworks is genetically related to the crystallographic symmetry of their precursory sulphide minerals. Blanchard and his co-workers[31] based their techniques of interpretation on the morphological characteristics of boxworks as viewed macroscopically in the field. It is now known, from mineragraphic examination of the more massive jasperoidal gossans without macroscopically visible boxworks, that the pseudomorphous preservation of supergene sulphide textures is not altogether uncommon[112,115]. In fact, we consider that a continuous spectrum of occurrence exists between gossans showing open cellular boxworks and those showing massive pseudomorphously preserved sulphide textures. Significantly, this expands the concept of textural interpretation of gossans well beyond that initially proposed.

Replica textures of individual minerals

Blanchard's studies of the morphology of cellular boxworks are described in detail in nine publications between 1925 and 1942[11,69,100–106], and summarized posthumously in 1968[31]. The descriptive work of

TABLE 5

DESCRIPTION OF DIAGNOSTIC REPLICA TEXTURES IN GOSSANS.

Mineral	Controlling Symmetry Planes	Diagnostic Replica Textures	Fig.
SULPHIDES			
Pyrite (FeS_2) Isometric	(100)	–Cubic pseudomorphs or cavities.	9(a)
	(100) (111)	–Concentric or nested cubic boxworks, commonly with octahedral modification.	9(c)
		–Concentrically zoned webworks or anhedra.	9(d) 11(b)
Marcasite (FeS_2)	(110)	–Radiating elongate boxworks with herring-bone cross walls.	
	(110)	–Pseudomorphs of stellate spear-head twin.	
Pyrrhotite ($Fe_{1-x}S$) Hexagonal or Monoclinic	(0001)	–Broadly hexagonal boxworks with hieroglyphic or hexagonal cells.	
	(001)	–Closely spaced, parallel, thin wall, cellular boxworks.	9(e)
		–Poorly connected hieroglyphic boxworks.	9(f) 11(d)
		–Curvi-lamellar to spheroidal forms: concentric shrinkage cracks.	3(a,e)
Chalcopyrite ($CuFeS_2$)	(111) (201)	–Rectangular boxworks with closely spaced, parallel cell walls and cross walls at 95-110°.	10(a,b,d)
	(111) (201)	–Coarse, quadrangular, equi-dimensional boxworks, often thick walled.	10(c) 11(d)
Bornite (Cu_5FeS_4) Isometric	(111)	–Spherical triangular boxworks.	
Tetrahedrite $[(Cu,Fe,Zn,Ag)_{12}Sb_4S_{13}]$ Isometric		–Sinuous, trench-like contour boxworks.	
Pentlandite $[(Fe,Ni)_9S_8]$ Isometric	(111)	–Delicate, thin walled, octahedral boxworks.	3(b,d)
		–Feathery margin developed after violarite along basal (0001) plane of original pyrrhotite.	3(a,b)
Galena (PbS) Isometric	(100)	–Regular, cubic, thin walled, concentric boxwork (more regular than pyrite).	10(e)
	(111)	–Stepped pyramidal boxwork.	
	(021)	–Rhombic, cerussite boxwork: thick walled, multiple parallel septa with acute angle ideally 62°.	
Sphalerite (ZnS) Isometric	(110) (111)	–Angular boxwork: well connected, oblique (45°C) cross walls and continuous, parallel long walls produce triangular sub-cells.	10(g,h) 11(f)
Molybdenite (MoS_2) Hexagonal		–Foliated, thin walled boxworks with smooth, rounded walls: casts of original interstices.	

Mineral	Controlling Symmetry Planes	Diagnostic Replica Textures
OXIDES, HYDROXIDES		
Goethite (FeOOH) Orthorhombic		–Spherical, cellular boxworks (gel texture). –Poorly connected hieroglyphic boxworks.
Magnetite (Fe_3O_4) Isometric	(111)	–Elongate external forms with closely spaced, parallel septa.
	(111)	–Octahedral pseudomorphs or boxworks.
	(110)	–Dodecahedral pseudomorphs or boxworks.
Ilmenomagnetite ($FeTiO_3$-Fe_3O_4) Hexagonal – Isometric	(111)	–Octahedral, trellis boxwork after leucoxene in magnetite.
CARBONATES, SILICATES AND GANGUE MINERALS		
Calcite ($CaCO_3$) hexagonal	(1011)	–Highly regular, rhombohedral boxworks or pseudomorphs: ideally acute angle 72°.
Siderite ($FeCO_3$) Hexagonal	(1011)	–Curved, rhombehedral boxworks or pseudomorphs: ideally acute angle 73°.
Fluorite (CaF_2) Isometric	(111)	–Very thin walled, delicate, octahedral carbonate boxwork: poorly silicified.
Serpentine Monoclinic		–Random meshwork (as olivine) with included residual magnetite.
Amphibole Monoclinic	(110)	–Fibro-lamellar, lath-like pseudomorphs.
Mica Monoclinic	(001)	–Decussate to foliated, flaky mats of goethite pseudomorphs.
Garnet Isometric	(110)	–Idiomorphic, dodecahedral forms with hieroglyphic internal texture.

Blanchard and his co-workers, reviewed in detail elsewhere[72,109], forms the basis of part of Table 5 in which diagnostic replica forms of individual sulphide minerals are summarized. In the present-day microscopic evaluation of gossans, Blanchard's pioneering studies continue to provide an elementary descriptive reference. Relatively little has been published on the mineragraphic study of gossans, illustrating individual and composite sulphide replica textures[112,115,116].

Boxworks after *pyrite* are very common and are not well described in the early literature, where emphasis is placed on the associated cellular sponges and accretionary exotic limonite[31]. Positive, and more commonly negative, pseudomorphs of cubic pyrite are ubiquitous in cellular pyritic gossans [Figure 9 (a-d)]. Nested cubes of thin-walled limonite are a second form of pyrite boxwork commonly expressing the octahedral (111) cleavage of the mineral in diagonal cell walls [Figure 9 (c)]. Cells within the broadly cubic boxwork may also reflect the poor cleavage of pyrite in irregular forms. Concentrically zoned anhedral-to-subhedral forms [Figure 9 (d)] are the microscopic expression of the cellular sponge of Blanchard[31] and are more readily identified in mineragraphic section.

Monoclinic *pyrrhotite*, the commoner of the two forms, is replicated, not only by a characteristic boxwork [Figure 9 (c)] (Table 5) but also by pseudomorphous replacement of translation twins by goethite, where the primary sulphide has been deformed. Blanchard[31] illustrates a polished surface of two-phase pyrrhotite from Broken Hill, N.S.W., which combines the closely spaced, parallel boxwork of monoclinic pyrrhotite with cells expressing the symmetry of the hexagonal form of the sulphide. The hieroglyphic pyrrhotite boxwork [Figure 9 (f)] shows no repeated cell morphology and was described by Blanchard[31] under the forms of 'flaky limonite crusts' and 'cellular sponge'. The 'pin-point' or 'pisolitic limonite' of Blanchard[31] is expressed in mineragraphic section as concentrically banded spheroids of silica-goethite [Figures 3(c), (d)], which pseudomorph marcasite or pyrite after pyrrhotite. The shrinkage that accompanies the alteration of pyrrhotite to marcasite is accommodated by concentric cracks that are generally centred along fractures in the altering pyrrhotite producing spheroidal forms.

[*Editor's Note:* Figures 9 and 10—color photomicrographs showing characteristic replica textures of pyrite, pyrrhotite, pentlandite, chalcopyrite, galena, and sphalerite in gossans—have been omitted.]

The sub-parallel ladderwork [Figures 10(a), (b)] and coarse quadrangular boxworks [Figures 10(c), (d)] of *chalcopyrite* are well described by Blanchard[69]. The controlling elements of the quadrangular cell configuration are thought by Whittle[113] to be the tetragonal symmetry planes related to the unit cell, rather than the poorly developed (201) cleavage. The prominent (111) twin plane guides the majority of replacement reactions in chalcopyrite[62,117] and possibly controls the formation of the long, continuous cell walls. The prismatic and fibrolamellar forms of amphibole replicas may superficially resemble the ladderwork form of chalcopyrite boxwork, but are much more regular.

The blocky octahedral (111) cleavages of *pentlandite*, expressed by symmetrically triangular patterns [Figures 3(g), (h)], are commonly preserved through progressive alteration and retained in the gossan [Figures 3(a) to (d), 9(g), (h)]. Another indirect indication of pentlandite is the replica of the feathery margin of secondary violarite that replaces adjacent pyrrhotite during supergene alteration[116] [Figures 3(a), (b)]. The bladed form of pentlandite flames in pyrrhotite is also commonly pseudomorphed in gossans [Figures 9(g), (h)]. The pentlandite alteration and boxwork textures are best discerned in mineragraphic section and were not described prior to 1973[112,116].

A higher degree of regularity generally distinguishes the cubic boxwork of *galena*[100,113] [Figures 10(e), (f)], from that of pyrite[31]. Rhombic cerussite boxwork[100] is a common associate recording the progressive replacement of galena by anglesite (no diagnostic replica), cerussite and goethite.

Idiomorphic forms of *sphalerite* are rare, but within the irregular external forms, the perfect dodecahedral cleavages (110) are portrayed as parallel, continuous cell walls and short oblique (45°) cross walls[100] [Figures 10(g), (h)]. This configuration produces typically 45° triangular cells.

The strong octahedral (111) parting and twin plane of *magnetite*[118] determines the form of the closely spaced, parallel cell divisions which are oblique to elongate grain axes. Pseudomorphs of idiomorphic grains are also commonly preserved.

Bornite has a characteristic habit of forming curved crystal faces that are preserved in the boxwork as a modification of the octahedral (111) cleavage to give a unique spherically triangular configuration[103]. *Tetrahedrite* exhibits another unique replica texture (Table 5) possibly reflecting its curvilamellar habit and lack of pronounced cleavage[103]. *Molybdenite* is readily weathered, leaving a foliated, rounded form represented by casts in the original sulphides[105].

Many gangue minerals are also expressed by boxworks in gossans. *Calcite* boxwork may be distinguished from the rhombic boxworks of other carbonates, such as siderite[119], by the precise replication of its perfect cleavage. The rock-forming silicates such as *olivine*, *serpentine*, *amphibole*, *garnet*, *mica*, and *quartz* form boxworks or replica textures determined by individual habit and crystal symmetry. Elucidation of the residual gangue textures may assist in the classification of an ore association in the case of a gossan, or in the recognition of lateritized silicates in the case of other ironstones.

Replica textures of principal ore types

By mineragraphic examination of samples from weathered ore profiles, the progressive alteration sequence through the vertical supergene zones to the surface gossan can be observed.

The recent studies of nickel–copper sulphide weathering profiles in Western Australia[12,13] describe in detail the supergene sulphide alteration of *pyrrhotite–pentlandite ore*. Many of the features of supergene alteration in these ores are preserved in jasperoidal gossans[116]. In the alteration sequence, two texturally distinct types of violarite (V_{pn} and V_{po}) are formed. The violarite replacing pentlandite (V_{pn}) retains the blocky octahedral partings of its host (Figure 3), whereas violarite replacing pyrrhotite (V_{po}) assumes marginal, feathery intergrowths controlled by the crystallographic (001) plane of former pyrrhotite (Figure 3). Because the amount of violarite that replaces pyrrhotite is genetically controlled by the amount of nickel released during violaritization of pentlandite (Figure 4), the proportions of the two texturally distinct violarite types are a measure of the original pyrrhotite:pentlandite ratio in the primary ore. Figure 4 shows that the molecular ratio of violarite replacing pyrrhotite is approximately one third of the original primary pentlandite. Thus, large feathery violarite margins are indicative of pentlandite-rich ores, whereas small margins are indicative of pentlandite-poor ores. These textures, commonly preserved in nickel gossans by goethite and silica replacement, are best identified by mineragraphic examination[112,116] (Figure 3). Despite the great chemical changes in gossan formation (Figure 8), the recognition of these residual textures in nickel gossans provides an indication of the ore grade in the underlying sulphide body.

Supergene marcasite (and pyrite), developed from the residual monoclinic pyrrhotite at the top of the transition zones, is commonly pseudomorphed by goethite-silica in the gossans. The colloform marcasite (or pyrite) textures, indicating a volume decrease in pyrrhotite replacement[116], (Figure 2), initially develop along incipient fractures. This aggregate of pisolitic spheroids ('bird's eye texture') may be retained in the gossan indicating a genetic link with the cores of primary pyrrhotite remaining after violaritization.

Many of the massive *copper–zinc sulphide ores* show recrystallized and annealed textures that result from diagenesis, metamorphism, or deformation. Annealing produces a coarse ore fabric comprising pyrite porphyroblasts in a poikiloblastic groundmass of (pyrrhotite)–chalcopyrite–sphalerite [Figure 11(a)]. The interstitial sulphides are more susceptible to weathering than the idiomorphic pyrite which almost outlasts the complete alteration of the matrix [Figure 11(f)]. Sphalerite is generally the first sulphide to be altered with goethite invading sub-parallel (110) partings [Figures 11(g), (h)]. The direct replacement of sphalerite by goethite occurs rapidly and only in coarse-grained aggregates does the diagnostic boxwork survive[115] (Table 5) [Figure 11(e)]. The common oxidate minerals derived from sphalerite (Table 4), themselves have no distinctive replica textures. For chalcopyrite, the initial oxidation product is normal covellite, which develops along one sub-parallel set of (111) or (201) planes. Enriched chalcopyrite ores also oxidize through covellite [Figure 11(c)]. The framework of the chalcopyrite boxwork is established by early oxidation and further alteration to goethite [Figure 11(d)], lepidocrocite, and malachite accompanies development of the boxwork.

Fine-grained portions of the matrix show little preservation of replica textures and are commonly occupied by exotic, colloform goethite. Elsewhere, a compositional range of goethitic jaspers may preserve detailed textural replicas of the original sulphides [Figures 11(a), (b)].

As pyrite is ultimately corroded, the cubic (100) and octahedral (111) cleavages of pyrite are invaded by silica-goethite, producing cubic and triangular prismatic cells within the replica form of the original grain. A range of textures (Table 5) may develop, thus preserving the original porphyroblastic texture of the primary sulphide [Figures 11(b), 11(e), 11(f)].

Pyrrhotite, where present, occurs in the matrix and, with sphalerite, undergoes early alteration. Coarse-grained aggregates of pyrrhotite commonly leave some form of residual texture (Table 5), whereas fine grains are replaced without trace.

The gross primary texture of pyrite–chalcopyrite–sphalerite ores is preserved macroscopically in gossan[115], but mineragraphic examination is necessary to elucidate the poorly preserved replica textures of the interstitial copper and zinc sulphides.

The replica textures of laminated, syndiagenetic *lead–zinc sulphide ores* are extremely difficult to evaluate because the fine laminations and finely colloform sulphides preclude the formation of substantial boxwork textures. Oxidate mineralogies, however, commonly reflect the primary mineralogies because the neutralizing effect of the host rocks inhibits leaching of the ore metals. In this ore type, sphalerite undergoes early supergene alteration[24] to a wide variety of oxidate minerals before hematite or goethite precipitate. Characteristic boxworks are associated only with the coarser phases of the sulphide (Table 5) [Figures 10(e) to (h)]. Lacking these, pseudomorphous replacement textures of sphalerite-derived oxidate minerals (Table 4) possess no diagnostic features.

As galena oxidizes, a protective coating of anglesite forms from which the carbonate, cerussite, eventually

[*Editor's Note:* Figure 11, color photomicrographs illustrating the incipient supergene alteration textures of pyrite—chalcopyrite—sphalerite ores and diagnostic textures of the derived gossans, has been omitted.]

develops[120,121]. Thus residual cores of galena may survive in gossans. Without these cores, two boxwork types may develop, which record the presence of galena: (i) rectilinear anglesite replacements along the cubic (100) cleavage of galena[100] (Table 5) [Figure 10(e), 10(f)], and (ii) rhombic mesh boxworks[100] (acute angle 62°) of goethite outlining the structure of oxidate cerussite. This orthorhombic cerussite boxwork is difficult to distinguish from the rhombic textures of trigonal carbonates (acute angles 73 to 75°). The composite boxwork form of a cubic core and rhombic envelope is rare but invariably present in gossans of this type and even in spatially discrete association, the two forms are wholly diagnostic of primary galena.

These examples illustrate the application of techniques in the interpretation of the textures of gossans that are derived from various ore types under a range of conditions. It should be borne in mind that in modern exploration, the interpreted characterization of gossans is based also on additional evidence, particularly geochemical data.

THE GEOCHEMISTRY OF GOSSANS

As sulphide minerals are weathered, a component (albeit variable) of the liberated ore metals, is generally retained within the gossan-forming system, although many other ironstones of non-sulphide derivation may also contain significant accumulations of trace elements. In mineral exploration, the objective of geochemical evaluation is to assess the significance of minor-and trace-element content, thus affording the genetic classification of gossans and ironstones.

The chemical factors controlling trace metal retention and accumulation in iron-rich rocks are complex in detail. There is little doubt that trace metals in exotic and indigenous limonites alike are largely adsorbed on, or coprecipitated with, finely crystalline goethite or amorphous, gel limonite. The 'scavenging' activity of hydrated manganese oxides in limonite appears to exceed that of the iron oxides alone (Table 6). Experimental simulation of the adsorption and coprecipitation mechanisms has shown that, at very low metal (sorbate) concentrations, the sorption capacities of hydrous iron[122,123] and manganese oxides[124,125] are high and strongly pH dependent, increasing with alkalinity in the range pH 3 to 8[122,124,125]. Adsorption is probably effected by surface ion exchange and perhaps, in the case of manganese, by the formation of a surface complex[125]. At high metal (sorbate) activities, dependence on pH decreases, commonly to zero, and continued high sorption capacity, at least for manganese[124], is variously attributed to mechanisms such as ion exchange in the lattice[125] or surface exchange with cation impurities[124]. In some gossans, microcrystalline disseminations of discrete ore-metal bearing oxysalts, oxides, and native grains of residual sulphides may also be sufficiently abundant to contribute to detectable trace element contents in gossans.

The great variability in content of single elements that exists between samples of the same genetic type (even on the scale of a hand specimen) has long been recognized as a hindrance to confident geochemical characterization of gossans. Geological factors controlling the ultimate content of, and distribution within, gossans include primary metal and sulphur contents, the tectonic, geomorphological, and climatic conditions of sulphide weathering, and the nature of the interactions between the host rock and sulphide weathering systems. Despite their variable geochemistry, the use of geochemical criteria to characterize gossans, in contrast to the use of textural criteria, does not necessarily prerequire *in situ* precipitation of iron (preserving texture), thus simplifying the inherent sampling limitations, although not necessarily the interpretational techniques, of evaluating gossans and ironstones.

The earliest geochemical comparison of gossans was published in 1944 by Blanchard[108], who listed major and minor element analyses (including Cu, Pb, Zn, Mn, and As) for twenty gossan limonites. Descriptions in the early literature of individual orebodies and their supergene alteration rarely included one partial gossan analysis. This no doubt largely reflected the time-consuming analytical methods then available. In present day exploration, the increased application of rapid, versatile, and precise methods of geochemical analysis, particularly atomic absorption spectrophotometry and to a lesser degree X-ray fluorescence and emission spectrography, has permitted the geochemical affinities of various gossan and ironstone types to be characterized in detail. Jasperoidal gossans are particularly resistant to chemical attack, but can generally be digested by some combination of hydrofluoric with perchloric or nitric acids. Losses of 'volatile' elements, notably arsenic, may be reduced by selective alkali fusion (low sample dilution) or, more versatile, sodium peroxide sintering[126] (high sample dilution). Precise characterization of a gossan commonly relies on the residual trace contents of immobile elements (As, Sn, Au, Pd, Pt, Ir, Se, Te, Mo, Ba) that, even with modern analytical technology, require costly determinations by non-routine methods. Thus, the problem in the exploration context is to distinguish between gossans and ironstones on the basis of elements that require only rapid, economical, routine analysis.

The geochemical evaluation of gossans has recently come to the fore. Since 1965 many of the major exploration companies have been actively compiling large volumes of data during the course of routine exploration. It is not within the scope of this paper to review the geochemistry of gossans from a wide range of environments, the data for which are mainly held in unpublished company files, but rather to discuss several well-documented examples of base-metal-sulphide gossan types and their associated ironstones. The interpretation of geochemical data on gossans and ironstone has matured on an empirical basis from detailed, comparative orientation at known deposits and mineral prospects. It still remains essential to evaluate the ironstone problem in mineral exploration within a well-founded metallogenetic, regional and climatic framework, since there is a dangerous tendency to over-extrapolate a locally diagnostic geochemical signature, even within a similar ore type.

The geochemistry of the massive *nickel–copper gossans* of Western Australia is well documented in several reviews[7,73,127–131] and a few specific orientation studies[8,132]. Diagnostic geochemical features of nickel copper gossans (Table 6) include high Ni–Cu, high Pd–Pt[8,133,134], variable Co-As and low Cr–Mn–Zn–Pb. Gossans capping non-nickeliferous sulphides, and ironstones representing lateritized ultramafics, can generally be distinguished on the basis of low Cu–Ni/high Zn, and high Ni–Cr–Mn/low Cu respectively (Table 6). Even highly siliceous nickel gossans generally retain the charac-

TABLE 6

GEOCHEMICAL COMPOSITION OF PRINCIPAL GOSSAN TYPES AND ASSOCIATED IRONSTONES.
THE DATA ARE FROM A NUMBER OF SOURCES (REFERENCE NUMBERS 7, 8, 73, 115).

NICKEL-COPPER GOSSANS AND ASSOCIATED IRONSTONES, WESTERN AUSTRALIA

COPPER-ZINC AND LEAD-ZINC GOSSANS AND ASSOCIATED IRONSTONES, SOUTHERN AFRICA

Note: Parameter values (p) are the ranges (r) at the 10 and 90 percentile limits, the median values (md) and means (m). The total number of samples (n) is also specified. Asterisks indicate samples from more than one locality.

	n	p	*Ni* ppm	*Co* ppm	*Cu* ppm	*Zn* ppm	*Fe* %	*Mn* ppm	*Cr* ppm	*As* ppm	*Te* ppm	*Pd* ppm	*Pt* ppm
Nickel Gossans	16	m	9200	300	2600	30	40	500	800	220		1.1	0.35
		r	900-24200	60-1050	900-6000	10-120	8,9-51,9	50-1700	60-3000	7-860		0,15-4,5	0,04-1,0
Nickel Gossans (Siliceous)	5	m	2600	110	580	54	5	110	580				
		r	1700-4400	22-100	490-760	42-55	4,4-5,7	225-735	320-1040				
Nickel Gossans*	33–47	md	7000	260	1500	85	17		1150	23	0.95		
		r	1130-25000	50-850	500-16000	50-260	4,3-57,0		430-2300	3-170	0,57-6,4		
Probable Nickel Gossans*	10–30	md	3200	740	600	50	55		900	30	1.6		
		r	450-22000	65-360	65-2500	25-470	5,8-60,0		230-2600	9-110	0,31-7,4		
Gossanous Tuffs in Ultramafic Rocks*	16–19	md	350	80	250	250	45		320	50	0.05		
		r	120-2500	55-280	45-690	100-720	3,7-57,0		60-1700	5-600	0,005-0,46		
Lateritized Ultramafic Rocks*	37–48	md	1300	140	100	90	8		2300	45	0.005		
		r	1200-3400	30-400	15-570	40-320	3,5-50,0		480-11000	5-350	0,004-0,10		
Iron Sulphide Gossans in Ultramafic Rocks*	11–16	md	1100	180	340	150	57		430	7	0.05		
		r	330-3200	75-1100	45-3100	70-360	10,0-60,0		150-5700	2-26	0,004-0,11		

	n	p	*Ni* ppm	*Co* ppm	*Cu* ppm	*Zn* ppm	*Pb* ppm	*Mn* ppm	*Cr* ppm	*Ba* ppm	*Ag* ppm	*Cd* ppm	
Copper-Zinc Gossans	50	m	27	70	1800	90	250	110		1800	8	2	
		r	25-50	40-120	800-3400	40-170	125-650	40-250		110-9400	2-12	1-4	
Copper-Zinc Gossans (Siliceous)	19	m	32	28	370	580	105	125	75	5900	2	7	
		r	25-40	18-48	150-960	160-1300	60-220	60-215	40-115	1300-18000	1-4	6-8	
Lead-Copper Gossans	5	m	30	40	330	800	3800	3300	50	720	9		
		r	19-35	20-60	75-15500	160-2840	480-7500	170-7000	20-90	380-1500	3-32		
Zinc-Copper Gossans	19	m	24	25	115	600	6200	360	65	260	7	7	
		r	13-30	10-35	40-220	210-1910	420-22500	110-760	40-100	55-660	3-17	5-10	
Stratiform Iron Sulphide Gossans (Minor Cu-Zn)	22	m	10	100	3900	920	390	1000		140	8	10	
		r	10-20	40-120	1100-8000	240-3500	100-850	200-2200		40-400	2-12	6-12	
Copper Gossans (Vein-type)	11	m	270	240	15200	50	220	330	45	50	3	6	
		r	80-750	60-460	8500-22800	30-90	150-315	90-920	25-80	20-90	2-4	5-8	
Pyrite Gossans (Vein-type)	12	m	25	20	45	45	30	70	25	30			
		r	20-40	12-30	18-75	20-80	15-40	20-270	12-35	15-40			
Ferricretes	32	m	75		15	105	65	1250				4	
		r	30-175		5-60	25-300	20-115	320-5000				1-8	

teristics of the type geochemical signature at diluted metal contents (Table 6).

Relatively few geochemical data have been published on massive *copper–zinc gossans*, but significant early work[135] established the importance of a high Ba–Au–Ag/low Mn association. Preliminary results from current research[115] (Table 6) on southern African gossans indicate that a high Cu–Pb–Ba/variable Zn/low Mn–Ni–Co–Cr association is diagnostic of certain volcanogenic ore types. Small, low grade copper–zinc bodies feature reduced Ba and increased Zn–Mn contents in the gossans in some cases. Silicification drastically reduces the absolute content of trace elements, with the possible exception of Ba. The geochemistry of the ore type is less consistent than that of nickel–copper ore; therefore, the derived gossans must be evaluated as independent sub-types having independent geochemical signatures. Gossan capping vein-type copper mineralization may display a distinct Cu–Ni–Co–Pb association, *Copper–lead–zinc gossans* generally have high contents of the constituent ore metals, broadly corresponding to original relative proportions. Published geochemical data are rare for any class of lead-zine gossan, probably because retained metal contents are commonly several per cent and, therefore, they present straightforward exploration targets. Lead–zinc gossans in carbonates and argillites[22,32,108] typically have an association of high Pb–Zn–Mn–Ag contents. In Australia, low-grade, stratiform copper and pyrite gossans[74,83,136], siderite-derived ironstones[83,137], ferruginized rock[74,83,128] and fault ironstone[74,83,128,136] display similar high Cu contents with variable Mn–Ag (Table 6). The gossan and oxide zones over pyrite bodies are commonly rich in selenium[138,139], but in volcanic sequences, high background selenium[140] may invalidate this criterion. Pyrite and pyrite-pyrrhotite gossans in southern Africa generally have very low-trace element signatures (Table 6), but in common with various types of ironstone, they may contain variably high Mn, Zn, and Cu. Iron-rich sedimentary rocks[137] principally bog ores, siderites, minettes, and chamosites, have sporadically high Mn (0,1 to 10%), As (100 to 800 p.p.m.), Zn (100 to 240), and P (0,1 to 2,6%).

The information presented in Table 6 illustrates how, within restricted geographical areas, the geochemical data on specific gossan and ironstone types can be regarded as discrete populations. The application of statistics is therefore an appropriate means of effecting an empirical classification of ironstone types within the exploration context. It is certainly expedient exploration practice to erect a classification through orientation surveys that discriminates between all known ironstone and gossan types within a search area. Once erected, this can be used to test the affinities of unknown sample types, and the classificatory criteria can be updated periodically and refined as new data come to hand.

Viewed as single-element populations, gossan and ironstone types may be characterized by simple statistical parameters. Measurements of central tendency (including the medians, the modes, and the arithmetic or geometric means) and measurements of variability (including the ranges, the ranges at the specific percentiles, and the standard deviations or coefficients of variation) can be applied to effect separation of large sample populations. Commonly, however, gossan geochemistry gives rise to variously skewed distributions of extreme variance that are not derived from sufficiently large sample populations to apply these parameters confidently. In these cases, the application of nonparametric statistical techniques such as the Mann–Whitney U test[141] may be more applicable than the parametric techniques. The shape of the distribution curve conveys the maximum information about single-element populations and the maximum departure of two curves may be gauged by the Kolmogorov–Smirnov test[142]. Involving the entire curve shape is considered to produce a more powerful test of difference between two distributions of gossan data than comparison of individual parameters.

The fact that gossan geochemistry is so characteristically variable generally necessitates a large number of samples to effect confident classification. Simple ratios of single elements are frequently used in the treatment of gossan geochemistry. Where two individual complex distributions are involved, this procedure tends to complicate, rather than simplify, discrimination between groups because the derived ratio distributions commonly have expanded, rather than reduced variances. It does appear possible, however, that the inherent single-element variability may be used to advantage, if the covariation of the elements is quantified and used as an empirical discriminant function. The effect of multiplying specific pairs or groups of highly correlated elements, that are differently associated within the two distributions, may cumulatively yield a total separation. This phenomenon has been noted by several workers[128]. The concept can be formally applied by the use of multivariate statistical techniques[143]. By considering all elements as variables, multivariate statistical techniques lend themselves to rediscribing the total variability of the data set in terms of the variability of associated elements.

Based upon Q mode principal components analysis on geochemical data for copper, nickel, lead, zinc, manganese, and chromium in a suite of gossans and ironstones from the Western Australian shield terrain, Joyce and Clema[9] showed that two factors, one approximating to log copper plus log nickel and the other to log chromium, account for as much as 67 per cent of the total variability in all the geochemical data between all the samples. Moreover, the first factor was shown to be related to nickel gossans whereas the second factor was shown to be related to nickel laterites. A simple practical scattergram consisting of chromium plotted against nickel times copper on log–log paper[9] serves to discriminate between lateritic ironstones and nickel gossans in the Western Australian terrain. It should be emphasized, however, that this type of scattergram is only applicable to classification of ironstones within their own restricted terrain; the same scattergram, for example, misclassifies many of the nickel gossans in Botswana. Principal components analysis is a useful tool to redescribe the total variability in terms of the main associations of elements. A much more satisfactory tool for classifying ironstones and gossans within the exploration context, however, is multivariate discriminant analysis.

In discriminant analysis, one or more discriminant functions are calculated to reflect the maximized difference(s) in the multi-element data between two or more sample populations. The advantage of discriminant analysis is that the sample populations can be chosen to represent known gossans and ironstones that typify a known classification. Once a file of data is accumulated, and the appropriate discriminant functions derived, a sample of previously unknown origin collected during the course of routine exploration can be tested by the discriminant function. It should be stressed, however, that a large number of samples is required to estimate the multi-element covariance matrix reliably. The covariance matrix forms the basis for the calculation of linear discriminant functions that may be used for classifying samples of unknown origin. Ideally, a sample size of at least ten times the number of variables per class is required for reliable estimation of the covariance matrix[143]. Reliably estimated discriminant functions can assign a probability of the sample being derived from each class. Samples given a favourable classification can thus guide subsequent evaluation. This type of methodology has been successfully applied to the classification of nickel gossans and ironstones in the Western Australian shield terrain[7]. Work in progress indicates that it can also be successfully applied elsewhere where the ironstone problem is differently manifest.

CONCLUSIONS

The problem of discerning gossans overlying economic sulphide mineralization within a background of other variously derived ironstones is seldom resolved conclusively in the field. Surface expression of massive sulphide bodies as gossans may be rendered extremely subtle by such processes as lateritization and silicification.

The iron oxide mineralogy of gossans is not universally characteristic, but metal oxidates and the residual sulphide-gangue assemblage may provide useful genetic criteria. Remanent textural evidence of sulphides, manifest as cellular or pseudomorphous replacement, may, under favourable conditions, indicate the original ore assemblage, fabric and even the grade. Gossan textures range from precise replicas to partially preserved, subtle or inconspicuous forms, or may be totally non-diagnostic. Trace-element geochemistry represents the most versatile and reliable criterion for gossan evaluation, being adversely affected only by extreme leaching or silicification.

In regional reconnaissance, the geochemical evaluation of ironstones, using the criteria of readily determined trace-element associations as discriminants, may indicate their exploration potential. In this way, large numbers of samples may be classified routinely. Confident classification of the samples having ambiguous trace-element affinities may be achieved by more costly analytical and mineragraphic interpretation. This type of information may guide geophysical surveys, ground acquisition, and drilling priorities. In target investigation, the integration of both geochemical and mineragraphic interpretation balanced with data from other exploration methods may be used to predict the economic significance of the underlying sulphide mineralization. This type of detailed interpretation may serve as a guide to prospect acquisition or optioning properties for further evaluation.

ACKNOWLEDGEMENTS

The authors wish to thank Professor G. R. Davis and other colleagues in the Mining Geology Division at the Royal School of Mines for their encouragement and advice. Financial assistance provided by NERC and several mining companies, notably RTZ, Anglovaal, and JCI, for research in this general field is gratefully acknowledged.

REFERENCES

1. FINLAYSON, A. M. Secondary enrichment in the copper deposits of Huelva, Spain. *Trans. Instn Min. Metall.*, vol. 20. 1910. pp. 61-72.
2. WILLIAMS, D. Gossanized breccia ores, jarosites and jaspers at Rio Tinto mines, Spain. *Bull. Instn Min. Metall.*, vol. 59. no. 526. 1950. pp. 1-12.
3. PRYCE, W. *Mineralogia Cornubiensis*, 1st ed. Bath, Pitman Press, 1778, reprinted 1972.
4. LOUGHLIN, G. F. The oxidized zinc ores of the Tintic District, Utah. *Econ. Geol.*, vol. 9. 1914. pp. 1-19.
5. SCHWARTZ, G. M. The nature of primary and secondary mineralization in porphyry copper deposits. *Geology of the porphyry copper deposits, Southwestern North America*. Titley, S. R. and Hicks, C. L. editors. Tuscon, Univ. of Arizona Press, 1966. pp. 41-50.
6. BATEMAN, A. M. *Economic mineral deposits*. 2nd ed. New York, Wiley, 1958.
7. BULL, A. J., and MAZZUCHELLI, R. H. Application of discriminant analysis to the geochemical evaluation of gossans. *Geochemical exploration 1974*, Elliott I. L. and Fletcher W. K. eds. Amsterdam, Elsevier, 1975. pp. 219-226.
8. WILMSHURST, J. R. The weathering products of nickeliferous sulphides and their associated rocks in Western Australia. *op. cit.* (7). pp. 417-436.
9. JOYCE, A. S., and CLEMA, J. M. An application of statistics to the chemical recognition of nickel gossans in the Yilgarn Block, Western Australia. *Proc. Australs. Inst Min. Metall.*, no. 252, 1974. pp. 21-24.
10. LOCKE, A. *Leached outcrops as guides to copper ore*, Baltimore, Williams and Wilkins., 1926.
11. BLANCHARD, R. Interpretation of leached outcrops. *J. Chem. Soc. S. Afr.*, vol. 39, 1939. pp. 344-372.
12. NICKEL, E. H., ROSS, J. R., and THORNBER, M. R. The supergene alteration of pyrrhotite – pentlandite ore at Kambalda, Western Australia. *Econ. Geol.*, vol. 69. 1974. pp. 93-107.
13. WATMUFF, I. G. Supergene alteration of the Mount Windarra nickel sulphide ore deposit, Western Australia. *Mineral. Deposita*, vol. 9. 1974. pp. 199-211.
14. BROTHERTON, R. L. The major element geochemistry of nickeliferous gossans. *Trans. Instn Min. Metall.*, 1977. (in press).
15. KEELE, R., and NICKEL, E. H. The geology of a primary millerite bearing sulfide assemblage and supergene alteration at the Otter Shoot, Kambalda, Western Australia. *Econ. Geol.*, vol. 69. 1974. pp. 1102-1117.
16. THORNBER, M. R. Supergene alteration of sulphides. III The composition of associated carbonates. *Chem. Geol.*, vol. 17. 1976. pp. 45-72.
17. WILLIAMS, D. The geology of the Rio Tinto Mines, Spain. *Bull. Instn Min. Metall.*, vol. 43. 1933-34. pp. 594-640.
18. EMMONS, W. H. The enrichment of ore deposits. *Bull. U.S. geol. Surv.* no. 625. 1917.
19. GILBERT, G. Oxidation and enrichment at Ducktown, Tennesee. *Trans. Am. Inst. Min. metall. Petrol. Engrs.*, vol. 70. 1924. pp. 998-1023.
20. MAGEE, M. The geology and ore deposits of the Ducktown District, Tennessee. *Ore deposits in the United States, 1933-1967*, Ridge, J. D., ed. New York, Inst Min. Metall. Engrs., vol. 1. 1968. pp. 207-241.
21. HILL, P. A. The gossans of Minas Carlota, Cuba. *Econ. Geol.*, vol. 57. 1962. pp. 168-194.
22. BENNETT, E. M. Lead-zinc-silver and copper deposits of Mount Isa. *Proceedings of the 8th Commonwealth Mining Metallurgical Congress*. vol. 1. *Geology of Australian ore deposits*. Melbourne, Australas. Inst. Min. Metall., 1966. pp. 233-246.
23. BLANCHARD, R. Significance of the iron oxide outcrop at Mount Oxide, Queensland. *Proc. Australas. Inst. Min. Metall.*, no. 114. pp. 21-50.
24. HEYL, A. V., and BOZION, C. N. Oxidized zinc deposits of the United States: Pt 1. General geology. *U.S. geol. Surv., Bull. 1135-A*, 1962. pp. A1-A52.
25. HEYL, A. V. The Upper Mississippi Valley Base-metal District. *op. cit*, (20). pp. 431-459.
26. TWETO, O. Leadville District, Colorado. *op. cit.* (20). pp. 681-705.
27. LEA, E. R., and DILL, D. D. Jr. Zinc deposits of the

Balmat-Edwards District, New York *op cit.* (20). pp. 20-48.
28. BROWN, J. S. Supergene sphalerite, galena and willemite at Balmat, New York. *Econ. Geol.*, vol. 31. 1936. pp. 331-354.
29. BROWN, W. H., and WEINBERG, E. L. Geology of the Austinville – Ivanhoe District, Virginia. *op. cit.* (20). pp. 169-186.
30. MOORE, J.McM. Supergene mineral deposits and physiographic development in southwest Sardinia, Italy. *Trans. Instn Min. Metall.* vol. 81. 1972. pp. B591B66.
31. BLANCHARD, R. Interpretation of leached outcrops, Nevada State Bur. Mines, *Bull. 66*, 1968.
32. JAQUET, J. B. Geology of the Broken Hill Lode and the Barrier Ranges Mineral Field, New South Wales. *Geol. surv. N.S.W.*, Mem. 5, 1894.
33. BOYLE, R. W. Geology, geochemistry and origin of the lead-zinc-silver deposits of the Keno Hill-Galena Hill area, Yukon Territory. *Geol. surv. Can. Bull. 111*, 1965. 302 pp.
34. BARSHAD, I. Weathering – chemical. *The encyclopaedia of geochemistry and the environmental Sciences*, Fairbridge, R. W. ed. New York, Van Nostrand-Rheinhold, 1972. pp. 1264-1269.
35. LOUGHNAN, F. C. *Chemical weathering of the silicate minerals*, New York, American Elsevier 1969.
36. HAWKES, H. E., and WEBB, J. S. *Geochemistry in mineral exploration*, New York, Harper and Row, 1962.
37. SATO, M., and MOONEY, H. M. The electrochemical mechanism of sulfide self-potentials. *Geophysica*, vol. 25. 1960. pp. 226-249.
38. THORNBER, M. R. Supergene alteration of sulphides. I. A chemical model based on massive nickel sulphide deposits at Kambalda, Western Australia. *Chem Geol.*, vol. 15. 1975. pp. 1-14.
39. GARRELS, R. M. *Mineral equilibria at low temperature and pressure*, New York, Harper, 1960.
40. GARRELS, R. M., and CHRIST, C. L. *Solutions, minerals, and equilibria*, New York, Harper and Row, 1965.
41. SATO, M. Oxidation of sulfide ore bodies: Pt 1. Geochemical environments in terms of Eh and pH *Econ. Geol.*, vol. 55. 1950. pp. 928-961.
42. THORNBER, M. R. Supergene attention of sulphides. II, A chemical study of the Kambalda nickel deposits. *Chem. Geol.*, vol. 15. 1975. pp. 117-144.
43. PETERS, E. Direct leaching of sulphides: chemistry and applications. *Metall. Trans. B*, vol. 7B. 1976. pp. 505-517.
44. FOX, R. W. On the electromagnetic properties of metalliferous veins in the mines of Cornwall. *Phil. Trans.*, Pt 2. 1830. 399 pp.
45. SKEY, W. On the electromotive power of metallic sulphides. *Trans. N.Z.* Inst. vol. 3. 1871. pp. 232-236.
46. BERNFIELD, I. D. Studien über Schwefelmetallelektroden. *Z. physik. Chem.*, vol. 25. 1898. pp. 46-78.
47. WELLS, R. C. Mineral electrodes. *Am. Electrochem. Soc. Trans.*, vol. 22. 1912. pp. 313-342.
48. WELLS, R. C. Electrical activity in ore deposits. U.S. geol. Surv., *Bull. 548*. 1914. 78 pp.
49. GOTTSCHALK, V. H., and BUEHLER, H. A. Oxidation of sulphides. *Econ. Geol.*, vol. 7. 1912. pp. 15-34.
50. RECHENBURG, H. The electrochemical potential series of ore minerals. *Neues Jb. Miner., Mh.*, 1951. pp. 88-91.
51. ROSETTI, V., and GESINI, A. M. Electrochemical effects between metal sulphides and mine water. *Periodico Mineral.*, vol. 26. 1957. pp. 63-102. (In Italian).
52. SVESHNIKOV, G. B. Role of natural electrochemical processes in the formation of oxide ore in sulphide deposits. *Vestn. Leningr. gos. Univ.*, vol. 18. 1963. *Ser. Geol.i Geogr.*, No. 4. 1963. pp. 54-61.
53. SVESHNIKOV, G. V., and RYSS, Y. S. Electrochemical processes in sulphide deposits and their geochemical significance. *Geochem. Int.*, no. 2. 1964. pp. 208-218.
54. NDDACK, W., and WRABETZ, K. Über das elektrochemische Verhalten einiger Schwermetallsulfide. *Z. Elektrochem.*, vol. 59. 1955. pp. 96-102.
55. WRABETZ, K. E. Über gevanische Zellen mit Sulfidelektroden. *Z. Elektrochem.*, vol. 60. 1956. pp. 722-731.
56. SATO, M. Oxidation of sulfide ore bodies, 2, Mechanisms of oxidation of sulfide minerals at 25°C. *Econ. Geol.*, vol. 55. 1960. pp. 1202-1231.
57. HABASHI, F. The mechanism of oxidation of sulfide ores in nature. *Econ. Geol.*, vol. 61. 1966. pp. 587-591.
58. BÖLVIKEN, B., and LOGN, Ö. An electrochemical model for element distribution around sulphide bodies. *op. cit.* (7). pp. 631-650.
59. THORNBER, M. R. Pyrrhotite – the matrix of nickel sulphide mineralization. *Newcastle Conference Australasian Institute Mining and Metallurgy*. Parkville, Vic., The Institute, 1972. pp. 51-58.
60. BLAIN, C. F., and BROTHERTON, R. L. Self potentials in relation to oxidation of nickel sulphide bodies within semi-arid climatic terrains. *Trans. Instn Min. Metall.* vol. 84. 1975. pp. B123-B127.
61. ANDERSON, C. A. Oxidation of copper sulfides and secondary sulfide enrichment. *Economic Geology*. 50th Anniversary vol. 1905. 1955. Bateman, A. M. ed. Lancaster, Penn., Economic Geology, pp. 324-340.
62. SILLITOE, R. H., and CLARK, A. H. Copper and copper-iron sulfides as the initial products of supergene oxidation, Copiapó Mining District, Northern Chile. *Am. Mineralogist*, vol. 54. 1969. pp. 1684-1710.
63. ZIES, E. G., ALLEN, E. T., and MERWIN, H. E. Some reactions involved in secondary copper sulfide enrichment. *Econ. Geol.*, vol. 11. 1916. pp. 407-503.
64. BLAIN, C. F. Chemical reactions in secondary enrichment of copper ores, 1977. *(in preparation)*.
65. TAKAHASHI, T. Supergene alteration of zinc and lead deposits in limestone. *Econ. Geol.*, vol. 55, 1960. pp. 1083-1115.
66. MORRISSEY, C. J., and WHITEHEAD, D. Origin of the Tynagh residual orebody, Ireland. *Proceedings 9th Commonwealth Mining and Metallurgical Congress London 1969*. Jones, M. J. ed. London, Instn. Min. Metall., vol. 2. 1970. pp. 131-146.
67. TAYLOR, J. H. Formation of supergene galena at Broken Hill, N. Rhodesia. *Mineral of Mag.*, vol. 31. 1958. pp. 908-913.
68. POSNJAK, E., and MERWIN, H. E. The system $Fe_2O_3-SO_3-H_2O$, *J. Am. Chem. Soc.*, vol. 44. 1922. pp. 1965-1994.
69. BLANCHARD, R., and BOSWELL, P. F. Notes on the oxidation products derived from chalcopyrite. *Econ. Geol.*, vol. 20. 1925. pp. 613-638.
70. DUTRIZIAC, J. E., and MacDONALD, R. J. C. Ferric ion as a leaching medium. *Minerals Sci. Engng.*, vol. 6. no. 2. 1974. pp. 59-100.
71. KRAUSKOPF, K. B. *Introduction to geochemistry*, New York, McGraw Hill, 1967.
72. WHITTEN, G. Leached outcrops. *Proceedings 8th Commonwealth Mining and Metallurgical Congress.*, vol. 2. 1966. pp. 193-204. *Exploration and Mining Geology*, Melbourne, Australas. Inst. Min. Metall.
73. COCHRANE, R. H. A. A guide to the geochemistry of nickeliferous gossans and related rocks from the Eastern Goldfields. *Rep. geol. Surv. W. Aust. 1972*, 1973. pp. 69-75.
74. BAMPTON, K., COLLINS, A., GLASSON, K., and GUY, B. Geochemical evidence of concealed mineralization in an area northwest of Mount Isa. *25th International Geological Congress*, Sydney 1976, Abstracts: vol. 2. sect. 10B. pp. 434-435.
75. STRAHLER, A. N. *Physical geography*, 3rd ed. New York, Wiley, 1969.
76. MAIGNIEN, R. *Review of research on laterites*. Paris, UNESCO, 1966.
77. MULCAHY, J. Laterites and lateritic soils in southwest Australia. *J. Soil. Sci.*, vol. 11. 1960. pp. 206-225.
78. SHARPE, J. W. The Empress nickel-copper deposit, Southern Rhodesia. *The geology of some ore deposits in southern Africa*, Haughton, S. H. ed. Johannesburg, Geol. Soc. S. Afr., vol. 2. 1964. pp. 497-508.
79. MIDDLETON, R. C. The geology of Prieska Copper Mines. *Econ. Geol.*, vol. 71. 1976. pp. 328-350.
80. VOET, H. W. The Areachap orebody. *16th Congress of the Geological Society of South Africa, Stellenbosch, 1975*. Excursion Guide Bk., No. 2. 1975. pp. 36-40.
81. COETZEE, C. B. Manganiferous iron ore, hematite, barite and sillimanite on Gams (Portion 1), Namaqualand. *Geol. Surv. S. Afr., Bull. 28*, 1958.
82. SOLOMON, M. Fossil gossans at Mt. Lyell, Tasmania. *Econ. Geol.*, vol. 62. 1967. pp. 757-772.
83. TAYLOR, G. F. The geochemistry of siderite in relation to ironstone in the Paradise Creek Formation, northwest Queensland. *J. geochem. Explor.*, vol. 2. 1973. pp. 367-382.
84. KELLY, W. C. Topical study of lead-zinc gossans. *New Mex. St. Bur. Mines*, Bull. no. 46, 1958.

85. MALONE, E. J., OLGERS, F., CUCCHI, F. G., NICHOLAS, T., and McKAY, W. J. Woodlawn copper-lead-zinc deposit. *Economic geology of Australia, Papua and New Guinea:* metals. Knight, C. L. ed. Parkville, Australas. Inst. Min. Metall., vol. 1. 1975. pp. 701-710.
86. STANTON, R. L. Petrochemical studies of the ore environment at Broken Hill, New South Wales: 3 – banded iron formations and sulphide orebodies: constitutional and genetic ties. *Trans Instn. Min. Metall.*, vol. 85. 1976. pp. B132-B141.
87. SCOTT, M. J. Case histories from a geochemical exploration programme – Windhoek district, South West Africa. *op. cit.* (7). pp. 481-492.
88. FURBISH, W. J. Geological implications of jarosite, pseudomorphic after pyrite. *Am. Mineralogist.*, vol. 48. 1963. pp. 703-706.
89. POSNJAK, E., and MERWIN, H. E. The hydrated ferric oxides. *Am. J. Sci.*, vol. 47. 1919. pp. 311-348.
90. KELLY, W. C. Mineralogy of limonite in lead-zinc gossans. *Econ. Geol.*, vol. 52. 1957. pp. 536-545.
91. MORRIS, R. C., and PEPPER, R. S. Magnesian gaspéite and nickeloan magnesites from Western Australia, W. Aust. Govt. Chem. Labs., *Rep. Invest. 1.* 1968.
92. JARRELL, O. W. Oxidation at Chuquicamata, Chile. *Econ. Geol.*, vol. 39. 1944. pp. 251-285.
93. ANDREWS, E. C. The geology of the Broken Hill District. *Geol. surv. N.S.W., Memo. 8.* 1922.
94. PAGE, D. C., and WATSON, M. D. The Pb-Zn deposit of Rosh Pinah Mine, South West Africa. *Econ. Geol.*, vol. 71. 1976. pp. 306-327.
95. BOYLE, R. W., and JONASSON, I. R. The geochemistry of arsenic and its use as an indicator element in geochemical prospecting. *J. geochem Explor.*, vol. 2. 1973. pp. 251-296.
96. VILJOEN, M. J., and MUFF, R. The geology of the United Jack Area – a proposed model for a mineralized complex on the antimony line. *16th Congress of the Geological Society of South Africa, Stellenbosch,* 1975 *Abstracts.* pp. 168-171.
97. SALES, R. H. Ore deposits at Butte, Montana. *Trans Am. Inst. Min. Engrs.*, vol. 46. 1913. pp. 3-109.
98. LINDGREN, W., and LAUGHLIN, G. F. Geology and ore deposits of the Tintic mining district, Utah. *Prof. Pap. 107*, 1919. pp. 1-282.
99. LOCKE, A., HALL, D. A. and SHORT, M. N. Role of secondary enrichment in genesis of the Butte Chalcocite. *Trans Am. Inst. Min. Metall. Engrs.*, vol. 70. 1924. pp. 933-963.
100. BOSWELL, P. F., and BLANCHARD, R. Oxidation products derived from sphalerite and galena. *Econ. Geol.*, vol. 22. 1927. pp. 419-453.
101. BLANCHARD, R., and BOSWELL, P. F. Status of leached outcrops investigation. *Engng. Min. J.*, vol. 125. 1928. pp. 280-285; 373-377.
102. BOSWELL, P. F., and BLANCHARD, R. Cellular structure in limonite. *Econ. Geol.*, vol. 24. 1929. pp. 791-796.
103. BLANCHARD, R., and BOSWELL, P. F. Limonite types derived from bornite and tetrahedrite. *Econ. Geol.*, vol. 25. 1930. pp. 557-580.
104. BLANCHARD, R., and BOSWELL, P. F. Additional limonite types of galena and sphalerite derivation. *Econ. Geol.*, vol. 29. 1934. pp. 671-690.
105. BLANCHARD. R., and BOSWELL, P. F. Limonite of molybdenite derivation. *Econ. Geol.*, vol. 30. 1935. pp. 313-319.
106. BLANCHARD, R. Leached derivatives of arsenopyrite. *Econ. Geol.* vol. 37. 1942. pp. 596-626.
107. GARRETY, M. D., and BLANCHARD, R. Post mine leaching of galena and marmatite at Broken Hill. *Econ. Geol.*, vol. 37. 1944. pp. 365-407.
108. BLANCHARD, R. Chemical and mineralogical composition of twenty typical limonites. *Am. Mineralogist*, vol. 29. 1944. pp. 111-114.
109. McKINSTRY, H. E. *Mining Geology*, New York, Prentice Hall, 1948. pp. 242-276.
110. KELLY, W. C., TAKAHASHI, T. BEHRE, C. H. Jr., DUREK, J. J., STOERTZ, G. E., and FELBER, B. E. Jr. Progress in leached outcrop studies (Abstr.). *Econ. Geol.*, vol. 53. 1958. p. 9211.
111. WHITTLE, A. W. G. Gossan interpretation as a method in exploration. *Jt. spec. group. meet., Canberra,* (Abstr.). Geol. Soc. Aust., 1972. pp. B1-B4.
112. ROBERTS, D. E., and TRAVIS, G. A. Textural evaluation of nickel sulphide gossans. *Western Australian Conference, Perth,* 1973. Abstracts. Parkville. Australas. Inst. Min. Metall. 1973. p. 97.
113. GROVES, D. I., and WHITTLE, A. W. G. Mineragraphy applied to mineral exploration and mining, Workshop course. (unpubl.), Adelaide, *Aust. Miner Fdn.* 1976. pp. 94-111.
114. NEWHOUSE, W. H. The paragenesis of marcasite. *Econ. Geol.*, vol. 20, 1925. pp. 54-66.
115. ANDREW, R. L. The geology, mineralogy and geochemistry of copper-zinc-lead gossans in southern Africa. *Unpubl. Ph.D. thesis* (In preparation), Univ. of London. 1977.
116. NICKEL, E. H. Violarite – a key mineral in the supergene alteration of nickel sulphide ores. *Western Australian Conference, Perth.* 1973. Abstracts. Parkville, Australas. pp. 111-116.
117. EDWARDS, A. B. *Textures of the ore minerals*, 2nd ed., Melbourne, Australas. Inst. Min. Metall. 1960.
118. DIMANCHE, F., and BARTHOLOME, P. The alteration of ilmenite in sediments. *Minerals Sci. Engng.*, vol. 8, 1976. pp. 1187-2000.
119. TRISCHKA, C., ROVE, O. N., and BARRINGER, D. M. Jr. Boxwork siderite. *Econ. Geol.*, vol. 24, 1919. pp. 677-686.
120. ANDERSON, A. L. The incipient oxidation of galena. *Econ- Geol.* vol. 25. 1930. pp. 528-542.
121. GOOSSENS, P. J. Geochemical behaviour of galena under semi-arid climatic conditions in western Africa, *op. cit.* (7). pp. 493-502.
122. DYCK, W. Adsorption and coprecipitation of silver on hydrous ferric oxide. *Can. J. Chem.*, vol. 46. 1968. pp. 1441-1444.
123. KURBATOV, J. D., KULP, J. L., and MACK, E. Jr. Adsorption of Sr and Ba ions and their exchange on hydrous ferric oxide. *J. Am. Chem. Soc.*, vol. 67. 1945. pp. 1923-9.
124. ANDERSON, B. J., JENNE, E. A., and CHAO, T. T. The adsorption of silver by poorly crystallized manganese oxides. *Geochim. cosmochim. Acta.* vol. 37, 1973. pp. 611-622.
125. MURRAY, D. J., HEALY, T. W., and FUERSTENAU, D. W. The adsorption of aqueous metal on colloidal hydrous manganese oxide. Adsorption from aqueous solution. Washington American Chemical Society. 1968. *Adv. Chem. Ser.*, vol. 79. pp. 74-81.
126. CORBETT, J. A., GODBEER, W. C., and WATSON, N. C. The application of sodium peroxide sintering techniques to the analysis of minerals and rocks by atomic absorption spectroscopy. *Proc. Australas. Inst. Min. Metall.*, no. 250. 1974. pp. 51-54.
127. MOESKOPS, P. G. Yilgarn nickel gossan geochemistry – a review including new data and considerations. *op cit.* (74) pp. 449-450.
128. CLEMA, J. M., and STEVENS-HOARE, N. P. A method of distinguishing nickel gossans from other ironstones on the Yilgarn Shield, Western Australia. *J. Geochem. Explor.* vol. 2. 1973. pp. 393-402.
129. HANCOCK, W., and WILMSHURST, J. R. Geochemical observation on some nickel-copper gossans. *Jt. spec. group. meet.*, Canberra, (abstr.) Geol. Soc. Aust. 1972. pp. B10-B11.
130. PONTIFEX, I. R. The mineralogy and geochemistry of some Western Australian gossans and pseudo gossans. *Jt. spec. group Meet. Canberra* (abstr). Geol. Soc. Aust. 1972. pp. B16-B17.
131. WILMSHURST, J. R. The recognition of gossans and related rocks. *op. cit.* (74). pp. 464-465.
132. DALGARNO, R. Geochemistry of the Redross nickel prospect, Widgiemooltha area, Western Australia. *Jt. spec. group Meet., Canberra,* (abstr.), Geol. Soc. Aust. 1972. pp. B12-B14.
133. KEAYS, R. R. Platinoids as discriminators of gossans derived from nickeliferous sulphides. *Jt. spec. group Meet., Canberra,* (abstr.) Geol. Soc. Aust. 1972. p. B11.
134. TRAVIS, G. A., KEAYS, R. R., and DAVISON, R. M. Palladium and iridium in the evaluation of Ni gossans in W. Australia. *Econ. Geol.*, vol. 71. 1976. pp. 1229-1243.
135. TURNER, H. W. Gossan outcrops of cupriferous pyrites.

Min. Mag. vol. 7. 1912. pp. 357-361.
136. GLASSON, K. R. A study of anomalous copper-lead-zinc values in relation to rock types and weathering profiles in an area north of Mount Isa. *op. cit.* (112). pp. 27-40.
137. JAMES, H. L. The chemistry of the iron rich sedimentary rocks, U.S. geol. surv., *Prof. Pap. 440W*. 1966.
138. SINDEYEVA, N. D. A geochemical indicator of pyrite deposits. *Akad. Nauk. S.S.S.R., Doklady*, vol. 104. 1955. p. 114 – in Russian.
139. PALEI, I. P. The concentration of native selenium in the oxidation zone of a pyrite deposit. *Geochemistry*, no. 7. 1957. pp. 640-641.
140. STANTON, R. L. *Ore petrology*, New York, McGraw – Hill. 1972.
141. SIEGEL, S. *Nonparametric statistics for the behavioural sciences*, New York, McGraw Hill. 1956.
142. STEPHENS, M. A. EDF statistics for goodness of fit and some comparions. *J.Am. statist. Ass.*, vol. 69. 1974. p. 730.
143. HOWARTH, R. J. The pattern recognition problem in applied geochemistry, *Geochemical exploration 1972*, Jones, M. J. ed., London, Institution of Mining and Metallurgy. 1973. pp. 259-273.

ERRATA

In Table 4: in column 5, row 3, "Regvesite" should read "Reevesite" and in row 4 of that column, "Heterogenite" should be deleted. In column 6, row 2, "Braunita" should read "Braunite" and in column 8, row 1, "Vivanite" should read Vivianite."

Editor's Comments on Papers 24 Through 28

24 SALES and MEYER
Wall Rock Alteration at Butte, Montana

25 LOVERING
Excerpts from *The Geochemistry of Argillic and Related Types of Rock Alteration*

26 SCHMITT
Excerpts from *the Fumarolic-Hot Spring and "Epithermal" Mineral Deposit Environment*

27 LOWELL and GUILBERT
Excerpts from *Lateral and Vertical Alteration-Mineralization Zoning in Porphyry Ore Deposits*

28 GUILBERT and LOWELL
Excerpts from *Variations in Zoning Patterns in Porphyry Ore Deposits*

WALLROCK ALTERATION

One of the critical qualities of an effective ore guide is that it yields a recognizable target larger than the ore deposit itself. Thus, an envelope of alteration surrounding an ore body may serve as an effective guide. Successive zones that indicate the direction of the ore body are particularly helpful. The paper by Sales and Meyer (Paper 24) was the first comprehensive study of the chemical and mineralogical changes that occur adjacent to a site of vein formation. It established the basis for later alteration studies illustrating progressive wallrock changes outward from mineralizing channel ways within a uniform rock environment. The nature of alteration was independent of time. Lovering (Paper 25), building on experience in the East Tintic District, Utah (Lovering, 1949), added another important step in our understanding of alteration effects in a discussion of the geochemistry and mechanics of alteration in both igneous and sedimentary rocks. Of considerable importance was the recognition of development of high permeability zones, as a consequence of alteration, that serve as important ore controls in metal deposition from late percolating fluids. This is particularly relevant in the localization of precious metal values. Lovering attributes development of alteration zoning to diffusion from channel ways under isothermal conditions. Involvement of meteoric waters in the alteration process, particularly in the epithermal environment, was recognized by Schmitt (Paper 26) in 1950. This added a new dimension to the concepts of mineralization and alteration, entailing the intermixing of hypogene and supergene fluids (mesogene) or alteration and redistribution of values in the host rock by meteoric waters with convection circulation.

On a broader scale, Lowell and Guibert (Papers 27 and 28) examine vertical and horizontal zoning on a district scale as reflected in the porphyry copper deposit at the San Manuel-Kalamazoo deposit and expand observations to a general alteration/mineralization model. Departures from this generalized model as a consequence of overprinting by supergene activity (Titley, 1969), meteoric water and hot spring effects, and variations in rock type were not considered in Paper 27. Some of these factors were clarified in Paper 28, which was intended to answer criticism to overgeneralization of the alteration/mineralization model. Gustafson and Hunt (1975) further demonstrate this overprinting of magmatic, hydrothermal, mesogene, and supergene effects upon alteration and mineralization in the El Salvador deposit.

REFERENCES

Gustafson, L. B., and J. P. Hunt, 1975, The Porphyry Copper Deposit at El Salvador, Chile, *Econ. Geology* **70:**857-912.

Lovering, T. S., 1949, Rock Alteration as a Guide to Ore, East Tintic District, Utah, *Econ. Geology. Monog. 1,* 65p.

Titley, S. R., 1972, Intrusion and Rock Porphyry Copper Deposits, *Econ. Geology* **67:**122-123.

24

Reprinted from *Am. Inst. Mining, Metall. and Petroleum Eng. Trans.* **178:**9–33 (1948)

Wall Rock Alteration at Butte, Montana

By Reno H. Sales* and Charles Meyer,* Members AIME

(New York Meeting, February 1948)

Abstract

At Butte, successive zones of sericitized and argillized quartz monzonite occur around every ore-bearing fracture regardless of its size, attitude, or relative age. The two types of alteration always occupy the same relative positions; sericite adjacent to the ore-bearing vein, clay minerals always between the sericitized rock and fresh quartz monzonite, except where overlap of alteration effects between adjacent fissures has eliminated the lower grade mineralogical products, as is the case in much of Butte's Central zone of pervasive sericitization. As long as active circulation continued in the channel, each zone migrated away from the fissure, that is, it grew at its outer edge and simultaneously receded at its veinward edge because of encroachment by the next innermost zone. The different mineralogical and chemical responses within the wall rocks to the attack by the ore fluid are dependent not on a drastic about-face change of composition of the fluid in the channel relative to its ability to alter the quartz monzonite but on continuously varying conditions of physicochemical environment within the wall rock outwardly from the vein.

Introduction

From the economic viewpoint it is highly desirable that much effort be expended for the purpose of increasing our knowledge of ores and their origin. It has been the observation of the senior writer, many times expressed, that the mining companies have spent too little money in research on the geology of nature's deposits of copper, lead and zinc, which we know to be of limited extent and which, when exhausted, must be replaced by new discoveries.

Experience at Butte and elsewhere has forced the conclusion that observation and mapping in the field must be supplemented by laboratory work if we are to arrive at a reasonably correct understanding of the geology of ore bodies and their environments. It is only through the medium of such correct understanding of geology that we shall be able to accomplish the maximum toward aiding mine operations and at the same time provide a proper background for field examination work whether it be geological or geophysical.

The Anaconda Copper Mining Company's answer to the need above expressed has been to equip a geological laboratory at Butte. The reason why Butte was selected is obvious, for situated there is one of the world's largest and geologically most interesting concentrations of copper and zinc. Furthermore, because of the great extent of underground workings both laterally and vertically, the opportunities for observation and study of ore occurrences probably are unequaled anywhere.

The laboratory was established with five principal objectives in view. These were:

1. As an aid to Anaconda Company mine operations at Butte and elsewhere, this work to supplement that of the mine geologists.

2. To aid in acquiring as complete an understanding as possible of the behavior of Butte veins and ore bodies laterally and in depth.

Manuscript received at the office of the Institute April 5, 1948.

* Chief Geologist and Research Geologist, respectively, Anaconda Copper Mining Co., Butte, Mont.

3. To render assistance to the metallurgist on problems of ore concentration.

4. To make rock, mineral and ore determinations for field geologists employed on examination work.

5. To contribute to the existing knowledge of ore deposits.

The work thus far has been concentrated mainly on the important Butte problem. It is obvious that any attempt made in a laboratory, or out of it, to attain objective number two, that is, determining the behavior of an ore body in depth beyond where it can be observed, will meet with many difficulties.

From the viewpoint of procedure there seemed to be two available points of attack on the Butte problem: (1) the vein minerals comprising the veins, and (2) the ever present wall rock alteration in association with the veins and fissures. Rock alteration was selected for the first line of attack for the reason that a study of wall rock effects might afford a clue as to the nature of the ore-bearing solutions passing a given point within a vein, at which point certain types or groups of minerals had been deposited. In other words a clue might be provided that would aid in the solution of the problems of mineral zoning in the Butte district.

Before offering some of the results of this rock-alteration study, it might be well to state that the geology of Butte was the subject of an extended investigation by W. H. Weed[1] during the years from 1901 to 1906. The results of his work were published in 1912, and in the same year a paper by Charles T. Kirk[2] appeared on mineralization conditions in the copper veins at Butte. A descriptive paper[3] on ore deposits at Butte was presented by the present senior author at the Butte Meeting of the AIME in 1913. A number of descriptions of sulphide relationships in the Butte veins appeared shortly thereafter, but in more recent years little concerning the geology of the district has been published.

Since 1913 there have been very extensive mine workings made in the Butte district both laterally and vertically. The present deepest mining level is the 4200 as compared with the 2800 in 1913. Around the fringes of the copper zone large zinc-ore developments have taken place, and in addition a substantial tonnage of manganese ore has been produced from the Emma mine. Manganese ore (rhodochrosite) was not recognized as an ore possibility in 1913.

As a reminder, it is in order to give a very brief résumé of Butte geology. Butte is an extensively fissured area. The country rock is quartz monzonite into which have been intruded a few narrow dikes of quartz porphyry. A later rhyolite breaking up through the western part of the district is generally thought to be post-vein in age.

The quartz porphyry was followed by extensive fracturing and mineralization resulting in what is known as the East-West system of veins. These veins are by far the most extensive and persistent of any in the district. Next, named in order of appearance, are the Blue faulting fissures, also heavily mineralized, and the northeast fissures, only slightly mineralized. Still later came the faults of Rarus and Middle ages, which cut and displaced all veins and fissures of the earlier systems. No ore bodies have been found in these later faults, yet the presence of wall rock alteration along them marks the time of their appearance as prior to the time mineralizing action had entirely ceased.

The ore and gangue mineral distribution within the veins has been previously described. They exhibit a broad mineral zoning within individual veins as well as within vein groups, arranged radially upward and outward from an undisclosed deeper central source. Nearest the source are the copper minerals; next above a mixture of zinc and copper sulphides, including chalcopyrite; then zinc and lead sulphides with manga-

[1] References are at the end of the paper.

nese as carbonate or silicate; then quartz with very limited sphalerite, galena, chalcopyrite and manganese. Yet outward from this economic type of mineralization is quartz and lime-rich carbonate with very little scattered pyrite.

Since 1913 deeper mine workings have greatly expanded the areal extent of the so-called "Central" area of quartz monzonite alteration. Increasing widths of altered quartz monzonite along vein walls with greater depth is the rule. Also, in certain parts of the district away from the Central zone, deep level workings have disclosed substantial areas of rock alteration, whereas but little or only a moderate thickness of alteration is present along vein walls vertically above at higher levels.

From the economic point of view the Butte problem involves the important question of the persistence of copper and zinc ores in depth. Already some reasons have been given why the study of rock alteration was undertaken as the first line of attack. Also, the idea had been expressed as to whether a full knowledge of the type and nature of Butte wall rock alteration in the higher levels, where many mineralized fissures are barren of copper, might enable a more confident prediction to be made in advance of exploration as to the presence of copper in the deeper portions of those veins. This, of course, was a big order, but the thought supplied a strong incentive to proceed with the study. Another important objective was to determine, if possible, what bearing, if any, a given type of alteration had on ore continuity at successively deeper levels.

We should add that in our attack on the problem of rock alteration we are not alone, for in recent years a number of geologists have given much time to the subject in the field and in the laboratory. Some of these investigations have proceeded on the broad theory that rock alteration might give clues that would prove helpful in the search for ores. These objectives, in effect, spelled attempts to determine something of the nature and composition of the ore-bearing solutions ascending from the depths. It is obvious that a sound theory of ore deposition must rest upon a foundation fashioned from facts. That carefully laid foundation should include a complete knowledge of the time relationships involved in the structural history of any area under consideration. It must incorporate also detailed information concerning the identity, stability ranges, and inter-relationships of the minerals involved. Errors of improper delineation of mineralogical stability ranges, or inadequate consideration of all the possibilities, may become greatly magnified in the finished theory, sometimes to the extent of invalidating the fundamental concepts of that theory. Interpretations about the white mica group of minerals, for example, may be cited as a case in point. Present knowledge of mineralogical stability ranges, though expanding rapidly thanks to many investigators concerned with the problems of mineral syntheses, is still very much of the permissive rather than of the compelling type. In fact, even the identity of several qualitatively and quantitatively important groups of alteration minerals is not adequately known.

These are mineralogical problems, therefore an investigation calculated to delineate a sound theory of ore deposition for a given deposit must begin with an intensive mineralogical study. The Butte laboratory is trying to meet this requirement and much time has been devoted to the study of minerals and mineral variations involved in the alteration of the wall rocks of Butte veins and fissures.

At the very outset it became necessary to choose a method of attack; in other words, to devise a coordinated field-laboratory technique. As already stated, the wall rocks of the veins and fissures of all ages exhibit various shades and degrees of intensity of alteration effected by mineralizing solutions. Obviously, alteration effects will be overlapping, or superimposed, wherever veins of different ages are closely spaced, in areas of fissure intersection, and where there has been fracturing or re-opening of a mineral-filled vein permitting a renewal of solution movements within or along the vein. In the so-called Central area it is extremely rare to find a piece of quartz monzonite or quartz porphyry not exhibiting intense alteration.

Selecting samples from that extensively altered area, or at random from vein walls at various elevations in different parts of the district, did not appear to be a sound procedure where superimposed alteration effects must be avoided to the greatest extent possible. It was thought highly desirable that the samples be chosen so as to provide a sound basis for: (1) a record of the results of continuity of action of a solution on the wall rock outward from the vein or fissure, and (2) for establishing the true nature of the gradational changes concurrently taking place in the minerals outward from any given point along the walls of the vein. And, conversely, it seemed that only from the results of a determination of the point-to-point chemical and mineralogical effects outwardly from the vein to fresh rock, or inwardly from fresh rock to the vein, all interpreted in the light of the complex structural pattern at Butte, would it be possible to arrive even at tentative conclusions such as are set forth later in this paper regarding the nature of the solutions effecting the alteration.

To meet these requirements it was necessary to select for sampling and study bands of alteration believed to have escaped the action of solutions passing through fissures other than the one directly under consideration. Samples were chosen representative of the alteration effects at various elevations in all parts of the district, and from wall rocks enclosing veins of different ages and different mineral compositions. More details of these procedures will appear in a later section of this paper.

And finally, because of the close genetic inter-relationship between wall rock alteration and mineral deposition in and along fissures, the story of Butte rock alteration cannot be completed until a full understanding is gained of the age relations of the vein minerals and their distribution in vertical and horizontal range throughout the district. The study of rock alteration will be continued with increasing emphasis on the relationship between the process of sulphide precipitation and the chemistry of alteration.

Distribution of Alteration Products

In the areas immediately surrounding the so-called "Central" zone of the Butte District the mineralized fissures are large persistent structures separated by relatively wide zones of subordinately fractured quartz monzonite. Within and along these major veins the wall rock is profoundly altered mineralogically and chemically. But between such major alteration envelopes most of the rock is fresh, showing alteration only where it is traversed by subsidiary channels. Each of these smaller faults or joints is bordered by its own alteration envelope—a complete miniature, in qualitative mineralogy, of the larger ones. Hence, a relatively small scale example (Fig 1) is adequate as an illustration for a preliminary definition of the mineralogical identity and spatial distribution of the alteration products adjacent to a mineralized fracture at Butte.

The alteration minerals are arranged in zones parallel to the fracture. These zones may be named according to their mineralogical content since by this criterion they are readily recognizable even megascopically. Beginning with the vein itself and proceeding outward toward unaltered quartz monzonite, one encounters successively Zone A (Fig 1), the sericitized zone, in which the predominant hydrothermal mineral is a white mica; and Zone B, where clay minerals, principally of the kaolinite and montmorillonite groups, are dominant. In published accounts of other deposits the useful term "argillized zone," proposed originally by T. S. Lovering[4], has been applied to units of hypogene alteration similar to Zone B.

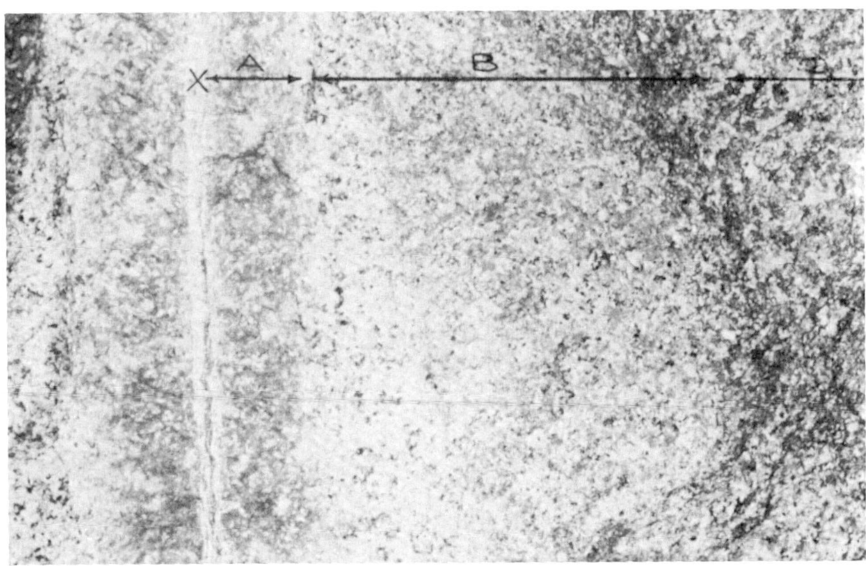

FIG 1—DISTRIBUTION OF ALTERATION PRODUCTS ADJACENT TO A VEINLET.
A. Sericitized zone. B. Argillized zone. C. Fresh quartz monzonite.
The argillized zone here shows three phases: the white kaolinite-rich portion in contact with the sericitized zone, the dark greenish band of montmorillonite-rich and nontronite plus allophane-hisingerite-bearing plagioclase and hornblende alteration adjacent to fresh quartz monzonite, and the lighter green "mixed clay" gradation between. Contacts between all three phases are intergradational.
Note the persistence of black biotite flakes up to the sericite front. Bright grains out in the darker colored portions of the argillized zone are light colored orthoclase crystals and direct reflections from cleavage surfaces. The veinlet (X) contains sphalerite, pyrite, quartz and rhodochrosite. X⅓.

In Fig 1 it is apparent that the white, bleached, veinward portion of the argillized zone, which is relatively rich in kaolinite, is easily distinguishable from the outer greenish part of that zone where montmorillonite-type clay is more abundant. These two phases of the argillized zone are actually intergradational in transition, yet for descriptive purposes it will be useful to refer to them independently as the kaolinite and montmorillonite subzones. Local areas of silicification in the sericitized zone merit similar evaluation.

All of these mineralogical zones are controlled in their identity and in their relative positions by the channel which now holds the vein. They are arranged in good bilateral symmetry around the fracture when the latter is simple and small and cuts otherwise unbroken quartz monzonite. Or, contrarily, they may be deformed in attitude and disposal to the extent that the fracture is intersected by other channels, or when it encounters a change in wall rock type. But even in such a maze of complexity of intersection as is the rule at Butte, considering each vein and its attending alteration suite independently, the same qualitative mineralogy prevails in the same relative distribution regardless of the size, attitude, or comparative age of the fracture. This factual premise, interpreted in the light of definite knowledge of the relative ages of fracturing and mineralization acquired from faulted intersections, will be used for formulating the hypotheses advanced in this paper: (1) that expansion of the total alteration envelope took place by continuous reaction between the wall rock and the ore-bearing solution in the channel; (2) that there was no fundamental qualitative change in the character of the original solution, relative to its capacity for producing the observed mineralogical changes in the quartz monzonite, over the entire period of active circulation in the channel; and (3) that as long as mineralization persisted on a significant scale in the district, each alteration zone maintained its identity and relative position by growing at its advancing edge while simultaneously receding at its inner edge because of the encroachment of the minerals of the next innermost zone. The varying mineralogical identity of each zone is thus visualized as merely the result of the attainment, at the advancing front of that zone, of any combination of conditions demanded by the true stability ranges of the minerals that are forming. Heat and some chemical ingredients came from the solution; other chemical requirements were met by the wall rock itself.

Following a major persistent vein from the peripheral area of the district toward the central zone of mineralization, or following it from high elevations in the mines into depth, one quickly perceives a widening of its alteration envelope. A width of 10 ft in the peripheral area may become 100 ft or more in the zone of intermediate mineralization intensity. This change is, of course, accompanied by a change in vein content following the district's zonal pattern of metal distribution as described by the senior writer in 1913.[3]

Belts of fresh rock between parallel major structures therefore become narrower going toward Butte's Central zone until finally the outermost alteration effects from two reference veins coalesce, and fresh quartz monzonite may be found only in rounded "residual boulders" ensheathed in concentric shells of the alteration products bordering the fractures which isolated them. With continued increase in intensity conditions, one by one of the mineralogical zones of alteration is flushed out by this progressive overlap. Fresh quartz monzonite disappears first, then the green montmorillonite subzone, and finally the kaolinite subzone until only sericitized rock remains. The logical conclusion of this gradual expansion of the sericite envelope around the veins—a more or less uniform zone of high intensity alteration extending from one vein to the next—is achieved in the Central zone at Butte. This zone has made Butte famous as an example of large scale sericitization.

Obviously, the foregoing generalized picture is vastly complicated in detail by variations in closeness of spacing of fracturing. The closer the spacing the more rapid the progress of elimination of fresh rock and low grade alteration products, both because of the greater channelizing effect on the migrating ore fluid and because of the fact that the advancing alteration zones did not have to go so far between fractures.

For the reasons cited earlier in this paper it was decided to begin determinative work on the alteration products around small veinlets in places where complete suites of alteration samples, including vein and fresh rock, could be obtained. Preliminary microscopical work demonstrated the then rather surprising fact that clay minerals of several groups were important constituents of certain of the alteration zones even on the deepest mine levels. Some of these clays were of the type susceptible to base exchange and rapid changes in hydration. Small quantities of lime-rich carbonates soluble in acid mine waters were found also, and rapid post-mine oxidation of sulphide alteration minerals was noted. It was then judged necessary, in order to raise reliability of the samples to the best possible level, to collect specimens as soon as the smoke was cleared from a blast in new country and seal them in paraffin on the spot. Large specimens were taken, overlapping when necessary to include all phases of all the alteration zones from veinlet to fresh wall rock. Their exact position and location were recorded on a sketch of the occurrence drawn to an appropriate scale.

Laboratory Procedure

Three sets of laboratory preparations were made from these samples; for chemical analyses, for examination in thin section and polished thin section, and for mineralogical determinative work.

The samples for chemical analyses were sawed from the large specimens into the form of rectangular blocks weighing at least 500 g each. When possible in these studies of narrow alteration envelopes, each block was cut to include the entire width of a single mineralogical alteration zone. Otherwise, if the zone was too wide, it was cut from a megascopically homogeneous area at the center of the zone. The sawed block was then again lightly coated with paraffin and its specific gravity measured. The paraffin was used in this determination to include voids and to prevent decrepitation of the water-swelling clays upon immersion. After removal of the paraffin by sawing off thin slices of the block, the sample was crushed in a case-hardened steel diamond mortar and a 50–75 g split pulverized to minus 100 mesh by hand in a mullite mortar. This split was submitted for chemical analysis. Several hundred analyses have been made in the course of this study.

Thin sections were made for conventional purposes, including quantitative estimates by the Rosiwal method of the mineralogical compositions of the various alteration zones. In narrow alteration envelopes, where a continuous sample was available, these sections were cut to overlap one another, in order to make possible a complete study of sequence relationships across individual alteration zones and zone transitions. Polished thin sections[5] were used to study the sulphidation of ferromagnesian iron.

Mineralogical determinative work, especially on the clay mineral groups, was done on fresh material by means of standard optical procedure including double variation methods, and by thermal analyses of centrifuged gravitationally separated samples. Supporting information was derived from chemical analyses of these separations, and from X-ray powder crystal patterns.

Samples from wide alteration envelopes around large veins were treated in the same way except that they were chosen underground as representative on a basis of megascopic appearance. They were then located on a sectional view of the vein and its attending alteration zones to the scale of 1 in. equals 10 ft. The underground note sheets, with their lateral control of distance from the vein, supplied the base for graphic presentation of the quantitative chemical and mineralogical data obtained during the investigation of a suite of samples. These cross-sectional views, with specimen locations, are transferred on the same scale to a large sheet of coordinate paper (Fig 6 to 10). On this sheet, above each sample location and on the vertical line which intersects it, are plotted the specific gravity of the sample as determined before it was pulverized and the mineralogical percentages obtained from Rosiwal analyses. Below the sectional view, also on the vertical ordinate of each sample, is plotted the chemical composition of the sample, each radical against its own base line.

The value of chemical analyses to this study consists principally of gaining an estimate of the quantitative interchange of chemical constituents between ore solution and quartz monzonite at different distances from the vein. This necessitates comparison of the amount of each radical in a given zone with the amount of that radical in the original rock and with the amounts in samples from other alteration zones of the same envelope. But the specific gravities of the altered wall rocks usually differ considerably from those of fresh quartz monzonite as well as among themselves. Therefore, unless there are proportionately compensating volume changes between fresh rock and altered, gravimetric percentages as reported in chemical analyses are not directly comparable for appraisal of gains and losses. But evidence obtainable on the matter indicates that no significant volume change has taken place during alteration. There are no megascopic nor microscopic compression nor slump fractures visible in areas of widespread alteration. Quartz and orthoclase surrounding completely argillized plagioclase grains (in which specific gravity decreased from 2.65 to 2.1) show no strain nor other disruption of original rock texture. And despite intense chemical and mineralogical change, the average distance between original rock quartz grains in altered quartz monzonite shows no increase over the average for fresh rock.

Proceeding then on the common assumption of constancy of volume comparative values of chemical content are most conveniently expressed in grams of each radical per cubic centimeter of rock. The gravimetric percentage of each radical quoted in the analyses, multiplied by the specific gravity of the reference sample in accordance with the foregoing discussion, is plotted against its own base line on the vertical ordinate drawn through the sample location. At the start of this study complete analyses were made, but after some experience was gained in evaluating the relative importance of the various radicals present, the group was cut down to the following: SiO_2, Al_2O_3, total Fe, FeO, MgO, CaO, Na_2O, K_2O, H_2O, CO_2, S.

In order to estimate the gains and losses for each alteration zone, and for the alteration envelope as a whole, the quantity of each radical in grams per cubic centimeter for fresh quartz monzonite is also drawn as a line across the suite of samples. Unless unusually good specimens of fresh rock were available from both sides of the vein, average values for composition of the fresh quartz monzonite were used for these "norm" lines. This is permissible because the wall rock is fortunately homogeneous over large areas of the camp. Evaluation of the areas between this norm line and a curve connecting the appropriate points of variation in altered rock permits the estimation of the gains and losses each zone has suffered.

Alteration Mineralogy

Montmorillonite Subzone

As fresh quartz monzonite out from the wall of a fissure first came under the influence of the advancing front of hydrothermal alteration, it encountered a solution whose original chemical character had been very drastically changed by interaction with the rock closer to the fissure. At this very outer edge of an alteration envelope, the ferromagnesian minerals hornblende and biotite were the first to show a response to the changing environment. Marginal transition to chlorite in the biotite, and to chlorite, carbonate and tiny specks of epidote in the hornblende are the first mineralogical changes observed. The solutions which accomplished this fringe alteration carried enough CO_2 to fix as carbonate part of the lime liberated from the hornblende. Except very occasionally, epidote is quantitatively subordinate to carbonate. A little iron separated out as magnetite from both biotite and hornblende during this process, and a grid of rutile needles formed in the altering biotite from TiO_2 displaced from the mica by base exchange. Some of the magnetite is frequently sulphidized in the fringe zone. This beginning of hydrothermal attack caused little or no megascopically visible change in appearance of the fresh quartz monzonite.

The first sign of alteration in the other rock minerals marks the beginning of the process of argillization. Tiny wisps of a montmorillonite-type clay appear along the cleavage zones of the plagioclase. This clay mineral imparts an olive-green or bluish-green tint to the quartz monzonite, the color deepening with increasing amounts of argillization. From the start, the attack on plagioclase is guided methodically by two loci of original plagioclase structure-cleavage and composition zoning. Calcic zones (An_{40} to An_{48}) are preferentially attacked first, whether they comprise the core of progressively zoned crystals or intermediate rings in grains where oscillatory zoning has occurred. Plagioclase cleavage generally also controls the orientation of the tiny new plates of montmorillonite with the result that beautiful plaid texture is frequently exhibited by the new clay growths.

By the time argillization of the plagioclase approaches completion hornblende has been entirely changed to carbonate and a nontronite-type clay. Alteration in biotite is still of the chlorite type, though usually it is far from complete and no longer very active. Orthoclase is unchanged.

The last residues of plagioclase in the network of montmorillonite-type alteration are frequently replaced by a series of amorphous clays having gradational properties even in small grains. The lower range of the index of refraction of this amorphous mineral is higher than that of ordinary allophane—about 1.50 in freshly broken samples—and it slowly rises to nearly 1.52 on exposure. Measurements of over 1.56 have been made on some samples of darker color. This allophane-hisingerite type mineral is often deep azure blue on freshly broken surfaces underground, but after a few hours of exposure the blue color fades to a dark olive drab or even a nearly black hue which is common in plagioclase grains in this zone of alteration. Probably both oxidation and changes in hydration influence the color change. The amorphous clay is always quantitatively subordinate to montmorillonite, even at the former mineral's peak development as a first product of residual plagioclase alteration. It decreases rapidly in relative proportion toward the vein, and its place is taken by more montmorillonite.

The montmorillonite-type mineral is the principal source of the greenish color of the so-called chloritic zone at Butte. It appears to be optically homogeneous but all of its measurable properties vary with respect to position in the montmorillonite subzone. All grains which were large enough to be tested were recognizably biaxial negative with a small 2V. The beta index of refraction is close to 1.53 for much of the material studied, though variations between 1.52 and 1.55 have been observed. Samples from near the center of the zone where this mineral is developed show the minimum index and also the strongest tendency for water swelling. The average index rises toward the vein with decrease in birefringence and lighter green color and also toward fresh rock with darkening of the green color and higher content of iron. Varying concommitantly with increasing refraction nearer the vein, however, is specific gravity of the clay mineral. The average gravity determined is about 2.20 with extremes of 2.07 and 2.40 as determined by bromoform separations.

Thermal analyses have provided a means for coordinating most of these variable properties into systematic groups and for suggesting interpretations of mineral identity. Samples prepared for thermal analyses were pulverized as needed to get reasonable separation when centrifuged and then held at standard relative humidity (31 pct) for 24 hr. A furnace heating rate of 10°C per minute was used on all runs.

Curve No. 1 (Fig 2), of a centrifuged gravity split between 2.08 and 2.16 from an argillized rock near the center of the greenish montmorillonite subzone, compares closely with data published by Grim and Rowland[6] for montmorillonite, having a low temperature doublet, unit endotheric peaks at 700 and 900°C, and a well-defined exothermic peak at about 950°C.

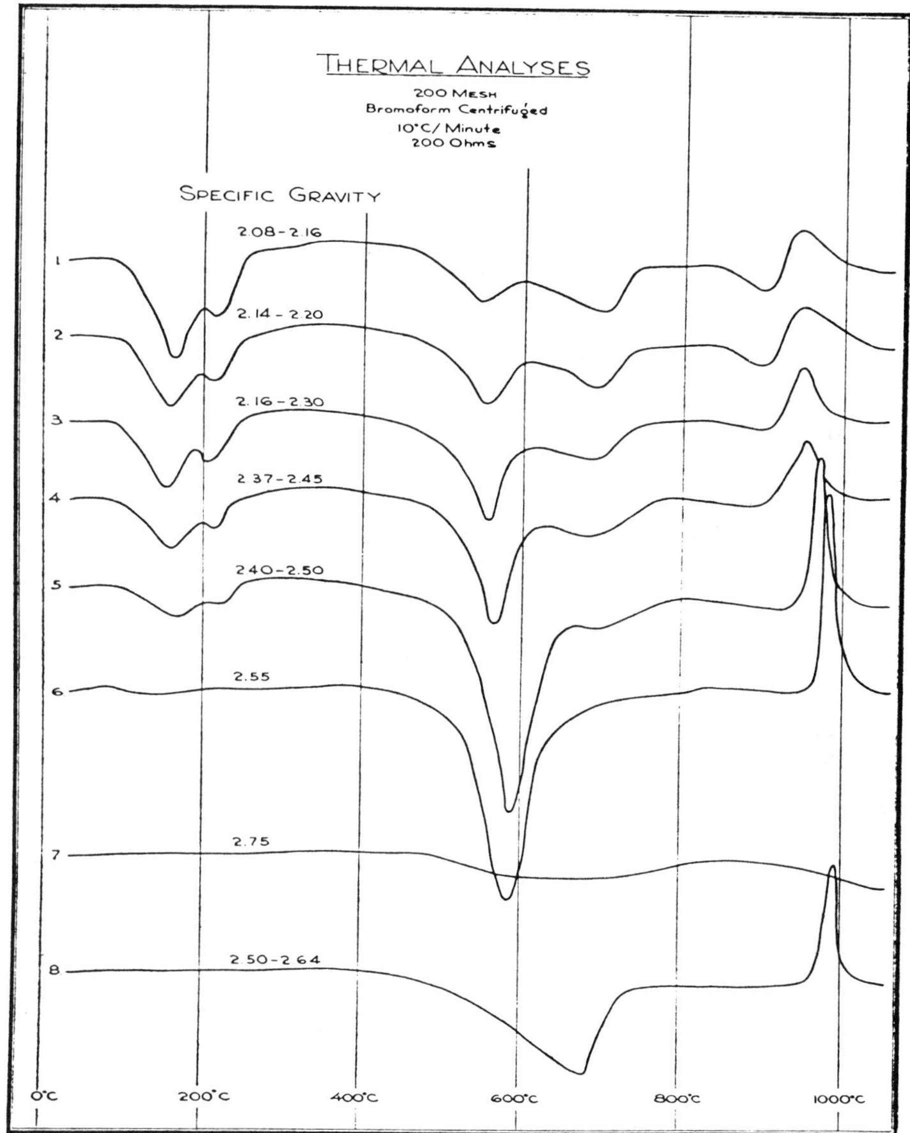

FIG 2—THERMAL ANALYSIS OF TYPICAL BUTTE ALTERATION MINERALS.

The endothermic peak at 550°, however, is not a montmorillonite peak. Following through curves 2, 3, 4 and 5 (Fig 2), taken in the same argillized zone of alteration from samples successively closer to the vein, one notes a gradual deepening of that peak and a progressive fading of the montmorillonite peaks. Also, there is a considerable sharpening and displacement toward higher temperature of the final exothermic peak, until in 5 the characteristics of the curve are those of *kaolinite* with a minor contamination of montmorillonite. X-ray powder patterns permissively accord in this estimate, and, as noted on the curves, the gravity has increased through this set of samples from 2.08 to 2.44 while the color has bleached from medium olive green to nearly pure white. This set of samples is typical of many that have been studied. Yet much of the mineral appears to be optically homogeneous. Hence, a multilayered repetitive alternation of the two lattice types, kaolinite and montmorillonite, in variable proportions to give the appropriate thermal reactions, seems to the writers to be the most promising possibility of identity in evidence at present. This mechanism has been proposed by Grim and Rowland[7] and by others to explain these anomalies for many clays of this type.

The foregoing details on mineralogy of this interesting green part of the argillized zone apply directly toward establishing certain tentative hypotheses which conclude this paper. A more complete account will follow at a later date.

Kaolinite Subzone

Very little of the montmorillonite-type thermal reaction shows up in curve 6. This curve is characteristic of nearly pure kaolinite. The alteration subzone characterized by this mineral is bleached white in color and speckled with black flakes of biotite, corroded pinkish or buff colored orthoclase residues, and grayish quartz. Even orthoclase sometimes shows noticeable attack at this stage, going to kaolinite more or less directly. This kaolinitic subzone, when developed, is always at the veinward edge of the argillized portion of the alteration envelope. Actually, it appears to be an end product of the gradual increase, toward the vein, of the kaolinite content of the mixed clay mineral in the argillized zone, for even the purest white kaolinite at Butte shows some small suggestion of the presence of montmorillonite by its thermal curve or low specific gravity. The zone is best developed in district-wise areas of intermediate mineralization intensity, and is sometimes nearly absent from alteration envelopes elsewhere, where a mixed clay mineral containing as much as 20 pct montmorillonite often lies in contact with the advancing sericite front.

Associated with this kaolinite in the kaolinite subzone, biotite becomes an important alteration mineral. In a foregoing comment on chloritization of rock biotite in the outer fringe of alteration, it was mentioned that the chloritization process was arrested when clay attack on plagioclase became intense. The unaltered biotite residuals are still visible as fairly large grains in the kaolinite alteration subzone, but the chlorite fringe is driven *back* to biotite. This alteration biotite is always fine-grained and frequently distributed as a halo around residuals of primary biotite (Fig 3). Fresh biotite is thus again available for attack at the sericite front.

Sericite Zone

From the viewpoint of mineralogical diversity, the sericitized zone is somewhat disarmingly less complicated than the argillized zone for it is composed chiefly of only three minerals; sericite, quartz and pyrite, in decreasing order of abundance—provided we admit certain rather wide deviations from true muscovite optics and composition within the term sericite.

White mica directly replaces all remaining silicate minerals in the argillized rock. It occurs as unit pseudomorphs after biotite, without the appearance of any chloritic intermediate phase, and also as aggregates of tiny plates, mixed with quartz, after the feldspars, whether or not the latter were first completely argillized. Microscopically, the encroachment of sericite on the clay minerals at the advancing front of the sericite zone is in evidence from textures which show sericite apparently cutting earlier clay (Fig 4).

FIG 3—ALTERATION BIOTITE.
Two residual grains of original rock biotite are present in this cluster of finely granular alteration biotite. Both are on the N-S cross hair, one at the top of the photograph and the other midway between the center and the bottom. Marginal and transectional corrosion are both apparent in the latter grain. Alteration biotite is the apparent replacing mineral but an intermediate stage of chloritization, here completely reconverted to biotite, probably actually accomplished the corrosion of the rock biotite. Plane polarized light. × 40.

Sericite rarely attacks the original quartz of the quartz monzonite, and in sericitized rock the total quartz content always exceeds the average amount of rock quartz. The intimate association of finely granular SiO_2 with sericite in feldspar areas suggests that the silica was, in part at least, coprecipitated with the sericite during the process of sericitization.

Locally, especially near quartz-rich veins in the district's Central zone, silicification becomes quite intense and silica assays of 80 pct are not uncommon in rocks still showing recognizable granitic texture. These rocks become very hard and are generally uniformly medium gray in color. Replacement vein material is the obvious end result of continued application of this silica invasion.

Microscopically, this silicification quartz commonly engulfs and replaces sericite, whose alignment may still preserve original plagioclase structures two generations removed (Fig 5). This is textural evidence of the aggression of silicification on sericitization, in the same way that sericite had advanced on the clay zone at the point of reference earlier in the mineralization process.

Because it involves no change in type of mineralogy, merely a change in relative proportions between two associates because of rather local conditions, these areas of silicification are properly regarded as sub-

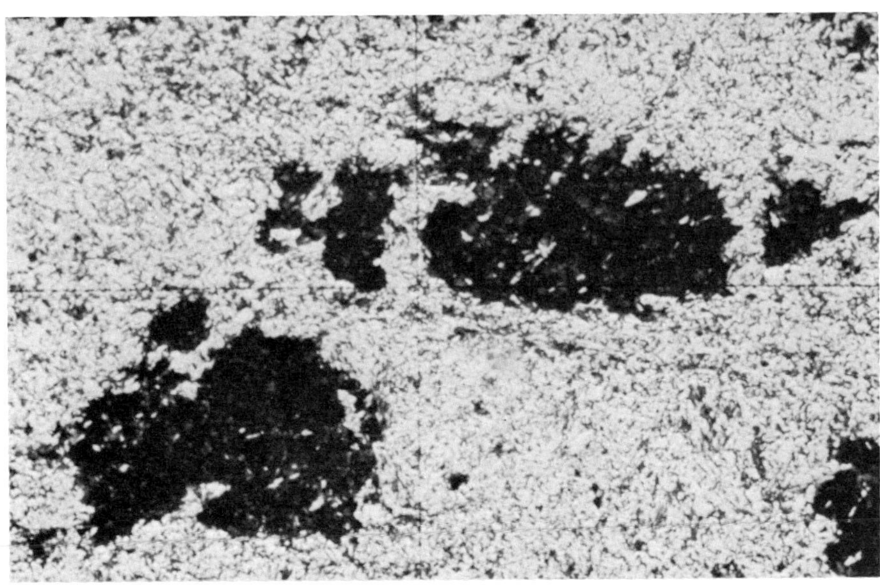

FIG 4—SERICITE APPARENTLY CUTTING KAOLINITE.

This space relationship is commonly shown in crystal areas which were once rock plagioclase in the zone of the active advance of sericitization on argillized quartz monzonite. Kaolinite appears dark and sericite light colored because of the difference in birefringence between the two minerals. Crossed nicols, × 100.

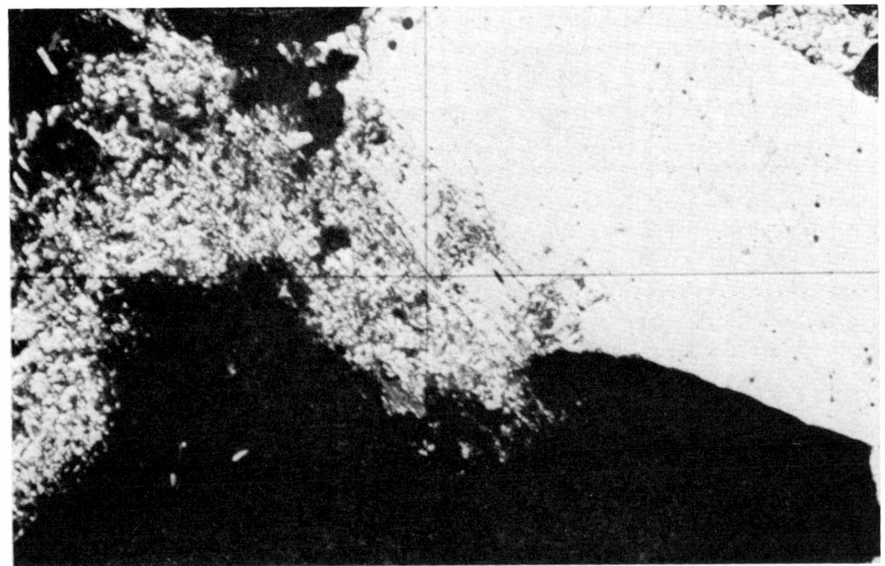

FIG 5—"GHOST" OUTLINE OF FELDSPAR EUHEDRON PRESERVED BY SERICITE.

It is noteworthy that only two quartz crystals (one at extinction position and one bright) make up the entire background of the photograph. In this manner sericite is engulfed and displaced by quartz during silicification. Crossed nicols, × 40.

zone members of the sericitized alteration zone, like the kaolinite subzone of the argillized zone.

Most of the sericite at Butte, including the greater part of that which comprises the pervasive sericitization of the district's Central zone, shows a fairly good approximation of the properties of muscovite except that it is somewhat low in alkali content, in indices of refraction, and in specific gravity. Its thermal curve (Fig 2, curve 7) is characterized in general by a broad, shallow trough centering at about 600 to 700°C, and a sharper endothermic peak at about 1040°C. The lower the alkali content, the narrower and sharper is the lower temperature endothermic peak, though the water content of the mineral does not go up more than a few tenths of a per cent. A pure sample of this latter mineral from a vug filling on a deep level in the Central zone shows a total of 8 pct alkali radical, including 1 pct Na_2O. The lowest index of refraction falls just below the higher index of quartz. This high-intensity low-alkali sericite is the only case of the fixation of Na_2O other than in montmorillonite observed to date in Butte alteration minerals. Low-alkali sericite is also an important constituent of the altered walls in deeper mine levels.

A common associate of this type of sericite, both as a vein constituent and in the wall rocks near the hot center of camp, is the high-intensity clay mineral dickite, identified optically, by means of its characteristic thermal curve (Fig 2, curve 8), and by X-ray powder pattern. Good pink crystalline alunite is also sometimes found in this association, though little of it penetrates very far into the wall rocks.

Both dickite and low-alkali sericite are commonly interlaminated with beautiful covellite plates and rich crystalline masses of digenite and chalcocite in the deep levels of exploration in the eastern part of the district.

Chemical and Mineralogical Changes in the Wall Rocks

As stated earlier in this paper, our attempt to evaluate the chemical interchange and mineralogical relationships between the ore fluid and the walls of the Butte veins has made use of detailed cross sectional diagrams like those of Fig 6 to 10. Several hundred such sections across minute veinlets, major structures, and all gradations between, have been studied. In general, as we have already pointed out, they all tell the same story, and the five sections from ore-bearing veins selected to accompany this discussion are representative of the types of activity that are characteristic of the Butte district.

The cross section of Fig 6 is from a 1900 level crosscut on a prominent East-West vein in a locality exhibiting near-peripheral type of mineralization intensity. The metallization is zinc ore composed of sphalerite, galena, pyrite and quartz. The section of Fig 7 is from the same vein farther west geographically but from a much deeper level, the 3800, where the metallization is copper ore with minor zinc. In each case the alteration envelope includes zones of sericite, kaolinite and montmorillonite, named in order of their position measured outward from the vein and generalized in the cross sections. Comparing the sections, it may be observed that in spite of the sharp differences in metallization of the vein at these widely separated localities, the chemical action on the wall rock and the mineralogical results effected by alteration processes working outward from the vein are qualitatively identical, although the total width of the alteration envelope at the 3800 level is about three times that at the 1900 level.

In both sections the lime, soda and silica curves show strong decline toward the vein across the argillized zone. The loss of magnesia by the walls is in the same direction but less rapid. Mineralogically, the

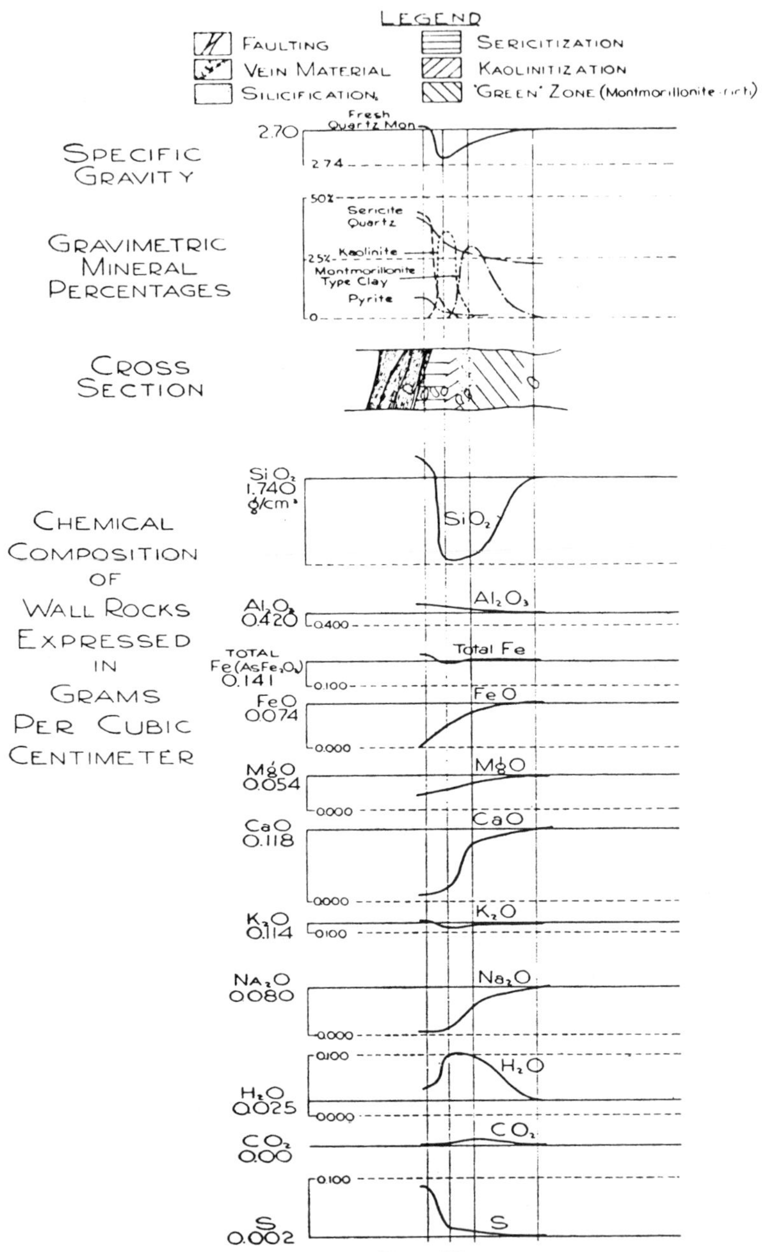

FIG 6—CROSS SECTION OF AN EAST-WEST VEIN, 1900 LEVEL.

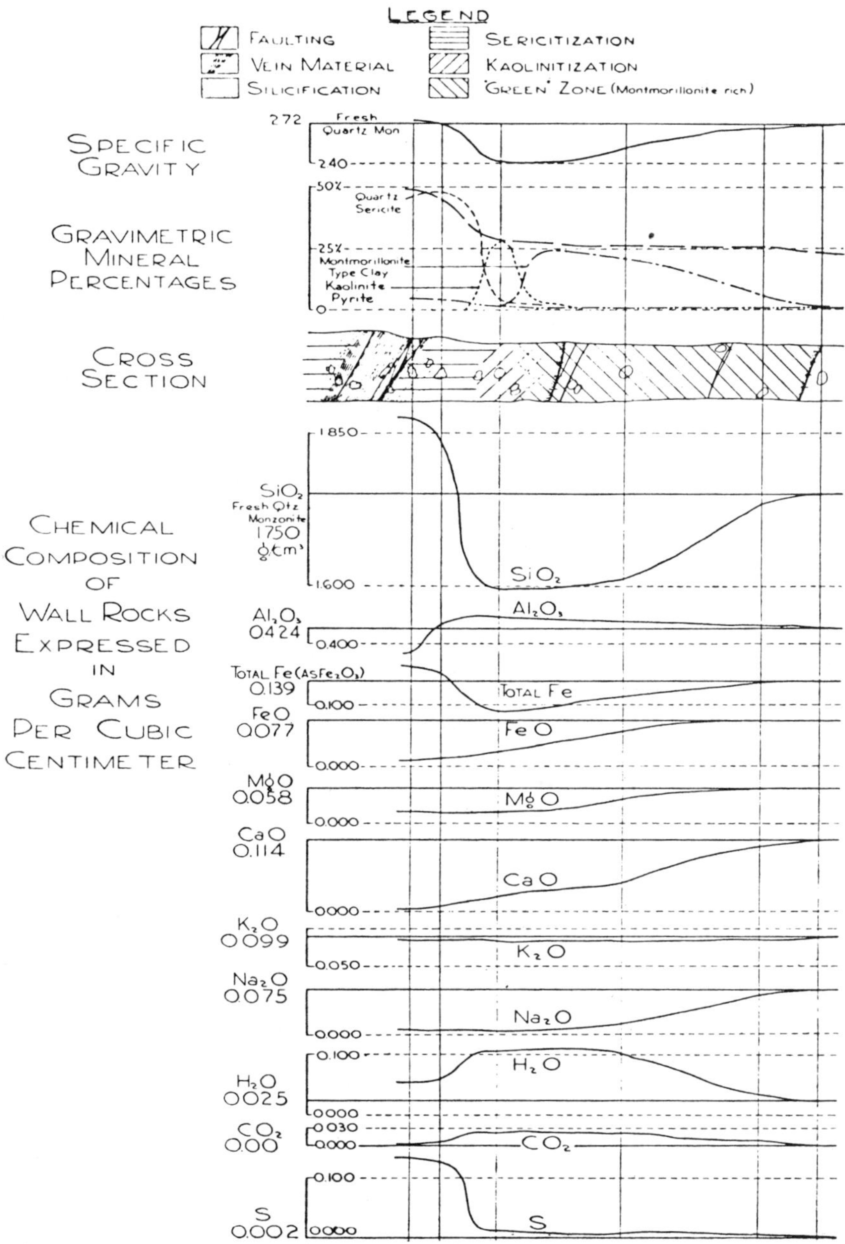

FIG 7—CROSS SECTION OF AN EAST-WEST VEIN, 3800 LEVEL.

losses of the above-mentioned radicals are evidenced by the progressive destruction of plagioclase with the temporary fixation of part of its constituents plus some magnesia and iron in the mixed clay mineral. The increasing proportion of kaolinite as compared to montmorillonite in the mixed clay, proceeding toward the vein through the argillized zone, is demanded by the continuing abstraction of the foregoing radicals during the gradual outward expansion of a progressively stronger leaching environment.

K_2O and Al_2O_3 remain significantly close to the lines defining the composition of fresh rock. K_2O is held up principally by the metastability of orthoclase in this environment, since the latter mineral shows only partial breakdown even in the kaolinite subzone. There is also a little potash feldspar in solid solution with the plagioclase and the potash from this source probably accounts for much of the decline that is shown by the K_2O curve. The proportion of alumina increases in the progressively changing clay minerals toward the vein, so, even though the specific gravity of the rock decreases, the alumina curve stays high.

The principal gains of constituents in the argillized zone are registered in water for hydration and in CO_2 for carbonates, plus a little sulphur for pyrite. Hence from the point of view of the destruction of rock minerals the significant process in the argillized zone is the increasingly intense selective removal of the base and alkali metals, plus their molecular equivalents in SiO_2, from plagioclase and hornblende.

Except for the lowest grade start toward chloritization it received at the outer edge of the alteration envelope, a member of the biotite group is within its range of true stability and the potash feldspar is metastable in the environment of the argillized zone.

The foregoing subtractive process operating in the argillized zone requires the presence of constituents in the fluid medium of transfer that could break down the silicate molecules of lime and soda and, to a lesser extent, magnesia, hold these constituents in solution, and avail their transport out of the zone. The increasing efficacy of this process toward the vein indicates increasing ability of the solution nearer the vein in the argillized zone to accomplish this removal and, at the same time, to press the abstraction process toward completion. The result over a period of time was a gradual pushing outward of the front of attack as the total envelope of alteration grew wider while maintaining the identity and positions relative to each other of its component zones.

Continuing supply of the ingredients required to push the alteration zones outward must have come from the channel, and spent products of reaction were removed to the channel. This two-way transfer could only have been accomplished by ionic diffusion. Because of the evident continuing reaction and solution of base and alkali silicates, illustrated by the gradual but persistent decline of the base and SiO_2 curves, and resulting in the gradual displacement of montmorillonite by kaolinite in the mixed clay mineral of this zone, there would be a tendency toward chemical neutralization of the solution progressively farther from the vein. Fulfillment of this tendency might be indicated by the low-intensity epidote and chlorite in the outermost alteration fringe.

Pertaining to the hypothesis mentioned earlier in this paper that sericitization and argillization are contemporaneous processes, Fig 9 illustrates a cross section from a Northwest vein which *offsets sulphide mineralization* on the East-West veins of Fig 6 and 7. The fundamental similarity of this diagram to those from the East-West vein is obvious both as to quantitative chemical variations and distribution and identity of mineralogical products. Argillization must have been accomplished on the

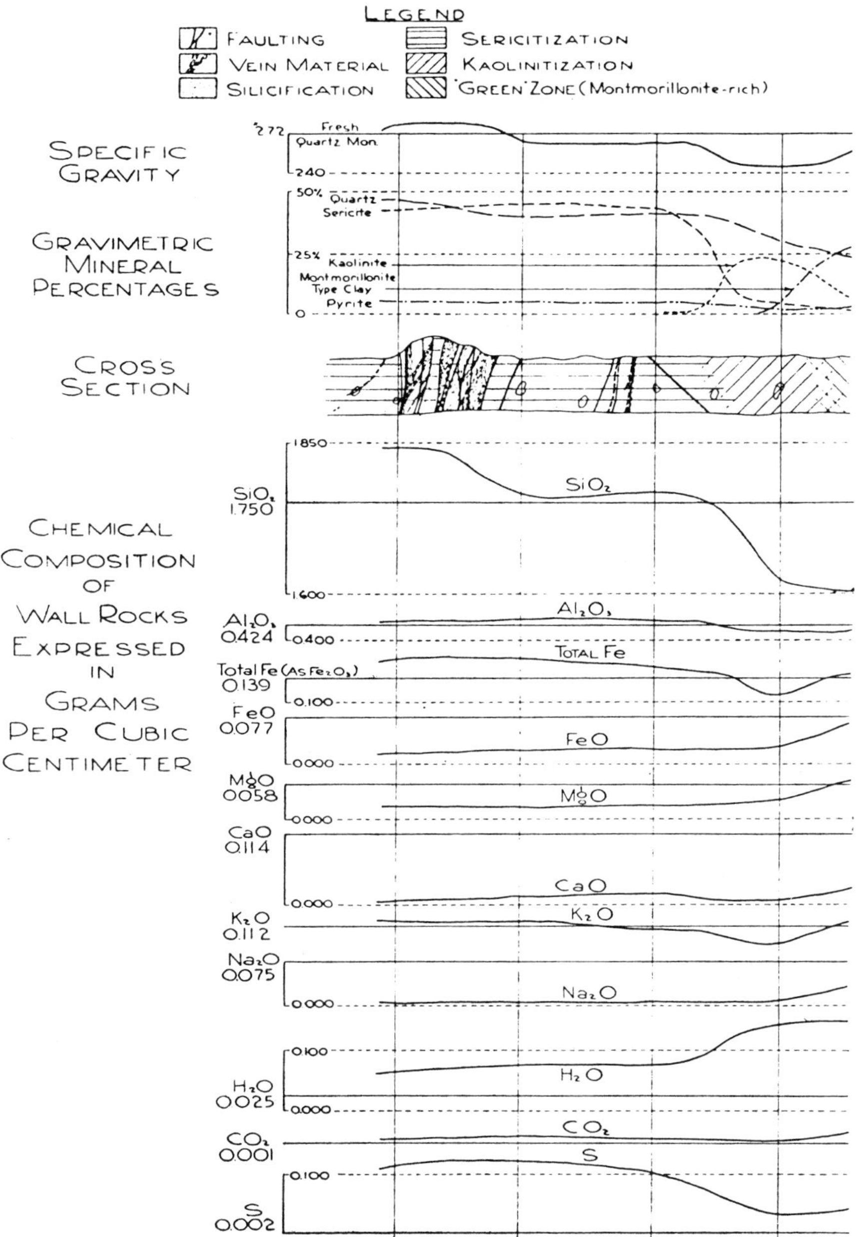

FIG 8—CROSS SECTION OF THE EAST-WEST VEIN, 4200 LEVEL.

Northwester after sulphide deposition had begun in the East-West vein which it offset. The ore-bearing solution, which then had free access to both structures, continued its work of deposition in the East-West fissure and began this process in the Northwest fault, contemporaneously attacking the walls of the latter according to the same chemical and mineralogical pattern as it had previously applied and was still applying to the walls of the earlier formed and partially metallized East-West fissure.

In every case the curves of base leaching from the walls by the solution continue their downward slope toward the vein *into* and *through* the zone of sericitization. The arguments of base-metal leaching applied to the argillized zone may be reapplied here, since the little remaining Na, Ca, and part of the Mg were still incapable of producing stable compounds in the environment of this zone. The soda and lime curves, in particular, drop to nearly zero. Potash and alumina came from the orthoclase and clay to make sericite, and some of the veinward migrating silica from the argillized zone was precipitated as quartz intermingled with the sericite, raising the silica percentage in the sericitized zone above the quantity present in fresh rock. Any excess of silica abstracted from the argillized zone was, of course, available for precipitation in the channel as vein quartz. Some quartz veins may grow at their outer surfaces by accretion of this veinward migrating silica.

The replacement of biotite by sericite is in the nature of a base-exchange phenomenon, for various stages of the change may be observed in some thin sections, and unit pseudomorphs of sericite after biotite are prevalent. Sulphidation of iron liberated from the biotite also takes place immediately and contemporaneously with the sericite-forming reaction, illustrating the close association and interdependency of sulphide formation with potash-mica crystallization.

Suggested Control Mechanism for the Sericite Zone Front

The big question now becomes: is the hypothesis of simultaneous sericitization and argillization at different distances from the vein compatible with what we actually know about the requirements for crystallization of the various minerals involved, particularly for sericite? Presenting permissive evidence from syntheses, it may be stated that several experimenters have formed sericite in basic or neutral solutions at temperatures as low as 200°C, and J. W. Gruner[8] has altered microcline and albite to sericite in an environment made acid with 0.1 N to 0.35 N HCl, though at temperatures of 350 to 525°C, and in the presence of excess KCl. Gruner also found that "as the acidity decreases the field of formation of muscovite probably extends to lower temperatures, but does not reach 300°C at 0.05 N." It is apparent from these two sets of data that the kindling temperature for sericite formation, assuming availability of appropriate constituents, takes an abrupt rise across the transition in pH from neutrality to slight acidity. But a point of significance is that if the temperature is high enough, that is, of the order of 350°C, sericite may form even in a decidedly acid solution (pH 1 at close of most of Gruner's experiments) where potassium is available. Below this kindling temperature only kaolinite was formed even in the presence of excess KCl.

In a neutral or slightly basic chemical environment, the position of the sericite alteration front might be determined by the passage of orthoclase from its state of metastability in the clay-forming zone into a range of chemical conditions where it could no longer maintain its identity. Its breakdown because of increasing capability of feldspar decomposition by the solution could then supply the potash for sericite. Except in a strictly permissive role, temperature (200°C requirement) may not be an important factor.

In an acid environment, however, *if* such be considered a requisite for the accomplishment of kaolinitization, two requirements especially pertinent in this instance must be met for the crystallization of sericite and for the consequent advance of the sericite front: availability of potash, and attainment of the required temperature. Much of the potash for sericite evidently comes from the orthoclase which has persisted through the zone of argillization up to the sericite front. As above, chemical attack by the sulphur-bearing ore fluid on the orthoclase, with consequent release of potash, may provide the critical control of sericite formation, assuming a sufficiently high temperature. Or, the temperature requirement itself may be the critical control, and the breakdown of orthoclase may be consequent from the vigorous crystallization of and attack by sericite when the kindling temperature of the latter is achieved. In this case, the position of the sericite front at any given time and its velocity of outward migration over a period of time may have been determined by the rate of supply of heat from the solution. Moreover, the rate of supply of chemical ingredients to accomplish the base leaching of the argillized zone, compared to rate of supply of heat, could influence the comparative widths of the sericitized and argillized zones of alteration. A relatively faster rate of heat supply would cause the front of sericitization to advance more rapidly on the argillized zone than the latter could progress against fresh wall rock, and a comparatively narrower argillized zone would result. Sericite might encroach on the clay mineralogy rapidly enough to prevent thorough base leaching from montmorillonite and the formation of a well-defined kaolinite subzone. This might be expected, for example, in areas of closely-spaced heavy fracturing where relatively rapid heating of the walls would be probable. In several such localities, particularly in and near the Central zone of the Butte district, this occurrence is well illustrated by the replacement by sericite of a clay mineral still relatively rich in the montmorillonite molecule.

Contrarily, broad areas of argillization cut by narrow bands of sericite around intersecting veinlets are common in the west central intermediate zone of the Butte district. In general, such areas resemble certain phases of some of the porphyry copper type of mineralization and attending alteration. The quantitative preponderance of clay over sericite at these places in Butte might have been accomplished, for example, by a chemically potent hydrothermal solution whose heat supply was relatively low; not necessarily by post-argillization veining by sericite deposited from later, radically different chemical solutions. The apparent age relationships, both field and microscopical, would be the same under each hypothesis. Merely the space-time concepts applied to their interpretation would be different. For Butte, structural proof is available for the district-wise contemporaneity of sericitization and argillization.

Adequacy of the heat supply in the hydrothermal solution at Butte for the operation of the temperature-front mechanism for sericitization is indicated by vein quartz crystallization temperatures as determined from observed temperatures of disappearance of the vapor phase in liquid inclusions. Applying Ingerson's[9] corrections for depth of burial of about 3 km, the observed temperatures range from 300 to about 380°C, bridging the temperature span of sericite formation in acid solution as determined by Gruner.

In the cross sections thus far illustrated by chemical diagrams, the losses by the argillized wall rock in silica, for example, are more than adequate to account for the silica gains of the sericitized zone of the same envelope. In the process of constant growth of the alteration zones visualized in the foregoing discussion, adjacent envelopes will eventually meet at their outer edges and overlap, with the complete eradication of fresh quartz monzonite. Meanwhile, sericitization continues to expand from the vein outward, and ultimately a position of the sericitized zone is reached where losses ascribable to a given argillized zone are balanced by gains in the widening sericitized zone at deeper levels. This condition is approximated in the cross section of Fig 8 which includes a section of wall rock measured outward from a large East-West vein on the 4200 level of the Mt. Con mine, at a point nearer the Central zone and deeper in the mines than the location of Fig 7. Here the vein carries copper but no zinc. No fresh quartz monzonite is present in the crosscut penetrating the walls of this vein because of the overlap of the alteration envelopes developed outward from the vein and from the next parallel structure 100 ft farther north. The lowest grade alteration present here is a light greenish montmorillonite-type clay intergrown with considerable kaolinite. However, a strong white bleached subzone containing nearly pure kaolinite is also present between the green clay and the sericitized zone.

Moving deeper into the district's Central zone along any of the major vein systems, chemical gains in certain radicals in the zone of sericitization begin to outweigh losses of those radicals in the argillized zone. This situation reaches its climax in the Central zone of mineralization where pervasive sericitization, increasingly wide at greater depth, plus or minus local silicification, occupies the entire distance from one vein or mineralized zone to the next. The cross section of Fig 10 from the Leonard vein deep in the Central altered zone of the Butte district illustrates the final sericitization phase of the alteration process. Here silica, alumina, potash and iron, in addition to much sulphur, have been added to and fixed within the walls so that the total amount of each radical now present exceeds the amount present in the originally fresh quartz monzonite. Moreover, the specific gravity of the rock is higher than it was originally, so quantitatively the total gains outweigh the total losses in other constituents. These materials must have been supplied to the walls from below by the hydrothermal fluid. How the fluid acquired these constituents is conjectural, since they could either be supplied from a deeper alteration zone through another change in mineralogical equilibrium conditions or they may have been integral to the ore fluid at the place of its accumulation. This matter need only concern us to the extent of observing that the latter method is not a requirement, and that a suggestion of depletion of potash supply *may* be indicated by the low-alkali sericite and dickite in the near-vein alteration in this cross section. These minerals show up in the chemical curves as a decline in the K_2O curve and a corresponding increase in the quantity of Na_2O, which radical, as pointed out earlier, attempts to fill part of the place occupied by potash in the mica molecule when the potash available for mica is low.

Conclusions

In the foregoing discussion, continuity and uniformity over the period of active circulation of the ore fluid have been the emphasized attributes of the alteration process as judged from the chemical action and mineralogical results of attack by the ore fluid on its channel walls. And the suggestion has been made that the mineralogical stability requirement of the hypotheses presented, as deduced from the evidence offered by mineral syntheses, are within the known ranges of fields of formation of the alteration minerals recognized at Butte. As a concluding argument we desire to emphasize once more the structural field evidence at Butte which, we believe, requires the operation of the hypothesis of the district-wise contem-

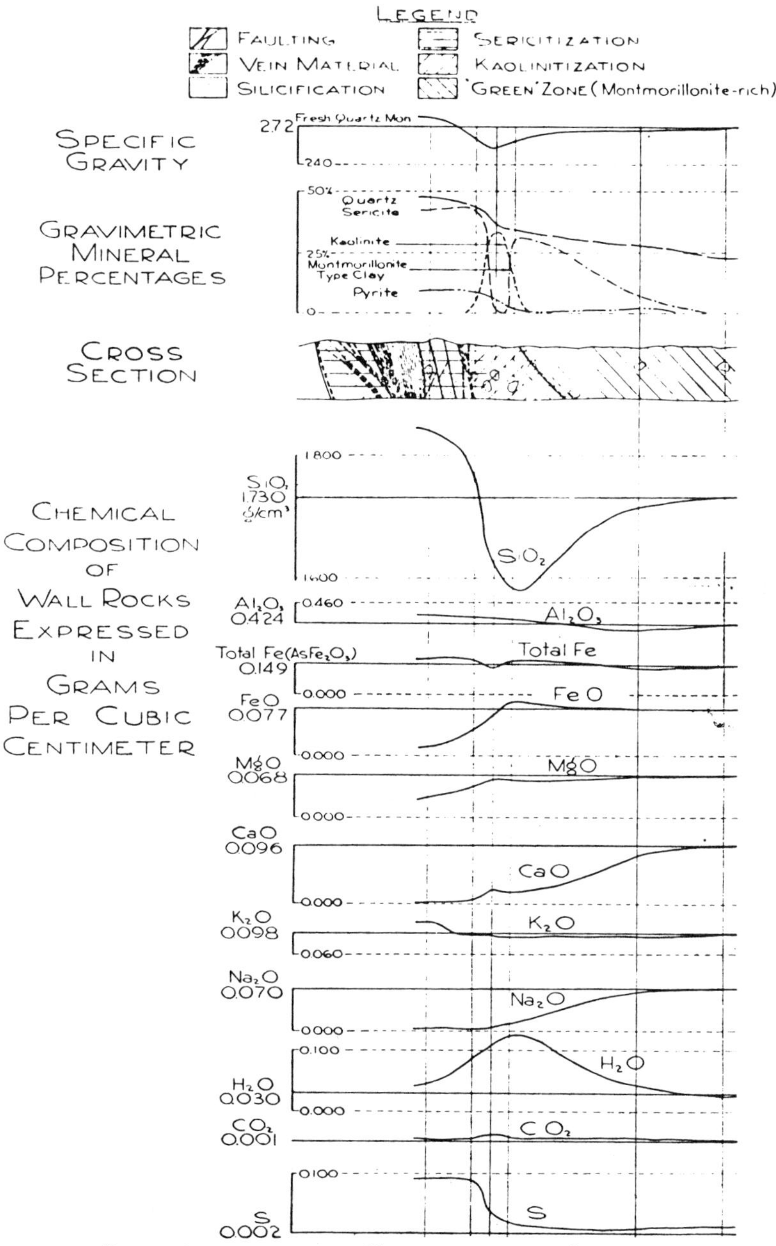

FIG 9—CROSS SECTION OF NORTHWEST VEIN, 2800 LEVEL.

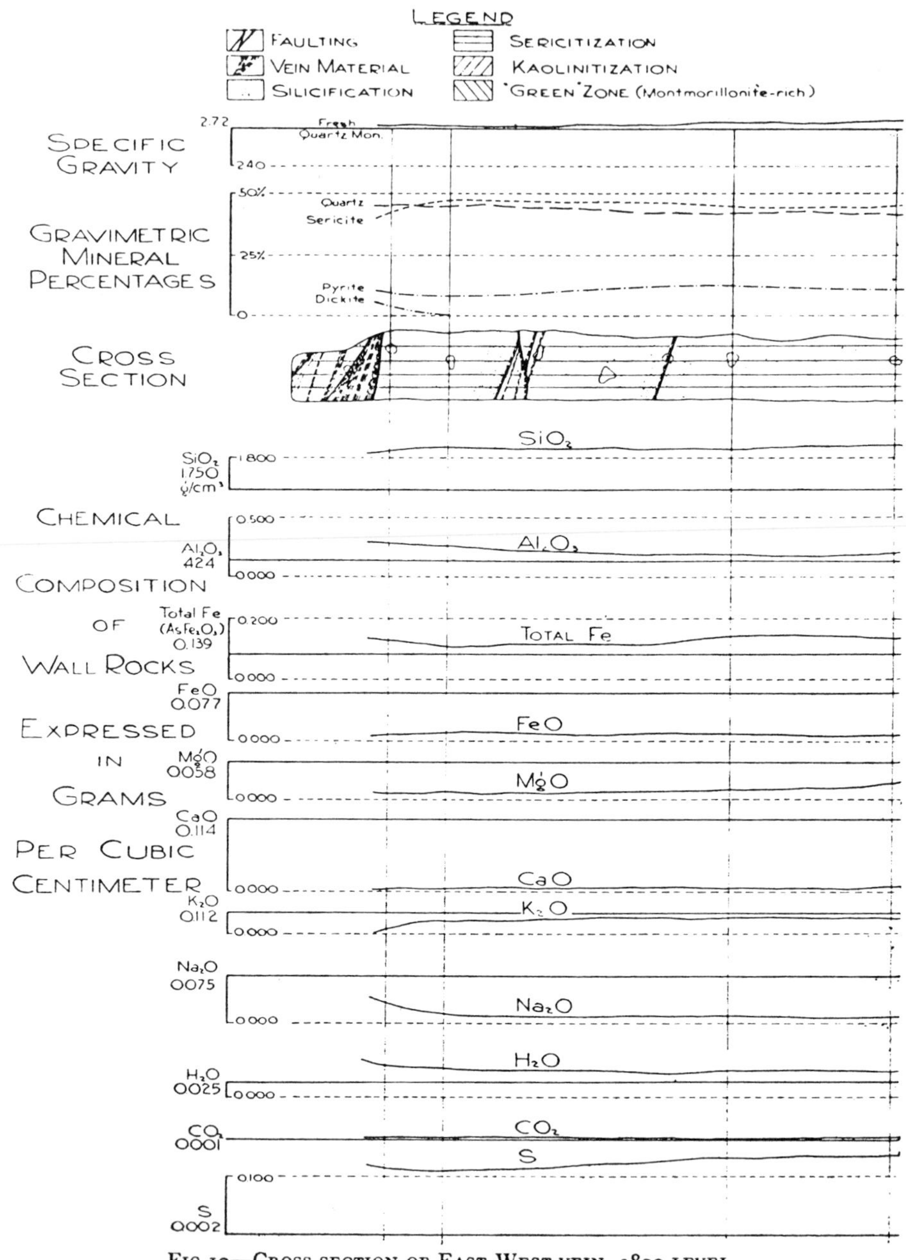

FIG 10—CROSS SECTION OF EAST-WEST VEIN, 3800 LEVEL.

poraneous formation of zones of sericitization and argillization.

It is obvious that any changes in fundamental character of the ore fluid, over the period of active circulation of that fluid through the succession of channels provided by intermittent faulting at Butte, should be recorded by the variable characters of the veins found in these faults and of the wall rock alteration adjacent to them. If an early period of alteration, characterized only by argillization, had been succeeded by one in which only sericite was formed in the walls by reason of a radical change in the composition of the ore fluid, that change should be recorded in the superimposition of sericitization on argillized rocks around early-formed fractures and by the total absence of clay minerals from the alteration zones around fractures which formed late in the period of metallization.

But, as we have stated, the observed facts of distribution of alteration products at Butte stand in sharp contradiction to this requirement. Successive zones of sericitized and argillized quartz monzonite occur around every ore-bearing fracture, regardless of size, attitude, or relative age; provided the lower intensity alteration zones are not flushed out by coalescence of the alteration envelopes of adjacent structures, as is the case in the Central zone. Furthermore, the two types of alteration always occupy the same relative positions: sericite adjacent to the ore-bearing vein; clay minerals always between sericitized rock and fresh quartz monzonite. And among themselves, the clay minerals always are arranged with the high kaolinite end of the series nearer the vein and the montmorillonite end toward fresh rock.

According to any theory based on the premise that an epoch of sericitization followed an epoch of argillization, the geometric pattern of alteration which envelopes every mineralized fissure in the Butte district would require that the invading zone of sericitization never advanced completely through the argillized zone to attack fresh rock directly. It would have had to advance only far enough out from every part of the respective channel walls to leave unchanged the characteristic band of argillization between the advancing sericite zone and unaltered rock. Furthermore, if the solutions which produced sericite were not capable of causing argillization, and if sericitization therefore wholly followed all argillization, sericitization could not have commenced until the very end of circulation in the channel maze after all mineralized fissures of all ages had been formed and sulphide deposition had all but ceased completely; for in areas where a complete structural sequence may be observed, argillization is always present, and in its customary zonal position, in the wall rocks of Northwest-faulting and Northeast-faulting fissures and even in the last formed fractures of the district along which there was any mineralizing activity. This would place sericite in the untenable position of post-sulphide age and run counter to the generally observed ubiquitous close association of sulphides and sericite at Butte and in other deposits of its type.

Structural evidence also indicates that many early East-West veins were completely "filled" with quartz and sulphides before they were offset along Northwest faults. Some of these early "sealed" veins should have escaped circulation of the postulated post-fracturing epoch of sericitization and be contained therefore, in an envelope of only argillized quartz monzonite without sericite. But, instead, the facts of occurrence are that the wall rocks around all early veins are abundantly sericitized. In fact, circulation along the East-West fracture system is chiefly responsible for the Central zone of intense sericitization which has widened rapidly with depth of exposure in the mines.

Logically, as well as geologically, only one alternative explanation of the ubiquity of argillization and sericitization around all ore-bearing fractures at Butte is available. In the opinion expressed here it fits the facts of distribution of alteration products better than an explanation involving two or more distinctly different chemical solutions operating at different times with different mineralogical results. This alternative proposition demands that sericitization and argillization are contemporaneous processes; that the frontal attacks of each zone on the next outermost alteration zone were made at the same time. As long as active circulation took place in the channel, each zone migrated away from the fissure, that is, it grew at its outer edge and simultaneously receded at its inner edge because of encroachment by the next innermost zone. And the different mineralogical and chemical responses to this attack are dependent not on a drastic about-face change of composition of the fluid in the channel but on smoothly varying conditions of physicochemical environment outwardly from the vein, demarcated into mineralogical zones by responses dictated by finite mineralogical true stability ranges.

This is the simplest geometric interpretation and the only one which fits all details of distribution, apparent age relationships, and true stability ranges of the minerals themselves. It became attractive at the beginning of this study as an explanation for distribution around small isolated veinlets cutting fresh quartz monzonite, where complete point to point observation of all mineralogical and chemical changes was facilitated. It was then applied on what might be termed an intermediate scale to large veins, where complications of intersection and bifurcation of fractures are prominent. And finally, viewing the district as a whole, it is clear that just as the sericite and argillite fronts were migrating outward from a vein during the period of active mineralization, so were they migrating district-wise radially upward and outward from a deeper centralized source. The pattern, when thus generalized, is the same for all three scales of application. The higher intensity alteration zone, where minerals of the sericite group predominate, is in each scale of application closer to the source; that is, to the supply of heat and necessary chemical ingredients, where rapid reaction to form sericite is more easily achieved.

References

1. W. H. Weed: Geology and Ore Deposits of the Butte District, Montana. U.S. Geol. Survey *Prof. Paper* No. 74 (1912).
2. Charles T. Kirk: Conditions of Mineralization in the Copper Veins at Butte, Montana. *Econ. Geol.* (1912) **7**, 35–82.
3. Reno H. Sales: Ore Deposits at Butte, Montana. *Trans.* AIME (1914) **46**, 3–106.
4. T. S. Lovering: The Origin of the Tungsten Ores of Boulder County, Colorado. *Econ. Geol.* (1941) **36**, 236.
5. Charles Meyer: Notes on the Cutting and Polishing of Thin Sections. *Econ. Geol.* (1946) **41**, 166–172.
6. R. E. Grim and R. A. Rowland: Differential Thermal Analyses of Clay Minerals and other Hydrous Materials. Part 1. *Amer. Min.* (1942) **27**, 755.
7. *Ibid.*, 801.
8. J. W. Gruner: Hydrothermal Alteration of Feldspars in Acid Solutions between 300° and 400°C. *Econ. Geol.* (1944) **39**, 578–589.
9. Earl Ingerson: Liquid Inclusions in Geologic Thermometry. *Amer. Min.* (1947) **32**, 375–388.

25

Reprinted by permission of The Colorado School of Mines from pages 231, 248–260 of *Colorado School of Mines Quart.* **45:**231–260 (1950)

THE GEOCHEMISTRY OF ARGILLIC AND RELATED TYPES OF ROCK ALTERATION *

By T. S. Lovering †

ABSTRACT

A zoning outward from many veins in many mining districts comprises: (1) a silicic zone; (2) a sericitic zone; (3) an argillic zone, with kaolin and montmorillonite minerals; (4) an allophane fringe zone transitional to (5) fresh rock. The alteration minerals characteristic of each zone are younger than those in zones farther away from the vein. The relative susceptibility of primary minerals to argillic alteration is similar in different deposits. A review of work on equilibria of the common igneous and alteration minerals in alkaline, neutral, and acid solutions shows that (1) solubility of hydrated silica in pure water increases rapidly with temperature and is almost unaffected by increasing acidity in the *p*H range of geologic interest in alteration studies, *p*H 7 to *p*H 0.1; the solubility is much greater in alkaline solutions than in acid solutions; (2) decomposition of silicates by acid solutions liberates hydrated silica and soluble cations (bases), continuing until the free acid present is neutralized. Feldspars are stable in weakly alkaline, neutral, and weakly acid solutions at elevated temperatures and pressures, if the solutions are saturated with silica and contain appropriate cations. In acid solutions containing potassium and aluminum the feldspars alter to sericite at high temperatures and to kaolinite at lower temperatures, but change to zeolites at temperatures from 60° to 550° C. in alkaline solutions undersaturated with respect to silica. Magnesian and ferromagnesian silicates of the pyroxene, amphibole, and chlorite groups are stable in alkaline solutions but are readily decomposed by hot acid solutions.

Alteration caused by present-day alkaline hot springs is characterized by decreased rather than increased porosity, by the absence of sericite and kaolin, and by the presence of zeolites, chlorite, and other minerals shown experimentally to be stable in alkaline solutions. Alteration by acid hot springs is attended by kaolin minerals, precipitated silica, and alunite, and by marked increase in porosity; the alkaline suite is conspicuously absent. It is concluded that solutions causing argillic and sericitic hydrothermal alteration are characterized by free acid.

The symmetrical zones of alteration in unfractured originally impermeable wall rock next to a fissure seem explainable only by the process of diffusion. The calculated rate of temperature rise in rock adjacent to a fissure is so fast as compared to that for diffusion of ions in unobstructed water that it is probable the reactions within all the zones of alteration adjacent to a hydrothermal conduit take place in an almost isothermal environment. Data from Solfatara, Italy, suggest that diffusion and decomposition of rock proceed at a rate only a few percent of that calculated for diffusion in pure water.

Diffusion always proceeds in the direction of lower concentration; leaching of constituents in outer zones and their addition close to a fissure indicate a change in the composition of the through-going solution at depth some time after alteration commenced.

The geochemistry of argillic alteration is similar in igneous and sedimentary rocks. The porosity of the argillic zone is usually much greater than that of the silicic zone, and some ore shoots seem to have been localized by the permeability developed in the argillized zones. The porosity and permeability caused by the leaching of acid hydrothermal solutions may prove as important as structural controls in determining the location of ore shoots in some districts.

* Published by permission of the Director, United States Geological Survey.

† Geologist, Mineral Deposits Branch, United States Geological Survey, Tucson, Arizona.

[*Editor's Note:* Material has been omitted at this point.]

GEOCHEMISTRY OF ARGILLIC ALTERATION

It seems evident that solutions containing free halogen acids or free sulphur acids in concentrations 0.01 to 0.10 N will quickly leach igneous rocks of Al_2O_3, the bases, and much of the SiO_2, leaving a porous residue of undigested silica in the zone of most intense alteration, an argillized zone adjacent to it, and a weakly altered fringe zone transitional to the fresh rock; in this fringe zone occurs a variety of minerals stable in a nearly neutral environment, such as alunite, gypsum, and epidote. The presence of carbonic acid in the altering solutions would change the mineral suite only in the fringe zone; so long as there was an appreciable amount of the stronger acids available, carbonic acid would be displaced from its salts, as with $CaCO_3$ and HCl, but in the outer zone where the pore solution was nearly neutral carbonic acid would still be an effective agent of attack. Bicarbonates of the bases would appear first, and might give way to carbonates in the outermost zone.

The activity of the bicarbonates is probably reflected by allophane and the montmorillonite group of clays. Calcite, chlorite, and uralite may form in the presence of alkali-carbonate solutions resulting from reaction of the bicarbonates with the cations liberated during allophanization of the silicates.

The innumerable tiny blebs of allophane that pockmark feldspars in the first stages of alteration deserve more attention than they have been accorded. These minute specks of alumina-silica gel that develop throughout the host mineral reflect the first loss of SiO_2 and the bases and a temporary concentration of alumina relative to silica. The feldspar matrix plus the allophane may subsequently react to form sericite, montmorillonite, or another feldspar such as orthoclase or albite; further leaching may convert the allophane to kaolinite, while the feldspar matrix continues to change first to allophane and subsequently to another clay. Such a change has been attributed by some to rising temperature, but the presence of allophane in partly replaced feldspars close to a vein as well as at a substantial distance from it, with argillic alteration intervening, suggests that temperature rise is not the primary cause of the change from allophane to kaolinite. The nearly isothermal condition that probably exists in the zone of hydrothermal alteration is discussed on page 251.

The alteration caused by neutral or slightly alkaline salt solutions will be largely that of ion exchange, the cations in the solution seeking equilibrium with those in the rocks. Calcic feldspars are commonly changed to albite or to orthoclase by this type of reaction.

Solutions containing free alkali dissolve silica and alumina but not the bases; and they usually cause the replacement of the feldspars by minerals having a higher ratio of base to silica. The relative concentrations of cations, aluminum, and silica in either acid or alkaline solutions are important in determining the alteration products formed. The relative concentration of cations in an alkaline solution deficient in silica may determine whether calcium, sodium, or potassium zeolites form; but with the same concentration of cations in a solution saturated with silica, the appropriate feldspar might appear instead of the zeolite. Similarly, at elevated temperatures an acid solution containing a suitably high concentration of potassium, aluminum, and silicon would form muscovite: if aluminum only were decreased, orthoclase might be stable; if potassium only were decreased, pyrophyllite might appear.

MECHANICS OF ALTERATION

Alteration involves a two-way process: bringing the reagent to the mineral and taking away the soluble products of reaction. This can be accomplished by the bodily transport of the solute or by the diffusion of material through a static fluid. Moving rather than static solutions are suggested by veins following fractures, especially those with relict open spaces, and by ready permeability in the direction followed by the altering solutions. Conversely, the absence of these features suggests static pore solutions and diffusion. Symmetrical zones of alteration in unfractured, originally impermeable rock adjacent to an open fracture or vein seem explainable only by diffusion of reactive ions into the walls from the fluid filling the fracture. A pervasive alteration in fractured rocks that shows no change in its intensity or mineralogic character along the fracture but does display an over-all major zoning with respect to some geologic feature suggests diffusion in a nearly stagnant pore solution. Rhythmic banding of the Liesegang type is also characteristic of diffusion, and its prominence in weathered rocks testifies to its importance in supergene processes.

Given a solution of electrolytes in prolonged contact with a slightly porous rock, it is impossible not to have them diffuse into the walls. The question is not whether they will diffuse into them, but only how far and how fast. In the argillized rocks adjacent to many veins there seems to be no agency that could remove the leached constituents of the rock other than diffusion, and the only avenue of escape is the vein. The ions and molecules abstracted from the wall rock must have moved back into the through-going solutions following the fissure, while the altering reagents diffused outward into the rock. Small-sized openings would not greatly impede diffusion, but the tortuous path to be followed in moving around solid grains might cause an ion to gain distance from the vein much more slowly than in open water.

Experimental work on diffusion of solutions through rocks has been carried on by Garrels and his associates at Northwestern University; and their report in advance of its publication, was very kindly made available.[20] In it the coefficients of diffusion in water are given for a number of ions; in complex solutions the coefficients change—the faster cations move faster, and the slow ones move slower, whereas the very fast anions are slowed down.[56] For a qualitative appraisal of the effect of diffusion, the writer has computed the change in concentration with diffusion for several different ions in pure water maintained at a constant concentration of 0.10 normality at the interface. In a porous rock the distances perpendicular to the vein traversed by the ions would probably be less than half those shown.

It is difficult to estimate the time that should be allowed for the operation of such a process in nature. The steam fumaroles at Solfatara, Italy, have been active for centuries, probably since the last eruption which took place A. D. 1198; the Plombieres hot springs were used by the Romans 2.000 years ago and were of much the same temperature and character then as now; the Yellowstone Park geysers have been active since Late Pleistocene time. Curves have therefore been drawn up for time intervals of 750 and 10,000 years, but these can be converted to other time periods by changing the x coordinates of the graph by the usual procedure:

$$\frac{\sqrt{t_2}}{\sqrt{t_1}} x_1 = x_2 \quad \text{since} \quad \frac{t_1}{t_2} = \frac{x_1^2}{x_2^2}$$

The equation of diffusion is identical in form with that for the heating of a body that has one wall maintained at constant temperature.

The equation of diffusion [20] is:

$$\frac{C_o - C}{C_o} = \phi\left(\frac{x}{\sqrt{4Dt}}\right) \quad \text{or} \quad C = C_o\left[1 - \phi\left(\frac{x}{\sqrt{4Dt}}\right)\right]$$

where C_o is the concentration at the interface; C is the concentration of the moving diffusion front; ϕ is the error function or probability integral of the argument enclosed in parentheses, for which tabulated values are readily available; x is the distance in centimeters from the interface (fissure); D is the diffusion coefficient in centimeters per day; and t is the time in days.[20]

The equation for temperatures in a body with one face maintained at the constant temperature T_o is:

$$T = T_o \left[1 - \phi\left(\frac{x}{\sqrt{4\alpha t}}\right) \right]$$

where T is the temperature at the point x centimeters from the hot face (vein fissure) after t seconds; and α is the thermal diffusity of the rock.[25] Because even the slow movement of water through a fissure would be sufficient to supply heat to a wall as rapidly as the heat was removed, it is of interest to compare the temperature distribution adjacent to the vein with the diffusion fronts. This is done in figure 4, and strongly suggests that diffusion takes place in an almost isothermal environment. The relatively rapid heating of the country rock would minimize or annul the importance of reactions due to temperature changes, as in the conversion of allophane to kaolinite or the change from kaolinite to sericite below 340° in acid potassic solutions, noted by Gruner.[22]

The alteration of the wall rock at the fumarole studied by Rittmann [47] 736 years after the eruption of the volcano gives the best data now available as to the probable rate of alteration by diffusion. The bodily movement of liquid under pressure in the wall rock of the steam vent since the volcanic eruption can be safely eliminated as an important factor in the alteration observed. The changes in the composition of the wall rock, as shown in figure 3, seem satisfactorily explained only by diffusion. The rate of decomposition and leaching would, of course, be slower than the rate of diffusion of the reagents, and the rate of diffusion in the walls of a steam vent might be appreciably slower than in a vein; this in turn would be slower than in unobstructed water.

The concentration of the soluble salts at the surface of the altered rock next to the superheated steam undoubtedly varies with the supply of rain. The H_2SO_4 supplied by the fumarolic gas itself should have a concentration of about 0.06 percent if all the H_2S were oxidized to acid; the pore fluid 30 centimeters from the fissure had a concentration of less than 0.04 percent and more than 0.017 percent, using as the upper limit the ratio of soluble sulphate to H_2O^- and for the lower limit the ratio of soluble sulphate to the amount of water that would be present if the pore space were saturated with water. These figures suggest that the concentration gradient at 30 centimeters was about one-third that at the fissure face; if alteration has been active here since the volcanic eruption in 1198, the rate of rock decomposition is such that the order of magnitude of sulphate diffusion is only about two percent as fast as that shown by the curve for $\frac{1}{2}SO_4^{--}$ in figure 4, p. 254.

The movement of material by diffusion depends on a concentration gradient. The material moving out of a rock towards the fissure must therefore have a higher concentration in the pore fluid than in the fissure, and its concentration from fringe zone to fissure must decrease progressively. In contrast, the leaching reagents such as HCl, H_2SO_4, or H_2CO_3 must continuously decrease in concentration from the fissure to the fringe zone.

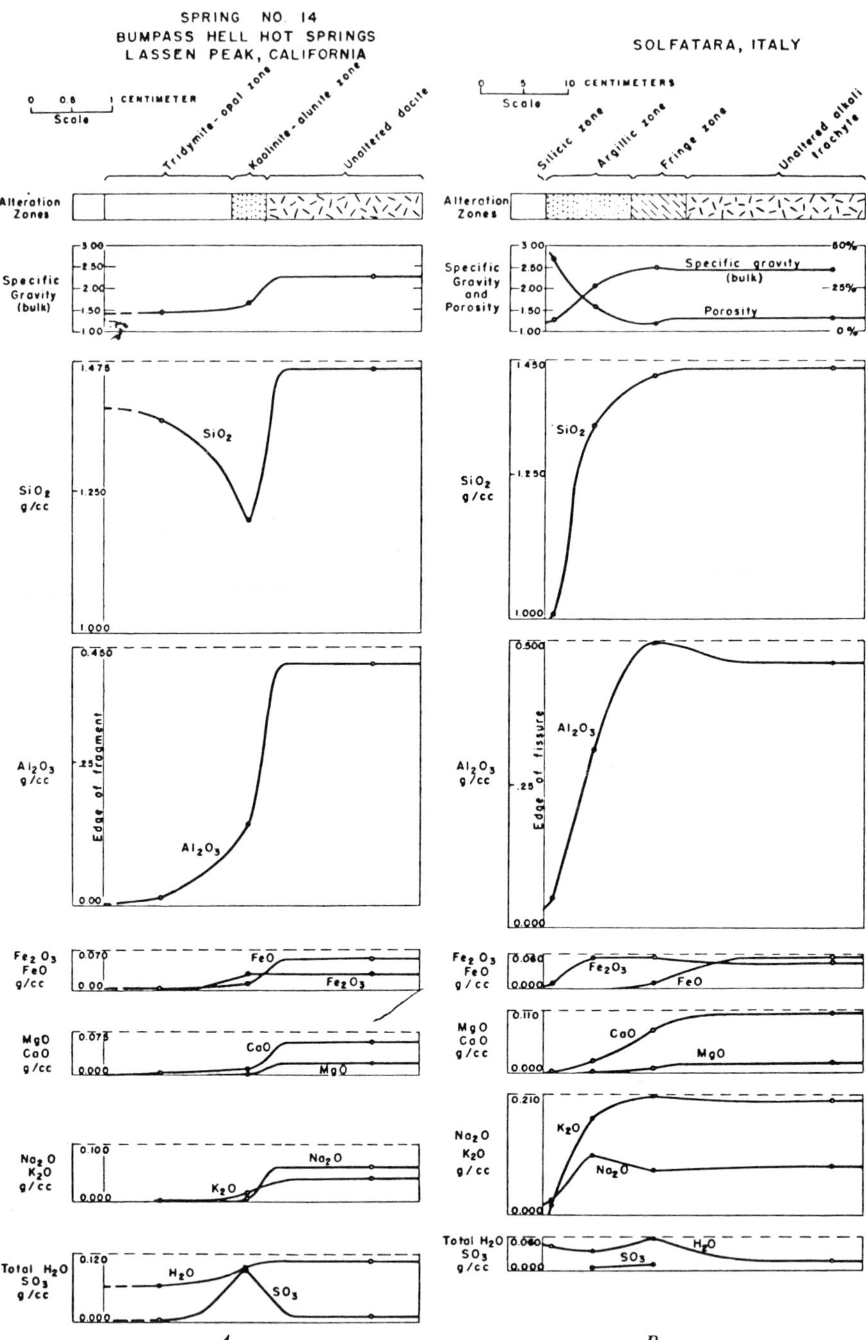

FIGURE 3. Chemical changes in dacite and trachyte caused by natural hot acid solutions. *A*. Chemical, physical, and mineralogical changes in a fragment of dacite altered by the weakly acid waters of Spring 14, Bumpass Hell Hot Springs, Lassen Peak, California (Anderson, 1935). *B*. Chemical, physical, and mineralogical changes in alkali trachyte wall of fumarole fissure, Solfatara, Italy (Rittmann, 1934).

Silica is but slightly ionized at ordinary temperatures and probably is present chiefly as molecules and molecular aggregates in hydrothermal solutions. In whatever form it is present, however, it readily moves through altered rock by diffusion. Its rate of diffusion is known to be slow as compared with such active ions as K^+, but quantitative data on its diffusion coefficients are not yet available, according to Professor Garrels. If silica were to move toward the fissure by diffusion, it would only be necessary that the pore solution be more nearly saturated with silica than the through-going solution in the fissure. If the solution in the fissure became saturated with silica, owing to a change in composition at depth after a time, the concentration gradient would be reversed next to the fissure, and the silica concentration in the pore solution would tend to equalize. Eventually the pore solution would all become saturated and the leaching of silica would cease, unless appreciable supersaturation were possible. This is not to say that destruction of silicates would also stop when silica was no longer leached; cations from the rock might still continue to diffuse out to the fissure even though the silica content of the altering rock remained "frozen" at the level it had reached just prior to saturation of the pore liquid, and continued decomposition of silicates would probably be accompanied by precipitation of silica. Molecules of SiO_2 in a supersaturated solution would, of course, move toward one of lower concentration even though that one were saturated, and precipitation would tend to spread outward from the fissure once the fluid began to deposit vein quartz. However, a thoroughly satisfactory understanding of the quartz envelope probably must await diffusion experiments with silica in both dilute and supersaturated solutions at high temperatures.

The same factors would also operate if the relative concentration of any other constituent in the through-going fluid increased beyond that in the pore fluid, even though saturation were not approached. The content of the cations in the pore fluid would be a function of the rates of decomposition of the wall-rock minerals except for those ions added by diffusion from the vein fluid. Decomposition, in turn, would depend largely on the *p*H of the solution and the kind and concentration of the ions or molecules brought in from the outside by diffusion. If the concentration of some cation in which the hydrothermal fluid was originally deficient increased sufficiently during some stage of alteration, movement of its ions by diffusion would be reversed, and the decomposition of a mineral supplying them in the wall rock might halt. The increasing concentration of cation might well be reflected in the conversion of an earlier-formed alteration mineral to a different kind, as in the change of kaolinite to sericite. Sericitic alteration might proceed concurrently with argillic alteration, however, if the original acid solutions contained a slight excess of potassium—enough to diffuse outward from the supply fissure slowly as compared with free acids, inhibiting the decomposition of potassic minerals and converting the earlier-formed kaolinite into sericite.

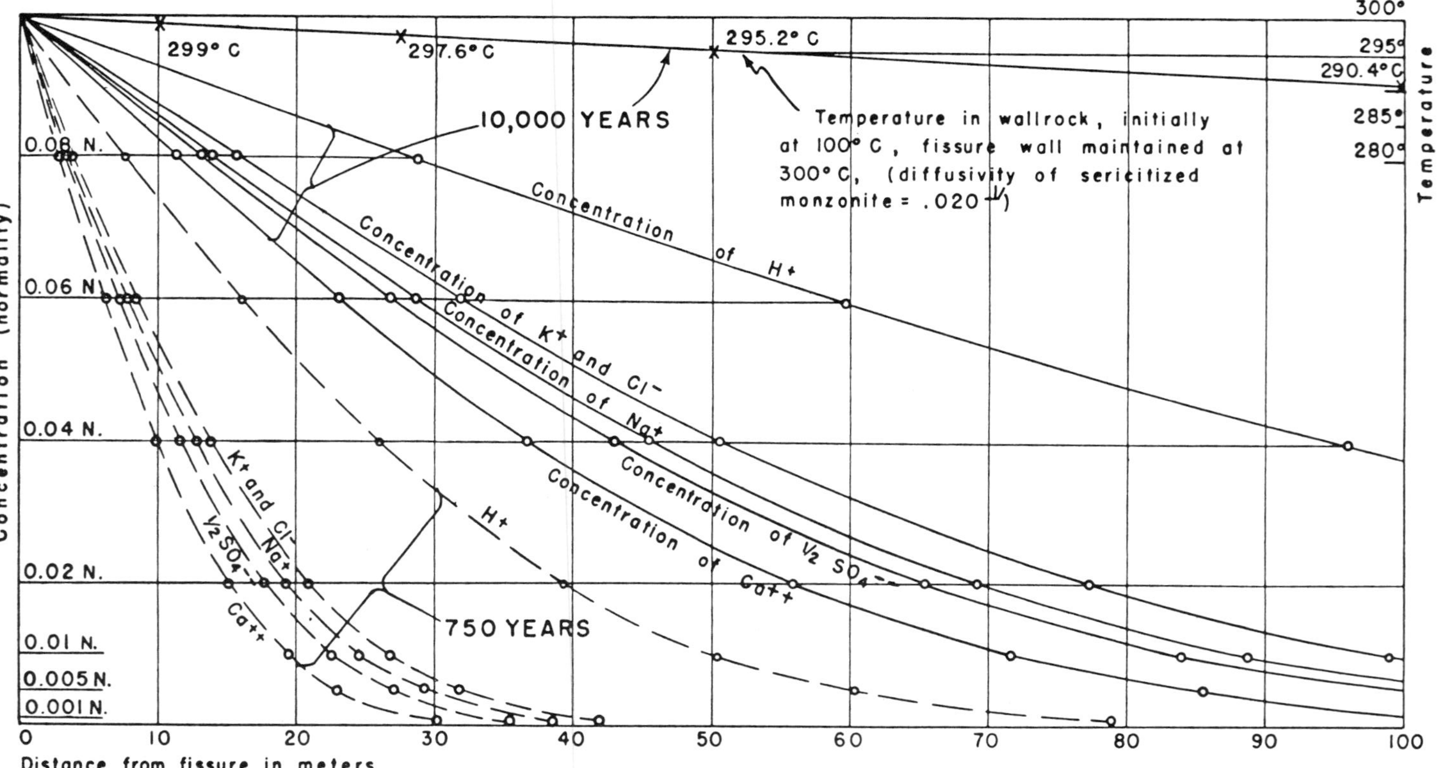

FIGURE 4. Temperature gradient in sericitized monzonite after 10,000 years, and concentration gradients calculated for diffusion of individual ions in pure water after 750 years and 10,000 years; diffusion coefficients from Garrels *et al.* (1949).[1] Lovering (1949), p. 5.

Diffusion for a short time is shown by relatively steep concentration gradients; a reversal in direction of diffusion from a vein fissure at a late stage in the period of alteration would therefore be characterized by a steeper composition curve than that which reflected an older diffusion source. The shape of variation diagrams for silica and other oxides in the walls of fissures should throw light on the processes and the time they operated, when more is known of the appropriate diffusion coefficients and the rates of decomposition of rocks under acid attack.

The curves showing the changes during alteration next to a tungsten vein and a copper vein (fig. 1) seem satisfactorily explained by the diffusion mechanism, and the silica concentration in the vein fluid evidently increased toward the end of the alteration period. The problem of whether the changes in concentration within the vein fluid with time were due to the contribution of material supplied by rock alteration at depth or to a change in the composition of the through-going solution at the source—hydrothermal differentiation [43]—or were the result of physical factors such as pressure and temperature changes or of a combination of these factors, must be considered in the light of the evidence available in the individual district. To the writer it seems plain that much alteration can have been accomplished only by diffusion; where diffusion alteration operated, zones must reflect concentration gradients, and it seems inescapable that reversals in chemical trends within wall rock must reflect changes in composition within the vein fluid.

ARGILLIZING SOLUTIONS IN VARIOUS ROCK ENVIRONMENTS

One of the universal features of argillic alteration is a notable increase in porosity attendant on the partial or complete leaching and removal of constituents of the wall rock. The instances cited have been purposely chosen to show the effect of argillizing solutions on igneous rocks of intermediate composition, rocks with minerals that are relatively insoluble in dilute acids or alkalies. To form the minerals characteristic of the argillized zone requires little but the molecular rearrangement of the unleached constituents in this type of rock. In the sericitized and silicified zones between the vein and the argillic zone, the porosity is less, and there is, of course, evidence of chemical addition; the minerals in these two zones are later than those of the argillized rock and replace or vein them.

Where the pre-ore silicic and potassic alteration materially reduced the permeability of the conduits, ore solutions may seek the more porous outer unclogged zone of argillic alteration. According to Kerr and his associates,[28] the ore zone in the Santa Rita copper porphyry shows a close association with argillized rock and shuns the quartz-sericite alteration zone; but the ore is later than the quartz and sericite, and these minerals in turn are later than the kaolin suite found in the argillized rock with the ore.

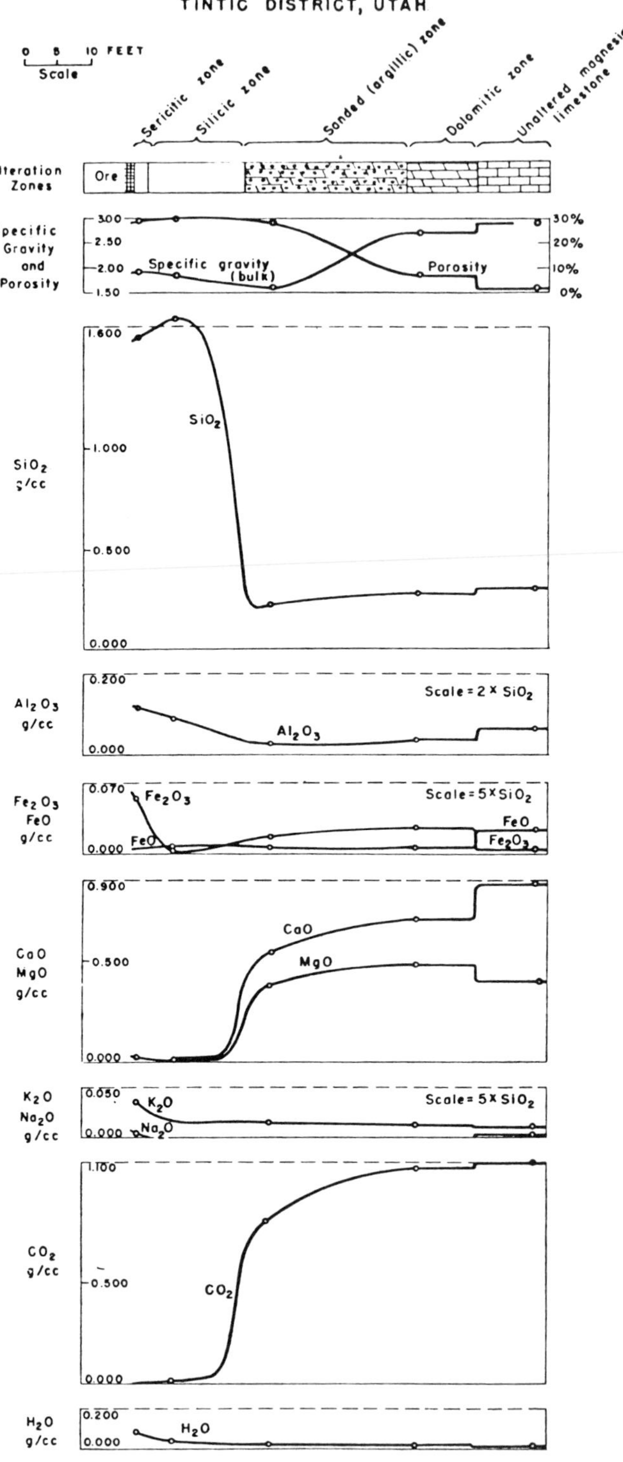

FIGURE 5. Chemical, physical, and litholic changes in dolomitic limestone adjacent to ore, Tintic Standard mine, Dividend, Utah (Lovering *et al.*, 1949).

The alteration zones, the paragenesis, and the porosity relations in argillized igneous rocks are common to so many mineralized areas that it is appropriate to ask what effect the solutions responsible for the alteration would have in completely dissimilar hosts.

In the Tintic district,[40] quartz latite lava lying above blind ore bodies in hydrothermally altered Paleozoic limestones shows argillic, sericitic, and silicic alteration. The alteration zones below the lava cap, surrounding a channel of mineralization in the limestones, are especially well displayed at the Tintic Standard mine. The lead-silver-zinc ore is in a zone of vuggy jasperioid that gives way upward to a porous sinterlike mass of silica. Close to the ore, sericite and hydrous mica are common. A short distance away from the ore the silicified rock is denser and contains little sericite; still farther from the ore the silica gives way abruptly to hydrothermal dolomite, which is commonly a porous, sandy-feeling, somewhat argillized rock at the contact. Clay minerals such as kaolinite and halloysite are locally abundant here at the outer edge of the jasperoid. The granular character of the dolomite is characteristic, and it is recognized by the textural term "sanded dolomite"; in many places the dolomite has been changed to an incoherent mass of small dolomite crystals or dolomite "sands." More commonly the delithifying process has not proceeded so far, but the crystals are so weakly bonded that it is easily possible to rub many dolomite grains loose from a specimen with the thumb. At different distances from the jasperoid, the "sanded dolomite" passes into a hard, relatively dense hydrothermal dolomite, which abruptly gives way to unaltered limestone.

The zone that is most obviously related to the argillic alteration described earlier is that in the dolomite adjacent to the jasperoid. Not only are clay minerals present here, but the marked porosity of the "sanded dolomite" bespeaks active leaching. The process of sanding was one in which the fine grains were dissolved and the bond between the larger crystals was weakened by intergranular solution. The hydrothermal solutions capable of such work must have carried free acid, not alkali.

Aluminum, silicon, and potassium were added in part by the bodily movement of solutions through the permeable "dolomite sand" and the leached dolomite along conduits, but in the large masses of rock adjacent to mineralizing channels, neither the argillizing nor the silicifying solutions apparently had any way of effecting alteration except by diffusion.

There is no chemical reason for a siliceous envelope always to intervene between argillized dolomite and ore. If the mineralizing solutions were low in silica, they might move through a substantial body of "sanded dolomite" before physical or chemical factors caused the silica to precipitate. If, as seems certain in many camps, the jasperoidizing solutions predated the ore solutions, there would be still less relation between jasperoid and ore. In the Tintic district ,[40] nearly all the ore has a siliceous casing; in the Red Cliff-Gilman district,[38] jasperoid commonly occurs some distance beyond the ore, toward the outlet side. In the Climax district,[8] on the other hand, silicified rock apparently occurs between the ore and its source.

At both Tintic and Gilman, however, and in many other districts, the effect of early argillizing solutions was an important factor in determining the channels followed by mineralizing solutions. This power of argillizing solutions to enlarge the openings followed by them and to render the adjacent rock more permeable, or at least more porous, may prove to be the most important aspect of argillic alteration for the mining geologist. It is certain that the permeability control exercised by argillizing solutions is a factor that has received insufficient attention in the past, and one that on further study may prove to be important as structural control in determining the location of ore shoots in some districts.

BIBLIOGRAPHY

1. Allen, E. T., and Day, A. L., *Hot Springs of the Yellowstone National Park:* Carnegie Inst. Washington Pub. 466, p. 112, 1935.
2. Anderson, C. A., *Alteration of the Lavas Surrounding the Hot Springs in Lassen Volcanic National Park:* Am. Mineralogist, vol. 20, no. 4, pp. 240-252, Apr. 1935.
3. Barshad, Isaac, *Vermiculite and Its Relation to Biotite as Revealed by Base Exchange Reactions, X-ray Analysis, Differential Thermal Analysis and Water Content:* Am. Mineralogist, vol. 33, pp. 655-678, 1948.
4. Blanch, E., and Passarge, S., *Die chemischen verwitterung in der agyptischen Wuste,* p. 75, Hamburg, 1925.
5. Brannock, W. W., Fix, P. F., Gianella, V. P., and White, D. E., *Preliminary Geochemical Results at Steamboat Springs, Nevada:* Am. Geophys. Union Trans., vol. 29, no. 2, pp. 211-226, Apr. 1948.
6. Burbank, W. S., *Geology and Ore Deposits of the Bonanza Mining District, Colorado:* U. S. Geol. Survey Prof. Paper 169, 166 pp., 1932.
7. Burbank, W. S., *Structural Control of Ore Deposition in the Red Mountains, Sneffels, and Telluride Districts, Colorado:* Colorado Sci. Soc. Proc., vol. 14, no. 5, pp. 141-261, 1941.
8. Butler, B. S., and Vanderwilt, J. W, *The Climax Molybdenum Deposits, Colorado:* U. S. Geol. Survey Bull. 846, pp. 195-237, 1933.
9. Chroustchoff, K. von, *Reproduction de l'amphibole:* Acad. sci. Paris Comptes rendus, tome 112, pp. 677-679, 1890.
10. Clarke, F. W., *The Data of Geochemistry,* 5th ed.: U. S. Geol. Survey Bull. 770, 841 pp., 1924.
11. Dana, E. S., *A System of Mineralogy,* 6th ed., New York, John Wiley & Sons, Inc., 1920.
12. Daubrée, A., *Memoire sur les depots mineraux formé par les source thermales de Plombieres, avant et pendant la periode actuelle:* Acad. sci. Paris Comptes rendus, tome 46, p. 1088, 1858.
13. Daubrée, A., *Études synthetique de geologie experimentales,* Paris, 1879.
14. Day, A. L., and Allen, E. T., *The Volcanic Activity and Hot Springs of Lassen Peak:* Carnegie Inst. Washington Pub. 360, 190 pp., Apr. 1925.
15. Doelter, C., *Ueber die kunstliche Darstellung und die chemische Constitution einiger Zeolithe:* Neues. Jahrb. Min. Geol., Bd. 1, pp. 118-139, 1890.
16. Duffel, Stanley, *Diffusion and Its Relation to Ore Deposition:* Econ. Geology, vol. 32, no. 4, pp. 494-510, June-July 1937.
17. Fenner, C. N., *Bore-Hole Investigations in Yellowstone Park:* Jour. Geology, vol. 44, no. 2, pt. 2, pp. 225-315, Feb.-Mar. 1936.
18. Friedel, C., and Sarasin, E., *Sur la reproduction de l'albite par voie aqueuse:* Acad. sci. Paris Comptes rendus, tome 97, pp. 290-294, 1883.

19. Friedel, C., and Grandjean, F., *Synthese de chlorite par action des solutions alkalines sur le pyroxene:* Bull. Soc. Min., vol. 32, pp. 150-154, 1909.
20. Garrels, R. M., Dreyer, R. M., and Howland, A. L., *Diffusion of Ions through Intergranular Spaces in Water Saturated Rocks:* Geol. Soc. America Bull., vol. 60, no. 12, pp. 1809-1828, Dec. 1949.
21. Grubenmann, Ulrich, and Niggli, Paul, *Die Gesteinsmetamorphose*, Borntrager, Berlin, 1924.
22. Gruner, J. W., *The Hydrothermal Alteration of Feldspars in Acid Solutions between 300° and 400° C.:* Econ. Geology, vol. 39, no. 8, pp. 578-589, Dec. 1944.
23. Harker, Alfred, *Metamorphism*, 360 pp., London, Methuen and Co., 1932.
24. Hitchens, C. S., *A Method for the Experimental Determination of Hydrothermal Solutions, with Notes on Its Application to the Solubility of Silica:* Inst. Min. Met. Bull., vol. 364, pp. 1-26, 1935.
25. Ingersoll, L. R., Zobel, O. J., and Ingersoll, A. C., *Heat Conduction*, New York, McGraw-Hill Book Co., Inc., 1948.
26. Johnstone, Alexander, *On the Action of Carbonic Acid Water on Minerals and Rocks:* Edinburgh Geol. Soc., vol. 5, pp. 282-285, 1886.
27. Johnstone, Alexander, *On the Action of Pure Water and Water Containing Carbonic Acid Gas on Minerals of the Mica Family:* Geol. Soc. London Quart. Jour., vol. 95, pp. 363-368, 1889.
28. Kerr, P. F., Kulp, J. L., Patterson, C. M., and Wright, R. J., *Alteration Studies, Santa Rita, New Mexico*, vol. 1 and 2, New York, Kings Crown Press, 1947.
29. Knopf, E. B., and Jonas, A. I., *Geology of the McCalls Ferry-Quarryville District, Pennsylvania:* U. S. Geol. Survey Bull. 799, 156 pp., 1929.
30. Lacroix, A., *La Montagne Pelée et ses Eruptions*, pp. 182-187, Paris, 1904.
31. Lemberg, J., *Ueber Silicatumwandlungen:* Zeitschr. geol. Gesell., vol. 28, pp. 519-621, 1876.
32. Lemberg, J., *Zur Kentnis der Bildung and Umbildung von silikaten:* Zeitschr. deutsch geol. Gesell., vol. 39, pp. 559-600, 1887.
33. Lenher, Victor, and Merrill, H. B., *The Solubility of Silica:* Am. Chem. Soc. Jour., vol. 39, p. 2630, 1917.
34. Leonard, R. J., *The Hydrothermal Alteration of Certain Silicate Minerals:* Econ. Geology, vol. 22, no. 1, pp. 18-43, Jan.-Feb. 1927.
35. Lindgren, Waldemar, *Mineral Deposits*, 4th ed., 930 pp., New York, McGraw-Hill Book Co., Inc., 1933.
36. Lovering, T. S., *The Leaching of Iron Protores; Solution and Precipitation of Silica in Cold Water:* Econ. Geology, vol. 18, no. 6, pp. 523-540, Sept. 1923.
37. Lovering, T. S., *The Origin of the Tungsten Ores of Boulder County, Colorado:* Econ. Geology, vol. 36, no. 3, pp. 229-279, May 1941.
38. Lovering, T. S., and Tweto, O. L., *The Geology and Ore Deposits of the Minturn Quadrangle, Colorado*, open-file report, U. S. Geol. Survey, Denver, Colorado, and Washington, D. C., 1943.
39. Lovering, T. S., *Geothermal Gradients, Recent Climatic Changes and Rate of Sulfide Oxidation in the San Manuel District, Arizona:* Econ. Geology, vol. 43, pp. 1-20, 1948.
40. Lovering, T. S., *et al.*, *Rock Alteration as a Guide to Ore—East Tintic District, Utah:* Econ. Geology Mon. 1, 64 pp., 1949.
41. Morey, G. W., and Ingerson, Earl, *The Pneumatolytic and Hydrothermal Alteration and Synthesis of Silicates:* Econ. Geology, vol. 32, no. 5 (supplement), Aug. 1937.
42. Moore, E. S., and Maynard, J. E., *Solution, Transportation, and Precipitation of Iron and Silica:* Econ. Geology, vol. 24, no. 4, pp. 365-402, June-July 1929.
43. Neumann, Henrich, *On Hydrothermal Differentiation:* Econ. Geology, vol. 43, pp. 77-83, 1948.
44. Noll, W., *Synthese von Montmorilloniten:* Chemie der Erde, vol. 10, pp. 129-153, Gustav Fischer, Jena, 1936.
45. Nutting, P. G., *The Action of Some Aqueous Solutions on Clays of the Montmorillonite Group:* U. S. Geol. Survey Prof. Paper 197-F, pp. 219-235, 1943.

46. Peterson, N. P., Gilbert, C. M., and Quick, G. L., *Hydrothermal Alteration in the Castle Dome Copper Deposits, Arizona:* Econ. Geology, vol. 41, no. 8, pp. 820-840, Dec. 1946.
47. Rittmann, A., *Der Alkalitrachyte der Solfatara und seine Zersetzung durch Fumarolengase:* Schweiz. Miner. und Petrogr. Mitt., Band 14, Heft 1, pp. 1-21, 1934.
48. Ross, C. S., and Kerr, P. F., *The Kaolin Minerals:* U. S. Geol. Survey Prof. Paper 165-F, pp. 151-180, 1930.
49. Ross, C. S., and Hendricks, S. B., *Minerals of the Montmorillonite Group, Their Origin and Relation to Soils and Clays:* U. S. Geol. Survey Prof. Paper 205-B, pp. 23-79, 1945.
50. Sales, R. H., and Meyer, Charles, *Wall Rock Alteration at Butte, Montana:* Am. Inst. Min. Met. Eng. Trans., vol. 178, pp. 9-35, 1948.
51. Schwartz, G. M., *Hydrothermal Alteration in the "Porphyry Copper" Deposits:* Econ. Geology, vol. 42, no. 4, pp. 319-352, June-July 1947.
52. Sestini, F., *Wirkung des Wassers auf naturliche Metasilicate:* Atti. Soc. Tosc. di Sc. Nat., Pisa. Proc. Verh., vol. 12, pp. 127-136, 1900. (Abstract in Zeit. Kryst., vol. 35, pp. 511-512, 1902.)
53. Smith, O. C., *Identification and Qualitative Chemical Analysis of Minerals,* 351 pp., New York, D. Van Nostrand Co., Inc., 1946.
54. Stephenson, E. A., *Studies in Hydrothermal Alteration:* Jour. Geology, vol. 24, pp. 180-199, 1916.
55. Sullivan, E. C., *The Interaction between Minerals and Water Solutions:* U. S. Geol. Survey Bull. 312, 69 pp., 1907.
56. Vinograd, J. R., and McBain, J. W., *Diffusion of Electrolytes and Their Mixtures:* Am. Chem. Soc. Jour., vol. 63, pp. 1008-1015, 1941.
57. White, D. E., and Thompson, George, *Thermal Springs and Their Possible Significance in the Future Discovery of Ore Deposits:* Am. Inst. Min. Met. Eng., Min. Engr., vol. 1, p. 37. See also: Brannock, Fix, Gianella, and White, (1948) 1949.
58. Wichers, Edward, Schlecht, W. G., and Gordon, C. L., *Preparing Refractory Oxides, Silicates, and Ceramic Materials for Analysis by Heating with Acids in Sealed Tubes at Elevated Temperatures:* U. S. Nat. Bur. Standards Jour. Research, vol. 33, pp. 451-456, 1944.
59. Zies, E. G., *The Valley of Ten Thousand Smokes:* Nat. Geog. Soc. Contr. Tech. Papers, vol. 1, no. 4, 79 pp., 1929.

26

Reprinted by permission of The Colorado School of Mines from pages 209–210 and 224–229 of *Colorado School of Mines Quart.* **45:**209–229 (1950)

THE FUMAROLIC-HOT SPRING AND "EPITHERMAL" MINERAL DEPOSIT ENVIRONMENT

By HARRISON SCHMITT *

ABSTRACT

The present data on fumaroles, hot springs, and "epithermal" mineral deposits are reviewed and discussed. They suggest a close relationship between the phenomena of volcanism and the "epithermal" mineral deposits, as many others have noted. If the conditions of origin of the "epithermal" deposits and of the fumaroles and hot springs are assumed to be nearly the same, the origin of the "epithermal" ore deposits differs from current theory. These deposits then are laid down in the meteoric-water shell, the silica is largely derived from wall-rock alterations, and the exotic, i.e., wholly foreign, elements are transferred to the meteoric-water shell by the vapor, gas, and sublimate phase, that is by volcanic emanations. The chief elements imported are carbon dioxide, sulphur, and the ore metals. The chemical field of the wall-rock alterations and of the fumarolic-hot springs and "epithermal" environments should correspond.

INTRODUCTION

During the first few and last few decades of the one hundred years just passed, volcanologists and geophysicists have carried out comprehensive researches on the phenomena of volcanism. This work seems to provide a basis for the understanding of fumaroles, hot springs, and geysers and when integrated with the present knowledge of "epithermal" mineral deposits, by a major assumption, appears to aid in the understanding of the wall-rock alteration and other phenomena associated with it.

The writer realizes the difficulties in the problem implied by the title. No easy solution is proposed. Data on the "epithermal" deposits pertinent to the subject are sparse. Furthermore, the writer's experience in this field is limited on the theoretical side, although professionally he has viewed many deposits of the "epithermal" type. Many of the data from which inferences are presently drawn have been in the literature up to three decades, but the inferences impelled by them do not seem generally to have been made. This may be in part the result of the difficulty of access to some of the publications.

The term "epithermal" must be used advisedly. It is evident that the "shallow" deposits were formed at high as well as low temperatures. Buddington's term "xenothermal" covers only the high-temperature possibility. There may be other variations; therefore, in this paper "epithermal" will be written with quotation marks. The "epithermal" deposits, in any case, are understood to be those formed at shallow depths. The maximum known depth is about 6,000 feet, but few exceed 3,000 feet.

* Mining Geologist, Silver City, New Mexico.

Certain commonly observed features include increasing structural complexity upward; crustified, vuggy, hackly, and colloform textures; and presence of the minerals adularia, alunite, amythestine quartz, barite, fluorite, stibnite, opal, and chalcedony, and of the elements mercury, tellurium, selenium, arsenic, bismuth, and antimony. Gold- and silver-bonanza ore bodies with complex antimony, arsenic, and tellurium minerals are characteristic. Pyrite and base-metal minerals are not abundant. Pending contrary proof at a few places such as the Casapalca (Peru), Fresnillo, Zacatecas, and Michigan copper deposits, the "epithermal" are not known to grade to the mesothermal type in depth. They are commonly associated with volcanic necks and areas of Tertiary volcanic activity.[28]

That the "epithermal" ore deposits originate or originated in the fumarolic-hot spring environment seems to lack proof except in the case of the mercury deposits. There are at least two examples of small mercury ore bodies associated with active hot springs, namely; in New Zealand and at Steamboat Springs, Nevada. At Huitzuco, Mexico, mercury ore bodies are associated with what are called "geyser tubes,"[33] and at Sulphur Banks, California, and at other places noncommercial mercury is found. The origin of other types of "epithermal" ore deposits is not so suggestive. At Roturoa, New Zealand, the Whakorewarewa hot springs have deposited sinter containing about $5.00 a ton in silver and gold, and a sludge in the springs at Steamboat Springs, Nevada, has a value of $12.00 in gold. The sinter contains cinnabar and stibnite.[9] Many hot springs and fumaroles are known to deposit small amounts of metals. Results of spectrographic studies by Fraser[20] indicate the close qualitative parallelism in the metals present in hot-spring and "epithermal" deposits. Although inconclusive, the evidence (many times reviewed, particularly in modern times by Lindgren) is so strongly inferential that we are justified in assuming that the "epithermal" deposits in most cases originated in the fumarolic-hot springs environment. In any case we shall see where this assumption may lead us.

[*Editor's Note:* Material has been omitted at this point.]

SUMMARY

1. A century of work by volcanologists and geophysicists has established that:

(*a*) The volcanic emanations are the universal motivation for volcanic, fumarolic, hot-spring, and geyser activity. They are characterized by superheated steam and much smaller quantities of CO_2, H_2S, NH_3, CH_4, HCl, HF, NaCl, and NH_4Cl. The ore metals accompany them in some cases. Temperatures may range up to incandescence; pressures up to hundreds of atmospheres.

(*b*) Meteoric water constitutes most of the water in the hot-spring and geyser systems and in places is important in fumaroles.

(*c*) The emanations may emerge at the surface nearly unchanged, but where meteoric water is dominant they are quenched and absorbed in depth.

(*d*) The steam from the time it gained its superheat could never have been in contact with an excess of liquid water.

(*e*) Fumarolic phenomena show that the ore metals are transferred in the gas-vapor-sublimate phase.

2. When the emanations are absorbed by the meteoric-water shell, there follows a more or less regular sequence of effects: alterations, leachings, and depositions accompanied by fumarolic, geyser, and hot-spring activity. The processes are dominated by carbon-dioxide and hydrogen-sulphide chemistry, even though these compounds are quantitatively small in the original emanations.

At Yellowstone Park, where the alteration of the rhyolite-dacite walls has been carefully studied by surface observation and core drilling, several types of alteration and their relationship in space and time are fairly well known. In general a beidellite-type clay results from approximately the first encounter of H_2CO_3-charged hot water with silicates of the walls. The tendency is to remove all the alkalies and alkaline earths and some of the silica, and thus to leave an "inert" clay composed of alumina, silica, and water. This is here called the "carbon-dioxide-leaching" phase. The effluent liquid from this reaction is charged predominantly with the alkali and alkaline-earth bicarbonates and colloidal silica. It seems to be oversaturated with potash and silica, and the next metasomatisms show an increment of potash, in part by base exchange in the feldspars, with a corresponding loss of soda to the liquid. This is here called the "potassic" phase. It is marked at Yellowstone Park by the development of a quartz-orthoclase alteration. The effluent liquid from this is saturated with soda and silica. The next phase then is sodic and still silicic. Orthoclase then gains soda by base exchange, and zeolites and calcite are stable and are deposited. This is here called the "sodic" phase.

Quartz seems to be stable in all the phases or zones beyond the beidellite-type clay phase with the metacolloids more prominent in the more distant zones. Calcite does not appear to be stable until much of the H_2CO_3 is neutralized by reactions or lost by lower pressure. At Yellowstone Park and at Steamboat Springs it is deposited characteristically under alkaline conditions, where it tends to choke up the near-surface openings.

3. It is well-established in the hot-springs areas that, where H_2S is in excess near the surface, oxidation may produce sufficient sulphuric acid to neutralize and make acid the normally alkaline spring water. This results in an alteration of the walls characterized by the kaolinite group of clays, the sulphates (particularly alunite), and often massive silicification. Like the beidellite-type clay alteration, the alkalies, and alkaline earths and much of the silica are taken into solution and "inert" clay left. Much silica seems to be redeposited as shallow silicification. Some of the "epithermal" deposits show a marked analogy to the foregoing system. Goldfield, Nevada, is a typical example.

4. The Yellowstone Park studies indicate a moderately complex circulation of meteoric water and volcanic emanations. The meteoric water appears to dominate the larger openings. The "gases" may be encountered near the surface even where the country rock is largely flooded by surface water. It is conceivable that the large variations in the structure of the country rock in "epithermal" districts may result in intricate and unique paths of circulation. The major fissures may serve both as an entrance and exit for the meteoric waters.

5. Ionic interchange must be important, as is demonstrated at Butte, but the convection process must often dominate. The alteration zones doubtless migrate, even if the areas dominated by meteoric water do not change in position, for when a clay zone, for example, approaches inertness the then-unmodified fresh solutions would attack previously, but differently, altered zones farther along.

6. Wall-rock alteration seems inadequately described for many important "epithermal" districts. (But see Lovering,[24] Burbank, [11 12] and Butler.[13]) It presents a confusing picture when compared with the clear-cut Yellowstone Park phenomena. Tentative conclusions are (1) that the "primary" clay alterations in some, perhaps many, cases must have been overlooked; (2) that the potassic and sodic alterations affect both the walls and selvages of the veins (their characteristic minerals are also deposited in the veins); and (3) that the primary argillization should progress forward and replace earlier zones of the potassic and sodic phases.

7. As for the ore-metal phase of the "epithermal" deposits, the volcanic gas-vapor sublimates or emanations seem wholly adequate as agents for the transfer of the metallic elements to the meteoric-water shell. These elements are known to be, and to have been, associated with the emanations at many places. The Yellowstone Park hot alkaline spring water with abundant NaCl and $NaHCO_3$ is a type common to volcanic

areas and in some regions contains small quantities of the ore metals. Similar waters at Steamboat Springs, Nevada, contain mercury, arsenic, and antimony in small quantities, and the older sinter contains cinnabar and stibnite. Sludges in these springs contain "commercial" gold with mercury, antimony, silver, and copper.

8. Sales and Meyer indicate that the various types of metasomatisms can be going on in the walls at the same time in successive zones. In the case of the metallic minerals in the veins, a requirement seems to be a change in the fluids at a given spot with time. Since the paragenesis is nearly constant from deposit to deposit, often starting with quartz, pyrite, arsenopyrite, and oxides, followed in order by the simple sulphides, the complex sulphosalts, the precious metals, and mercury, with the gradation in order of decreasing solubility, increasing softness, etc., the final explanation must fit these established facts. Brown [10] thinks that a progressive change in the source of the metallic emanation is the explanation.* In the system outlined a change of the ore liquid could occur, particularly in the amount and proportion of the principal salts and colloidal silica due to changes in the character of the walls through progressive alteration. The relative concentration of the chloride ion in the metal-bearing water should progressively rise. Newhouse shows that primary liquid inclusions in sulphides have a high saline content. The abundant early quartz in the vein may be primarily related to the loss of silica during argillization. A decrease in silica is to be expected with time.

9. The postulated formation of the "epithermal" ore bodies in the meteoric-water shell appears to simplify the explanation of several features of these deposits:

(*a*) They show a nearly universal shallow bottoming. This bottom often may well be at the base of the vadose-water zone. In some districts a belt of primary ore bodies of a given vein parallels the present topography.

(*b*) The delicate layering of the vein minerals may be due to seasonal and longer climatic changes, i.e., changes largely in the quantity of water entering the meteoric-water shell.

(*c*) The difficulty of classifying many "epithermal" minerals as hypogene or supergene is explained as the result of the intimate intermingling of the hypogene and supergene environments.

(*d*) "Epithermal" ore bodies commonly do not outcrop, but the apex of the ore is at shallow depth. This is possibly explained by the rapid change in conditions near the original surface. Such a change is evident in the hot-springs areas.

---

* J. W. Gruner in his criticism here states: "Is this not a matter of mass? Pyrite is millions of times as abundant and should be deposited immediately and all along. Gold while being present in very small amounts will be deposited when the proper saturation is reached." Gruner believes that geologists have neglected the implications of mass action (molecular concentration) in puzzling over paragenesis. It must be said that pyrite does not seem very abundant in "epithermal" deposits but is nearly always present to some degree.

SUGGESTED GENERAL IMPLICATIONS

1. The free silica in the "epithermal" deposits, particularly if these belong to the environment postulated, can be accounted for as derived from alterations of the walls. Schwartz [29] concluded that the silica in wall-rock alterations, not necessarily "epithermal," however, about balanced. This implies that new or "exotic" silica was not needed. Lindgren's examples of wall-rock gains and losses [23] show silica loss in the altered wall and gain in the vein. Lovering's data [24] and those of Sales and Meyer [26] show the same thing. It seems unnecessary to call on the magma for silica for the "epithermal" deposits. Knopf [21] and Schmitt [27] think that gains of silica in "mesothermal" quartz veins and in garnet zones can be accounted for by losses in the walls due to alteration.

2. Such a source for the silica appears to make it less necessary on an *a priori* basis to postulate a magma as the source for the "epithermal" deposits. The volcanic emanations can be derived by the heating alone of many kinds of rocks. In depth this could be made effective by release of pressure. The deeper earth shells should contain progressively more of the metals. The principles demonstrated by Zies [34] and by Liddell [22] of the chloridizing roast may be adequate to account for the collection and transfer of the metals.

3. The many resemblances of the "epithermal" deposits to those of other classifications, particularly the mesothermal and hypothermal in noncarbonate rocks, suggest that all fit into one eclectic frame. The evidence is weak, however, that the "epithermal" deposits grade into mesothermal deposits in depth, for the "epithermal" deposits usually show true bottoms such as wholly barren quartz and barren breccia. Schmitt [28] suggests that "epithermal" deposits may be unique to the late age of the earth.

4. A possible outcome of the verification of a close tie in origin of the "epithermal" deposits and the hot springs would be the development of guides to likely conditions in depth. It may be found that "epithermal" ore deposition is usually just above, or to one side of, the clay zone and associated with a sericitic-quartz or potassic phase. If a given exposure was in the zeolitic-calcite phase of wall alteration, one might expect most of the ore shoots to be largely intact in depth, or if in the clay zone one might interpret this to mean one was near the bottom of the zone of ore deposition. The foregoing frame may not be the general one at all, but it does seem as though this particular phase of investigation could have results of considerable utility in the exploration for ore. A systematic field and laboratory study of the possibilities should be made. The data in the literature are inadequate.

ACKNOWLEDGMENTS

The writer is greatly indebted to J. W. Gruner, Augustus Locke, and C. N. Fenner for their careful criticism of this paper.

REFERENCES

1. Allen, E. T., *Chemical Aspects of Volcanism with a Collection of the Analyses of Volcanic Gases:* Franklin Inst. Jour., vol. 193, no. 1, pp. 29-80, Jan. 1922.
2. Allen, E. T., *Geyser Basins and Igneous Emanations:* Econ. Geology, vol. 30, no. 2, pp. 1-13, Jan.-Feb. 1935.
3. Allen, E. T., and Day, A. L., *The Sources of the Heat and the Source of the Water in the Hot Springs of the Lassen National Park:* Jour. Geology, vol. 32, no. 3, pp. 178-190, Apr.-May 1924.
4. Allen, E. T., and Day, A. L., *Steam Wells and Other Thermal Activity at "The Geysers," California:* Carnegie Inst. Washington Pub. 378, 106 pp., 1927.
5. Allen, E. T., and Day, A. L., *Natural Steam Power in California:* Nature, vol. 122, pp. 17-18, 27-28, July 7, 1928.
6. Allen, E. T., and Day, A. L., *Hot Springs of the Yellowstone National Park:* Carnegie Inst. Washington Pub. 466, 525 pp., 1935.
7. Allen, E. T., and Zies, E. G., *A Chemical Study of the Fumaroles of the Katmai Region:* Nat. Geog. Soc. Contr. Tech. Papers, Katmai Ser., no. 2, pp. 75-155, 1923.
8. Bandy, M. C., *A Theory of Mineral Sequence in Hypogene Ore Deposits:* Econ. Geology, vol. 35, no. 3, pp. 359-381, May 1940; no. 4, pp. 546-570, June-July 1940.
9. Brannock, W. W., Fix, P. F., Gianella, V. P., and White, D. E., *Preliminary Geochemical Results at Steamboat Springs, Nevada:* Am. Geophys. Union Trans., vol. 29, no. 2, pp. 211-226, Apr. 1948.
10. Brown, J. S., *Ore Genesis,* 204 pp., Hopewell, N. J., The Hopewell Press, 1948.
11. Burbank, W. S., *Geology and Ore Deposits of the Bonanza Mining District, Colorado:* U. S. Geol. Survey Prof. Paper 169, 166 pp., 1932.
12. Burbank, W. S., *Structural Control of Ore Deposition in the Red Mountains, Sneffels, and Telluride Districts of the San Juan Mountains, Colorado:* Colorado Sci. Soc. Proc., vol. 14, no. 5, pp. 141-261, 1941.
13. Butler, B. S., *et al., The Copper Deposits of Michigan:* U. S. Geol. Survey Prof. Paper 144, 238 pp., 1929.
14. Clarke, F. W., *The Data of Geochemistry,* 5th ed., Chap. 8, *Volcanic Gases and Sublimates:* U. S. Geol. Survey Bull. 770, pp. 261-292, 1924.
15. Coats, R. R., *Propylitization and Related Types of Alteration on the Comstock Lode:* Econ. Geology, vol. 35, no. 1, pp. 1-16, Jan.-Feb. 1940.
16. Dahlblom, Th., *The Permeability of Rocks:* Econ. Geology, vol. 19, no. 4, pp. 389-392, June-July 1924.
17. Day, A. L., *The Hot Springs Problem:* Geol. Soc. America Bull., vol. 50, no. 3, pp. 317-336, Mar. 1, 1939.
18. Fenner, C. N., *Bore-Hole Investigations in Yellowstone Park:* Jour. Geology, vol. 44, no. 2, pt. 2, pp. 225-315, Feb.-Mar. 1936.
19. Fenner, C. N., *The Nature of the Ore-Forming Fluid; A Discussion:* Econ. Geology, vol. 35, no. 7, pp. 883-904, Nov. 1940.
20. Fraser, H. J., *Genetic Correlation of Cinnabar and Some Gold-Silver Deposits* (abstract) : Econ. Geology, vol. 34, no. 4, p. 468, June-July 1939.
21. Knopf, Adolph, *The Mother Lode System of California:* U. S. Geol. Survey Prof. Paper 157, 88 pp., 1929.
22. Liddell, D. M., *Handbook of Non-Ferrous Metallurgy,* Vol. 2, New York, McGraw-Hill Book Co., Inc., 1926.
23. Lindgren, Waldemar, *Mineral Deposits,* 4th ed., 930 pp., New York, McGraw-Hill Book Co., Inc., 1933.
24. Lovering, T. S., *The Origin of the Tungsten Ores of Boulder County, Colorado:* Econ. Geology, vol. 36, no. 3, pp. 229-279, May 1941.
25. Newhouse, W. H., *The Composition of Vein Solutions as Shown by Liquid Inclusions in Minerals:* Econ. Geology, vol. 27, no. 5. pp. 419-436, Aug. 1932.
26. Sales, R. H., and Meyer, Charles, *Wall Rock Alteration at Butte, Montana:* Am. Inst. Min. Met. Eng. Trans., vol. 178, pp. 9-35, 1948.
27. Schmitt, H. A., *The Pewabic Mine:* Geol. Soc. America Bull., vol. 50, no. 5, pp. 777-818, May 1, 1939.

28. Schmitt, H. A., *Uniformitarianism and the Ideal Vein*, presented Soc. Econ. Geologists, San Francisco, Calif., Feb. 1949.
29. Schwartz, G. M., *Hydrothermal Alteration of Igneous Rocks:* Geol. Soc. America Bull., vol. 50, no. 2, pp. 181-237, Feb. 1, 1939.
30. Shepherd, E. S., *Kilauea Gases, 1919:* Hawaiian Volcano Obs. Bull., vol. 9, no. 5, May 1921.
31. Shepherd, E. S., *The Analysis of Gases Obtained from Volcanoes and from Rocks:* Jour. Geology, vol. 33, no. 3 (supplement), pp. 289-370, Apr.-May 1925.
32. Shepherd, E. S., and Merwin, H. E., *Gases at the Mt. Pelée Lavas of 1902:* Jour. Geology, vol. 35, no. 2, pp. 97-116, Feb.-Mar. 1927.
33. Vaupell, C. W., *Mercury Deposits at Huitzuco, Guerrero, Mexico:* Am. Inst. Min. Met. Eng. Tech. Pub. 842, 14 pp., 1937.
34. Zies, E. G., *The Valley of Ten Thousand Smokes:* Nat. Geog. Soc. Contr. Tech. Papers, vol. 1, no. 4, 79 pp., 1929.

27

Reprinted from pages 373-379, 381, 383, 385-386, and 404-408 of *Econ. Geology*
65:373-408 (1970)

Lateral and Vertical Alteration-Mineralization Zoning in Porphyry Ore Deposits

J. DAVID LOWELL AND JOHN M. GUILBERT

Abstract

The geologic history of the San Manuel–Kalamazoo deposit has provided an opportunity for the examination of vertical and horizontal zoning relationships in a porphyry copper system. Precambrian Oracle "granite," a Laramide monzonite porphyry, and a Laramide dacite porphyry are hosts to zones of potassic, phyllic, argillic, and propylitic assemblages shown to be coaxially arranged outward from a potassic core through phyllic, argillic, and propylitic zones. Alteration zones at depth comprise an outer chlorite–sericite–epidote–magnetite assemblage yielding to an inner zone of quartz–K-feldspar–sericite–chlorite. Mineralization zones are conformable to the alteration zones, the ore zone (with a 0.5% Cu cutoff) overlapping the potassic and phyllic zones. Occurrence of sulfides changes upward and outward from dissemination at the low-grade core of the deposit through microveinlet to veinlet and finally vein occurrence indicating the progressively increasing effect of structural control.

Several aspects of San Manuel–Kalamazoo geology suggest that it is exemplary of the porphyry copper deposit group. To test that idea and to evolve three-dimensional aspects of these deposits, a table of geologic characteristics of 27 major porphyry deposits is presented. Consideration of the table indicates that the "typical" porphyry copper deposit is emplaced in late Cretaceous sediments and metasediments and is associated with a Laramide (65 m.y.) quartz monzonite stock. Its host intrusive rock is elongate-irregular, 4,000 × 6,000 feet in outcrop, and is progressively differentiated from quartz diorite to quartz monzonite in composition. The host is more like a stock than a dike and is controlled by regional-scale faulting. The orebody is oval to pipelike, with dimensions of 3,500 × 6,000 feet and gradational boundaries. Seventy percent of the 140 million tons of ore occurs in the igneous host rocks, 30 percent in preore rocks. Metal values include 0.45% hypogene Cu with 0.35% supergene Cu, and 0.015% Mo. Alteration is zoned from potassic at the core (and earliest) outward through phyllic (quartz–sericite–pyrite), argillic (quartz–kaolin–montmorillonite), and propylitic (epidote–calcite–chlorite), the propylitic zone extending 2,500 feet beyond the copper ore zone. Over the same interval, sulfide species vary from chalcopyrite–molybdenite–pyrite through successive assemblages to an assemblage of galena–sphalerite with minor gold and silver values in solid solution, as metals, and as sulfosalts. Occurrence characteristics shift from disseminations through respective zones of microveinlets (crackle fillings), veinlets, veins, and finally to individual structures on the periphery which may contain high-grade mineralization. Breccia pipes with attendant crackle zones are common.

Expression of zoning is affected by exposure, structural and compositional homogeneity, and postore faulting or intrusive activity. Vertical dimensions can reach 10,000 feet, with the upper reaches of the porphyry environment perhaps only at subvolcanic depths of a few thousand feet. The vertical and lateral zoning described is repeated with sufficient constancy that depths of exposure at many deposits can be cited against the model of San Manuel–Kalamazoo.

Several lines of evidence suggest relatively shallow depths of formation and significant variations in water content in the porphyry environment. Shallow emplacement is consistent with the appearance of breccia pipes associated with ring and radial diking and with vertically telescoped zoning. Models of the source of altering-mineralizing fluids are considered.

Contents

Introduction

EXPLORATION of the Kalamazoo portion of the San Manuel–Kalamazoo district, Pinal County, Arizona, has presented an unparalleled opportunity for the study of a porphyry copper deposit in three dimensions. The coaxial symmetry of alteration and mineralization zones which was the basis of the exploration model has been verified in the exploratory drilling (Lowell, 1968) of the Kalamazoo portion of the district and in exploitation of the San Manuel portion. As exploration proceeded, it became increasingly apparent that many elements of mineralogy, occurrence, and geometry of other porphyry copper deposits were explicitly represented at San Manuel–Kalamazoo. Zoning patterns there can be considered a refined base for the study of mineralization and alteration relationships in other porphyry copper deposits, and this is the subject of the study reported here, with compilation of data from 27 major porphyry copper and molybdenum deposits in North and South America. Most significant is the emergence from the many descriptions of a more generally applicable unifying theme of large-scale alteration-mineralization zoning in these large deposits than has generally been recognized. Stringham (1953, p. 990) stated that "a review of hydrothermal studies of porphyry copper deposits shows as many dissimilarities as similarities to the hydrothermal features at Bingham Canyon." We now take the opposite position that there are many characteristics which link Bingham Canyon and many other deposits to the general porphyry copper deposit type. There appears to have been little published effort specifically to compare and contrast the porphyry deposits as a group.

The first portion of this paper describes both lateral and vertical alteration-mineralization relationships at San Manuel–Kalamazoo. The exploration model included and substantiated approximately 70 degrees of postmineralization tilting. Thus this geologic system provides information concerning both vertical and horizontal axes of a porphyry deposit. A three-dimensional synthesis is given of hydrothermal alteration mineralogy and assemblages, of the distribution and quantitative aspects of sulfides, and of the structural occurrence of sulfide and oxide minerals. Vertical treatment of alteration and mineralization geometry is still tentative, but some vertical zoning changes can be identified.

Comparison of other major porphyry base-metal deposits to San Manuel–Kalamazoo by means of published data assembled in Table 1 permits development of a generalized lateral and vertical zonation model for the deposit group. Finally, that model is used to examine the genesis and environment of formation of the porphyry deposits. The data suggest that it is sometimes possible to estimate the position of the present erosion surfaces of other porphyry deposits with respect to their original columns of mineralization. Depth parameters have been assigned to nine deposits, and it is hoped that both scientific and explorational use can be made of three-dimensional alteration-mineralization zoning.

The porphyry copper and molybdenum deposits, hereafter called "porphyries," must first be defined. A necessarily flexible definition emerges from consideration of many deposits and descriptions of a "typical" one.

A porphyry deposit is here defined as a copper and/or molybdenum sulfide deposit consisting of disseminated and stockwork veinlet sulfide mineralization emplaced in various host rocks that have been altered by hydrothermal solutions into roughly concentric zonal patterns. The deposit is generally large, on the scale of several thousands of feet, although smaller occurrences are recognized. The relatively homogeneous and commonly roughly equidimensional deposit is associated with a complex, passively emplaced stock of intermediate composition including porphyry units. It contains significant amounts of pyrite, chalcopyrite, molybdenite, quartz, and sericite associated with other alteration, gangue, and ore minerals and metals including minor lead, zinc, gold, and silver. Mineralization and alteration suggest a late magmatic-mesothermal temperature range. The deposit is generally associated with breccia pipes, usually with a large crackle brecciation zone, and is surrounded by peripheral mineral deposits suggestive of lower temperature mineralization.

The grade of primary mineralization in typical porphyry copper deposits ranges up to 0.8% Cu and 0.02% Mo, and porphyry deposits in which molybdenite is the chief economic mineral have grades ranging up to 0.6% Mo and 0.05% Cu. All porphyry copper deposits contain at least traces of molybdenite, and all porphyry molybdenum deposits contain some chalcopyrite. Many deposits contain recoverable quantities of both minerals, either in separate orebodies or in ore with approximately equal copper and molybdenum dollar values. Although typical porphyry copper deposits differ from typical molybdenum deposits in some respects, the existence of gradational characteristics in metallization suggests a common origin.

This definition is somewhat generalized because it must permit consideration of many deposits whose local geologic circumstances vary as expressed by their geometries and physical characteristics. We believe the porphyry deposits to be a petrological-mineralizational class, and individual porphyry deposits are best interpreted as greater or lesser departures from the unifying model of the above definition as elaborated upon below.

Genetic Models of Porphyry Deposits

Several genetic models have been proposed to relate the characteristics of porphyry copper and molybdenum deposits. All of the models recognize the important involvement of porphyritic intrusive rocks with ore deposition, and all are fundamentally magmatic-hydrothermal, differing in the sequences of events, depths of intrusion, the timing of derivation of fluids, and the source of fluids. The models considered here are the orthomagmatic model, Fournier's model of intrusion of a water undersaturated melt, and the White model of multilevel circulation of brines adjacent to a heat source.

The orthomagmatic model has been best described in the recent writings of Burnham (1967) and Nielsen (1968). It is the genetic model tacitly adopted in most deposit descriptions, as for example, those described in Titley and Hicks (1966). It sometimes involves penetration of the source to levels as shallow at 1,500 feet (Nielsen, 1968), but more commonly to depths apparently on the order of 3,000–5,000 feet. The model depends on a melt derived at some greater depth, probably near the mantle-crust boundary, which becomes saturated with water as it approaches the upper surface. Release of that water may occur when internal vapor pressure developed by supersaturation exceeds the lithostatic load pressure or when the intrusive system is rent by external stresses. Crystallization then proceeds presumably along the lines of Emmons' (1933) cupola or R. H. Sales's sub-hood cupola development.

As described by Nielsen (1968), the sequence of events can be paraphrased as intrusion, early marginal crystallization which produces a solid shell, and rupture of that shell to produce porphyritic-aphanitic textures in subsequently crystallized rocks. Volatiles released by the quenching migrate outward through crackle, stockwork, and brecciated zones in the cooler margins where, augmented by diffusion effects, alteration and mineralization occur in response to gradients "from near magmatic temperatures at the center of the stock to relatively cool temperatures in the wall rocks" (p. 37). Silicate sulfide reactions of the type described by Hemley and Jones (1964) prevail. Other authors would not necessarily limit the separation of volatiles to the period of quenching, but rather would consider evolution of the hydrothermal fraction a quasi-continuous separation of volatiles in response to the many variables related to temperature and pressure. The loss of volatiles from near-surface portions of a melt may permit the upward and outward replenishment of mineralizers from greater depths.

Fournier (1968) suggests that the initial deep porphyry copper melt was unsaturated with water at one to three percent, that it was intruded to depths of less than about 4,500 feet, and that rupture by faulting would cause sudden, even explosive loss of water and supercooling of the silicate melt. Crystallization would then abruptly halt the upward progress of the now dry melt. Subsequent "extensive argillic alteration shown by most porphyry copper deposits is probably due to a superimposed circulating hot-spring system, fed mainly by meteoric and connate water" (p. 101).

White (1968) in a particularly stimulating paper suggests that circulation of sulfur-deficient Na–Ca–Cl brines, with salt contents generally equivalent to 5% to 40% NaCl, are responsible for many base-metal deposits. Such brines may be produced in porphyry systems by deuteric reaction of residual liquids with earlier formed plagioclase and ferromagnesian minerals to achieve high contents of calcium and base metals. Although White in his paper does not develop a specific space-time model for the porphyry deposits, he implicitly develops a model involving multilevel circulation of deuterically metal-enriched or connate-meteoric sulfur-deficient metallizing solutions under the influence of thermal gradients established by an adjacent or subjacent magmatic heat source. The model differs importantly from the orthomagmatic model in that the source of the solutions, and perhaps the metals, is almost completely external to the magmatic system, with convective overturn of circulating solutions producing alteration-mineralization envelopes and zones.

Geology of the San Manuel-Kalamazoo Deposit

The San Manuel–Kalamazoo deposit (Lowell, 1968), located in Pinal County, Arizona, is here accepted as the type porphyry copper deposit, and its geology and other characteristics are presented for comparison and contrast with others (Table 1).

Precambrian quartz monzonite of the Oracle Granite batholith in the San Manuel area was intruded in Laramide time by swarms of monzonite porphyry dikes and irregular masses of monzonite porphyry, more properly termed biotite latite porphyries, although long-established "monzonite porphyry" terminology will be followed here. Closely related in time and space to the activity was a porphyry copper mineralization event that produced the San Manuel–Kalamazoo orebody and its associated concentric alteration zones. The hydrothermal system appears to have been centered in the middle of the monzonite porphyry dike swarm, and metallization is almost equally distributed between the monzonite porphyry and the Oracle Granite host rocks (Fig. 1).

Following hydrothermal mineralization and alteration (Fig. 1a), the whole district was tilted to the northeast, and the block including the San Manuel–Kalamazoo orebody was probably relatively elevated. Erosion of this block exposed the top of the orebody, and supergene activity formed a thin chalcocite enrichment blanket. At this time, the long axis of the orebody may have plunged at about 65°SW. Shortly thereafter, terrestrial sediments began to cover the deposit.

Further tilting, perhaps 15°, followed deposition of the lowermost Cloudburst Conglomerate. An erosion surface formed on the Cloudburst sediments was later covered by the Gila Conglomerate. A third-stage tilt of about 30° gave the Gila Conglomerate its present inclination and brought the originally vertical axis of the San Manuel–Kalamazoo orebody into a 20° southwest-plunging attitude. The San Manuel fault then diagonally offset the original, nearly cylindrical orebody into two roughly equal-sized pieces, the San Manuel and the Kalamazoo portions. The upper Kalamazoo portion moved about 8,000 feet in a down-dip, S55°W direction.

Small, high-angle, northwest-trending normal faults later displaced both halves of the original orebody, and erosion stripped most of the Gila Conglomerate from the east end of the present San Manuel orebody (Fig. 1b).

The original, unfaulted orebody, as defined by a 0.5% copper limit, formed a slightly flattened or elliptical cylinder which was at least 7,700 feet long and from 2,500 to 5,000 feet in diameter. The top of the cylinder, at the east end after tilting, may have been rounded, with the bottom, at the west, having an irregular shape. The center of the orebody is poorly metallized, so that ore actually forms a hollow cylinder or cylindrical shell. The shell surrounding the low-grade center varies from about 100 to 1,000 feet in thickness. Mineralization and alteration zones are approximately coaxial.

The alteration assemblages in the San Manuel–Kalamazoo deposit form regular, smoothly bounded zones, which, as in most porphyries, are locally gradational and difficult to place within a hundred feet, although they are well defined on a broad scale. The boundaries are more clearly defined than they are in most porphyry deposits, presumably because the mineralizing fluids affected intrusive, essentially homogeneous, isotropic plutonic and hypabyssal host rocks of intermediate composition. These rocks responded to the indicated alkali chemistry without important gains or losses. No marginal sediments, compositionally contrasting intrusive rocks, planar rock fabrics, or prominent tectonic elements produced steep physical or chemical gradients to influence the uniform zoning and symmetry.

Mineralogic zoning at Kalamazoo and elsewhere suggests that at least four alteration assemblages are easily discernible in the porphyry copper and molybdenum deposits. The terms potassic, phyllic, argillic, and propylitic have been adapted or adopted from the literature (Burnham, 1962; Creasey, 1966; Meyer and Hemley, 1968) to describe the four principal assemblages. The terms "argillic" and "propylitic" are well known and widely accepted, broadly describing quartz-kaolin-montmorillonite-chlorite-biotite and chlorite-calcite-epidote-adularia-albite alteration assemblages, respectively. "Phyllic" is here applied to the assemblage quartz-sericite-pyrite with less than 5% kaolin, biotite, or K-feldspar, and "potassic" is suggested (Guilbert and Lowell, 1968) to include introduced or recrystallized K-feldspar and biotite, with minor sericite and highly variable but persistent and generally minor amounts of anhydrite. Each of these assemblages will be more fully described below, especially as they occur at San Manuel–Kalamazoo. Other assemblages

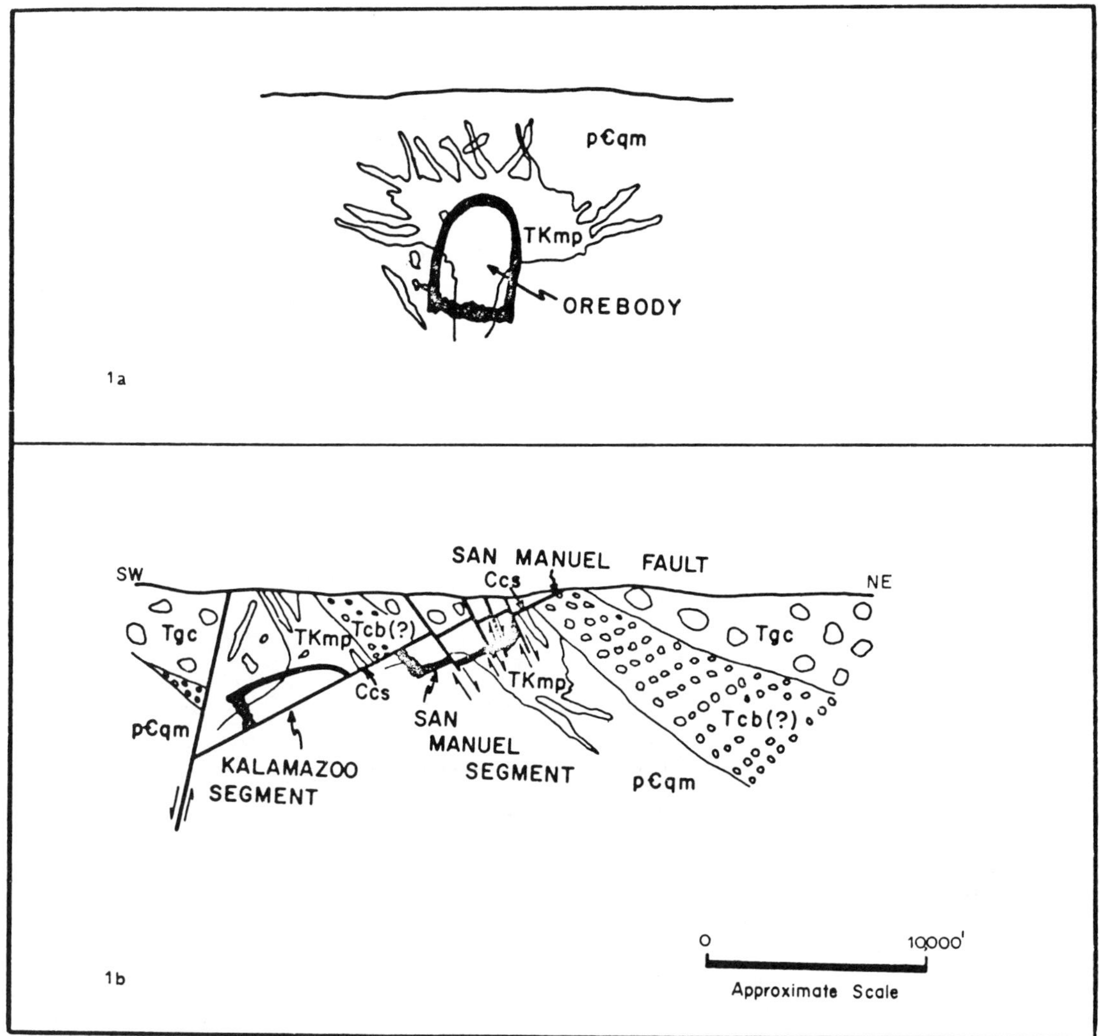

FIG. 1. Schematic drawing of structural history of San Manuel-Kalamazoo deposit. (a) at time of emplacement and (b) at present. Note the umbrella-like flare of dike swarm and the chalcocite enrichment zone (CCa). p€qm = Oracle Granite, TKmp = monzonite porphyry, Tcb = Cloudburst Formation, Tgc = Gila Conglomerate.

rarely encountered in the porphyry environment are the advanced argillic (Meyer and Hemley, 1968) and pegmatoid, respectively involving quartz and pyrophyllite, with traces of dickite or kaolinite, topaz, and zunyite, and quartz–coarse sericite–K-feldspar, with or without carbonate, anhydrite, and apatite.

Hydrothermal alteration assemblages in the San Manuel–Kalamazoo deposit are summarized in Figure 2, which shows alteration changes mineral by mineral and assemblages on AKF–ACF diagrams. Supergene activity is limited to a 200-foot thick zone near the top of the deposit.

The alteration zones were separated during Kalamazoo exploration as follows. The inner limit of the propylitic zone was placed where the total quartz–montmorillonite, quartz–kaolin, or quartz–sericite content in plagioclase sites exceeds the total of chlorite and epidote replacing mafic minerals; here the color usually changes from green to light gray. The argillic zone, in which kaolin or montmorillonite predominates in plagioclase sites and chlorite replaces biotite, was not generally mapped separately and is least significant quantitatively. The inner limit of propylitic alteration is locally the outer

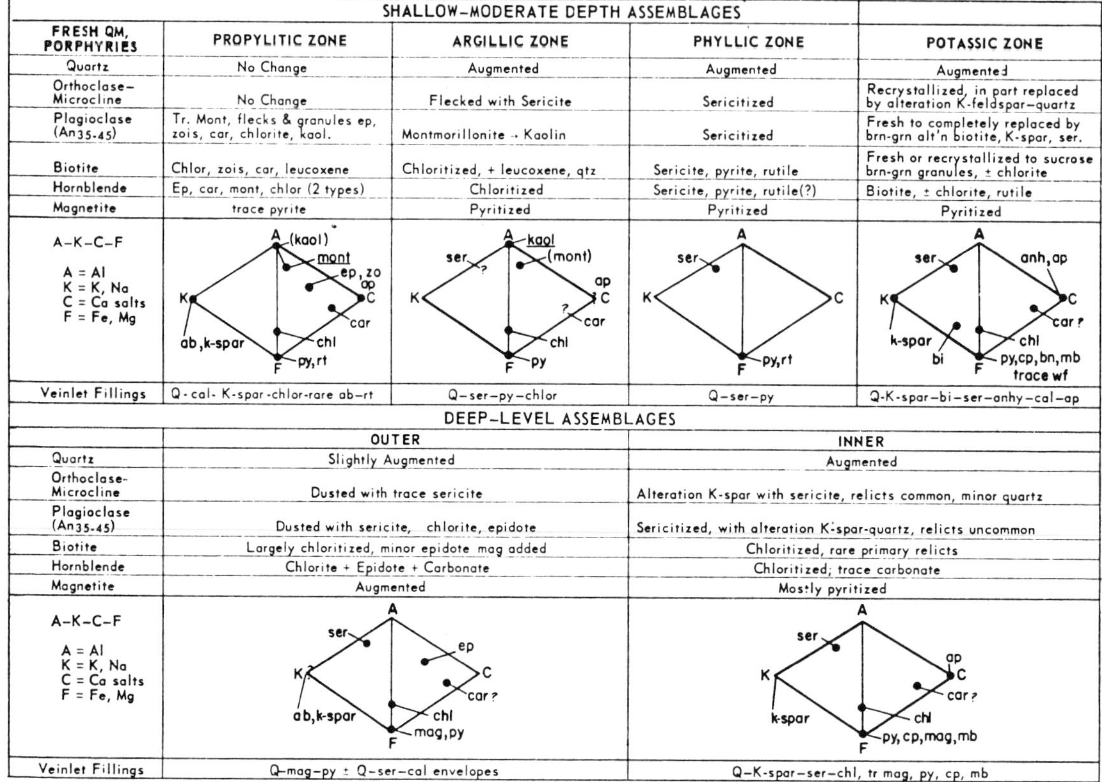

	SHALLOW–MODERATE DEPTH ASSEMBLAGES			
FRESH QM, PORPHYRIES	PROPYLITIC ZONE	ARGILLIC ZONE	PHYLLIC ZONE	POTASSIC ZONE
Quartz	No Change	Augmented	Augmented	Augmented
Orthoclase–Microcline	No Change	Flecked with Sericite	Sericitized	Recrystallized, in part replaced by alteration K-feldspar–quartz
Plagioclase (An35-45)	Tr. Mont, flecks & granules ep, zois, car, chlorite, kaol.	Montmorillonite → Kaolin	Sericitized	Fresh to completely replaced by brn-grn alt'n biotite, K-spar, ser.
Biotite	Chlor, zois, car, leucoxene	Chloritized, + leucoxene, qtz	Sericite, pyrite, rutile	Fresh or recrystallized to sucrose brn-grn granules, ± chlorite
Hornblende	Ep, car, mont, chlor (2 types)	Chloritized	Sericite, pyrite, rutile(?)	Biotite, ± chlorite, rutile
Magnetite	trace pyrite	Pyritized	Pyritized	Pyritized
A–K–C–F A = Al K = K, Na C = Ca salts F = Fe, Mg	A (kaol), mont, ep, zo, ap, C, car, K, ab, k-spar, chl, F, py, rt	A, kaol, (mont), ser, ?, ap, C, ?, car, K, chl, F, py	A, ser, K, C, F, py, rt	A, ser, anh, ap, K, C, car?, k-spar, bi, chl, F, py, cp, bn, mb, trace wf
Veinlet Fillings	Q-cal-K-spar-chlor-rare ab–rt	Q–ser–py–chlor	Q–ser–py	Q-K-spar–bi–ser–anhy–cal–ap

	DEEP–LEVEL ASSEMBLAGES	
	OUTER	INNER
Quartz	Slightly Augmented	Augmented
Orthoclase-Microcline	Dusted with trace sericite	Alteration K-spar with sericite, relicts common, minor quartz
Plagioclase (An35-45)	Dusted with sericite, chlorite, epidote	Sericitized, with alteration K-spar-quartz, relicts uncommon
Biotite	Largely chloritized, minor epidote mag added	Chloritized, rare primary relicts
Hornblende	Chlorite + Epidote + Carbonate	Chloritized; trace carbonate
Magnetite	Augmented	Mostly pyritized
A–K–C–F A = Al K = K, Na C = Ca salts F = Fe, Mg	A, ser, ep, K, C, car?, ab, k-spar, chl, F, mag, py	A, ser, ap, K, C, car?, k-spar, chl, F, py, cp, mag, mb
Veinlet Fillings	Q–mag–py ± Q–ser–cal envelopes	Q–K-spar–ser–chl, tr mag, py, cp, mb

FIG. 2. Summary of hydrothermal alteration assemblages at San Manuel–Kalamazoo.

limit of either the argillic or the phyllic zone of pervasive conversion to quartz, sericite, and pyrite. The inner limit of the phyllic zone is the outer limit of the first continuous section of secondary K-feldspar and secondary biotite, even though the total quartz and sericite content here ordinarily exceeds the total K-feldspar plus biotite content. The zoning patterns and intercepts can be projected remarkably well from hole to hole. Subsequent petrographic study has contributed to these descriptions of the zones, and subsequent publications by J. M. Guilbert describing the chemical and structural mineralogy and physical geochemistry of the alteration-mineralization processes are planned.

In the following sections, the fresh rocks at San Manuel–Kalamazoo are first discussed and alteration zones exposed on a horizontal plane at moderate depth are described successively outward from the center. Alteration and mineralization changes with depth are discussed last and are summarized schematically in Figure 3a.

Fresh Rocks

The unaltered rocks at San Manuel–Kalamazoo include Precambrian Oracle porphyritic quartz monzonite and two varieties of much younger biotite porphyries. The Oracle "granite" is coarse grained (Fig. 4) with anhedral subrounded quartz units about a centimeter across and commonly tangential to their nearest neighbors, rectangular to irregular plagioclase tablets (An_{35-45}), and interstitial quartz and K-feldspar. K-feldspar species include microcline, orthoclase, and microperthite. Several authors, especially Banerjee (1959) have considered the rock palingenic, although many other workers accept its orthomagmatic origin. Accessory minerals include biotite and hornblende, with trace amounts of zircon, apatite, sphene, magnetite, and very sparse monazite.

The porphyries are of at least two types. One (here called Type A) is a quartz monzonite porphyry distinguished by its zoned and twinned oligoclase-andesine phenocrysts which average about 5 mm and range up to 15 mm across (Fig. 5), its quartz–K-feldspar groundmass commonly containing

[*Editor's Note:* Figures 4 through 12 have been omitted.]

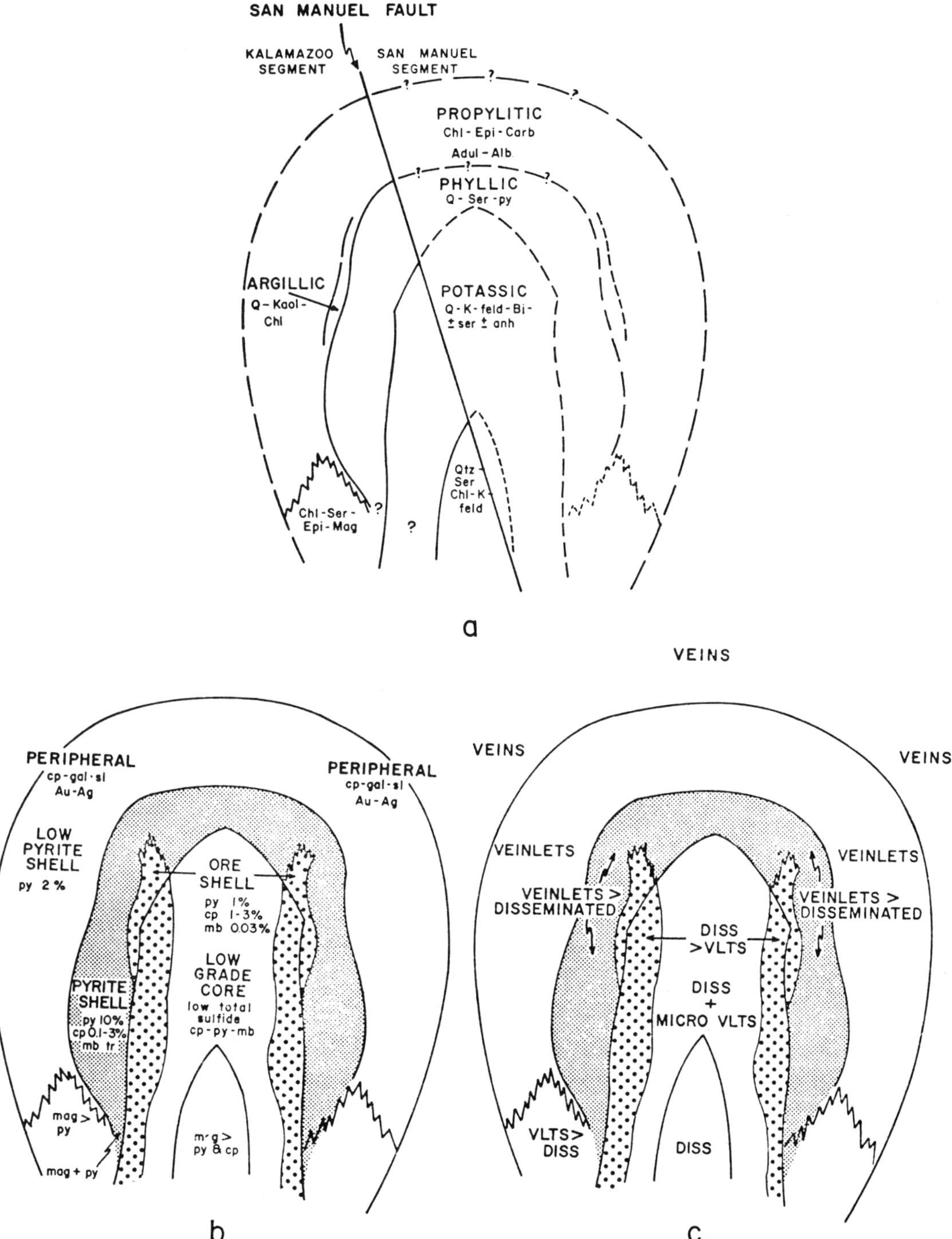

FIG. 3. Concentric alteration-mineralization zones at San Manuel–Kalamazoo. (a) schematic drawing of alteration zones. Broken lines on Kalamazoo side indicate uncertain continuity or location and on San Manuel side extrapolation from Kalamazoo. (b) schematic drawing of mineralization zones. (c) schematic drawing of the occurrence of sulfides.

fine-grained, embayed quartz "eyes" with stippled overgrowth rims. Few quartz eyes exceed 1 mm in diameter. Accessory biotite, hornblende, apatite, rutile, zircon, and minor magnetite are generally euhedral, the first two reaching 5 mm in length. No K-feldspar phenocrysts were observed. The groundmass is that of the widespread quartz latite porphyry and quartz monzonite porphyry of the porphyry deposits. Its grain size averages 0.1 mm and its texture is granular sucrose. Though locally variable, it averages 55 percent quartz and 45 percent K-feldspar, so that the overall whole rock feldspar composition averages about 35 percent plagioclase and 25 percent K-feldspar. The K-feldspars are anhedral, granular, and mutually intergrown with quartz; granular, often euhedral apatite and rutile and shreds of mafic minerals are sparse.

A second porphyry (here called Type B) is a biotite dacite. Plagioclase phenocrysts in Type B are generally roughly square to rectangular or even trapezoidal in cross section (Fig. 6) rather than compound and zoned as in Type A. Rarely do they exceed 5 mm on a side. Biotite phenocrysts up to 3–5 mm are prominent. Quartz phenocrysts are absent, and the biotite-to-amphibole ratio is slightly greater than that of Type A. The groundmass is composed of intergrown microcrystals of sparsely twinned plagioclase with quartz, apparently slightly later, and sparse K-feldspar. Rutile and apatite accessory minerals are rare.

It is difficult to estimate from drill core the relative abundances of the two varieties of porphyry. Type A predominates along the core of the San Manuel–Kalamazoo system. Porphyry units form an umbrella or mushroom-shaped outward expansion of diking at higher levels (Fig. 1a). Although porphyry-quartz monzonite contacts are predominantly sharp, they may in some cases appear gradational in diamond drill core, and the porphyry "dikes" must be highly sinuous and variable in attitude, especially at greater depths. Indeed, an approach to wholesale mobilization of porphyry concurrent with the potassic alteration is suggested by coarsely vermicular and diffuse contacts between quartz monzonite and porphyry seen in drill core from deep within the orebody.

Alteration Zones

Alteration zone boundaries are not affected by rock type interfaces, at least at the scale of study to date. Systematic comparisons of fresh and altered rocks on either side of a particular contact have not yet been made, but the various starting material compositions, structural characteristics, and fabrics seem to have responded nearly identically to alteration processes.

Potassic Zone.—Several authors, especially Hemley and Jones (1964), Creasey (1966), and Meyer and Hemley (1968), have discussed the potassic alteration environment. Hemley and Jones have delimited an environmental interface between K-feldspar and sericite stabilities, the latter with higher HCl/KCl ratios at a given temperature, an environment consistent with late magmatic or early hydrothermal conditions in the K-feldspar–sericite–kaolin (pyrophyllite) system. Inclusion of iron and magnesium should bring biotite or chlorite into consideration with K-feldspar, sericite, and quartz, an assemblage increasingly noted in porphyry copper deposits (Creasey, 1966) and assignable to a late magmatic-early hydrothermal "deuteric" environment. Such a biotite–K-feldspar alteration assemblage with quartz, sericite, anhydrite, pyrite, chalcopyrite, molybdenite, and traces of bornite generally constitutes the low-grade center and part of the ore shell of the Kalamazoo deposit (Figs. 7, 8, 9).

This innermost alteration zone (Fig. 3a) involves pervasive and veinlet replacement of primary minerals by secondary biotite, K-feldspar, quartz, sericite, and to a lesser extent anhydrite (Fig. 9). K-feldspar occurs with quartz as microveinlet fillings that heal minute stockwork-like fractures in the primary rocks and also replace original feldspars to varying degrees. "Rock" orthoclase is flesh colored when fresh, turning slightly orange where extensively replaced by alteration K-feldspar. Typically, quartz heals quartz grains, and K-feldspar heals orthoclase, with K-feldspar also commonly replacing andesine plagioclase extensively, either by rimming or by advance along twin planes. K-feldspar also locally replaces plagioclase in the porphyry groundmass. No albitization has been found, although preliminary examination of alteration K-feldspar indicates it to be more sodic than the primary orthoclase.

Alteration biotite occurs in four important modes: (1) as hairline veinlet fillings along with chalcopyrite, alteration silicates, and anhydrite; (2) as sparse to massive replacement of plagioclase phenocrysts; (3) as bright black euhedral units megascopically nearly identical to primary rock biotite; and (4) as locally pervasive replacements of groundmass feldspars (Fig. 8). Alteration biotite is recognizable both by its fine-grained, sucrose, subhedral to euhedral form and by the coexistence of two distinctive color variants, one a light tan to brown which mostly predominates, the other a light apple green. Shagreen is not present, and birefringence is slightly lower than that of the rock biotite. Chlorite intergrown with biotite is common.

The altered rocks, especially the porphyries, are distinctively pigmented by groundmass biotitization. Porphyries megascopically showing the smoky gray color normally found with advanced potassic alteration generally also carry significant base metal values. Such rocks also show K-feldspar-rich veinlets up to ¼ inch wide (Fig. 9).

The potassic alteration assemblage generally involves sparse to trace amounts of anhydrite, carbonates, and apatite. Rutile and wolframite have been observed in several veinlets. Anhydrite, not previously reported as a widespread alteration mineral, commonly forms granules in the quartz–K-feldspar-rich gash veinlets and in microveinlets which cut individual rock feldspar grains. Unlike biotite, it is not generally a replacement mineral. It is widespread but rarely abundant. Carbonates occur both in veinlets and as bits and shreds dispersed through the entire rock. Apatite, though not yet well studied, occurs both as a veinlet mineral and as minor but pervasively distributed anhedral units.

Phyllic Zone.—Surrounding and to some extent overlapping the biotite–K-feldspar zone is a zone in which alteration minerals include quartz, sericite, pyrite, hydromica, minor chlorite, and traces of rutile. This zone (Figs. 3a and 3b) generally includes part of the ore zone and all of the marginally mineralized and pyritic zones and is nearly coextensive with strong pyrite mineralization. Sericite predominates in the inner part of this zone, clay minerals and hydromica in the outer margins. The most distinctive assemblage, both megascopically and petrographically (Fig. 10), is that of complete sericitization of all silicates except quartz. Original rock plagioclase and orthoclase are both pervasively replaced by a felted mat of fine-grained muscovite with abundant ultrafine granular quartz. Vestiges of cleavage, zoning, and twin planes of plagioclase are retained in most instances in preferred orientations of sericite flecks. Original biotite sites can be identified by relatively well-oriented alteration sericite flecks, by less abundant alteration quartz, and by either anhedral or sagenitic rutile or leucoxene, presumably representing titanium from the original biotite. Primary quartz is unaffected but generally overgrown.

K-feldspar is totally sericitized in the innermost phyllic zone, but shreds and scraps of K-feldspar persist in the outer part. Pyrite is abundant; chalcopyrite is variable, generally occurring as disseminated grains, commonly in sericitized sites. Pyrite forms veinlets and generally granular disseminations in the pervasively phyllic-altered material. Pyrite content ranges from 2–30 percent by weight, averaging 5–10 percent. Apatite and rutile again appear to have been recrystallized and redistributed. Silicification well beyond that expected from the breakdown of feldspars to sericite plus quartz plus alkali ion appears common. Neither carbonates nor anhydrite were identified in the phyllic zone assemblage.

The phyllic assemblage at San Manuel–Kalamazoo closely resembles the quartz-sericite-pyrite alteration at Butte (Sales and Meyer, 1951), at Morenci (Moolick and Durek, 1966), and at many other southwestern North American porphyry deposits.

Contacts of the phyllic zone with the potassic zone have been described above; they are generally gradational over a hundred feet or so. Contacts of the phyllic zone with the next outer most, argillic zone are less definite.

Argillic Zone.—The argillic zone at San Manuel–Kalamazoo is least well understood at this stage, both mineralogically and distributionally. It is the least well developed and is the most likely to be absent in any given penetration of the ore deposit symmetry. It is characterized by the conversion of plagioclase to either kaolin nearer the orebody or montmorillonite farther away from the orebody center (Fig. 11). Kaolin is the more common reaction product, grading outward to sparse outlying montmorillonite. Pyrite is common but much less abundant than in the phyllic zone. It is generally distinctly veinlet controlled rather than disseminated. Primary biotite may be essentially unaffected, persisting as shiny black megascopic flecks in a white, earthy rock, or it may be in part converted along cleavage to chlorite. The compositional characteristics of this chlorite have not yet been compared with those of the chlorite of the potassic and deep zones. K-feldspar shows minor flecking with sericite and dusting with kaolin, but it is generally not extensively affected.

Propylitic Zone.—This zone contains the most widely distributed and least distinctive of the alteration assemblages. Plagioclase generally remains fresh (Fig. 12), although it is locally ribbed with either montmorillonite, kaolin, or an apparent mixture of the two minerals. Amorphous mineraloid clouding the plagioclases was not conclusively identified but is suspected in small amounts. Biotite is replaced along cleavage by both chlorite and carbonate, which generally decrease in abundance outwardly. Epidote and calcite are common as fine granules in plagioclase and as coarser aggregates with montmorillonite in amphibole sites. Both albite and veinlet K-feldspar with minor carbonate, quartz, and epidote are rare. Rock quartz is unaffected. Chalcopyrite is rare, but pyrite constitutes one to three percent by volume of the rock. The propylitic assemblage grades into argillic or phyllic phases at the inner side over an interval of from 10 to 100 feet and is presumed to fade over perhaps thousands of feet in the outer reaches, although this has not been proved.

Deep Zones.—The deep zones at San Manuel–Kalamazoo cannot be described with certainty. Our findings are based on only a few drill intercepts and may be modified in detail by further work and better exposures. The gross relationships are shown schematically in Figure 2. A slight uncertainty, especially with respect to chlorite–biotite relationships, is introduced by the nearness of the post-Laramide San Manuel fault and its possible effects.

As shown in Figure 2, the propylitic assemblage which rims the deposit at moderate depths grades downward from propylitized Oracle quartz monzonite into a zone in which both rock feldspars are dusted with sericite. Biotite is largely chloritized, and chlorite and epidote replace amphibole. Quartz–magnetite–minor pyrite veinlets up to $\frac{1}{2}$ inch wide are common and generally have narrow quartz–sericite–chlorite selvages. The rocks are greenish and free of alteration K-feldspar and biotite.

The phyllic zone is widest, possibly with some repetition by steep faulting just below the midpoint of the orebody on the Kalamazoo side, but is virtually absent from the deepest levels (Fig. 2). Moving laterally toward the center of the deposit at depth (Fig. 3a), sericite content in altered plagioclase sites increases; magnetite content of the zone in veinlets and as disseminations diminishes but does not disappear. Narrow veinlets of chalcopyrite and pyrite occur which lack anhydrite but have selvages of nearly normal potassic alteration but without biotite. Neither typical argillic nor phyllic assemblages are discernible. The deepest penetration into the core zone shows an assemblage in which K-feldspar and sericite dust both primary plagioclase and orthoclase; in which veinlets of quartz–K-feldspar are flanked and intergrown with selvages of sericite after biotite and plagioclase; and in which magnetite, chalcopyrite, pyrite, and trace molybdenite occur as disseminations and microveinlets. This deep-level aspect of the symmetry and character of the Kalamazoo assemblage resembles that at Butte where alteration envelopes flanking Main Stage veins decrease in width at deepest levels with increasingly common quartz–K-feldspar–sericite assemblages and with chlorite replacing biotite (Meyer et al., 1969). No real argillization is present in this deep zone at San Manuel–Kalamazoo.

Mineralization Zones

Concentric mineralization zones are coaxial with the alteration zones as shown in Figures 3a and 3b. A plane normal to the axis of the deposit at a moderate depth shows the following zones of mineralization.

Potassic Zone.—An inner zone entirely within the potassic alteration zone averages about 2,600 feet in diameter and contains about 0.3% Cu almost totally as chalcopyrite. Total sulfide content is low and pyrite to chalcopyrite ratio is about 1:2; magnetite is rare or absent. Most sulfides are disseminated grains. Surrounding this zone (Fig. 3b) is the ore shell as defined by a 0.5% Cu cutoff lying in the potassic zone but also overlapping into the phyllic zone. This ore shell averages about 600 feet in thickness and ranges from 0.5%–1.0% Cu in grade with a pyrite-to-chalcopyrite ratio of 1:1. Pyrite generally forms stockwork veinlets; chalcopyrite occurs in disseminated grains.

Phyllic and Argillic Zones.—There are three rather distinct types of "ore" mineralization in the phyllic alteration zone. The outer portion of the ore shell, as just mentioned, lies in the phyllic zone. Surrounding the ore shell and entirely within the phyllic zone is a zone about 200 feet thick in which copper mineralization ranges from 0.1%–0.5% Cu, with a pyrite-to-chalcopyrite ratio of about 10:1. Most of both the pyrite and chalcopyrite forms veinlets. Surrounding this zone of marginal mineralization but still entirely within the phyllic and argillic zones is a zone of pyrite mineralization which ranges from 1,000–1,500 feet in width and contains 6%–25% pyrite by weight. Pyrite occurs with quartz in veinlets ranging up to $\frac{1}{2}$ inch thick.

Propylitic Zone.—Mineralization in the propylitic zone consists of a few small, high-grade silver, gold, chalcopyrite veins, and pervasive pyrite in veinlets which constitutes 2%–6% by weight of the rock. Since the outer edge of the propylitic zone does not crop out it is uncertain whether disseminated pyrite is coextensive with propylitic alteration. The area of pervasive pyrite veinlets contains 100–500 ppm copper which is apparently included in the pyrite since discrete primary copper minerals have not been found in this material.

Vertical Changes in Mineralization.—Total sulfide content and copper content in the low-grade portion of the phyllic zone decrease with depth. The character of the mineralization appears also to change with depth from finer grained disseminated grains to coarser grained blebs. In the ore shell, there is remarkably little change in copper grade with depth, but the chalcopyrite again changes downward to a predominant bleb-type disseminated occurrence. As shown on Figure 3b, a progressively greater portion of the ore shell occurs in the potassic alteration zone as depth increases. Little change with depth is noted in the marginal zone except that magnetite substitutes for much of the pyrite near the bottom of the orebody. Similarly, magnetite substitutes for most of the pyrite in the zone of peripheral pyrite mineralization near the bottom of the orebody. These relationships are also shown in Figure 3a.

Molybdenite and Bornite Distribution.—Molybdenum shows a tendency to be concentrated in the middle two thirds of the ore shell (Fig. 3b) with lower grade zones at the upper and lower portions of the ore shell. Within the 0.5% Cu zone, molybdenum grade in individual drill intercepts tends to increase with copper grade and with thickness of the ore shell. Higher grade molybdenum occurs in both the potassic and phyllic zones and does not seem to be controlled by lateral alteration zoning, but from the standpoint of vertical zoning, molybdenum content drops off at about the same level that deep alteration assemblages become important (Fig. 3a).

Although only a dozen or so bornite identifications were recorded in drill hole logging, bornite also tends to occur in a short vertical column less than half the total length of the column of copper mineralization and nearly centered between top and bottom of the ore-grade copper interval. Most bornite is found with potassic alteration but it occasionally occurs also in phyllic and propylitic zones.

[*Editor's Note:* Material has been omitted at this point.]

Conclusions

The foregoing summary forcefully demonstrates that the porphyry copper-molybdenum deposits display important unifying geologic characteristics including various lateral and vertical zones. The fact of zoning is not new, but several important aspects, such as sulfide species, detailed alteration assemblages, and the characteristic occurrences of the sulfides, is far more widespread than has previously been realized. Indeed, a "typical" porphyry copper deposit can be hypothesized from Table 1 and is included along the bottom of the table.

It is especially noteworthy that many, and perhaps most, porphyry deposits have coaxially cylindrical alteration zones. Factors that limit the development of discernible symmetry in porphyry deposits include the following:

1. Regional or local structural fabric that may produce asymmetry in alteration and mineral ore zones.
2. Heterogeneous and contrasting composition of preore rocks, especially the presence of sedimentary "screens."
3. Dislocations of the original geometry by fault displacement or by postore intrusions.
4. Exposure of the porphyry system laterally and at depth.

The vertical dimension interpreted for the San Manuel–Kalamazoo system is on the order of 8,000–10,000 feet. No definite evidence suggests that this vertical dimension is either typical or normal, but the mineral assemblages typical of different vertical zones in San Manuel–Kalamazoo appear to be useful in estimating the depth of formation of several deposits. These "depth levels" of present exposure surfaces for several porphyry copper deposits are shown in Figure 13. Morenci is placed high in the hypothetical vertical section because of the wide exposure of the phyllic zone without exposure of potassic assemblages. Several aspects of Morenci geology—breccia zones, the broad-scale alteration symmetry, and the occurrence and distribution of sulfides—suggest that potassic alteration will be encountered at depth under the existing open pit.

It is also noteworthy here that phyllic zone alteration assemblages, with their high pyrite content and their profusion of veinlets and microveinlets, are chiefly responsible for the extensive development of supergene oxidation, leaching, and enrichment of southwestern North American deposits. This high level of exposure appears to be the most common, especially in supergene-enriched deposits.

Recent publications on the Chino deposit at Santa Rita, New Mexico, show that an island of low-grade material is being left in the center of the northern portion of the pit area. This island of low grade is symmetrically and centrally disposed with respect to secondary K-feldspar, chalcopyrite, and pyrite distribution as reported by Nielson (1968, Figs. 6, 7, and 9). This "low grade island" may represent the cropping out of a low-grade barren zone analogous to the central core at San Manuel–Kalamazoo. Lastly, Gilluly's (1946) description of the Ajo deposit involves much the same K-feldspar–biotite–chlorite–sericite and magnetite–chalcopyrite assemblages and zonal characteristics as those encountered at depth in San Manuel–Kalamazoo. It appears possible, therefore, to assign a third dimension to at least several deposits, and many others may be assigned depth parameters as further information develops. For example, brecciation and ring diking may have significance in regard to depth of formation.

It also appears significant that the major porphyry deposits of British Columbia (for example, the Bethlehem and Lornex deposits) occur in quartz diorite, and the K-feldspathic rock types reported at Ajo yield outward to a quartz diorite composition (Wadsworth, 1968). The evidence concerning large-scale metasomatism of rocks, generally with attendant enrichment in K-feldspar and quartz as described at Bingham Canyon by Stringham (1956), may well prove to be more general than is now realized. The alteration assemblages, mineralization characteristics, and occurrence of sulfides at Bethlehem and Lornex are consistent with deep exposure, and we may see now exposed a relatively deep-seated porphyry environment. The fact that these deposits also involve quartz diorites rather than granodiorite or quartz monzonites may be another manifestation of the vertical dimension in porphyry deposit genesis.

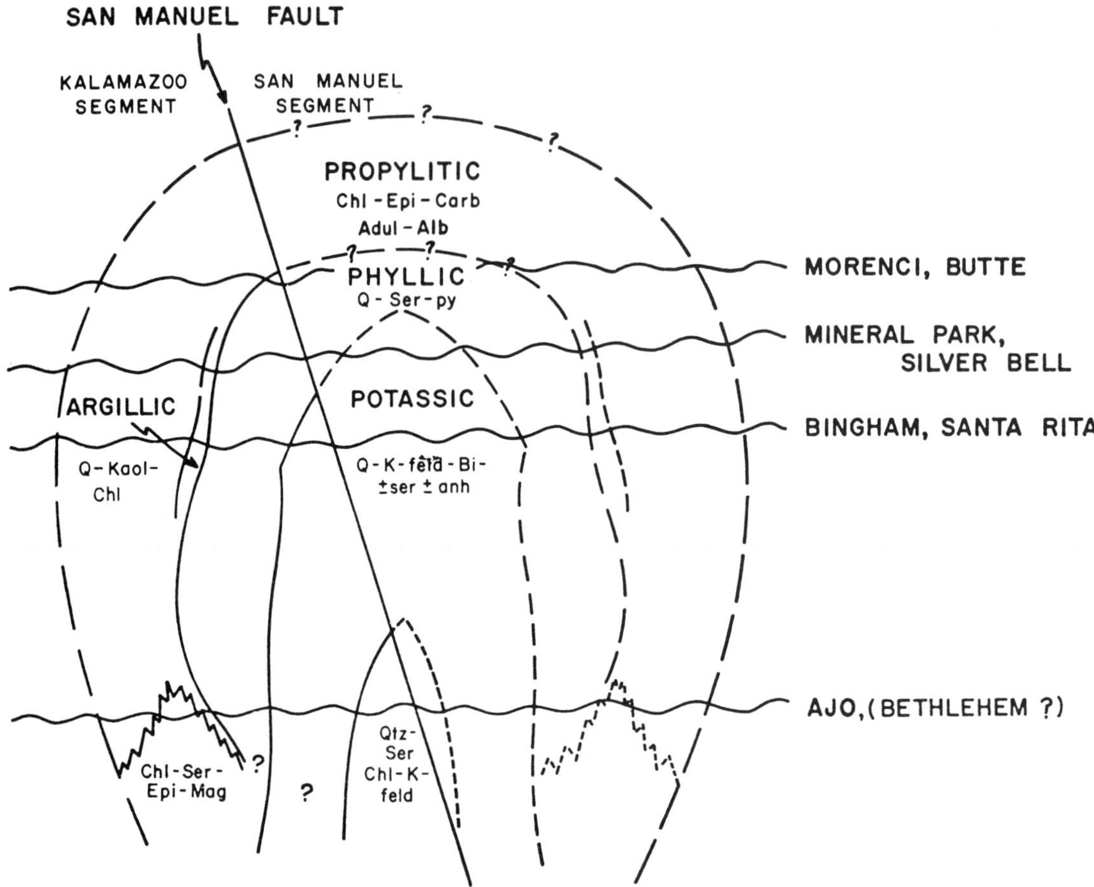

FIG. 13. Schematic drawing of San Manuel-Kalamazoo showing exposure levels of several porphyry copper deposits. Other deposits could be added, but these few serve to show a vertically developed dimension.

A growing body of data indicates that the porphyry deposit minerals may form at depths as shallow as 5,000–10,000 feet. Facts supporting this conclusion are (1) the occurrence of porphyry rocks in all 27 deposits of Table 1; (2) the cutting of all deposits by postore erosion surfaces; (3) the widespread occurrence of brecciation (even though the host intrusions are usually passively emplaced); (4) the location of 14 deposits in Cretaceous or younger preore rocks while the intrusions themselves are of late Cretaceous or younger age; (5) regional structural-stratigraphic considerations; and (6) the common occurrence of porphyry-ore-forming environments in cupola-like stocks less than one square mile in area at the ore-forming elevation.

Deposits seem to range from "wet" types having high pyrite-to-chalcopyrite ratios and surrounded by enormous halos of pyrite–sericite–quartz hydrothermal alteration to "dry" deposits with relatively low sericite–pyrite content. Although perhaps the terms are too casual, "wet" and "dry" refer to the net apparent abundance, involvement, and permeation of a mineralizing-altering fluid. Concentric zoning is also present in "dry" deposits, but it is telescoped laterally into a small fraction of the halo thickness of the "wet" type. The "wet" type is represented by most of the Southwest deposits, such as Bingham and Morenci, and includes most of the large porphyry copper deposits. The "dry" type is represented by many of the British Columbia deposits, such as Bethlehem, and includes many of the hypogene ore-grade porphyry coppers.

The most distinctive feature of the porphyry deposits is simply their huge size as compared with other hydrothermal ore deposits. Including ore-grade mineralization and surrounding alteration and mineralization, they assume dimensions more commonly associated with stocks than with ore deposits.

The bulk shape of porphyry deposits reflects large-scale structural control of mineralization and may also be related to the original depth of formation. Flat-tabular, cone, and flat-dipping tube-type deposits may represent relatively shallow depth of formation where steep environmental gradients prevail. Steep, columnar deposits with long vertical dimensions and little brecciation seem to indicate relatively great depth of formation and gentle environmental gradients.

Breccia pipe deposits, such as Toquepala, with only thin alteration halos and with evidence of violent emplacement, are clearly representative of a different genesis in which the mineralizers may have evolved suddenly in a more or less open vent with relatively steep pressure gradients. Examples of blind mineralized breccia pipes at Cananea, Pilares, and elsewhere indicate, however, that mineralized breccia pipes need not necessarily be either open to the surface or emplaced at shallow depth.

Porphyry deposits tend to have either elongate, vertical, columnar shapes (San Manuel–Kalamazoo and Bingham) or foreshortened columnar, almost discoid shapes (Climax or Ray). These shapes suggest that migration of hydrothermal fluids was controlled by nearly vertical gradients and that fluids, however derived, migrated upward across large areas, up to tens of thousands of feet in diameter. It appears likely that the mineralizers originated as a separation of fluids at the point of crystallization of the "host intrusive body." It should be noted that the "overhang" effect or beet shape of San Manuel–Kalamazoo could also be consistent with an influx of deeply circulating, externally derived, perhaps cooler water, although both the depth and wall rocks involved make this explanation seem unlikely.

Many characteristics described in Table 1 and systematized in Figure 13 are consistent with one another. Variations in the character of sulfide occurrence, for example, appear best explained by considering that dissemination textures are compatible with a model involving crystallization of rock-forming silicates (i.e., the potassic zone) such that the sulfides, which are really igneous accessory minerals, are deposited either as truly included minerals or in fractures and microfractures in newly competent rocks which are subsequently healed by local crystallization.

The San Manuel–Kalamazoo deposit thus appears to be typical and illustrative of porphyry copper and molybdenum deposits. We suggest in conclusion that the integrated model of vertical and lateral silicate-oxide alteration, sulfide mineralization, and sulfide occurrence characteristics in the porphyry deposits may be useful to economic geologists both explorationally and scientifically.

Acknowledgments

The writers wish to express their appreciation to their many colleagues whose discussion and comments have contributed to this paper, to L. B. Gustafson, James Gilluly, and T. W. Mitcham who reviewed the manuscript, and to H. R. Hauck, M. T. Wolf, and B. Townsend for their assistance in preparing the manuscript and illustrations.

5211 N. Oracle,
Tucson, Arizona,
and
Department of Geology,
University of Arizona,
Tucson, Arizona,

December 19, 1969; March 16, 1970

REFERENCES

Anderson, C. A., 1968, Arizona and adjacent New Mexico, *in* Ore Deposits of the United States, 1933–1968 (Graton-Sales Volume), John D. Ridge, ed., Vol. 2: New York, A.I.M.E., p. 1163–1190.

——, Scholz, E. A., and Strobell, J. D., Jr., 1955, Geology and ore deposits of the Bagdad area, Yavapai County, Arizona: U. S. Geol. Survey Prof. Paper 278, 102 p.

Banerjee, A. K., 1957, Structure and petrology of the Oracle granite, Pinal County, Arizona: Unpub. Ph.D. thesis, Univ. of Arizona, Tucson, Arizona.

Bauer, H. L., Jr., Breitrick, R. A., Cooper, J. J., and Anderson, J. A., 1966, Porphyry copper deposits in the Robinson mining district, Nevada, *in* Geology of the Porphyry Copper Deposits, Southwestern North America, S. R. Titley and C. L. Hicks (eds.): Tucson, Arizona, The University of Arizona Press, p. 233–244.

Beall, J. V., 1965, Southwest copper—a position survey: Min. Engr., v. 17, p. 77–92.

Bryant, D. G., 1968, Intrusive breccias associated with ore, Warren (Bisbee) mining district, Arizona: Econ. Geol., v. 63, p. 1–12.

——, and Metz, H. E., 1966, Geology and ore deposits of the Warren mining district, *op. cit.* (4): p. 189–204.

Buddington, A. F., 1933, Correlation of kinds of igneous rocks with kinds of mineralization, *in* Ore Deposits of the Western States (Lindgren Volume): New York, A.I.M.E., p. 350–385.

Burnham, C. W., 1962, Facies and types of hydrothermal alteration: Econ. Geol., v. 57, p. 768–784.

——, 1967, Hydrothermal fluids at the magmatic stage, *in* Geochemistry of Hydrothermal Ore Deposits, H. L. Barnes (ed.): New York, Holt, Rinehart and Winston, Inc., p. 166–235.

Carpenter, R. H., 1968, Geology and ore deposits of the Questa molybdenum mine area, Taos County, New Mexico, *op. cit.* (1), Vol. 2: p. 1328–1350.

Carr, J. M., 1960, Porphyries, breccias, and copper mineralization in Highland Valley, B.C.: Can. Min. Journal, Nov., p. 71–73.

——, and Lee, R., 1966, Geological map of the Highland Valley area, British Columbia: Dept. of Mines and Petrol. Res., Vancouver, B.C., Prelim. map, British Columbia.

Cooper, J. R., 1960, Some geologic features of the Pima

mining district, Pima County, Arizona: U. S. Geol. Survey Bull. 1112-C, p. 63-103.

Creasey, S. C., 1965, Geology of the San Manuel area, Pinal County, Arizona: U. S. Geol. Survey Prof. Paper 471, 64 p.

——, 1966, Hydrothermal alteration, *op. cit.* (*4*): p. 51-74.

Damon, P. E., and Mauger, R. L., 1966, Epeirogeny-orogeny viewed from the Basin and Range province: A.I.M.E. Trans. v. 235, p. 99-112.

Dixon, D. W., 1966, Geology of the New Cornelia mine, Ajo, Arizona, *op. cit.* (*4*): p. 123-132.

Drummond, A. D., and Kimura, E. T., 1968, Hydrothermal alteration at Endako mines—a comparison to experimental studies: Presented at Canadian Inst. Mining Annual Meeting, Vancouver, B.C., April 24, 1968, 28 p.

Eidel, J. J., Frost, J. E., and Clippinger, D. M., 1968, Copper-molybdenum mineralization at Mineral Park, Mohave County, Arizona, *op. cit.* (*1*), Vol. 2, p. 1258-1281.

Emmons, W. H., 1933, On the mechanism of the deposition of certain metalliferous lode systems associated with granitic batholiths, *op. cit.* (*8*): p. 327-349.

Ettlinger, I. A., 1928, Ore deposits support hypothesis of a central Arizona batholith: A.I.M.E. Tech. Pub. 63, 23 p.

Fournier, R. O., 1967, Syngenetic formation and subsequent alteration of the porphyry copper deposit near Ely, Neveda: Econ. Geol., v. 62, p. 57-81, 207-227.

——, 1968, Depths of intrusion and conditions of hydrothermal alteration in porphyry copper deposits: Geol. Soc. America Program Annual Meetings, p. 101.

Gilluly, J., 1946, The Ajo mining district, Arizona: U. S. Geol. Survey Prof. Paper 209, 112 p.

Gott, G. B., and McCarthy, J. H., Jr., 1966, Distribution of gold, silver, tellurium, and mercury in the Ely mining district, White Pine County, Nevada: U. S. Geol. Survey Circ. 535, 5 p.

Guilbert, J. M., and Lowell, J. D., 1968, Potassic alteration in porphyry copper deposits: Geol. Soc. America Program Annual Meetings, p. 120.

——, and Sumner, J. S., 1968, Distribution of porphyry copper deposits in the light of recent tectonic advances, *in* Southern Arizona Guidebook III, S. R. Titley (ed.): Tucson Arizona, Arizona Geological Society, p. 97-112.

Hemley, J. J., and Jones, W. R., 1964, Chemical aspects of hydrothermal alteration with emphasis on hydrogen metasomatism: Econ. Geol., v. 59, p. 538-569.

Howell, F. H., and Molloy, J. S., 1960, Geology of the Braden orebody, Chile, South America: Econ. Geol., v. 55, p. 863-905.

Jarrell, O. W., 1944, Oxidation at Chuquicamata, Chile: Econ. Geol., v. 39, p. 251-286.

Jones, W. R., Hernon, R. M., and Moore, S. L., 1967, General geology of the Santa Rita quadrangle, Grant County, New Mexico: U. S. Geol. Survey Prof. Paper 555, 144 p.

Kimura, E. T., and Drummond, A. D., 1966, Geology of the Endako molybdenum deposit: Presented at Canadian Inst. Mining Section Meeting, Victoria, B.C., 10 p.

Kinnison, J. E., 1966, The Mission copper deposit, *op. cit.* (*4*): p. 281-287.

Laughlin, A. W., Damon, P. E., and Watson, B. N., 1968, Potassium-argon dates from Toquepala and Michiquillay, Peru: Econ. Geol., v. 63, p. 166-168.

Lindgren, W., 1905, Copper deposits of the Clifton-Morenci district, Arizona: U. S. Geol. Survey Prof. Paper 43, 412 p.

Lowell, J. D., 1968, Geology of the Kalamazoo orebody, San Manuel district, Arizona: Econ. Geol., v. 63, p. 645-654.

Lutton, R. J., 1959, Pegmatites as a link between magma and copper-molybdenum ore: The Mines Magazine, Dec., p. 15-24.

Lynch, D. W., 1966, The economic geology of the Esperanza mine and vicinity, *op. cit.* (*4*): p. 267-280.

Mayo, E. B., 1958, Lineament tectonics and some ore districts of the Southwest: Min. Engr., v. 10, p. 1169-1175.

Metz, R. A., and Rose, A. W., 1966, Geology of the Ray copper deposit, Ray, Arizona, *op. cit.* (*4*): p. 177-188.

Meyer, C., and Hemley, J. J., 1967, Wall rock alteration, *op. cit.* (*10*): p. 166-235.

——, Shea, E. P., Goddards, C. C., Jr., Staff, 1968, Ore deposits at Butte, Montana, *op. cit.* (*1*), Vol. 2: p. 1373-1417.

Moolick, R. T., and Durek, J. J., 1966, The Morenci district, *op. cit.* (*4*): p. 221-232.

Moore, W. J., Lanphere, M. A., and Obradovich, J. D., 1968, Chronology of intrusion, volcanism, and ore deposition at Bingham, Utah: Econ. Geol., v. 63, p. 612-622.

Nielsen, R. L., 1968, Hypogene texture and mineral zoning in a copper-bearing granodiorite porphyry stock, Santa Rita, New Mexico: Econ. Geol., v. 63, p. 37-50.

Olmstead, H. W., and Johnson, D. W., 1966, Inspiration geology, *op. cit.* (*4*): p. 143-150.

Perry, V. D., 1933, Applied geology at Cananea, Sonora, *op. city* (*8*): p. 701-709.

——, 1952, Geology of the Chuquicamata orebody: Min. Engr., v. 4, p. 1166-1169.

——, 1961, The significance of mineralized breccia pipes: A.I.M.E. Trans., v. 220, p. 216-226.

Peters, W. C., James, A. H., and Field, C. W., 1966, Geology of the Bingham Canyon porphyry copper deposit, Utah, *op. cit.* (*4*): p. 165-175.

Peterson, N. P., 1954, Copper Cities copper deposits, Globe-Miami district, Arizona: Econ. Geol., v. 49, p. 362-377.

——, 1962, Geology and ore deposits of the Globe-Miami district, Arizona: U. S. Geol. Survey Prof. Paper 342, 151 p.

——, Gilbert, C. M., and Quick, G. L., 1951, Geology and ore deposits of the Castle Dome area, Gila County, Arizona: U. S. Geol. Survey Bull. 971, 134 p.

Ransome, F. L., 1904, Geology and ore deposits of the Bisbee quadrangle, Arizona: U. S. Geol. Survey Prof. Paper 21, 168 p.

——, 1919, The copper deposits of Ray and Miami, Arizona: U. S. Geol. Survey Prof. Paper, 115, 192 p.

Renzetti, B. L., 1957, Geology and petrogenesis at Chuquicamata, Chile: Unpubl. Ph.D. dissertation, University of Indiana.

Richard, K., and Courtright, J. H., 1958, Geology of Toquepala, Peru: Min. Engr., v. 10, p. 262-266.

——, and ——, 1966, Structure and mineralization at Silver Bell, Arizona, *op. cit.* (*4*): p. 157-163.

Ringwood, A. E., 1955, Principles governing trace element behavior during magmatic crystallization: Geochim. et Cosmochim. Acta, v. 7, p. 242-254.

Robertson, F., 1962, Crystallization sequence of minerals leading to formation of ore deposits in quartz monzonitic rocks in the northwestern part of the Boulder batholith, Montana: Geol. Soc. America Bull., v. 73, p. 1257-1276.

Robinson, R. F., and Cook, A., 1966, The Safford copper deposit, Lone Star mining district, Graham County, Arizona, *op. cit* (*4*): p. 251-266.

Rose, A. W., and Baltosser, W. W., 1966, The porphyry copper deposit at Santa Rita, New Mexico, *op. cit.* (*4*): p. 205-220.

Sales, R. H., and Meyer, C., 1948, Wall-rock alteration at Butte, Montana: A.I.M.E. Trans. 178, p. 9-35.

Schmitt, H. A., 1966, Porphyry copper deposits in their regional setting, *op. cit.* (*4*): p. 17-34.

Schwartz, G. M., 1953, Geology of the San Manuel copper deposit, Arizona: U. S. Geol. Survey Prof. Paper 256, 63 p.

Simmons, W. W., and Fowells, J. E., 1966, Geology of the Copper Cities mine, *op. cit.* (*4*): p. 151-156.

Spencer, A. C., 1917, Geology and ore deposits of Ely, Nevada: U. S. Geol. Survey Prof. Paper 96, 189 p.

Spurr, J. E., 1925, Basic dike injections in magmatic sequences: Geol. Soc. America Bull., v. 36, p. 545-582.

Steele, H. J., and Rubly, G. R., 1957, San Manuel prospect: A.I.M.E. Tech. Pub. 2255, Mining Technology, v. 11, no. 5, 12 p.

Still, A. R., 1962, Uranium at Copper Cities and other porphyry copper deposits, Miami district, Arizona: Unpub. Ph.D. thesis, Harvard University.

Stringham, B., 1953, Granitization and hydrothermal alteration at Bingham, Utah: Geol. Soc. America Bull., v. 64, p. 945–991.

——, 1966, Igneous rock types and host rocks associated with porphyry copper deposits, *op. cit.* (*4*): p. 35–40.

Swayne, W. H., and Trask, F., 1960, Geology of El Salvador: Min. Engr., v. 12, 344–348.

Thomas, L. A., 1966, Geology of the San Manuel ore body, *op. cit.* (*4*): p. 133–142.

Velasco, J. R., 1966, Geology of the Cananea district, *op. cit.* (*4*): p. 245–250.

Wadsworth, W. B., 1968, The Cornelia pluton, Ajo, Arizona: Econ. Geol., v. 63, p. 101–115.

Wallace, S. R., et al., 1960, Geology of the Climax molybdenum deposit: a progress report: Geol. Soc. America Guidebook for Field Trips (Guide to the geology of Colorado), Field Trip B3, p. 238–252.

——, Muncaster, N. K., Jonson, D. C., Mackenzie, W. B., Bookstrom, A. A., and Surface, V. E., 1968, Multiple intrusion and mineralization at Climax, Colorado, *op. cit.* (*1*), Vol. 1: p. 605–640.

White, D. E., 1968, Environments of generation of some base-metal ore deposits: Econ. Geol., v. 63, p. 301–335.

White, W. H., Harakal, J. E., and Carter, N. C., 1968, Potassium-argon ages of some ore deposits in British Columbia: Can. Inst. Mining Bull. v. 61, p. 1326–1334.

——, Thompson, R. M., and McTaggart, 1957, The geology and mineral deposits of Highland Valley, B.C.; Can. Min. and Met. Trans., v. 60, p. 273–289.

28

Reprinted from pages 99–104 and 108–109 of *Canadian Inst. Mining and Metallurgy Bull.* **67**:99–109 (1974)

Variations in Zoning Patterns In Porphyry Ore Deposits

John M. Guilbert, Professor, Economic Geology, Department of Geosciences, University of Arizona

J. David Lowell, Consulting Geologist, Tucson, Arizona

Abstract

Models of alteration and mineralization of porphyry copper deposits have been useful in both scientific investigation of the deposits and in exploration activities, but many uncertainties remain. It is commonly difficult geologically to interpret and economically to evaluate individual occurrences and assemblages that depart from the norm.

In this paper, departures from the "typical" porphyry copper zoning pattern and assemblages are described, and the functions of several variable conditions in the porphyry copper environment are discussed. Depth of exposure, composition of pre-mineral wall rocks, pre-ore structural controls on intrusion, variations in composition of both igneous host rock and mineralizing fluids, size of the mineralizing system, variations in development of contemporaneous controls on mineralization (veinlets, veins and brecciation), and the breadth of stability fields of important silicate alteration minerals such as sericite and K-feldspar all affect the size, shape, mineralogy, symmetry, intensiveness and extensiveness of porphyry copper alteration-mineralization. Study of these variables can help in predicting the three-dimensional geologic-geometric characteristics of individual deposits.

Examples described to demonstrate the effects of these variables include the Valley Copper, JA and Bethlehem orebodies in British Columbia, Marcopper and Atlas in the Philippines, and the Ajo orebody in Arizona.

Introduction

SEVERAL RECENT PUBLICATIONS (Lowell & Guilbert, 1970; Rose, 1970; James, 1971; De Geoffroy & Wignall, 1972) have developed the concept that there are unifying geologic characteristics which are displayed to varying degrees by porphyry copper and molybdenum deposits. The first-named of these papers involved a tabulation of published data on the geologic features of 27 of the best-known deposits. It elicited from them, by comparison with the San Manuel - Kalamazoo model, a definition of porphyry copper-molybdenum deposits, a description of a "typical" deposit and a summary of the "normal" characteristics of porphyry ore deposits. Nineteen of those deposits are in southwestern North America, the remaining 8 being chiefly South American and Canadian deposits. It was the intent of that paper to isolate unifying geologic characteristics and to demonstrate that porphyry coppers have an interpretable lateral and vertical zoning. It was not proposed, however, that all deposits are identical. This paper will consider the nature of departures from idealized zoning patterns, reasons for those departures, and examples of variant assemblages at Bethlehem, Valley Copper and the JA orebody, B.C., Marinduque (Marcopper) and Atlas in the Philippine Islands, and Ajo, Arizona.

We are deeply indebted to the Bethlehem Copper Corporation, Ltd. for access to specimens, thin sections and deposits, and to our colleagues in Canada and Arizona — notably H. G. Ewanchuk and S. R. Titley — for stimulating discussion and critical comment. Portions of this paper have been published in Spanish (Guilbert & Lowell, 1971).

Before considering departures, the norm must be established. The "typical" southwestern North American porphyry copper deposit is emplaced in late Cretaceous sediments and metasediments and is associated with a Laramide (65 m.y.) quartz monzonite porphyry stock. Its host intrusive rock is elongate-irregular, 4,000 by 6,000 ft in outcrop, and is progressively differentiated from quartz diorite to quartz monzonite in composition. The host is more like a stock than a dyke and is controlled by regional-scale faulting. The orebody is oval to pipelike, with dimensions of 3,500 by 6,000 ft and gradational boundaries. Seventy per cent of the 140 million tons of ore occurs in the igneous host rocks, thirty per cent in pre-ore rocks. Metal values include 0.45% hypogene Cu, 0.015% Mo and 0.35% supergene Cu. Alteration (Fig. 1) is zoned from potassic (earliest and at the core) outward through phyllic (quartz-sericite-pyrite), argillic (quartz-kaolin-montmorillonite) and propylitic (epidote-calcite-chlorite) zones, with visible alteration commonly extending 2,500 feet beyond the copper ore zone. Over the same interval, sulphide species (Fig. 2) vary from chalcopyrite-molybdenite-pyrite through successive assemblages to an assemblage of galena and sphalerite with minor gold and silver in solid solution in sulphides, as metals, and as sulphosalts. Sulphide occurrence characteristics (Fig. 3) shift from disseminations through respective zones of microveinlets (crackle fillings), veinlets and veins, and finally to discrete peripheral structures that may contain high-grade mineralization. Breccia pipes with attendant crackle zones are common.

De Geoffroy and Wignall (1972) more than doubled Lowell and Guilbert's control group to 58 deposits, chiefly by the addition of 23 British Columbia occurrences, for a computerized statistical study of porphyry base-metal deposits. Their results, summarized in their Tables 4 and 5, permit their description (Table 6, p. 666) of a "composite geological model of the most typical ore-bearing porphyry intrusive complex", which is in very close agreement with the Lowell-Guilbert conclusions. The addition of a large number of examples from British Columbia has not appreciably modified the characteristics of the "average" deposit.

Variations from the alteration model are obvious in some deposits, however. In an attempt to unravel some of the causes of departure from an ideal model, this paper will consider variations imposed by and related to depth of exposure, composition of pre-ore wall rocks, pre-ore structures, source and chemistry of igneous host rocks, chemistry and volume of mineralizers, deposit size, brecciation and contemporaneous structures, and extent of both hypogene and supergene alteration. Several of these variables, especially the effects of supergene alteration, have also been considered by Titley (1972).

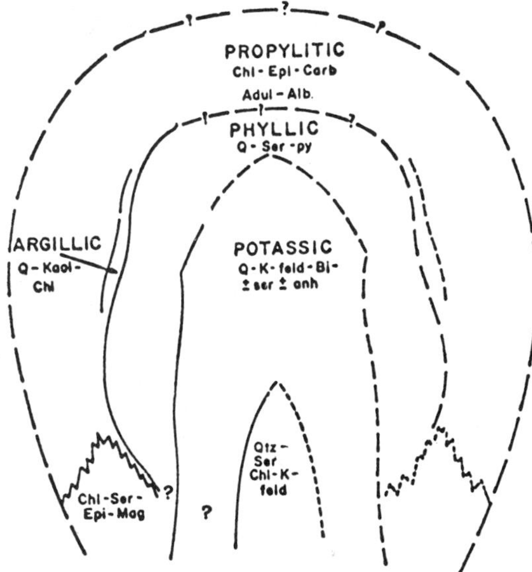

FIGURE 1 — Schematic drawing of alteration zoning in a typical porphyry ore deposit.

Variations Related to Depth of Exposure

Although several authors have considered changes in porphyry copper deposits in the vertical dimension, important uncertainties in both concept and detail still exist. De Geoffroy and Wignall (1972) show that deposits fall most commonly in moderate to shallow depth exposure level categories, with great depth only rarely reported. Variations in clay mineralogy, calcian silicate equilibria (Greenwood & Gordon, 1971) and other mineralogical-textural relationships are doubtless affected by increasing temperature and pressure at depth, but the details of these variations are far from understood on either an observational or experimental basis. We can report on only a few known deep assemblages, and infer other deep occurrences based on those assemblages. Modern studies (Roedder, 1971; Nash & Theodore, 1972) suggest temperatures up to 600-700°C and pressure to 2 kb as reasonable upper limits, with the bulk of porphyry copper mineralization formed at temperatures from 300-400°C. Porphyry molybdenum deposits may have formed at slightly higher temperatures.

At San Manuel - Kalamazoo, near the center of the axis and at depth, a quartz-K-feldspar-chlorite-sericite assemblage is stable, probably grading into fresh rock at greater depth (Fig. 1). At Butte, alteration envelopes in quartz monzonite characterized by quartz-sericite-pyrite at intermediate depths grade downward into the same assemblage described for San Manuel - Kalamazoo. Laterally outside the deep San Manuel - Kalamazoo assemblage is a quartz-chlorite-sericite-epidote-magnetite zone. The downward narrowing of the phyllic zone toward that zone is probably real, but may be in part a structural effect (Lowell, 1968). Beyond these scant observations concerning the deeper reaches of porphyry deposits, little is known. A peripheral quartz-K-feldspar (salmon-colored)-epidote-chlorite propylitic assemblage is described from the Sierrita deposit, from the Bethlehem, B.C., deposit, and from several others. It

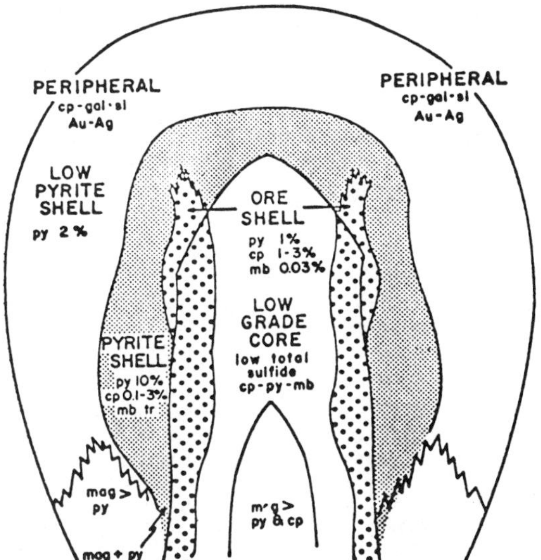

FIGURE 2 — Schematic drawing of mineralization zoning in a typical porphyry ore deposit.

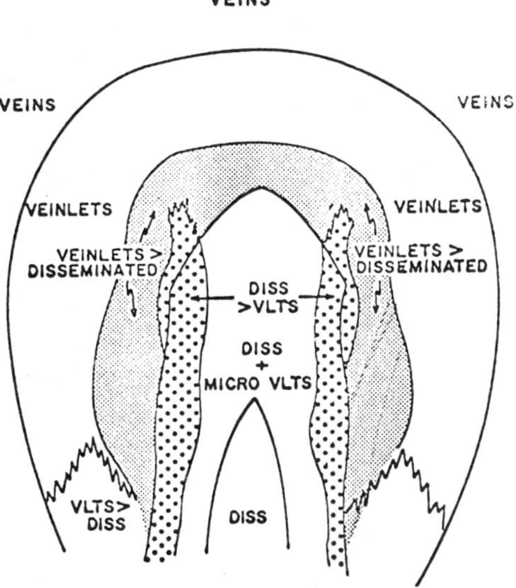

FIGURE 3 — Schematic drawing of sulphide-occurrence zoning in a typical porphyry ore deposit.

is reasonable to expect that a calcian silicate, such as epidote or zoisite, would be present in a deep outermost assemblage with either sericite or K-feldspar, as will be shown. As yet, however, the transition downward from altered rocks to fresh pyrogenic ones is not well understood.

Hydrothermal alteration minerals may also grade outward into metamorphic, deuteric or even supergene assemblages with which they can be confused. The several possible variant assemblages will be discussed below. Lastly, the environment most difficult to interpret is the shallow, volcanic-subvolcanic one, where complications of alteration-mineralization patterns result from contrasting adjacent rock types, structural anisotropism, and steep physical and chemical environmental gradients. Telescoping of assemblages (Titley, 1972), the so-called 'solfataric' alteration, and short-range variations in alteration intensity and extensiveness are to be expected.

Variations Related to Composition of Pre-Ore Wall Rocks

One of the more obvious obstacles to the development of alteration symmetry by a mineralizing system is the presence of chemically heterogeneous wall rocks, especially screens of compositionally divergent rocks (Fig. 4). Limestones lead to the production of calcian, magnesian and ferrian silicates such as tremolite, diopside, hedenbergite and garnet. Chemical and probably thermal gradients are so telescoped at contacts with carbonate sections that sudden precipitation of metal values commonly produces high-grade but narrow skarn or tactite zones in silicate-oxide assemblages that are quite diagnostic both mineralogically and spatially. Christmas and Morenci, Arizona, and Santa Rita, New Mexico, are examples. At the Christmas mine near Winkelman, Arizona, potassic, phyllic and propylitic alteration zones are well developed in an elongate quartz monzonite porphyry stock, but the zones are sharply cut off on either side by substantially richer skarn zones in upper Paleozoic carbonate sediments into which the stock has intruded. Rarely, widespread skarn mineralogy is produced hydrothermally (rather than essentially just thermally) in carbonate rocks, as at Mission-Pima and Twin Buttes, Arizona. These deposits are characterized by low pyrite content (Titley, 1969), and the action of the carbonate wall rocks as buffers of CO_2-O_2-S_2 relationships in hydrothermal fluids (Burt, 1971) is certainly important. Low alumina and silica starting compositions predispose carbonate wall rocks toward skarn rather than toward clay-mica mineralogy.

Quartzite is a relatively poor indicator of alteration zoning, because quartz, the only abundant material present, is essentially nonreactive. In the inner alteration zones of several deposits, quartz from quartzite can be shown to have been mobilized into quartz metacrysts and quartz veinlets and combined with other elements to form feldspar metablasts, so that the resulting rocks appear transitional into true igneous rocks. They thus have become metasomatic felsites, quartz porphyries or aplites. Arkose tends to reflect alteration zoning in a manner similar to that of intermediate and felsic intrusive rocks, although zoning in arkose is generally less perfectly developed, perhaps due to the planar fabric of the rock.

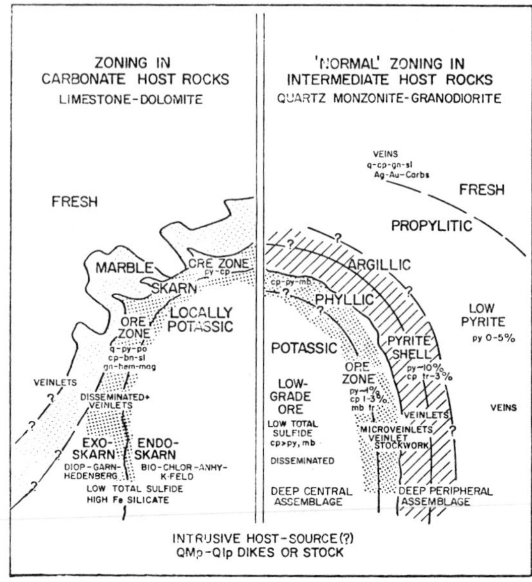

FIGURE 4 — Schematic comparison of alteration and mineralization in carbonates and intermediate rocks.

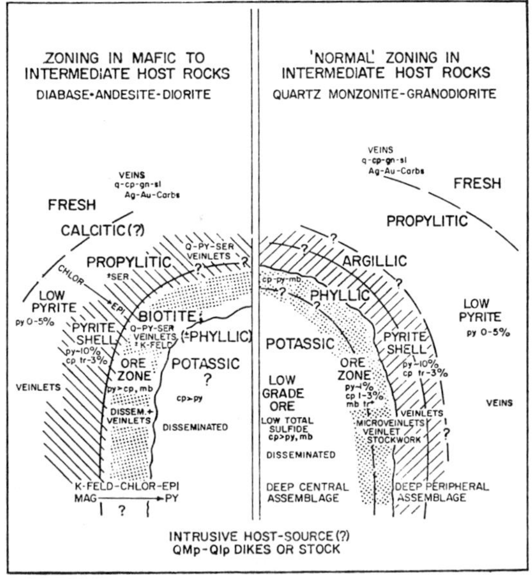

FIGURE 5 — Schematic comparison of alteration and mineralization in mafic and intermediate rocks.

Mafic volcanic rocks such as basalts, diabases and andesites (Fig. 5) are relatively rich in iron and magnesium, with moderate amounts of calcium. Where these rocks are the dominant pre-ore wall rocks, their compositions affect the bulk composition of the ore-forming environment and bias that system toward the formation of more mafic alteration minerals. These reactant materials, with sufficient alumina and silica, favour the development of wide propylitic zones (epidote-chlorite-carbonate) and expand the stability field of biotite in part because of the relatively higher activity of iron, magnesium or both.

Biotite differs from sericite only in the composition of the octahedral layer, biotite having $3(Mg + Fe)^{+2}$, sericite having $2Al^{+3}$. They are structurally virtually identical. Biotite in effect takes over from sericite in a high-iron-magnesium environment, and wide biotitic "potassic zones" with narrow phyllic zones, as at Safford and Ray, Arizona, normally result. Another excellent example of this expanded biotite zone is at El Salvador, Chile, where biotite is abundant and the calcium component of the andesite host rocks is represented by abundant anhydrite. The Ingerbelle and Babine Lake deposits in British Columbia may also be examples of this effect. The presence of widespread alteration biotite, with variable epidote, chlorite, and sometimes even amphibole and K-feldspar, is thus common in mafic volcanic and plutonic sections, even into the quartz diorite-dacite categories. It should thus be anticipated and accounted for in exploration; for example, in southwest Pacific terrains and in the more mafic rocks in British Columbia.

Metamorphic rocks are probably the most difficult in which to discern hydrothermal effects. Albitization, chloritization and biotitization are common and can usually be distinguished petrographically by the determination of the general lack of silicification and hydrothermal accessory minerals (Guilbert & Lowell, 1970).

Intrusive rocks of intermediate composition register alteration mineralogy well, and it is in these rocks that zoning is most extensively and symmetrically developed, as at San Manuel - Kalamazoo (Lowell, 1968). Mafic plutonic rocks behave chemically like the mafic volcanic rocks previously described; more felsic and aluminous rocks favour the formation of more highly aluminous and siliceous alteration minerals such as corundum, quartz, topaz, tourmaline and pyrophyllite. All of these variations must be dealt with both in exploration and in attempting to apply alteration models to individual deposits.

Variations Caused by Pre-Ore Structures

The shape of both the intrusive igneous host-rock mass and the resultant ore deposit and its related alteration zones are demonstrably modified by structural control. Although large orebodies and even districts tend to be circular to oval in over-all plan shape, the elongate nature of some deposits almost certainly represents the effects of pre-ore structural grain, as at Silver Bell (northwest), Chuquicamata (north-south) and the Safford district (northwest). Local ore-controlling structures are probably more readily obliterated by ore-forming processes than are regional effects. Structural features must be considered in the analysis of alteration-mineralization zoning patterns at a mine or prospect, the most difficult aspect being the isolation and consideration of only those structures that were effective at the time of intrusion-mineralization.

Variations in Source and Chemistry of Igneous Host Rock

The discovery in the southwest Pacific and Puerto Rico of porphyry copper deposits associated predominantly with diorites and quartz diorites has caused re-examination of the association between copper deposits and igneous rock types in southwestern North America. It appears that the more felsic igneous source rocks generally associated with thicker continental crustal rocks characterize the Cu-Mo association, but that Cu-Au is the more common association in deposits associated with more mafic rocks and oceanic crust in the southwest Pacific (Kesler, 1971). Alteration envelopes associated with diorites and quartz diorites in many of these occurrences (e.g., Marcopper) display broadened envelopes of calcian, magnesian and ferrian minerals, with or without potassium, so that propylitization, with or without K-feldspar, is well developed. Mafic rocks intrusive into andesites and diorites in island-arc terranes, as along modern subduction systems in the southwest Pacific, generate alteration zoning generally enriched in calcium, magnesium and iron alteration minerals, probably also with abundate carbonate minerals. Hybridization of mafic intrusive and felsic host rocks should favour more siliceous-aluminous alteration minerals. Felsic melts, produced either by differentiation or by fusion of continental crust, intruded into sialic host rocks favour silicification and silica-alumina-rich alteration products such as sericite, K-feldspar, albite and even peraluminous phases such as corundum. It also appears that the more primitive starting material of porphyry copper intrusive complexes with low initial strontium 87/86 ratios in southwestern North America may be more dioritic than monzonitic, quartz monzonitic or granitic. Igneous rock zoning over a distance of several kilometers from more mafic rocks on the margins to more felsic rocks near the center of Cordilleran porphyry ore deposit districts is reported from several deposits, including Bingham Canyon, Utah, and Ajo, Mineral Park, and Sierrita, Arizona. There is strong evidence for hydrothermal metasomatic "upgrading" in felsic constituents like quartz, K-feldspar, muscovite and even biotite toward the centers of porphyry districts.

Variations Relative to Chemistry and Volume of Mineralizers

This variable is almost certainly a key one. Without separation of a quantitatively significant aqueous phase from the intrusive material (Burnham, 1967; Holland, 1972) or without the circulation of meteoric water to some extent (Sheppard, Nielsen & Taylor, 1971), neither concentration and transportation of metals nor hydrolysis of preexisting silicates can occur. If the H^+ ion is not appropriately abundant at various pressure-temperature conditions, apparently neither alteration nor mineralization can proceed to economically interesting levels. The H^+ ion is provided by dissociation of either H_2O or of HCl within it, probably one of the reasons that chlorine has been found to be abundant in most districts where the geochemistry of hydrothermal fluids has been carefully scrutinized. Water and hydrochloric acid, sulphuric acid, and various acid constituents involving phosphorus, carbon dioxide, fluorine and others are all donors, by dissociation, of H^+ (hydronium or hydrogen ion), the chief instigator of hydrolytic alteration and a determinant of its extensiveness. These constituents are also donors of the radicals which form many alteration accessories — chlorine and chlorides in fluid inclusions, apatite and zunyite, sulphate in anhydrite and gypsum, phosphorus in apatite, carbon dioxide in calcite, siderite, magnesite and dolomite, and fluorine in fluorite and apatite. The presence of these trace minerals helps distinguish hydrothermal from metamorphic events.

The intensity of alteration, which, like metamorphic rank, is strictly a temperature-pressure index, is not necessarily linked to extensiveness — the degree or extent to which susceptible minerals are converted to their alteration products. One aspect is intensive, like temperature; the other extensive, like heat content. Rocks can be brought to appropriate temperature-pressure conditions, but major ore-grade mineralization as part of a well-established zoning pattern appears impossible unless H^+ activities are maintained at sufficient levels to produce pervasive, extensive, thorough alteration. Variations in amount of these constituents and in the abundance of water itself must control the efficacy of H^+ and thus the dimensions and pervasiveness of alteration zones. Also, the activity of oxygen will control the type of biotite (Yoder & Eugster, 1954) and the distribution of hematite and magnetite present (Buddington & Lindsley, 1964). The ratio of activities of sulphur and oxygen will determine sulphide/oxide species (Holland, 1965) or the distribution of pyrrhotite, pyrite, anhydrite, magnetite, hematite, chalcopyrite, bornite and so on (Barnes & Czamanske, 1967; Barton, 1970). Sulphur availability almost certainly affects silicate reactions as well. The activity of silica in the aqueous phase at the sites of alteration (Fournier, 1967) may drastically modify relative stabilities of feldspars and micas and the distribution of resultant alteration assemblages, as at Ely and perhaps at Valley Copper, B.C. Lastly, an element vital to the distribution of alteration assemblages is potassium. Without it, sericite, K-feldspar and biotite stability fields are drastically restricted; with abundant potassium, the stability fields of both sericite and K-feldspar are extended, and these minerals are widely distributed, especially in the absence of iron and magnesium.

In summary, the volume, efficacy and composition of mineralizing solutions — in turn dependent upon their meteoric or juvenile source, the types of paths followed through host rocks, and dissociation-association coefficients of both solvent and solute — can all be expected to affect the relative abundance of H_2O, sulphur, oxygen, potassium, anion groups and metal values. Each of these items in turn affects the distribution of alteration-mineralization assemblages in a manner in part predictable for individual deposits the geologic parameters of which are known or can be mapped. Some individual deposits will be described and discussed below.

Variations Relative to Size of Occurrence

Experienced students of porphyry base metal deposits have observed that the larger the occurrence the more regular and well developed the zoning in terms of both mineralization and alteration. The reverse is equally true — that smaller deposits *even of ore grade* are inclined to show significantly less regular alteration effects. The smaller occurrences typically are erratic in terms of intensity and extensiveness factors, and even assemblages and rock-type responses are commonly irregular.

Variations in Brecciation and Structures Contemporaneous with Mineralization

A popular model for the genesis of porphyry ore deposits involves periodic buildup of the vapour pressure of volatile components, their release by various mechanisms, sealing or healing, and re-establishment of growing pressures. The location and nature of conduits will affect both the distribution of alteration and mineralization and their zoning symmetry. La Caridad shows zoning related to broad and complex brecciation events. Butte shows fault-controlled leakage or bleed-off of volatiles such that the pressures of mineralizers may never have built up to the explosive breccia-forming stage. San Manuel - Kalamazoo appears neither to have been cut by through-going structures nor to have discharged its volatiles through breccia columns. The distribution of alteration-mineralization zoning around each of these districts is consistent. Orebody zoning and occasionally district zoning (Park & McDiarmid, 1970) occur around breccia pipes and internal fault vein structures that tend to destroy district-scale zoning symmetry.

Breadth of Stability Fields of Alteration Silicates

The most troublesome mineral in alteration assemblages to the field geologist and economic petrologist is certainly sericite. "Sericite", like "kaolin", is more a family name than a species description. Sericites vary in alumina-to-silica ratio, in total alkali content, in K^+/Na^+ ratio (possibly even $K + Na/Ca$), in water or hydroxyl content and in structural or polymorphic state. Thus it is no surprise to find a fundamentally mica-like mineral of several different morphological-optical characteristics extending from the pegmatite environment possibly to the supergene environment. Coarse, well-crystallized sericite-muscovite is typical of the pegmatitic to potassic alteration stages, such as at Valley Copper, Butte and San Manuel. Finer-grained sericite of several descriptive types is normal in phyllic zones, as at Morenci, Chuquicamata, La Caridad, Cananea and virtually all other deposits. Poorly ordered, extremely fine grained hydromicas and "sericites" commonly extend well out into propylitic zones and create problems in the interpretation of alteration zoning.

As shown by Hemley and Jones' 1964 diagram (Fig. 6), both sericite and K-feldspar can form at high and low temperatures. Both form at the pegmatitic stage and both can form in the alkaline, low-temperature conditions of the sea floor. Both can also form, if temperature, pressure and composition are appropriate, at the low temperatures and pressures presumed for the propylitic assemblage. K-feldspar, albite and sericite can remain stable into outer alteration assemblages and, although they are not normal or typical in that assemblage, their presence must be watched for. Central potassic zones and external K-feldspathic propylitic zones can thus be confused, and careful attention must be paid to accessory mineralogy, geochemistry and the factors described above in their interpretation. K-feldspar-epidote-chlorite ± biotite propylitic assemblages appear not uncommon at deposits such as Bethlehem and Sierrita-Esperanza.

A detailed discussion of supergene complications will not be included here, but their importance as modifiers of hypogene relationships, and therefore to interpretation of hypogene zoning, is widely recognized (Titley, 1972). The supergene development of clay minerals, especially the kaolin family, is well established, but the growth of sericites and montmorillonites in supergene cappings over orebodies is not yet well documented. Epidote and the zoisites can occur only as metamorphic or hydrothermal minerals (Ellis, 1967), and chlorite is of the same general origin, apparently only very rarely occurring as a supergene phase. A better understanding of the role and history of pH and Eh in capping evolution is required along with better data on the distribution, polytypism and order-disorder relationships in definite examples of supergene and hypogene alteration products.

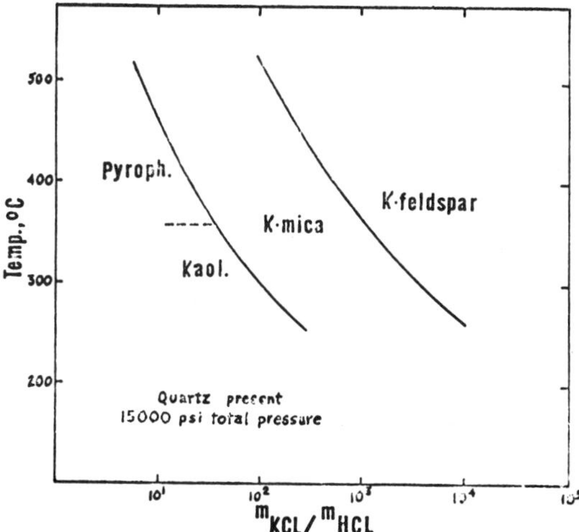

FIGURE 6 — Stability fields of K-feldspar, mica and kaolin-pyrophyllite, after Hemley and Jones (1964). A similar diagram relates sodium feldspar (albite) and sodium mica (paragonite) stabilities to temperature and alkali-to-hydrogen ion ratios. Clearly, both feldspars and micas can be stable over a wide temperature range, probably even down to surface temperatures in alkaline environments. Also, one can cross stability fields without necessarily changing temperature.

[*Editor's Note:* Material has been omitted at this point.]

Conclusions

A large variety is possible in both the assemblages of minerals making up the various porphyry alteration zones and in the size and configuration of the zones themselves. Well-developed zoning at San Manuel - Kalamazoo, Morenci, Bingham Canyon and Valley Copper suggest that the larger and higher-grade porphyry deposits show better developed and more consistent hypogene alteration-mineralization zoning. Smaller and lower-grade deposits exhibit a range of alteration patterns and assemblages, but are united by more similarities than differences. It is the thesis of this report that the variations are generally predictable and interpretable on the basis of local geologic variables. Continued careful geologic, petrographic and X-ray study of alteration envelopes is indicated. Further careful reporting of such phenomena as net abundance of alteration minerals, mineral ratios, veinlet widths and sequences, veinlet spacings and pertinent geologic variables in individual deposit descriptions is urged. The application of new tools such as mica polytypism, isotope studies and minor-element ratio analysis offers powerful insights into variations in zoning patterns in porphyry ore deposits and prospects.

Bibliography

Allen, J. M., and Richardson, J., 1970, The geologic setting of the Valley Copper orebody; CIM Annual Meeting, Toronto, April, 1970.

Barnes, H. L., and Czamanske, G. K., 1967, Solubilities and transport of ore minerals, chapter 8 (p. 334-381) *in* Geochemistry of Hydrothermal Ore Deposits; Holt, Rinehart and Winston, New York, 670 p.

Barton, P. B., Jr., 1970, Sulfide petrology; Min. Soc. Amer. Spec. Pap. 3, p. 187-198.

Bethlehem Copper Corp., 1972, Annual Report.

Buddington, A. F., and Lindsley, D. H., 1964, Iron-titanium minerals, oxide minerals and synthetic equivalents; Journ. Petrology, vol. 5, p. 310-357.

Burnham, C. W., 1967, Hydrothermal fluids at the magmatic stage, chapter 2 *in* Geochemistry of Hydrothermal Ore Deposits, H. L. Barnes (ed); Holt, Rinehart and Winston, New York, 670 p.

Burt, D. M., 1971, Some phase equilibria in the system Ca-Fe-Si-C-O; Carnegie Institution of Washington Year Book 70, p. 178-184.

Cheney, E. S., and Trammell, J. W., 1972, Inside-out hydrothermal alteration in a porphyroid Cu/Mo deposit; Geol. Soc. Amer. Ann. Mtgs. Abstracts, p. 471.

De Geoffroy, J., and Wignall, T. K., 1972, A statistical study of geological characteristics of porphyry copper-molybdenum deposits in the Cordilleran belt — application to the rating of porphyry prospects; Econ. Geol., vol. 67, p. 656-668.

Ellis, A. J., 1967, The chemistry of some explored geothermal systems, chapter 11 (p. 465-514) *in* Geochemistry of Hydrothermal Ore Deposits; Holt, Rinehart and Winston, New York, 670 p.

Ewanchuk, H. G., and Anderson, R. F., 1972, Bethlehem Copper Corporation Annual Report.

Fournier, R. O., 1967, The porphyry copper deposit exposed in the Liberty open-pit mine near Ely, Nevada; Econ. Geol., vol. 62, p. 57-81, 207-227.

Gervacio, F. C., 1971, Ore deposits of the Philippine mobile belt; unpublished report, National Research Council of the Philippines.

Gilluly, J., 1946, The Ajo mining district, Arizona; USGS Prof. Paper 209, 112 p.

Gordon, T. M., and Greenwood, H. J., 1971, The stability of grossularite in H_2O-CO_2 mixtures; Am. Miner., vol. 56, p. 1674-1688.

Guilbert, J. M., and Lowell, J. D., 1968, Potassic alteration in porphyry copper deposits; Geol. Soc. Amer. Prog., Ann. Mtgs., p. 120.

Guilbert, J. M., and Lowell, J. D., 1971, Variaciones en zonacion de yacimientos porfidicos; IX Convencion Nacional, AIMMGM, Hermasillo, Sonora, Mexico, in press.

Hall, D. K., 1972, Hydrothermal alteration and mineraization in the East Camp of the Turquoise District, San Bernadino County, Calif.; unpublished M.S. thesis, Univ. of Arizona, 137 p.

Hemley, J. J., and Jones, W. R., 1964, Chemical aspects of hydrothermal alteration with emphasis on hydrogen metasomatism; Econ. Geol., Vol. 59, p. 538-569.

Holland, H. D., 1965, Some applications of thermochemical data to problems of ore deposits; Econ. Geol., Vol. 60, p. 1101-1166.

Holland, H. D., 1972, Granites, solutions and base metal deposits; Econ. Geol., Vol. 67, p. 287-301.

Hylands, J. J., 1972, Porphyry copper deposits of the Guichon Creek Batholith, B.C.; 24th Internat. Geol. Cong., Mineral Deposits, Section 4, p. 241-251.

James, A. L., 1971, Hypothetical diagrams of several porphyry copper deposits; Econ. Geol., Vol. 66, p. 43-47.

Kesler, S. E., 1971, Cu-Mo-Au abundances in island arc and continental porphyry copper deposits; Geol. Soc. Amer. Ann. Mtg., Wash., D.C. (abs.), p. 623.

Langlois, J. D., 1971, Hydrothermal alteration of intrusive igneous rocks in the Eureka mining district, Nevada; unpub. M.S. thesis, Univ. of Arizona, 112 p.

Loudon, A. G., 1972, Marcopper disseminated copper deposit, Philippines; AIME Preprint, Tokyo, May 24, 1972.

Lowell, J. D., 1968, Geology of the Kalamazoo orebody, San Manuel district, Arizona; Econ. Geol., Vol. 63, p. 645-654.

Lowell, J. D., and Guilbert, J. M., 1970, Lateral and vertical alteration-mineralization zoning in porphyry ore deposits; Econ. Geol., vol. 65, p. 373-408.

Madamba, F. A., 1972, Geology and mineralization of the Atlas-Cebu disseminated copper deposits; AIME Preprint, Tokyo, May 24, 1972.

McMillan, W. J., 1972, The Highland Valley porphyry copper district; 24th Internat. Geol. Cong., Field Excursion AO9-CO9 (Copper and Molybdenum Deposits of the Western Cordillera), p. 53-69.

Meyer, C., and Hemley, J. J., 1967, Wall rock alteration, chap. 6 *in* Geochemistry of Hydrothermal Ore Deposits, H. L. Barnes (ed.), Holt, Rinehart and Winston, New York, 670 p.

Nash, J. T., and Theodore, T. G., 1971, Ore fluids in the porphyry copper deposit at Copper Canyon, Nev.; Econ. Geol., vol. 66, p. 385-399.

Northcote, K. E., 1969, Geology and geochronology of the Guichon Creek batholith; B.C. Dept. of Mines, Bull. No. 56, 73 p.

Park, C. F., Jr., and McDiarmid, R. A., 1970, Ore Deposits; W. H. Freeman & Co., 522 p.

Roedder, E., 1971, Fluid inclusion studies on the porphyry-type ore deposits at Bingham, Utah, Butte, Montana, and Climax, Colorado; Econ. Geol., vol. 66, p. 98-120.

Rose, A. W., 1970, Zonal relations of wallrock alteration and sulfide distribution at porphyry copper deposits; Econ. Geol., vol. 65, p. 920-936.

Sheppard, S. M. F., Nielsen, R. L., and Taylor, H. P., Jr., 1971, Hydrogen and oxygen isotope ratios in minerals from porphyry copper deposits; Econ. Geol., vol. 66, p. 515-542.

Titley, S. R., and Hicks, C. L. (eds.), 1966, Geology of the porphyry copper deposits, Southwestern North America; University of Arizona Press, Tucson, Ariz., 287 p.

Titley, S. R., 1968, Environment of deposition of the hydrothermal metamorphic (pyrometasomatic) deposits; Geol. Soc. Amer. Ann. Mtgs. Abs., p. 297-298.

Titley, S. R., 1972, Intrusion and wall rock, porphyry copper deposits; Econ. Geol., vol. 67, p. 122-123.

Wadsworth, W. B. 1968, The Cornelia pluton, Ajo, Arizona; Econ. Geol., vol. 63, p. 101-115.

White, W. H., Thompson, R. M., and McTaggart, K. C., 1957, The geology and mineral deposits of Highland Valley, B.C.: CIM Bulletin, vol. 50, p. 487-503, Aug. 1957.

Wolfe, J. A., 1972, Setting of porphyry copper deposits in the Philippines; AIME Preprint, Tokyo, May 24, 1972.

Worley, Birl, 1971, personal communication.

Yoder, H. S., and Eugster, H. P., 1954, Phlogopite synthesis and stability range; Geochim., Cosmochim. Acta, vol. 6, p. 157-185.

Part V

EVALUATION

Editor's Comments on Papers 29 Through 34

PROSPECT EVALUATION

On the one hand, great deal of exploration funds is wasted on continued search in deposits or districts where it should be apparent that even successful discovery would not meet minimum economic limits; on the other hand, a geologist is not quickly forgiven abandonment of a prospect that a competitor develops into a profitable mine. Thus, a balance must be achieved in an examination between rapid screening and exhaustive search. Company policy will dictate which end of the spectrum is expected, dependent upon staffing and funding. This in turn determines techniques utilized and the detail of search.

An economic measure of justifiable expenditure may be indicated as:

$$\text{EXPENDITURE} \simeq \frac{\text{PRESENT VALUE OF REWARD} \times \text{RISK.}}{\text{FACTOR}}$$

(Factor is equal to one for governments, two for large companies, and three for small companies.)

As economic conditions change and as additional geological knowledge becomes available, the attractiveness of various prospects changes and reevaluation may be justified.

Joralemon (Paper 29), one of the early highly successful mine/prospect evaluation consultants, applies the principle of the weakest link in the rapid screening of prospects or operations that are available for purchase. The principle applies as well to delimiting countries, regions, or specific zones that justify exploration expenditure. In contrast, Sales (Paper 30), in response to a paper by Joralemon reiterating the thoughts presented in Paper 29, cautions against haste in the screening process. A sound approach in mineral search is to back up examinations with detailed geological mapping to avoid overlooking major deposits. However, much depends upon experience—for example, an experienced geologist may in a few hours delimit evidence of a successful or unsuccessful venture that would take a less experienced geologist many weeks to piece together. Dr. Harrison Schmitt, for example, in a few hours outlined the Esperanza porphyry copper deposit on the basis of wallrock alteration and a single adit exposure. Months of detailed mapping, geochemical and geophysical surveys, and drilling only served to confirm his initial analysis.

In an evaluation, the geologist is confronted with a decision as to the geometry of the mineral deposit relative to the erosion surface: Is he looking at the top, the central portion, or the bottom? Mines have been abandoned as having bottomed only to be reopened and the so-called bottom found to be only a structural interruption. Schmitt (Paper 31) deals with this problem as applied to hydrothermal vein and replacement deposits in the southwestern United States and Mexico. Sillitoe (Paper 32) considers the problem relative to porphyry copper deposits.

Following World War II came a great surge in organized mineral exploration to replenish depleted reserves, and great hopes developed that geophysical and/or geochemical techniques would serve as a panacea in the location of new deposits. Geophysical techniques in particular were often utilized under situations for which they were not suited and developed a reputation of uncertainty and unreliability. Brant (Paper 33) placed the utilization of geophysics in perspective by defining its applications and limitations. Each type of deposit develops a characteristic signature in terms of its geochemical, geophysical, and geological characteristics. This signature has been best illustrated and documented in the porphyry copper exploration programs that dominated the search following World War II. Jerome (Paper 34) outlines the geological, geophysical, and geochemical signature of the typical porphyry copper deposit.

Broad scale usage of geochemistry as an exploration search

technique evolved in Russia. Practical utilization of this technique, based upon the Russian experience, is summarized by Beus and Grigorian (1977).

A comprehensive review of the current status of mineral exploration philosophy, techniques, and future challenge is presented by Kužvart and Böhmer (1978) and by Peters (1978).

REFERENCES

Beus, A. A., and S. V. Grigorian, 1977, *Geochemical Exploration Methods for Mineral Deposits,* R. Teteruk-Schneider, trans., A. A. Levinson, ed., Applied Publishing Co., Wilmette, Ill., 287p.

Kužvart, M., and M. Böhmer, 1978, *Prospecting and Exploration of Mineral Deposits,* Elsevier, New York, 431p.

Peters, W. C., 1978, *Exploration and Mining Geology,* Wiley, New York, 696p.

29

Reprinted from *Eng. and Mining Jour.* **125:**536–540 (1928)

The Weakest Link; or,
Saving Time in a Mine Examination

BY IRA B. JORALEMON
Consulting Geologist, San Francisco

PLATO was one of the first to remark that the strength of a chain is the strength of the weakest link. It is easy to forget this obvious bit of philosophy in many of the affairs of busy lives; and in none more easily than in examining mines. If engineers more generally took Plato's saying as their guide, the time spent in examinations would be greatly reduced, and, also, the features essential to making a correct appraisal would be unencumbered by details which may be true but which have no real bearing on the question.

Ira B. Joralemon

The problem of how thorough an examination should be is a puzzling one. Owners of properties nearly always think the engineer is too hasty. They bitterly, and now and then justly, criticize the examiner who makes a decision after an inspection of a few minutes or a few hours. In the minds of most of them is a necessary connection between physical exertion and correct mental conclusions. Few prospectors or mine operators will complain of even an adverse report if the engineer has climbed every raise and mountain top in sight so rapidly that he and his guides have their tongues hanging out and their minds numb with exhaustion. Those for whom the report is made are less anxious for a long examination or a detailed report, for every day means more dollars from their pockets. But it is to be feared that even many prospective buyers value the looks and size of a report and the pleasing color scheme of the maps that accompany it more than they do the brief paragraph that settles the matter.

A mine examination is simply a process of testing the successive links of a chain of observed facts. As soon as any link is found to be too weak for the purpose it must serve, the examination is over. Any further time spent confuses the issue and wastes the money of the client.

Of the many findings that should cut short an examination, inadaptability of the property to the needs of the client is one of the most obvious. If the examination is made for a company so big that only a profit of half a million a year or more will have an appreciable effect on dividends, a 3-ft. vein of $10 ore can be eliminated at a glance. The fact that careful operation might yield a profit of a few thousand dollars a month has no bearing on the subject. A tactful or kind-hearted engineer may suggest to the owners of the property the names of people who may be interested in their mine; but in fairness to his clients he must not waste any time sampling or studying costs or geology. If the clients have only a limited amount to spend, and so must try several properties that can be proven for a small sum, any expensive gamble should be discarded off-hand, even if it looks attractive. A week's geological work might have disclosed the possibility of finding a faulted orebody by sinking a 1,000-ft. shaft, but this does not interest such clients, unless they want to play the part of middlemen. To protect those who employ him, an engineer must in all cases know what they want and what they can afford to take. Usually, a short talk with the owners of a property will show whether or not it is of a type that may suit his clients, often making the examination unnecessary. If the answer is negative, as soon as he sees that the property will not meet the requirements, the engineer cannot honestly look further.

The price or terms asked for a property often make it advisable to cut short an examination. Frequently the price of a developed mine is asked for a raw prospect. This variety of shortsightedness on the part of property owners has been notably common in the Mother Lode section of California, and is probably one of the most important reasons for the decline of mining in that state. A demand for a large cash payment for a slightly developed property will in most cases make it inadvisable to proceed further. There are exceptions to this rule, when the history of near-by mines or some unusual geological features make it safe to count on much additional ore. The time required before payments must be made should of course vary with the amount of development that must be done to throw light on the future of the property. Usually, the specified time is too short. Even with apparently simple geological and operating conditions, unexpected delays are likely to occur, and leave the optionees in the unenviable position of begging for an extension of time when the optionors know that money already spent makes it almsot impossible to withdraw. As soon as an engineer sees that the time offered in an option is not sufficient for the necessary work, a few hours' study will tell him whether he should ask for better terms or definitely give up the examination. Occasionally the total price asked for a property is so high that further consideration is not worth while. If a Uath Copper price is asked for a nice little lease, nothing is to be gained by finding out just how nice the lease may be. But if sufficient time is given before payments are required, a seemingly high price is seldom cause for cutting short an examination. If the owners are reasonable enough to grant ample time, they can usually be persuaded to cut their price to one that is fair.

A brief trip underground and a study of mine maps often will reveal the fact that, even if the owner's assay value and estimated cost are accepted, it has cost a dollar and a half to find every dollar's worth of ore. In such cases re-sampling and careful study are not justified. The owner's estimates are seldom overly conservative, and may be accepted for the time as giving the property all it deserves. The problem then becomes one of geology. If a preliminary study of the structure and ore occurrence gives the impression that the early development was not well planned, or that it has not progressed far enough to yield representative results, the property may merit more careful geological study. If the ore is apparently bunchy, selective mining may turn failure into success. But too much optimism is dangerous in considering a mine in which previous operators have not found enough ore to pay for the hunting. Even the prospector who most scorns geology and all varieties of science seems to have an uncanny faculty for doing his development in the best place. The road of the geologist who tries to "show up" the old-time miners is indeed rocky.

THE problem of how thorough an examination should be is a puzzling one. Owners of properties nearly always think the engineer is too hasty. They bitterly, and now and then justly, criticise the examiner who makes a decision after an inspection of a few minutes or a few hours. . . . A mine examination is simply a process of testing the successive links of a chain of observed facts. As soon as any link is found to be too weak for the purpose it must serve, the examination is over. Any further time spent confuses the issue and wastes the money of the client.

The history of the district in which a mine is situated may help an engineer to reach an early decision as well as the history of the mine under consideration. Orebodies seldom grow alone. If there are no other prospects, and if no obvious general mineralization occurs in the neighborhood, there is little chance of developing even a moderately valuable mine. If other mines near by have played out at shallow depth or have passed into lean material, perhaps below a zone of surface enrichment, the chances are that the new mine will have a future as disappointing. A geological survey or other report that tells about the district is, therefore, of the greatest value. But except for the history and general geology, it is not safe to put too much reliance on such reports. New developments may have shown some reason why the mine in question may differ from surrounding mines; or in a general examination, perhaps only a reconnaissance, the earlier observers may have overlooked important possibilities. Nearly all the properties an engineer now visits have been condemned by half a dozen others. Unless he keeps an open mind and relies on his own observation he will seldom find a mine.

Transportation Often Decisive Factor

The relation of character of ore to transportation facilities may quickly convince an engineer that he would be wasting his time on a long examination. Ore is often much too lean to stand the cost of hauling to a custom treatment plant. Then if proved or probable reserves do not justify erecting a plant at the mine, it has no value. If the mine is far from a railroad, the cost of hauling supplies and metal or concentrates may be too great, even if a plant is built at the property. Sometimes a high-grade product near a railroad has no value, owing to market conditions and the location of fuels, fluxes, or factories for making finished products. For instance, a fairly rich zinc-carbonate ore is valuable only in certain parts of the country. High-grade iron ore is in many places worthless, and may continue to be so for a dozen years or more. Coal can be profitably exploited only under relatively favorable conditions. However, before definitely condemning a mine because of the character of the ore or the locality, an engineer should make allowances for future developments in metallurgy and transportation. In doing this he should bear in mind the fact that carrying charges soon double the investment. Unless it is likely that an ore will become valuable within a few years, it is usually cheaper to pay a higher price for the mine after it is certain that a profit can be made.

An engineer often sees at once that 100 ft. of development work will be much more useful than a thorough examination. A short crosscut may develop a vein beyond a fault that had been overlooked in old workings. A drill hole or two may show whether or not the sulphide zone under an iron-stained porphyry outcrop contains copper in paying quantity. Surface trenching may prove or disprove the continuity of mineralization along a shear zone or vein. Or a test pit may reveal the fact that a limonite outcrop in limestone passes in a few feet into a lead carbonate orebody.

In many such circumstances careful geological work might give nearly as reliable information at far greater profit to the engineer. Detailed mapping often makes a suspected fault a certain one. Microscopic study of an outcrop may prove to the satisfaction of the examiner that lead or copper sulphides, as well as iron, were present in the primary material that was oxidized to an iron capping. Slight alteration of wall rocks may show that a vein or mineralized shear zone continues even though the outcrop is covered. Painstaking geological study that reveals such facts quite frequently is of the greatest possible value.

But if the same information can be had at approximately the same cost by underground work, the engineer must regretfully tell his clients that what they need is a hole in the ground and not an elaborate report. He may seem to be making a casual examination, but the interests of his clients forbid more thorough work. A report based on a month's detailed study will convince only those who can follow the technical reasoning, whereas a vein exposed by a crosscut is patent to anyone who can swing a pick. The links in the chain of evidence thus far discussed can usually be tested quickly. If they are satisfactory, the more difficult work begins. The time necessary for sampling, geological mapping, study of methods of mining and treatment, and probable costs varies with every property visited. Each of these phases of the examination must be taken up in turn, in just sufficient detail to give the information desired. Too much time spent on any one item wastes the money of the client and makes a correct conclusion more difficult, by exaggerating the importance of one link in the chain.

In nine out of ten examinations only a few rough check samples, or no samples at all, need be taken. If underground work has not progressed to a point where ore developed is likely to be representative of new ore that will be found, it is foolish to waste a lot of time and money in ascertaining the grade of the exposed ore. Even if this ore is likely to be representative, it is seldom worth while to do careful sampling unless the quantity of ore developed is reasonably great. In most such cases grade is not the link of the chain that needs testing, but tonnage. If it is uncertain whether there will be 1,000 or 1,000,000 tons, the question whether the ore is worth $8 or $9 per ton is purely academic.

Often the record of shipments is a much safer index of grade than any amount of sampling. A few samples in a bottom of a drift to prove the ore has not been cut off, together with smelter returns from a stope above the drift, are just as informing as a long series of channel samples.

Sampling that is too careful often given a wrong impression of the degree of accuracy of the examination as a whole. When the grade of ore is established to a tenth of a per cent, uncertainty as to whether dilution in mining will be 5 or 10 per cent, and whether the determined grade represents 10,000 or 20,000 tons, is often lost sight of. Accordingly, time spent in making an estimate of grade far more accurate than estimates of other factors in the examination is totally wasted.

WITHOUT a thorough knowledge of the size and grade of the deposit, the geological conditions, and the physical character of the ore, an intelligent study of mining methods is impossible. Some of the worst mistakes in the history of mining have resulted from accepting false premises on these points. Several huge gold mining enterprises have foundered because of the assumption that bulk mining on a great scale would yield the same grade as the smaller stopes of former operations. Failure to foresee the effect of a well-known fault on a caving system has cost millions of dollars. The selection of open-pit mining where the orebody was better suited to underground mining nearly wrecked a great corporation; and an underestimate of the cost of drilling and breaking an ore postponed the success of another for at least a score of years, after millions had been spent in equipment.

There are of course instances in which the greatest possible care is necessary in sampling. If a great tonnage of low-grade ore is developed, on which the margin of profit will surely be low, a slight mistake in the estimate of grade may be fatal. In such a property there are uncertainties enough at the best. Mining grade is one thing that can be accurately determined, and this should be done at any expense. If a payment must be made for a property, or a large sum appropriated for a road or treatment plant, the engineer should make sure that the value of ore developed justifies the expenditure. He must then adopt all possible safeguards in his sampling.

If even a small amount of already developed ore is likely to be fairly typical, careful sampling of this ore may be necessary to show whether or not further exploration is advisable. In short, when the grade of the ore that can be seen is judged to be a vital factor in arriving at a decision, sampling becomes the most necessary part of the examination. When the contrary is true, little time need be spent on sampling.

As always, the interpretation of samples in making an estimate of developed ore requires more skill than the mechanical operation of sampling. An understanding of the geology and a general idea of the probable method of mining must precede the estimate. Assays of all the material that will be mined must be included in calculating the grade, and must be weighted in proportion to the quantity of material they represent. A simple mathematical computation will give the cubic contents with an accuracy far greater than that of other factors in the estimate. Specific gravity must be investigated, with the realization that even if the cubic contents is exactly figured, the tonnage estimate is no more accurate than the specific gravity determination. With all the care that can be exercised, experience and a knowledge of the quirk in psychology or fate that always places irregularly spaced workings in those parts of an orebody that contain the most gold or silver or copper are indispensable.

A study of detailed geology is more often necessary in planning development than in an examination for purchase. Generally , a brief reconnaissance will solve the simple geological problems involved in deciding whether or not to buy or option a property. There are, of course, exceptions. These exceptions are becoming more common as easily discovered, shallow orebodies are exhausted and more expensive deep exploration must be undertaken to find new mines.

If complex faulting makes the correlation of exposures underground or on the surface uncertain, the engineer must devote care and time to solving the problems. Often a new property is in an old district, but on account of property lines the proposed work is far from developed ore. In such circumstances a detailed geological study is necessary to fix the depth and position of strata of veins. Surface exposure in beds which do not contain ore may throw light on deeper mineralization. The recent discovery of the North Lily orebody in the Tintic district in Utah is an outstanding example of successful long-range ore finding by thorough geological work.

It sometimes happens that the developed ore in a district is associated with a certain type of surface alteration. Then an accurate mapping of areas having similar alteration is worth while. Frequently a very slight structural deformation, that can be discovered only by a careful geological survey, is of vital importance. In certain oil fields a dip of only a few feet in a mile is the only guide to the location of productive wells. Accordingly there are many cases in which an attempt to cut short the geological features of an examination would be fatal.

The newly developed geophysical methods often help to solve uncertainties in the geological aspects of an examination. In areas covered by a shallow, fairly uniform mantle of soil or glacial drift, an examination without the use of such methods has already become an anachronism. With increasing experience, the field of geophysical methods is constantly widening. As yet they are of little benefit in most of the Western districts.

Generally, the ore is too deep to effect the instruments as much as various extraneous influences, such as power lines, pipe lines, flow of underground water, or irregular surface contours. Even where ore comes near the surface, rock exposures are so common that a study of outcrops will usually give more information than a geophysical survey. But though these new aids of geology are thus far strictly limited in their application, they must always be considered as possible aids to an examination.

Mining Method and Metallurgy

A study of the mining method that is most economical need be undertaken only after the engineer is reasonably certain that there will be a fair amount of ore of workable grade. Then the system of mining will have a very important effect on the rate of production, the grade of ore that will be mined, and the cost.

In most small mines the proper method of mining is obvious. A more or less offhand estimate of the possible rate of production, mining cost, and dilution by waste from the walls is usually much more accurate than many other factors determined in the examination. The most doubtful question is likely to be whether or not sorting and selective mining may make it possible to improve on results obtained by former operators. In many instances this change has transformed an operating loss into a good profit. A study of stope assays, where these have been systematically recorded, often throws much light on this question. Sometimes special sampling, supplemented by experiments in sorting, must be resorted to. Selective mining will nearly always be met by violent opposition on the part of foremen and superintendents, who have been trained to regard per-ton costs, and not costs per unit of metal, as the measure of efficiency.

Features That Determine Operating Detail

In larger orebodies the influence of mining methods on the result of the examination is more likely to be decisive. The rate of production often determines success or failure. If a large tonnage can be mined daily, the resulting reduction of the overhead expense per ton may more than atone for the low grade of the ore. The adjustment of the overhead expense to size of orebody; the selection of open-cut, caving, or other plan of mining; the amount of dilution that will result; the cost of preliminary development and the amount of capital required for this development; the question of selective mining versus bulk mining—all of these problems must be decided before the future of a large low-grade mine can be judged.

Without a thorough knowledge of size and grade of the deposit, the geological conditions, and the physical character of the ore, an intelligent study of mining methods is impossible. Some of the worst mistakes in the history of mining have resulted from accepting false premises on these points. Several huge gold-mining enterprises have foundered because of the assumption that bulk mining on a great scale would yield the same grade as the smaller stopes of former operations. Failure to foresee the effect of a well-known fault on a caving system has cost millions of dollars. The selection of open-pit mining where the orebody was better suited to underground mining nearly wrecked a great corporation, and an underestimate of the cost of drilling and breaking an ore postponed the success of another for at least a score of years, after several millions had been spent in equipment.

Seldom can large-scale experiments in mining be carried on during the examination. The engineer must exhaust every line of attack in learning all the facts about the orebody, and then after planning various systems of mining in detail, must make the best guess he can as to the probable results of the system chosen. If these results agree fairly well with those at operating mines of similar character, he can be sure he has made no great mistake.

Unless enough ore is developed to be typical, a study of the probable method of treatment of the ore is not worth while. If a considerable tonnage has been developed, or if new ore is likely to be like that already developed, metallurgical study becomes a vital part of the examination.

Value of Former Metallurgical Records

If there has been a mill on the property, or if ore has been shipped to a custom mill or smelter, the problem is simple. A study of records will give a definite figure to start from. The old results may be improved on by a change in the shipping product brought about by selective mining or a change in character of ore in depth. The treatment in use may obviously be out of date or poorly adapted to the ore. But if actual results of past operations are known, there is a sure foundation on which improvements may be built. Elimination of any uncertainty should make the engineer happy.

If there have been no past operations to help in a study of ore treatment, this subject must be approached warily. It must be preceded by sufficient sampling to give a fair idea of the average quality and of the quantity of ore available. Then the cost of metallurgical tests must be in proportion to the size of the mine. If the orebody is small, it cannot bear the expense of elaborate experiments. The choice will then lie between shipping to a custom plant and putting up a simple mill, using a well-known process. Estimates of cost of treatment by the custom plant, or estimates of cost of construction and operation by some metallurgical laboratory which makes tests cheaply, must be accepted, subject to the general information of the engineer. If there is tonnage enough to justify greater expense, more difficult problems may be attacked. Then, in addition to asking the regular metallurgical laboratories to make suggestions, it often pays to employ independent metallurgists for the job, and to build small test plants. Such proceedings of course require a long option, or a price so low that it pays to buy a property on the chance that eventually the ore may be successfully treated.

It is a tribute to the metallurgists of today that if there exists a large tonnage of ore with a reasonable metal content, with practicable transportation facilities, it is usually assumed that a way will be found to treat the ore successfully. The experiments may be long and expensive, but finally the ore will yield to treatment.

Cost the Decisive Factor

If one link in a chain can be more important than the others, the most vital link in a mine examination is that of costs. The whole object of the work has been to find out whether a mine will make money. Yet the engineer seldom arrives at a point where careful consideration of this question is necessary. In most cases there is a factor so obviously weak that a detailed study of costs is unnecessary. A very rough approximation, based on experience with similar properties, will show that there is no chance of a successful outcome. Then a more meticulous cost estimate would be foolish. Even in favorable reports, the best that can usually be said is that further development may find ore. Until this development is done, an estimate of costs or value is a pure guess. All that an engineer can do is to let his imagination loose, guided only by his experience, and paint a picture of the sort of mine he hopes to find. If this possible prize, in the event of success, is great enough to justify risking the money required for development, he can recommend the venture. In simpler terms, the pot is big enough to justify the ante.

If the method of treatment is not well established, an accurate cost estimate is, of course, impossible. Here the successful outcome depends on metallurgical instead of geological development. The chance of success in metallurgical development is usually better, and the justified outlay for purchase and experimentation is greater in proportion to the profit hoped for. But here too the engineer can only say that he thinks the prize is worth the risk.

Accurate Cost Data Depend on Well-Established Process

Once in a while an examination must be made of a mine in which a large tonnage of representative ore has been developed; in which future developments are likely to be commensurate with the price asked; and in which the probable method of treatment can be well established. Such a property gives the engineer a chance to use all his skill. He must sample and estimate the ore accurately, foretell the future by geological study, plan the method of mining and treatment, and finally make an estimate of the costs and value. Only after the first three stages of the examination are complete can the last one be attacked intelligently. Then the whole process of development, mining, and treatment must be followed through in detail, and a probable cost estimated for every step. The safe way is to make a cost sheet just like the one that will be used when the property is operating. If possible, an actual cost sheet of a similar mine should be used as a guide. By doing this, the chance of overlooking any step is minimized and estimated costs are checked against actual costs. In every item the probable error should be weighed, and a maximum and minimum figure set down, as well as the figure accepted as most likely. The total of these figures will show the probable degree of accuracy as well as the accepted estimate.

The last step is to calculate the present value of the property under consideration. With the costs, the rate of production, and the probable life determined, this is just a matter of applying the Hoover or Findlay formulas. The only new uncertainty is the rate of profit that should be expected—or the risk rate, as it is called by the Bureau of Internal Revenue. The engineer can either set a rate which in his opinion will make up for the degree of risk involved, owing to fluctuating metal prices or other doubtful features, or can set a low rate of 6 or 8 per cent, and tell his clients in some other way about what risk they will be compelled to run in the enterprise.

The examination has been a process of elimination, testing in turn the various points in the order of weakness as seen in the first few hours of work. The interests of the client have determined every step. He is entitled to the facts necessary for a correct decision arrived at with the least possible expense. As soon as there is sufficient evidence on which to base a definite recommendation, the examination is finished.

30

Reprinted from pages 42–43, and 44 of *Mining Cong. Jour.* **27**:42–44 (1941)

MINING GEOLOGY TODAY and TOMORROW

By RENO H. SALES
Chief Geologist
Anaconda Copper Mining Company

I HAVE read with much interest Mr. Joralemon's paper entitled "Mining Geology Today," presented at the Colorado Springs meeting of the American Mining Congress. I am happy to be afforded the privilege of discussing certain features of it, and I shall take advantage of the opportunity to call attention to what I consider important phases of the subject not touched upon by the author. Although I find myself in general agreement with the substance of the paper, there are many statements which do not seem to harmonize with my own field experience.

Primarily in our consideration of the subject of mining geology, we should assume that we are dealing with trained men and geologists competent to exercise proper judgment in passing on the common types of geological problems confronting the prospector, small mine operator or the client. Under such reasonable assumption many of the difficulties cited by Joralemon disappear automatically. For example, there can be no excuse for a geologist spending months, or even weeks, on the job of mapping a small mine or prospect. A competent geologist does not make expensive laboratory studies covering phenomena unrelated to the problem in hand. But, I might add, even the experienced geologist should not permit himself to pass final judgment on matters of geology vital to his problem without adequate mapping and study of the geological facts available at the surface and in accessible mine workings.

Editor's Note: Mr. Joralemon's paper which is discussed by Mr. Sales appeared in the December, 1940, issue of MINING CONGRESS JOURNAL.

Urges Geology Be Built on Facts

Always there will arise the question as to what constitutes adequate mapping. It is undoubtedly true that the experienced geologist will be expected to more quickly form his opinion, but even he should not do a hurry up, half job, of mapping against time in order that he free himself to run to other prospects and mines, with assurances to his client that, allied with Lady Luck, he will meet up with a good one sometime, somewhere, somehow. Actually, good judgment dictates to him which properties are worthy of careful study and which are not, and he should act accordingly. I am firm in my belief that the geologist should build up geology, geological theory, and geological opinion, on facts, and not on hastily conceived notions, or on hunches.

Cautions Against Haste in Mine Examinations

In fact, throughout Joralemon's paper there runs the suggestion of hurry and haste. "Haste becomes a virtue," he says, "and in most cases decision must be adverse before geology starts," and further, "there is seldom need for anything save a very hasty and casual geological mapping." Worst of all we are told, "where the preliminary view shows that geology will play an important part in an examination, detailed work must be looked upon with suspicion," and later, "too detailed a study of phenomena that will probably prove not to be related to ore deposition will not only waste time, but often will give a wrong idea as to the occurrence." This is laying it on pretty thick to the discomfiture of the mining geologist. Here the value of careful geological work in connection with mine examination work is severely depreciated, and Joralemon does it on the assumption that the examining geologist is not competent to appraise the importance of the respective geological factors involved. Otherwise, be he competent, no part of his work will be looked on with suspicion, or will he waste time on geology unrelated to the problem in hand.

Mr. Joralemon injects a peculiar bit of reasoning in stating that when a small mine hires a geologist who spends "many months in a detailed survey of the mine and district, the ore found is likely not to pay for the finding," and on a later page he adds, and I quote:

"In making decisions based on the examinations of a day or two that are all the smaller companies can afford, the geologist runs a risk of missing some hidden factor that might profoundly influence his conclusions. Occasional ore bodies may escape him, or he may spend too much money to find ore. He may even cause the mine to shut down when a more detailed study might find profitable ore. This is a risk that he and his clients must run. If he has had a fair amount of experience, added to the intuition that no amount of detailed drudgery can entirely replace, the chance of making such mistakes is slight."

From which reasoning one must conclude that the geologist who does careful detailed mapping, or the "detailed drudgery" of an examination, finds only non-commercial ore, whereas the experinced man acting hastily with "intuition" and "chance" as "allies," plops into the big and valuable ones. Here again is unwarranted depreciation of the value of careful geological mapping, and applause for the "stop and run" school. But Joralemon reverses himself and supplies a partial answer when he says, "the most insignificant detail of today may lead to an entirely new outlook in a great district tomorrow." We might well ask, could not this be just as true, even more so, when we are dealing with prospects or small mines where every scrap of evidence is needed to enable the geologist to form intelligent conclusions? I think Joralemon will agree that the less the information in hand, the greater the difficulty in arriving at an opinion. Which, I repeat, does not mean wasting time on non-essentials.

Day of Simple and Easy Discovery of Mines Is Past

It is not exactly clear why the author suggests that the present day mining geologist should adopt some of the methods used by early day prospectors and mining scouts, to whom is credited, and rightly so, centuries of success in finding mines. I am certain no one realizes better than Joralemon that the day of the simple and easy discovery of mines has passed. There are no rich veins or deposits sticking out of the ground ready to be found by merely chipping off a sample or digging a 10-ft. hole. Geological gambles, so-called, are now the order of the day. How far would the early day prospector get today pawing over what now remains for our consideration? How many new mines are currently being found by the prospector of today who not only uses the methods of the early prospector, but in addition has the advantage of much published geological and mine information covering most of the country? Any mining man can supply the answer.

[*Editor's Note:* Material has been omitted at this point.]

Cites Examples in Support of Need for Careful Geological Studies

Let us pause to consider a few examples. Had the Butte central copper area been covered by the 500 to 1,000 feet of the extrusive rhyolite which spreads out over the northwestern part of the district, the picture there today might have been one of worked out, abandoned silver-lead-zinc mines such as the Bluebird, Nettie, Alice and Black Rock, with the presence of valuable copper ores unsuspected. A similar situation would have existed had there been 1,000 to 1,500 feet less erosion of the upper portions of the copper veins. When the first prospector arrived, he would have found only silver-lead-zinc ore above the copper area, and perhaps worthless outcrops above the Rainbow and other silver-lead-zinc veins.

In the Southwest, if Bisbee gulch and the original small Copper Queen outcrop had been hidden under a covering of Gila conglomerate or other deposit of relatively recent origin, leaving only the Sacramento Hill with its leached outcropping visible to the early prospector in that immediate area, one could readily visualize the Bisbee camp today as an abandoned porphyry ore pit with but little positive evidence indicating the existence of the tremendous high grade copper ore deposits which have made that district famous. In Utah, only slightly less erosion of rhyolite would have kept in reserve the important Tintic Standard and North Lily ore bodies, and in the Black Hills, S. Dak., erosion may have played a friendly role by removing enough Cambrian rocks to bring to light the outcrop of the famous Homestake ore body.

In the case of porphyry or other developed ore bodies, it may be that little or no geological mapping is needed to aid the day to day operation. But what of the day of reckoning when the end is near? Could there be a better place to explore for new or additional ore than one with existing up-to-date mine and metallurgical plants, and with established homes and communities depending upon the continuance of the enterprise? And could there be a sadder story for the management than a last minute discovery of evidence that valuable ore concentrations might exist in the immediate vicinity, only to find no record of pertinent geological facts upon which to build a proper prospecting program? Large sums are spent in metallurgical research to save pennies per ton in costs, or the last ounce of metal in recoveries, but little or nothing to make as certain as possible that millions in ore reserve are not being overlooked.

31

Reprinted from *Eng. and Mining Jour.* **134:**52–54 (1933)

DETERMINATION OF ORESHOOT BOTTOMS

Harrison Schmitt

THE subject of the bottoming of oreshoots is of predominant concern to the mine operator when he seems to have reached the downward end of his ore. He wishes to know if the pinching is true or apparent, and whether or not he should risk sinking the shaft another level, possibly against large volumes of water. That a problem usually so critical should be so difficult is unfortunate, but the best efforts of the mining geologist are therefore challenged. It seems worthy of a program of organized research.

This paper presents descriptions of the bottoms of several ore deposits studied by me in northern Mexico and in the Southwestern United States, concluding with tentative generalizations.

The oreshoots at Santa Barbara, Chih., Mexico, have been described by Spurr[1] and Schmitt[7]. Within the last six years two different types of bottoms have been reached in the district. One was found between the 8th and 10th levels of the Hidalgo mine south of Arroyo de las Vacas, where the average ore above the 8th level had approximately the following composition: Quartz, 60 per cent by weight; fluorite, 15; galena, 12; sphalerite, 10; chalcopyrite, 2; arsenopyrite, less than 1 per cent. A specimen selected as typical, from the 10th level, when examined with the microscope showed the following approximate composition: Quartz, 85 per cent by weight; fluorite, 10; pyrite, 2; chalcopyrite, less than 1; sphalerite, less than 1; galena, less than 0.1; arsenopyrite, less than 0.1 per cent. Minor primary carbonate was also found, associated with the quartz. Apparently the essential change is increase in quartz, decrease in sulphides, and appearance of primary carbonate. But for this latter *quality* change, the bottoming of the shoot could be assigned to lack of sufficient brecciation (Bain[5]; Hulin[6]) of the early barren quartz, and lack of sufficient openings for the deposition of commercial quantities of the sulphides. But the carbonate is probably indicative of a more fundamental change, unless it also existed in the upper part of the shoot and was replaced by the later metallic minerals.

Another type of bottom reached in the same district with more definite indications than the one just described is about 50 meters (160 ft.) below the surface in La Hundida mine, about 1 km. (0.6 mile) south of San Francisco del Oro. This seems to be a true bottom, for fundamental changes occur downward. Near the outcrop the shoot is similar to the Hidalgo vein above the 8th level; but the lead decreases rapidly downward, and the conditions on the lowest level are shown by Figs. 1 and 2. Here the minerals are coarse textured, with sphalerite and a few galena crystals up to 6 in. diameter, and chalcopyrite up to 2 in. diameter in stringers that cut coarse calcite and quartz (Fig. 2) and that contain no admixed gangue. The vein at this place is analogous to a pegmatite from the standpoint of grain size. The sulphide stringers characteristically appear lenslike in plan (Fig. 1), but rootlike in longitudinal vertical projection.

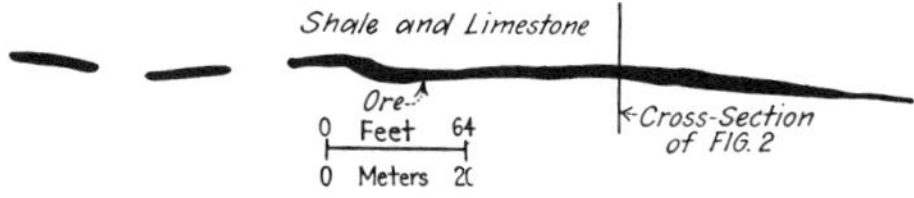

Fig. 1—Plan of second level, La Hundida mine, Santa Barbara district, Chihuahua. Shows lenslike horizontal section of the root of the main oreshoot.

A variation of this second type of bottom was reached at the 12th level of the San Albino vein at Santa Barbara, according to T. P. Clendenin, where the sides of the vein are composed of coarse pyroxene and other silicates, and the center is of coarse sulphides. The minerals were much finer grained above this level.

The resemblance of the texture of these bottoms to that of pegmatites is suggestive, particularly because orthoclase—with quartz, calcite, and silicates such as pyroxene and garnet—is found in veins in this district (Schmitt[7]). The geneticists seem to have a possible case here of a connection of veins with pegmatites. No granitoid intrusive is known in the vicinity of the veins.

The majority of the known oreshoots of the Parral district of Chihuahua, Mexico, except a few in the vicinity of the city of Parral, have been explored to and below their bottoms, so much information about them is available. They seldom extend below an elevation of 1,700 meters (5,450 ft., sea-level datum), according to John Barry, which means 50 to 300 meters (160 to 960 ft.) below their outcrops. They have the form of V's, and bottom in one or more roots (Fig. 3 and Schmitt,[7] pp. 284-85), which appear lenticular in plan on the 5½ and 6 levels of the Sierra Plata mine (Fig. 4), and on the lower levels of the Plata Verde mine, and are enclosed by a gangue of silicified breccia. At places in the Plata Verde mine, development below the silicification and sulphide roots disclosed a coarse unmineralized breccia the interspaces of which are voids. This mine made over 3,000 g.p.m. of water, and loose filling, such as gouge, in the interstices, may have been washed out with the rush of the water when the level was opened. I was told that the change, from massive silicified vein with sulphides to this breccia, took place within a few meters. The fluorite gangue was coarser at the bottom of the mineralized section. This breaking up into roots and change of mineral character in the Sierra Plata and Plata Verde shoots, I believe, represents a true bottom. The root of the Palmilla oreshoot (Fig. 3) on the 7th level showed a horizontal section of silicified rock about 50 meters (160 ft.) diameter, cut by veinlets of sulphides up to 3 cm. wide. The diminution in size, structural relationships, and change in character of this root also indicate a true bottom.

It was previously stated (Schmitt,[8] p. 283) that the shoots of the district show an increase in lead and zinc in depth. This may not represent a hypogene condition, for much zinc was leached out in the near surface zone; and in the early days the silver-rich lead may have been selectively mined near the outcrop. But if the down-

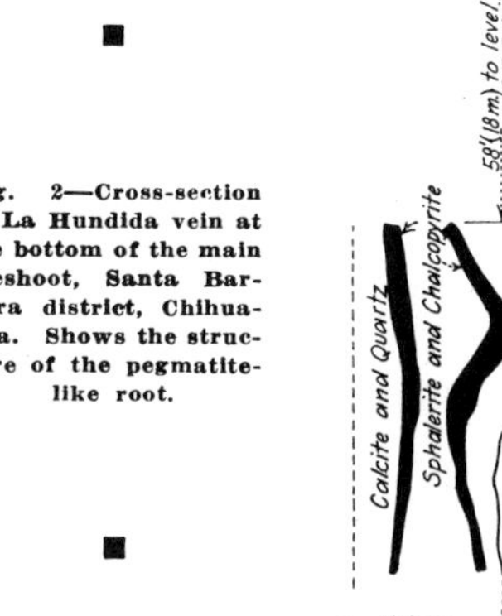

Fig. 2—Cross-section of La Hundida vein at the bottom of the main oreshoot, Santa Barbara district, Chihuahua. Shows the structure of the pegmatite-like root.

ward increase in base metals does represent a hypogene condition, it is evidence of approach of the bottom of the shoots.

Summarizing, evidence indicating bottoming of shoots in the Parral district includes: (1) diminution in size and tendency to extend downward as roots; (2) position relative to the 1,700 meters (5,450 ft.) above-sea-level elevation; (3) change to coarse fluorite; (4) change to unmineralized breccia; (5) increase in lead and zinc.

Most of the sphalerite ore at Hanover, N. M., occurs as replacement bodies localized by fissures that cut a 120-ft. white, pure crinoidal limestone bed near the contact of an intrusive "stock," and is associated with andradite, hedenbergite, and other silicates. An 18-ft. shale bed superjacent to the crinoidal limestone appears to have acted as an "impervious" bed, because the ore in many places spreads out

Plan of Ore on 550 Level

Plan of Ore on 600 Level

0 Feet 100

0 Meters 30

Fig. 4—Plan of the bottom levels of Sierra Plata oreshoot on the sixth level, Parral district, Chihuahua.

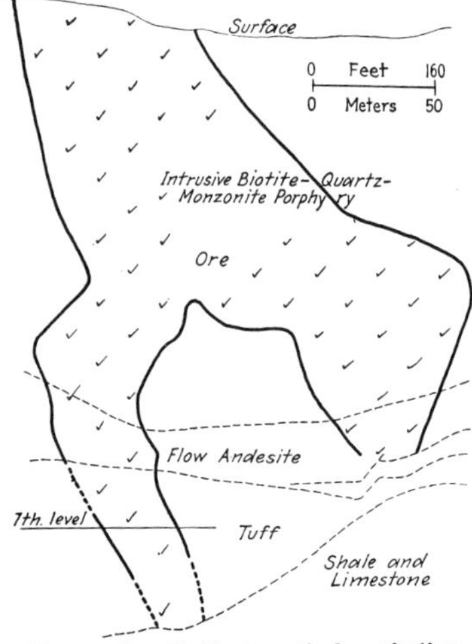

Fig. 3—Longitudinal vertical projection of the Palmilla vein and main oreshoot, Parral district, Chihuahua. After F. W. Smith.

along its lower contact. A 200-ft. gray limestone bed, 200 ft. of fissile shale, and finally a thick series of dolomites and magnesian limestones occur in succession below the crinoidal limestone.

The largest of the oreshoots in the crinoidal limestone, particularly when localized by crosscutting faults, and when extending downward through the entire 120-ft. section, have a bottom of sphalerite (6 to 10 per cent Zn) in the 200-ft. gray limestone. Here the sphalerite and associated gangue silicates have a much finer grain, apparently caused by the finer texture of the host rock. When the localizing fissures are followed downward still farther, through and below the 200-ft. shale, they are found to contain pyrite with no sphalerite. This is believed to represent a fundamental change and to indicate the true bottom of the zinc ore.

The characteristics of bottoming of the manto-chimney type of deposit within the area selected for this discussion are unique and should be mentioned. The downward and lateral connection of the mantos with chimneys at Erupcion-Ahumada and Santa Eulalia, Chihuahua, and at Gilman, Colo., seems to present an indication of value, because no extension has ever been found below the bottoms of the chimneys. Moreover, the change from manto to chimney is commonly accompanied by a mineralization change. For example, at Erupcion-Ahumada, Chihuahua, the oxidized lead mineralization of the manto becomes high in vanadium and molybdenum minerals (vanadinite, descloizite, and wulfenite) when it connects with and turns down into the chimney. At Santa Eulalia, Chihuahua, the iron silicates first appear in the bottom of the Mina Vieja chimney; and at Gilman, Colo., the chimneys are of massive pyrite, with minor chalcopyrite and tetrahedrite, whereas the connected mantos carry oxidized and unoxidized lead and zinc minerals. In this connection it is interesting to note that the lead-zinc sulphide replacement orebodies in the upper levels of the Nitt mine, Magdalena district, N. M., turn to roots of massive pyritic copper ore in depth and pinch out quickly downward, although the distinctive manto-chimney form is not developed there.

Silver-manganese-quartz deposits below the Devonian Percha shale in southwestern New Mexico have been described (Lindgren,[10]). The ore is localized by fissures that cut the shale and replace the magnesian limestone just under the shale; it never extends more than 100 ft. below the contact. The topography is rugged in the six districts that exemplify this type; and, although no ore outcrops in the limestone beds exposed for hundreds of feet stratigraphically below the shale, fruitless deep development has still been done on the ore-bearing fissures, with the idea that they are veins. If no deep development had been done at all, it should have been obvious from a statistical point of view that the fissure ore would die out within ± 100 ft. below the shale contact.

The "funnel" ore of Utah and Nevada, described to me by Augustus Locke and Paul Billingsley, seems to be a distinct type. It is well developed at Tintic, Park City, Ely, and Eureka, and has affiliations with the manto-chimney type; indeed, it merges with the latter and similarly is essentially the convergence downward of horizontal elements into a vertical element. The convergence is funnel-like, and not marked by a sharp break from manto to chimney. This is probably explained by the fact that the true manto-chimneys occur in flat-lying massive undisturbed limestones, whereas the "funnel" is found characteristically in intensely thrust folded and faulted zones. Many of the "funnels" bottom in coarse breccias, and this phenomenon appears to be an important indication of bottom, along with mineral changes such as a decrease in base-metal content.

A distinction should be made between the true and the apparent bottoms of oreshoots. The apparent bottom may have no more significance than the ending of a shoot horizontally, and should show no fundamental mineral change. Similar shoots may exist below. The true bottom is critical, and represents the change from ore stability to non-stability. How may these types be distinguished?

Our present knowledge suggests that in the province under discussion, the position of ore in relation to the surface of the earth is the product of three factors: (1) position of ground preparation in relation to the surface of the earth; (2) position of the zone of ore precipitation in relation to the surface of the earth; and (3) effect of erratic chemically and/or texturally reactive bodies on ore precipitation. In other words, given a crosscutting structure—such as a fault, fracture, dike contact, intrusive contact, breccia chimney, or breccia cone—the locus of the ore depends on the vertical range along this structure in which the normal factors of temperature, pressure, and concentration exert an ore-precipitating effect.

Therefore, if the zone of the factors that exert a positive action on ore deposition (zone of precipitation) is higher than the bottom of a given fault or mass of breccia (zone of ground preparation), the ore will end downward in unmineralized broken ground, and will have a true bottom such as has been described. If the zone of precipitation is lower than the zone of ground preparation, the ore will end downward without fundamental mineral change, and will have an apparent bottom. In the second case, prediction usually must be indefinite, because the possibility of the ground opening up again in depth is usually indefinite; in the first case, we may be able to develop reliable indications.

The problem is complicated by the possibility that the stability zones of the gangue minerals may be higher or lower than those of the ore minerals; and still more complicated when the gangue minerals, particularly quartz or silicates, are older than the ore minerals. In this circumstance the introduction of the ore minerals must be preceded by shattering, unless the openings were not previously completely filled by the gangue. Therefore, if the evidence indicates the succession: early quartz, shattering, and ore minerals, the early gangue is literally wall rock, and the shattering and ore mineralization are the immediate factors to consider.

At this point the problems of structure deserve attention. Experience indicates that the chances are fewer than one in twenty that a given oreshoot will extend more than 2,000 ft. (Hoover,[1] p. 32) vertically below the surface; and, say, one in a thousand that it will extend more than 5,000 ft. Hoover says (p. 31) that the average ratio of outcrop length to depth for a large number of veins is near 2 : 1 and calculations by Schmitt ([9], p. 324) indicate that this ratio is the average of eight important veins in Utah. Observation suggests that the broken ground in many mineralized districts decreases in form and rate comparable to a cone whose sides converge at 45 deg. Many hanging-wall fracture systems, "funnel" systems, and so-called stocks converge downward at about this rate. The normal section of this hypothetical cone would have a ratio of base to depth of 2 : 1—Hoover's ratio for oreshoots or ore zones in veins. This suggests that, on the average, oreshoot extension in depth may be closely related to the depth of preparation of the ground.

The many factors that may cause ground preparation to depart from the ideal downward diminution just outlined include those that tend to cause horizontal restriction, such as vertically fractured, faulted, or schistose rocks, or any vertical structural expression, and those that cause horizontal expansion, such as massive horizontal limestones, unconformities, impervious beds, and thrust faults. A critical plane commonly present in the West is the contact between the basement, in which the vertical factors are likely to predominate, and the higher disturbed or undisturbed bedded rocks. This plane often marks the bottom of a manto-chimney or funnel system.

So much for the broad relationships of structure to depth extension. When a specific orebody or vein is under consideration, reliable relationship between outcrop length or area and depth is obviously not known, except possibly such as may be determined and applied in a given district. For some districts a level below which few orebodies extend may be ascertained. Such broader data all enter into and strengthen assumptions or speculations as to the bottom of a shoot. Probable definite indications for true bottoms include: (1) unmineralized pre-ore breccia extending below ore; (2) coarsening of texture of gangue and ore minerals up to that of a pegmatite; (3) gradation of ore minerals to contemporaneous gangue minerals; (4) change to chimney form in the manto-chimney or related deposits, especially if accompanied by a change of mineralization, particularly from base-metal sulphides to pyrite or iron silicates; (5) tapering of orebodies to distinct roots; this usually must be judged with other evidence. Unreliable indications (they suggest a chance for new ore below) include: (1) pinch in ore without textural or mineral changes; (2) change from ore to pre-ore mineral gangue; (3) changes of the zoning type; these can sometimes be judged, however, by the horizontal zoning.

■

Acknowledgment

I am greatly indebted to F. F. Grout and Augustus Locke for criticism of the manuscript of this paper.

References

1. Hoover, H. C.: "Principles of Mining," McGraw-Hill Book Co., New York, 1909.

2. Prescott, B.: "The Underlying Principles of the Limestone Replacement Deposits of the Mexican Province," *Eng. & Min. Jour.*, Vol. 122, pp. 246-53 and pp. 289-96, 1926.

3. Hulin, C. D.: "Structural Control of Ore Deposition," *Econ. Geol.*, Vol. 24, pp. 15-49, 1929. Discussion: Koeberlin, F. R., Vol. 24, pp. 657-63; Porter, C. A., Vol. 24, pp. 866-69.

4. Bain, G. W.: "Structure of Gold-Bearing Quartz in Northern Ontario and Quebec," A.I.M.E. Tech. Paper 327, 1930.

5. Hayward, M. W. and Triplett, W. H.: "Occurrence of Lead-Zinc Ores in Dolomitic Limestones in Northern Mexico," A.I.M.E. Tech. Paper 442, 1931.

6. Spurr, J. E.: "The Ore Magmas," McGraw-Hill Book Co., New York, 1923.

7. Schmitt, H.: "Geologic Notes on the Santa Barbara Area," *Eng. & Min. Jour.*, Vol. 126, pp. 407-11, 1928.

8. Schmitt, H.: "Geology of the Parral Area," *Trans.* A.I.M.E., pp. 268-89, 1931. Discussion by H. E. McKinstry, pp. 289-90.

9. Schmitt, H.: "The Extension of Oreshoots," *Trans.* A.I.M.E., pp. 318-324, 1929.

10. Lindgren, W., Graton, L. C., and Gordon, C. H.: "The Ore Deposits of New Mexico," U. S. Geological Survey, Prof. Paper 68, 1910.

32

Reprinted from pages 799 and 813-815 of *Econ. Geology* **68:**799-815 (1973)

The Tops and Bottoms of Porphyry Copper Deposits

RICHARD H. SILLITOE

Abstract

Although it is now widely accepted that porphyry copper *deposits* consist of zonally arranged shells of alteration and mineralization centered on high-level, calc-alkaline stocks, the nature of their uneconomic upward and downward extensions remains undocumented. This paper attempts to characterize these upward and downward extensions and to integrate the resulting concepts into a hypothetical model for complete porphyry copper *systems*. Examples from Chile, Argentina, and elsewhere are used to aid in the substantiation of the model. Programs of exploration for porphyry ore deposits can clearly benefit from the application of a model of this sort.

A typical porphyry copper-bearing stock is inferred to grade downward into stockwork mineralization and potassium silicate alteration in a phaneritic intrusive, which in turn is transitional downward to an essentially unaltered pluton of considerably larger dimensions than the stock. Porphyry copper deposits are normally located in the basement beneath a comagmatic volcanic pile, which is transected by a column of hydrothermal alteration representing the upper parts of the porphyry copper system. This alteration consists of propylitic and argillic types with localized patches of silicification and advanced argillic alteration. The volcanic pile is thought to constitute a stratovolcano which possesses large native sulfur deposits and small quantities of base metals, particularly copper, in sublimates at high-temperature fumaroles in the vicinity of its central vent; these surficial deposits are considered as the effluent products of active porphyry copper systems.

The available evidence favors the emplacement of the tops of typical porphyry copper *deposits* at depths of 1.5–3 km beneath the summits of stratovolcanoes and suggests that entire porphyry copper *systems* possess vertical extensions as great as 8 km.

At Chuquicamata, Chile, a major high-angle fault may have cut the porphyry copper deposit, and subsequent erosion has removed the portion of the deposit that was situated in the upthrown block. The hydrothermal alteration pattern in the remaining part of the ore body is incomplete and terminates abruptly against the fault. The unaltered, phaneritic granodiorite, containing minor veins and pegmatitic bodies, in the upthrown block is interpreted as the root zone of the Chuquicamata porphyry copper system.

The lower most, *mineralized* part of a porphyry system is believed to be exposed at Los Loros, Chile. There a zone of molybdenum-rich and copper-poor potassium silicate alteration carrying abundant K-feldspar occupies an area in the interior of a relatively large pluton of phaneritic granite.

At Farallón Negro, northwest Argentina, several small porphyry copper deposits pierce the infrastructure of a temporally related, andesitic stratovolcano. This unusual locus of the deposits above the subvolcanic basement enables it to be determined that porphyry copper emplacement was a late event in the construction of the stratovolcano, succeeded only by the formation of minor rhyolite intrusives and "epithermal" veins.

Extensive zones of pyritic alteration including widespread silicification, in which intrusive rocks are virtually absent, are visible in the centers of eroded stratovolcanoes, as at Cerro Marquéz in northern Chile. Such zones are interpreted as the columns of alteration spanning the vertical interval between porphyry copper deposits and the vent areas of uneroded volcanoes. At Cerro Queva in northwest Argentina, lead-silver mineralization associated with advanced argillic alteration is located in an alteration zone beneath the summit regions of a stratovolcano.

It may be concluded that during the final stages of construction of stratovolcanoes, fumarolic and hot-spring activity are the surficial manifestations of the efflux of metal-bearing magmatic fluids from magma chambers during retrograde boiling, the interaction of these fluids with the groundwater system and the consequent formation of alteration and mineralization. The proposed model implies that porphyry copper systems effectively span the boundary between the plutonic and volcanic environments.

[*Editor's Note:* Only the abstract and reference list have been reproduced here.]

REFERENCES

Aguirre, L. L., and Egert, R. E., 1970, Cuadrángulo Lambert (La Serena), Provincia de Coquimbo: Inst. Inv. Geol. [Santiago] Carta no. 23, 28 p.

Bamford, R. W., 1972, The Mount Fubilan (Ok Tedi) porphyry copper deposit, Territory of Papua and New Guinea: ECON. GEOL., v. 67, p. 1019–1033.

Bazin, D., and Hübner, H., 1969, Copper deposits in Iran: Iran Geol. Survey Rept. 13, 232 p.

Brown, A. S., 1969, Mineralization in British Columbia and the copper and molybdenum deposits: Canadian Mining Metall. Bull., v. 62, no. 681, p. 26–40.

Burbank, W. S., and Luedke, R. G., 1969, Geology and ore deposits of the Eureka and adjoining districts, San Juan Mountains, Colorado: U. S. Geol. Survey Prof. Paper 535, 73 p.

Caelles, J. C., Clark, A. H., Farrar, E., McBride, S. L., and Quirt, S., 1971, Potassium-argon ages of porphyry copper deposits and associated rocks in the Farallón Negro-Capillitas district, Catamarca, Argentina: ECON. GEOL., v. 66, p. 961–964.

Carter, N. C., 1970, Copper and molybdenum porphyry deposits in central British Columbia: Canadian Mining Jour., v. 91, no. 4, p. 74–76.

Casertano, L., 1963, General characteristics of active Andean volcanoes and summary of their activities during recent centuries: Seismol. Soc. America Bull., v. 53, p. 1415–1433.

Clark, A. H., 1970, An occurrence of the assemblage, native sulfur-covellite-"$Cu_{5.5x}Fe_xS_{6.5x}$," Aucanquilcha, Chile: Am. Mineralogist, v. 55, p. 913–918.

Emmons, W. H., 1927, Relations of the disseminated copper ores in porphyry to igneous intrusives: Am. Inst. Mining Metall. Engineers Trans., v. 75, p. 797–815.

García, H. H., 1969, Cansideraciones sobre algunas alineaciones de desarrollos hidrotermales tipo pórfiro cuprífero en el noroeste Argentino: Dirección Nac. Geología Minería Rev. [Buenos Aires], no. 18, p. 55–66.

—— 1971, Geología del pórfiro cuprífero Bajo de San Lucas, Depo. Andalgalá, Prov. de Catamarca, República Argentina: Assoc. geol. Argentina Rev., v. 26, p. 327–342.

González Bonorino, F., 1950, Geología y petrografía de las Hojas 12d (Capillitas) y 13d (Andalgalá), Provincia de Catamarca: Dirección gen. Industrias Minas [Buenos Aires] Bol. 70.

Guest, J. E., 1969, Upper Tertiary ignimbrites in the Andean Cordillera of part of the Antofagasta Province, northern Chile: Geol. Soc. America Bull., v. 80, p. 337–362.

Gustafson, L. B., and Hunt, J. P., 1971, Evolution of mineralization at El Salvador, Chile [abs.]: ECON. GEOL., v. 66, p. 1266–1267.

Hamilton, W., 1969, The volcanic central Andes—a modern model for the Cretaceous batholiths and tectonics of western North America: Oregon Dept. Geology Mineral Industries Bull. 65, p. 175–184.

Hemley, J. J., Hostetler, P. B., Gude, A. J., and Mountjoy, W. T., 1969, Some stability relations of alunite: ECON. GEOL., v. 64, p. 599–612.

Howell, F. H., and Molloy, J. S., 1960, Geology of the Braden orebody, Chile, South America: ECON. GEOL., v. 55, p. 863–905.

Ivanov, V. V., 1959, Present-day hydrothermal activity within the Kurile-Kamchatka island arc and its relation to volcanicity: Bull. Volcanol., ser 2, v. 2., p. 137–154.

James, A. H., 1971, Hypothetical diagram of several porphyry copper deposits: ECON. GEOL., v. 66, p. 43–47.

Jarrell, O. W., 1944, Oxidation at Chuquicamata, Chile: ECON. GEOL., v. 39, p. 251–286.

Katsui, Y., and González, Ferrán, O., 1968, Geología del área neovolcánica de los Nevados de Payachata; con consideraciones acerca del volcanismo cenozoico superior en los Andes chilenos; provincia de Tarapacá, departamento de Arica: Univ. Chile Dept. Geología [Santiago] Pub. 29, 61 p.

Kilinc, I. A., and Burnham, C. W., 1972, Partitioning of chloride between a silicate melt and coexisting aqueous phase from 2 to 8 kilobars: ECON. GEOL., v. 67, p. 231–235.

Kinkel, A. R., Jr., 1966, Massive pyritic deposits related to volcanism and possible methods of emplacement: ECON. GEOL., v. 61, p. 673–694.

Llambías, E. J., 1970, Geología de los yacimientos mineros de Agua de Dionisio: Asoc. Argentina Minería, Petrología Sedimentología, v. 1, no. 1–2, p. 2–32.

—— 1972, Estructura del grupo volcánico Farallón Negro, Catamarca, República Argentina: Asoc. geol. Argentina Rev., v. 27, p. 161–169.

López, V. M., 1939, The primary mineralization at Chuquicamata, Chile, S. A.: ECON. GEOL., v. 34, p. 674–711.

Lowell, J. D., 1968, Geology of the Kalamazoo orebody, San Manuel district, Arizona: ECON. GEOL., v. 63, p. 645–654.

—— and Guilbert, J. M., 1970, Lateral and vertical alteration-mineralization zoning in porphyry ore deposits: ECON. GEOL., v. 65, p. 373–408.

Meyer, C., and Hemley, J. J., 1967, Wall rock alteration, *in* Barnes, H. L., ed., Geochemistry of hydrothermal ore deposits: New York, Holt, Rinehart and Winston, Inc., p. 166–235.

Mitchell, A. H. G., and Garson, M. S., 1972, Relationship of porphyry copper and circum-Pacific tin deposits to palaeo-Benioff zones: Inst. Mining Metallurgy Trans., sec. B, v. 81, p. B10–B25.

Mizutani, Y., 1970, Copper and zinc in fumarolic gases of Showashinzan volcano, Hokkaido, Japan: Geochem. Jour., v. 4, p. 87–91.

Moore, W. J., 1973, A summary of radiometric ages of igneous rocks in the Oquirrh Mountains, north-central Utah: ECON. GEOL., v. 68, p. 97–101.

—— Lanphere, M. A., and Obradovich, J. D., 1968, Chronology of intrusion, volcanism, and ore deposition at Bingham, Utah: ECON. GEOL., v. 63, p. 612–621.

Naboko, S. I., 1959, Volcanic exhalations and products of their reactions as exemplified by Kamchatka-Kuriles volcanoes: Bull. Volcanol., ser. 2, v. 20, p. 121–136.

Nielsen, R. L., 1968, Hypogene texture and mineral zoning in a copper-bearing granodiorite porphyry stock, Santa Rita, New Mexico: ECON. GEOL., v. 63, p. 37–50.

Perry, V. D., 1952, Geology of the Chuquicamata orebody: Mining Eng., v. 4, p. 1166–1168.

Portigliati, C. N., 1971, Distribución zonal de los minerales de ganga en la mina El Teniente: Memoria de Prueba, Univ. de Chile, Santiago, 92 p.

Quirt, S., Clark, A. H., Farrar, E., and Sillitoe, R. H., 1971, Potassium-argon ages of porphyry copper deposits in northern and central Chile: Geol. Soc. America Mtgs., Abs. with Programs, v. 3. no. 7, p. 676–677.

Robinson, R. F., and Cook, A., 1966, The Stafford copper deposit, Lone Star mining district, Graham County, Arizona, *in* Titley, S. R., and Hicks, C. L., eds., Geology of the porphyry copper deposits, southwestern North America: Tucson, Univ. Arizona Press. p. 251–266.

Roedder, E., 1971, Fluid inclusion studies on the porphyry-type ore deposits at Bingham, Utah, Butte, Montana, and Climax, Colorado: ECON. GEOL., v. 66, p. 98–120.

Rose, A. W., 1970, Zonal relations of wallrock alteration and sulfide distribution at porphyry copper deposits: ECON. GEOL., v. 65, p. 920–936.

Ruiz, F. C., Aguirre, L., Corvalán, J., Klohn, C., Klohn, E., and Levi, B., 1965, Geología y yacimientos metalíferos de Chile: Inst. Invest. Geol., Santiago, Chile, 385 p.
Salas, O. R., Kast, R. F., Montecinos, P. F., and Salas, Y. I., 1966, Geología y recursos minerales del departamento de Arica, provincia de Tarapacá: Inst. Inv. Geol. [Santiago] Bol. 21, 114 p.
Sales, R. H., 1954, Genetic relations between granites, porphyries, and associated copper deposits: Mining Eng., v. 6, p. 499–505.
Sheppard, S. M. F., Nielsen, R. L., and Taylor, H. P., Jr., 1971, Hydrogen and oxygen isotope ratios in minerals from porphyry copper deposits: Econ. Geol., v. 66, p. 515–542.
Sillitoe, R. H., 1972a, A plate tectonic model for the origin of porphyry copper deposits: Econ. Geol., v. 67, p. 184–197.
— 1972b, Formation of certain massive sulphide deposits at sites of sea-floor spreading: Inst. Mining Metallurgy Trans., sec. B, v. 81, p. B141–B148.
—— 1973, Geology of the Los Pelambres porphyry copper deposit, Chile: Econ. Geol., v. 68, p. 1–10.
Stoiber, R. E., and Rose, W. I., Jr., 1970, The geochemistry of Central American volcanic gas condensates: Geol. Soc. America Bull., v. 81, p. 2891–2911.
—— 1971, Sublimates at volcanic fumaroles of Cerro Negro volcano, Nicaragua: 3rd Reunión de Geólogos de America Central, Actividades generales y resumenes, San José, Costa Rica, p. 10–11.
Swayne, W. H., and Trask, F., 1960, Geology of El Salvador: Mining Eng., v. 12, p. 344–348.
Takeuchi, T., Takahashi, I., and Abe, H., 1966, Wall-rock alteration and genesis of sulphur and iron sulphide deposits in northern Japan: Tohoku Univ. Sci. Repts., 3rd ser., v. 9, no. 3, p. 381–483.
Taylor, A. V., Jr., 1935, Ore deposits at Chuquicamata, Chile *in* Copper resources of the world: Internat. Geol. Cong., 16th, Washington, v. 2, p. 473–484.
Thomas, N. A., 1970, Beitrag zur Tektonik Nordchiles: Geol. Rundschau, v. 59, p. 1013–1027.
United Nations, 1971, Investigación detallada de minerales en áreas seleccionadas de las provincias de Atacama y Coquimbo: Los Loros: Informe tecnico no. 2, Programa de las Naciones Unidas para el Desarrollo—Empresa Nacional de Minería, New York, 100 p.
Vilela, C. R., 1969, Descripción geológica de la Hoja 6c, San Antonio de los Cobres, Provincias de Salta y Jujuy: Argentina, Dirección Nac. Geologia Minería Bol. 110, 66 p.
White, D. E., Muffler, L. J. P., and Truesdell, A. H., 1971, Vapor-dominated hydrothermal systems compared with hot-water systems: Econ. Geol., v. 66, p. 75–97.
White, D. E., and Waring, G. A., 1963, Data of geochemistry, 6th ed., chap. K: Volcanic emanations: U. S. Geol. Survey Prof. Paper 440-K, p. K1-K29.
White, W. H., Thompson, R. M., and McTaggart, K. C., 1957, The geology and mineral deposits of Highland Valley, B.C.: Canadian Mining Metall. Bull., v. 50, p. 487–503.

33

Reprinted from *Eng. and Mining Jour.* **156:**25-32 (1955)

THE ROLE OF GEOPHYSICAL METHODS IN MODERN EXPLORATION PROGRAMS

DR. ARTHUR A. BRANT
Newmont Exploration, Ltd.
Jerome, Ariz.

I. Introduction

WE, IN MINING, frequently wonder why geophysical methods, particularly seismic, have been so singularly successful as to return oil in 1 out of 5 holes drilled, even to depths of several thousand feet, while mining geophysical approaches have continued to be relatively uncertain, incapable generally of detection below a few hundred feet of depth, or of diagnosis as to nature or concentration of mineralization.

A list of mining geophysical methods are:

Self Potential	Geothermal
Resistivity	Gravitational
Electromagnetic	Magnetic
Radioactive	Seismic
Geochemical	

We wish briefly to discuss these methods and their limitations, to indicate new developments and suggest the future.

For all these methods, except the seismic, the response recorded varies in some manner inversely with the distance away from the body or condition causing the anomaly.

In general we may write:

Response R is proportional

$$(P_2 - P_1) \iiint \frac{dv}{r^n}$$

where $(P_2 - P_1)$ is the physical property difference between the volume causing the anomaly and its surroundings, and $\frac{dv}{r^n}$ is a portion of the volume at distance r from the point of observation.

n is some integral power that varies with the method used, the factor measured and the shape of the anomalous volume—

Certain aspects are immediately clear.

(1) Unless a physical property difference is present, no anomaly results. Thus there may well be geological differences without physical property differences. For example, in two copper-bearing areas, Jerome and Southern Peru, there is little magnetic contrast between flows, porphyries and diorites so that these formations cannot be differentiated magnetically.

At Bisbee, the porphyry and limestones are essentially the same density, but this is useful as the only density contrast is then with the massive sulphides which may thus be sought gravitationally (or with the solution caverns in the limestone: and underground, a dense body below you is the same as a cavern above).

(2) The response varies inversely as the distance to the causative body. Thus, for a given volume, the nearer the cause, the much stronger the anomaly. Hence beware of surface material and features. The soil and drift cover usually has contrasting physical properties from the rock below and the air above. Hence anomalies may readily result from changes in the thickness or nature of the cover and from topographic features. This statement applies to most methods and necessitates either satisfactory corrections or limits the use of the method geographically. This will be apparent in what follows.

(3) The response varies with the volume and shape of the contrasting body. For a given breadth or thickness of the zone causing the anomaly, the response will be greater for a dike or manto, than for a cylinder or pencil, than for a sphere or lens.

Actually a volume can be expressed as some linear dimension cubed, so that the response involves the ratio of said linear dimension to distance from the station. Thus the response from a 1,000-ft wide tabular zone under 1,000 ft of cover is about the same as from a 10-ft wide similar zone under 10 ft of cover.

For a given size and shape of body, the response decreases rapidly with distance away and provides the serious limitation of all mining geophysical methods in detecting to depth.

For most conditions a lens-shaped body cannot be detected farther away than its diameter, a pencil farther than twice its diameter and a dike or manto farther than 10 times its width or thickness. These estimates are actually a bit hopeful.

II. Discussion of Methods in Use

The methods now will be discussed in the order previously listed.

Self Potential. The self potential method responds to sulphides undergoing oxidation. It does not respond to sulphides not undergoing oxidation such as sulphides below the water table. Topographic ridges, particularly if outcrop, or thinly covered, will give anomalies similar to sulphides. (Fig. 1.) A good conducting dipping strata or zone bounded by poorer conductors will also give a characteristic negative anomaly. (Fig. 2.) The former is due to local atmospheric electrical potential variations at relatively poor conducting surface ridges; the second is due to potential differences arising from horizontally flowing telluric or ground currents.

The self potential method is also seriously limited for work to depth since essentially only the oxidizing portion of the sulphides are effective, thus limiting the volume contribution to the response.

Nevertheless if the anomaly can be recognized as due to sulphides, the method can be said to be direct in evincing mineralization, even if scattered. Absence of anomaly does not eliminate sulphides and spurious sulphide-appearing anomalies may result as mentioned above. An example of a successful use of the method is S. F. Kelley's contribution to the Noranda C orebody discovery. Recent Phelps Dodge work at Tyrone has also been helpful.

Resistivity. The crux to the re-

Acknowledgments

To Newmont Mining Corp. for permission to publish this paper and to
Mr. E. Brown—Sherritt Gordon Mining Co.
Dr. G. Herzog—Texas Oil Co.
Dr. E. E. Maillot—Phelps Dodge Corp.
Magma Copper Co.
O'okiep Copper Co.
Tsumeb Corp.
Aero Service Corp.
Aeromagnetic Surveys
Newmont Exploration Ltd.
for items of information included.

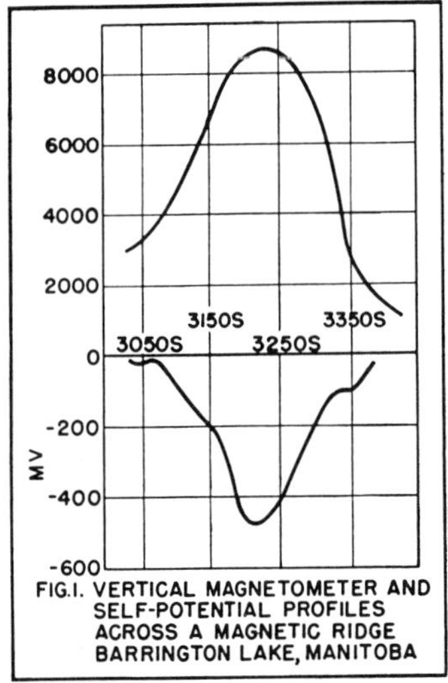

FIG. 1. Topographic ridges often give anomalies similar to sulphides.

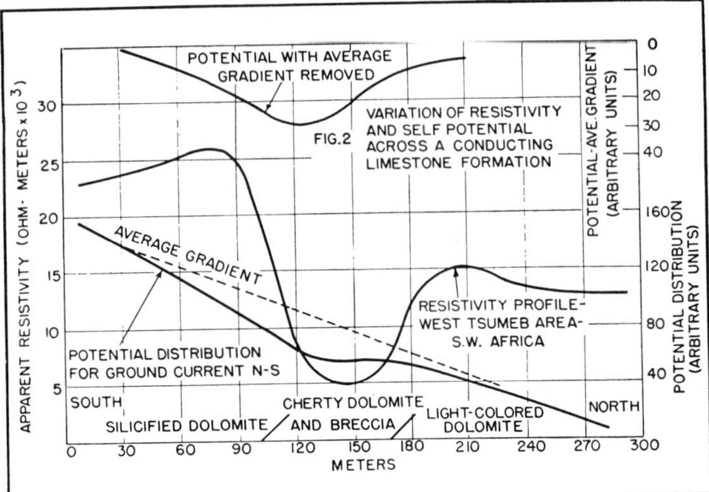

FIG. 2. A good conducting dipping strata bounded by poorer conductors will also give a characteristic negative anomaly.

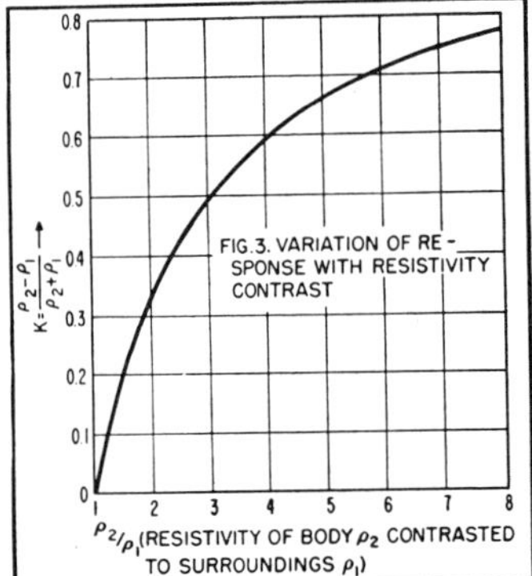

FIG. 3. Note that small resistivity contrasts cause relatively large anomalies.

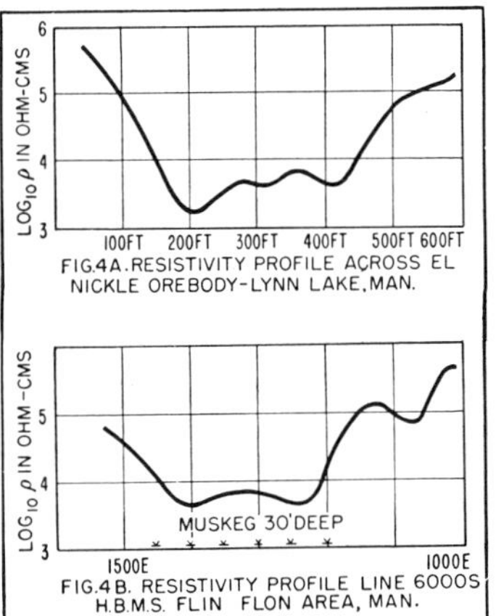

FIG. 4A, B. Resistivity surveys reveal depth of cover and fracturing.

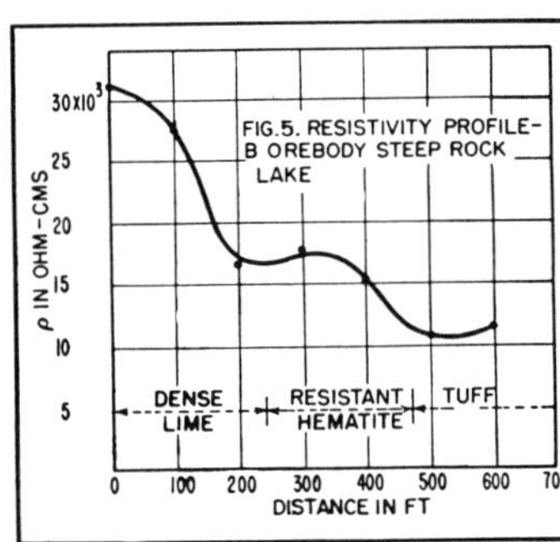

FIG. 5. Note how resistivity drops over hematite orebody.

sistivity method is the fact that the response due to a volume of contrasting resistivity ρ_2, in surroundings ρ_1, is governed by the ratio $\frac{\rho_2 - \rho_1}{\rho_2 + \rho_1}$. From Fig. 3 we see that this ratio rises rapidly for low values of ρ_2/ρ_1 and flattens for large. Thus the resistivity response is proportionately large for low resistivity contrast, and does not increase greatly for large resistivity contrasts. Thus small resistivity contrasts cause relatively large anomalies.

Hence anomalies readily occur with changes in surface material, e.g. muskegs and Gila conglomerate valleys are good conductors, with topographic ridges which actually simulate good conductors and with varying thickness of cover. Anomalies also result from minor changes of rock material, rock fracturing, etc. Sulphides when present then need not give rise to exceptional anomalies, and the resistivity method therefore is generally not direct, any presence of mineralization having to be inferred from relative juxtaposition of anomalies and known geology. Frequently it is difficult to determine whether anomalies result from the bedrock or cover conditions.

Nevertheless resistivity methods have been extremely useful, particularly in areas of the younger formations, to give depth of cover, evidences of fracturing, and other factors.

To illustrate the above, Fig. 4a and 4B show a resistivity profile over the Lynn Lake L orebody and a Flin Flon muskeg. They agree within 10%.

Fig. 5 is one of the original resistivity profiles across the tuff-limestone contact at Steep Rock Lake B orebody, evincing the more resistant hematite.

The Buchan's ore zone was extended by the equipotential line technique—a form of resistivity procedure.

Electromagnetic. The understanding of this method is readily derived from the work of J. R. Wait, Geophysics, Vol. 4, No. 6, October 1951, who (Fig. 6) shows how, for any conducting body, the in-phase and quadrature response components vary with the conductivity σ, size of the anomalous body r, and frequency f used accord-

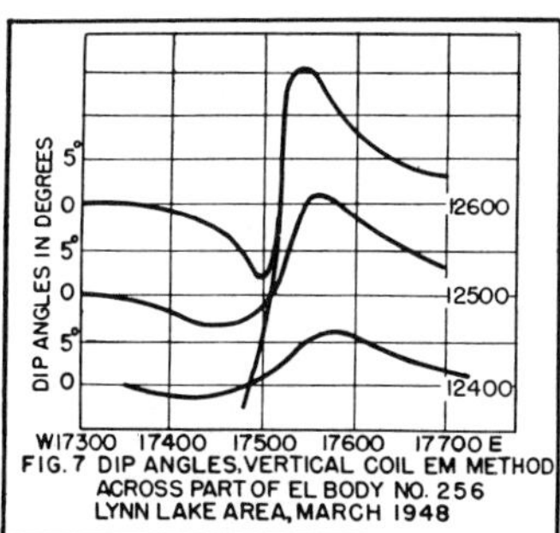

FIG. 7. Response obtained with conventional vertical-coil equipment.

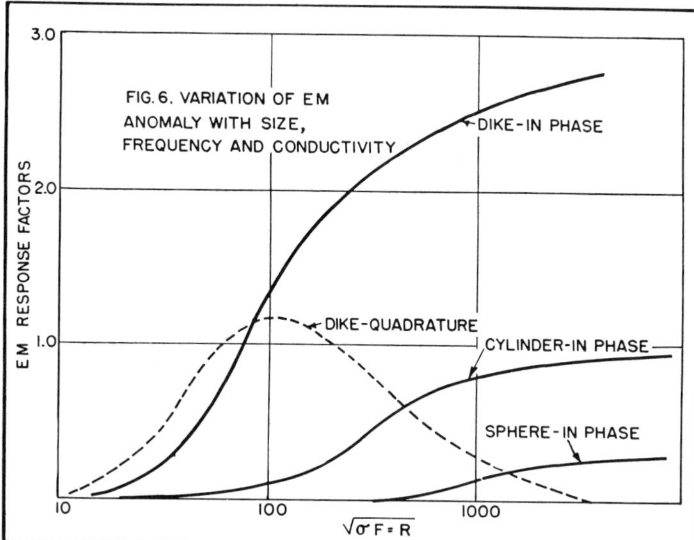

FIG. 6. In-phase and quadriture response components vary with conductivity, size of the anomalous body, and frequency used.

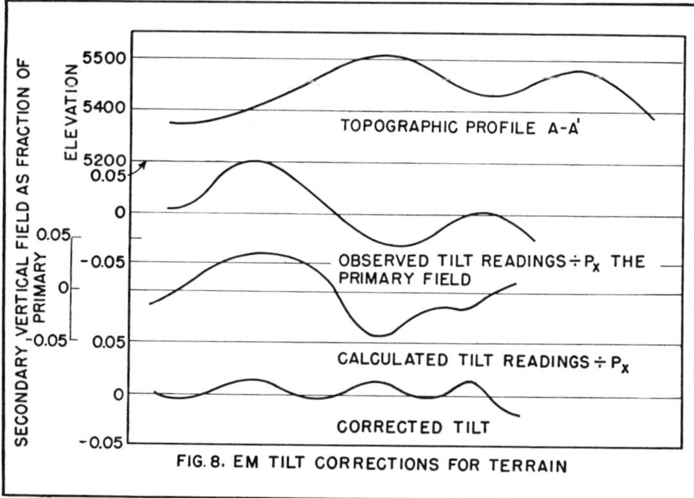

FIG. 8. Any component of terrain slope not perpendicular to the vertical generating coil can cause a response due to good soil cover.

ing to the relation $\sqrt{\sigma f} \cdot r$, see Fig. 6.

Usually a fixed frequency between 400 and 1200 cycles is used whence from Fig. 6 it may be shown that for the average conductivity 10^2 of pyrite, chalco, or galena, a truly disseminated deposit with particles of 10^{-2} meters size cannot be detected. Also it can be shown that broad areas of relatively low resistivity ($\rho = 1/\sigma = 10 -$ 30 ohm-meters), e.g. late volcanics as in southwest United States, salt water clays as in Australia, and the dry, evaporatively-salted cover of arid South Africa, Southwest U. S. and South America can give anomalies of a magnitude comparable to many massive sulphide occurrences.

This is why EM work has been largely concentrated in Canada, Sweden and Finland, lands of shallow cover and, strangely enough, higher resistive cover (50 ohm-meters or more).

Fig. 7 shows the response across the EL orebody at Lynn Lake by the conventional vertical coil, 1,000-cycle, ground method used in Canada.

Fig. 8 shows results of the same ground method in exploration using 800 cycles south of the United Verde mine with best orientation of coil relative to terrain. Here any component of terrain slope not perpendicular to the vertical generating coil can cause a response due to the good soil cover. Knowing the terrain, the form of the response can be calculated and here agrees with the response profile, so it is apparent that the response came from the terrain.

To date, the ground electromagnetic method has been effective in exploration for essentially massive or interconnected sulphides as witness the recent aid rendered in the New Brunswick discoveries.

The method has been extended to continuous recording from aircraft by the International Nickel Co. in their recent patent 2,623,924, work on which was actually initiated by S. Davidson, A. Brant and the present McPhar group, supported by Sherritt Gordon, Falconbridge, and Ventures. Another airborne magnetic device is being operated by Aeromagnetic Surveys of Toronto, Canada using 2 parallel sets operating at 400 and 2300 cycles and recording as a phase angle the total direct plus anomalous signal versus the direct. Other airborne devices are in development stimulated by Amco's base metal discovery in New Brunswick in connection with one of a series of airborne EM anomalies resulting from their use of the Inco apparatus.

Fig. 9 is a test flight by Aeromagnetic Surveys over the Amco discovery.

Radioactive The key point here is the absorption of the various radiations by the earth and rock materials. Gam-

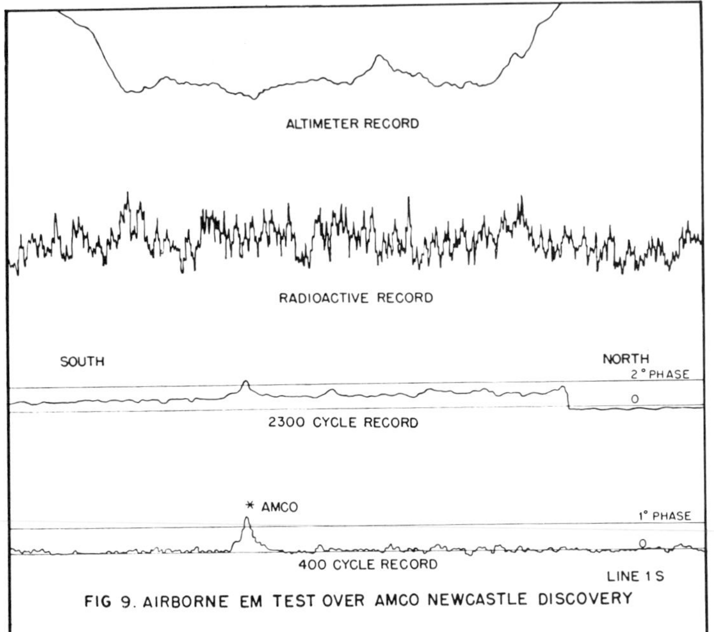

FIG. 9. Results of a test flight over the Amco Newcastle discovery. Note how blips differ in the 2300 and 400-cycle records.

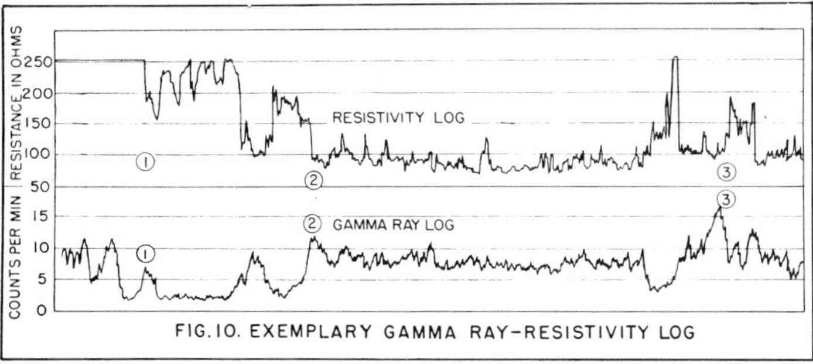

FIG. 10. Change of formations is readily recognizable at points 1, 2, and 3. From a survey on the Colorado Plateau.

ma, the most penetrating of the emissions is reduced to about 1/40 of its initial intensity by 5 ft of average soil cover and to 1/1600 by 10 ft of cover.

Neutron radiations arising underground are readily absorbed in 10 ft or so of water which limits distance for any drill hole work while there is a considerable scattered background intensity over a 45 deg solid angle cone in the atmosphere from cosmic ray showers.

Regarding gamma rays, the research department of the Texas Oil Co. at Houston, who have run tests over many U. S. mineralized areas, feel that there is a perhaps 70% correlation between change of gamma intensity and one time mineralization action. Any such change is, however, readily blanketed by shallow soil cover or later formations.

The early work at Great Bear Lake showed that changes in intensity resulted from change of formation and topography as well as change in thickness of the soil cover.

Thus it is apparent that radioactive prospecting on the ground or in the air will detect only essentially those deposits or areas of increased radioactivity under shallow cover. Nevertheless this has sufficed to aid in the rapid exploration of U. S. uranium resources.

The useful application of the possible gamma ray change with mineralization might well be in drill holes as an indication of nearby associated mineralization much as alteration indications are now used. More data is necessary here before warranting conclusions.

Due to the impetus of uranium prospecting in the Colorado Plateau and elsewhere, Geiger counters, and scintillometers are in commercial production for surface, drill hole and airborne use.

It is rather tragic that the radioactive emanations should be largely stopped by such short sections of earth or rock. Theoretically the radon should diffuse through great thickness or normal shale and sandstone but the limitation here is the slow rate of radon formation.

Fig. 10 is a modern radioactive log with associated resistivity recording from the Colorado Plateau. Change of formation is readily recognizable at the 3 positions indicated. Sandy beds show as high resistivity, low radioactivity; shaly beds as low resistivity, higher radioactivity.

Geochemical This is a present day refinement of the old visual trace or float prospecting and may be applied in a series of forms, although most work done now is along or parallel to the soil sampling techniques and treatments developed by the USGS.

It is imperative that the samples taken contain in some way material derived from the mineral source. In general a mineral source breaks down by solution, (essentially oxidation action) and decrepitation. Thus source material may be carried by actual water transport, ion diffusion in ground moisture, or remain as a soil residue.

The above would preclude detection of a source that has never emerged above the water table and thus never experienced solution, or of a source covered by impervious layers, or by later formations essentially following the original mineral source decrepitation.

Thus at the Bluebird property in Idaho, the residual soils intermixed with transported materials had to be recognized and sampled. Soil sampling has been effective in the shallow cover of New Brunswick but not so in all muskeg areas.

Again the source material spreads out with distance or thickness of cover. This gives rise to down slope or glacial direction fan spreading.

Some host rocks hold the metal solution products better than others: zinc ions in themselves are more mobile

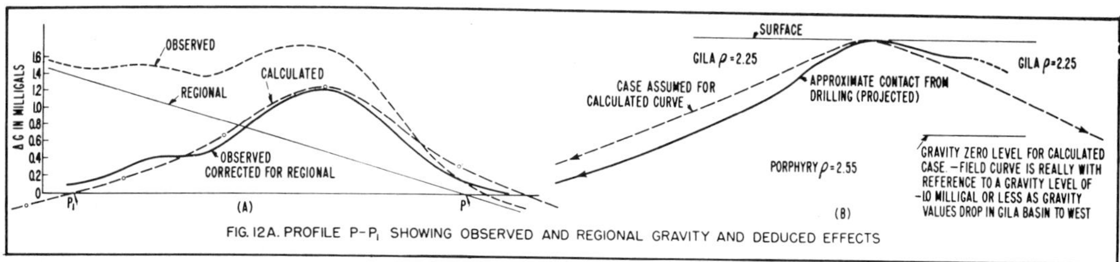

FIG. 12A. Readings shown in Fig. 12 below were corrected for regional intensity as shown.

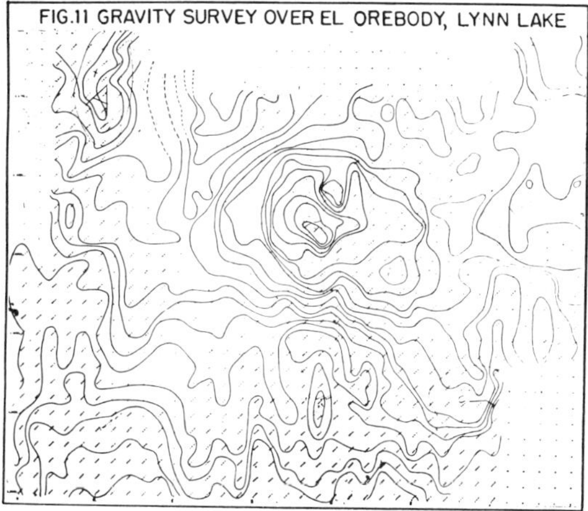

FIG. 11. Gravity anomaly over 6-million ton L orebody.

FIG. 12 GRAVITY SURVEY SAN MANUEL COPPER CORP. (CORRECTED FOR REGIONAL INTENSITY)

—.2— GRAVITATIONAL INTENSITY IN MILLIGALS
CONTOUR INTERVAL IS 0.2 MILLIGALS
0 1000 2000 3000
SCALE IN FT

FIG. 12. This anomaly is a reflection of bedrock.

than lead which in turn is more mobile than copper. On the Rhodesian Copperbelt, soils above quartzitic copper bearing rocks may show only 150-200 parts per million while above argillites evince 500-1000 parts per million for about the same grade, against a 75 ppm background. In New Brunswick, watch out for excess zinc in small drainage traps.

At this point it might seem that geochemistry works best in shallow covered areas of residual soil, or for general broad area indications by stream, water or transport fan analysis. The important fact is it can indicate the mineral present (but not necessarily its absence). Stringers, disseminations and contaminations may act like better concentrations. It looks like a powerful tool for rapid reconnaissance.

Geothermal Fluctuations of surface temperature here practically eliminate the use of any quick, simple, surface technique. Also the thermal conductivity contrast between the metallic minerals and rock is only a factor of about 5, whereas electrical contrasts may readily be 100. There is some possibility that the radioactive heat generated by (e.g.) a uranium ore could be used to detect a near miss in a drill hole, but this has not been tried.

Earth temperature measurements and studies on regional scales should give us some better and useful idea of geologic processes initiating folding and intrusion but this is in the future.

Gravity Here a considerable volume and density difference are necessary to produce an anomaly. Thus massive sulphides or iron are about the only occurrences of ore that could possibly be detected.

Corrections have to be made for topography, and for station elevation. This however does not eliminate anomalies that can arise at the bedrock surface because of density differences between the cover and bedrock.

Fig. 11 shows a gravity survey over the L orebody of some 6-million tons at Lynn Lake. Average density of the body was 3.3 g per cc. The numerous lesser anomalies which still might represent sulphide bodies of appreciable size are actually due to bedrock ridges under the muskeg.

Fig. 12 is a gravity coverage at San Manuel. The anomaly is a reflection of the bedrock as an outcropping ridge dipping off under Gila conglomerates.

At Labrador the dense chert iron formations with 20-30% iron gave the maximum anomalies of about 0.5 milligal. Lower more irregular anomalies generally on strike were actually indicative of higher iron content.

Difficulties in mining application are the relatively large volumes required to give measurable responses above errors arising from generally rough terrain, and the anomalies, difficult to correct, due to bedrock topography and varying thicknesses of cover.

Gravity work has been useful in iron country, both in Labrador and

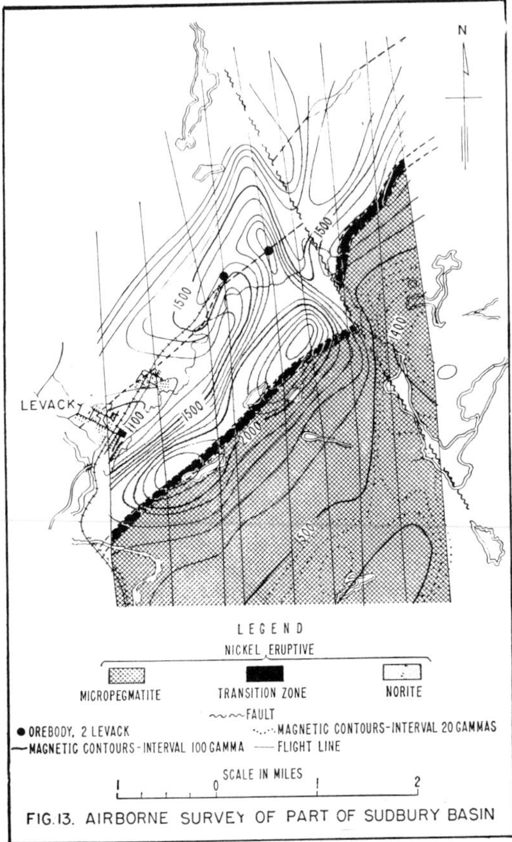

FIG.13. AIRBORNE SURVEY OF PART OF SUDBURY BASIN

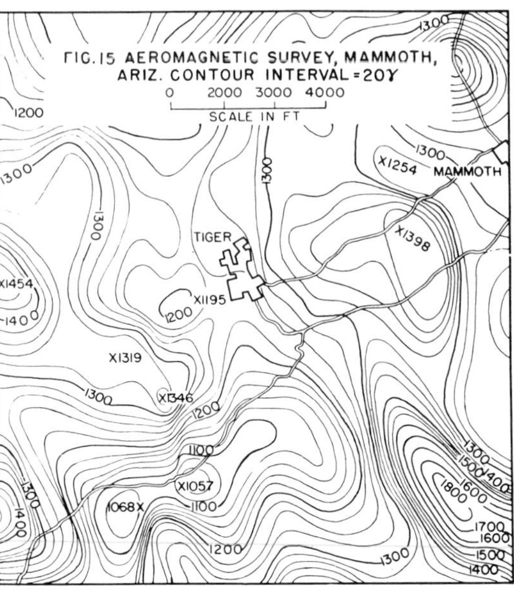

FIG. 15. (ABOVE) The aeromagnetic pattern here is essentially due to varying thicknesses of Gila conglomerate. See also Fig. 16.

FIG. 13. (LEFT) Anomalies in the vicinity of the orebody are due to magnetite in volcanics and norite.

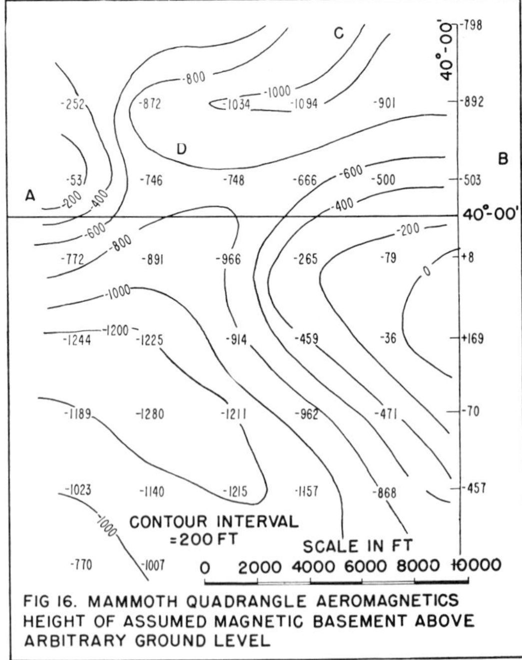

FIG. 16. Thicknesses of the Gila conglomerate have been mathematically derived.

Minnesota. It has been used for massive base metal exploration in the Noranda, Sudbury, and Bisbee areas. However, the main applications have been limited and in general, indirect, as, for example, outlining the ore controlling denser diorite at Jerome, the less dense areas of increased fracturing or solution alteration as in the Tri-State, and the thin portions of the covering volcanics on the Rand extension into the Orange Free State.

Test work has shown that large scale geologic folds, grabens, etc. frequently show up gravitationally. This may have eventual useful regional application.

Successful work has been done underground at Bisbee in detecting massive sulphides in the limestone-porphyry complex which fortunately is of consistent density.

Magnetic. World wide, this has been one of the most useful mining geophysical methods even though it is indirect, magnetic anomalies generally being referable to formations rather than to mineral bodies except in the

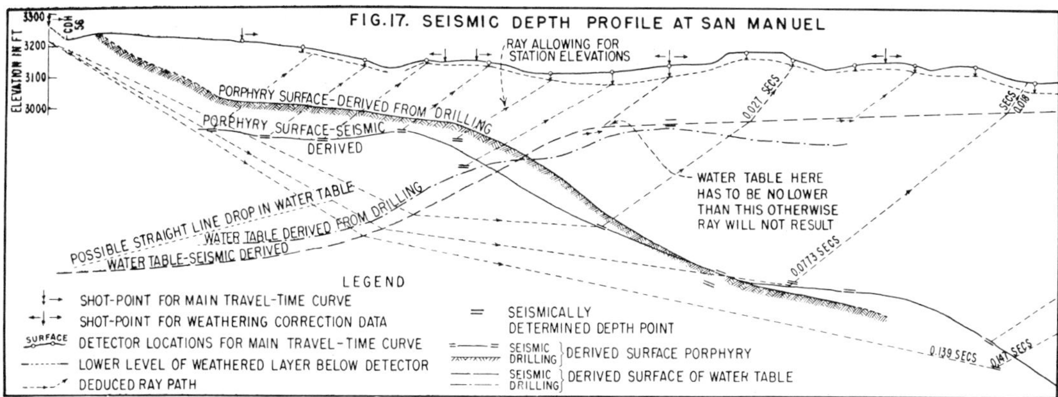

FIG. 17. This seismic profile at San Manuel shows how the refraction method may be used to determine depth of cover. Short surface shots were used to determine near-surface velocities and layerings.

case of iron, nickel and titanium illmenites. However, mineralization and alteration are not infrequently accompanied by changes in magnetic property.

Basic rocks are generally more magnetic than acidic, plutonic more magnetic than sedimentary. Most rocks are of relatively low magnetic susceptibility, so that topography and cover considerations can frequently be neglected.

However the broad late Cretaceous and Tertiary areas from southern Utah into South America present a series of covering lavas that are frequently magnetic and effectively blanket much of the area to magnetic exploration. Also the broad Gila filled valleys are of such dimensions as to cause marked magnetic expression and again make (e.g.) airborne work, of very limited value.

The more extensive and useful application of magnetics then, has been in the older formation areas of the Canadian and Scandinavian shields and the South African plateau.

As examples, Fig. 13 shows an airborne survey section of part of the Sudbury Basin. Represented is an orebody area and anomalies due to scattered magnetite in the volcanics and norites.

Fig. 15 shows airborne results over San Manuel. That these are essentially due to the varying thickness of Gila is clear from Fig. 16, where the necessary thickness of Gila to account for the magnetics have been mathematically derived by an extension of the continuation downwards procedure of Peters, *Geophysics*, Vol. XIV, No. 3, 1948.

The usefulness of magnetic methods, despite these limitations, are exhibited by O. Weiss' work tracing the iron formations into the Rand Extension, the extension of the Chesterfield Inlet Nickel zone, the discovery of the first Lynn Lake nickel bodies Canada, the outlining of the ilmenites at Allard Lake, Quebec, and the discovery of magnetite near Morgantown, Pa., and Marmora, Ont., the last three resulting from airborne work by Aero Service Corp.

Seismic. The seismic method has been seldom used in mining work.

One of the main applications is for determining depth of cover, and Fig. 17 shows a cross section derived southward from the North San Manuel shaft. The procedure was a series of refraction shots with short surface shots to determine near surface layerings and velocities. Knowing these and the emergence angle, the ray can be drawn back downwards from the pick up. The velocity in the bedrock is apparent from the refraction shots and the only unknown is the thickness of the lower layer of cover. Knowing the total travel time to the station permits the depth to bedrock to be determined and the dip of the bedrock.

Orebody detection was tried at the Chief Consolidated mine, Tintic, by H. H. Pratley of National Geophysics, using the slower velocity of the brecciated and crumbly ore area to provide a diffracting shadow defined from the emergence angles. Here the smallest size that can be detected is given by

$$\sqrt{\frac{dv}{f}} \text{ where } d = \text{distance from body,}$$

v = wave velocity and f = frequency. Only a low frequency, about 60 cycle per second was tried.

This possibility, and checking on ore continuity, exists seismically with more hope at higher frequencies, although above a few thousand cycles, dispersive effects at small fissures, etc., become extreme.

III. Present Developments

So much for the past and much of the present. Recent developments are the airborne magnetic, electromagnetic and scintillometer devices previously mentioned.

In connection with magnetic work, the mathematical means whereby results are converted to those at higher or lower elevational levels, as long as outside the anomalous volume, is of great value. A case of this was the derivation of thickness of cover from the San Manuel airborne results.

Some interesting electrical developments are taking place. In resistivity, tests have been made to roughly determine the size and continuity of conducting sulphide orebodies. In both cases the orebody is contacted by the one current electrode and a nearby potential. Measurements are made to the far potential, on line of strike or in the other showing, if continuity is desired, or in other underground workings or drill holes, if size is desired.

Fig. 18 shows the potential thus derived relative to an ore intersection of some 10 ft in a drill hole. The size of the intersection is readily computed from

$$P_1 \overset{\triangle V}{\quad} P_2 = \frac{\rho\,\text{ore}}{4\pi} I \left\{\frac{1}{r_1} - \frac{1}{R\,\text{ore}}\right) + \frac{\rho\,\text{rock}}{4\pi} I \left\{\frac{1}{R\,\text{ore}} - \frac{1}{r_2}\right)$$

where R is the radius of the equivalent spherical orebody, r_1 and r_2 are the distances to the potential electrode

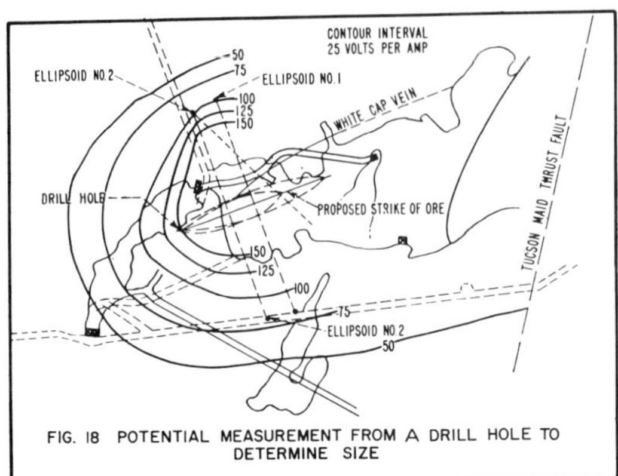

FIG. 18. Here the potential method was used to determine the size of an intersected orebody.

pickups in ore and rock from the current electrode in the ore, and ρ^{ore} and ρ^{rock} are the known resistivities of ore and rock. In this case size worked out to 3,000-4,000 tons. The latter was obtained.

Also it is possible to predict the azimuth of a conducting body near a drill hole or drift. Current electrodes are placed on a line in a plane at right angles to the hole or drift so that the hole or drift is midway between them. The potential is then measured between a movable pickup in the hole and one far away. Fig. 19 is the result. If the polarity of the current electrodes is known, the azimuth of the conducting body can be deduced, for the potential will be more positive opposite the body if the body is on the positive current side.

Work has been going on with methods to detect scattered sulphide mineralization, most methods heretofore being adapted to massive sulphide bodies, or being dependent on some indirect association as (e.g.) associated magnetite. Fig. 20 shows the deduced percent by volume of sulphide mineralization and depths to sulphide at a porphyry copper prospect in Southern Peru.

Parallel developments are taking place in electromagnetic prospecting for massive sulphide bodies. Size and azimuth determinations are possible in drill holes. There are at least two ways of accomplishing the former and several for the latter.

IV Future Possibilities

The future is not going to be marked by the sudden development of exceptional methods. Nothing on the immediate horizon indicates an approach to mining geophysical exploration that will solve the present existent depth limitations. Laboratory phenomena do, however, give some hope that other methods besides geochemical may eventually indicate the nature of the material or ore present.

All in all, it looks like continuing gradual development on all fronts; in equipment, interpretation and better understanding of cases, rather than the sudden burgeoning of any exceptional techniques.

Continuing attempts will be made to make the various methods airborne essentially with helicopters to permit low "grouse" flying. Rapid reconnaissance application will be stressed. We will work more and more from the outside in, as the oil people do. Within the Americas, Europe, and Africa, a slow breakdown of political barriers should evolve with slowly increased ease and interest in "foreign" exploration by American companies.

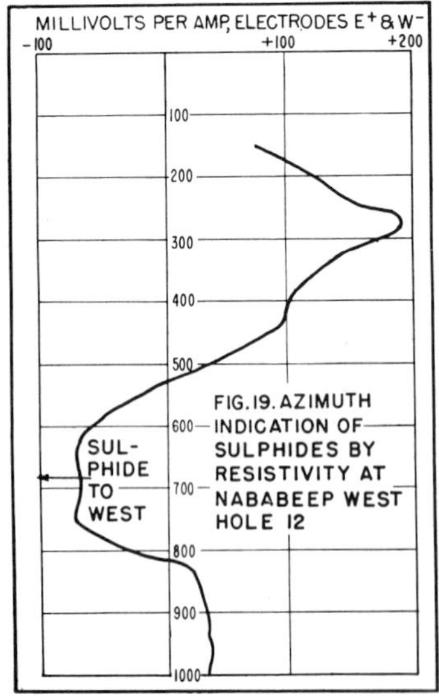

FIG. 19. Current electrodes were used to predict azimuth of orebody.

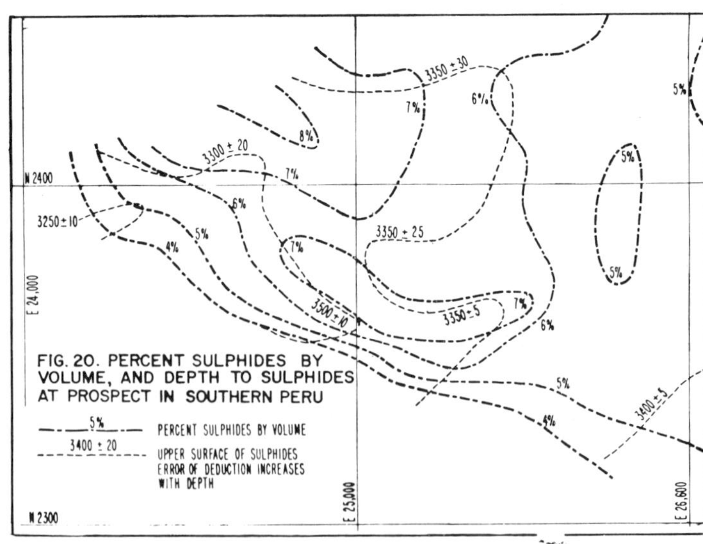

FIG. 20. Electrical methods were used in southern Peru to deduce percent by volume of scattered sulphides.

34

Reprinted by permission from pages 75-85 of *Geology of the Porphyry Copper Deposits, Southwestern North America,* S. R. Titley and C. L. Hicks, eds., University of Arizona Press, Tucson, Ariz., 1966, 306p.

◊ ◊ ◊ SOME FEATURES PERTINENT IN EXPLORATION OF PORPHYRY COPPER DEPOSITS*

BY S. E. JEROME

INTRODUCTION

This paper discusses some geological, geochemical, and geophysical characteristics of porphyry copper districts, calls attention to some areas of ignorance concerning them, and suggests some research that might yield fruitful results in exploration for them. Certain facts personally observed and widely distributed in the literature have been assembled in a generalized and somewhat speculative fashion on the accompanying figure. The assemblage is examined in terms of the writer's experience, particularly in the southwest porphyry copper province of Arizona and New Mexico, gained while with several exploration companies and universities, including the New Jersey Zinc Co., University of Utah, Gulf Minerals Co., Bear Creek Mining Co., Hunting Geophysical Services, Inc., and the University of Nevada (Nevada Bureau of Mines).

Contributions to the writer's thinking made over the years by a host of associates too numerous to mention individually are gratefully acknowledged, but the contributors are not to be blamed for such odd twists as this thinking may seem to take from time to time. Reference material, credited at appropriate places, is listed at the conclusion.

Director Vernon E. Scheid of the Nevada Bureau of Mines allowed the time for preparation of the paper. Barbara Webb's skillful drafting of the figure is appreciated, as are the constructive comments of two of the writer's colleagues, Laurence Beal and John Schilling. Helen Mossman has done a remarkable job in converting the rough draft to final copy.

GEOLOGICAL SETTING

The geologic events involved in a "typical" porphyry copper district, with discussion of some significant departures from the "typical," are developed from the idealized cross section at the bottom of figure 1.

The bedded rocks shown include carbonates, shales, sandstone, and, possibly, volcanics. Actually, a striking aspect of the porphyry copper deposits is the variety of lithologies and environments in which they occur. Regionally, they are found in shield, shelf, and geosyncline; locally, the ore bodies and the intrusives with which they commonly are associated occur in schists, gneisses, granite, foliated and nonfoliated eugeosynclinal assemblages, nonfoliated miogeosynclinal assemblages, sediments of nongeosynclinal origin, and salic to intermediate volcanic rocks; these same generalizations have been made by Jerome and Cook (10) for most of the base-metal and precious-metal mining districts of the West, and they apply elsewhere in the world. The environmental varieties indicate that stratigraphic conditions are not prime determinants in a regional sense in positioning porphyry copper districts, although in a local sense stratigraphy does influence the disposition of both primary and secondary mineralization. The lack of preference for a particular lithologic environment and the tendency for ore bodies to occur in clusters, as in the Southwestern United States, may reflect concentrations of metalliferous material deep within the crust or in the mantle, the positions of which probably were established during the cooling history of the earth. A similar view has been expressed by others.

As the cross section indicates, the bedded rocks were folded and faulted. The importance of anticlinal folds and domes in localizing mining districts, as expounded by Wisser (31), is one of the popular thoughts of the day. Although such folding may not be present or recognized in all cases, it is conceivable that pressure relaxation on material at depth, occasioned by folding and faulting, might promote lateral and upward migration of materials, much as in the formation of salt domes.

The deformed rocks were invaded by an intrusive. Ordinarily, such intrusives are of the dimensions of stocks rather than those of batholiths. Some have com-

* Reprinted by permission of "Western Miner and Oil Review."

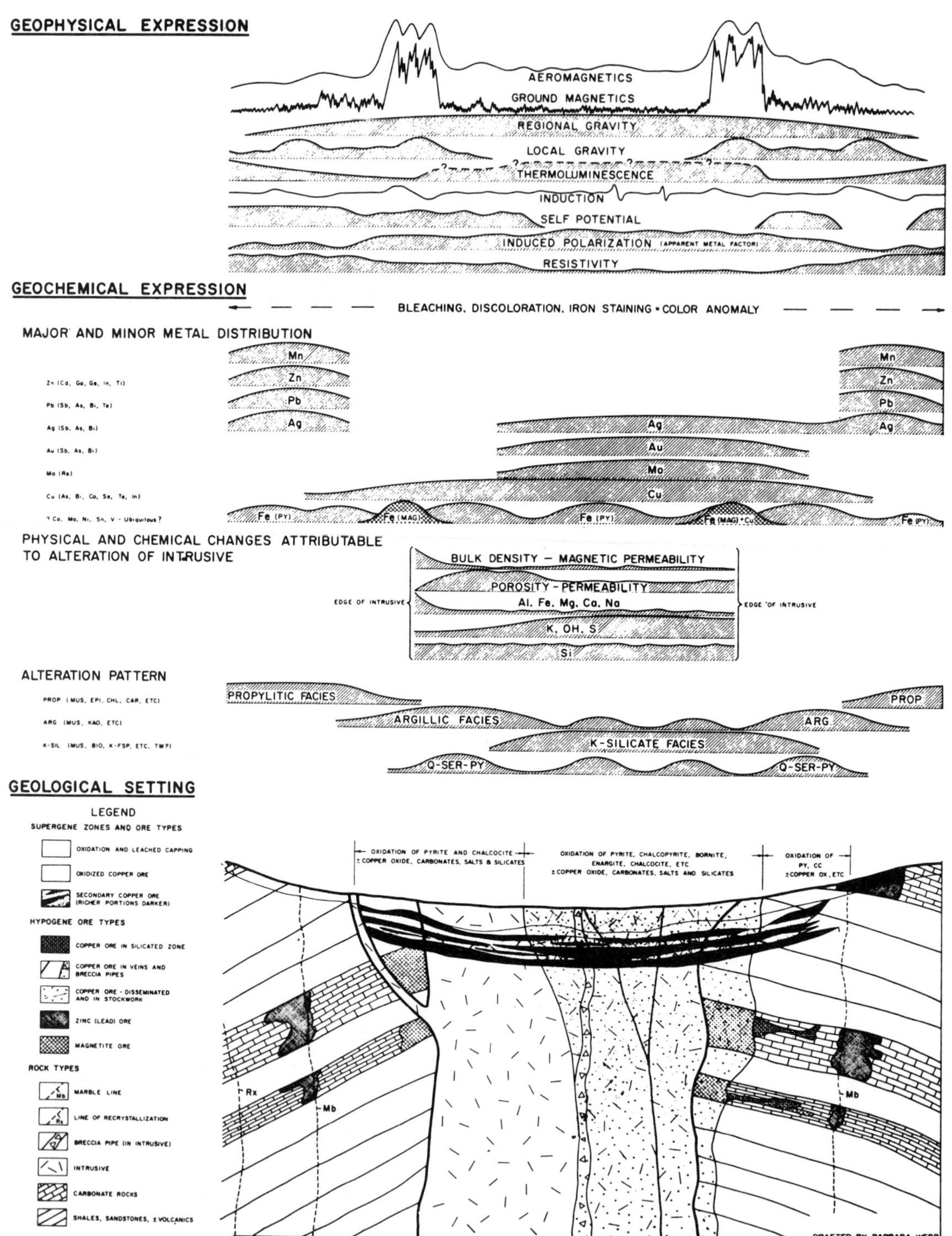

FIGURE 1.—Some generalizations and speculations, porphyry copper districts.

plex or turnip shapes, but in most districts drilling has not progressed sufficiently to establish either the shape or the continuity in depth. Monzonite, quartz monzonite porphyry, and granodiorite are the intrusive rocks most commonly associated with porphyry copper deposits, although granite and more salic intrusive rocks may be present occasionally. Considering the Cordilleran mountain chain from Alaska to the southern tip of South America, the ages of most of these intrusives range from late Mesozoic to middle Tertiary. It does not appear that a significant number of such deposits was generated during the Nevadan period of isoclinal folding, metamorphism, and batholithic invasion (granitization[?]). Copper Mountain, British Columbia; Ely, Nevada; Ajo, Arizona; and Bisbee, Arizona may be among the exceptions, but field relations often are so ambiguous that ages have been established only by analogies between districts. More comprehensive knowledge of ages of western intrusives should expedite fitting known ore bodies into the proper set of geologic events and should contribute to additional discoveries. The Bethlehem porphyry copper ore bodies at Highland Valley, British Columbia, illustrate the point. These postdate but appear closely related to Late Cretaceous mid-Tertiary(?) intrusives that cut the Jurassic-Cretaceous(?) Guichon batholith (30). In attempting to duplicate the Bethlehem type of deposit, prospecting the younger intrusives and their walls is likely to be more productive than concentrating on the older Guichon batholith. Some reservation must be held concerning the particular example, however, for when experience indicates that prizes such as Craigmont may be found on the margins of the older intrusive, it obviously would be unwise to restrict one's exploration efforts to only one type of ore body associated only with rocks of a certain age.

The thermal and chemical effects of the intrusion result in characteristic contact silicate minerals in the carbonate rocks, shales, lithic sandstones, and volcanics; silicification of the shales; recrystallization of the carbonate rocks; and some recrystallization of the sandstones. The results depend on the raw materials available for the change and the physical-chemical conditions superimposed. If the intrusive enters granitic or silica-rich volcanic terrane, the changes that can be directly attributed to intrusion will not be nearly so obvious as in the mixed rock environment being described. All the pyrometasomatic effects "fade" into fresh rock away from the intrusive. The position of the abrupt change of silicated carbonate beds to marble is referred to as the "marble line." Before giving way to normal rock, the zone of marbleization usually varies substantially in width within specific beds, depending on their physical and chemical natures.

The next stage visualized is the revival of faulting and the formation of breccia pipes and shattered zones (stockworks) in the solidified intrusive and invaded rock. The breccia pipes and faults may be much more abundant and occur at greater distances from the intrusive than the section indicates. Where the intrusive tops out at depth, its position may be revealed by breccia pipes observed at higher levels; the likelihood of an associated porphyry copper ore body at depth may be suggested by primary copper mineralization in the pipes. See Perry's observations (20).

Ground preparation may involve the entire mass of small intrusives, but in the large ones the breaking may be limited to a part, especially the edges. The late structural pattern could just as well be superimposed entirely on country rocks away from the intrusive, although when this actually is the case and the mass of broken rock is mineralized, many geologists would not be disposed to call the result a porphyry copper. Their conventional use of the term requires that most of the copper mineralization be superimposed on a porphyritic intrusive rock. In any event, many of the exploration problems in connection with large low-grade deposits are similar regardless of the host. The superimposed structural features perhaps were related to adjustments along the same master fault that admitted the intrusive, to tectonic forces acting in new directions, or to resurgence of withdrawal of the intrusive magma.

It is suggested that the intrusive rock and adjacent country rocks, not chemically insulated from the attack of fluids, were hydrothermally altered ahead of, and to a lesser extent contemporaneous with, sulfide deposition. Evidence for alteration being pre-sulfide is set forth by Lovering et al. (16) for the East Tintic district, Utah; the case for alteration contemporaneous with sulfide deposition is presented by Sales and Meyer (22) in a classic paper on Butte, Montana. The porphyry copper districts have been the particular beneficiaries of these two fundamental studies, as indicated by the large number of papers and theses they stimulated in the past 13 years on porphyry copper alteration. In reviewing this information, the writer has been impressed that, although many of the same alteration minerals appear in the descriptions of one district after another, the suites reported have been of such great variety that mineralogical comparisons and interpretations between districts have become increasingly confusing, to this writer at least, and more difficult to use with any confidence in exploration work. The same impression apparently was held by Creasey (5), who made a major contribution in the application of ACF and AKF diagrams to the problem. He assigns porphyry copper alteration to one or more of three principal facies:

Potassium silicate facies—characterized by muscovite (sericite), biotite, potash feldspar, and including other nondiagnostic minerals.

Argillic facies—characterized by muscovite (sericite), kaolinite, and related clay minerals such as dickite, montmorillonite, etc., and including certain other nondiagnostic minerals.

Propylitic facies—characterized by muscovite (sericite), epidote, chlorite, carbonate minerals, etc.

Creasey notes that the mineralogical and chemical characteristics of the first two facies are well described in the literature, but the same detail is lacking for the propylitic facies. Burnham (4) applauds Creasey's approach but believes that his facies should be reduced to two, argillic and phyllic; but Burnham then proposes subdividing argillic into three and possibly four types and subdividing phyllic into two types. The types all are realistic and usable but are subfacies of local application. This writer arranges Creasey's three facies in a manner that can be useful in exploration and discusses the arrangement under the heading "Geochemical Expression."

The position of magnetite in the sequence of events often is debatable, but it usually appears to have been deposited during the transition from the higher temperature stage of pyrometasomatism to the lower temperature stage of sulfide deposition. Where carbonate rocks are present, substantial magnetite ore bodies may form; where such rocks are absent, it may not be readily evident that magnetite has been added. The close spatial relation of magnetite to the intrusive tempts the belief that magnetite ore bodies in carbonate rocks are due to fluids emanating directly from the intrusive and attacking the reactive rocks, but where magnetite occurs as clots in breccia pipes or as veins in intrusives, it clearly is late. Because chalcopyrite frequently occurs near or with magnetite in appreciable quantities, effort always should be made to understand the age and structural relations between the two. Craigmont is an outstanding example of the association of the two minerals. Magnetite can be a guide to porphyry and other types of copper ores.

Principal minerals deposited during the hydrothermal sulfide stage include pyrite, molybdenite, chalcopyrite (plus bornite, enargite, and other copper sulfides on occasion), sphalerite, and galena. Gold and silver may occur with several of the above and certain other minerals. The types of primary ore bodies that may form are:

1. The primary porphyry copper ore body includes copper, iron, and related sulfides and precious metals as disseminations and fillings of previously opened fissures, stockwork, and breccia pipes in and adjacent to the intrusive. Rare veinlets of sphalerite and galena may cut chalcopyrite. This has been called protore, but the designation is inappropriate if economic conditions permit its mining. Several major copper producers now are largely or entirely dependent on values in the primary zone, i.e., Bingham, Utah; Ely, Nevada; and Ajo, Arizona. The Bethlehem enterprise was initiated on primary values that certainly would have been protore in the early history of porphyry coppers.

In some porphyry copper districts high-grade fissure fillings of chalcopyrite, and locally bornite, were exploited long before the true nature of the total deposit was realized and, of course, before the total deposit could be developed economically. Today one occasionally sees in open-pit walls remnants of drifts and timbered stopes on these veins. The Banner-Pima-Mission ore body near Tucson, Arizona, illustrates that even in this day of supposed sophistication in porphyry copper geology and exploration methods, it may be an agonizingly long time before it is appreciated that a mass of sulfide, obscured by 200 feet of alluvium, discovered by geophysical methods, and mined for some time only as a vein deposit, actually is only a small part of a great low-grade copper ore body. Those to whom the light finally dawned deserve much credit for their astuteness. It is pleasant to see the reverse situation at Bethlehem where discovery of higher grade veinlets of bornite followed aggressive development of apparently marginal porphyry copper ore bodies.

It was customary for some to think of porphyry copper deposits as "disseminated" ore bodies. To others the term implies that the sulfides are, in effect, late stage rock-forming minerals. This over-simplifies the situation and can confuse the approach in exploration. The possible variety of textures represented in the primary ore body and the fact that porphyry copper-type ores may occur beyond the intrusive support the idea of sulfide mineralization after the intrusive is in place, solidified, and broken. The actual position of the primary ore body depends on a source capable of supplying the tremendous volume of fluids this type of ore body represents, a plumbing system adequate to transmit these fluids, and a broken "sponge" physically and chemically receptive to the deposition of sulfides. Some might like to draw an analogy here with petroleum where economic concentrations depend on source rock, migration, reservoir rock, and a trap.

2. Replacements, fissure fillings, and disseminations of primary copper minerals may occur in the silicated zone surrounding the intrusive. The greatest number and size of such deposits may be on the side of the intrusive where the primary porphyry copper ore body forms, although this relation is by no means consistent. Copper mineralization in the silicated zone may terminate quite abruptly at the marble line; in some districts, although not necessarily porphyry copper zones, the commercial concentrations are restricted to its vicinity. San Pedro, New Mexico, is a good example. The writer believes there were no significant pressure and temperature differences between the deposition of copper in silicated zones and in the primary ore body; the only real differences are in the nature of the ground preparation and the host rock upon which mineralization was superimposed.

Copper ore may occur in silicated rock independent of a porphyry copper deposit and vice versa, but when one type of deposit is known and a geological possibility exists that the other could occur, a search for it is merited.

Where carbonate rocks are not present, replacement copper ore bodies in silicated rock may be absent, and the metal values either may be confined to the primary ore body or may be deposited in outlying veins and breccia pipes.

3. Zinc-lead replacement ore bodies may form on either side of the marble line. Such ore bodies also may occur randomly in the marbleized zone and for some distance out into unaltered carbonate rocks where structural conditions were favorable during mineralization. If rocks susceptible to replacement apparently lack important zinc and lead ore bodies, as at Ely, Nevada, such ore bodies may remain to be discovered; or, at the depths being considered, the metals may have been deposited in vein zones or have been widely dispersed in minute amounts; or they may have traveled to higher levels where either they were dissipated or formed ore bodies that subsequently were eroded; or the source may have been deficient in the missing elements, an idea at which the exploration geologist may arrive quite reluctantly.

The pattern of distribution of major metals and associated trace metals is discussed further under the heading, "Geochemical Expression."

The final stage in the geologic evolution of the district involved erosion, usually with an attendant fluctuation of the water table. Oxygenated ground water coming in contact with pyrite became rich in ferric sulfate and sulfuric acid. These strong oxidizers dissolved copper from its primary sulfides above the water table and precipitated it at or below the water table where the solutions became dilute. The importance this process attains depends on climatic conditions and rates of erosion. In steep terrane or humid climate, or any combination of the two, the enrichment process may be of little importance, and the primary porphyry copper ore body and satellitic ore bodies, must be of sufficient grade to justify development.

Assuming arid conditions and a moderate rate of erosion, the water table descends and the process of solution and deposition of copper is repeated until a subhorizontal blanket of secondary copper sulfides that usually represents a substantial upgrading of primary material is formed, hence the designation "zone of secondary enrichment." This zone is shown on the figure as occurring irrespective of structure, rock type, and alteration. Actually, all three profoundly affect the distribution of values in this and other zones, and the tops and bottoms of the zones are not as regular as many think. Detailed mapping of the three is absolutely essential for a proper understanding of the geology for intelligent pit development and for use in exploration.

The enriched zone may show a subhorizontal banding of secondary values; such grade bands possibly represent varied positions of the water table. The enriched zone commonly is overlain by an irregular zone in which secondary copper carbonates, salts, and silicates with local cuprite and native copper are spectacularly displayed; this grades upward into completely leached and oxidized capping. Over the secondary ore body the leached zone will show limonitic products derived from original pyrite and from oxidation and reoxidation of the chalcocite blanket. Over the combined primary and secondary ore bodies oxidation products of pyrite, primary and secondary copper sulfides, and molybdenite may be recognized. Much work has been done on such secondary materials by Locke (15), Blanchard (2), and others in attempts to correlate them with the nature of the sulfides and the copper values to be anticipated at depth. Some experienced observers claim proficiency in making accurate forecasts of grade at depth, but the writer suspects such claimants have had several unpleasant surprises in their careers. This suspicion is bolstered by Locke's (14) study, which demonstrated great inconsistency in such forecasts between six experienced people using identical samples and repeating their own work. It would appear that greater success might attend predictions if outcrop observations were supported by thorough geochemical studies.

Although the leached capping may show brilliant blue, green, and brown patches and smears of secondary copper minerals and iron oxides, such striking coloration does not necessarily lie over the best part of the ore body, nor does it invariably mean that a commercial ore body exists. However, although it sometimes is said facetiously that "a copper penny can stain a mountain," the exploration man will demonstrate wisdom by thoroughly appraising the mountain before he attributes what he sees only to a penny.

In some districts the secondary ore body was complicated during its formation by faulting, tilting, fluctuations of the water table, burial by volcanics and (or) alluvium, and in some cases by partial or total reexposure by erosion. All these contribute to the confusion of the observer, and may delay discovery, but they do not change the fundamental concepts of the evolution of the district.

Before proceeding it may be appropriate to summarize some of the main characteristics of the porphyry copper ore body. It is a large, low-grade ore body dominated by primary and secondary copper minerals in disseminated grains, stockwork veinlets, and breccia pipe and vein fillings superimposed on an intermediate to salic intrusive rock and its wall rocks. The ore body may be a steeply dipping pipe of primary mineralization, or it may be an asymmetrically shaped mushroom with the cap representing the secondary ore body and the stem the primary ore body. The size of the combined ore bodies in terms of today's economics ranges from 20 million to over 4 billion tons, with the value of several of the largest districts exceeding $4 billion; continued exploration and development and more economic methods of mining and milling will expand these figures. Worthy exploration targets indeed! Also, it should not be forgotten that the central ore bodies often are surrounded by satellites of copper, magnetite, lead, and zinc, some with precious metal values. These latter ore bodies, although greatly overshadowed by their big brothers, contribute substantially to the tonnage of metal and the dollar value of such districts.

Grade of deposits currently being exploited ranges between 0.7 and 2.5 percent copper; today's mill feed from pits in the United States averages between 0.7 and 1.0 percent copper. Gold content normally is under 0.004 ounce, silver under 0.1 ounce, and molybdenite under 0.1 percent (28, p. 350, 738, 540).

Because of the nature of porphyry copper ore bodies, published dimensions may not be too meaningful, and the sizes of the primary zones may be impossible to identify, but the overall range in plan seems to be between 1,000 by 1,200 feet and 3,500 by 9,100 feet (18, 19). The thicknesses of the secondary zones are about as follows: capping 0 to 1,000 feet, averaging about 150 feet; oxidized ore zone too variable to generalize; enriched ore 100 to 1,000 feet, averaging about 300 feet. The vertical extensions of primary zones usually are impossible to establish from published information, at least for those ore bodies in the United States.

GEOCHEMICAL EXPRESSION

Alteration Pattern

This writer has arranged Creasey's (5) three alteration facies on the figure in a manner that accords with but may overgeneralize experience and the facts in the literature. The K-silicate facies is correlated with the position of the primary ore body based on the fact that pits being excavated in the primary ore zone have minerals of this facies well expressed. Examples include Ely, Nevada; Ajo, Arizona; and Bingham, Utah. Tourmaline is placed on a questionable basis with the K-silicate facies. The argillic facies may margin the K-silicate facies as Schwartz (25) and Creasey (5) report for San Manuel, Arizona, but the curve is drawn to show that argillized rock also does occur within the area of the K-silicate facies. The presence of substantial argillic alteration poses the problem as to how much of it actually is hypogene. Probably more of it is supergene than has been reported, a thought suggested by the prominent display of argillic alteration in those pits where secondary ore is mined. Santa Rita, New Mexico, is an example. In cases where paragenetic studies show kaolinite and related minerals older than sericite, biotite, etc., it can be concluded that at least part of the argillic alteration is hypogene. Also, where drilling encounters argillized rock below the zone of secondary effects, argillization can be viewed as hypogene.

Although the propylitic facies may occur with either or both of the other facies, its proper position appears to be on the fringes of the central ore and alteration zone, and it may extend out for considerable distances from the main areas of mineralization.

Obviously, in positioning these three facies, the district rock types must be considered; in some rocks, particularly multimineralic ones, the presence of all three facies is readily detectable, whereas in carbonate rocks and relative nonreactive rocks, such as sandstones, the three facies either never formed or are only poorly expressed. The most favorable situation for expressing the pattern is a single large intrusive, intermediate in composition, upon which alteration and mineralization have been superimposed.

The frequently encountered combination of quartz, sericite, and pyrite is shown to occur in variable amounts in the K-silicate and argillic facies and to extend weakly into the propylitic facies. This may disturb some accustomed to thinking of quartz-sericite-pyrite as most closely associated in time and space with the valuable hypogene sulfides. The mineral combination actually may be a facies in its own right.

The zonal pattern depicted poses some questions important in exploration. What are the third and fourth dimensional implications of the pattern; if propylitic alteration is exposed at the surface, might it grade into argillic and finally into K-silicate alteration at depth, possibly with associated primary ore? Do the alteration facies develop simultaneously with ore deposition as suggested by the studies at Butte, or in pre-ore stages as suggested by the Tintic work, or both ways? Some probably feel that the evidence usually is so ambiguous or too little understood to permit defendable conclusions. Does the propylitic facies result from products flushed from rocks that were argillized or converted to the K-silicate facies? These and related questions merit the continued attention of field and laboratory men.

Physical and Chemical Changes Attributable to Alteration of Intrusive

Alteration of the intrusive produced substantial physical and chemical changes. Undoubtedly, similar ones were produced in reactive rocks adjacent to the intrusive, but since the changes in such rocks have not been reported in any comprehensive manner, except for the zone of silication (as for example Schmitt's work on the Pewabic mine, 23), the data diagrammed are restricted to the intrusive. Chemical and physical studies need to be extended beyond the walls of altered intrusives to the total environment.

Porosity and permeability may be increased due to leaching in the argillic facies, and this may account for occasional concentration of higher grade secondary copper values in argillized rock (13). The curve is drawn to suggest that these two may decrease as the rock is freshened and reconstituted in the K-silicate facies, although even in this facies they may be greater than in unaltered rock. Bulk density, closely related to porosity, is reduced in argillized rock and probably increased in that converted to K-silicate minerals. Magnetic susceptibility may be reduced to nothing due to the destruction of magnetite during alteration, but the picture can be confused somewhat by the magnetite introduced during mineralization.

The chemical curves, based on data presented by Anderson (1) and Creasey (5), show an increase in K, OH, and S in the argillic facies and further increase in the K-silicate facies. Al, Fe, Mg, Ca, and Na all are reduced substantially in both facies. Si remains fairly constant in both, although Si was extracted from various minerals in the original rock and reconstituted into quartz that does occur in concentrated patches (the quartz-sericite-pyrite curve is drawn with fluctuations to suggest this). The silicification that sometimes is present evidently does not represent any great flood of hydrothermal silica.

Major and Minor Metal Distribution

The distribution pattern of major and minor metals in a porphyry copper district can be established by systematic field and laboratory studies of carefully selected ores, limonites in stringers and gossans, and fresh and altered rocks including jasperoids. All too few comprehensive studies of primary metal dispersion patterns have been made on a district-wide basis, and published results are scanty. This is an intriguing field of research for the geochemist and the exploration geologist that should yield results pertinent in the interpretation of drill holes into covered targets.

The writer's concept of the distribution of major metals in a porphyry copper district has been described above in the discussion of types of ore bodies, but it also is expressed by diagrammatic curves on the figure. The arrangement is somewhat idealized and obviously is influenced by the concept of zoning. This concept has good geographic basis and can be useful in exploration, whether or not the reasons for it are understood. It is fully realized that structures and rock types present will cause significant deviations from the curve patterns shown.

Iron in the form of magnetite is concentrated near the borders of the intrusive where carbonate or other reactive rocks are in the section; where nonreactive rocks are present, magnetite, if it actually was part of the mineralization story, is too widely dispersed to be recognized. Magnetite, as mentioned before, may occur as fillings in breccia pipes or veins, normally fairly close to the intrusive. Magnetite, which occurs as an accessory mineral in the intrusive, commonly is destroyed during hypogene and supergene alteration; therefore, iron in that form is shown as deficient over the intrusive.

Iron in the form of pyrite commonly is widespread and spans a variety of rock types. The undulations shown in this curve only indicate the variability in distribution of pyrite, not concentrations and deficiencies directly correlatable with the geological setting. Widespread pyrite implies that a large volume of hydrothermal fluid moved through the country and where present invites intensive prospecting.

The copper distribution curve, unlike any of the other metal distribution curves, is a composite of both primary and secondary metal dispersion patterns, and because of secondary enrichment shows a greater span than any of the other curves, except that for pyrite. Although this and other metal curves are turned down at the edges of the figure (which also suggest the edges of the district), copper and other metals may have geochemical expression in float, stream sediments, surface water, and ground water well beyond the limits of the district. It is not the writer's intention, however, to discuss geochemical prospecting beyond the district limits; attention is restricted here to what might be termed "anomalies observable in place." The pattern of secondary copper distribution may conceal the position of the primary ore body from the casual observer, but it is pertinent in planning a drilling program to make every effort to distinguish the two. It also is important to observe what types of copper minerals favor particular alteration types. The relation shown on the figure suggests that the better primary values correlate with the K-silicate facies and that commercial secondary values occur in both the K-silicate and argillic facies.

While molybdenum is mobile under certain circumstances, a qualification that also can be applied to the other metals, it usually is not extracted from the primary zone nor spread through the secondary copper blanket. Its position, therefore, is shown to correspond to that of the primary ore body. This is substantiated by the presence of molybdenite in all those pits where ore is being extracted from the primary zone. According to Waterman (29) the best molybdenum values in outcrop at Chuquicamata, Chile, lie directly over the best molybdenum values at depth. The same observation has been made by Schmitt (24) for the Esperanza ore body near Tucson, Arizona.

Gold is interpreted as largely restricted to the primary zone and may be residually enriched over it. However, gold may occur in veins at some distance from the main ore body.

Silver is shown related to the primary porphyry copper ore body and present in much greater amounts in satellitic ore bodies, usually with lead and zinc. It appears to be one of the more mobile primary metals, that is, it may be a "far-traveler" and span quite a variety of environments and even extend well beyond the satellitic lead-zinc ore bodies. Silver alone does not appear useful in distinguishing the center of a district from its fringe, but combined with lead and zinc values it suggests the fringe. Both gold and silver are recovered in porphyry copper concentrates; in fact a large percentage of the gold and silver produced annually in the United States is from porphyry copper ores.

Although veinlets of lead and zinc minerals in several areas cut primary porphyry copper ore, these metals characterize the fringe zone. In most districts zinc and lead occur together, although commonly differing in relative ages and amounts. Lead may be found somewhat farther out on the district fringes than is zinc.

Manganese, although rarely a major metal, occurs often enough to warrant consideration with the above. Its minerals, rhodochrosite and manganosiderite, may occur in the fringe zone by themselves or associated with galena and sphalerite.

Summarizing the above: the major metals of the central copper zone include copper, molybdenum, gold, and silver; the major metals of the fringe zone are silver, lead, zinc, and manganese; iron in the form of pyrite is abundant in both zones. The metal distribution patterns pose some of the same time and space problems as suggested by the alteration patterns.

It is most likely that geochemical exploration for in-place anomalies initially should be concentrated on the major metals, but in any total geochemical study of the porphyry copper environment the minor metals are not to be ignored. Not only can they be important in helping to establish whether or not a really interesting metalliferous target exists, but certain of them that occur with the major metals may lag in the zone of oxidation and suggest the former presence, and possible presence in depth, of the major metals. Cadmium in greenockite as indicative of the former presence of sphalerite is a good example.

Some of the major and minor metal associations are shown on the figure; most of these are described in standard references on ore deposits. They also are nicely summarized in Hawkes and Webb's recent book on geochemical exploration (7, p. 19, 359–377) and are reported in the discussions of various commodities in "Mineral Facts and Problems" (28). In reviewing the data it appears that the combinations most definitive of particular zones are the molybdenum-rhenium association and possibly the copper-selenium association in the central ore zone and the unique association of the minor metals cadmium, gallium, germanium, indium, and thallium, with zinc in the fringe zone. According to Hawkes and Webb (7, p. 19, 369) rhenium content is important in molybdenites associated with porphyry copper deposits. Although antimony, arsenic, bismuth, and tellurium are found in both zones, because of their association with most of the major metals cited, their quantitative distribution patterns may help in distinguishing the two zones.

The occurrence of cobalt, molybdenum, nickel, tin, and vanadium in minor amounts in the major metals presents some interesting problems and implications. During World War II, Kaiser, et al. (11) collected a wide variety of mineral materials originating in North and South America for spectrographic determinations of the trace metal content. Their objective was to locate unsuspected sources of strategic metals that might be expanded for use in the war effort. The mineral materials included gossans, capping, varied ore types, concentrates, tailings, and all varieties of smelter products relating to copper (including porphyry coppers), lead, zinc, gold, mixed base and precious metals, pyrite-pyrrhotite, antimony, arsenic, mercury, molybdenum, and manganese. Although they noted some interesting combinations, these workers did not attempt a thorough evaluation of their results.

This writer classified their data according to metal and type of material, took the number of samples involved, and without regard to the actual spectrographic amounts of the metal oxides reported, calculated the percentages of the samples in which the various metal oxides were reported. The deficiencies in this qualitative approach are recognized, but the results substantiated the association of the major and minor metals shown in figure 1, and unexpectedly showed that cobalt, molybdenum, nickel, tin, and vanadium occur a high percentage of the time in practically all these types of materials. Hence, the five elements are shown on the figure as ubiquitous in occurrence.

The implications of these results are somewhat speculative, but some kinship of origin of these materials is suggested. Although some of these metals are in more concentrated amounts—though still minor in some ore deposits—than in others, the results suggest these metals do not show unique associations that will distinguish the ore types of the central zone from those of the fringe. The molybdenum in molybdenite as discussed above is a different matter. Each district must be approached with the realization that absolute amounts of these metals will vary between districts, and what may be a suitable guide in one district may not be definitive in another. The spectrographic work by Burnham (3) on sphalerite and chalcopyrite, from which he developed his metallogenic provinces of the Southwestern United States, lends considerable support to the above thoughts on widespread occurrence, especially for cobalt, molybdenum, and tin. He does not report vanadium.

Although this writer has not seen the results of mercury-distribution studies in porphyry copper districts, Hawkes (8) has suggested that anomalous amounts of mercury may be indicative of base metal districts. Lemaire Instruments (12) in an advertising brochure states: "Dispersion coronas around base and precious metal deposits are generally weaker [than those around mercury deposits] but are generally strong enough to be useful exploration aids."

Color Anomalies

The combination of pyrometasomatism, alteration, and mineralization effects, after exposure to arid weathering conditions, produces a large, sometimes miles square, color anomaly distinguished by brownish to greenish coloration, bleaching, and variegated iron staining. The extent and intensity of the anomaly usually are not fully appreciated until viewed and photographed in color from the air. It is the writer's impression that in some areas in the Southwest a crude correlation appears to exist between the size and intensity of the color anomaly and the size and grade of the contained porphyry copper ore body, but there are too many possible complications for this to be other than a suggestive working hypothesis. The nature and intensity of the color depend on the rock types present, whether or not a pyrometasomatic halo is present, the degree of oxidation, the intensity of alteration, whether the ore body is primary or both primary and secondary, the ratio and total percentages of pyrite and chalcopyrite, the degree of exposure, and probably a few other conditions.

In southern Arizona several porphyry copper districts have large, distinctive, reddish areas near the ore bodies. Examples include Bisbee, San Manuel, and Banner-Pima-Mission, but not all porphyries are distinguished by such "red thumbs." Usually, these monuments are due to nothing more than oxidation of barren pyrite abundantly disseminated in and veining a siliceous matrix. Sometimes these have distracted from the main objective, but wherever they occur in partly covered inadequately explored ground favorable for porphyry copper ore bodies, intensive geological, geochemical, and geophysical work around them is indicated. Such work should be supported by, and in some cases supplanted by, an adequate number of drill holes so spaced that significant ore bodies could not be missed.

In a humid area with much vegetation cover, the broad color anomaly may not develop, or may not be well exposed, or may not be particularly striking where it is exposed, but whether the colored outcrops are large or small, if the necessary geological elements are present, they have the same implications in exploration as more striking color anomalies in an arid climate.

GEOPHYSICAL EXPRESSION

Some geophysical approaches used in combination with geology and geochemistry may yield results suggestive of the porphyry copper environment. The basis for the curves on the figure are standard references on geophysical prospecting, which describe the various methods and permit predictions of possible results for the geological setting described. Published case histories also are cited.

The resistivity curve shows decreasing resistance or increasing conductance, depending on how one prefers to view it, toward the primary ore body. This is intended to correlate with an increase in conducting sulfides and an increase in electrolytes in more porous, shattered, altered rock. The curve is rippled somewhat to suggest abrupt local changes caused by irregular masses of conducting material.

The induced polarization curve in which observed geophysical measurements are converted to an "apparent metal factor" is based on data in a brochure distributed by McPhar (17) on a disseminated ore body in Arizona. While the induced polarization results do not distinguish the type of sulfide, they do suggest the percentage of conducting sulfides in both the primary and secondary ore bodies. The curve is drawn to suggest a possible correlation of the higher apparent metal factor with the primary sulfide zone. Where induced polarization surveys are made of covered areas, results may be confused by salt-bearing ground water or other nonsulfide conducting material, but, of course, the same reservation applies to other electrical prospecting methods.

The self-potential curve shows a major negative potential center over the oxidizing primary sulfide zone and a lesser but equally sharp negative center over the oxidizing zinc ore body on the right side of the cross section. Self-potential anomalies occur under such a variety of conditions that they are not definitive by themselves unless of especially large size; they are produced by electrochemical reactions at quite shallow depths.

While the induction method normally is not regarded as applicable in porphyry copper exploration, it can be useful in detecting satellitic ore bodies, veins, and richer sulfide concentrations in the primary zone. Such was the case at the Pima mine, Arizona. In spite of varied field approaches and frequencies used the method lacks substantial depth penetration; most geophysicists interpreting the results are reluctant to make predictions beyond 300 or 400 feet, and these are maximums. In drawing the curve it has been assumed conveniently that the veins and massive ore bodies are within the range of detection and have not been destroyed by oxidation. The massive ore bodies are expressed by the broad deflections in the line, so drawn to suggest moderate depths of burial and moderate widths. The split vein in the primary ore body is shown by two sharp deflections; the left deflection is shifted to the right to approximate the position in the inclined vein of sulfides not destroyed by oxidation.

Too few data are available on thermoluminescence to present other than a suggestive questionable curve. Ingerson (9) discusses the effect in limestones and mentions it in potash feldspars. Hot intrusives should destroy the thermoluminescence of invaded limestones; they gradually should regain this property with increasing distance from intrusives. Propylitic and argillic alteration should destroy the original feldspars and their attendant thermoluminescence, but it seems reasonable to anticipate the effect in the K-silicate alteration facies where the rock is reconstituted.

According to Sables (26) the thermoluminescence of any carrier is activated by the presence of copper, silver, and manganese, and he observes that feldspars are ideal materials on which to work.

The local gravity curve shows four highs that express the positions of the zinc and iron ore bodies. Of some interest in this regard is the Houston-Thomas property in New Mexico (21, 27) where drilling of a pronounced local gravity high revealed a massive zinc-lead-pyrite replacement ore body. Another survey by the curious operator, R. Mathis, after he had mined the ore body, disclosed an anomaly in the same position, but no additional ore was found. This happy, although incompletely understood, coincidence suggests a note of caution in blindly applying the one-to-one correlation the curve shows. The central gravity low correlates with the zone of greatest alteration in the intrusive and reflects decreased bulk density and increased porosity. Possibly this is overidealized, for rock reconstituted in the K-silicate facies actually may show a higher bulk density than in the argillic facies. More detailed gravity studies of altered zones are needed to permit refinements in interpretation.

Superimposing maps showing the distribution of mining districts on those with regional gravity contours shows coincidence of districts with gravity highs frequently enough to suggest that major districts may occur where the mantle is somewhat closer to the surface than normal.

Ground magnetic-survey results will correlate quite well with those of in situ or laboratory susceptibility measurements except that the former provide a continuous rather than a spot picture of the variations. Analyzing the ground magnetics curve from left to right: the gentle low-level oscillations correlate with relatively nonmagnetic rock. The first sharper oscillations may reflect the zinc ore body at the marble line, but the correlation will be poor to nonexistent unless the zinc minerals are accompanied by magnetic accessories. Increasing irregularity, or contrast, is caused by the ferromagnetic minerals in the silicated zone. The magnetite ore body produces the sharply higher ragged curves. The situation shown is a symmetrical one with magnetite ore bodies on both sides of the intrusive, but one easily can conceive of a variety of geological conditions that would be nonsymmetrical magnetically and might even lack expression if magnetite were not present. Between the sharp magnetic peaks, irregularity in the curve correlates with altered rock. Figure space was not available to express this as an extreme low-level effect, frequently equivalent to or even less than that observable over unaltered sediments.

The airborne magnetometer "smooths out" many of the local anomalies obtained from ground magnetic work. The sharp deflections over the magnetite ore bodies are superimposed on a broad general high that reflects the magnetic contrast of the intrusive with its sedimentary environment, the striking "thumb print" anomaly often seen on aeromagnetic maps. The general high may be confused or cluttered with irregularities due to alteration. Airborne surveys usually are made at too high a speed for the alteration low that exists between the two magnetic peaks to be well expressed, but this can be compensated for by the use of a helicopter-borne magnetometer.

Other data curves could be presented. Some approaches that might yield information significant in porphyry copper exploration include radioactivity measurements that would contribute to the understanding of both metal dispersion and potassic alteration patterns, AFMAG, infrared aerial photography, and total chemical studies of various metalliferous environments. A highly speculative possibility of airborne geochemical surveys is suggested in a recent article by Feder (6) on possible applications of radar in detecting physical and chemical variations on the land surface. It does appear, however, that opportunities for rewarding research in some of the conventional fields are as great or greater than in the more exotic ones.

CONCLUSION

In a fully exposed porphyry copper district, the geological setting, alteration pattern, and metal distribution pattern can be established by direct observation supported by chemical and instrumental analyses. With these data available results of geophysical surveys can be interpreted more precisely. The combination of methods should make it reasonably easy to locate the ore bodies if prospectors have not done so already. If, instead, the ore bodies are totally obscured by post-ore alluvium or volcanics, the problem is much more difficult. Investigation may be prompted in the first place by projective geology, geochemical prospecting based on soils, vegetation and water, and regional aeromagnetic and gravity surveys. Once the area of interest is selected, more detailed geophysical surveys might be completed, but evidence for the existence of a major porphyry copper district still would be indirect.

However, the geophysical curves depicted on the figure, excluding that for thermoluminescence which depends on direct measurements, do bolster each other since the electrical, gravity, and magnetic curve groups depend on different physical properties. With curve combinations such as those pictured, it should be possible to plan a drill grid with a minimum number of holes which would insure that a porphyry copper ore body would not be missed. Each hole drilled adds supporting direct evidence, not visible in the totally covered area, on the geological setting, alteration pattern, and metal distribution pattern that should cause the explorer to move from the "outer target rings" closer to the "bull's-eye."

LITERATURE CITED

1. Anderson, C. A., 1950, Alteration and metallization in the Bagdad porphyry copper deposit: Econ. Geology, v. 45, no. 7, p. 609–628.
2. Blanchard, R., 1965, Interpretation of leached outcrops: Nevada Bur. Mines. (In press.)
3. Burnham, C. W., 1959, Metallogenic provinces of the Southwestern United States and northern Mexico: New Mexico Bur. Mines Min. Resources Bull. 65, 76 p.
4. ——— 1962, Facies and types of hydrothermal alteration: Econ. Geology, v. 57, no. 5, p. 768–784.
5. Creasey, S. C., 1959, Some phase relations in the hydrothermally altered rocks of porphyry copper deposits: Econ. Geology, v. 54, no. 3, p. 351–373.
6. Feder, A. M., 1962, Radar geology can aid regional oil exploration: World Oil, v. 155, no. 1, p. 130–138.
7. Hawkes, H. E., and Webb, J. S., 1962, Geochemistry in mineral exploration: Harper and Row, 415 p.
8. Hawkes, H. E., Personal communication.
9. Ingerson, E., 1955, Methods and problems of geologic thermometry: Econ. Geology, 50th Anniversary Volume, p. 382–383.
10. Jerome, S. E., and Cook, D. R., 1958, Relation of western hypogene mining districts to regional tectonics, with special reference to southern Arizona and southwestern New Mexico (abs.): Mining Eng., v. 10, no. 1, p. 54–55.
11. Kaiser, E. P., Herring, B. F., and Rabbitt, J. C., 1954, Minor elements in some rocks, ores, and mill and smelter products: U.S. Atomic Energy Comm., TEI-415, 119 p.
12. Lemaire Instruments (no date), Advertising brochure, Type S 1 Mercury detector: Reno, Nevada, Lemaire Instruments, 1 p.
13. Leroy, P. G., 1954, Correlation of copper mineralization with hydrothermal alteration in the Santa Rita porphyry copper deposit, New Mexico: Geol. Soc. America Bull., v. 5, no. 8, p. 739–768.
14. Locke, A., 1922, Experiment in ore hunting geology: Mining Metallurgy, no. 184, p. 27–29.
15. ——— 1926, Leached outcrops as guides to copper ore: Williams and Wilkins Co., 175 p.
16. Lovering, T. S., and others, 1949, Rock alteration as a guide to ore—East Tintic, Utah: Econ. Geology, Mon. 1, 64 p.
17. McPhar Geophysics Ltd. (no date), The uses of induced polarization in mining exploration: McPhar Geophysics Ltd., 9 p.
18. Parsons, A. B., 1933, The porphyry coppers: Am. Inst. Mining Metall. Petroleum Engineers, p. 344–345.
19. ——— 1957, The porphyry coppers in 1956: Am. Inst. Mining Metall. Petroleum Engineers, 270 p.
20. Perry, V. D., 1961, The significance of mineralized breccia pipes: Mining Eng., v. 13, no. 4, p. 367–376.
21. Romberg, F., 1949, Gravity meter survey leads to ore discovery: Eng. Mining Jour., v. 150, no. 3, p. 52–55.
22. Sales, R. H., and Meyer, C., 1948, Wall rock alteration of Butte Montana: Am. Inst. Mining Metall. Petroleum Engineers Trans., v. 178, p. 9–35.
23. Schmitt, H. A., 1937, The Pewabic mine: Geol. Soc. America Bull., v. 50, no. 5, p. 777–818.
24. ——— Personal communication.
25. Schwartz, G. M., 1953, Geology of the San Manuel copper deposit, Arizona: U.S. Geol. Survey Prof. Paper 256, 65 p.
26. Sables, B. E., Personal communication.
27. Soule, J. H., 1948, West Pinos Altos zinc-lead deposit, Grant County, New Mexico: U.S. Bur. Mines Rept. Inv. 4237, 8 p.
28. U.S. Bureau of Mines, 1960, Mineral facts and problems: U.S. Bur. Mines Bull. 585, 1015 p.
29. Waterman, G., Personal communication.
30. White, W. H., 1956, Bethlehem copper—a porphyry copper deposit: Canadian Min. Jour., v. 77, no. 4, p. 76–77.
31. Wisser, E., 1960, Relation of ore deposition to doming in the North American cordillera: Geol. Soc. America Memoir 77, 117 p.

Editor's Comments on Papers 35, 36, and 37

DRILLING

The principal objective of drilling is to obtain maximum geological and grade information, sufficiently accurate to satisfy both present and anticipated requirements, within the limits of available finance. The major cost is normally the actual drilling. Additional costs to insure good recovery and proper logging, sample preparation, and assaying are sometimes begrudged, which represents poor economy. Too often situations occur where drilling programs have to be repeated or where properties may be unwisely developed or abandoned because of improper sampling and logging. The hole by itself is of little value; its value lies only in the information it yields.

Drilling technology is continuing to advance, producing greater penetration rates, improved recoveries, and reduction of salting or dilution effects. Logging techniques have not developed as rapidly. No substitute yet exists for painstaking meter-by-meter examination of core or cuttings. Neither geophysical, geochemical, nor photographic techniques will supply the character or detail of data required.

Diamond drilling is usually regarded as being the most useful technique for obtaining geological information and for obtaining samples for metallurgical testing. It is satisfactory for obtaining assay information where core recovery is good. Special precautions are required in highly fractured ground and where metal values are contained in highly friable minerals. Development of "wireline" drilling techniques and the use of drilling muds to stabilize hole walls and to prevent water loss have greatly improved reliability of assay results and reduced the necessity for reliance upon sludge samples.

Rotary drilling is faster and cheaper than core drilling, sampling of the cuttings is simple, and the sample can be split at the drill rig. It provides an excellent sample for assay where values are disseminated but dilutes values found in narrow structures. It is usually unsatisfactory for detailed geological information, but general rock type, mineralization, and alteration information can be obtained from examination of the chips.

Rotary-percussion drilling has gained favor in the sampling of bulk low-grade deposits because of its speed and mobility. Where reversed fluid circulation is utilized, excellent recovery of values is obtained even for gold values. It has the same drawbacks as rotary drilling in that much geological information is lost.

Percussion rock drills are often used for short holes to obtain low-cost samples. Churn drilling is seldom used any more in the testing of mineral deposits in rock but is still commonly used in the sampling of alluvial deposits.

Exploration drilling has a twofold purpose: first, to determine in a preliminary fashion the presence or absence, size, and grade of a mineralized zone; and second, to determine whether the geological environment is such that ore mineralization could be in the vicinity. Emphasis in the early stages is on obtaining rapid and low-cost information in sufficient detail that evidences of significant mineralization would not be overlooked. This emphasis will influence the selection of the initial drilling method. When probability of discovery has been increased, greater expense can be justified. This is not an excuse, however, for careless logging or sampling in early exploration holes. Care in logging and sampling adds little to the cost of the program and helps to insure that important data are not lost. It also enables this early drilling information to be incorporated with confidence into later planning and development.

Most exploration programs are organized in a series of stages. Each stage culminates in a decision point to decide whether probability of discovery has been sufficiently improved to justify further expenditure.

Kursell (Paper 35) is dated in his description of churn drilling sampling procedures in use by the Cerro Corporation and the use of standard rather than wireline diamond drilling. However, the procedures and precautions are applicable. Of particular importance is the emphasis on care in obtaining and handling the sample. A careful record of all procedures utilized must be maintained, which is essential not only to insure uniformity but will be demanded when finance is sought.

Cook (Paper 36), in another company instruction manual, sets out procedures for core logging and sample preparation. Few companies prepare such procedural manuals, often leaving decisions to an inexperienced geologist who is unaware of the many pitfalls. Instruction manuals as illustrated by Papers 35 and 36 should not be followed blindly but should serve as a guide to be constantly checked to establish the production of optimum results.

Logging of drill core can be monotonous and uninspiring and is often relegated to the most junior geologist on a project, yet it is the most important output from a drilling program. Proper logging will often suggest a near miss of an ore intercept; may show where assay results are in error or where samples have been mixed, or salted; gives information of value to ore reserve estimation and to mine planning ; and may indicate where problems will be encountered in extraction or beneficiation of the ore. Ekström, Wirstam, and Larsson (Paper 37) outline a data system for recording drill hole information that simplifies the task of recording and storing relevant geological data so that it can be readily manipulated by the computer. Unfortunately, it is often very difficult initially to recognize which geological parameters are important and relevant to a particular operation. The same features are not necessarily relevant in different types of deposits or in deposits formed in different geological environments. Paper 37 thus serves as an example, not a fixed template.

35

Reprinted from pages 19 and 22-24 of *Eng. and Mining Jour.* **156:**19-24 (1955)

CERRO DE PASCO DRILLERS' MANUAL DEFINES EXPLORATION PROCEDURES

H. A. Kursell

THE DRILLING of a porphyry copper deposit requires utmost care and accuracy, but it must also be done economically. Standardization of drilling procedure is the most effective means of obtaining the correct end result.

The drill hole, whether churn or diamond drill, is, for all practical purposes, a sample. An accurate sample at a permissible cost is the primary objective. Speed of drilling is of importance in the cost of sampling, but is of secondary consideration in the overall objective.

Essential steps for obtaining accurate churn drill samples are these:

a. Casing must be carried as close as possible to bottom of the hole.

b. The sample must be bailed from the bottom of the hole as cleanly as possible until no appreciable quantity of cuttings is found. The same applies to cuttings from reaming operations.

c. Sample equipment at the drill rig must be thoroughly cleaned before finishing each sample.

d. Unintentional salting must be avoided at all points from the drill hole through the assay laboratory.

e. The sample drying process must be carefully controlled to prevent overheating and resulting loss of combined water and sulphur.

f. All records must be complete and accurate. Too much information is better than too little.

Essentials for obtaining accurate diamond drill samples are, in principle, the same as those for churn drilling. Those essentials peculiar to diamond drilling may be summarized as follows:

a. The primary objective is to obtain as high a core recovery as possible.

b. Accurate sludge sampling operations should be maintained at all times, even though core recovery is high.

c. All possible measures must be taken to ensure that at least 80% of the water used for drilling is returned to the surface, and that the tanks and equipment used to measure the water are properly maintained.

What Drilling Must Yield

When exploring a porphyry copper deposit by drilling, it is the aim to achieve greatest possible accuracy of sampling compatible with efficient work and at a cost that is economically permissible.

The following procedures, based on experience and long observation of such work, are recommended for the Cuajone deposit. Improvements of methods are of course, possible and the engineer in charge of an exploration campaign should always be on the alert for such improvements. However, before changes and amendments are made, extensive comparative tests made by others should be consulted. Before being put into regular use, such changes must have the approval of the supervisory and managerial staff.

Standardization, establishment and follow-up of a strict routine to be used from the start of the campaign will insure accuracy of the end results, and reduce operating expense.

Drilling and Casing

Churn drill holes are normally collared and drilled 13 in. dia through the surface wash and soft capping 12-¾ in. O.D. surface pipe is installed and drilling is continued with a 12-in. bit through the volcanic capping, leached capping and oxide ore until the enriched zone is reached, or until caving and protection of the oxide ore samples call for installation of 10-¾ in. O.D. casing before reaching the enriched zone.

The 10-¾ in. casing is then set at the top of the enriched zone and drilling continued. In the enriched zone the casing will be carried down with the drilling not more than 15 m from the bottom of the hole, either by straight reaming or underreaming. Under normal conditions at Cuajone, in which the enriched zone is of substantial thickness and of considerably higher grade than primary ore, the casing will be set at the bottom of the enriched zone, and will remain in place until the hole is completed. In case the enriched zone proves to be relatively thin and low grade, the casing may be run to depths below the enriched zone.

The primary ore zone will then be drilled and cased as dictated by the condition of the hole walls in order to minimize salting of samples and to protect the tools, in accordance with the judgment of the project superintendent. When it is determined that a column of 15 m of protore (copper content below assumed cut-off grade) has been cut, the hole will normally be cased to the top of the protore before proceeding with the drilling, unless in the opinion of the superintendent the condition of the ground and location of casing previously set in the ore above does not warrant such practice.

Sampling at the Drill Rig

Whenever drilling water has to be put into the hole, the driller must lower the water to the bottom of the hole in the bailer, to avoid washing the sides of the hole. Standard short dart-valve bailers or flat-valve bailers are used to remove the sludge from the hole at the established sampling interval. Choice of the bailer will depend on drilling and sludge conditions encountered.

[*Editor's Note:* Material has been omitted at this point.]

The Authors

The Manual in its present form represents the combined efforts of the following:

H. A. Kursell, consulting mining engineer, was author of the major part of the Manual; also F. N. Spencer Jr., resident mining engineer; R. S. Bemis, general superintendent of mining; W. C. Lacy, chief geologist; G. T. Vandel, general superintendent of concentrating; T. McPardee, superintendent of the Cuajone Development Project; A. Davidson, assistant project superintendent; J. D. Hamilton, geologist of the project, and M. de Losada, ore dressing technician.

GEOLOGIC LOG

Cerro de Pasco Corporation

File No. ____________

Diamond Drill Hole No. ____________ **Date** ____________

Total Depth of Hole ____________ **Property** ____________ **Log By** ____________

Depth		Rock Classification	MINERALIZATION			Alteration	HEAVY MINERALS IN CONCENTRATES	REMARKS
From	To		OXIDE	SULPHIDE	Silicif.			

Property: ____________ **Area:** ____________ **COMPOSITE DIAMOND DRILL LOG** **File No.** ____________

Collar - Grid Coordinates N. ________ **E.** ________ **Cerro de Pasco Corporation** **D. Drill Hole No.** ____________

Collar Elevation ____________ **m.** **Scale 1 cm. = 4/m.** **Logged by:** ______ **Date Started:** ____________

Depth of Hole ____________ **m.** **Bearing** ________ **No. of Sheets, this log** ________ **Drawn by:** ______ **Date Finished:** ____________

Water Level ________ **m.** **Inclination** ________ **EXPLANATION** **Date:** ________ **No. of 8 Hour Shifts:** ____________

Ore	Rocks		Mineralization		Alteration	Hardness
O = Oxide Ore E = Enriched Sulphide S = Primary Sulphide P = Protore	t = Talus, Overburden v = Post Mineral Volcanics c = Leached Capping mp = Monzonite Porphyry qmp = Quartz Monzonite Porphyry	di = Diorite bt = Basalt rp = Rhyolite Porphyry bx : Breccia	cry = Chrysocolla cu = Native Copper li = Limonite cc = Chalcocite bn = Bornite py = Pyrite	sl = Sphalerite mo = Molybdenite cp = Chalcopyrite	St = Strong F = Fair W = Weak	H = Hard M = Medium So = Soft

Elevation	Sample Depth Recalculated to Every 2m. intervals	Final Copper Assay From Assay Log Recalc. for 2m. Sample Interv.				Classification					% Recovery		Casing				Remarks	Copper Assay Graph
		Total	Non Sulphide	Average	Composite Assays	Ore	Rocks	Mineralization	Alteration	Hardness	Core	Total	Hole Drilled	Hole Reamed	Hole Cased	Bit Size		% Cu: - .5, - 1.0, - 1.5, - 2.0, - 2.5, - 3.0
1	2	3	4	5	6	7	8	9	10	11	12	13	14	15	16	17	18	19

Assaying

A 2-½-g sample is used for assaying. Each sample is assayed for total copper and non-sulphide copper content. The standard iodide method is used for the total copper determinations and the sulphur dioxide method for the non-sulphide copper determinations. Determinations are carried to the third decimal. Until such time that checks within 2% of contained copper are obtained with Oroya (where also electrolytic check determinations should be made from time to time), every sample in ore will be checked. Later it will be adequate to check in Oroya every fifth sample (10 m) of better grade ore, every tenth sample (20 m) of medium grade ore and every twentieth sample of material less than the assumed cut off.

Samples from non-commercial low-grade runs in protore or waste may be combined for longer columns up to 30 m. Precious metal content and molybdenum should be determined only for the composite samples in Oroya. Spectographic analyses should be made from time to time, on not more than one composite for each hole, for Pb, Zn, W, Mo, Bi, Se, Te, Ge, and whatever other determinations are normally made at Oroya.

Churn Drill Records

(1) A form of a suitable report is attached (see p 21).

(2) The form of the geologic log attached (see top p 20), is prepared for recording megascopic and microscopic examination of the samples. The microscopic examination of the samples is made with a binocular

DIAMOND DRILL ASSAY LOG
Cerro de Pasco Corporation

Property: CUAJONE

File No. ____________
D. Drill Hole No. ____________
Date: ____________

Depth		Hole Size	Sp. Grav.	Core			Combined Core 2 Met. Run		Sludge 2 Met. Run			Recovery %		Average For 2 Meter Run							Oxide Co. Assay			Remarks
																Adjusted Assays								
From	To			Wgt.	Cuajone Assay	Oroya Assay	Wgt.	Assay	Wgt.	Cuajone Assay	Oroya Assay	Core	Sludge	From	To	Wgt:	Core	Wgt.	Sludge	Final Comb. Assay	Core	Sludge	Comb.	

microscope. All the minerals indentified and their relative amounts are recorded on a log, also the rock formation penetrated.

(3) The composite log for each completed drill hole forms the permanent record and is a condensation of all useful information pertaining to the hole. This log is compiled from the daily shift reports, geologic and assay logs.

(4) The sludge board is prepared from a part of the divided sample by a suitable trained workman. A description of the board follows:

The sludge or sample board affords at a glance a detailed picture of results obtained in the drill hole. The board is 1.60 m long, 0.12 m wide, and 0.01 m thick. It is made of seasoned lumber (Grade A), or its equivalent, has a hole at the top for filing in the sludgeboard rack, and is painted white.

The sample (200 g) is washed free of slime and a small portion of the larger rock fragments are saved. The rest of the sample is then panned till a concentrate is formed. Rock fragments and panned concentrates are dried and glued on the board opposite the sample interval marked in black India ink on the left side of the board. An allowance of 1 cm for each sample is made vertically and 5 cm for the coarse fragments and 3 cm for the concentrates horizontally. Coarse fragments are glued on the left side and concentrates adjoining them to the right. The assay value of each sample is recorded to the right of the concentrates. Each band of 1 cm represents a 2 m sampling interval. The hole number, its coordinates and elevation of the collar of the hole are recorded at the top of the sludge board. Should a hole exceed 150 m in depth, two or more sludge boards represent one hole.

(*Please turn page*)

DIAMOND DRILL SHIFT REPORT
Cerro de Pasco Corporation

File No. ________

Property ____________________

Hole No. ______________ Date ______________ 19____

Drill No. ______________ Shift ______________

Size of Bit - Number - From - To - Accuml. Total Driller ______________

______________ Helper ______________

______________ Water - Depth Encountered __________ Feet

Casing Lowered - Size - From - To Lost - Water At __________ Feet

______________ Feet ______________

Casing Removed ______________ Feet

EMPLOYMENT OF TIME

Moving and Setting Up ______________ Cement Used - Type ________ Quantity ______

Drilling ______________ Drilling Cement - From ________ To ________

Setting Casing ______________ Cementing Hole - From ________ To ________

Equipment Repair ______________ Drilling Mud - From ________ To ________

Other Delays ______________ Reaming Hole - From ________ To ________

SAMPLES

Depth		Recoveries			Hardness of Rock V.H. Very Hard H-Hard M-Medium S-Soft V.S. Very Soft	Condition of Hole	Color of Sludge	Dry Weight		Sp. Gr.	Remarks
From	To	% Core	% Sludge	% Water				Core	Sludge	Core	

GENERAL REMARKS

Depth of Hole at Beginning of Shift ________ Samples Left in Sludge Box ________

Depth of Hole at End of Shift ________ Samples Canned ________

Total Loads of Water Hauled During Shift ____ Sampler ________

Water on Hand ______________ Gallons

REPORTS FORMS appearing on this and preceding pages were designed to assemble vital exploration and geologic information obtained by churn drilling and diamond drilling at Cerro de Pasco's Cuajone project in southern Peru.

Diamond Drilling

Powerful diamond drills, capable of drilling a "BX" size hole to 2,000 ft depth, are required if diamond drills are to be used at Cuajone. Drill derricks should be quickly dismountable, lightweight, (aluminum pipe), strong, and of sufficient height to pull at least 40 ft of rods.

Drilling Procedure

NX (3-in. hole dia, 2⅛-in. core) BX (2⅜-in. hole dia, 1⅝-in. core), AX (1⅞-in. hole dia, 1⅛-in. core) and eventually EX (1½-in. hole dia, 13/16-in. core) bit sizes will be used with non-rotating, self sealing double-tube core barrels. Casing of the holes is essential and the casing should follow the drill closely (within 10 m). The casing is lowered after reaming the hole, and underreaming should not be relied upon or attempted. Drilling with non-rotating double-tube barrels is slow and requires frequent pulls. Should ordinary double-tube core barrels be used, sludge samples become of increasing importance. The greatest emphasis is placed on the necessity to return at least 80% to 85% of the water used for a run. Therefore, adequate tanks with the necessary measuring scales should be provided so that the driller may know at all times how much water he has pumped into the hole during a run, and how much sludge has actually been returned. No return water must be taken for drilling from the overflow of a sludge-catching device before a run is completed. Adequate tanks and a proper set up must be available. If necessary an additional pump must be provided. There are many sludge-catching devices used in diamond drilling. Of the several devices tried, a cylindrical tank, 1.6 m high, 1.22 m dia with a conical bottom and a spigot in the bottom to draw off the sludge, have been proved to be satisfactory. Two such tanks can be provided for each drill for alternate 2 m runs. The clear water is syphoned off into a reserve tank and at the same time the sludge and wash water are caught in a 10-gal milk can and sent to the sample room. Here the sample undergoes the same treatment as the churn drill samples, with the exception that samples are cut only for assaying and composite assays.

Core and Sludge Sampling

Core recovered in drilling is carefully placed into core boxes with markers of hole depth, which are locked and sent to the sample room. Cores for each run as pulled are dried and weighed to establish core recovery. Only a small specimen from each run to indicate the nature of the rock penetrated and for microscopic studies, etc., is kept as a geological sample and stored in a suitable core box. The balance of the core is crushed and split on a Jones splitter. One part of the crushed core is stored in glass jars with hermetical metal seals, for reference, metallurgical tests, specific gravity determination, etc. The balance of the core is pulverized for assaying, check assaying, and composite samples. Core weight and assays of the several runs as pulled are first combined for one 2 m run, corresponding to the 2 m sludge run and core, and sludge samples for a 2 m run are then combined in proportion of the weight of the core actually recovered to the theoretical weight of the sludge, adjusted for sludge recovery. Overage is deducted from 100% to establish percentage of sludge recovery.

The adjusted assay for a 2-m interval drilled equals actual weight of core × core assay plus theoretical weight of sludge × sludge recovery × sludge assay, divided by actual weight of core plus theoretical weight of sludge × sludge recovery. For example, if the actual weight of core is X and the core assay is 1.25% Cu, the theoretical weight of sludge is Y, sludge recovery 80%, and sludge assay 1.33% Cu, then the equation would look as follows:

$$\frac{X \times 1.25 + Y \times .80 \times 1.33}{X + Y \times .80}$$

In practice it has been found that sludge actually recovered under most perfect running conditions may vary from 80% to 120% of calculated sludge.

Assaying

Assaying of diamond drill cores and sludge samples does not differ from assaying churn drill hole samples as previously described.

Surveying

Deep diamond drill holes should be surveyed for deflection and the results recorded in the shift reports and composite drill holes log under "Remarks." A Maas compass and glass tube of suitable composition for etching with hydrofluoric acid is sufficiently accurate for holes up to 300 m in depth. Diameter of the tubes will be the maximum commensurate with the diameter of the part of hole being surveyed.

Keeping of Diamond Drill Records

(1) The attached diamond drill shift reports are a more or less standard form. (See p 23.) Depths of holes are measured in feet and decimals, and later converted by the engineer compiling the composite drill hole log into meters, as it is not practicable to change the habits of American diamond drillers.

(2) The geologic log (see top p 22) is based on examination of the core megascopically and if need be, microscopically. All minerals recognized are recorded, as also the nature of the rock penetrated and any other observations obtainable from core study.

(3) The assay log (see p 23) contains all individual assays of short lengths of core obtained from several runs, usually required to drill the standard 2 m intervals and combines the results of core assays and sludge assays into final figures to be used for the permanent record.

(4) The composite log of the completed drill hole (see p 22) combines all pertinent information on the hole as a permanent record. It is compiled from shift reports, the geologic and assay logs.

Storage of Core Specimens

Each specimen of diamond drill core retained from every 2-m run is to be spotted with white paint and labeled with hole number and depth represented. Specimens are then to be placed in a heavy manila sample envelope (10.5 cm x 18 cm) with metal closing tab, labeled as follows:

CERRO de PASCO CORP.
Diamond Drill Core
Hole ....
.... feet to .... feet
........meters to ....meters

Specimens in the labeled envelopes are placed in open boxes, fabricated of construction lumber with inside dimensions 11.5 cm wide x 84 cm long x 12.5 cm deep. Each box will hold about 50 specimens. The boxes should be labeled on the ends as to hole number and depth, and placed in a rack constructed so that any box may be removed independently.

Editor's Note: In addition to the foregoing information, the Cerro de Pasco Drillers' Manual contains instructions pertaining to: The point at which to stop a wildcat drill hole; mapping and engineering drill grids, standards, scales, sections; details to be included in regular monthly progress reports; interpretation of results.

36

Unpublished report, Kennecott Corporation, Salt Lake City, Utah, 1972, 16p.

NOTES ON CORE LOGGING, SAMPLE PREPARATION AND ORE RESERVES

Annan Cook

1. CORE LOGGING

The objective of core logging is to obtain a written "foot by foot" description on a standard form, of the rock types being drilled, with accompanying notes on the presence or absence of mineralization, rock alteration or structure. This is done before splitting but, as some features show up best on a broken surface, core should also be examined after splitting. Usually the texture of the rock shows up better by wetting it.

In addition, core split for assay must be recorded, so appropriate columns in any drill log are needed for all this information as well as for assay data. Additional columns are also required to enter some of the information listed below. In order to have room enough for all the information needed, a two-page log is necessary.

1.1 General

At the top of the log on the left hand side is shown the name of the project or examination with the appropriate drill hole number, coordinates, elevation and angle of inclination and bearing. Other useful information concerning the hole, such as the date started, the date completed, the name of the drilling company and final depth of the hole, should also be shown here. If the hole is deep and down-the-hole surveys have been made, these readings should also be tabulated. Core size or sizes should also be shown and the type of drilling described (e.g. rock-bitting or diamond drilling). It is also useful to know what make of drill rig was used and this should be stated. Casing depths should also be recorded.

1.2 Geological Data

This information is shown on the right hand page of the log. Rock types are shown by footage ("From" and "To") in the columns provided. Footages should be shown in feet and tenths of a foot (instead of in feet and inches). A column for naming the rock type should follow, with an adjoining column which can be colored or marked with symbols representing the specific rock type. Faulting or shearing (described in separate columns) can also be shown diagramatically in this column. Larger columns are also needed for the detailed geological description of the length of core being described. Other columns for describing alteration and mineralization should also be available.

In general, it will help if the spacing on the log is done to a given scale. This is, however, not essential—particularly if hundreds of feet of barren rock overlie and ore body and only a few sections of core are split for assay in the barren portion of the hole. It should be borne in mind, however, that in a case like this, two different scales could be used in the same log, to give a clearer picture of the position of any ore that has been intersected.

Most of the engineering data (mainly assays) are usually shown on the left hand side of the two-page drill log. Sample lengths are, as before, described by footage ("From"and "To"). The actual sample length is then shown, plus the length recovered. The percentage recovery and sample number follow.

The assay length should normally never go beyond a given pull, because different core recoveries may have been obtained from adjoining pulls. Core size may also have been reduced when the hole has been cased. When this happens, the sample taken should never include two sizes of core, otherwise incorrect weighting of results will result.

Adjacent to the column for sample numbers are columns in which the assay values for the metals being assayed, weighted average of a given group of assays can be entered.

Space on the log should also be made for composite analyses, in case ore-grade material has been intersected. Composites are made up from individual sample rejects, a standard weight per foot of sample being taken from each reject. The sample is then mixed carefully and assayed for the same metals that were run before plus any others, such as cobalt, chromium, silver, gold, lead, zinc, etc. Assays from composites should naturally check the average of the weighted assays of the individual samples that make up the composite.

Additional engineering data that should be recorded are the amount of fracturing present (faults, shears, gouge, zones, broken zones, etc.) and the character, or strength of the rock—especially if ore has been intersected. At four porphyry copper deposits we have used the categories of strong, moderately strong, moderately weak and weak to describe the character of the rock. In a future underground mine this breakdown will help to determine the best mining method that will eventually be used.

The drillers log, which is filled out at the end of each shift, usually contains a record of the hardness or softness of the ground. This information, as well as the rate of advance per shift, will help the geologist to describe the strength of the rock.

Intersection, angles made on drill core by bedding, joints, cleavage, schistosity, shearing or faulting should be recorded in a special column. Specific gravity measurements should also be made routinely on any ore intersection. These must be made on dry samples after heating, say to 105°C. If the sample is porous, it *must* be coated with paraffin or shellac before being weighed in air and in water.

1.3 General Comments on Drill Hole Logging

Drill logs can be designed specifically for a given type of ore being sought.

Obviously when rotary-drilling is being done, all you have to log is drill cuttings. Accordingly, items like core recovery, fracturing, character of the rock and angle of intersection (with structure or bedding) are impossible to obtain by standard methods. However, if a sample-run is weighed, it can always be compared with the theoretical weight of sample that should have been obtained had the recovery been 100%.

Before wire-line drilling was introduced, core recoveries used to be poor, particularly in well-mineralized rock. Sludges used to be recovered and assayed separately from the core. By a process of weighting the sludge and core assays (based on recoveries), a combined core-sludge assay was obtained. Unless the core recovery was high, the assay obtained was of dubious value. Thanks to the vastly improved core recovery now generally obtained by wire-line drilling, such practices are no longer necessary.

With regard to copper analyses—particularly if copper is your main target—an extra assay column *must* be included in the log to show the nonsulfide (or oxide) copper analyses. The total copper analyses minus the nonsulfide analyses, give the amount of copper present as sulfide. In most cases we are looking for sulfide deposits, so the amount of copper present as sulfide must be known.

When submitting individual assay pulps from ore intersections to an assay office, it is routine practice to send a duplicate of every tenth or twentieth sample to another assay office (if available) as a check. Checks can also be made by sending a duplicate pulp to the *same* assay office under a different number and comparing the results.

Core logging always requires a hand lens, a pocket knife, a tape (marked in feet and tenths of a foot) and a clino-rule (to measure intersection angles). The job can naturally be improved by using a binocular microscope and having available suitable reagents for microchemical qualitative analysis for one or more elements. As a minimum you will need the standard acids, ammonia, 2% dimethyl glyoxime solution and 3% potassium mercuric thiocyanate solution—for nickel and copper tests.

Microchemical tests for ore minerals are described in U.S.G.S. Bulletin 914.

Thin and polished sections are naturally needed from time to time from drill core.

When coring, routine scanning of the core should be done with a fluorescent lamp and a scintillometer or Geiger counter.

Core in ore bearing zones or country rock can be photographed. Such photographs provide an excellent record of just how solid or broken the ore really is. For future mining purposes this becomes very valuable information.

Deep drill holes are routinely surveyed. Where the rock is magnetic, a gyro-compass may be used to obtain the correct bearing.

Core should always be carefully marked and stored for future reference. The cost of doing this is usually minor, compared with the actual cost of drilling.

Frequently a specimen core log from a given drill hole is found to be very useful. This can be made up by removing pieces of core at specific intervals from the drill hole and filing them in ordinary core boxes. Needless to say, every core sample will be labelled with the drill hole number as well as the depth from which it was taken.

2. SAMPLE PREPARATION

2.1 Marking the Sample

After logging, the drill core is marked off by the geologist in the intervals decided upon for sampling. Whereas in a porphyry copper deposit a ten foot sample interval is satisfactory (or even 20′ in a marginal zone), generally in smaller types of deposit smaller sample intervals are required. The smaller sample intervals (say 5′) obviously give a better control of where the mineralization occurs.

Sample tickets can be made up in booklet form. These describe the location and number of the drill hole, the footage sampled, the date, the name of the sampler and the list of metals to be assayed. On the right hand side of the ticket a duplicate (or duplicates) of the sample number appears, showing the date, the sampler, the metals to be assayed. This right hand portion of the sample book is made so that it can be torn out and put in the core box at the end of each sample length. This ticket goes with the split core sample.

If desired, enough space can be provided on the portion of the sample ticket that is removed to show the drill hole number and footage—as well as the sample number. Normally, however, when sending samples to an outside assay office, only the sample number, date, and metals to be assayed should be shown for security reasons.

While drillers normally show the end of each run with the footage written on a block of wood (or cardboard), it may be necessary to split a given pull into one or more samples. Additional wooden blocks should be kept on hand to show where the sample is taken, for future reference.

2.2 Splitting the Core

After marking the core, it is ready for splitting. For this work, either a core splitter or a diamond saw is used. The core splitter has been in use for many years and if *exact* splitting of the core is done, satisfactory results usually result. More recently the diamond saw has been being used for core splitting. In spite of the fact that fine material from sawing the core is lost, a better representative sample is obtained by using this method of core splitting.

During the process of splitting, attempts must be made to split core at right angles to bedding or structure. When using a core splitter, all the core and fine particles from the splitting that fall into one of the trays provided, must be included in the sample for assay. All the core and fine material from the other half of the core must be returned to the core tray. On one occasion I saw the nicely split core returned to the core box, while the fines from both sides of the splitter were put into the sample for assay! This was found (by assaying the fines separately) to high-grade the sample.

I want to emphasize the care that needs to be given to core splitting. Like underground mine sampling, it needs to be done by responsible people. Normally it is regarded as the lowliest job around a core laboratory, instead of being one of the most important.

2.3 Preparation of the Sample

In the early stages of drilling project (if no crushing and grinding equipment is available), split cores from individual samples are handed over to an assay office for the job of both preparing the sample pulp and assaying it. This is also the routine at most operation mines. In exploration however, it is fundamentally the task of the project manager to *prepare a correct sample for assay.* This task must therefore be done correctly.

Drill hole samples may be any size from 1 lb. to 30 lbs. and after splitting, may consist of pieces from 3'' or more in length to the finest powder. The assayer requires about a pound ground to a fineness which will permit free play to the chemical processes involved, and which will give concordant results. Between these two positions lies the work of reducing the sample bulk and the particle size, in a manner and with a care which will yield a real sample representative of the costly taken diamond (or rotary) drill hole sample.

The principle involved is that the bulk should not be reduced more rapidly than the size (or particle dimension). Two tabulations below give you an idea of the minimum weight of a representative sample needed for the size of the material.

2.4 Smallest Permissible Weight to Yield a Representative Sample. (From U.S. Steel - through C. Dahl)

	Mesh Size	*Min. Wt.(lbs.)*	*Kg.*
	6''	10,000	4500
	5''	7000	3200
	4''	4500	2000
	3''	2800	1300
	2''	1000	450
	1''	300	135
	¾''	150	70
	½''	75	35
	3 Mesh	20	10
	4 Mesh	10	5
			Grams
	6 Mesh	4.4	2000
3 mm	8 Mesh	2.2	1000
2 mm	10 Mesh	1.1	500
	14 Mesh	0.66	300
1 mm	20 Mesh	0.33	150
	28 Mesh	0.165	75
	35 Mesh	0.077	35
	48 Mesh	0.044	20
0.010'' or .25 mm	65 Mesh	0.022	10

2.5 Smallest Permissible Weight of Sample to Yield a Representative Sample for Gold Ores
(From "Mining Engineers Handbook" - Peele)

Size, diam. in.	*Lbs.*
2	10,000
1½	5000
1	2000
¾	1000
½	400
⅜	300
¼	200
3/16	100
⅛	75
6 Mesh	50
10 Mesh	25
18 Mesh	10
30 Mesh	4
50 Mesh	1

Reduction is achieved in stages and embraces crushing, mixing and dividing—in this order. Though bulk may not be reduced more rapidly than size, size, may be reduced as rapidly as the crushing equipment allows. In that way, two stages of bulk reduction may be and often are, included within one stage of size reduction. Size being taken to be diameter, equal reduction-ratios for bulk and for size connotes an increased number of particles in the retained portion, since particle number increases as the cube of the diameter decreases; it is this increase in particle number as the bulk gets smaller which secures real representation of the original core sample. In any finally retained portion, the largest particles should not be of such a size that one such particle, however rich, would sensibly affect the value of the assay portion.

2.6 Crushing and Grinding

First dry and then crush the entire sample in a jaw crusher. For a couple of dozen samples a day a 2'' X 3'' crusher is satisfactory (like the 3'' Van Gelder crusher). If however, as in any sizable drilling program, the machines are to be worked for hours on end, use about a 3½'' X 4'' Denver Type 4 Laboratory Jaw Crusher or 4'' Van Gelder crusher. This will do about four times the work of the smaller crusher.

Clean the jaw crusher with a brush and with compressed air after each sample run. Generally the jaw crusher has a free discharge and offers little chance for lodgement of the sample. Pieces are, however, apt to fly out of the mouth.

At this time the material is usually −¼''. Put all the sample through a coffee-grinder or rolls—which, like the jaw crusher, discharges freely, and crush to 95%-20 mesh (0.033'') with no particles greater than 10 mesh (0.07''). Clean the coffee-grinder (or rolls) as before. The sample is now fine enough to mix and roll. Follow this up by splitting with a Jones splitter until about 2 lbs. (900 grams) is left. (Actually 500 grams (1.1 lbs.) is enough, but is safer to use the larger size).

The sample is now pulverized to 95%-100 mesh, with nothing greater than 80 mesh. If necessary this can be done in two stages. This constitutes the assay pulp, which is sent to the assay office. The balance of the sample reject is bagged and stored in a dry place.

Assay laboratories use anything from 0.4 grams to 2.0 grams for determinations so it will be obvious why the sample has to be carefully prepared. If the original sample weighed 20 lbs. (9072 grams), the amount of material actually used for assay (say 2.0 grams) is only 0.022% of this.

[*Editor's Note:* Material on computation of ore reserves has been omitted.]

3.3 Erratic High Values

When an abnormally high assay value has been obtained in an ore intersection, the question arises as to whether or not it should be included at its full value in averaging the ore-grade interval.

First of all the possibility must be eliminated that the high assay is simply a mistake caused by faulty splitting of the core, through faulty preparation of the sample or through accidental salting in the laboratory. If the core has been satisfactorily split, careful reduction and reassay of the sample reject should establish this.

If the value remains high, is the sample erratic or is it simply high? If a series of high values have been obtained they are obviously not erratic. In your drilling for nickel deposits, you might obtain high values in a massive sulfide intersection, which would be quite normal. The problem is thus basically geological rather than purely mathematical.

If numerous ore intersections have been made—particularly through mine development, the grade of blocks of ore can be calculated by the frequency-weighting method. At the moment we are, however, only discussing the exploration stage, so no details of this method will be discussed now.

If there are no good geological reasons to explain a high value, there remain arbitrary methods to take care of it:

(1) Average the high value with the two samples adjoining it and substitute the average of the three samples. An alternative method is to average the high value with the four adjoining samples and substitute the average of the five samples.
(2) Calculate the mean of the whole ore intersection and substitute the high value (or values) with the mean value. The intersection is then averaged again using the substituted value.
(3) Cut all highs to an arbitrary value.
(4) Apply a percentage correction to the average grade of the ore intersection.

The first two methods are more suitable to exploration drilling. The last two methods can really only be used at an operating mine, where experience has been obtained in comparing ore reserve grades with the actual head values of the mined ore.

It must always be remembered that the second half of the split drill core is still available for further testing. This can always be quartered and the assays compared with those originally obtained.

Additional ways of checking an ore intersection are to wedge the drill hole and obtain a second intersection. Finally, if necessary, a check drill hole can be put down adjoining the original hole.

3.4 Core Loss

The percentage recovery of drill core is extremely important. The more core recovered, the more accurate the sample. If part of an ore intersection is missing, the question arises as to what to do about it.

Probably the best method (already mentioned above) is to wedge the drill hole and try and obtain a second intersection. Alternatively, the hole can be redrilled.

If only a few feet are missing, the problem of assigning a grade to it exists. As also mentioned above, about all one can do is to average the grade of the adjoining two or three samples to the missing core and substitute this value. Whatever is done, however, should be mentioned in the drill log or in a report.

REFERENCES

"Mine Economics" by S. J. Truscott, Mining Publications Ltd., 1937.

"Mine Examination and Valuation" by Charles H. Baxter and Roland D. Parks, Michigan College of Mining and Technology, 1939.

"Mining Geology" by H. E. McKinstry, Prentice Hall, 1948.

"Engineering and Mining Journal," March, 1971.

37

Reprinted from *Econ. Geology* **70:**359-368 (1975)

COREMAP—A Data System for Drill Cores and Boreholes

Tom K. Ekström, Åke Wirstam, and Lars-Elve Larsson

Abstract

COREMAP is a data system for recording and processing information from drill cores and boreholes. The system may also be used in tunnels, adits, etc. The standard sheets include general, geological, chemical, joint, and borehole deviation data, but these may be extended to include any other type of information. In COREMAP it is also possible to computerize notes.

The processing routines of COREMAP include decoded lists, lists of any specified parameter or parameters, brief descriptions of the core by histograms, and a number of plotted maps. Different types of planar and section maps are plotted on a drum or a microfilm plotter. The projections are corrected for borehole deviation and the strike and the dip of the rocks. The length of each analyzed interval is plotted together with one chemical element. COREMAP is a sister system of GEOMAP, the field data system for hard rocks. The two systems comprise a valuable tool in prospecting and geological mapping.

Introduction

Manual processing of data is often time-consuming, dull, and expensive, because of the sheer bulk of the various types of information from drill cores and boreholes. This means that a rational retrieval and storage of data is necessary. During the last decade computer techniques have been applied to many types of geological problems. Most of the systems developed were originally designed for special problems and are therefore of limited interest.

When the present project started in the beginning of 1971, no general data management system for boreholes and drill cores was available. Some systems dealing with borehole and drill core data have since been published. Oil geology was probably the first field of geoscience to make use of data systems for storage and retrieval of data from boreholes. Here only the system developed by Buller (1972) will be mentioned. Gover et al. (1971) have presented a pilot system for geological information from cored boreholes in Scotland. Look and Vinken (1971) report from Germany on present activities with borehole data, and in England Farmer (1973) has presented a number of programs for plotting boreholes. The "Geolog System" presented by Blanchet and Godwin (1972) is of special interest. This system was originally designed for porphyry deposits, and the standard formula is mainly restricted to geological information. The method of collecting and coding data in the "Geolog System" is similar to some field systems published by Wynne-Edwards et al. (1970) and Berner et al. (1971, 1972).

Although a data management system for drill cores and boreholes may treat different types of data, it should be possible to apply it to other types of information—in other words, the system should have high flexibility. As a result of favorable experience with the field data system GEOMAP in Scandinavia (Berner et al. 1971, 1972), its geological parameters, codes, etc., are also used in COREMAP, the new system for boreholes and drill cores. The principles of the COREMAP system are shown in Figure 1. The lefthand side of the figure shows the input documents, including general, geological, chemical, joint, and borehole deviation data. Note the possibility of adding new data to the system in the future. All data are punched on cards, sorted, tested for errors, and stored on magnetic tape. The righthand side of the figure shows the major routines of COREMAP, namely, listing of decoded sheets, listing of any

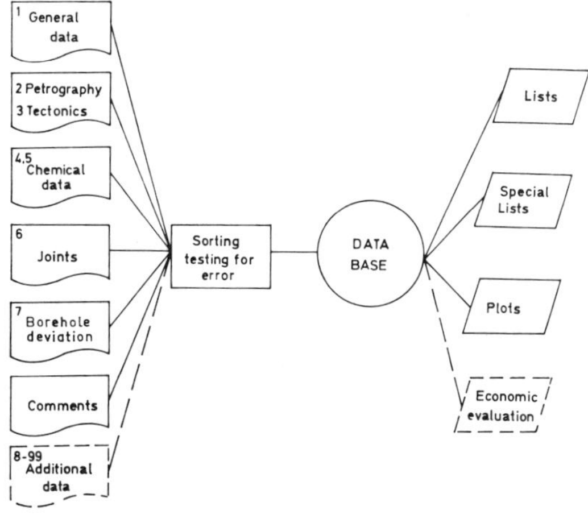

Fig. 1. The principles of the COREMAP system.

parameter or parameters, and the plotting of different types of maps.

Data sheets

The COREMAP system has at present seven different types of data sheet. These include (1) general data, (2) petrography, (3) tectonics, (4, 5) chemical data, (6) joint, and (7) borehole deviation data. A special sheet is also used for comments or notes. As stated before, COREMAP could be applied to any type of drill core or borehole data.

Six of the forms are shown in Figures 2 and 3. The user has the option of reducing the parameters of the original sheets and of giving them a personal design. However, the numbering and the order of the parameters must be retained. A detailed description of the different input documents of COREMAP and instructions on recording the data, together with codes, are given in a special handbook. Only the principles and the most important features of COREMAP will be treated in the following pages.

1. General data: The form named "General data" is the only one which is essential in COREMAP. This sheet includes information which is registered only once for each borehole, i.e., starting coordinates, soil depth, total length, geologist, etc. The code that identifies a certain borehole or a core is found in the entries 6–16. This consists of locality, hole number, and type of data. "Type of data" is a numerical value found in columns 6 and 7. "General data" has number 1, "petrography" number 2, "tectonics" number 3, etc. COREMAP at present includes seven different types of input form apart from the comment sheet, but the system may be expanded to include 99 different forms. In cases where there is not sufficient space to annotate the information from one interval in 80 columns, column number 8 is used for an internal numbering or continuation of the sheets. The three parameters in columns 6–16 must always be given. This also applies to the starting coordinates and the azimuth and the dip of the borehole. All positions in the borehole or at the core are given with accuracy of 0.1 meter and directions with 0.1 of a degree. As most of the other parameters are self-explanatory they will not be discussed here.

2. Petrography: The geological information is noted on two different sheets, one for petrography (2) and one for tectonics (3). The method of mapping (logging) is very similar to that of GEOMAP, the field data system of hard rocks developed by Berner et al. (1971). Parameters, codes, etc., tend to be the same in the two systems. A few differences, however, will be pointed out. Because of the continuation of the core, each mapped interval is assigned an interval number, and the beginning and the end of each interval are also given. It is then possible to give a petrological description of two different rock types by means of the parameters, structure, color, grain size, and texture. Megagrains, e.g., porphyroblasts, may be described in one rock type, as well as the name and quantity of four major or minor minerals. The names of the major (I) and subordinate (II) rocks are then coded. These names are often combined with a prefix which may be an abbreviation of a local name, a stratigraphical or a lithological term. Furthermore, it is possible to add one type of alteration or one characteristic mineral to the major rock. The aim of coding two different rock types in each interval is to describe a mineralization as well as the host rock. Samples examined by chemical analyses, thin sections, etc., are coded in the last six columns of each row.

In the GEOMAP and COREMAP systems the codes of structures, minerals, rock types, etc., may be complemented or exchanged according to the user's desire.

3. Tectonics: Tectonics, dikes, and the location of certain samples are given on form number 3 (Fig. 2). When the orientation of a piece of core is missing (which is the normal case), only a few measurements of structures are possible. COREMAP may also be used in tunnels, pits, etc., to complement borehole data. In such cases careful tectonic measurements are of great help in interpreting borehole information. There is space for recording two planar structures and one linear, one fold, and one dike on each row of form 3. The position of a dike or a collected sample may also be given and the loss of core specified at the end of each row.

4 and 5. Chemical data: Form number 4, which is not illustrated, is intended for use in the chemical laboratory and includes such parameters as analytical methods, number of account, type of notes, etc. The last code mentioned gives the sequence of seven chemical elements to be chosen by the user. Number one, e.g., is the following row of elements: Au (ppm), Ag (ppm), Cu %, Pb %, Zn %, As %, and S %, which are registered on form number 5. Each interval of the core to be analyzed is defined by the parameters, type of data, locality, and, usually, an interval number. If there is a standard analysis, the current sequence of elements is preprinted on form 5 (Fig. 3). Unwanted elements are marked with a cross, and the sheet serves as an order form for the laboratory. Ten rows at most may be used for each interval; in other words the upper limit is 70 chemical elements in the COREMAP system.

6. Joint data: A special sheet (no. 6) was also developed for joint data. The purpose of such a logging is to arrive at a relative measure of the rock

COREMAP

1. GENERAL DATA

Field		Card columns
INTERNAL CODE		1–5
TYPE OF DATA		6–8 (1)
LOCALITY		9–11
HOLE NUMBER		12–16
MAP UNIT		17–21
LOCALITY	TYPE	22
	M/NO	23–26
PROFILE		27–30
CO-ORDINATES	BASE	31
	+ N - S X	32–37
	+ N - S Y	38–43
	+ - Z	44–49
AZIMUTH		50–53
DIP		54–57
DIAMETER		58–59
ABOVE GROUND		60–61
SOIL DEPTH		62–64
TOTAL LENGTH OF HOLE		65–69
DRILLING	STARTED	70–73
	ENDED	74–77
GEOLOGIST		78–79
STATUS		80

2. PETROGRAPHY

Field		Card columns
INTERNAL CODE		1–5
TYPE OF DATA		6–8 (2)
LOCALITY		9–11
HOLE NUMBER		12–16
INTERVAL NO.		17–19
INTERVAL	FROM	20–24
	TO	25–29
ROCK I	STRUCTURE	30–31
	COLOUR	32–33
	GRAIN SIZE	34
	TEXTURE	35
ROCK II	STRUCTURE	36–37
	COLOUR	38–39
	GRAIN SIZE	40
	TEXTURE	41
THICKNESS		42
MEGA GRAINS	MINERAL	43–44
	HABIT	45
	SIZE	46
	%	47
MINERALOGY	MINERAL	48–49
	%	50
	MINERAL	51–52
	%	53
	MINERAL	54–55
	%	56
	MINERAL	57–58
	%	59
ROCK I	PREFIX	60–61
	NAME	62–65
	ALT	66–67
ROCK II	PREFIX	68–69
	NAME	70–73
	%	74
SAMPLES		75–80

3. TECTONICS

Field		Card columns
INTERNAL CODE		1–5
TYPE OF DATA		6–8 (3)
LOCALITY		9–11
HOLE NUMBER		12–16
INTERVAL NO.		17–19
INTERVAL	FROM	20–24
	TO	25–29
PLANAR STRUCTURES	TYPE	30
	STRIKE	31–33
	DIP	34–35
	PHASE	36
	STRAT.	37
	TYPE	38
	STRIKE	39–41
	DIP	42–43
	PHASE	44
LINEAR. ST.	TYPE	45
	STRIKE	46–48
	PLUNGE	49–50
	PHASE	51
FOLDS	TYPE	52
	λ	53
	AMPLITUDE	54
	SYMMETRY	55
DYKES ETC.	TYPE	56
	STRIKE	57–59
	DIP	60–61
	PHASE	62
	FILLING	63
	THICKN.	64
SAMPLE	TYPE	65
	POSITION	66–70
	CORE LOSS	71–72
SAMPLE	TYPE	73
	POSITION	74–78
	CORE LOSS	79–80

FIG. 2. Standard sheets including general data, petrography, tectonics.

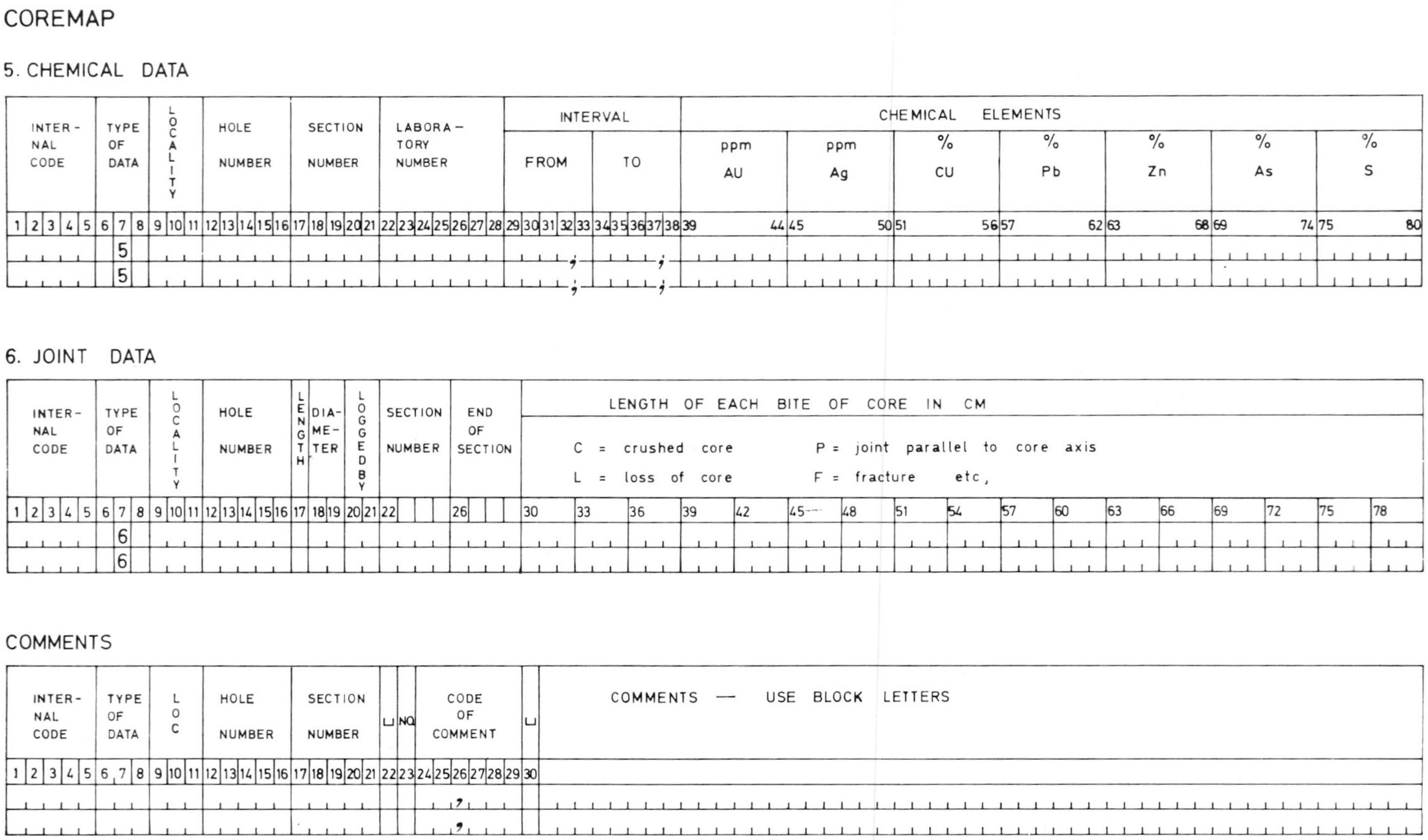

COREMAP

5. CHEMICAL DATA

INTER-NAL CODE	TYPE OF DATA	LOCALITY	HOLE NUMBER	SECTION NUMBER	LABORA-TORY NUMBER	INTERVAL FROM	INTERVAL TO	CHEMICAL ELEMENTS ppm AU	ppm Ag	% CU	% Pb	% Zn	% As	% S
1 2 3 4 5	6 7 8	9 10 11	12 13 14 15 16	17 18 19 20 21	22 23 24 25 26 27 28	29 30 31 32 33	34 35 36 37 38	39 44	45 50	51 56	57 62	63 68	69 74	75 80
	5					;	;							
	5					;	;							

6. JOINT DATA

INTER-NAL CODE	TYPE OF DATA	LOCALITY	HOLE NUMBER	LENGTH	DIA-METER	LOGGED BY	SECTION NUMBER	END OF SECTION	LENGTH OF EACH BITE OF CORE IN CM C = crushed core P = joint parallel to core axis L = loss of core F = fracture etc,																
1 2 3 4 5	6 7 8	9 10 11	12 13 14 15 16	17	18 19	20 21	22	26	30	33	36	39	42	45	48	51	54	57	60	63	66	69	72	75	78
	6																								
	6																								

COMMENTS

INTER-NAL CODE	TYPE OF DATA	LOC	HOLE NUMBER	SECTION NUMBER	⊔	NO.	CODE OF COMMENT	⊔	COMMENTS — USE BLOCK LETTERS
1 2 3 4 5	6 7 8	9 10 11	12 13 14 15 16	17 18 19 20 21	22	23	24 25 26 27 28 29	30	
							,		
							,		

FIG. 3. Standard sheets of chemical data, joint data, and comments.

strength. The principle of the method is to measure the length of each piece of core within a constant interval. By using the Kiruna method (Hansagi, 1972), it is possible to calculate a joint factor, which varies from 0.0 to 1.0. The higher the value the better the quality of the rock.

The length of each interval may be 1, 2, 3 m, etc., and it is also necessary to register the diameter of the core. The lengths of the pieces are recorded in centimeters in the second and third positions of each triple column. The first position is reserved for such information as "crushed core (c)", "loss of core (1)", "parallel joint (p)", etc. Although this method is time-consuming it gives a continuous description of the rock strength of the core.

7. *Bore-hole deviation:* There is also a special form for borehole deviation data but it is not shown here. The distance from the collar, the dip, and the azimuth are registered for each observation point. If borehole deviation data are missing, the starting values from sheet number 1 are used. The dip and azimuth of the borehole are recorded to 0.1 degrees and the distance from the collar in meters.

Comments: It is not possible to prepare input forms covering all parameters of a special object. As notes or comments are always required to complement the data on the different sheets, a special technique for computerizing notes was developed in the GEOMAP system. This method is also applied to the COREMAP system, and a form including all types of comment is shown at the bottom of Figure 3. When a comment is recorded, the parameters, type of data, locality, and hole number are always given, and normally an interval number as well. A comment code refers to the annotated column number, including the code which enables the computer to retrieve the note. In such a case the text must be restricted to one data card. There is also a common note which may be extended to 10 cards. As this type of comment has no reference code it cannot be computerized, and therefore an active retrieval by computer is not possible.

This presentation of the standard sheets of COREMAP gives only a brief outline of the possibilities of recording data with this system. Each user of the COREMAP system may design his own sheets, reduce or increase the variables, and determine the combination of the forms.

The COREMAP System

The COREMAP system is a set of modular routines connected with a general data processing system and statistical package called INTEREST II (Haglund and Edblad, 1970). The programming language is FORTRAN IV and the system is now running on an IBM 370/155.

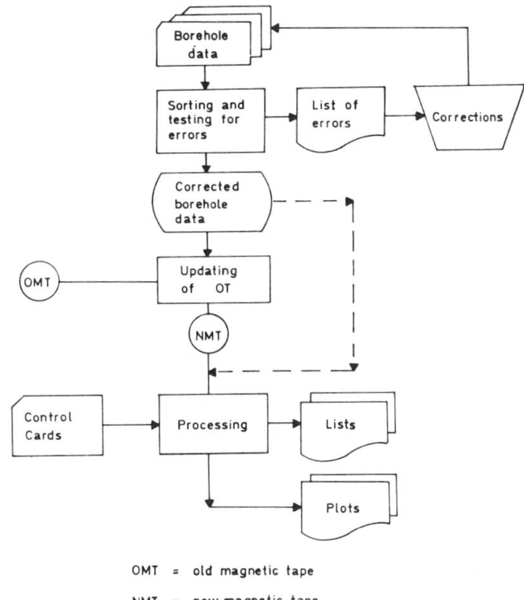

FIG. 4. Simplified flow chart of the COREMAP system.

The system is organized in two major modules: (1) a module for sorting, testing, and updating, and (2) a module for retrieval and display. In the latter module data may be updated, corrected, or deleted. Input is given on punched cards, and the new file is obtained by merging the new and the old data. This file may be stored on magnetic tape or temporarily on a disc.

COREMAP has a simple retrieval language. A short description of the control statements will be given here. For a deeper analysis the reader is referred to the INTEREST II manual (Haglund and Edblad, 1970). Instruction consists of standard functions and conditions. For example, the function GEOLST means a geological list, VERPLT means a cross section plot, etc. The example GEOLST (VI EQ KED AND V2 EQ 1415) means: print a list of the geology of borehole 1415 from the Kedträsk mine.

The number of functions is now 13 and they are based on four components:

Observation references, e.g., V1, V3, V9
Constants, e.g., 9.6, 91A
Logical operators NOT, AND, OR
Relational operators, e.g., GT = greater than, EQ = equal to, etc.
In every analysis each parameter is assigned a variable name V1, V2, etc.

```
GEOLOGY

MAP UNIT  23J2I        LOCALITY  KED        BH NO  125G
LEVEL     220          PROFILE   165        LOCAL  CO-ORD
CO-ORD    X = 1436.6 Y =         -1625.9    Z =    219.0
AZIMUTH   0.4          DIP       -0.8
HOLE DIAM 36 MM        LENGTH    123.8
STARTED   APR 70       FINISHED  APR 70
LOGGED BY AW           CORE      STORED

NOTE CHECKED BY SCINT. NEGATIVE RESULT.
++++++++++++++++++++++++++++++++++++++++++++++++++++++++++++++

FROM - TO    SAMPLE    GEOLOGICAL DESCRIPTION

 0.0 -  1.6  A, R      PYRITE-ORE(93A),MASSIVE,0.3-1.0 MM,
                         EQUIGRANULAR,PYRITE,90 PCT,
                         SPHALERITE,STRONG IMPREG.
 1.6 -  2.4  A         SERICITE-QUARTZITE(91A),SCHISTOUS,GRAY(75),
                         0.05-0.3 MM.EQUIGRANULAR,PYRITE; STRONG
                         IMPREG.,CHALCOPYRITE,TRACE
                       NOTE 49,PY BANDS OF A FEW CM
 2.4 - 15.3            TUFFITE(92A),SERICITIZED,SCHISTOUS,GRAY(75),
                         0.05-0.3 MM,EQUIGRANULAR,
                         DYKE,QUARTZ, > 2 CM, AT 11.6 M
                         S-SURFACE, 75
15.3 - 18.9  A         PYRITE-ORE(93A),MASSIVE BANKS,INEQUIGRANULAR,
                         PYRITE, 50 PCT.,SPHALERITE, IMPREG.,
                         CHALCOPYRITE,TRACE
                         SERICITE-QUARTZITE(91A),30-40 PCT., WEAK
                         SCHISTOSITY,LIGHT GRAY(76),0.05-0.3 MM,
                         EQUIGRANULAR
                         S-SURFACE,70
18.9 - 24.7  T         SERICITE-QUARTZITE(91A),SCHISTOUS,GRAY(75),
                         0.05-0.3 MM,EQUIGRANULAR,ANDALUSITE,
                         ANGULAR ELONGATED,2-5 MM, 10 PCT.,PYRITE,
                         STRONG IMPREG.
                         THIN SECTION,AT 22.4 M
```

FIG. 5. Borehole log produced by line-printer with decoded geological information.

Output

The system has a number of standard lists including, for example, geological, chemical, joint and borehole deviation data. All these lists start with general data from sheet number 1. A list of geological data is shown in Figure 5. This is the information from sheets 1, 2, and 3 in a decoded form. The description of the core gives the beginning and the end of each interval, samples collected, and a petrological tectonical description. The petrography starts with a full name of the major rock type and goes on to structure, color, grain size, texture, megagrain, and the name and quantity of four different minerals. A subordinate rock type is described directly after the major rock. Any comments are found at the end of the description of each interval.

The list of joints (Fig. 6) may serve as another example of output. Crushed core (CC), "sköl" (SK), parallel joint (PJ), core loss (CL), fractures (FR), and the C-factor are given from each interval together with the name of the major rock. If the rock strength is determined, this may be multiplied by the C-factor to give the rock stability. This value is printed on the righthand side of the list.

Special lists

The COREMAP system allows the selection of any type of data and its printing on special lists. The instruction to the computer is given with the function SPCLST, combined with the desired conditions. A maximum of eight variables may be obtained on a single list. The headlines of the special lists are determined by the user. If, for example, the user wishes to know all intervals with Au greater than 3 ppm from the Kedträsk mine, we give the instruction

SPCLST (V9 GT 3, V3, V4, V6, V9)

and we obtain a list as shown in Figure 7A.

The instruction may also be more complicated. For example, the user is interested in all intervals with sericite-quartzite from Kedträsk which are carrying more than 0.80 percent Cu. Furthermore, the area is restricted to the Y-coordinates 1600 to 2000. The instruction

SPCLST (V1, EQ 91A AND V4 GT 0.8 AND V6 GT 1600 AND V6 LT 20000, V2, V3, V7, V4, V9)

then gives the result shown in Figure 7B.

Printed description of the core

A simplified representation of the borehole may be obtained on the line-printer. This description includes a variable scale, the rock code, and two different chemical parameters in the form of histograms. For example, if we want to visualize the relationship between the content of lead and zinc from a certain core, we can produce a line plot as shown in Figure 8. The contacts between different types of rocks

```
ROCK STABILITY

MAP UNIT  23J2I          LOCALITY  KED        BH NO  123G
LEVEL     220            PROFILE   165        LOCAL  CO-ORD
CO-ORD    X = 1314.5     Y =       1639.8     Z =    221.3
AZIMUTH   271.4          DIP       -2.3

+++++++++++++++++++++++++++++++++++++++++++++++++++++++++++++++++++++++++++++++++++++++
```

SNO	FROM - TO	SL	CD	NN	CC	SK	PJ	CL	NJ	C-FACTOR	ROCK TYPE	ROCK STABILITY
		M	MM		CM	CM	CM	CM				
1	0 - 1	1	22	TE	5	0	0	0	14	0.44	PYRITE-ORE(93A)	
2	1 - 2				0	0	8	0	14	0.43	PYRITE-ORE(93A)	
3	2 - 3				0	0	0	0	12	0.46	SERICITE-QUARTZITE(91A)	
4	3 - 4				0	0	0	9	7	0.51	PYRITE-ORE(93A)	
5	4 - 5				13	0	5	0	15	0.29	SERICITE-QUARTZITE(91A)	
6	5 - 6				0	15	0	0	10	0.40	SERICITE-QUARTZITE(91A)	
7	6 - 7				0	0	0	0	7	0.63	SERICITE-QUARTZITE(91A)	
8	7 - 8				8	0	0	0	9	0.53	PYRITE-ORE(93A)	
9	8 - 9				0	0	9	0	6	0.54	PYRITE-ORE(93A)	
10	9 - 10				0	0	0	10	7	0.48	PYRITE-ORE(93A)	

FIG. 6. List of rock stability data.

```
KEDTRASK. INTERVALS WITH AU > 3 PPM.

BH NO   INT NO   FROM - TO      AU PPM

125G    1008      3.3 -  4.6      5
125G    1010     14.5 - 16.3     14
125G    1013     23.4 - 28.3     10
130G    4031     46.4 - 52.3      8
130G    4036     75.2 - 79.4     15          A
```

```
KEDTRASK. CU > 0.8%.ROCK TYPE SERICITE-QUARTZITE. Y-CO-ORD.
1600-2000.

BH NO   INT NO   FROM  - TO      CU %

156     3425     124.6 - 134.5   1.56
156     3428     167.8 - 173.8   0.89
156     3432     208.3 - 213.5   1.08
178     2214      34.9 -  45.5   0.81
178     2216      52.7 -  55.9   2.08
178     2218      60.7 -  67.9   1.68          B
```

FIG. 7A. List of sampling intervals with Au > 3 ppm from the Kedträsk mine.
FIG. 7B. List of sampling intervals with Cu > 0.8% from a restricted area of the Kedträsk mine.

are approximate, and the scale of the histogram may be defined in each run. If a more exact and detailed picture is required, the user is referred to the different types of plots in the COREMAP system.

Plotting

The following types of plotted maps are included in the COREMAP system:

(1) Map of borehole deviation
(2) Planar map (type 1)
(3) Planar map (type 2)
(4) Vertical section

The different maps are plotted on a drum (CALCOMP) or microfilm (BENSON) plotter.

```
MAP UNIT  23J2I         LOCALITY  KED        BH NO   127G
LEVEL     220           PROFILE   166        LOCAL   CO-ORD.
CO-ORD    X = 1439.9    Y =       1660.8     Z =     220.1
AZIMUTH   397.8         DIP       - 4.2
HOLEDIAM  36 MM         LENGTH    43.9 M
STARTED   MAY 69        FINISHED  JUNE 69
LOGGED BY AW            CORE      STORED

VARIABLES: ZN = 0.4 %, PB = 0.2 %
+++++++++++++++++++++++++++++++++++++++++++++++++++++++++++++++++++++

M     ROCK CODE        VARIABLES

0     KED  92A
      KED  92A
      KED  92A         ZNZN                       PB
           93A         ZNZNZNZNZNZNZNZ            PBPBPBPBPBPBPB
10         93A         ZNZNZNZNZNZN               PBPBPBPBPB
           93A         ZNZNZNZNZNZNZNZNZNZN       PBPBPBPBPBPBPBP
           93A         ZNZNZNZNZNZNZN             PBPBPBP
           91A         ZNZN                       PB
           91A         ZNZNZN                     PBPBPB
20         91A         ZN
           93A         ZNZNZNZNZNZN               PBPBPBPB
           93A         ZNZNZNZNZNZNZNZNZNZN       PBPBP
           93A         ZNZNZNZNZNZNZN ZNZN        PBPBPBPPBPBPB
      KED  92A         ZNZNZNZN                   PBPBPBPBPBPBPBP
30    KED  92A         ZN                         PBPBPB
      KED  92A
           91A
           91A
           91A         ZN                         PBPB
40         91A
           91A
           91A
```

FIG. 8. A brief core log by line-printer. Two chemical elements are represented in the form of histograms.

The plots shown here were redrawn to alter the original Swedish text.

(1) The deviation of a borehole with respect to azimuth and dip is plotted on two different maps, one corresponding to the vertical plane, the other to the horizontal (Fig. 9). The heads of all plots are titled with some general data from sheet number 1. The starting coordinates are marked with a ring and the end with a triangle. The different measured points are plotted to enable the continuous deviation to be drawn by hand. The computer determines the scale so that is is possible to plot in a standard format (A-4). In all other plots the scale and size are determined by the user. It may be pointed out that the effects of borehole deviation are always included with the other plots.

(2) There are two types of planar maps. The first one is mainly intended for steep boreholes. In this case the user has to define a certain z-coordinate or a level, a scale, and the starting coordinates. Such a plotted map is shown in Figure 10. All boreholes which cut the defined surface are plotted as circles. Furthermore, the number of the borehole and the content of *one* chemical element from the appropriate level are plotted.

(3) In underground mines, tunnels, etc., horizontal or low-angle boreholes are often drilled to determine the boundaries of an ore lode, etc. In this

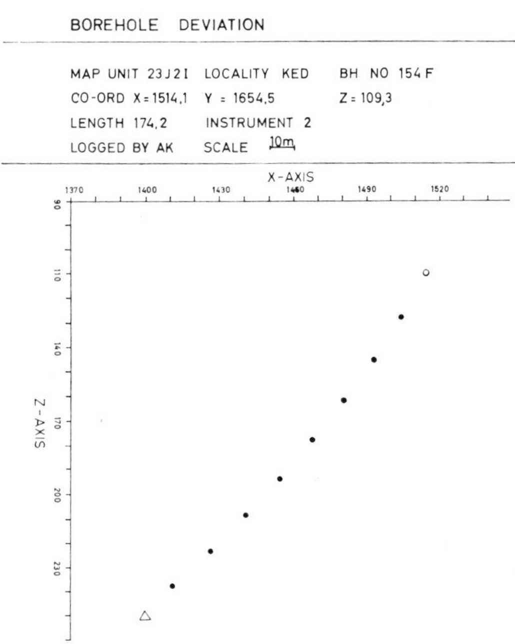

FIG. 9. Plot of borehole deviation on the vertical plane.

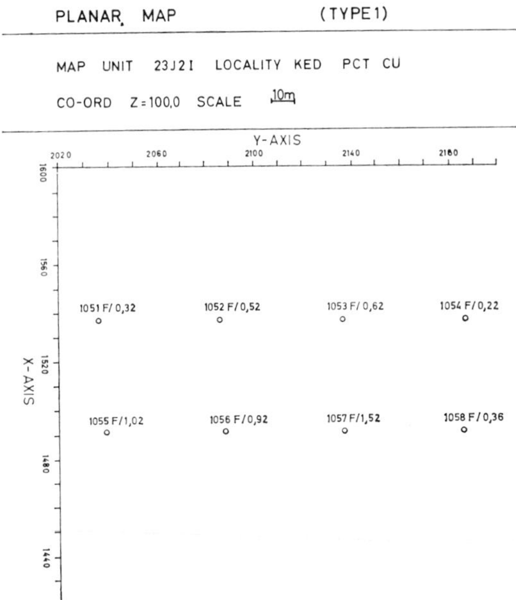

Fig. 10. A plot of borehole locations and the content of Cu on a specified level. Type refers to steep boreholes.

case, a projection of the whole borehole onto a certain level is wanted. If the geology is complicated, projection of the borehole over too great distances is not satisfactory. This explains why it is possible to give a projection interval in this type of plot. If, for example, z = 220 ± 5 m, the computer will select all boreholes in the interval 215 to 225 m and project them on the 220-m level. The horizontal extension of the "slice" is defined by the given scale and starting coordinates. The correct length of the intervals is marked on the projected boreholes, and the content of *one* chosen chemical element is plotted in conjunction with each interval analyzed. Such a plotted map is shown in Figure 11. This plot also demonstrates the possibility of complementing borehole data with information from adits, tunnels, etc. The boreholes no. 159 G, 160 G, 132 G, and 133 G are drilled from an adit in the ore. Between these four boreholes one adit was cut. By mapping one of the walls as a simulated borehole, important geological, chemical, etc., data could be added to the data bank, thus complementing the picture of the orebody.

(4) The vertical section maps are at present the most advanced plots of the COREMAP system. An interval or a slice has to be defined as in the case of the level maps. By determining the starting coordinates and the scale of the maps the plotted volume is limited. Any borehole exceeding this defined volume will be projected normally onto the center-plane. If the strike of the rock is not at right angles, or nearly parallel to the x- or y-axes, a considerable error of projection will arise. However, this may be corrected in COREMAP by defining the angle between the strike and the x- or y-axes. Before the plotting it is also possible to make a similar correction for the dip.

A map of a vertical section is shown in Figure 12. The borehole to the left has no end symbol, which means that it exceeds the specified volume. On the other hand, hole no. 212 F comes from the outside but stops within the defined interval. Thus, by combining a number of cross sections (parallel to x or y), it is possible to construct a three-dimensional picture of the geological object. As has been stated previously, the borehole deviation data, if present, are used in all plots. A contouring program may also be added so that it is possible to distinguish areas with the same content of a certain chemical element.

Cost of output

When using a computer-oriented system cost is often the determining factor. From a general point of view, the cost of such a system should be lower than the traditional method of processing the data.

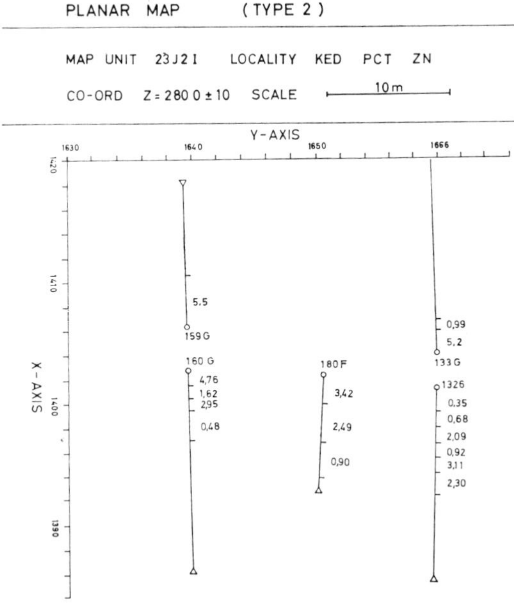

Fig. 11. Boreholes projected onto a specified level. The length of the sampling intervals and the content of one element are also plotted. Type 2 refers to low-angle boreholes.

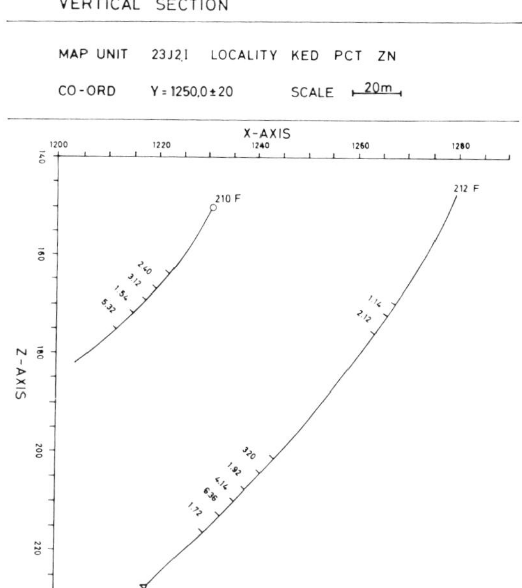

FIG. 12. Boreholes projected onto a vertical section.

Otherwise, the saving in time or the possibility of making more advanced calculations should offset the higher expenditure.

The costs of punching the cards vary, and no estimation is given here. Moreover, we are now testing a method of recording the data directly onto magnetic tape, which may be routine in future. In the following discussion the costs of COREMAP refer to an updated file.

The price of output here is based on an IBM 370/155 computer costing \$600 per hour. The approximate CPU time is also given for comparison. All types of listing, including specified lists and the brief description of the core on a line-printer, amount to \$0.30 per page. Each extra copy costs \$0.06. The CPU time of a page is approximately 1 second.

The microfilm plots are more expensive. An ordinary plot as shown in this paper costs \$1.00 to \$2.00. The electronic method offers great advantages over a pen-plotter, as the negative is easy to copy and store. The use of COREMAP is much cheaper compared with the traditional method of typing the geological description of the core. The selection of specified information from a large amount of data by hand is time-consuming and expensive. The special list of the COREMAP system solves this problem at a low cost. The gain is still greater for the different plots. Last but not least, there is a great advantage in storing the data on magnetic tape, which will make advanced retrieval possible in the future.

Range of application

A small sulfide orebody in the Skellefte field (Sweden) was used as a testing area. The COREMAP system was later applied to the quarrying of limestones. A large number of vertical boreholes were drilled in this area according to a system. Every third meter of the core was analyzed for SiO_2, $CaCO_3$, Al_2O_3, etc. By using different conditions, e.g., $CaCO_3 > 95\%$, and $SiO_2 < 2\%$, etc., it was possible to select certain intervals and to plot them for each level. These maps were then a valuable tool for the planning of the quarry. Special interest pertains to a copper mineralization in northern Sweden, which is being explored by the Swedish Geological Survey. This project is using COREMAP for the cores and GEOMAP for the geological field data. The similar method of collecting data in the two systems allows retrieval of all the information in a uniform manner. The COREMAP system is also used by some Swedish mining companies, such as Boliden AB, Luossavaara-Kiirunavaara AB, Stora Kopparbergs Bergslags AB. Further development of the COREMAP system includes a program for ore evaluation.

Conclusions

COREMAP is a data system for recording and processing data from drill cores and boreholes. The system may also be used in adits, tunnels, etc. When using COREMAP it is possible to record general, geological, chemical, joint, and borehole deviation data in standard formats, but the system may also be extended to other types of data. A method of processing notes is also included. General borehole data may be combined with any other type of collected information.

COREMAP produces decoded lists, lists of any specified parameter or parameters, and a simplified description of the core on a line-printer. COREMAP plots the deviation of boreholes, data locations on chosen levels, and vertical sections parallel to the x- or y-axes. Furthermore, COREMAP plots the length of each analyzed interval together with one optional chemical element. All boreholes running through a defined slice of the rock are projected on a given plane. The scale of the plot is determined by the user, and it is also possible to correct the error of striking and dip during the projection.

COREMAP is a sister system to GEOMAP which stores and retrieves field data of hard rocks. It is easy to combine the two systems by using the same parameters and the same codes. This gives the user a valuable tool for prospecting and geological mapping.

Acknowledgments

Many persons were consulted during the development of the COREMAP system; to all these we should like to express our gratitude. The authors are very grateful to Mr. J. Boman and Dr. E. Grip for their support. Special thanks are due to Drs. O. Stephansson and H. Berner for critical review of the manuscript.

This project was sponsored by the Swedish Board for Technical Development grant 72-881/U737.

T. K. E.
Institute of Geology
University of Uppsala
Box 555
S-751 22 Uppsala, Sweden

Å. W.
Boliden AB
S-930 50 Boliden, Sweden

L-E. L.
UDAC
Box 2103
S-750 02 Uppsala, Sweden

November 19, 1973; June 24, 1974

REFERENCES

Berner, H. Ekström, T., Lilljequist, R., Stephansson, O., and Wikström, A., 1971, Data storage and processing in geological mapping. I. Field data sheet: II. Data file: Geol. Fören. Stockholm Förh., v. 93, p. 85–101, 693–705.

Berner, H., Ekström, T., Lilljequist, R., Stephansson, O., and Wikström, A., 1972, GEOMAP—a data system for geological mapping: Internat. Geol. Cong., 24th, Montreal 1972, sec. 16, p. 3–11.

Blanchet, P. H., and Godwin, C. I., 1972, "Geolog system" for computer and manual analysis of geologic data from porphyry and other deposits: Econ. Geol., v. 67, p. 796–813.

Buller, H. V., 1972, The Saskatchewan government computerized well information system: Dept. Mineral Resources Rept. 162.

Farmer, D. G., 1973, A suite of Fortran programs for stratigraphic and cartographic display using an IBM 1130 computer and CIL drum plotter: Great Britain, Inst. Geol. Sci. Computer Unit Bull. 73/2.

Gover, T. N., Read, W. A., and Rowson, A. G., 1971, A pilot project on the storage and retrieval by computer of geological information from cored boreholes in central Scotland: Great Britain, Inst. Geol. Sci. Rept. 71/13, 30 p.

Haglund, A., and Edblad, L., 1970, Interest II manual: IMB A.B., Sweden.

Hansagi, I., 1972, Accordance between results of determination of the rock masses and the real facts in Kiruna: IVA-rapport 45, p. 15–22. (in Swedish)

Look, E.-R., and Vinken, R., 1971, Elektronische Datenverarbeitung bei der Aufnahme und Herstellung von geologischen Karten: Hannover, 29 p.

Wynne-Edwards, H. R., Laurin, A. F., Sharma, K. N. M., Nandi, A., Kehlenbeck, M. M., and Franconi, A., 1970, Computerized geological mapping in the Grenville Province, Quebec, Canada: Canadian Jour. Earth Sci., v. 7, p. 1357–1373.

Part VI

REPORTING

Editor's Comments on Papers 38 and 39

38 LOUGHLIN
Excerpts from *Reports on Mining Districts*

39 BEAVIS
Excerpts from *Minerals Exploration: Mapping and Drafting Procedures*

The art of presentation of geological observations, analysis of observation, and interpretations or conclusions has not changed appreciably during the past century though terminology and techniques have expanded. The U.S. Geological Survey set a high standard for the preparation of geological reports that has served as a pattern throughout the world. Suggestions to authors, prepared over a half century ago, are relevant today. Loughlin (Paper 38) refers principally to government reports, but advice given in both organization and method of presentation is relevant to all reports on mining districts, mines, or prospects.

Many reports fail to convey an accurate picture of geological relations or to reveal potential exploration targets because of poor illustrations and map presentations. The geologist may be required to serve as his own draftsman or may be fortunate in having a draftsman's services available to him. Beavis (Paper 39) outlines suggestions to the geologist and the draftsman regarding the proper preparation of minerals exploration maps.

38

Reprinted from pages 31-35, 45, 46, 47, and 49-55 of *Suggestions to Authors: Papers Submitted for Publication by the United States Geological Survey*, 4th ed., B. H. Lane, ed., U.S. Government Printing Office, Washington, D. C., 1935, 126p.

Reports on mining districts

By G. F. Loughlin

General suggestions

Reports on mining districts differ considerably in scope. Professional papers on large, extensively developed districts, which involve a vast amount of detailed study, are the most comprehensive and should be authoritative sources of information on all branches of geology that are conspicuously represented in the districts. Reports on small or only slightly developed districts are necessarily based on much less comprehensive information, and reports on certain special features may be much more restricted in scope. No matter what the scope may be, however, the purpose of the report is economic, and all parts of it therefore should be focused on those features that may be of direct value in the development of the district described. Too many authors tend to give equal emphasis to all sections of the report. The detailed study essential to the more comprehensive reports is likely to disclose valuable information on the more fundamental branches of geology, and the treatment of this information calls for careful consideration. It may be so closely related to economic problems that rather full treatment of it is essential to its practical application, or it may be more appropriately presented as a separate paper and only summarized in the economic report.

The report as a whole, as well as its major divisions, should be planned to give the reader a general idea at the outset, so that he may more readily understand the detailed descriptions that follow and appreciate their bearing on the author's conclusions. Although the author is treating a technical subject, he should also bear in mind that many of his readers will not be geologists. He should therefore avoid unnecessary technical terms and endeavor to express himself in simple, plain English. Certain technical terms that have become well established in the district being described and certain general terms that cannot be avoided without much circumlocution may be appropriately used, but even these terms may need definitions for the benefit of readers not familiar with the district or with the science of geology. There is a tendency among authors who are writing their first reports to continue the practice of their student days and attempt to display their command of technical terms and flowery expressions, but such writing is likely to give the reader who seeks practical information the impression that the report is impractical and only of academic interest.

Subject order

General order.—More elaborate reports should be planned under the following main divisions: Introduction, geography, geology, and mineral or ore deposits, in that order. The order of treatment of other reports depends somewhat upon their scope and object but ordinarily should follow the same plan so far as it is applicable to the facts presented. The author should carefully consider this order of treatment and should depart from it only for the sake of leading his readers in methodical fashion to a realization of the main economic contributions of the report. Any departures from the usual order of treatment should be explained either in the general introduction or in the introductory statements at the beginnings of the major parts of the report. Every report, whatever its scope, should be preceded by an abstract that outlines the basic facts and states the author's conclusions. (See pp. 43–45.)

Introduction.—The introduction may comprise a statement of the purpose of the investigation and the events leading up to it, the conditions under which the work was done, acknowledgment of cooperation and favors, a summary of previous work in the same field, and an annotated bibliography. It should call attention to the author's outstanding conclusions regarding local or regional problems or on the further development of current theories and to noteworthy differences between his conclusions and those expressed in earlier publications.

Geography.—The attention given to geography should depend upon the district studied. If the district is little known, relatively complete information on location, routes of approach, topography, climate, vegetation, and other geographic features may be justified; if it is well known and its geographic features have been adequately treated elsewhere, only a brief statement emphasizing those features that affect the local mining industry is necessary. The origin of topographic features, as they are the latest results of geologic processes, should generally not be treated until the geologic features have been described.

Geology.—The discussion of geology should describe the distribution, character, and composition of the original and altered facies of the different rock formations, in order of age, their structural relations, their influence on topographic development, and their economic aspects. Descriptions of the formations should be followed by a section on geologic structure that interprets the origin of the different structural features to the extent that information permits and points out the extent to which structure has controlled the distribution and form of intrusive igneous masses and the courses followed by ore-forming solutions. The interpretation of topographic features should follow, and the entire section on general geology may be appropriately concluded with a summary of the geologic history.

Ore deposits.—The part on mineral deposits should aim to present historical descriptive data that will aid the operators in appreciating the factors on which to plan both the more thorough development of opened ground and the extension of exploration both horizontally and downward, not only within their own properties, but in undeveloped areas. Its general order of subjects treated should be history and production, general character and classification of deposits, mineralogy, description of each class of deposits, genesis of original deposits, processes of alteration and enrichment, practical applications, and detailed mine descriptions. This sequence may be changed if orderly presentation can be improved thereby.

In reports on the older and larger districts the part on ore deposits should contain an adequate history of discovery, mining and cultural development, and production. The amount of information on the leading districts may even entitle this history to equal rank with the parts on general geology and ore deposits; on the other hand, the information on small districts may be so scant that it will be most conveniently appended to the general introduction, where it will not break the thread of consistent geologic treatment. The review of production should cite the geologic, technologic, and economic conditions that have accounted for fluctuations or changes in the rate of production, and it should conclude by calling attention to information in subsequent sections that may have a bearing on ore reserves and future development.

The section on mineralogy should logically precede the general description and classification of the mineral deposits, for it is at least a partial basis for classification. The more elaborate reports have contained alphabetic lists of minerals with their chemical formulas, and if the number of minerals is great, such a list may be a convenient reference, but the author should consider whether it is any more than a duplication of headings to the mineral descriptions. There may be some difficulty in deciding how much information should be included in the section on mineralogy and how much in that on the ores. Conciseness and clarity of treatment should govern the decision. Where preference is given to the section on mineralogy, this section should contain adequate descriptions of (1) the different ore minerals, grouped according to (*a*) their principal metal contents and (*b*) their hypogene or supergene origin; (2) the gangue minerals, appropriately subdivided. Any significant features regarding temperature of deposition, effect on metallurgical processes, and other matters may be appropriately included, thereby increasing the practical value of the presentation. The section may end with a review of paragenesis, which may be largely a review of pertinent parts of the preceding description and which should serve as a basis for any discussion of the classification, origin, and zoning of ore deposits.

The description of the general character of the deposits should serve as an introduction to subsequent sections and should present any classification of deposits that the author may devise for convenience and clarity of description. Much has been written about the classification of mineral deposits, and most writers today favor a genetic classification when considering mineral deposits in general; but, as all the deposits of a mining district may fall within two or three subdivisions of such a classification, although they differ markedly in form, structure, and metal contents, a special classification based on these differences is likely to be most practical in a report on the district. The order in this classification should be followed in the detailed description of each class and subclass.

Descriptions of the different classes and subclasses of ores or mineral deposits should follow, including distribution of the main deposits and the ore shoots in them, controlling geologic features, structural relations, processes of ore deposition, and related alteration or replacement of wall rock. If the classes of deposits are few they may all be conveniently described before their genesis is considered; but if they are too diversified for such simple treatment, the original or hypogene deposits may be described and their genesis considered before the altered or supergene ores and the processes of alteration are described.

The subject of rock alteration also may be so complex as to deserve a special section enumerating the processes of alteration which have been active at different and perhaps widely separated times and some of which have particular significance with respect to the distribution and origin of mineral deposits, whereas others have no such significance. The most appropriate position of a special section on rock alteration depends upon its contents. It may follow the treatment of intrusive rocks or may be more closely associated with the discussion of mineral deposition. The author in his discussions of alteration and genesis should recapitulate the essential facts brought out in the descriptions, show their bearing on the problems under consideration, and deduce such theoretical conclusions as they may warrant. If the descriptive matter has been properly focused on these problems and has brought out the significance of the different items, the reader will be so well prepared for these interpretive discussions that they may be brief and do little more than bring together and confirm the minor interpretations that have been made throughout the report.

The general part of the report should conclude with a section on practical applications in which the author points out how his work may aid in the development of ore bodies and the finding of new ones and expresses his conclusions regarding ore reserves and the economic future of the district.

The detailed descriptions of mines should contain as full information as is available on history, development, production, and geologic features of each mine and should be adequately illustrated. The general reader may have little interest in these detailed descriptions, but operators and others who may become interested in a mine even many years later will look to these detailed descriptions for information and may value them more than the general part of the report. Their preparation should not, therefore, be regarded as an irksome duty but as the preservation of important geologic records. The more nearly complete they are the better is the author's basis for his general conclusions; in fact, it is advisable to prepare rough drafts of the mine descriptions before completing the general sections of the report, as this practice gives a strong appreciation of the practical significance of different geologic features and prepares the author for emphasizing these features throughout the report. Care should be taken that the final drafts of mine descriptions do not include or needlessly repeat data that properly belong in the general sections of the report.

[*Editor's Note:* Material has been omitted at this point.]

Illustrations

Publications of any class may be illustrated, but illustrations may not be used for mere embellishment; each one should serve a definite scientific or practical purpose and should illustrate some feature described in the text.

[*Editor's Note:* Material has been omitted at this point.]

The author should give each illustration a distinctive number—plates in one series, figures in another. The numbers should be assigned in the order in which references to the illustrations appear in the text. It should be noted, however, that an incidental mention of an illustration need not determine its number, which should be governed by the place where it is principally mentioned or discussed.

[*Editor's Note:* Material has been omitted at this point.]

Plates are numbered in arabic (1, 2, 3, etc.). If two or more subjects are combined to make a single plate they should generally be designated by italic capitals, as "plate 4, *A*"; but in plates showing a group of fossils the individual subjects are numbered in arabic and called "figures", as "plate 4, figure 7." For details of a single subject that are too long to be lettered directly on the plate, italic lower-case letters (*a*, *b*, *c*, etc., or initials, as *q*, for quartz; *g*, for galena; *cp*, for chalcopyrite) are used and the explanation is given in the plate description.

[*Editor's Note:* Material has been omitted at this point.]

Full descriptions of the plates should be furnished if the brief titles given in the list of illustrations are not sufficient. It is desirable that the full descriptions be attached to the illustrations, especially photographs, as the length of the description may govern the amount of reduction necessary to bring an illustration within the page limits, but a separate list comprising the full descriptions should always be furnished, and a carbon of this list may be cut apart for attachment to the illustrations. The descriptions should, if possible, be kept short enough to be printed on the plates themselves. For example, on a plate showing fossils it is usually sufficient to give the name of each subject with a reference to the page on which it is described, as "3. *Sequoia concinna* Heer (p. 57)." Descriptions that are too long to be printed on the plates must be printed on facing pages, which adds to the expense. Such long descriptions can sometimes be avoided by incorporating the details in the text, but if unavoidable they should be inserted in the manuscript at the proper places, the description of each plate written on a separate page. Otherwise the titles of plates should not be written in the text, as they are set up at a different time.

Figures are numbered in arabic (1, 2, 3, etc.); details of a figure are lettered in lower-case italic (*a*, *b*, *c*, etc.). Symbols or letters that are used on the figures must be explained, preferably in the titles, but if the explanation is long and complex it may be incorporated in the text. The titles of the figures, including all necessary explanations, are set up at the same time as the text and therefore should be written in the manuscript at the places where the figures are to appear.

[*Editor's Note:* Material has been omitted at this point.]

Suggestions as to expression

"Writing maketh an exact man. He who writes little had need have a good memory."—Bacon.

General counsel

An eminent critic has said that the ideal of literary style is "the speech of the people in the mouth of the scholar", but this ideal cannot be consistently maintained throughout a series of reports that are in large part technical. Some of the reports of the United States Geological Survey are technical contributions to that universal encyclopedia that is being compiled by scientific men in all parts of the world. Of the immediate or ultimate practical utility of much of this material there can be no question, but a part of it is of necessity unintelligible to the ordinary reader. Many of the Survey's reports, however, such as those on placer mining, coal beds, or water-bearing beds, may be written in a simple and somewhat popular style, and simplicity and clearness in both technical and untechnical matter may profitably be cultivated by all writers.

The following extracts from an article written by a former Director of the Survey[14] are suggestive and highly significant:

[14] Smith, G. O., Plain writing: Science, new ser., vol. 42, pp. 630–632, Nov. 5, 1915.

At its best, science is simple; for science is not much more than arranging facts so as to set forth the truth. Scientific thought is exact and direct, and scientific writing must therefore be accurate and to the point. The scientist should think directly and with the precision of one of the instruments of his trade, and above all his language must present that thought exactly. * * *

Of course, any writer's first duty is to be intelligible. Choice of language thus resolves itself largely into an understanding of the audience. If a scientific investigator desires to announce his discovery to his fellow workers, he does well to use those exact terms that carry the same shade of meaning the world over and indeed may have the same form in several languages; if, on the other hand, his results have immediate value for the mine operator or the prospector, the geologist does not and cannot accomplish his purpose unless he writes in plain language, using words possibly less exact but surely more understandable. * * *

It is not a coincidence that some of the deepest thinkers in geological science have also possessed a literary style conspicuous for clarity of expression. On the other hand, some authors whose English needs the most editing are equally careless in the quotation of facts determined by others and, indeed, in the statement of their own observations. I mention this simply to show that I am strong in my belief that plain writing is not something beneath the plane of endeavor of the scientific investigator—indeed, it is something so hard to attain that the most of us need to aim high, to raise our standards of scientific thinking. The use of common words is worthy of any writer if his purpose is to transmit thought.

In a later article [15] Dr. Smith wrote:

The geology I plead for is that which states facts in plain words—in language understood by the many rather than only by the few. Plain geology needs little defining, and I may state my case best by trying to set forth the reasons why we have strayed so far away from the simple type.

First of all, I suppose we may as well admit a certain liking for the sound of words, and the longer the word the more sound it has. Especially enjoyable is this mild form of hypnotism if both ideas and words are such as to make us feel that we are moving in the highest circles. At the meeting of the British Association this year one physicist frankly explained that the idea of relativity is popular because to most people it is "pleasantly incomprehensible." * * *

Again, I fear lest in our writing we lose sight of our audience, if, indeed, some of us ever see at all the audience to whom we address our written reports. The chief purpose of words is to convey thoughts, and unless the wave lengths of the words are right the receiving apparatus will utterly fail to pick up the thoughts. * * *

This brings me to a third reason for our use of highly technical language: we too often try to overdress our thoughts. Just as there is a somewhat prevalent notion that clothes make the man, so we subconsciously believe that words make the idea. We follow the precept "To be scientific, use scientific terms", and in so doing we deceive ourselves. I do not wish to be unduly autobiographic in this analysis, but to show my true sympathy for those whose practices I denounce, I confess that I, too, have had the unhappy experience of stripping the technical words from what looked like a good-sized geological deduction only to find that the naked idea was rather small and not my own. It is also a common experience to make the sad discovery that a piece of involved and obscure writing is simply the product of roundabout reasoning or twisted thinking. Our own words fool us, and unconsciously we cover up with long words or tangled rhetoric our lack of plain thinking. * * *

There is really somewhat of an obligation upon us, both as scientists and as partners in the world's business, to show the world that geology is not mystery or magic but only

[15] Smith, G. O., Plain geology: Econ. Geology, vol. 17, pp. 34–39, 1922.

common sense. I have told practical men of business that they should give little credence to the geologist who cannot tell his story in common language. The world has a right to discount our usefulness and even to distrust our honesty if we persist in concealing our thoughts, or lack of thoughts, behind a mask of professional jargon.

Much of the popular ridicule excited by some scientific reports is justly chargeable to the authors' indifference and neglect in the matter of expression. An eminent critic,[16] who is also a man of science, writes:

It is strange that scientific men who habitually work in dimensions of a ten-thousandth of an inch are either blind to gross confusions of argument and to false refractions of meaning or regard them with indifference. * * * A writer who wishes to convince must lay his mind alongside that of the reader, who must be carried along in a quick and equable current.

Another English critic [17] says:

You must first be your own reader, chiseling out the thought definitely for yourself, and after that must carve out the intaglio yet more sharply and neatly, if you would impress its image accurately upon the wax of other men's minds.

An English instructor in a New York City college said recently, "Clarity is not infantile, nor is complex opacity profound."

If judged by their literary merit many scientific papers would not deserve publication; only the facts they relate give them value; but unless the facts are stated plainly, in proper order, and with skill to carry conviction the purpose of their publication will not be achieved.

Most of the masters of literature devoted great time and care to the task of writing and rewriting their works. The labor of correction and revision that Fontaine gave to his fables is almost incredible. Buffon rewrote his matter several times and then had it read aloud to him, that he might note where the reader halted or failed to catch the meaning at once, and revised or rewrote accordingly. Macaulay repeatedly rewrote and revised his manuscript, and Robert Louis Stevenson, one of the masters of modern English, who devoted a large part of his life to the study of style, revised and polished his matter until his critical fancy was fully satisfied. The writers of the Geological Survey's reports are not so fortunate in having the privilege of thoroughly revising their manuscripts; the demand for the printed results of their scientific work is immediate and insistent, and the transition from field notes and sketches to finished maps and drawings and complete manuscripts, ready for printing, must be made in the shortest time possible. It is true that the manuscript is read critically by the Survey's editorial staff, but the editorial work is done under the same stress and is necessarily somewhat superficial. Many of the refinements of style that characterize the best literature therefore cannot reasonably be expected in a Government scientific report, where the chief object should be precision or clearness of statement.

---

[16] Allbutt, Clifford, Notes on the composition of scientific papers, 2d ed., p. 30, 1905.

[17] Quiller-Couch, Arthur, On the art of writing, p. 162, 1916.

Adams Sherman Hill,[18] professor of rhetoric at Harvard University from 1876 to 1904 and afterward professor emeritus, gives the three following rules for good writing:

1. The rule of precision: Of two forms of expression which may be used in the same sense that one should be chosen which is susceptible of but one interpretation. Observance of this rule tends to give each word a meaning of its own.

2. The rule of simplicity: Of two forms of expression which may be used in the same sense the simpler should be chosen. The simpler a word or phrase the more likely it is to be understood, and simplicity in language, like simplicity in dress or manners, belongs to the best society.

3. The rule of euphony: Of two forms of expression which may be used in the same sense that one should be chosen which is the more agreeable to the ear. It is of course wrong to give undue weight to considerations of euphony, but when no sacrifice is involved it is desirable to avoid an expression that is unusually difficult to pronounce or to substitute for an extremely disagreeable word one that is agreeable to the ear.

Correctness, clearness, and conciseness are ideal qualities of good scientific writing. Clearness alone is not sufficient, for a statement that is entirely clear may contain grammatical errors or may be expressed in terms that are not well adapted to a scientific report; and conciseness may be gained at the expense of both clearness and correctness. However, all abstract good counsel as to forms of expression is suggestive only, and its ready acceptance by a writer does not imply that he has the detailed knowledge or the skill to apply it in practice; and without that knowledge and skill all faith in good rhetorical doctrine becomes "faith without works."

The technical student's need of better training in English has been recognized by many professional and scientific men. In his preliminary analysis of replies received from 23,000 circular letters sent to engineers to ascertain the needs of the profession Dr. Mann,[19] of the Carnegie Foundation for the Advancement of Teaching, wrote:

> In questioning the efficiency of the engineering schools at the present time there are four conspicuous things in which the professional men show a fair degree of unanimity. The first and most important is English. A large majority of the letters received mention the absolute necessity for higher efficiency in the training in English.

In a lecture on the teaching of English at the universities Sir Stanley Leathers said:

> Education without English is like the wisdom of the dumb: it lacks the one thing which gives outward value to the whole. And what sort of professional training is that which has not made the man fit to put his meaning on paper that others can grasp and respect it?

Fortunately for the scientific writer the reader expects far less from him than from the writer of "polite literature", whether of prose or poetry. He looks for no feats of imagination, no flights of fancy, no rhyme or rhythm, and few rhetorical figures of any kind—the fewer the

[18] Hill, A. S., The principles of rhetoric, pp. 18–22, 1895.

[19] Mann, C. R., Soc. Promotion Eng. Education Bull., vol. 6, p. 100, October 1915.

better. All he asks is a plain description of something seen or inferred—of observations or experiments made and conclusions reached—and the plainer the tale the better he will like it. But this ideal simplicity is hard to attain. A French master of style [20] writes:

> A simple style is like white light. It is complex, but not to outward seeming. In language a beautiful and desirable simplicity is but an appearance, and it results only from the good order and sovereign economy of the various parts of speech.

Economy and efficiency in expression may well be studied by the man of science, not only for his own credit but in the interest of science itself.

The dictionaries must be used with vivid realization that as a rule they do not and cannot prescribe usage—they only record it, and they record both good and bad usage. One of the few scientific men who have also undertaken literary criticism [21] writes:

> Certain critics have argued that as such and such a use, which I deprecate, is quoted by the New English Dictionary therefore the use is justified against me. I suspect that no one would be more taken aback by such a protest than the editor of that great work. A dictionary may give select uses or all uses; the editor of the New English Dictionary decided—wisely, in my opinion—to give all uses and to leave to the inquirer the advantage of comparing them and their sources for himself. The dictionary "sanctions" nothing of its contents, but it enables us by consultation of its stores to compare and choose for ourselves. In using this liberty we shall neither be subservient to the prescriptions of age nor scornful of modern freedom; in every use we shall be guided by historical growth, the example of the best authors, and our present necessities.

The scientific writer, above all others, should choose words that have precise meanings or to which precise meanings may be given. He knows how dictionaries are made—he may have helped to make them—and he should use them with discrimination and never forget that English is a live and growing language.

Probably every writer in the Survey has been put through the usual course in rhetoric and has learned some of the elementary principles of writing; he has been taught, for example, that the emphatic phrase or word should stand at the beginning or, better, at the end of the sentence, and that emphasis may also be gained by other devices, but the reports of many an author display here and there not his schoolboy knowledge but his adult forgetfulness of the elements of force in expression. So much of that earlier study was fruitless of practical results that many of its supposed lessons might be forgotten without serious loss. Thorough studies of Chaucer, of the obsolete style of the Spectator or of the sounding periods of Burke's address on conciliation with the Colonies are doubtless valuable, but if long continued they diminish the time available for the study and practice of present-day English—the sort of English that should be used by the author of a report on mineral deposits or geologic structure and processes.

---

[20] France, Anatole, On life and letters, trans. by A. W. Evans, 1914.

[21] Allbutt, Clifford, Notes on the composition of scientific papers, 2d ed., pp. x–xi, 1905

If the author wishes his message to have its best chance of being effective, it is well worth his while to burden himself with all the labor of making it clear and forceful. The author's perspiration will lead to the reader's inspiration.

First or third person

An author should determine at the outset of his work on a report whether he will write it in the first or in the third person. Both "I" and "the writer" should not be used indiscriminately. The first person is more straightforward than the third and if not overused is no less modest or becoming; moreover, it prevents the ambiguity that may occasionally arise where "the writer" may be taken to refer to some other writer who has just been mentioned, as in the following example:

These analyses, according to a letter received from Mr. Ellicott, were made * * *
Though the samples analyzed were collected from localities unknown to the writer * * *

Order within the sentence

Essentials of a good sentence

An adequate vocabulary and a keen discrimination of the best meanings and proper uses of words are essential to all good writing, but still more essential is the ability to arrange words properly in sentences that will convey at once, clearly and forcefully, the ideas to be expressed. Among the indispensable requisites are the choice of the best subject nominative, the determination of its proper place, and the selection of an effective closing phrase or word. Other points that must be considered are consistency of construction, relation of all parts of the sentence to the subject nominative, and continuity of thought. Order of statement or of arrangement is of primary importance, and related words and phrases should be kept together. The obvious lack of proper order in the following example is corrected in the rewritten sentence:

His interpretation differs from that of the writer for the ancient glaciation of the region here described in several particulars. (His interpretation of the ancient glaciation of the region here described differs in several particulars from that of the writer [or from mine].)

Brevity is, of course, always desirable, but brevity should not be gained at the expense of clearness or correctness. The following sentence of 130 words shows not only lack of brevity but numerous other faults. The labor of untangling such a muddle of ideas should have been assumed by the author before he submitted his manuscript for publication.

The view, looking west, shows a portion of the Park City mining district and in the distant background a peak, Clayton (elevation 10,728), formed by an extensive stock of diorite, cutting and metamorphosing sediments adjoining on north as the argillite composed Jupiter Peak which are traversed by master northeast fissures carrying extensive veins of valuable silver-lead ore as the Crescent vein on Crescent Ridge; which faults Jurassic down 1,000 feet against Triassic; and in near foreground a ridge marking a major overthrust fault, overriding two formations; and in immediate foreground the mouth of

Ontario tunnel (5 mi. in length) of Ontario mine, which produced, between 1872 and 1896 silver lead ores which yielded over $60,000,000 worth of silver, surrounded by the plant of the Park Utah Consolidated Mining Co.

Sentences badly begun

Some writers unconsciously get into the habit of beginning sentences with "there is", "there are", "it is"—a trick of construction that may not only multiply words but may have the effect of putting in an inferior place a subject nominative that should preferably stand at or near the beginning of the sentence. "There is no quartz in the ash and it is probably andesitic in composition" could be profitably rewritten "The ash contains no quartz and is probably andesitic", with 9 words instead of 14 and with a gain in force and clearness. Some other sentences of this kind are corrected below.[22]

There are many other primary minerals contain*ing* phosphorus.

There are [At] some places *where* lignite beds are exposed. (Beds of lignite are exposed at some places.)

There has been some faulting [occurred] subsequent to the deposition of the ore.

It is believed that these vugs probably represent openings *which were* formed by recent faulting.

It is the belief of the miners [believe] that the ground now worked may be a slide.

There is a probability that some of the veins may have had their gold content increased by enrichment. (Better: The gold content of some of the veins may have been increased by enrichment.)

"There is some stibnite in the ore" is not so good as "The ore contains some stibnite": the second sentence is not only briefer than the first but more normal and more forcible; it begins with a concrete term, the proper subject, and it ends with the term that should stand at the end of the sentence, in the place of superior emphasis.

"There is little direct evidence from outcrops of the faulting" was written to mean "The outcrops afford little direct evidence of the faulting." "There is also a difference in the grade of the valleys" was replaced with advantage by "The valleys differ also in grade", for the context showed that "grade" was the term to be emphasized.

An initial "there is" or "there are" may also undesirably detach a sentence from one that precedes it, as in the following example: "The Niagara is mainly a light-gray to light-buff fine- to medium-grained dolomite. *There are* [It contains] both thick and thin beds, and at certain horizons *there is* considerable chert."

The phrases "There are", "There were", "There have been", "It is", "It was", etc., may, of course, properly and preferably begin many sentences, but the writer who is about to use one of these phrases should consider whether he cannot express his thought more forcibly in some other way.

[22] Corrections in the examples quoted in this pamphlet are indicated by italicizing the words that should be omitted and placing in brackets the words that should be added.

39

Reprinted from pages 21-26 and 28-38 of *Minerals Exploration: Mapping and Drafting Procedures*, unpublished report, Broken Hill Proprietary Co. Ltd., Melbourne, Aust., 1979

MINERALS EXPLORATION: MAPPING AND DRAFTING PROCEDURES

G. Beavis

[*Editor's Note:* In the original, material precedes and follows these excerpts.]

GEOLOGISTS DRAFT

It is suggested the geologist, when compiling his draft map, used a colour or a rock name abbreviation in preference to the tedious job of drawing symbols.

Note:- It is not advisable to leave blank areas on a map which represents rock types. Add either colour, pattern symbol or the rock type abbreviation (lower case lettering: gr, ms).

Sample:-

THE GEOLOGISTS RESPONSIBILITY

The geologists draft or manuscript copy should not be cluttered up with unnecessary details nor lacking in relevant information for orientation purposes.

To ensure satisfactory results from the draftsman, the geologist must compile his draft in a clear complete form.

We realize not all geologists are fair draftsmen, but they must make some effort to eliminate unwanted linework and ensure that all lettering is clear and legible.

It is not our responsibility to compile a map from a series of partially drawn maps presented piecemeal.

It is unsatisfactory for the draftsman to have to make repeated alterations to maps because the geologist was not sure of details on maps when presented for fair drawing.

Some of the difficulties encountered in producing accurate and intelligent drawings from the drafts received from field parties are: -

> Drafts are illegible by reason of cramped, smudged or ill-formed lettering, or because of faulty erasure of unwanted lines and super-position of corrected information.

Accuracy of geological boundaries (See page 24.)

Titles for map, not sufficient thought; the title should explain and name the map in simple terms.

Plans received have no reference to recognisable features. This is particularly true of plans showing stream sampling where the only feature may be a crooked line depicting an unnamed creek in an undisclosed locality.

Location of survey:- some known identifiable ground feature and/or geographical or A.M.G. coordinate intersections to be shown where scale permits, applicable north points.

Grids, traverses and drill locations to be tied into named features or transferred on to aerial photos or mosaic as an aid for the draftsman to relate and draw in on final map. All maps should be self explanatory, located and aptly titled.

Two, three or more plans may be received which on careful study (in the absence of proper references) are found to be overlapping or adjacent parts of one area and are better shown on one plan.

Titles given to field sheets are not adequate. Map scales are sometimes given as a proportion e.g. "1:50,000" or "1cm = 500 metres". On photocopying such scales may become misleading. Graphic scales should be shown.

There are problems which we often encounter from the report writer.

The geologist or geophysicist compiles his report and maps then presents them for fair drawing. It is often obvious that some maps could be combined or even deleted, thus eliminating some Figures from the report. "A4 maps" are best distributed through the text - the writer ought to refer to the figures "on page opposite".

Maps presented before his report is written, could well be changed, deleted or may have to be redrawn because when writing the report he had overlooked some details.

Before finalising his report the geologist or geophysicist should confer with the supervising draftsman to discuss what maps are to be drawn for his report.

Please ensure:-

- All references to figures are included in text.

- Correct title as on map should show in "List of Maps" and drawing number to be included.

Points to look for are:

1. Composite maps or combining geology with geochemical etc.

2. Reduction of scales (maps for reports are normally drawn at a suitable (round figures) scale to fit the report and not intentionally for field use).

3. Drawing sections in available space on the geological map or a series of sections on one map.

4. Keeping number of maps to a minimum (refer to rule 23 C.R. 1401 - C.P. Taylor).

5. Present all figures together - not piecemeal.

6. Titles are brief and to the point.

7. Before submitting his final draft maps for fair drawing the writer should check through his report to ensure all maps are quite legible, clear, precise and contain all and only relevant information mentioned.

   Ensure that all maps are referred to in text and a list of maps with titles as stated on final maps are listed in report.

If the geologist studied the Suggested Rules 1 to 16 as set out in "Presentation of Field Data" C.P. Taylor 1973 and "Drafting Principles" G. Beavis; most of the above problems would be eliminated.

The standard of drafts prepared by field staff is generally poor and not up to the standards required. All geologists and geophysicists etc must make a more determined effort to improve his own standard of map presentation.

It is my opinion that the geologist ought to depict or relate his geological formations etc,in a colour rather than by a symbol;- bore logging is the exception as these sheets are usually acceptable in the geologist's own form providing care is taken.

[*Editor's Note:* Figures 1 through 9 have been omitted.]

The important aspect of geological map presentation is to display the accuracy of the mapping.

In order to represent the boundaries of rock types the following standards should be applied.

Geological Boundaries

Established boundary - position accurate

Established boundary - position approximate or interpreted

Inferred boundary

Outcrop boundary

Faults

Established fault - position accurate

Established fault - position approximate or interpreted

Inferred fault

Unconformities

Plan (U opens to younger rocks)

Section

Draft presentation of mapping is often misinterpreted by the draftsman when fair drawing a map; so for clarity of geological map presentation could we adopt a standard procedure.

The majority of mapping and photo interpretation is concerned with available outcrops.

symbol for outcrop

The stratigraphic sequence is then defined within these outcrop areas. You may map or interpret one or more geological rock types in the one outcrop. In this case we can show the geologic contact within the outcrop (Fig. 1a)

mapped within outcrop

approx.

In the case of mapping along a traverse or stream the geologist should always attempt to draw in the outcrop boundary; at least across the traverse or stream (Fig. 2).

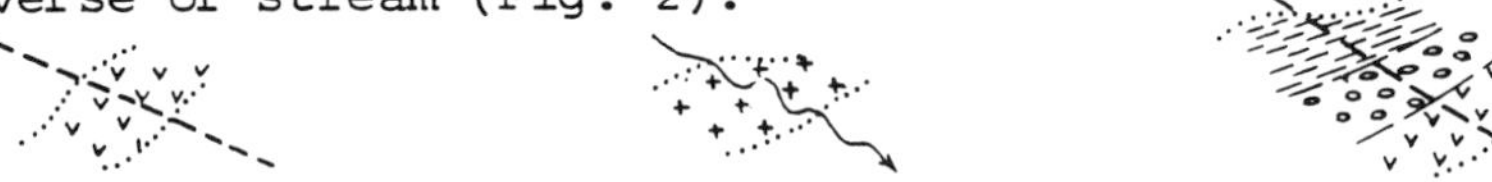

It often happens that the geologist draws in a series of rock types not divided by a boundary. It is left to the draftsman to distinguish where one rock type ends and the other starts. Surely the geologist is the best judge where the boundaries are and should show the approximate boundaries on his map.

If the geologist (he being the most efficient) wishes to add his interpretation then this can be added by drawing a finer line for his interpreted boundaries. The map when being coloured would show the outcrops darker than the interpretation
Figures 1 a-b and 2 illustrate standard types of map presentation.

MAPPING SCALES

In order to standardize and minimise the number of mapping scales the following list has been recommended.

MAPPING		PRESENTATION
1:2500 (1cm = 25m) 1:5000 (1cm = 50m)	Prospect detail drill holes 100 m apart grid lines 200 m apart sample sites 50 m apart	1:2500 1:5000
1:10,000 (1cm = 100m)	Prospect detail traversing 200 m grid semi regional mapping	1:10,000
1:25,000 (1cm = 250m)	Semi regional mapping 400 m x 400 m grid	1:50,000
1:50,000 (1cm = 250m)	Regional mapping	1:50,000 or 1:100,000
1:80,000 (1cm = 800m)	Photo scale mapping	1:100,000
1:250,000 1:1,000,000	Broad geological outlines - interpretations Location map Regional sampling Reconnaissance drilling	

Tenement considered too large to show (hatched)
(A0) (A3) Sheet size

	N.T. E.L.	QLD. A.to P.	N.S.W. E.L.	VIC. E.L.	TAS. E.L.	S.A. E.L.	W.A. T.R.-M.C.	P.N.G. P.A.
	MAX. AREA 500 sq. m 1300 sq. km (44km × 30km)	MAX. AREA 100 Sub-blocks (14 × 7) (1.7km × 1.7km each)	MAX. AREA 256 sq. km (16km × 16km)	MAX. AREA 860 sq. km 12 Sub-blocks (9km × 8km each)	NO LIMIT (NOT DEFINED)	MAX. AREA 2500 sq. km (60km × 40km)	MAX. AREA T.R. 200 sq. km (10km × 20km) M.C. (1·5 × 0·8) km	MAX. AREA 10 000 sq. km (120km × 85km)
SCALES	SIZE – cm	SIZE – cm	SIZE – cm	SIZE – cm	SIZE – cm	SIZE – cm	SIZE – cm	SIZE – cm
1 : 2 500 (1cm = 25m)							M.C. 60 × 32 (A1)	
1 : 5 000 (1cm = 50m)							M.C. 30 × 16 (A3)	
1 : 10 000 (1cm = 100m)							T.R. 200 × 100 (4 × A0) M.C. 15 × 8 (A4)	
1 : 25 000 (1cm = 250m)	176 × 120 (2 × A0)	95 × 48 (A1)	64 × 64 (A0) (2 × A1)	128 × 108 (2 × A0)		240 × 160 (4 × A0)	T.R. 80 × 40 (2 × A1)	
1 : 50 000 (1cm = 500m)	88 × 60 (2 × A1)	48 × 24 (A2)	32 × 32 (A2)	64 × 54 (A0) (2 × A1)		120 × 80 (A0) (2 × A1)	40 × 20 (A2)	240 × 170 (SERIES – MIN 9 × A1)
1 : 80 000 (1cm = 800m) PHOTO SCALE	55 × 38 (A1)	30 × 15 (A3)	19 × 19 (A3)	40 × 34 (A2)		75 × 50 (A1)	25 × 12·5 (A3)	150 × 106 (2 × A0) (4 × A1)
1 : 100 000 (1cm = 1km)	44 × 30 (A2)	24 × 12 (A3)	16 × 16 (A3)	32 × 27 (A2)		60 × 40 (A1)	20 × 10 (A4 – A3)	120 × 85 (4 × A1)
1 : 250 000 (1cm = 2·5km)	18 × 12 (A4)	10 × 5 (A4)	6·5 × 6·5 (A4)	13 × 11 (A4)		24 × 16 (A3)	8 × 4 (A4)	48 × 34 (A1)

COMPARISONS OF RECOMMENDED SCALES, SHEET AND TENEMENT SIZE

STANDARD MAP REQUIREMENTS

Maps compiled should include:-

(a) Some known and identifiable ground features - roads, fences, creeks (named if known), stations, tracks, buildings and wells.

(b) True North and/or Magnetic North point showing relationship to grid north, if local grid used.

(c) Geographical and/or A.M.G. co-ordinates where possible.

(d) Graphic scale (see below).

(e) A complete title:- e.g. T.R. 4367 COOBINA, W.A.
DETAIL GEOLOGICAL MAP
SHEET 24

<u>or</u>

P.A. 105 NEW BRITAIN T.P.N.G.
RAMANDU RIVER AREA
STREAM SEDIMENT SURVEY
RESULTS - COPPER P.P.M.

<u>or</u>

T.R. 3697 MT. MONGER, W.A.
GROUND MAGNETIC SURVEY
MAGNETIC PROFILES - SHEET 7

(f) A locality map (in top right-hand corner).

(g) A note explaining how or where map details were obtained, e.g. photo interpretation or traced from Roy Hill 1:250,000 Geol. map, B.M.R.

(h) 1. The datum point or the field origin of local exploration grids should be a well known and identifiable ground features. These features should be shown and described in terms of geographical latitude and longitude or Transverse Mercator or Universal Transverse Mercator grids. The co-ordinates of the datum point of local grids should be chosen to ensure that the grid and extensions lie in the north east quadrant.

2. Method of showing grids on maps. It is more convenient and practical to show grid lines in a fine full line forming a 10 cm grid over working area of 50 cm x 60 cm or 50 cm x 50 cm.

(i) Locations and identification mark of all samples, costeans, pits, drillholes, roads, bulldozed tracks and traverses must be marked on plans.

(Show bar scale in multiples of 50, 100, 500, 1000. Add also the Ratio scale, e.g. 1:10,000)

Scale 1: 10,000

0 200 400 600 800 1000 Metres

DRAFTSMAN'S RESPONSIBILITY

A map presented by the geologist for drafting is usually a record of his field observations and interpretations.

Because the geologist does not realise nor take sufficient care with his draft, additional information which is necessary for the reader to locate and orient the survey and fully understand the map are neglected.

The draftsman's responsibility is to ensure his map is a true record of the geologist's work and to give assistance to the reader at all times.

Ensure necessary information is shown to geographically locate and orient the survey and understand the map without making unnecessary assumptions.

For example, readers are not expected to -

a. make assumptions regarding information on the map.
b. depart from the map in hand to locate and orient the survey.
c. extend base lines or long lines across maps and sections to join scales in order to measure off information.
d. assume the relative base line for each section or profile.
e. assume grid graticules are in metres or feet.
f. assume the accuracy of the mapping - add source of data.
g. decipher difference between figures because of poor titling.

The final appearance of the map needs to be aesthetically pleasing but should clearly portray and display the priorities of the collected data which the geologist, geochemist, geophysicist wants to present (see Fig. 3, 8.).

When a symbol on the map is drawn (out of character - too large and bold or too similar to other symbols - Figs. 7b) and if the priorities of values are misinterpreted by the draftsman then the map can impart the wrong impression to the reader.

The drawing of the symbol is left to the discretion of the draftsman. In the majority of cases there are no defined dimensions for symbols so the visual effect on the map will depend on the experience of the draftsman and his ability to read and present a proper and pleasing result of the field personnel's work.

DRAFTSMAN PRESENTATION

The fair drawing of the map is usually left to the draftsman and he or she has to decide which is the best method of displaying the geologist's work.

Place yourself in the reader's position, give the reader all the help necessary to make the map simple, clear and easy to read. Know your work, know your author, bring what is important to the reader's attention by high-lighting those important features.

When the reader has difficulty defining detail on a map or has to assume certain detail then the draftsman has failed to do his job.

Cartography is not art in the sense that one may have complete freedom with techniques and media. In the majority of cases there are standard symbols for most items. This is apparent in topographic , survey, geology and its associated surveys, plans and maps.

The aim of the cartographer (meaning all draftsmen) is to present the map data in such a fashion that the map as a whole, appears as an integrated unit and so that each item included is clear, legible and neither more or less prominent than it should be.

In some instances, certain detail by its importance needs to be high-lighted. Only experience will give you the ability to read into and present a proper and pleasing result of the author's work (see Fig. 8).

A good designed map is pleasing to look at, but it is a fact map and requires to be located. Therefore it doesn't matter how well it is drawn and presented if it is not located (geographic position by localities or geographic co-ordinates) then it doesn't mean much. However, in minerals exploration topography is important but it is secondary to the showing of the collected data.

Titles are important - by reading the title you should straight away know what the map is all about.

Irrelevant data is often drawn in by many draftsmen; get rid of it - it is time consuming and doesn't help the reader (see Fig. 6a-6b.)

Large symbols drawn on a map convey the impression of a large scale; small symbols give the impression of a small scale. Therefore it is important that the draftsman when drawing a map understands its purpose and the impact it will have on the reader. (Fig. 7a-7b).

DISTINGUISHING UNITS ON THE MAP

The best way to differentiate between rock units on a map is to use colours. However many maps have to be produced in black and white because of the work involved in hand colouring a large number of maps.

METHOD A. Colour and/or Letter Symbol

When it is decided to use colour to denote a rock type it is advantageous also to designate all rock type areas shown on the map a letter symbol as well as a colour.

Sample:-

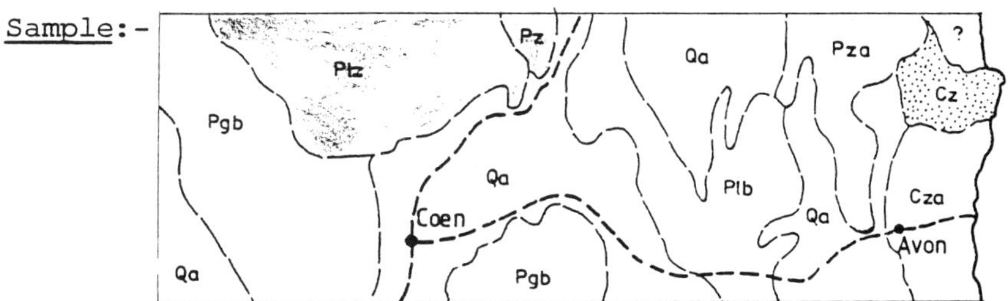

The orginal coloured map may be lost and if all areas on the map are denoted with a letter symbol, it is a simple matter to make another coloured copy.

METHOD B. Black and White Patterns

Using symbols (or patterns) the contrast in the adjoining drawn symbols is the most important aspect for definition.

The IMPORTANT ASPECT of black and white patterns is the degree of contrast and definition for ease of reading.

Sample:-

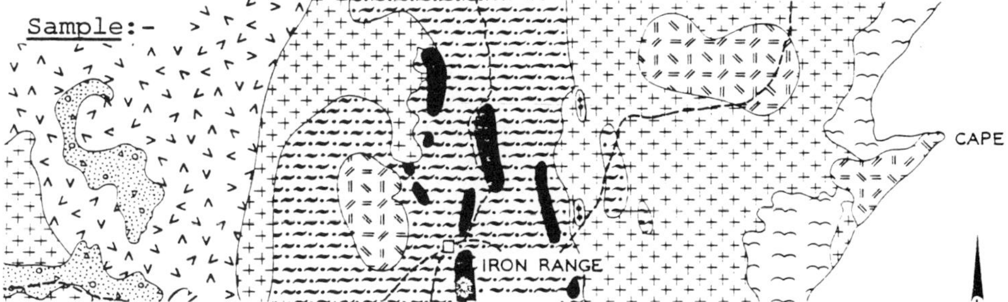

The size and density of the patterns set out in this Exploration Manual are only a guide. When fair drawing the patterns, the size and density of the symbol should be adjusted for clarity of reading (contrast) and recognition.

[*Editor's Note:* In the original, the grey areas on the sample map appear in various colors.]

STANDARD MAP REQUIREMENTS

Maps compiled should include:-

(a) Some known and identifiable ground features - roads, fences, creeks (named if known), stations, tracks, buildings and wells.

(b) True North and/or Magnetic North point showing relationship to grid north, if local grid used.

(c) Geographical and/or A.M.G. co-ordinates where possible.

(d) Graphic scale (see below).

(e) A complete title:- e.g. T.R. 4367 COOBINA, W.A.
DETAIL GEOLOGICAL MAP
SHEET 24

or

P.A. 105 NEW BRITAIN T.P.N.G.
RAMANDU RIVER AREA
STREAM SEDIMENT SURVEY
RESULTS - COPPER P.P.M.

or

T.R. 3697 MT. MONGER, W.A.
GROUND MAGNETIC SURVEY
MAGNETIC PROFILES

(f) A locality map (in top right-hand corner).

(g) A note explaining how or where map details were obtained, e.g. photo interpretation or traced from Roy Hill 1:250,000 Geol. map, B.M.R.

(h) 1.The datum point or the field origin of local exploration grids should be a well known and identifiable ground features. These features should be shown and described in terms of geographical latitude and longitude or Transverse Mercator or Universal Transverse Mercator grids. The co-ordinates of the datum point of local grids should be chosen to ensure that the grid and extensions lie in the north east quadrant.

2.Method of showing grids on maps. It is more convenient and practical to show grid lines in a fine full line forming a 10 cm grid over working area of 50 cm x 60 cm or 50 cm x 50 cm.

(i) Locations and identification mark of all samples, costeans, pits, drillholes, roads, bulldozed tracks and traverses must be marked on plans.

(Show bar scale in multiples of 50, 100, 500, 1000. Add also the Ratio scale, e.g. 1:10,000)

Scale 1: 10,000

0 200 400 600 800 1000 Metres

STAGES IN THE DRAWING OF GEOLOGICAL MAPS

1. Perusal of geologists field map to be fair drawn and selection of main features for highlight.

2. Selection of standard sheet size.

3. If the final map base is to be gridded, draw grid graticules (spacing 5" x 5" or 4" x 4" British Standard or 10 cm x 10 cm metric) in fine lines on back of tracing. Grid north parallel to sheet borders.

4. Drawing area of A1 sheet should be:-

   50 cm x 60 cm or 50 cm x 50 cm (metric).

   This leaves the right hand sections of your map vacant to add locality map (top right corner), legend, scale and notes.

5. Secure geologists draft copy to drawing board, place standard sheet in position for copying detail and secure sheet to board with tape.

6. If detailed geological data, trace off creeks, strikes and dips, faults and structure, at this early stage.

7. Topographic features such as roads, railways, towns, exploration developments (costeans, drill holes, sample location etc.) can now be added to specified standards.

8. Plan your sizes for all lettering and add notes and names with the aid of stencils or other means (Note: ensure that all lettering can be read from bottom and right hand edges of map).

9. Trace off carefully and accurately geological boundaries and add letter symbol for rock types where applicable.

10. If it is decided to use a black and white pattern symbol, then the pattern is best drawn on the reverse side of the tracing.

11. Where topographic contours are to be shown, they may also be drawn (in fine lines) on the reverse of the tracing.

12. Add legend, locality map, scale and title block details.

GUIDELINES FOR DRAFTSMEN
DRAFTING NOMENCLATURE, NOTES AND TITLES ON MAPS

1. Spacing of Lettering and Words:-

Spacing of words should not be greater than an average letter width:-

eg. Correct	Poor Spacing
▲ Mt. Lockeridge	▲ Mt. Lockeridge
☐ Aeolian sands.	☐ Aeolian sands
☐ Saline lake deposits.	☐ Saline lake deposits

2. Use of Arrows:-

The drawing of arrows on geological maps, should be the exception rather than the rule.

Arrows are generally used where there is one note depicting two or more locations or where space is insufficient to show notes clearly in appropriate area.

The geologist when drafting his map may use as many arrows as he wishes to ensure clarity and definitions of his work. BUT the draftsman should in the main be able to depict the geologist's work without using arrows.

eg. 1. General note describing rock type and sample site.(2)

A dot can be used instead of the arrow drawn by the geologist and the note written close to the dot.

2.

AF 2741
x AF 2741

Geologist. Draftsman

- 2 -

Recommended Use of Arrows:-

Strong
Copper

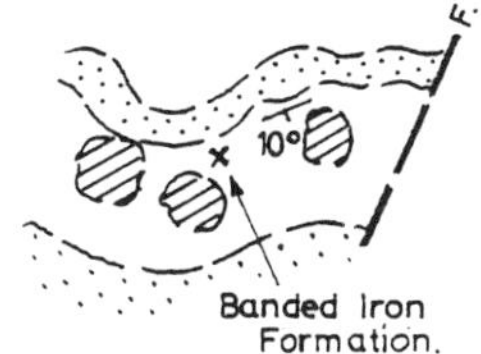

3. Sizes of Lettering:-

Lettering and numbering on maps must be drawn to a size and weight where detail can be clearly read when reduced by half scale. (Especially important on A3, A2, A1, A0 sheet sizes).

Recommended Heights of Lettering:-

Sheet size A4 - not greater than 2.5 but usually 2mm or 1.6mm

A3 - not less than 2mm (unless unduly crowded with - assay results or detail descriptions of rock types - strike and dip readings - then down to 1.6mm)

A2 - not less than 2mm - preferably 2.5mm

A1 - not less than 2.5mm - unless unduly crowded - (2mm)

Lettering on maps is read with the north point pointing directly away from the reader or facing to the NW or NE quadrant.

Lettering of legend details should not be written in larger than general note sizes on map.

4. Titling of Maps:-

Map Sizes:-	Lettering Height
A4) A3)	2.5 to 3mm
A2	3mm to 3.5mm
A1	3mm to 4mm

-3-

Titles should be apt and simple, eg:

1. Geological Map
   Nymagee Mine Area, N. S. W.

2. E. L. 71, Finniss River, N. T.
   Magnetic Profiles, South Grid

Poor Title Layout

Geological Plan Nymagee
Mine Area, N. S. W. Data From
Cyprus Mines Drawing No. 7-168

NOTE: (Data from etc. should be written outside of title block - preferably under the scale or below legend details.)

The source of map data should always be shown and written outside of title block, preferably below legend or scale.

The draftsman's job is:

a) Confer with and advise the geologist on the best method of depicting his work in order to create a clear picture of what the maps represent.

b) Minimise the work and paper bulk by drawing composite maps and reducing large maps to A3 and A4 where practicable, providing no relevant information is lost.

c) The reader should not have the problem of assuming any details; the map should be laid out clearly and neatly with just sufficient information to portray the geologist's work.

d) Reproduction of maps. It is the draftsman's responsibility to ensure that the printer supplies the best possible copy.

   This is important in the case of sepia transparencies, especially when printing from a base map where copies from this base will later be used as final maps.

   Ensure that inkwork on base is black (greys do not reproduce) and that the printer supplies good dark line reproductions. It is best to have a slight background on sepia copies.

The scale on all maps must be drawn in bar or graphic form and located in either of the applicable positions.

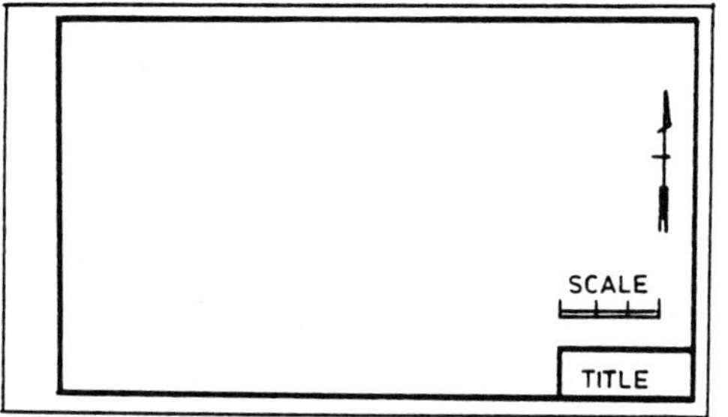

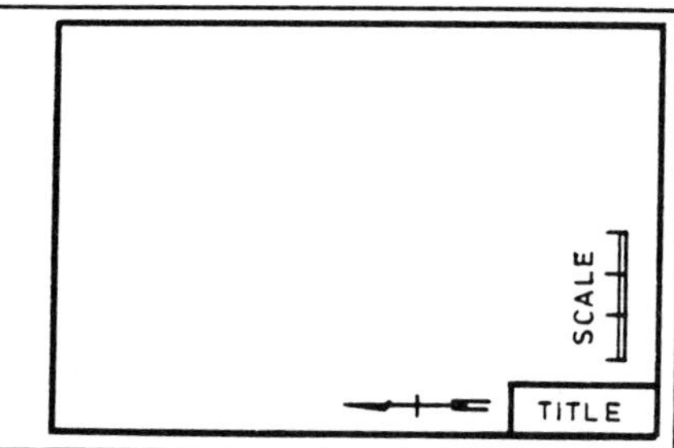

The ratio scale (e.g. Scale 1:2,000) should also be written above the bar scale.

Divisions of the bar scale should finish on the even hundred or thousand, (fifty on occasions).

e.g.

Scale 1:1,000

0	20	40	60	80	100	metres
0	200	400	600	800	1000	metres
0	50	100	150	200		
0	10	20	30	40	50	kilometres

ATTITUDE AND STANDARDS OF DRAFTING

Work is not set out exactly as the geologist has done on his rough copy.

Common sense prevails in most jobs, therefore planning of lettering, titles, setting out, (showing main features a little bolder than the rest), locating the survey are things which should be considered before starting the job.

There are certain standards and symbols to be adhered to. (Refer to previous sections of this manual).

Put yourself in the reader's position; give the reader all the help necessary to make the map simple and clear to read.

Bring important features to the reader's eye; scales, north points, legends should be clear but apart from the actual map area and not to dominate the map itself.

Plan your lettering to be in keeping with plan size. On larger maps; it may be necessary to reduce in size; so lettering and line work needs to be drawn to a size where at least a half scale reduction can be obtained without loss of detail.

When a reader has difficulty defining detail on a plan, then the draftsman has failed to do his job.

If work is too fine to print black, or the ink work is grey, then we must strive to correct these faults.

Present details clearly to the reader; relate to the reader what the plan represents, and don't leave him in the position where he has to assume certain details.

Always check lettering for errors in spelling and blackness.

Titles should be apt and should convey to the reader where the locality is and exactly what the plan represents.

(Area) - TR 4386 - KOOLYANOBBING, W.A.
(Type) - Detailed Geological Map
(Locality) - Sheet 26 - North Range

Geophysical and geochemical profiles - draw in profile line to highlight profiles above other linework.

On Locality Map draw to the attention of the reader the location in question. (Slightly larger printing for locality name.) (Fig. 3)

On small maps (A4-A3) keep lettering down to a size in keeping with plan size.

LEGENDS - Do not make the legend blocks and notes more pronounced than plan itself.

For clear black lettering, pens and guides should be kept clean.

Good shapes in hand lettering are not accidental, they are formed deliberately and only become automatic when practised regularly - Time has been offered for practise; do take advantage of this time.

Quality of Work

Each of us should work to our highest standard - self satisfying is a job well done. However quality has to be scarificed at times because of the urgency in reporting to managers, directors or government authorities concerned. This is no reason to disregard good design and clarity.

Good work is only produced by keenness and if the keenness is not there, then carelessness and laziness takes over.

Nothing satisfies more than a job well done. This applies to all aspects of this office:-

1. Painting on a wash
2. Folding of plans, even trimming
3. Flush joining maps
4. Lettering
5. Ruling lines and making a perfect intersection.

If you don't apply this technique to your work, then work will become a drag. You should work to the standard you would expect if you were paying for labour.

When you are keen, you will also remember the plans you have drawn. If you care, you will remember.

Lettering

Sloping lettering for all water features, ——— U/L
for Creeks - Gullies - Lakes. —— *e.g. Lake Eyre.*
CAPITALS FOR BAYS - RIVERS - OCEANS. *e.g. DOVER RIVER.*

Vertical lettering for all land features, including coastal names. e.g. Ayres Rock - Rocky Point.

Part VII

ETHICS

Editor's Comments on Papers 40 and 41

40 FLEMING
The Professional Engineer in the Mineral Industry

41 THE ANACONDA COMPANY NATURAL RESOURCES GROUP
Environmental Awareness in Mineral Exploration

The minerals industry does not stand high in public opinion in spite of the great contributions it has made, and continues to make, to the basic needs and comforts of society. The industry has a reputation of exploitation, pollution, and questionable stock promotion programs. During periods of mineral booms, with severe competition for staff and information, epidemics of company jumping with confidential information, of reports written to mislead and to encourage speculators, and of speculation by geologists on the basis of privileged information have occurred. Many articles have been written on professional ethics within the geological profession, but Fleming (Paper 40) covers not only aspects of ethical responsibilities during exploration but also continuing responsibilities through production and restoration.

Since the exploration geologist is usually the first contact that a land holder has with a mining company, it is essential that this contact leave a favorable impression. The Anaconda Company (Paper 41) has taken the lead in issuing to their exploration staff specific instructions regarding individual responsibilities in respecting land ownership and preservation of the environment. Attention to such details goes a long way toward improvement of the public image.

40

Reprinted from *Inst. Mining and Metallurgy Trans.* **80:**24–30 (1971)

THE PROFESSIONAL ENGINEER IN THE MINERAL INDUSTRY

M. G. Fleming

The professional mineral engineer is an individual of personal integrity who has undergone rigorous intellectual instruction in the scientific disciplines fundamental to his vocation; he has learnt to apply these disciplines to the discovery, mining and treatment of mineral resources and, through practical training and experience, has acquired competence and judgment to fulfil his undertakings. In the course of this development he has recognized and accepted a threefold responsibility to society, to his employer and to his profession.*

The definition contains some dozen elements, and this list is not exclusive: (1) individuality; (2) personal integrity; (3) intellectual discipline; (4) application of science to the arts, i.e. technology; (5) identification with the mineral industry; (6) practical training; (7) experience; (8) competence; (9) judgment; (10) responsibility to society; (11) responsibility to the client or employer; and (12) responsibility to the profession.

I shall consider some of these elements in more detail, but what I should like to stress is not the components but the unity of the whole, for this is not just a definition—it is a doctrinal system which has evolved over two centuries, its principles being incorporated in the Bye-laws of our Institution. This is not to say that the doctrine is universally accepted or even approved. However, this address is not about the defence, or even the justification, of our professional organization: it is about understanding ourselves.

In speaking so, I run the risk of sounding both pompous and patronizing, because most of you need no aid to understanding of this subject—it is built into your characters. Furthermore, my exposition must be largely subjective because, although I have started with a definition, the subject is too subtle to be defined,† and one can be no more precise than to say that a professional mineral engineer is the product of some function of each of the twelve factors listed, with each term modified in some variable degree by factors of time and place. This means only that we are all different and we all change, but that in no case can any one of the twelve elemental factors have a zero value.

I would justify myself on three grounds. First, young people entering our profession are entitled to know what they are letting themselves in for and noviciates to know what is expected of them. Secondly, many of the elements in my definition are changing very rapidly—science, education, industry and society itself. We can neither ignore these changes nor allow ourselves unwittingly to be swept up in them. Thirdly, all doctrines, including those based on intellect, contain an element of faith and, if they are proved satisfying and worthy, then those who hold to them should be prepared to declare their faith and to promote them by discussion and interpretation.

The mineral industry

Let us start with the mineral industry itself, because it is a striking and unusual feature of our particular profession that we serve a single industry. It is very old, very important and very self-contained. It is a platitude to say that it is expanding because civilization is based on minerals so, unless there is some catastrophic series of events that reverses the population trend and suppresses industrialization and technological development, the industry must keep pace. It has always been international—for the simple reason that geological patterns were established before political frontiers; it is becoming rapidly more so because of the uneven industrial development which led to Britain changing from a net exporter in 1928 to our present unfavourable balance of mineral imports and which, in the United States, has caused the net value of mineral imports over exports to treble in the last decade.[9]

The industry has frequently influenced the economics of individual countries, but it is itself dominated by the overall economic pattern of a capitalist society. Therefore, although on a broad time-base it continues to develop at an ever-increasing rate, it is subject to all the fluctuations of unstable world economics. It is an

*In this phrase I would embrace the disciplines of mineral exploration, mining geology, mining, mineral technology and metallurgy, which the new Bye-laws of the Institution identify as our particular concern. These fields were recognized by the 'Brighton Conference'[1] as comprising the core of technological manpower in the mineral industry. In the United States, too, employers in the mineral industry have expressed a strong preference for men trained in these specialities.[2]

†Most authors have avoided the difficulty of defining a profession. Even Carr-Saunders and Wilson[3] in their classic and eminently readable work *The professions* did not try to distil their exhaustive analysis of the subject. Reader[4] preferred 'to let the definition emerge rather hazily—as it does in real life—from the discussion of particular cases'. Lewis and Maude[5] attempted no hard and fast definition, and Millerson,[6] having moved towards it stealthily, proposed that 'It is a type of higher-grade, non-manual occupation, with both subjectively and objectively recognized occupational status, possessing a well-defined area of study or concern and providing a definite service, after advanced training and education'. Bennion[7] and others have adopted the evasion of enumerating qualities which they say a body must possess before it can justifiably call itself professional. Even so, they do not always demand a perfect score. These tactics are exemplified by the Monopolies Commission,[8] who, in their recent examination of professionalism, stated that no occupation was regarded by them as falling within the ambit of their enquiry unless it had some of the following characteristics. (i) Practitioners apply a specialized skill enabling them to offer a specialized service. (ii) The skill has been acquired by intellectual and practical training in a well defined area of study. (iii) The service calls for a high degree of detachment and integrity on the part of the practitioner in exercising his personal judgment on behalf of his client. (iv) The service involves direct, personal and fiduciary relations with the client. (v) The practitioners collectively have a particular sense of responsibility for maintaining the competence and integrity of the occupation as a whole. (vi) The practitioners tend or are required to avoid certain methods of attracting business. (vii) The practitioners are organized in bodies which, with or without State intervention, are concerned to provide machinery for testing competence and regulating standards of competence and conduct.

The further reservation is then applied that occupations having characteristics (i) and (vii), together with some at least of characteristics (ii)–(vi) would, as a matter of convenience, be referred to as 'professions'.

industry that lives with risk—the risk of flooding, cave-in, fire, political changes, labour disputes, fluctuating prices, product substitution, changing tax structures and public condemnation. And yet it is organically incapable of quick adaptation, because it may take ten years and £40 000 000 to bring a new discovery into production. Its present trends have a sort of inevitability about them: big new discoveries are of lower grade, requiring big production units which involve a switch from labour-intensive to capital-intensive operations, which, in turn, need big international companies to finance them.

Where does the professional mineral engineer stand in relation to this juggernaut? The short answer is that he is a part of it—for better or for worse. Unquestionably, it is a vital part, because the planning, design and control of modern mining operations has so far outstripped the art as to be conceivable only under the guidance of highly trained engineers. Nor could the exploration, research and development that are essential prerequisites for expanding production be effective without their specialist skills. There are, however, other facets to the industry's structure, as can be seen from the functional diagram below. Each demands its particular skills and each is inexorably linked together:

EDUCATION
AND TRAINING

OPERATION
AND PRODUCTION
|
MANAGEMENT
AND ADMINISTRATION

EXPLORATION, RESEARCH
AND DEVELOPMENT
|
EQUIPMENT AND
SUPPLIES

MARKETING
AND FINANCE

hence, none is significant in isolation and no five can continue to prosper without the sixth. It is salutary to consider that even in Canada mineral engineers account for less than $2\frac{1}{2}$ per cent of the manpower employed by mining companies; in Africa this figure would be very much smaller. They neither dominate nor control the industry; they are an essential part of a diverse and complex organism in which technology must be matched by capital and labour, and the whole must be in harmony with legal, market and social restraints.

Nevertheless, the influence and authority of this small proportion of professional mineral engineers permeates the entire structure. In addition to the esoteric nature of their specialist skills, they have by training and experience a necessary knowledge of the related technologies. A concentrator superintendent must understand the mining problems that limit his supply of ore, just as he must appreciate the requirements of the smelter to which he delivers his products. Similarly, his colleagues, the mine and smelter superintendents, share a breadth of interest and of technical knowledge which can extend to every facet of the industrial structure, qualifying them to act as interpreters to others and putting these men in a favourable position, if they possess them, to exploit those qualities of ambition, vitality, personality, foresight and exceptional ability that are required for successful general and executive management.

The financial rewards are usually good.* However, for his skills and the right to practise them, for his influence, authority and opportunity, the professional mineral engineer has a high price to pay, because in a very real sense he is the conscience of the industry. This role—which is always difficult and sometimes seemingly impossible—devolves on him because of his identification with the industry. It is not assigned to him and, indeed, it is a service which industry would sometimes willingly forgo; but it cannot have his competence without his integrity and responsibility because it is the essential combination of these three qualities that make him a professional engineer.

Professional ethics

Professional ethics are the acknowledgement of our responsibility towards the society we serve and the philosophy we represent. The concept is not new. In law and in medicine it goes back to the Middle Ages; in the church, much further than that.† Professional associations are the guardians of the ethic. They arise when they are needed, that is, when the corpus of knowledge is substantial and when its application becomes vital to society. Thus, although non-military engineering was a distinguishable activity in the middle of the eighteenth century,‡ the first professional engineering institution (the Institution of Civil Engineers) was not founded until 1818, and our own Institution was not established until 1892. Professional ethics are neither absolute nor static. Since the body of knowledge in the science and practice of mineral engineering is continually expanding, and since our society is changing at an alarming rate, it follows that our responsibilities too must change, and, this being so, we are obliged to reconsider them from time to time.§

*A survey, conducted by Dr. E. M. Goodger, based on the replies of 6098 people who graduated from Imperial College, London, between 1929 and 1963 showed that past students of the Royal School of Mines had a markedly higher median income throughout their careers than graduates in other engineering fields or in science. This advantage, which amounted to about £800 per annum between the ages of 25–50, corresponded to a similarly higher level of responsibility for the R.S.M. graduates.

†For a good description of the evolution of the professions see Reader.[4]

‡Armytage[10] quoted Campbell in the *London Tradesman* of 1747 in a passage which anticipates the views of some of our leading industrialists: 'The Engineer makes Engines for raising Water by Fire, either for supplying Reservoirs or draining Mines . . . The Engineer requires a very mechanically turned head . . . He employs Smiths of various sorts . . . He requires a large stock . . . to set up with, and a considerable Acquaintance among the Gentry . . . He ought to have a solid, not a flighty Head, otherwise his Business will tempt him to many useless and expensive Projects'.

§This fluidity discounts the value of the vow or oath which is often proposed as a means of establishing a specific moral obligation on the part of members of a profession. There is a danger that, instead of fulfilling its original purpose, it may become an issue of contention—dividing instead of unifying. Having survived for nearly 2300 years, the oath of Hippocrates now seems to conflict with the need to reduce the birth rate. That any such vow should be closely examined is demonstrated by a recent proposal from a London professor that all applied scientists and engineers should swear an oath which included the solemn promise to minimize destruction of mineral resources![11] The Iron Ring Ritual, prepared by Rudyard Kipling for Canadian engineering graduates, includes an expression of intent, but not an oath.

For convenience, I have considered three aspects of responsibility, although these overlap widely and, so far as they relate to our professional ethic, it is my personal opinion that they are indivisible.

(1) *Responsibility to society*

First under this heading I would place the obligation to supply man's mineral needs; for, although the engineer is far from self-sufficient, his is the leading role, and, without him, that obligation cannot be met. Linked with this is the aim of optimum utilization of mineral resources and the avoidance of waste, and this is a technological challenge worthy of the best scientists and engineers in our ranks. It involves not only the reduction of losses in pillars and tailings but production, within the economic framework, of useful by-products so that, in a perfect world, we should recover from the pegmatite ore not only high-grade cassiterite for the tin smelter but spodumene and beryl, apatite for the fertilizer manufacturer, pure quartz for the glassmaker, feldspar for the potteries and mica for the building trade.

Minerals cannot be produced without disturbing the environment, but for too long many of us were content to leave nature to heal the scars of essential mining operations. In this we did a disservice to our industry by exposing it to the vilification of those who take its benefactions for granted. 'There is a tendency, even a perverse willingness, to suppose that the despoliation sometimes produced by technology is an inevitable and irremediable process, a trampling down of nature by the big machine. Of course it is nothing of the kind. The deterioration of the environment produced by technology is a technological problem for which technology has found, is finding and will continue to find solutions'.* Today few of us can question our responsibility in this respect; for only we have the knowledge to inform the public of what solutions are available and what they must pay to have them.

Protection of the community from fraud or from its own gullibility is not just a matter of personal integrity—it requires vigilance from those best qualified to recognize dubious practices. Legal action provides redress (after the event), but the public's best safeguard is the availability of honest and competent advice on matters which it is incapable of judging for itself. This is one of the basic reasons for the association of professional men. The Institution gives rules for guidance on reports in connexion with commercial enterprises and recommends standards of practice which the public can recognize. However, it goes much further than this in that it assumes a collective responsibility and, in effect, it guarantees the honour of its members.

Protection of members of the community, including those who work with him and under him, from risks to life, health and safety is a responsibility that clearly rests with the engineer and which, in fact, he can share with no one but his professional colleagues. All too frequently this is the young engineer's first experience with the grave implications of his professional status. Consideration for the general welfare of his employees and their families, including the security of their employment, is a later stage of social awareness.

*This quotation is from Sir Peter Medawar's Presidential Address to the British Association for the Advancement of Science.[12]

(2) *Responsibility to client or employer*

The most clearly defined of the professional responsibilities are those of the consultant to his client.* In our particular profession the freelance consultant is fast becoming a rarity. To those who remain and to those privileged few whose occupations permit them to act as consultants in their spare time, these rules of conduct clearly apply. However, the great majority of our members is in salaried employment† where, quite obviously, some of the restrictions imposed on the independent consultant are irrelevant. Does this relieve them of their other obligations and leave them with a duty only to satisfy requirements of their employers? On this issue I, personally, have no doubts, and I would say (i) that a man's first responsibility is to his profession; (ii) that the characteristics of the professional mineral engineer which we are considering are so clearly beneficial to the legitimate aims of any employer that there should be no conflict of duty; and (iii) that where the actions of the employer are incompatible with the engineer's professional standards, he should, with patience, commonsense and considered judgment, first make sure that he fully understands the situation, then try to influence it (and this may mean waiting until he has acquired the authority to do so) and, finally, if he is unsuccessful, seek other employment.

I would suggest that the specific responsibilities of the professional engineer to an employer do not differ radically from those which a client is entitled to expect. Basically, they are (i) to serve his employer well and faithfully; (ii) to give only his best service without regard for what he will be paid and to subordinate his personal interests to the needs of his employer—which is only to say that he should take 'professional pride' in his work; (iii) to advance the technical efficiency and the economic capability of his employer's business, to help it to meet competition, increase profit and to anticipate future needs; (iv) not to disclose the confidential affairs of his employer; and (v) to avoid financial involvement with anything outside his employment, but related to it, without the knowledge and consent of his employer.

(3) *Responsibility to the profession*

Professional associations, as I have said, historically arose when they were needed. Starting as social gatherings—often dining clubs—of people having a

*See *Rulings and Recommendations* made by the Council of the Institution on 11 May, 1961, for publication to members.
†The Pilkington Report[13] showed that in 1958 only 2 per cent of all professional engineers were principals in private practice.

common interest, they developed into societies for the exchange of ideas, the discussion of research and, by publication, the dissemination, advancement and preservation of their intellectual discipline. Having discovered themselves, they sought public recognition and status within their social framework, and the crowning success of these efforts was the Royal Charter. If the fruits of this specialized knowledge directly affected the community—and this was always the case with science-based disciplines—some standard of practice was desirable. At the same time, the pride which professional men took in the quality of their work made them want to see it developed by education, recognized by formal qualification and protected from disparagement by the actions of incompetents.* The development of an interest in the qualifications of its members was the main feature that distinguished a professional institution from a learned society or a study group; it enabled the institution to assert a special competence in its own field, based on intellectual training, and to guarantee the competence of its members to a public unable to judge for itself. Concomitant with the guarantee of technical competence was a guarantee of probity, for the same considerations that led to the first led inevitably to the second: the client required security that the advice he received was financially disinterested, public confidence in honest practitioners was not to be discounted by the behaviour of scoundrels, and honourable men were unwilling to associate with either charlatans or knaves.

The responsibilities which the professional engineer owes to his institution derive from these traditions. In brief they are (i) to contribute to the advancement of the science and practice of his discipline; (ii) to repay his predecessors by passing his knowledge and experience to his successors; (iii) to uphold the professional standards of which his institution is both the guardian and the guarantor; and (iv) to protect and enhance the professional reputation of both his institution and of his fellow members.

'I hold every man a debtor to his profession; from the which as men of course do seek to receive countenance and profit, so ought they of duty to endeavour themselves, by way of amends, to be a help and ornament thereunto.'[15]

Conflicts

In the changing scene, conflicts develop between the three broad areas of responsibility to society, the employer and the profession.* Many of these are more apparent than real, but some are real enough to cause distress and frustration to engineers who accept their threefold responsibility. If these conflicts are to be resolved, they must be discussed and understood. Let me mention some which seem to cut across and divide our loyalties.

'Pollution' is a word that drives a wedge between society and our employers. No one who has visited the Sudbury area of Canada or the old coal fields of Britain can question the past depredations of the industry, and the monuments they provide have made a society newly awakened to the horrors of overcrowding antagonistic to the whole mineral industry.† It was not always so. I remember my father telling me of the excitement and enthusiasm that greeted the discovery of the world's richest nickel deposit, and the stimulus which its exploitation gave to the opening up of northern Canada and the development of the entire country was incalculable. The Industrial Revolution was nourished on coal, and a country on the road to becoming the richest and most powerful in the world had little time to tidy up behind it. It is ironic that, if the spoliation is picturesque enough or even big enough, it may hold a charm of its own: one of the main tourist attractions in the U.S.A. is 'the biggest man-made hole on earth', the Bingham open-pit. However, there is no charm in the public's present dismal view of mining in this country or in North America, for the mineral industry is out of phase with the temper of the time.

How is this conflict to be resolved? Perhaps the first move is to recognize that, up to a point, society is right and that future civilization will need all the natural amenity that can be preserved for it. It is our business to be alert and sensitive to social attitudes, just as it is our business to be sensitive to changing material requirements. I think we must now agree that mining operations should be designed from the outset to minimize damage to the environment and that rehabilitation of essential damage should be incorporated in the original plans for any new project. We shall have to accept the fact that this is as much a part of technical efficiency as are adequate safety precautions. Pioneering operations have shown that such planning is practicable and effective. These examples can help in the second move, which is to convince the public

*Reader[14] quoted two emotional outbursts that well illustrate this attitude.

'This is the rule in every other learned profession, and there can be no reason why it should not be adopted by the engineers ... at present ... any man without business, competent or not, dubs himself engineer, starts a project, well or ill founded ... *generally the latter*, and issues a prospectus to the public ... (From Sir John Rennie, 1875.)

'Our engineers have no real scientific instruction, and we let them learn their business at our expense, by rule of thumb; but it is a ruinous system of blunder and plunder. A man without the requisite scientific knowledge undertakes to build a difficult bridge; he builds three which tumble down, and so learns to build a fourth which stands; but somebody pays for the three failures. In France or Switzerland he would not have been suffered to build his first until he had satisfied competent persons that he knew how to build it, because abroad they cannot afford our extravagance.' (From Matthew Arnold, 1868.)

*Some of these problems, seen through American eyes, are given in question and answer form in a book sponsored by the American Society for Engineering Education, the Engineer's Council for Professional Development and the National Society of Professional Engineers.[16]

†In an address given at the annual dinner of the Royal School of Mines Association in 1970 Mr. Ian MacGregor quoted the results of a recent nationwide public opinion poll in the United States in which the public was asked to give its general impression of eight major industries.

Industry	Per cent having favourable impression
Food and grocery products	59
Automobiles	47
Airlines	41
Textiles	31
Oil	18
Soap and detergents	16
Railroads	13
Mining	6

that vandalism is no part of the intention of the mineral industry, now or in the future. The second will be the more difficult for an industry that has always been better at technical achievement than public relations; but it is important that it should succeed because, in the words of Maurice Bowra,[17] 'The trouble with a society which does not believe in engineering or see its possibilities is not that it cannot be civilized, but that it is not likely to survive'.

There is also room for conflict between social consciousness and professional responsibility. The professions have always been exposed to calumny by the press, the public, by government and its agencies, sometimes by the very organizations they serve. They have been seen as assuming privilege, authority and status on an exclusive basis; their restriction of membership, of title, of charges and of commercial activity have been criticized; the self-discipline and the sanctions they impose have been suspect. Where, as is the case with the engineering institutions, a professional body assumes no responsibility for the *personal* interests of its members, including the terms of employment or remuneration and where the right to practise is not exclusive, such arguments are really very flimsy. Nevertheless, a voluntary corporate body is an easy scapegoat. There is a conspicuous tendency for professional people to undervalue themselves and to accept passively attacks from those who through jealousy, malice or misunderstanding have misrepresented their motives. Although most such attacks are ephemeral and best ignored, we must be vigilant for misrepresentation that can do us permanent damage. In this respect the Council of Engineering Institutions can provide a bulwark which our own resources could not equal.

The third area of potential conflict, between the responsibility due to the employer and that to the profession, is to my mind the most difficult. I should like to mention two aspects which I, personally, have found very worrying. The first is the danger that the professional man may be swept up in the powerful and inexorable trade union movement. The pressures are very strong and the arguments can sound attractive to someone who has not absorbed the principles of professionalism or who feels, in any case, that these are discounted by an employer who neither appreciates his standards nor respects his independence. Professional institutions should appreciate that there is a real and recurrent danger not only to their corporate freedom but also to their corporate unity in that young engineers may accept the inducements of a militant organization that emphasizes their 'rights' in preference to the stern prescriptions of one that emphasizes service and responsibility.* For these reasons it is important to discriminate between (*a*) trade unions, which are political, self-motivated and concerned primarily with the remuneration and conditions of employment of their members, and (*b*) the professions, which are non-political, disinterested and essentially concerned with competence, ethics and bringing knowledge to the service of society.

This is not to say that professional associations are or should be precluded from protecting the individual interests of their members. Most unchartered associations do,* and one might say that it is an accident of incorporation that excludes the engineering institutions from this function. With the great majority of its members in salaried employment, and with the increasing pressure of inflation, the professional institution (however much it might hope that the status and position of corporate membership would make such action unnecessary) cannot take exception to its members associating themselves to protect their interests. Nevertheless, although this is the main and proper function of a trade union, those other functions of a professional institution which we have been considering are superior and must not be confused or diluted. In particular, it is obvious that the standard weapons of trade unionism, the withdrawal of labour and the 'go slow', are completely incompatible with the obligation to serve and to give only the best service. This puts a limitation on the type of protective association that engineers may feel free to join and makes it necessary that it should be composed of professional men with standards similar to their own. It would not prevent collective bargaining by such an association if this were felt to be desirable, because collective bargaining can be effective without the sanction of the strike if it is conducted regularly, publicly and with all cards on the table.

The opposite side of this coin is the attitude of mining companies to professional engineers in their employ. If they demand subservience and blind obedience, a professional man cannot work for them and retain his self-respect—for independence is the essence of professionalism and the professional must be free. Not 'free to do his own thing', in the current jargon, but free to perform the special service for which he has been trained, disciplined and indoctrinated, free to exercise his judgment, and free to maintain the habit of integrity in both thought and action which is as indispensable to the engineer as is the habit of cleanliness to the surgeon.

The mineral industry's problems are seldom routine, and dealing with the changing forces of nature is no

*This month the members of the Association of University Teachers will vote to determine whether or not the Association is to affiliate with the Trade Union Congress. At Imperial College the Association of Scientific, Technical and Managerial Staffs, an aggressive trade union, has energetically solicited membership from postgraduate students and junior members of academic staff.

*Millerson[18] described the structure, functions and limitations of the qualifying associations. He commented: 'Incorporation by Royal Charter brings disadvantages, as well as advantages of increased status. The Privy Council must approve the Bye-laws and any changes in Bye-laws; a major change demands the extra expense of petitioning for a Supplemental Charter. Great care must be taken to avoid any purely selfish objects, for example, directly negotiating over members' remuneration, or working conditions. A Royal Charter binds the association more strongly to serve the public good. Restrictions may be imposed on the value of property held by the association, or on some other aspect of structure.'

job for a subservient engineer who waits for orders. But let me emphasize that none of this is incompatible with teamwork, and the huge scale of today's mining operations makes cooperative effort more than ever necessary. Provided that his colleagues and employers are prepared to act within the professional code, then freedom and independence of discussion and judgment can be maintained, together with a willingness to loyally accept decisions arrived at in this open way.

The pressures on a senior engineer, in an employed situation, to withhold professional opinions which may run counter to his company's short-term wishes are very great. They increase with the size of his salary and the period of his pension rights. Companies, too, have responsibilities, and one of these is to ensure that its engineers feel free to express unbiassed opinions without incurring personal loss. The lapse into expediency on the part of an engineer is a strong temptation, but there is no doubt that both sides are the losers if standards are lowered because of the employee–employer relationship.

In this respect the position of the consultant is obviously to be envied, and it can be argued that only he enjoys to the full the professional qualities that originally came into being because the freedom and independence of private practice encouraged their growth. Once brought into being, however, these qualities can flourish in employed service if the atmosphere is right and ... a man who belongs to a profession which has won for itself prestige and a position of dignity, may pass from the service of one organization to that of another. Though he remains salaried all his life, he takes his stand upon his proved competence and experience; he serves one client after another much as does a free-lance worker. He is attached primarily to his profession, whence he goes out, as occasion may offer, to render his services in some co-operative organization, and whither he returns. The difference between the position of such a man and that of an ordinary salaried worker is not to be found in their respective legal situations; the difference is subtle, but it is vital; and the position of the former is as enviable as that of the free-lance man'.[19]

The making of the professional engineer

Is the professional mineral engineer an anachronism in 1971? Has he perhaps served a useful purpose and should he now disappear? Is he merely a residue of middle-class elitism as obsolescent as patriarchal professors? There are employers who seem to hold this view and there are vociferous sections of society who would support them. I think they must be mad, because the best investment that any mining company could make—at any price—would be in a staff of ideal professional engineers, complete with integrity, competence, responsibility *and* independence. Perhaps they have not succeeded in finding these paragons and it might be as well to consider where they come from and how they are produced.

The development of character is a complicated process. Obviously, the inheritance, upbringing and early education of the child are of great importance, and it is the combined result of these formative influences that the university and mining school endeavour to assess in their selection of candidates. Thereafter, university education, practical training, experience and the example of teachers, colleagues, associates and employers are the factors that determine in what degree and proportion the professional qualities, about which I have been speaking, are finally matured. Recruiting is the exploratory stage, and this, together with the development, extraction, concentration and refining of these professional qualities, is the job of the whole mineral industry. I am afraid that we have not taken it as seriously as we should. We are inclined to treat it as a side issue instead of giving it the priority it deserves. Until very recently we did not even provide central R and D facilities.*

The capabilities we seek will not arise spontaneously. If we require high standards in the product, the quality of the university education and of the vacation and graduate training given in industry must be high as well. Since, as Sir Eric Mensforth[20] has said, 'Judgment based on experience remains as important a factor as technical knowledge in (the engineer's) ability to serve the community, there must also be a real opportunity in the early years for him to apply his skills and to bite into challenging problems that will give him varied and substantial experience. In effective career development the role of the university teacher is of great importance, but so is the role of the manager and of the engineer employer. Men are always the most important factor in any enterprise, but too often their development is haphazard, unplanned and almost casual. The professional mineral engineer deserves better than this. If he is to recognize his responsibilities, he must see them in the example of his seniors; if he is to accept his responsibilities, he is justified in expecting a satisfying career with commensurate rewards and, above all, respect for his competence, achievement and independence.

I am afraid that, having been promised an address, you have been given a sermon. I have said that this is a personal view, and it must be apparent that it could not be otherwise. I am quite aware that I have spoken in an idealistic vein and that neither I nor many of my senior colleagues could sustain this ideal by our personal histories. Most of us have trod a winding path through our professional careers; never quite abandoning our principles, but sometimes leaving them to look after themselves. It is easier for the engineer than for most professions to fall from grace, since his discipline is not to seek perfection but to find solutions that work. But our tenets are not invalidated by our frailty and if, as I believe, professional standards are

*The Mineral Industry Manpower and Careers Unit was established in July, 1970

worth preserving, they are worth discussing, they are worth modifying and bringing into line with the current situation, and, so modified, they are worth passing on.

References

1. *The technologist in the mineral industry of the future* (London: Overseas Mining Association, 1969), 20 p.
2. *Mineral science and technology—needs, challenges and opportunities: a report ...* (Washington, D.C.: National Academy of Sciences, 1969), 129p. (p. 9)
3. **Carr-Saunders A. M. and Wilson P. A.** *The professions* (Oxford: Clarendon Press, 1933), 536 p.
4. **Reader W. J.** *Professional men* (London: Weidenfeld and Nicolson, 1966), 248 p. (p. 1)
5. **Lewis R. and Maude A.** *Professional people* (London: Phoenix House Ltd., 1952), 284 p.
6. **Millerson G.** *The qualifying associations, a study in professionalization* (London: Routledge and Kegan Paul, 1964), 306 p. (p. 10)
7. **Bennion F. A. R.** *Professional ethics* (London: Charles Knight, 1969), 278 p. (p. 15)
8. *The Monopolies Commission. A report on ... professional services* (*Sir Ashton Roskill chairman*) (London: H.M.S.O., 1970), 88p. (p. 5) (*Cmnd.* 4463)
9. Reference 2, p. 26.
10. **Armytage W. H. G.** *A social history of engineering* (London: Faber and Faber, 1961), 378 p. (p. 81)
11. **Thring M.** A Hippocratic oath for applied scientists. *New Scientist*, **49,** 7 Jan. 1971, 25–6.
12. **Medawar *Sir* P.** The effecting of all things possible. *The Listener*, **82,** 2 Oct. 1969, 437–42.
13. *Royal Commission on doctors' and dentists' remuneration 1957–60. Report.* (*Sir Harry Pilkington chairman*) (London: H.M.S.O., 1960), 346 p.
14. Reference 4, p. 124–5.
15. **Bacon F.** Maxims of the law. In *The works of Francis Bacon* **Spedding J. Ellis R. L. and Heath D. D. eds** (London: Longman, Green and Co., etc., 1859), vol. VII, 307–87. (p. 319)
16. **Alger P. L. Christensen N. A. and Olmsted S. P.** *Ethical problems in engineering* (New York: Wiley, 1965), 299 p.
17. **Bowra *Sir* M.** The impact of engineering on society. *Proc. Instn mech. Engrs*, **170,** 1956, 57–63. (p. 58)
18. Reference 6, p. 91.
19. Reference 3, pp. 502–3.
20. **Mensforth *Sir* E.** The basis of our man-made world. *The Guardian*, 6 Nov. 1970, p. 11.

41

Brochure, The Anaconda Company, Tucson, Ariz., 1971

ENVIRONMENTAL AWARENESS IN MINERAL EXPLORATION

The Anaconda Company Natural Resources Group

The Anaconda Company's Geology Department indirectly participates in all of the Company's environmental programs. As geologists, we also have individual responsibilities for the protection of the surface environment. These efforts require a minimum of time and expense in contrast to the extensive pollution control programs sponsored by the mining industry.

The disciplined use of our resources will allow us to maintain our environment for future generations. We are in accord with Federal and State regulations providing for multiple use of public lands, and will make all reasonable efforts to insure that our activities cause as little disturbance as possible.

There are three basic R's which we should all consider in environmental control procedures during the exploration programs:

RESPECT the environment we work in; it's the only one we have.

RESTRAIN from unnecessarily disturbing nature's balance.

RESTORE the natural setting of areas where we have worked.

The following practices are considered standard procedures for Anaconda exploration personnel:

FIELD RECONNAISSANCE

1. Be familiar with regulations governing mineral exploration in the location you are working. Consult with appropriate Federal, State, and/or local agencies for required permits, bonds, road standards, and environmental control procedures. Seek assistance from The Anaconda Company's Legal and Environmental Engineering personnel as appropriate. Allow sufficient time to satisfy administrative requirements without causing undue delay in subsequent field work.
   Investigate and study land ownership for local requirements; for example, some states have specific land owner notification laws which may be applicable. In the event of conflict over land ownership or rights, contact The Anaconda Company's Land Records and/or Legal Departments for consultation.
2. Respect fences and gates to minimize disturbance of livestock and crops. Observe signs posted by land owners.
3. Avoid unnecessary disturbance of vegetation and wildlife with procedures such as keeping vehicles on established roads whenever possible. Obtain permission from land owner when travel to off-road sites is necessary.
4. Carry out what you carry in. Use proper receptacles for refuse and attempt to leave an area cleaner than when you arrived.
5. Check local agencies for fire laws, as some areas prohibit fires at certain times. Carry appropriate fire fighting equipment; this is required by some land management agencies during seasonal fire periods.

PROPERTY ACQUISITION

1. Investigate the status of both surface and mineral rights before taking action to acquire a property position. Know the regulations and mining laws relating to property acquisition and exploration procedures for the particular region in which you are working.
2. After mineral rights have been acquired, discuss Anaconda's presence in the area with surface rights owners to avoid misunderstandings. Maintain cordial relations with local residents during the course of the program.
3. Avoid the use of excessive excavations for discovery purposes. Whenever possible, utilize drill holes to satisfy location work requirements rather than pits or cuts. Keep surface disturbance of vegetation to a minimum.
4. Provide for the grading of discovery pit overburden and revegetation of discovery-related excavations as soon as practical after excavation.

ROAD BUILDING AND ACCESS

1. Check with local administering agencies for road and trail development requirements, including permit acquisitions when necessary.
2. Plan and locate roads and trails to serve the maximum purpose with the least amount of construction and disturbance. Take advantage of landform and vegetation to screen exploration activity whenever possible.
3. Clear right-of-way timber and debris in an orderly manner. Dispose of debris by burning if permissible, or by burying in the road fill whenever practical.
4. Provide for erosion and surface water control by installing culverts, bridges, water bars, ditches, or other means whenever necessary. Keep insloping to a minimum. If insloping is necessary, provide for frequent water diversions off of road. Remove all unnecessary berms from road edge.
5. Plan and accomplish revegetation of roads and trails as soon as practical, and definitely upon abandonment.

GEOPHYSICAL AND GEOCHEMICAL SURVEYS

1. Avoid unnecessary brush cutting and blazing of trees in the course of laying out lines.
2. Remove all wire, flagging, stakes, and other extraneous material upon completion of survey.
3. Avoid contamination of soil and water in the use of chemicals and of other material necessary to accomplish your function.
4. Backfill, cover, and revegetate all excavations, drill holes, and other surface disturbances.

DRILLING PROGRAMS

1. Emphasize to all contractors involved in the exploration program that they represent Anaconda while under contract, and must abide by our environmental control regulations. Include appropriate environmental specifications in contracts.

2. Keep noise pollution produced by equipment at a minimum by specifying in contracts that equipment meet existing Federal, State, and local noise control requirements.
3. Choose the location of drill sites in such a manner so that both the access and the actual site will result in a minimum of disturbance. Take advantage of local land forms and vegetation to screen drilling activities whenever possible.
4. Use minimum size drilling equipment and drill holes whenever feasible to reduce the need for elaborate access roads.
5. Insure that proper receptacles are provided and utilized for debris at the drill sites.
6. Do not allow drilling wastes to run uncontrolled over the land surface. Recycle with the use of tanks or sumps as much drilling mud and water as possible.
7. Do not allow water lines to run free. Install and use shut-off valves when water is not required.

COMPLETION OR ABANDONMENT OF EXPLORATION ACTIVITY

1. Outslope and remove all berms from roads no longer needed, and install appropriate water barriers and other erosion control structures. Remove culverts, low bridges and other structures which may cause blockage and unwanted diversions of water courses.
2. Backfill and grade as close to original contour as practical all open excavations such as exploration trenches, adit and shaft waste dumps, etc. Grade to contour or smooth out drill sites, building sites, road, or other excessive construction disturbances.
3. Cap all drill holes and close, or otherwise make safe from unauthorized entry, all tunnels, shafts, and other openings.
4. Remove all trash and other foreign material from the exploration area. Inert, non-polluting wastes should be buried at least two (2) feet deep in an area which will not be affected by, or where it will not affect, either surface or ground waters.
5. Revegetate all surface disturbances with plant/tree species suited for the area at the proper time of the year in order to be effective. Check with local agencies and The Anaconda Company Environmental Engineering Department for assistance.
6. Inspect gates and fences used for access to ascertain that no damage exists.
7. If an area acquired by the location of claims is abandoned, remove physical evidence of location such as posts, etc.

These procedures should be followed in all exploration projects. Each project is unique in some respect and consequently we must adjust accordingly. As the various state and federal agencies adopt new mine regulation and reclamation laws, we must become cognizant and adjust to these changes accordingly. The Anaconda Company Legal and Environmental Engineering Departments at General Mining Division Headquarters will provide assistance as required.

Editor's Comments on Paper 42

42 **TURNER and COFFMAN**
Excerpts from *Geology for Planning: A Review of Environmental Geology*

ENVIRONMENTAL IMPACT STUDIES

Evaluation of mines and prospects is neither limited to the determination of size and metal concentrations nor to technological considerations of mineral processing. Increasingly, the geologist is expected to record the character of the environment and to include an environmental impact study showing the effect of mining and beneficiation on the environment. Certain areas may be excluded from exploration or development for environmental reasons, or the mining or metallurgical technology may be modified to minimize impact effects.

Turner and Coffman (Paper 42) summarize mapping procedures and established environmental resource units that would assist in presentation of a cost/benefit analysis.

42

Reprinted by permission of The Colorado School of Mines from pages 27-30, 43-51, and 125-127 of *Colorado School of Mines Quart.* **68:**1-127 (1973)

GEOLOGY FOR PLANNING: A REVIEW OF ENVIRONMENTAL GEOLOGY

A. K. Turner and D. M. Coffman

[*Editor's Note:* In the original, material precedes and follows these excerpts.]

REVIEW OF SOME CURRENT ENVIRONMENTAL GEOLOGY MAPPING PROCEDURES

Planners have turned increasingly toward state and federal geological agencies for assistance in developing regional plans having due regard for the natural resources and natural hazards of an area. In Chapter 2, we discussed the trends and rationale underlying the growing concern for the environment. These concerns and the resulting demands for natural resources information prompted the start of many specialized mapping projects.

The objective of these special maps was to supply information on geological factors that was immediately usable by the planner. Several approaches have been tried, but none has been entirely successful. This chapter will review a selection of these techniques in order to outline the major advantages and disadvantages of the various competing systems.

Goal of Environmental Geology Mapping Programs

The goal of all the environmental geology mapping programs was assistance to the planner. Most studies came to the same conclusions:

(1) the geologist can aid the planner to anticipate the effects of seven criteria that are essential in the planning analysis; and
(2) an environmental geology map should show at least six classes of information.

The criteria considered most appropriate included:

(1) foundation conditions and excavation costs;
(2) seismic activity and presence of active faults;
(3) slope stability;
(4) the presence of building materials;
(5) the ground water and surface water resources before and after development;
(6) the danger of flooding;
(7) the acceptability of certain areas as waste disposal sites.

The important classes of information included:

(1) the physical properties of the soil and rock units, including seismic stability, potential as building materials, and acceptability for waste disposal;
(2) slope stability, including potential for landslides, rockfalls, soil creep, and avalanche;
(3) presence of active or possibly active faults, and epicenters of historical earthquakes;
(4) the contours of the ground water table or the depth-to-ground water contours;
(5) thickness of the soil or depth to bedrock;
(6) the extent of any floodplains present.

Naturally, the most appropriate classes of information vary with the conditions within the area mapped. For example, in a seismically inactive area, epicenters of historic earthquakes and active faults would not appear on the map. Inactive faults would still be mapped, however, since they might affect foundation stability. The effect of certain geological conditions on land use is shown in figure 7.

X = PRIMARY IMPORTANCE

O = SECONDARY IMPORTANCE

	TYPICAL LAND-USES						
GEOLOGIC CONDITIONS	LIGHT STRUCTURE CONSTRUCTION	HEAVY STRUCTURE CONSTRUCTION	WASTE DISPOSAL	BUILDING MATERIAL RESOURCES	EASE OF EXCAVATION	ROAD PERFORMANCE	AGRICULTURE
PHYSICAL PROPERTIES OF SOILS AND ROCKS	X	X	X	X	X	X	X
SLOPE STABILITY	X	X	O	O	O	X	O
THICKNESS OF SURFICIAL MATERIALS	X	X	O	X	O	O	O
DEPTH TO GROUND WATER	X	X	X	X	O	O	X
SUPPLY OF SURFACE WATER	O	O	O	X	O	O	X
DANGER OF FLOODING	X	X	X	X	X	X	X

Figure 7.—Geologic conditions that affect land use.

Classification of Environmental Geology Mapping Procedures

Two general concepts have been used in environmental mapping programs. The first involves the use of traditional geologic maps; the second involves the creation of special format maps.

Traditional Maps

The traditional map method establishes the engineering properties of already mapped geologic map units, and publishes these properties along with the map. It is fast and inexpensive where geologic maps already exist; however, it is often ineffective in terms of planning. Geologic map units are often defined on stratigraphic terms. Such units do not necessarily reflect consistent lithologic or engineering properties, but reflect only a difference in geologic age or a minor variation in mineralogy, color, or fossil content. Thus the planner, when trying to use a traditional geological map, may be faced with two shales, identical in engineering properties, differentiated in age because of the presence of foraminifers. Similarly, this same planner may find a thick sequence of clay-rich shales capped by sandstones mapped as a single unit.

Another drawback of traditional geologic maps is their tendency to show bedrock, whether overburden is 1 foot thick or 30 feet thick. Alluvial deposits are usually lumped as a single unit, whether they consist of fine-grained floodplain deposits or coarse-grained river channel deposits. This creates such a wide range of values in the engineering properties of these materials as to render them virtually useless to the planner or engineer. To some extent, this problem is being eased by modern geological procedures which study and map surficial deposits in reasonable detail in areas where they are significant. Such mapping is especially common in areas subjected to continental glaciation, where bedrock and surficial geology maps are often prepared on separate sheets.

Special Format Maps

Special format maps have been developed by many environmental mapping projects. In some cases only a single map sheet is prepared (Gardner and Hart 1971; Lutzen 1968; Lutzen and Rockaway 1971), but in many cases multiple maps are developed (Scott 1972a; 1972b; 1972c; 1972d; 1972e; 1972f). In some cases the maps are combined with texts and bound to form a single atlas or folio (Christiansen 1970; Yon 1972).

In developing these special format maps, the map units are chosen on the basis of their observed, or in some cases inferred, engineering and physical properties. With these maps the planner has only the information that is important to him, eliminating confusing or extraneous data. Where complex factors occur, multiple maps are often used with special purpose maps delineating specific hazards, such as landsliding or flooding.

Several attempts have been made to supply both traditional and interpreted data on a wide variety of environmental conditions within a single atlas publication. A recent Canadian example is the "Physical Environment of Saskatoon, Canada" (Christiansen 1970). This publication includes some 54 maps and diagrams covering such diverse fields as bedrock and surficial geology, pedology, climate, land use, land capability for recreation and agriculture, and engineering applications (including soil mechanics, groundwater resources, and slope stability). Such wide-ranging compilations are at the moment experimental and are somewhat controversial. There is little doubt, however, that a single source document containing all available data is useful to the planning agency.

[*Editor's Note:* Material has been omitted at this point.]

ENVIRONMENTAL RESOURCE UNITS: A METHOD OF MAPPING THE TOTAL ENVIRONMENT

Proper management of our natural resources and planned growth of our society must incorporate an understanding of environmental principles. Planners, managers, developers, and legislators must consider an area's capabilities and sensitivities relative to human use before pursuing any specific course of action. However, sound environmental policy requires sound environmental information.

Communication of the information required to plan the use and conservation of resources is best accomplished through environmental maps. A map is an ideal format to present data which are dependent on location, elevation, areal extent, and proximity to other parameters. The synthesis and graphical presentation of environmental information are feasible because of the systematic nature of the environment.

The Environment as a System

The environment is a well-structured system of components—rocks, soil, water, biota, climate. These components are complexly interrelated and controlled by processes operating within the system. For example, weathering of rock as controlled by the climate will determine many of the physical characteristics of a region. These physical characteristics influence the vegetation types and their successional rates and trends. Vegetation, climate, and topography combine to control wildlife types and dynamics. In turn, the combined physical and biological properties determine the suitability of the land for human habitation and use. Only human activity is likely to be poorly correlated with respect to other portions of the environmental system.

It must be emphasized that the environment of any area is a function of both structure and process. While the structure, or components, of an environment are the most readily identified and mapped units, the processes which have determined the forms and materials comprising these components provide the key to understanding and properly managing the environment.

In most environmental investigations, artificial separation of environmental components is imposed. As a result, geology, soils, vegetation, and land use maps are produced by independent professional groups. This procedure ignores the basic concept that the environment is an entity in which all components are intimately interrelated. The failure of conventional mapping practices to use this interrelationship has prohibited effective extrapolation of information from intensively studied areas to similar but less studied regions.

Recent attempts to surmount this problem and effectively map a region's environment have tried to deal with the natural systems (Brown et al. 1971; 1972.) These efforts, however, have suffered from two problems. They have emphasized structure rather than process, and they have lacked a truly interdisciplinary approach in the selection of the primary map units.

For example, environmental capability maps have been produced whose basic units are derived from geologic or soil associations. Such efforts have produced excellent displays of engineering and agricultural properties of the physical environment. However, it is unrealistic to expect that a map, compiled without any input from the social or biological sciences, can be interpreted to explain the complete spectrum of an area's environmental sensitivities and capabilities.

Environmental Resource Units

In the selection of primary map units for environmental data compilation, we should consider process equally with structure. We should also consider biological and cultural factors equally with physical factors.

The environment is naturally divided into units characterized by specific assemblages or patterns of both structural components (rocks, water, soils, biota, climate, etc.) and natural processes (erosion, runoff, pedogenesis, productivity, wildlife behavior, etc.). A portion of the environment with a similar suite of characteristics is called an "environmental resource unit" or ERU. Every ERU is the sum product of present and past processes, and in turn determines the structural characteristics and rates of ongoing natural processes. An ERU is a method for broadening the biological concept of an ecosystem to include the physical and cultural portions of the environment.

Environmental resource units and subunits are significant because the map boundaries are identical for physical, potential biological, and potential socioeconomic elements. It is therefore possible to display meaningful management information on a single map in terms of the capabilities, sensitivities, and hazards of each ERU. For special purposes, derivative maps showing only physical, biological, or cultural ERU properties may be produced.

Once the ERU units are selected in an interdisciplinary manner, the physical characteristics of an ERU may be presented independently to provide an environmental geology map. The major unit boundaries will reflect geomorphic, pedogenic, and hydrologic processes. Subunits are typically differentiated on the basis of lithology, geologic structure, and topography.

The physical ERU map is based on those elements of the environment which are not subject to rapid change. It will provide, like conventional applied geology maps, information on engineering properties of materials and locations of natural hazards. Additionally, it can be used to predict the potential ecosystems of a region, and to determine probable land uses and values, aesthetic characteristics, and potential locations of urban areas.

An actual biological ERU map can be prepared to show existing distributions of the more sensitive ecologic communities. In a climax ecosystem, the actual biological unit boundaries will match the boundaries of the physical ERU map. The disparity between the actual biological ERU and the potential biological ERU derived from the physical ERU map provides a quantitative measure of the degree of alteration an environment has undergone. This procedure permits a realistic evaluation to be made concerning the quality of the existing environment of a region.

Care must be taken, however, in such evaluations, since natural processes can occasionally cause disparities between the physical and biological ERU boundaries. Where such natural disparities exist, they generally provide valuable insight to a region's history.

The biological ERU maps may be used to extrapolate the natural stratification and original composition of biological communities within a study area. With these potentials clearly understood, the state of such processes as productivity, succession and disturbance responses can be determined. Decisions as to which areas can be returned to a natural condition, how best to manage these systems as a resource, and how best to apply appropriate reclamation procedures can be made with reasonable assurance.

The environmental impact of any activity has a character determined by the nature of the activity, and by the dynamic structure of the environment upon which the activity is superimposed. Impact assessment can effectively be carried out only on a systems basis, by determining the basic changes in structure and dynamics resulting from proposed actions. An ERU map is ideally suited to the efficient and accurate evaluation of system impacts.

Once the system impacts are understood, it is then a straightforward matter to itemize specific resource losses, and otherwise to meet the reporting requirements of the National Environmental Policy Act of 1969 or of individual state environmental legislation.

Methodology of Environmental Resource Unit Mapping

The key to mapping environmental resource units is an interdisciplinary approach. The mapping requires close cooperation among experienced scientists representing several disciplines, with each scientist being able to communicate across disciplinary boundaries. A project team might include specialists in soils, geology, hydrology, limnology/oceanography, vegetation ecology, animal ecology, social science, and land use. The team works together to determine patterns of structure and process which represent environmental entities. For example, vegetation mapping is not completed by the plant ecologist working independently of the others. Rather, vegetation patterns are related to soils, geology, water supply, coastal influences, animal patterns, and human responses in order to develop an ERU classification for mapping which is of direct and immediate value in establishing land capabilities and sensitivities.

Two major steps are involved in preparation of an ERU map: selection of primary units and the sampling of these units. Determination of units requires the consensus of a multidisciplinary team, and will depend on the scale of mapping to be done. On a continental scale, ERU units would be approximately equivalent to physiographic or vegetation provinces. A strong point of this methodology, however, is that units can be defined for any given detail of study.

ERU units usually are initially identified from remotely sensed imagery. A combination of high altitude or space imagery and medium altitude color infrared photography has proven most satisfactory for the diverse needs of a multidisciplined team.

Sampling of ERU units seeks to identify three types of areas: typical, transitional, and anomalous. The typical characteristics are most important in identification of resource capabilities; the transitional properties are indicative of many environmental dynamics; and the anomalous characteristics are often sources of sensitivities and hazards.

Example of Environmental Resource Unit Mapping

The concepts of ERU mapping are best shown by an actual example. An ERU map for a site in western Indiana is shown in figure 13. The

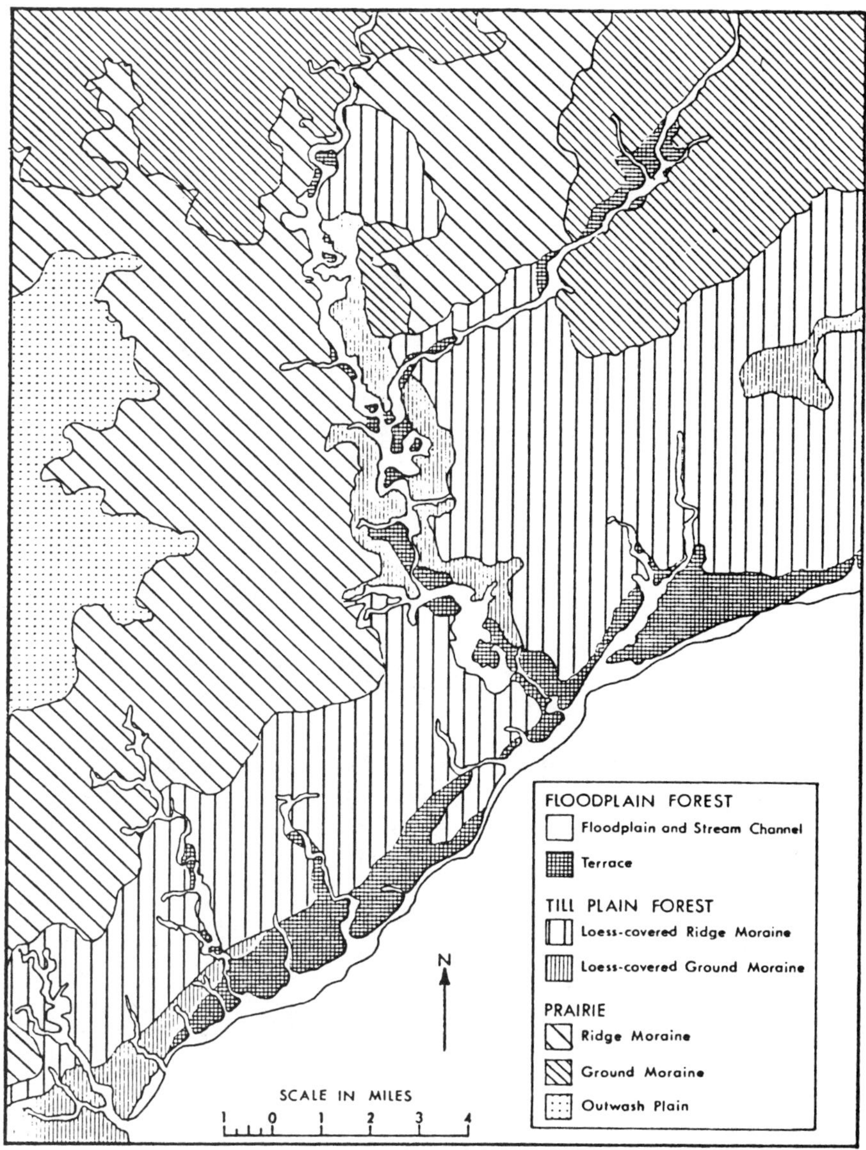

Figure 13.—Environmental resource units mapped in western Indiana.

region is typical of the Till Plains Section of the Interior Lowlands Physiographic Province (Fenneman 1938).

TABLE 5.—*Environmental resource units—western Indiana area*

Unit Name	Floodplain Forest	Till Plain Forest	Till Plain Prairie
Location and Size	About 2100 acres located along Big Pine Creek and lower Fall Creek	About 10,230 acres contiguous to and eastward from Big Pine Creek	About 30,000 acres located west and north of Big Pine Creek
Physical Subunits	Stream channel and floodplain; Pleistocene Terraces	Loess covered ground moraine; loess covered ridge moraine	Ground moraine; ridge moraine; outwash plain
Geology	Recent alluvial floodplain deposits and Pleistocene stratified sands and gravels; sedimentary rock in valley walls includes sandstones and shales of Devonian and Mississippian age.	Predominantly glacial till; near surface sedimentary rocks including sandstone and shale of Devonian and Mississippian age; localized occurrences of coal	Predominantly glacial till; stratified sand, silt, and gravel as outwash deposits
Landforms	Steep walled, level floored floodplain; minor terraces and rock benches; V-shaped, steep-sloped tributary ravines; pool-riffle channel	Rolling to hilly ridge moraine; relatively level till plain	Rolling to hilly ridge moraine; relatively level till plain; extremely flat outwash plain
Landscape Dynamics	Fluvial processes, particularly channel and overbank flow, are dominant; mass wasting occurs on valley walls	Fluvial processes, especially overland flow and sheetwash, are dominant; erosion rates are high; inherited landforms are still prevalent	An inherited landscape from Pleistocene glaciation; internal drainage is characteristic with little natural stream dissection; ditching to improve poor drainage is common
Soils	Sandy clay developed over lenticular deposits of sand, silt, and clay	Sandy clay, clay loams, and silty loams develop on loess and till; ridge tops are excessively drained resulting in acid soils	Silty clay loam developed on till
Vegetation	Silver maple-American elm forest	Sugar maple-beech forest on cool slopes and wet depressions; oak-hickory forest on well drained ground	Naturally occurring blue stem and other prairie grasses have been com-completely replaced by farmland
Wildlife and Fish	Characteristic wildlife include the eastern gray squirrel, raccoon, hairy woodpecker, barred owl, warblers, red-backed salamander, smallmouth and rock bass	Characteristic wildlife include deer, turkey, wood frog, slimy salamander, and acadian flycatcher	No characteristic large mammals remain; farmland species (e.g. deer, ring-necked pheasant, plains pocket gopher) now are common

TABLE 5.–(*Cont'd.*)

Ecosystem Dynamics	The high groundwater table controls vegetation patterns; goldenrod and sycamore-shrub successional series occur on abandoned fields	Drainage differences and insolation regimes control vegetation; oak-hickory is succeeding towards maple-beech forest except where adaphically maintained as climax; shrub and thicket series occur in abandoned fields	No natural systems remain
Land Use	About 50% of the unit is cultivated, 25% is in forest, and the remainder is in pasture or abandoned fields	About 40% of the unit is cultivated, 31% is in forest, and the remainder is in pasture or abandoned fields	100% of the unit is in farmland
Transitions	Steep valley walls provide abrupt unit borders with the upland ERUs	The boundary with the Floodplain ERU is abrupt due to steep valley walls; the boundary with the Prairie ERU is gradational, determined by degree loess cover and vegetation changes	The boundary within the Floodplain ERU is abrupt due to steep valley walls; the boundary with the Till Plain Forest ERU is gradational, determined by degree of loess cover and vegetation changes
Anomalies	The wide floodplain area in southern Big Pine Creek results from excavation of a preglacial buried valley trending north-south; Springs occur at the intersection of the present and buried valleys	Kettles occur within the moraines providing peat deposits and organic soils	None recorded in study area
Capabilities	The land is ideally suited to agriculture, recreation, and forest management. This unit is unusually scenic for the physiographic region. Construction materials including quarry stone, sand, and gravel aggregate are available	The land is naturally a hardwood forest suitable for lumbering; once cleared, this unit provides rich agricultural land	The land is highly valued agricultural property especially suited to large single crop farms
Sensitivities	Much of the unit is subject to periodic flooding; stream water quality is degraded from agricultural nutrients; valley walls are subject to slumping	Poor agricultural practices can quickly lead to accelerated erosion and severe soil losses	This unit is characterized by poor drainage; low areas remain covered by ponded water for considerable periods during the wet seasons

Three primary assemblages of environmental elements and processes were identified in the study area, resulting in three ERUs—floodplain forest, till plain forest, and till plain prairie. See table 5 for abbreviated description of the chief characteristics of each unit.

The data in table 5 is summarized in figure 14. It clearly demonstrates that each ERU includes unique geologic and topographic combinations. These in turn cause the development of characteristic soil associations, which in turn produce characteristic flora and fauna assemblages. Accordingly, each ERU has unique environmental characteristics and sensitivities.

[*Editor's Note:* Figure 14 follows.]

ERU	FLOODPLAIN FOREST	TILL PLAIN FOREST		PRAIRIE
TOPOGRAPHY				
SURFACE GEOLOGY	Floodplain and Terrace Alluvium	Loess Covered Glacial Till Of Ground and Ridge Moraines		Glacial Till from Moraines, Sands & Silts of Outwash Plains
SOIL ASSOCIATIONS	Genesee-Shoals-Eel Ockley-Fox	Miami-Hennepin Raub-Ragsdale	Miami-Russell-Fincastle	Sidell-Parr Parr-Corwin Odell-Chalmers Rensselaer-Darroch
SOIL TYPES	Sandy Clay	Sandy Clay	Clay	Silt Clay Loam
CHARACTERISTIC VEGETATION	Silver Maple-American Elm Forest	Maple-Beech Forest on Slopes and In Depressions	Oak-Hickory Forest on Knolls & High Ground	Prairie Grasses: Now Completely Removed & Replaced by Cultivated Farmland.
CHARACTERISTIC ANIMALS	Hairy Woodpecker Squirrel Cerulean Warbler Raccoon Prothonotary Warbler Skunk Red-backed Salamander Owl	Acadian Flycatcher Wood Frog Slimy Salamander	Wild Turkey Ruffed Grouse	No Prairie Is Left To Support Naturally Occurring Species.

Figure 14.—Principal characteristics of the environmental resource units of western Indiana.

REFERENCES

Anderson, J. R., and others, 1972, A land-use classification system for use with remote sensor data: U.S. Geol. Survey Circ. 671.

Beazley, R., 1967, Conservation decision-making: A Rationalization: Natural Resources Jour., v. 7, no. 3, p. 345-360.

Bickmore, D. P., 1968, Maps for the computer age: Geog. Mag., no. 14, p. 221-227.

Brown, L. F., and others, 1971, Resource capability units—their utility in land- and water-use management with examples from the Texas coastal zone: Bur. Econ. Geol., Texas Univ., Austin, Geol. Circ. 71-1, 22 p.

Brown, L. F., and others, 1972, Environmental geologic atlas of the Texas coastal zone-Galveston-Houston area: Bur. Econ. Geol., Texas Univ., Austin, 94 p.

Bryant, B., and Reed, J. C., Jr., 1972, Map showing approximate density of houses in the Evergreen quadrangle, Jefferson County, Colorado: U.S. Geol. Survey Misc. Inv. Map I-786-D.

Claire, W. H., (ed), 1969, ASCE urban planning guide: Am. Soc. Civil Engs., Manuals and Reports on Engineering Practice, no. 49, 299 p.

Christiansen, E. A. (ed.), 1970, Physical environment of Saskatoon, Canada: Sask. Research Council and Natl. Res. Council Canada, Ottawa.

Colton, R. B., 1960, Surficial geology of the Windsor Locks quadrangle, Connecticut: U.S. Geol. Survey Geol. Quad. Map GQ-137.

Davis, J., 1972, Environmental policy needed: a sense of perspective: Eng. Digest, v. 18, no. 1, p. 29-30 (Jan.).

Dole, H. M., 1971, The federal government and urban geology: *in* Environmental geology and planning: U.S. Govt. Printing Office, p. 60-69.

Emmons, S. F., Cross, Whitman, and Eldridge, G. H., 1896, Geology of the Denver Basin, Colorado: U.S. Geol. Survey Mono. 27, 556 p.

Federal Register, 1970, Interim guidelines from the President's Council on Environmental Quality: Federal Register, v. 35, p. 7890 (April 30).

———— 1971, Statements on proposed federal actions affecting the environment: Federal Register, v. 36, no. 19, p. 1398-1402, *and* no. 79, p. 7724-7729.

Fenneman, N. M., 1931, Physiography of Western United States: McGraw-Hill, 534 p.

———— 1938, Physiography of Eastern United States: McGraw-Hill.

Gardner, M. E., 1968, Preliminary report on the engineering geology of the Boulder quadrangle, Boulder County, Colorado: U.S. Geol. Survey Open-file Rept.

———— 1969, Preliminary report on the engineering geology of the Eldorado Springs quadrangle, Boulder and Jefferson Counties, Colorado: U.S. Geol. Survey Open-file Rept.

Gardner, M. E. and Hart, S. S., 1971, Preliminary engineering geology map of the Golden quadrangle, Jefferson County, Colorado: U.S. Geol. Survey Misc. Field Studies Map MF-308.

Gardner, M. E., and Johnson, C. G., 1971, Engineering geologic maps for regional planning: *in* Environmental Planning and Geology, U.S. Govt. Printing Office, p. 154-169.

Gross, D. L., 1970, Geology for planning in De Kalb County, Illinois: Environmental Geology Note no. 33, Illinois State Geol. Survey, 26 p.

Hackett, J. E., and McComas, M. R., 1969, Geology for planning in McHenry County, Illinois: Illinois State Geol. Survey Circ. 438, 29 p.

Hachman, F. C., and others, 1971, Report on the implementation of composite computer mapping: Bur. Economic and Business Research, Center for Economic and Community Development, Utah Univ., Salt Lake City, Utah.

Hayes, W. C., and Vineyard, J. D., 1969, Environmental geology in Towne and Country: Missouri Geol. Survey and Water Res., Rolla, Mo., Educational Series no. 2, 42 p.

Hubert, J. F., 1960, Petrology of the Fountain and Lyons formations, Front Range, Colorado: Colorado School Mines Quart., v. 55, no. 1, 242 p.

Hyatt, David E., 1973, The environmental impact statement: a new requirement in planning the mining operation: Colorado School of Mines Mineral Industries Bulletin, v. 16, no. 3, 10 p.

Legget, R. F., 1973, Cities and geology: McGraw-Hill, 624 p.

Leopold, L. B., and others, 1971, A procedure for evaluating environmental impact: U.S. Geol. Survey Circ. 645, 13 p.

LeRoy, L. W., 1946, Stratigraphy of the Golden-Morrison area, Jefferson County, Colorado: Colorado School Mines Quart. v. 41, no. 2, 115 p. (April).

Lutzen, E. E., 1968, Engineering geology of the Maxville quadrangle, Jefferson and St. Louis Counties, Missouri: Eng. Geol. series no. 1, Missouri Geol. Survey and Water Res., Rolla, Mo.

Lutzen, E. E., and Rockaway, J. D., 1971, Engineering geology of St. Louis County, Missouri: Eng. Geol. series no. 4, Missouri Geol. Survey and Water Res., Rolla, Mo.

Maberry, J. O., 1972, Slope map of the Parker quadrangle, Arapahoe and Douglas Counties, Colorado: U.S. Geol. Survey Misc. Geol. Inv. Map I-770-F.

Macleod, M. H., 1973, Semi-automated large scale map production: *in* Proc. NATO Advanced Study Institute "Computer Systems for Highway Design," Copenhagen, Denmark, p. 213-223.

Manheim, M. L., 1966, Hierarchical structure: a model of planning and design processes: M.I.T. Press, Cambridge, Massachusetts Inst. of Technology Rept. 7.

McHarg, I., 1967, Where should highways go?: Landscape Architecture, p. 179-181.

——— 1968, A comprehensive highway route selection method: Highway Research Record 246, p. 1-15.

———, 1969, Design with nature: The National History Press, Garden City, New York, 197 p.

——— 1970, Ecological values and regional planning: Civil Eng. (Am. Soc. Civil Engs.), v. 40, no. 8, p. 40-44 (Aug.).

Newville, J., 1970, Comment on McHarg's paper: Civil Eng. (Am. Soc. Civil Engs.), v. 40, no. 8, p. 44 (Aug.).

Nez, G., 1972, Toward a common system of land-use classification: Tech. Paper 17, Federation Rocky Mtn. States, Denver, Colo.

Pollard, W. S. and Moore, D. W., 1969, The state of the art of planning: Proc. Am. Soc. Civil Engs., Urban Planning Div., 95, UP1, paper 6517, p. 27-41.

Quay, J. R., 1966, Use of soil surveys in subdivision design: *in* Soil surveys and land use planning: Soil Sci. Soc. of America and Am. Soc. Agronomy, p. 76-87.

Radbruch, D. H., 1957, Areal and engineering geology of the Oakland west quadrangle, California: U.S. Geol. Survey Misc. Geol. Inv. Map I-239.

——— 1969, Areal and engineering geology of the Oakland east quadrangle, California: U.S. Geol. Survey Geol. Quad. Map GQ-769.

Rhind, D. W., 1971, Towards instant and efficient maps: the work of the Experimental Cartography unit: Rev. Geogr. Montr., v. XXV, no. 4, p. 391-410.

Rockaway, J. D., and Lutzen, E. E., 1970, Engineering geology of the Creve Coeur quadrangle, St. Louis County, Missouri: Eng. Geol. series no. 2, Missouri Geol. Survey and Water Resources, Rolla, Mo.

Schmidt, P. W., 1972, Slope map of the Evergreen quadrangle, Jefferson County Colorado: U.S. Geol. Survey Misc. Inv. Map I-786-C.

Scott, G. R., 1972a, Geologic map of the Morrison quadrangle, Jefferson County, Colorado: U.S. Geol. Survey Misc. Geol. Inv. Map I-790-A.

——— 1972b, Map showing landslides and areas susceptible to landsliding in the Morrison quadrangle, Jefferson County, Colorado: U.S. Geol. Survey Misc. Geol. Inv. Map I-790-B.

——— 1972c, Map showing areas containing swelling clay in the Morrison quadrangle, Jefferson County, Colorado: U.S. Geol. Survey Misc. Geol. Inv. Map I-790-C.

———1972d, Map showing potential source areas for non-metallic mineral resources, Morrison quadrangle, Jefferson County, Colorado: U.S. Geol. Survey Misc. Geol. Inv. Map I-790-D.

——— 1972e, Map showing points of geologic interest, Morrison quadrangle, Jefferson County, Colorado: U.S. Geol. Survey Misc. Geol. Inv. Map I-790-E.

——— 1972f, Map showing floods and water courses, Morrison quadrangle, Jefferson County, Colorado: U.S. Geol. Survey Misc. Geol. Inv. Map I-790-F.

Smith, H. J., 1964, Geology of the sedimentary rocks of the Morrison quadrangle, Colorado: U.S. Geol. Survey Misc. Geol. Inv. Map I-428.

Smith, J. A., Miller, L. D., and Ells, T., 1972, Pattern recognition routines for graduate training in the automatic analysis of remote sensing imagery-RECOG: Science series no. 3A, Dept. Watershed Science, Colorado State Univ., Fort Collins, Colo., 86 p.

Tanguay, M. G., 1969, Aerial photography and multispectral remote sensing for engineering soils mapping: Ph.D. Thesis (unpubl.), School of Civil Engnieering, Purdue Univ., Lafayette, Ind.

Tanguay, M. G., and Miles, R. D., 1970, Multispectral data interpretation for engineering soils mapping: Highway Research Record 319, p. 58-77.

Turner, A. K. and Miles, R. D., 1971, The GCARS system: a computer assisted method of regional route location: Highway Research Record 348, p. 1-15.

U.S. Government, Department of Defense, 1953, The unified soil classification: U.S. Army, Corps of Engineers, Technical Mem. no. 3-357.

——— Department of Interior, 1971, Statement of environmental impact: Dept. Interior Manual Part 516, Chapter 2; *in* Federal Register v. 36, no. 192, p. 19340-19347.

——— 1972, Environmental quality—preservation and enhancement: Bur. Reclamation, General Instructions, series 350, Part 376, Release 376-4, Jan. 12, Washington, D.C.

——— Department of Transportation, 1969, Public hearings and location approval: Federal Highway Admin. Policy and Procedure Mem. 20-8, Washington, D.C.

Waage, K. M., 1955, Dakota group in northern Front Range foothills, Colorado: U.S. Geol. Survey Prof. Paper 274-B, p. 15-51.

Waldschmidt, W. A. and LeRoy, L. W., 1944, Reconsideration of the Morrison Formation in the type area, Jefferson County, Colorado: Geol. Soc. America Bull. v. 55, p. 1097-1113.

Yon, J. W. (ed.), 1972, Environmental geology and hydrology, Tallahassee area, Florida: Bur. Geology, Florida Dept. of Nat. Resources, 61 p.

AUTHOR CITATION INDEX

SUBJECT INDEX

About the Editor

WILLARD C. LACY, Professor Emeritus, was Foundation Professor of Geology at James Cook University of North Queensland. Professor Lacy received the B.Sc. from DePauw University, the M.Sc. from the University of Illinois, and the Ph.D. from Harvard University. He served as geologist for Titanium Alloy Manufacturing Company, as petrologist to chief geologist for the Cerro Corporation in Peru, as visiting professor in economic geology at Harvard, then as professor of mining and geological engineering at the University of Arizona before taking the position in Australia.

Professor Lacy has served as consultant for many major companies in areas of economic geology and geotechnics, is author of many papers, and is a frequent speaker at professional meetings.